THE ROUTLEDGE HANDBOOK OF ANIMAL ETHICS

There isn't one conversation about animal ethics. Instead, there are several important ones that are scattered across many disciplines. This volume both surveys the field of animal ethics and draws professional philosophers, graduate students, and undergraduates more deeply into the discussions that are happening outside of philosophy departments. To that end, the volume contains more nonphilosophers than philosophers, explicitly inviting scholars from other fields—such as animal science, ecology, economics, psychology, law, environmental science, and applied biology, among others—to bring their own disciplinary resources to bear on matters that affect animals.

The Routledge Handbook of Animal Ethics is composed of 44 chapters, all appearing in print here for the first time, and organized into the following six sections:

I. Thinking About Animals
II. Animal Agriculture and Hunting
III. Animal Research and Genetic Engineering
IV. Companion Animals
V. Wild Animals: Conservation, Management, and Ethics
VI. Animal Activism

The chapters are brief, and they have been written in a way that is accessible to serious undergraduate students, regardless of their field of study. The volume covers everything from animal cognition to the state of current fisheries, from genetic modification to intersection animal activism. It is a resource designed for anyone interested in the moral issues that emerge from human interactions with animals.

Bob Fischer is Associate Professor of Philosophy at Texas State University. He is the author of *The Ethics of Eating Animals* (Routledge, 2019) and *Animal Ethics: A Contemporary Introduction* (Routledge, 2020).

ROUTLEDGE HANDBOOKS IN APPLIED ETHICS

Applied ethics is one of the largest and most diverse fields in philosophy and is closely related to many other disciplines across the humanities, sciences and social sciences. *Routledge Handbooks in Applied Ethics* are state-of-the-art surveys of important and emerging topics in applied ethics, providing accessible yet thorough assessments of key fields, themes, thinkers, and recent developments in research.

All chapters for each volume are specially commissioned, and written by leading scholars in the field. Carefully edited and organized, *Routledge Handbooks in Applied Ethics* provide indispensable reference tools for students and researchers seeking a comprehensive overview of new and exciting topics in applied ethics and related disciplines. They are also valuable teaching resources as accompaniments to textbooks, anthologies, and research-orientated publications.

ALSO AVAILABLE:

THE ROUTLEDGE HANDBOOK OF NEUROETHICS
Edited by L. Syd M Johnson and Karen S. Rommelfanger

THE ROUTLEDGE HANDBOOK OF THE ETHICS OF DISCRIMINATION
Edited by Kasper Lippert-Rasmussen

THE ROUTLEDGE HANDBOOK OF THE PHILOSOPHY OF PATERNALISM
Edited by Kalle Grill and Jason Hanna

THE ROUTLEDGE HANDBOOK OF THE ETHICS OF CONSENT
Edited by Peter Schaber and Andreas Müller

THE ROUTLEDGE HANDBOOK OF ETHICS AND PUBLIC POLICY
Edited by Annabelle Lever and Andrei Poama

THE ROUTLEDGE HANDBOOK OF ANIMAL ETHICS
Edited by Bob Fischer

For more information about this series, please visit: www.routledge.com/Routledge-Handbooks-in-Applied-Ethics/book-series/RHAE

THE ROUTLEDGE HANDBOOK OF ANIMAL ETHICS

Edited by Bob Fischer

Routledge
Taylor & Francis Group
NEW YORK AND LONDON

First published 2020
by Routledge
605 Third Avenue, New York, NY 10017
4 Park Square, Milton Park, Abingdon, Oxon OX14 4RN

First issued in paperback 2022

Routledge is an imprint of the Taylor & Francis Group, an informa business

Publisher's Note
The publisher has gone to great lengths to ensure the quality of this reprint but points out that some imperfections in the original copies may be apparent.

Library of Congress Cataloging-in-Publication Data
A catalog record for this book has been requested

ISBN 13: 978-1-03-247495-3 (hbk)
ISBN 13: 978-1-138-09506-9 (pbk)

DOI: 10.4324/9781315105840

Typeset in Bembo
by Apex CoVantage, LLC

CONTENTS

FIGURES

TABLES

BOXES

NOTES ON CONTRIBUTORS

Cheryl Abbate is Assistant Professor in the Philosophy Department at the University of Nevada, Las Vegas. She specializes in nonhuman animal ethics. Her publications include "A Defense of Free-Roaming Cats from a Hedonist Account of Feline Well-Being" in *Acta Analytica* (2019), "How to Help when it Hurts: The Problem of Assisting Victims of Injustice" in *Journal of Social Philosophy* (2016), and "Adventures in Moral Consistency: How to Develop an Abortion Ethic from an Animal Rights Framework" in *Ethical Theory and Moral Practice* (2015).

Elan Abrell is Senior Regulatory Specialist at The Good Food Institute and Visiting Assistant Professor in the Animal Studies Program at Wesleyan University. He is the author of many essays on animal sanctuaries.

Nekeisha Alayna Alexis is an independent scholar with wide-ranging interests in race/racism, human and animal oppression, intersectionality, co-liberation with other animals, and Christian ethics and theology concerning other animals. She is also Intercultural Competence and Undoing Racism coordinator at Anabaptist Mennonite Biblical Seminary in Elkhart, Indiana. She is involved in Vegan Michiana and organizing around other local social justice concerns.

Catherine E. Amiot is Full Professor of Psychology at the Université du Québec à Montréal. Her research is in the field of social psychology and human–animal relations. Her work has been funded by the Social Sciences and Humanities Research Council of Canada, the Australian Research Council, and the Québec Funds for Research on Health and Culture.

Jonathan Anomaly is Assistant Professor of Philosophy in the Department of Philosophy at the University of San Diego.

Raymond Anthony is Professor of Philosophy at University of Alaska–Anchorage. His publications are at the intersection of environmental, animal, climate, and food ethics and the philosophy of technology. He serves on the American Veterinary Medical Association's (AVMA) Animal Welfare Committee and is a co-author on the AVMA's Euthanasia, Humane Slaughter and Depopulation Guidelines.

Brock Bastian is Associate Professor in the School of Psychological Sciences at the University of Melbourne. He is trained as a social psychologist, and his research broadly focuses on the topics of ethics and well-being.

Maria Botero is Associate Professor of Philosophy at Sam Houston State University. Her recent publications focus in animal cognition, primatology, and ethics of animal experimentation.

Bernice Bovenkerk is Associate Professor of Animal and Environmental Ethics at Wageningen University. Her work focuses on animal agency, moral status of animals, animal experimentation, ethics of biotechnology, domestication, climate ethics, and deliberative democracy. She has been a member of an animal experimentation committee, an alternative to animal testing committee, and a scientific assistant of the Dutch Committee for Animal Biotechnology. She is editor of *Animal Ethics in the Age of Humans* (2016).

Donald W. Bruckner is Associate Professor of Philosophy at Penn State University, New Kensington. His recent publications not only focus on human well-being but also include contributions to the applied ethics literature, especially food and animal ethics.

Mark Budolfson is Assistant Professor at the University of Vermont, and works on interdisciplinary issues in public policy, economics, and ethics, especially in connection with sustainable development, planetary health, and collective action problems such as climate change and other dilemmas that arise in connection with common resources and public goods. He is coauthor of *Food, Ethics, and Society* (2016) and co-editor of *The Oxford Handbook of Food Ethics* (2018) and *Philosophy and Climate Change* (forthcoming).

Larry Carbone is an independent scholar of animal ethics and welfare. Prior to his retirement, he was a veterinarian and Director of the Animal Care and Use Program Office at the University of California San Francisco. He is the author of *What Animals Want: Advocacy and Expertise in Laboratory Animal Welfare Policy* (2004).

Janet M. Davis is Professor of American Studies and History at the University of Texas at Austin. Her most recent book is *The Gospel of Kindness: Animal Welfare and the Making of Modern America* (2016).

David DeGrazia is Elton Professor of Philosophy at George Washington University and Senior Research Fellow in the Department of Bioethics at the National Institutes of Health. His animal-related books are *Taking Animals Seriously: Mental Life and Moral Status* (1996), *Animal Rights: A Very Short Introduction* (2002), and, with Tom Beauchamp, *Principles of Animal Research Ethics* (2019).

Tim S. Doherty is Alfred Deakin Post-doctoral Research Fellow at Deakin University, Australia. He conducts applied research in the fields of wildlife ecology and conservation biology, aiming to inform environmental policy and management.

Sara Dubois is Adjunct Professor in Applied Biology at the University of British Colubmia and serves as Chief Scientific Officer at the BC Society for the Protection of Cruelty of Animals (SPCA), the largest animal welfare organization of its kind in North America. She is the lead author among 20 esteemed coauthors of the *Conservation Biology* (2017) article "International Consensus Principles for Ethical Wildlife Control" and contributed a chapter on bear tourism to the book *Tourism and Animal Welfare* (2018).

Max Elder is Research Director in the Food Futures Lab at the Institute for the Future, the world's leading futures-thinking organization. He is also a fellow at the Oxford Centre for Animal Ethics, the world's first academy dedicated to the ethical enhancement of the status of animals.

Karen S. Emmerman is an independent scholar and part-time faculty in the Department of Philosophy and the Comparative History of Ideas program at the University of Washington. She is also Philosopher-in-Residence at John Muir Elementary School in Seattle. Karen writes on ecofeminism, animals ethics, and philosophy for children.

Erin M. Evans is Assistant Professor of Sociology at San Diego Mesa College. She specializes in social movements and institutionalization, interrogating incrementalist approaches to policy reform and identifying the effects of institutionalizing activists' demands through laws and policies. She is currently Chair-Elect of the Animals & Society Section of the American Sociological Association, and her work can be found in *Social Movement Studies* (2016, 2019), *Sociological Perspectives* (2016), *Society & Animals* (2010), and edited volumes such as *The Handbook for Political Citizenship* (2014), and *Animals in Human Society* (2015).

Catia Faria is Postdoctoral Researcher at the Centre for Ethics, Politics and Society at the University of Minho. She works in normative and applied ethics, in particular in how they relate to the consideration of nonhuman animals, discrimination, gender, and sexuality.

Katherine C. Grier is Professor of History at the University of Delaware. She is the author of *Pets in America: A History* (2006). A specialist in American material culture and the history of animal–human interaction, her current research is on fancy breeding and showing of small animals.

Joanna Grossman works as a lobbyist for and leads the equine program at the Animal Welfare Institute in Washington, DC. She was previously the organization's federal policy advisor, advocating for a wide range of animal protection issues on Capitol Hill. Her writings and work have been featured in or covered by numerous media outlets including The Hill, Roll Call, Politico, NBC News, Newsweek, the Daily Caller, and the Washington Examiner.

Markus Gusset served as Chief Conservation Officer for the World Association of Zoos and Aquariums (WAZA) and published extensively on wildlife conservation, population management, environmental education, and animal welfare. He is now with the International Affairs, Research and Innovation Directorate at the Swiss Federal Office for Agriculture.

John Hadley is Senior Lecturer in Philosophy in the School of Humanities and Communication Arts at Western Sydney University. He is the author of *Animal Neopragmatism: From Welfare to Rights* (2019) and *Animal Property Rights: A Theory of Habitat Rights for Wild Animals* (2015).

Kristina M. Horback is Assistant Professor in the Animal Science Department of University of California, Davis. Her research combines methods of experimental psychology, comparative cognition, and applied ethology to identify personality traits and enhance assessment of affective states among nonhuman animals (domestic livestock and captive wildlife).

Oscar Horta is Professor of Philosophy at the University of Santiago de Compostela, in Spain, and a founding member of Animal Ethics. His research and advocacy focus on speciesism and wild animal suffering.

Lisa Kemmerer is Professor of Philosophy and Religion at Montana State University Billings and the founder of Tapestry: Women's Institute of Integrated Justice. She is an award-winning author who has written nine books, including *Eating Earth* (2014), *Animals and World Religions* (2012), and *Sister Species* (2011).

Mimi E. Lam is Marie Curie Fellow at the Centre for the Study of the Sciences and the Humanities, University of Bergen in Norway. She specializes in the human dimensions of fisheries, with emphasis on marine resource values, policy, and governance. Seafood ethics is the topic of her current Marie Curie fellowship and forthcoming monograph.

Shelby Elaine McDonald is Assistant Professor in the Virginia Commonwealth University School of Social Work and a research affiliate with the VCU School of Medicine Center for Human–animal Interaction. Dr. McDonald's program of research centers on understanding the potential risk and protective effects of human–animal interaction on relations between childhood adversity (i.e., abuse, neglect) and child health and development.

Steven McMullen is Associate Professor of Economics at Hope College in Holland, Michigan. He is the author of *Animals and the Economy* (2016).

James McWilliams is Professor of Practice in the Department of History. His most recent book is *Eating Promiscuously: Adventures in the Future of Food* (2017).

Joy A. Mench is Emeritus Professor of Animal Science at the University of California, Davis. She is a zoologist whose work focuses on the behavior and welfare of animals in captive and managed settings. She has authored or edited several books, most recently *Animal Welfare* (2011), *Advances in Poultry Welfare* (2018), and *Advances in Agricultural Animal Welfare* (2018).

Hamish Morrin is Lecturer in Animal Science at Capel Manor College, London. His interests are in the welfare and ethics of wild animals in captivity.

F. Bailey Norwood is Barry Pollard, MD/P&K Equipment Professor of Agribusiness in the Department of Agricultural Economics at Oklahoma State University. His most recent book is *Meet the Food Radicals* (2019).

Ben J. Novak is a researcher and advisor for Revive & Restore, a conservation non-profit organization dedicated to advancing innovation and adoption of biotechnologies for genetic rescue applications for wildlife and ecosystems.

Christine Overall is Professor Emerita and holds a University Research Chair at Queen's University in Kingston, Ontario. She is the editor of *Pets and People: The Ethics of Our Relationships with Companion Animals* (2017).

Clare Palmer is Professor of Philosophy at Texas A&M University. She is the author of *Animal Ethics in Context* (2010) and coauthor of *Companion Animal Ethics* (2015)

John M. Pearce is Professor at the School of Psychology, Cardiff University, and Honorary Professor at the Department of Psychology, University of Sydney. He is the author of *Animal Learning and Cognition: An Introduction* (2008).

Govind Persad is Assistant Professor of Health Law at the Sturm College of Law at the University of Denver. His research interests center on the legal and ethical dimensions of health insurance, health care financing (both domestic and international), and markets in health care services, as well as professional ethics and the regulation of medical research.

Jessica Pierce is a bioethicist and Affiliate Faculty at the Center for Bioethics and Humanities, University of Colorado Anschutz Medical School. She is the author of 10 books on animal ethics and environmental bioethics, including *Run, Spot, Run* (2016) and *The Last Walk* (2012).

Lonneke Poort is Assistant Professor at Erasmus School of Law, working in the fields of legal theory, legal ethics, and methodology. Her research deals with processes of decision-making and lawmaking around controversial issues with a strong moral impact, such as animal biotechnology. She is coeditor of *Symbolic Legislation Theory and Developments in Biolaw* (2016) and is a member of the subcommittee on ethical and societal aspects of genetic modification of the Netherlands Committee on Genetic Modification.

Pandora Pound is Research Consultant at Safer Medicines Trust, United Kingdom. Her work explores evidence relating to the clinical translation of animal research, and the emergence of pre-clinical, human-focused technologies. She has recently contributed a chapter on the clinical translation of animal models of traumatic brain injury to *Neuroethics and Nonhuman Animals* (forthcoming).

Mark Rowlands is Professor of Philosophy at the University of Miami. His most recent publication, *Can Animals Be Persons?* (2019), explores the issue of animal personhood. He has written extensively in the areas of philosophy of mind, moral psychology and ethical areas pertaining to nonhuman animals and the environment.

James C. Russell is Associator Professor in Biological Sciences and Statistics at the University of Auckland, New Zealand. He works throughout the world on remote islands to provide conservation solutions by applying a combination of scientific methods.

Peter Sandøe is Head of Section, Consumption, Health and Ethics in the Department of Food and Resource Economics at the University of Copenhagen. He has published more than 140 papers in international journals with peer review, six international books, more than 300 other academic contributions, and more than 230 contributions aimed at a broader audience.

Jeff Sebo is Clinical Assistant Professor of Environmental Studies, Affiliated Professor of Bioethics, Medical Ethics, and Philosophy, and Director of the Animal Studies MA Program at New York University. He works primarily on bioethics, animal ethics, and environmental ethics. He is coauthor of *Chimpanzee Rights* (2018) and *Food, Animals, and the Environment* (2018), and is working on *Why Animals Matter for Climate Change* (forthcoming).

Adam Shriver is Research Fellow at the Oxford Uehiro Centre for Practical Ethics and the Wellcome Centre for Ethics and Humanities. He works at the intersection of ethics and the cognitive sciences and is a coauthor of the forthcoming book Neuroethics and Nonhuman Animals.

Dean Spears is Assistant Professor of Economics at the University of Texas at Austin and a visiting economist at the Indian Statistical Institute. He is the author of *Where India Goes: Abandoned Toilets, Stunted Development, and the Costs of Caste* (2017), winner of the Joseph Elder Prize in the

Indian Social Sciences, and of *Air: Pollution, Climate Change, and India's Choice between Policy and Pretence* (2019).

Nancy Staus is Senior Researcher at the Center for Research on Lifelong STEM Learning at Oregon State University. Formerly a conservation biologist, she now studies how, why, and where people learn science, and is particularly interested in the role of emotion in science learning and understanding. She has contributed chapters to *Large Mammal Restoration* (2001) and *International Handbook on Ecotourism* (2013).

Jessica du Toit is a PhD student in the Philosophy Department at Western University in Ontario, Canada. Her primary research interests are in Animal Ethics, Medical Ethics, and the intersection of these two areas of philosophy.

Gary Varner is Professor of Philosophy at Texas A&M University. His publications include *Personhood, Ethics, and Animal Cognition: Situating Animals in Hare's Two-Level Utilitarianism* (2012) and *In Nature's Interests?: Interests, Animal Rights, and Environmental Ethics* (1998).

INTRODUCTION

Reflecting on Animals

We eat animals. We wear their skins. We do experiments on them. We put them in zoos. We breed them to be pets. We regard them as nuisances that need to be managed.

Maybe all this is fine. But it's worth noting that in each of the cases just mentioned, animal interests are being sacrificed in service of human interests. At least on the face of it, it isn't good for animals to be eaten, or to be turned into belts, or to be the subject of experiments, and so on. Indeed, it often seems to be pretty bad for them. But since it's good for us to use them in these ways, we do.

Serious moral reflection about animals should begin by acknowledging this mismatch between our interests and the interests of animals. It may turn out, of course, that there is nothing wrong about the way that we use nonhuman animals. (I say "nonhuman" animals because we, of course, are animals too.) But we shouldn't pretend that we are living in a world where animals are generally better off as a result of their interactions with us. Most the time, this isn't because we are *trying* to be cruel. It usually isn't because we don't care at all about animal welfare. Rather, it's because we're trying to address more or less pressing human interests—economic, nutritional, aesthetic, medical, and so on—which happen to conflict in significant ways with what's good for animals.

This book explores these conflicts. There are essays on what it would take to run more ethical zoos, the moral complexities of bringing woolly mammoths back from the grave, surveys of the environmental problems associated with intensive animal agriculture, and discussions of the ethics of domestication. And it doesn't explore them in the usual way. If you were to pick up a different handbook of animal ethics—which you should, as there are others that are very useful—you would largely find the work of philosophers. In this book, however, you'll find at least as many historians, sociologists, and animal scientists. That's because I don't think that philosophers have a corner on the animal ethics conversation. Ethics is really hard, and when philosophers do it, we tend to abstract away from some of that complexity, thinking about simplified, toy examples, or getting lost in debates about how things ought to be in an ideal world, rather than on the incremental changes that we need to make in the short run. That's well and good, and philosophers should keep doing it. I'll be the last to say that those conversations are unimportant. However, those conversations aren't the only ones worth having, and my hope is that by bringing together scholars from diverse fields, the discussion can be richer for all involved. Moreover, different fields take different things for granted when they discuss animals, and it's valuable to put those differnces side by side. We can learn something about the shape of each discipline by seeing how its representatives frame the moral problems associated with animals.

You'll need a bit of background to engage with the contributions to this book: many of them presuppose that you have some familiarity with the major theoretical issues in animal ethics. However, this may be your first time thinking about animals in a systematic way. So my aim in this introduction is to give you a quick overview of the ideas, arguments, and major moral theories that have shaped discussions about animal ethics for the last 45 years. Before doing that, however, I need to say just a bit about what ethics is and isn't. So, let's get to it.

Ethics: What It Isn't; What It Is

It's really hard to give a definition of "ethics." It's so hard, in fact, that I don't even try to do it when I teach ethics courses. Instead, I just say something like this. We all know that there are some things that you shouldn't do. You shouldn't lie. You shouldn't steal. You shouldn't kill people, and so on. After just a bit of reflection, however, it's obvious that these principles admit exceptions. Sure, you shouldn't lie. But if you're in Germany in 1942, and you're hiding Jews in your attic, and the Nazis are at your door asking whether you have seen any Jews, then you should definitely lie. Obviously, you shouldn't kill people. But if someone attacks you with no provocation, and your only means of self-defense is to use a deadly weapon, then use it you may. Once we recognize this, we quickly realize that the principles of common sense—don't lie, don't steal, don't kill people—aren't even meant to be exceptionless principles. Instead, they are more like rules of thumb: they are very good guides to what you ought to do most of the time but very bad guides to what you should do all the time.

This invites us to ask: Why should we do the things that we ought to do? In other words, what are the deeper principles that explain why the right things are right and the wrong things are wrong? These questions invite us to do some moral theorizing. That is, they invite us to start developing hypotheses about those deeper principles. Is the ultimate truth about morality that we ought to make the world the best place we can? Is it that we ought to do our best to respect others? Is it that we ought to cultivate character traits that allow us to live well? Maybe ethics is ultimately about caring for others or pursuing the end of oppressive structures or enabling people to live out their own visions of the good life? And, of course, if we find ourselves tempted by any one of these ideas, how do we make it precise? How do we state, in unambiguous terms, exactly what we mean by an idea like, "We should make the world the best place we can"?

As we pursue this project, we also have to recognize that history is littered with examples of mistaken moral beliefs. You don't have to go very far back in time to find people who thought that slavery was not just morally okay but that it's actually mandated by God. Or people who thought that women shouldn't be able to choose whether to marry—their families should make those choices for them. Or people who thought that children ought to be beaten severely for what we would now regard as minor infractions. Or people who thought that animals exist purely for our pleasure, and so if people enjoy seeing them tear one another apart—as in the cases of bearbaiting and cockfighting—then there's nothing wrong with setting up such competitions.

There are two ways of reacting to this kind of radical moral disagreement. One is to get a sense of vertigo, accompanied by the thought that perhaps moral norms have no objective basis and are simply fabricated by particular cultures at particular times. This isn't the place to assess this reaction in any detail, but let me simply make the following observation about it. Even if this were correct, it wouldn't show that there is no cause for moral theorizing. Regardless of whether morality is somehow written into the fabric of the cosmos, it remains the case that we use moral norms to govern our interactions with one another. And since the world is becoming more interconnected, it's increasingly the case that we can't avoid those interactions. We have to find ways of balancing the interests of diverse groups, and that requires finding norms that people can accept despite their many differences. Those norms are going to have to be impartial, at least to some degree. They can't simply say that you matter if and only if you are a white landowning male. People simply won't accept a principle of that kind. So we need to find rules to live by that don't just serve your interests or mine. Instead, they need to be designed to be compelling to anyone who's willing to think hard about ethics for a diverse world. This obviously isn't easy, and I'm certainly not saying that we'll get there without an enormous number of false starts. But the take away here—the really crucial point—is that even if you're skeptical about *objectivity* in ethics, in the sense that you don't think that there are any transcendent, mind-independent facts about what's morally right and wrong, you can still think that ethics is very important and that we've got very good reasons to keep doing it. If it turns out that we invented right and wrong, then it's worth asking why we did it. Odds are, the answer will help explain why ethics is still worth thinking about today.

But let's set that skeptical view aside. Instead, let's think about another way you might react to that massive difference between the views we have now and the views of our ancestors: namely, you might become a bit humbler. Human beings have made some pretty big mistakes when it comes to morality: we have believed many things that weren't simply incorrect; they were horrifically and terribly incorrect, not least because those false beliefs are responsible for untold amounts of suffering. So as we do our moral theorizing, we should be very sensitive to the possibility that we've made very significant errors in our moral judgments. It does seem very plausible that human lives are more important than animal lives, basically across the board. But it also once seemed plausible that white lives were uniformly more important in black lives, that gay sex was an abomination, and that husbands had every right to force their wives to have sex with them. The fact that something seems true to you means it's a good place to start in our moral theorizing. It doesn't mean that it's always a good place to end. We've made mistakes in the past, and we're probably making mistakes now. One of the goals of moral theorizing is to figure out where those mistakes are. Which of our moral beliefs can survive scrutiny? Which ones can't? Difficult questions, of course, but not ones to avoid asking.

Now we come to the methodology that many ethicists employ. We start with some ordinary observations about what seems true, morally speaking. As I mentioned before, it seems pretty clear that you shouldn't lie, steal, or kill people. It also seems pretty clear that there are some exceptions. And so we take these facts and start generating hypotheses about the moral principles that would, if true, explain them. We then apply those principles to new cases. What do those principles say about the rights of women or the legitimacy of different sexual practices or about how the powerful can treat the weak? And then we begin a complicated process of (1) revising principles, whether by refining them or generating new ones, when they conflict with our considered moral judgments and (2) revising our moral judgments when the principles seem more plausible.

This is a difficult process, and there is an unavoidable risk that our biases will lead us to endorse the conclusions that we want to be true, rather than those for which we can offer the best arguments, all things considered. But that's why moral theorizing isn't something we do alone, in the privacy of our homes, cut off from the rest of the world. Instead, it's something we do together, in public, with as many different sorts of people as possible. We need to have our biases checked, and the best people to do that are the ones who aren't like us in various ways. (Not incidentally, this is another reason why I think it's important to have psychologists and conservation biologists in the same book with philosophers.)

What's the upshot of all this? Here's one way to think about it. Ethics—at least as philosophers conceive of it—isn't just what your grandma says you should do, or what the law happens to say right now, or what most people in the society happen to believe, or what some people used to believe at one time or another. Instead, all these things are resources for doing ethics. Ethics is the business of thinking hard about how to fit together all our ideas about what's good and bad, right and wrong, morally important and morally insignificant. It's an attempt to come up with principles that make sense of our moral judgments. Inevitably, our views won't fit together perfectly, and so we will have to do a lot of work, figuring out exactly which should change and why. This is something we do in conversation with others, checking our biases and recognizing that the aim is rational persuasion: we are looking for conclusions that others can accept and take seriously, not just ones that we happen to like, given our own position in the world and its accompanying privileges.

As you might imagine, there is so much more to be said here. In a different book, I might well say it. But for now, this very brief primer will have to do. In the next section, I start to explore animal ethics specifically. How have people thought about animals in the past, and what are some of the challenges for those views? And now, on reflection, what sorts of principles should guide our interactions with animals? As we will soon see, these turn out to be questions with fascinating answers.

The Moral Community

Which beings matter morally? In other words, of all the things there are in the world, which ones are such that we ought to factor them into our moral deliberations, whether we want to or not? Equivalently, which beings *deserve* moral consideration, quite independently of whether we have some desire to protect their interests? These are all different ways of asking about the scope of the moral community—the group of beings who count, morally speaking.

It may help to think about the legal parallel here (although we will need to note some differences as well). The law recognizes a difference between *citizens* and *noncitizens*, and it assigns different rights to them accordingly. Citizens get some very significant rights that noncitizens don't. Moreover, the citizen/noncitizen distinction isn't arbitrary: there's some explanation for why any particular person is or isn't a citizen. The US, for instance, has birthright citizenship: if you were born in the US, then you're a citizen. This means that your being born here is what *explains* why you're a citizen, or your *not* being born here (and your not having been naturalized, etc.) explains why you aren't one.

Membership in the moral community is a bit like citizenship. If you are a citizen of a country, then you have a certain privileged status in that country. If you are a member of the moral community, then others ought to factor your interests into their deliberations. If you *aren't* a citizen of a country, you still have some rights, although not as many. However, if a thing isn't a member of the moral community, it just doesn't matter at all. For instance, you, dear reader, definitely have moral value and so are a member of the moral community. By contrast, it seems very plausible that grass clippings aren't members of the moral community: they have no moral value in and of themselves. (Some unusual person might be obsessed with grass clippings, but it's hard to believe that you'd be making a mistake if you didn't give the clippings a second thought.) Finally, just as there is some explanation for your being or not being a citizen, there is an explanation for membership or nonmembership in the moral community. The fact that you matter morally isn't arbitrary; there's something about you that explains why you deserve moral consideration. Likewise, we should be able to explain why grass clippings don't matter. Presumably, grass clippings lack some important feature that's shared by beings who deserve moral consideration. We will consider some candidates for that feature shortly, but before we do, we need to make sure that we're on the same page about the idea of the moral community.

We can draw a distinction between having obligations *to* something and having obligations *regarding* it. Here's the idea. If I own some firewood, then I can do whatever I want to it. For instance, if I want to burn it, I'm well within my rights to do so. But I could also paint pictures of owls on it or carve wooden toys out of it or sell it to someone in Missouri. But just because *I* can burn it, it doesn't follow that you can. There are constraints on how you relate to that firewood. Unless I give you permission, you can't burn it—or paint it or carve wooden toys out of it or truck it off to a different state. Obviously, there is nothing intrinsically wrong with any of these actions. Painting, carving, and moving firewood are all perfectly fine things to do, as long as you own the firewood in question. So neither you nor I have any obligations *to* the firewood. It's the kind of thing that we can, in principle, treat in any way we like, up to and including destroying it. But you have some obligations *to me* that *regard* that firewood. You're obligated *to me* not to sell it to anyone in Missouri, since you don't own it. Your obligation *pertains* to the wood, but it isn't *to* the wood.

Questions about the scope of the moral community are questions about the list of beings *to whom* we have obligations—not the list of beings *regarding which* we have obligations. We can have obligations regarding all sorts of things that don't matter in and of themselves, instead mattering only because someone else owns them or cares about them or has some interest in their protection. But the members of the moral community? Those beings matter in and of themselves.

So which beings are those? Here's one thought: you matter morally if you're a human being; otherwise, you don't. But this thought isn't terribly plausible. Imagine a stray dog, one that no one cares about. You come home from work after a hard day and decide that you need to take a walk to clear your head. As you're out there, fuming about something that happened in the office, this stray comes up to you. And to relieve some frustration, you give him a solid kick in the ribs, taking some pleasure in hearing them crack. Have you done anything wrong?

Surely you have. But the question is, Why? Nobody cares about that particular dog. It doesn't matter to anyone whether that dog lives or dies. So why shouldn't you break his ribs if that makes you feel better? Why shouldn't you take out your frustration on him? A very plausible answer is you shouldn't do it because *it's very bad for the dog*; it causes him pain. It's almost too obvious to say, but let's be explicit: getting kicked would *hurt* the dog, and that's the reason not to kick it.

If you agree, then you seem to be committed to saying that the stray dog is a member of the moral community. After all, you are saying that it doesn't matter whether anyone cares about the stray dog: you've got a good reason not to kick him simply because it would be bad for the animal were he to be kicked. The dog matters in and of himself. We have, then, a simple argument for expanding the moral community beyond the human. Are human beings the only members of the moral community? No: because if that were true, dogs would be out. But dogs are in, so the moral community is larger.

You might be perfectly satisfied with the idea of recognizing a larger moral community, one including both human and nonhuman animals. But historically, many people have balked at this idea. They've thought that people matter in and of themselves, but animals matter only insofar as we vest them with importance. Dogs? Those are important because we (Westerners) think of them as pets, and pets shouldn't be injured just to relieve some stress. Chickens? Those are for eating, and so it's fine to kill them to satisfy our preferences.

The basic problem for this view goes as follows. If we are going to say that all human beings are members of the moral community and that the moral community is limited to human beings (i.e., we are the only members), then we need to find some feature (1) that we all share, since all humans are in; (2) that no animal possesses, since all animals are out; and (3) that has the ability to explain why some beings matter. It's hard to satisfy the first requirement because human beings differ so much. Think about the really important differences between second-trimester fetuses, normal adults, dementia patients, and those in permanent vegetative states. It's hard to satisfy the second requirement because there's so much overlap between species in their characteristics. Human beings aren't the only rational, tool-using beings; we aren't the only ones with linguistic abilities, complex social orders, and rituals for mourning the dead. It's hard to satisfy the third requirement because the properties that we *do* seem to share, and that don't seem to be shared by any other animal, don't seem to be up to the task of explaining why we have this special status. Granted, all and only human beings have a certain kind of genome, but that isn't saying much: it's roughly equivalent to "All and only human beings are human beings." Moreover, having a certain kind of genome doesn't seem like a good explanation for being a member of the moral community. It seems that what matters are the characteristics that certain genomes produce — such as having the capacity to feel pain, or being an experiencing subject of a life — rather than the kind of genome per se.

So let's run with the idea that we shouldn't try to draw the boundaries of the moral community so narrowly; animals ought to be included in it. If we go this route, then we have to ask, Why not expand the moral community even further? Why, for instance, shouldn't we include all living things? To answer this question, imagine a weed growing up in the middle of an empty lot that happens to be owned by the city. Again, you've had a long day, you're really frustrated, and you go for a walk. As you pass by the lot, you noticed the weed and decide to pull it up, tossing it in the trash afterward. Did you anything objectionable? I think most people would say: No, you didn't. That weed isn't a member of the moral community. It's the kind of thing where you might have obligations *regarding* it—if, for instance, someone had owned the lot, as opposed to it being vacant public property—but in this case, you have neither obligation regarding it nor to it. It doesn't need to factor into your moral deliberations.

So far, so good. The next step is to ask why both people and dogs are in the moral community while weeds are out. Jeremy Bentham offered a famous hypothesis in 1789, when he wrote this:

> The day may come, when the rest of the animal creation may acquire those rights which never could have been withholden from them but by the hand of tyranny. The French have already discovered that the blackness of the skin is no reason why a human being should be abandoned without redress to the caprice of a tormentor. It may come one day to be recognized, that the number of the legs, the villosity of the skin, or the termination of the os sacrum, are reasons equally insufficient for abandoning a sensitive being to the same fate. What else is it that should trace the insuperable line? Is it the faculty of reason, or, perhaps, the faculty of discourse? But a full-grown horse or dog is beyond comparison a more rational, as well as a more conversable animal, than an infant of a day, or a week, or even a month, old. But suppose the case were otherwise, what would it avail? the question is not, Can they *reason*? nor, Can they *talk*? but, Can they *suffer*?[1]

According to Bentham, what distinguishes the members of the moral community from all other beings? Answer: they are *sensitive* beings—that is, they're *sentient*, having the capacity to feel pain. In other words, when beings don't like being harmed, we ought to consider their interests before harming them; when beings don't care whether they are harmed or not—because they don't have cares at all—we may still have obligations regarding them but don't have any obligations to them.

Some people will object to this view: they'll say that we were too quick when we were thinking about the weed. It isn't that the weed doesn't matter at all: it's that the weed doesn't matter very much, to the point that it's easy to think it's morally irrelevant. The idea here is that moral *relevance*—that is, membership in the moral community—is one thing; moral *importance* is another. The morally relevant beings are just the ones that you ought to think about, whether or not you care about them. Moral importance, by contrast, is a property had in varying degrees by beings who are morally relevant. Some morally relevant beings are not very important at all: it doesn't take much to justify harming them. Others are extremely important, and it takes a great deal to justify harming them. Obviously, we need to say a lot more about those factors that influence moral importance, and we will explore that later. For now, however, the point to note is that someone could have a thought along these lines: the weed matters in and of itself—it's a member of the moral community—but it isn't a terribly important member of the moral community.

And why might we think that the weed has this status? One way of reaching this conclusion is by thinking about other nonsentient organisms that seem to be valuable, like redwoods, hyacinths, and Venus flytraps. These remarkable organisms all seem to be morally important—you shouldn't destroy them for fun—and if that's right, then maybe they're members of the moral community too. But they're also *less* important than a human being—if it's a stranger or the hyacinth, please pick the former—and *more* important than mere weeds. So they provide further confirmation of the "relevance is one thing, importance is another" hypothesis. So maybe it was a mistake to draw the line at sentience, and we should instead draw the line at life. Perhaps all living things are members of the moral community, and nonliving things are the outsiders. Then, we should say that not everything matters as much as everything else: the moral community is *hierarchical.*

Let's set aside the hierarchy issue for a bit: we'll return to it when we discuss moral theories. For now, let's focus on the argument that got plants into the moral community. One problem is that it's easy to generalize that way of thinking. Google "Antelope Canyon," a stunning slot canyon in Page, Arizona, and scroll through a few pictures. These remarkable formations were created by ancient rivers, now long gone, and they are undeniably beautiful. Wouldn't it be wrong to destroy Antelope Canyon for fun? Of course. But is it because the canyon is a member of the moral community? That's less obvious. Maybe this is one of those cases where we have obligations *regarding* something but not *to* it. After all, the canyon doesn't care what we do to it!

Recall, though, that plants don't care what we do to them either. Granted, they *react* to what we do to them, but they don't have minds; they don't get upset that they're being harmed, even if, as self-regulating systems, they respond to damage. By contrast, conscious beings can suffer; they care about how things go for them in a way that plants simply don't. Again, plants have interests—things go well for them when they have enough water and badly when they don't, and so forth—but they don't feel anything when these interests go unsatisfied. There isn't anything that it's like to be a plant that doesn't get enough water, whereas there is something that it's like to be a dog that doesn't get enough water. That difference seems crucial, and it makes the presence or absence of consciousness seem exceptionally important.

Moreover, the view that life is sufficient for membership in the moral community—*biocentrism*—faces another problem. What, exactly, makes something alive? Crystals, for instance, can grow, respond to stimuli, and achieve a kind of internal equilibrium. Do they have moral standing too? What about ecosystems or other ecological complexes? There are senses in which they, too, can act in these ways. Obviously, there are ways of restricting the definition of life, but then there will just be examples of things that *seem* to be alive—like viruses—that may not satisfy our more stringent criteria. Based on such problems, some biologists actually argue that it's a mistake to try to define "life," and that we should accept that there are simply different kinds of living things, with no features common to all. That may be good biology, but it makes for messy morals, as we're left without a clean way of determining which beings matter in and of themselves. Instead, we just have to look to see what biologists say for their own purposes—which almost certainly aren't the ethical purposes that drive our inquiry.

For these reasons, many people have concluded that consciousness is indeed the line between those beings who are and aren't members of the moral community. If we go this route, then we are immediately faced with the difficult question of determining which beings, exactly, are conscious. I can't pursue this question in any detail here, but I can suggest how we might begin to answer it.

Which Beings Are Conscious?

At this point, most people take for granted that most animals are conscious. Cats and camels and crows? All beings who can feel, all beings who are aware of the world around them. But René Descartes (1596–1650), a French philosopher and scientist, thought that animals are automatons, unconscious machines that behaved as though they were in pain, although, in fact, they felt nothing. And even now, there are those who think that consciousness requires cognitive sophistication that very few nonhuman animals have—if any at all. Indeed, one of the contributors to this book, John Pearce, an expert on animal intelligence, believes that almost all animal behavior can be explained via associative mechanisms, requiring no conscious processing of information.

However, this is certainly the minority view, and most people take the crucial question to be about *which* nonhuman animals are conscious rather than *whether* most are. How do we go about answering this question? There is no simple or straightforward test that we can employ. It isn't as though we know that consciousness is produced by one particular pattern of neural activation, so the question of whether a being is conscious simply becomes the question of whether they have the relevant neural structures and activation patterns. Thankfully, there are other strategies available to us.

The first is to look for behavioral evidence. Does this organism behave in ways that would be difficult to explain without consciousness? For instance, is there evidence that the organism is forming representations of the world around it—perhaps using them to navigate, or catch prey, or what have you? Second, we can look for morphological similarities between human beings—critters that, presumably, we know to be conscious—and various nonhuman animals. For instance, do they have brains that are organized more or less like ours? Or, failing that, do they have structures that seem to perform the same functions, integrating and processing information from different sensory modalities? Third, we can think about various evolutionary considerations. For instance, what's our evolutionary relationship to the organism in question? Are we fairly closely related, or was our last common ancestor much further back? Also, we might wonder about the selection pressures on a given type of organism. How would consciousness benefit the members of the species, and what would the cost be of having it? Could organisms like this navigate their environment using simpler forms of information processing, in the way that current artificial intelligence operates? Or not?

Obviously enough, some of these questions are much easier to answer than others, and there is bound to be reasonable disagreement in some cases. At this stage, it's deeply implausible that chimps aren't conscious: the evidence of their sophistication is extraordinary. When it comes to cockroaches, however, things are much dicier. They can, for instance, learn to avoid potential sources of damage, but it isn't clear what to make of this. In any case, there is plenty of evidence that the beings that will largely be the focus of this book—the animals we farm for food, the animals we keep in zoos, the animals we keep our homes—are indeed conscious. So, we will have to deliberate carefully about how we ought to relate to them.

Moral Theory

This brings us to the final topic that we need to cover in this introduction—although, I hasten to add, it isn't one that we will be able to get through quickly. In short, the issue is this. Suppose we agree that consciousness is the dividing line between those beings who are and aren't members of the moral community. And suppose we agree, at least in the majority of cases, about those beings who are and aren't conscious. Still, we haven't said much at all about how we actually ought to act. Granted, chickens and chimpanzees are morally important. But how important? And what principles should guide our interactions with them? There are many proposals that we could discuss here, and I can't do justice to them all in the space available. This isn't a general introduction to ethics, which would have to be far more comprehensive. Instead, I'm only going to be able to sketch the outlines of the most prominent theories that animal ethicists have developed over the last 50 years. (For some others, see the Suggested Readings section.)

Consequentialism

Perhaps the most famous and familiar proposal is *consequentialism*. According to this moral theory, action is morally right—it's morally mandatory or morally obligatory; it's what you must do, morally speaking—just in case it produces the most good. It's wrong otherwise. On this view, then, the right course of action is the one with the best effects, the one that produces the best results. But which results count? Consequentialists don't agree about this, but one tribe within consequentialism gives a particularly notable answer: *utilitarians* say that an action is morally required if it produces the most utility overall. We can think of utility as representing a unit of happiness, a way of measuring the amount of happiness that an action produces. To be clear, "happiness" in this context doesn't simply refer to the feeling you get when you taste good chocolate; it's supposed to be a much broader notion, encompassing all the ways that you can be psychologically better off. So there is a kind of happiness associated with eating chocolate, but there are also the various kinds associated with playing with your child, finishing a particularly complex math problem, and taking a quiet walk in the woods. These are all very different pleasures, but they are no less real for that. (Also, and as you might imagine, there are lots of philosophical complexities associated with measuring happiness, but we aren't going to worry about them here. There is an entire field of research—positive psychology—that deals with such puzzles, many of which can be addressed.)

For present purposes, what matters is that utilitarianism says we shouldn't care about anything *other* than maximizing happiness. Notice that there is no mention here about whose happiness we ought to maximize: it's just happiness impartially considered. On this sort of view, your happiness is no more important than a mouse's happiness, which is no more important than an elephant's happiness. The theory is thoroughly egalitarian in this respect.

Utilitarianism has many attractive features:

1. As just mentioned, the theory is thoroughly egalitarian. Everyone counts for one, and not more than one. This, of course, leads to some radical consequences, but in a world full of deeply partial beings, an impartial moral view might be particularly attractive: it corrects our tendency to favor the near and dear, insisting that no one matters more than anyone else. (Recalling the discussion from earlier, utilitarianism rejects any hierarchy in the moral community. Plants and slot canyons are out, animals are in, and all animals are equal.)
2. In principle, there's a straightforward way to assess whether a particular course of action is justified. Should you adopt a dog from your local animal shelter? Well, consider how things are likely to go—for you, the dog, and anyone else who might be affected—if you do and if you don't decide to adopt. It certainly takes some work to run this calculation, not least because it's difficult to be sure that you've considered all the potentially affected parties. But in principle, this is work we can do, and so the answers to our moral questions are within reach.
3. The theory provides a natural explanation of what's wrong about many practices that are increasingly seen as objectionable. For instance, what's wrong with doing painful research on (literal, actual) guinea pigs? In a great many cases, the odds of success are extremely low, the potential benefit to human beings would not be significant, and the suffering of the animals is extraordinary. In other words, it looks like the cost outweighs the benefits, and those are exactly the circumstances in which utilitarianism condemns an action.

Of course, the view also has its liabilities. For instance, many have objected that utilitarianism is an incredibly *demanding* moral theory. It's easiest to see this by noticing the category of action that doesn't figure at all in utilitarian thinking, namely, the category of *permissible but suboptimal* behavior. According to utilitarianism, you either do the best or you act wrongly, and the best thing you can do in a situation may often be a lot. Suppose, for instance, that things would be best overall if you, personally, were to stop eating animals. Admittedly, that probably wouldn't be best for you: animal products are tasty and, in moderation, nutritious; they are culturally important to many groups; and at least in the US, it's also much more convenient to consume them. However, it may well work out that any costs you incur are outweighed by the great benefit to the animals who would otherwise be harmed. If utilitarianism is true and the calculation does work out that way, then it's wrong for you to eat animals.

Many people would regard this as a demanding result. And you might think that that's a reason to reject utilitarianism. Before we accept that conclusion, however, we should note that our judgments about what's burdensome are going to be relative to what we regard as normal. In a world where you were used to stealing to get whatever you want, following the "Don't steal" rule would feel like it was asking a lot of you: all that stuff used to be free (to you, since you were just taking it), and now you have to pay! No fun. But obviously, it doesn't seem so demanding to us, who follow the "Don't steal" rule, to follow it. We've built lives that simply take for granted that it's wrong to run off with other people's things. Likewise, we might think that when utilitarianism tells us that we have "extreme" obligations, the problem isn't with the obligations, but with us. We're used to downplaying the interests of animals, so any pressure to play them up seems radical. If we are living in a morally broken world, however, maybe radical conclusions are what we need.

The point here isn't that we should be utilitarians or that we should endorse any particular conclusion about the ethics of eating animals. Those are important issues, of course, but not ones I can weigh in on here. Instead, the point is that we need to scrutinize the intuitions that we use both to support and criticize moral theories. None of them gets a free pass. Granted, if moral theories answer—at least in part—our intuitions about what is and isn't morally required, then utilitarians will need to do some work: they will need to explain why the tensions between utilitarianism and those intuitions are either (1) merely apparent or (2) such that the intuitions should give. But that may well be work that utilitarians can do.

The Rights View

A prominent alternative to utilitarianism is the rights view, which has been most thoroughly developed by Tom Regan. According to this theory, utilitarianism is objectionable because it regards individuals as mere "receptacles of value." In other words, according to utilitarianism, what's valuable about you is your experiential states, or your pleasures and pains. Those are the things that can be summed alongside the pleasure and pain of all other beings, thereby revealing what ought to be done. It's almost as though you—the individual, the container for your experiences—aren't yourself valuable. The rights view sees this as violating our intuitions about the worth of individuals. What matters, according to the rights view, is that individuals be recognized as having inherent worth, valuable in themselves, regardless of their usefulness to others. And because they have inherent value, they're entitled to a certain kind of treatment—namely, respectful treatment, which acknowledges their independent worth, and which forbids their being used merely for others' purposes. As Regan (2004, *xvi–xvii*) summarizes it:

> Some nonhuman animals resemble normal humans in morally relevant ways. In particular, they bring the mystery of a unified psychological presence of the world. Like us, they possess a variety of sensory, cognitive, cognitive, and volitional capacities. They see and hear, believe and desire, remember and anticipate, plan and intend. Moreover, what happens to them matters to them. Physical pleasure and pain—these they share with us. But also fear and contentment, anger and loneliness, frustration and satisfaction, calming and imprudence. These and a host of other psychological states and dispositions collectively help define the mental life and relative well-being of those (in my terminology) subjects-of-a-life we know better as raccoons and rabbits, beaver and bison chipmunks and chimpanzees, you and I.
>
> The basic moral right to respectful treatment places strict limits on how subjects-of-a-life may be treated. Individuals who possess this right are never to be treated as if they exist as resources for others; in particular, harms intentionally done to anyone subject cannot be justified by aggregating benefits derived by others . . . The rights view recognizes the equal inherent value of all subjects-of-a-life, including those who lack the capacity necessary for moral agency.

It's important to recognize that the rights view isn't committed to saying that nonhuman animals have every right that humans have. Suppose that human beings have a right to a basic education: it doesn't follow from this that monkeys have a right to a basic education, as they aren't harmed by not receiving the kind of education we offer. The idea is that each subject-of-a-life has a right to be respected as the kind of thing it is. Since all subjects-of-a-life can, by definition, experience pain, and they are harmed by that experience, they have a right not to have pain inflicted on them to benefit others. But not all subjects-of-a-life have an interest in voting (skunks come to mind), and so not all subjects-of-a-life need the right to vote.

The rights view also has a number of attractive features:

1. It captures the intuition that there are some bright moral lines that we just shouldn't cross. Suppose that we could make many people slightly better off by torturing one animal—for instance, we could bring some short-term pleasure to many people by letting them watch a cockfight, which usually ends with one rooster bleeding to death from an array of injuries. As long as there are enough people on one side of this moral equation, the utilitarian has to say that such an action isn't just morally permissible, but is rather morally required: it's what we ought to do, morally speaking. This is a hard pill to swallow. The rights view explains why such actions aren't permissible: they fail to respect the individual who would be tortured, using that one as a mere means to benefit others. That seems like the correct result.
2. The rights view seems to allow us to draw a distinction between those actions that we absolutely *ought* to do, morally speaking, and those actions that it would be *good* to do but that aren't morally required. According to the rights view, as long as you aren't violating anyone's rights, you have some freedom: you don't have to spend all your time trying to make the world the best place it can possibly be; you can have your own projects and pursue your own good. (The rights view does insist that you ought to help those whose rights have been violated, but it doesn't necessarily require you to give up everything for them. By contrast, if utilitarianism allows you to hold anything back, it's only because if you were to try to do any additional good, you would burn out and stop caring about acting rightly. It isn't that you actually have any license to do what doesn't maximize the good.)
3. The rights view is based on the powerful thought that there is something about the *subjectivity* of nonhuman animals that requires our moral response. It's the fact that they are, as Regan says, "subjects-of-a-life" that explains why we have reason to be concerned about how their lives go for them. We might think that one of the first steps in moral thinking is to recognize that you are neither unique nor alone in the world: it's full of other creatures with their own vantage points on reality, their own beliefs and desires, and so forth. Perhaps, then, we might want to say that what matters fundamentally isn't pain per se but subjectivity—and that's what the rights view calls us to recognize.

All that said, we shouldn't understate the ways in which this too is a radical moral theory, at least if judged in terms of its assessment of the status quo. As Regan makes amply clear, the rights view calls for the complete and immediate end of animal agriculture—industrial or small scale, humane or cruel. There are no permissible forms of animal farming according to the rights view: they all invariably involve using others as a mere means to our own ends, and that, Regan argues, is very seriously wrong. Likewise, the rights view calls for an end to animal experimentation, as well as keeping animals captive in circuses and most zoos. (Zoos that function as animal sanctuaries are the likely exception.) It also has significant implications for conservation policy. For instance, most conservationists see no problem with killing invasive species to save noninvasive ones or culling some members of a species to improve the welfare of others. But if the rights view is correct, then these actions involve using some just to benefit others, and that's wrong.

Virtue Theory

We might worry that utilitarianism and the rights view don't capture everything that matters to us in the moral domain. Isn't there a lot more to being a good person than these theories discuss? What about the internal life of the person? What about your feelings and dispositions and character? If you find yourself having this reaction, then virtue theory may be of interest. Virtue theory denies that we're going to be able to assess what we should do without thinking about the kinds of people we want to become. And what kind of people are those? Presumably, they are ones who exhibit some key virtues: honesty, compassion, courage, diligence, humility, and so on. According to virtue theory, there may not always be one right thing to do. But when there is, it's because that's the action that the virtues favor. If a particular action is what kind, honest, temperate, judicious (etc.) people would do, then it's the right thing to do.

There are lots of ways of developing virtue theory, and this isn't the space to survey them. For present purposes, what matters is just the way that the virtue ethicist encourages us to think about moral problems. According to utilitarianism, even very bad people can know what they ought to do. To answer that question, they just need to know how to deploy the principle of utility. But according to virtue theory, bad people may *not* be able to know what they ought to do. It's good people—the morally wise—who are in a position to tell what's best. So instead of having a simple principle, such as "Maximize utility," virtue theory tells us that it's only the virtuous who are going to know how to balance these very important considerations—such as being just and merciful, kind and honest—in a particular situation. There isn't going to be a shortcut, where we can enter the information about a situation into an algorithm and have it spit out the correct answer. Instead, we have to do the hard work of becoming people with virtuous characters, and only then can we be trusted to know what to do in the hardest situations.

This is both quite compelling and remarkably disappointing. It's compelling insofar as it's a refusal to give pat answers in ethics. If nothing else, we can safely say that ethics is hard. In the kinds of cases that we bother to discuss, we discuss them precisely because there is so much disagreement about what ought to be done. And virtue theory says that this is exactly what you'd expect. It's only the morally wise who are going to be able to figure out how to balance the virtues, which often pull in different directions. (Sometimes, the kind thing isn't fair, or the honest thing isn't compassionate.) And since lots of us aren't morally wise, at least if we're honest with ourselves, confusion and disagreement are exactly what we should expect.

At the same time, there's a sense in which the virtue ethicist is punting. What should we do? Who knows! Ask the virtuous! But we often *identify* the virtuous by the courses of action that they recommend. And if we don't know what's right, then we don't know who to count as virtuous. This sort of problem is especially serious when it comes to issues relating to animals, where seemingly good people differ so radically in their assessment of our obligations. On the face of it, then, the virtue theorist gives us little advice at all.

Virtue theorists say different things at this juncture, but one attractive maneuver is to dismiss this worry as too removed from the reality of virtue-based decision-making. Consider, for instance, the way that Rosalind Hursthouse—a prominent virtue ethicist—discusses the ethics of eating animals:

> Can I, in all honesty, deny the ongoing existence of [the suffering of farmed animals]? No, I can't. I know perfectly well that although there have been some improvements in the regulation of factory farming, what is going on is still terrible. Can I think it is anything but callous to shrug this off and say it doesn't matter? No, I can't. Can I deny that the practices are cruel? No, I can't. Then what am I doing being party to them? It won't do for me to say that I am not actually engaging in the cruelty myself. There is a large gap between not being cruel and being truly compassionate, and the virtue of compassion is what I am

> supposed to be acquiring and exercising. I can no more think of myself as compassionate while I am party to such cruelty than I could think of myself as just if, scrupulously avoiding owning slaves, I still enjoyed the fruits of slave labor . . . The practices that bring cheap meat to our tables are cruel, so we shouldn't be party to them.
>
> *(2006, 141–143)*

The point here isn't that Hursthouse's conclusion is correct—though the argument is certainly worth considering—but rather that her reasoning is familiar. Granted, there may be large, unanswered questions about how human beings should relate to nonhuman animals, and there is massive disagreement about how to balance the various character traits that we judge to be morally valuable. Still, when you are doing your best to be morally serious, you probably have a good sense, at least in many cases, about what is and isn't virtuous. Sure, you can be vulnerable to self-deception, you can do your best to excuse your own behavior, and you can try hard to suppress negative interpretations of your behavior. But you can also be intellectually honest and appropriately humble, and see your behavior as either admirable or objectionable, depending on the case. Admittedly, this doesn't make collective moral deliberation particularly easy, as we may have a very hard time convincing one another to adopt our particular ideas about what cruelty is and isn't, or what honesty does and doesn't require. But the virtue ethicist will just say that as we all know, ethics is challenging. We will have to do our best, learning what we can from those who seem to be more virtuous than we are.

Care Ethics

Like virtue ethics, we can think of care ethics as a reaction against utilitarianism and the rights view (or, rather, deontology, which is a catch-all term for non-consequentialist, rule-based moral theories, of which the rights view is an example). However, the reaction is importantly different. Virtue ethicists reject utilitarianism and the rights view because they want to put the focus on our characters. At least when it was first developed, care ethicists rejected these approaches because they saw them as overly patriarchal ways of understanding the ethical project. However, the more important reaction is against the traditional goal of moral theory: namely, providing a theory of value (what's good and bad) plus a theory of the right (why every right action is right and every wrong one is wrong). Care ethics is, in this sense, a "critical" theory of ethics, regarding other tasks as more important. But arguably, traditional versions of care ethics didn't quite make it all the way to this critical position—a point to which I'll return in a moment.

We can distinguish three important aspects of traditional care ethics. We can call the first *particularism*. Care ethics began as a feminist critique of traditional moral theories, such as utilitarianism and Kantianism. Both those theories say that if you have all the facts about a situation, then you can simply apply their preferred principle to find out what people should and shouldn't do in that situation. Carol Gilligan and Nel Noddings were some of the first to argue that this amounts to privileging overly rational, "male" ways of thinking. On their view, there isn't any one principle that explains why all right actions are morally right, and all wrong ones are morally wrong. Actions are still right and wrong—they aren't denying that. They are saying, however, that you can't just feed the details of a situation into some "tell me the morally correct thing to do" machine, turn the crank, and get an answer.

We might worry that this makes it difficult to know what we should do. But care ethicists think that these worries are overblown. This is because they offer us two different tools for assessing our obligations, which brings us to the second aspect of care ethics, namely, the importance of care itself.

Here's one way of thinking about it. Care has many dimensions. It can involve feeling certain emotions: being happy when a friend is happy and being sad when she's sad. It can involve empathizing with her: recognizing her as a distinct individual, being attentive to her situation, and attempting to understand her on her own terms. It involves wanting what's good for her. It involves being responsive to her needs—sometimes at considerable cost to yourself. Actions are good, on this view, if they flow from this kind of rich, multidimensional care; they are bad if they flow from various attitudes that are incompatible with care, such as neglect or hostility.

Of course, there are plenty of cases where it's difficult to know exactly what caring requires of you. Am I enabling this person or simply being supportive? Am I caring for someone the appropriate amount, or am I caring too much and so letting others down? But regardless of our moral outlook, these kinds of cases are often tricky, so maybe it's no fault of care ethics that it doesn't give precise guidance on them.

This brings us to the third aspect of care ethics, which concerns the importance of relationships. Care isn't some abstract idea: it's a way of feeling and relating to another person. And the nature of that care, and the responsibilities it generates, are affected by the kind of relationship in which it occurs. My care for my parents is different than my care for my wife, which is different from my care for my children, which is different from my care for my colleagues at work. We can't separate a discussion of care from a discussion of the the special obligations that we have as parents, partners, and so on—not to mention the really complicated details of individual relationships. So the claim isn't just that we need to act out of care. Rather, it's that we need to care in ways that are sensitive to the special responsibilities we have given particular lives we lead.

There is something very attractive about this picture. It's hard to find a single principle that sums up our ethical obligations. Plainly, caring is very important to being a good person and to acting well in so many situations. And it obviously matters that we are sisters and children and parents and employees; any ethic that overlooks these relationships is missing something important. However, we might have two worries.

The first is that care ethics is a bit too anthropocentric: in its early days, it was focused on relations of care between humans *and other humans*. This was partly because those who developed it thought that caring relationships required a kind of reciprocity, and they didn't think you could have that with the citizens of the nonhuman world.

The second worry is that care isn't enough to address all the moral questions we face. It seems right to pay closer attention to our emotions and relationships, but a mistake to limit ourselves to caring. What about the various character traits that can help us rein in our tendency to care too much, such as honesty and civic responsibility and a sense of justice? Think about the mother who is convinced that she ought to lie to cover up her son's crime. She might be caring excellently and yet still acting wrongly. The care ethicist might agree and say that the problem here is that she doesn't care enough about victims, or that she cares for her son in the wrong way. But how are we going to explain why she isn't caring enough about the victims or the sense in which this is the wrong way to care? What explains why she's striking the wrong balance? It would be helpful to have other values to which to appeal—something else to tip the scales in favor of truth telling and explain why it's the right choice in this case.

Care ethicists make various moves here. For instance, some simply say that care is only part of the moral story: we need an ethic of care *and* an ethic of justice. But some have thought that this isn't a game worth playing: there will always be another puzzle to address, and the back and forth only distracts us from more important issues, such as real-world oppression and marginalization. When you combine this with the worry about anthropocentrism, you get the basic motivations for ecofeminism.

Ecofeminism

Like care ethics, ecofeminism is committed to particularism, the importance of caring, and the significance of relationships, but it completes the critical turn: it goes much further than care ethics in rejecting the traditional conception of moral theory. It isn't that ecofeminists have nothing to say about right and wrong. Rather, it's that ecofeminists don't think that questions about right and wrong are the most important questions to ask, and may even distract us from more significant features of the ethical landscape. To explain the basic idea, Carol Adams and Lori Gruen—two ecofeminists—write this:

> Ecofeminism addresses the various ways that sexism, heteronormativity, racism, colonialism, and ableism are informed by and support speciesism and how analyzing the ways these forces intersect can produce less violent, more just practices. In the 1990s, ecofeminists worked to remedy a perceived problem in feminist theory, animal advocacy, and environmentalism, namely, a lack of attention to the intersecting structures of power that reinforce the "othering" of women and animals, and contribute to the increasing destruction of the environment. Though sometimes called "utopian" or "concerned with too many issues," ecofeminist theory exposes and opposes intersecting forces of oppression, showing how problematic it is when these issues are considered separate from one another.

The crucial idea is that *power relations* deserve close attention, as the structural features of the social world are often as important—if not even more important—than whether any particular individual acts well or badly, rightly or wrongly. Ecofeminists are concerned, first and foremost, with structural critiques of the world in which we find ourselves, which they see as organized by various binaries that preserve positions of privilege: human/animal, male/female, black/white, high class/low class, cis/trans, and so forth. So rather than worrying about whether going vegan maximizes utility, ecofeminists think of veganism as part of a political movement, an attempt to resist the social realities that frame animals as food rather than as individuals, as consumable rather than as having their own lives and interests. Instead of asking whether you should avoid wearing fur, you might note—as Marti Kheel does—that women who wear fur are situated in a culture that "robs women of their own self-image and then sells it back to them in distorted form."[2] This doesn't necessarily excuse fur-wearing, behind which there's an enormous amount of suffering. But it keeps the context in view: this isn't *just* an unwillingness to attend to animals, even if it's that too.

None of this is to suggest that ecofeminists don't care about individual action: nothing could be further from the truth. It is, however, to say that ecofeminists frame the significance of individual action very differently, and so they're less concerned to criticize individuals than they are to criticize the social conditions within which individuals make choices. In nonideal circumstances—namely, our actual circumstances—all the options that an individual has are bad. In such circumstances, there may be little point to criticizing, say, a researcher who has to use mouse models to get funding for cancer research. Still, there are good reasons to be worried about the environment within which this choice makes sense.

It's hard to compare ecofeminism to traditional moral theories precisely because it's engaged in a slightly different project. It doesn't answer many of the questions that we expect other theories to answer: there is no account of membership in the moral community; there is no theory of value or criterion of right action; there is no explanation for why individuals have the particular rights that they have, or how, exactly, those rights should be traded off against others. This isn't to say that ecofeminists can't or don't answer these questions; they can and do. Rather, it's just to say that when they answer these questions, they aren't answering them in a distinctive way. Instead, they often borrow from more traditional moral theories to fill in details, or they develop novel answers that aren't shared by other ecofeminists. This isn't a criticism. Rather, it's a way of highlighting an important difference in what ecofeminist think ethics is about. For ecofeminists, the main goal is to diagnose—and ultimately address—real-world oppression. It's all well and good to explore the precise boundaries of the moral community, and such projects have their place within the ecofeminist tradition. But they aren't primary, and it would be a mistake to expect the theory to deliver what it isn't designed to provide.

Contractualism

All the theories that we've considered thus far lead to some radical conclusions, at least relative to what many people believe, about our obligation to and regarding nonhuman animals. Is there a moral theory that *isn't* so revisionary?

Perhaps. At least with respect to animals, contractualism is usually framed as the more modest alternative. There are different versions of contractualism, but the basic idea is that morality is a set of rules to which we would consent—*if* some other condition were satisfied. So, for instance, perhaps morality is the system of rules to which we'd consent *if we were perfectly self-interested and rational*, or *if we were free and equal*, or *if we were determined to live in accord with rules that others couldn't reasonably reject*. Whatever the condition, the idea is that morality is a system of rules that's created by agents, by beings with the capacities of normal adults. Unsurprisingly, most versions of contractualism imply that other beings like that—other agents—have various important rights. If we were perfectly self-interested and rational, we would see that it's in our interest to live in accord with rules like "Don't murder (other agents)." By agreeing to live by such rules, we give up the freedom to use lethal violence against others, but we effectively get the right not to have lethal violence used against us. It's rational to make this trade, and so all agents have that right. But you can't make such deals with tigers: they won't abide by them. So, it isn't in our self-interest, or rational, to give up the freedom to use violence against such animals.

What about young children and those with severe cognitive disabilities? You can't make deals with them, either. So do contractualists have to say that they don't have rights?

They usually don't think so. Some argue, for instance, that it wouldn't be rational for us to agree to arrangements that don't protect the vulnerable beings about whom we care deeply, such as young children and those with severe cognitive disabilities. But since humans generally don't have such strong attachments to animals, it can be rational for us to agree to arrangements that don't extend even further.

We might object that morality isn't about enlightened self-interest. Indeed, we might think that morality is *essentially* other-regarding; the whole idea is that it serves as a *corrective* to self-interest. And when we switch to other contractualist frameworks, it's less clear that animals won't end up with rights. Suppose we think that we ought to act in accord with the rules to which we'd all agree if we were free and equal. If "we" here means *everyone*, animals included, then animals will end up with rights. And if the "we" here *doesn't* mean everyone, then we need some story as to why not.

But let's set this objection aside. Even if it doesn't work, we should note that contractualism doesn't imply that anything goes with respect to animals. It's just that animals matter *indirectly*; we may have obligations regarding them, though not to them. This is true in two ways. First, it might matter how we treat animals because they're valuable to us in many ways, and treating them poorly sets back our own interests. This is, very roughly, the way that many producers talk about animal welfare in intensive agricultural systems: stressed pigs are unproductive pigs, so stress should be reduced. Stress isn't *intrinsically* bad from the perspective of such producers: it's bad because it makes pigs less productive. Still, that's a way of caring about animal welfare, and it places some significant constraints on agricultural practices.

Second, it might matter how we treat animals because of links between such treatment and our characters. To appreciate this point, consider someone who, after a long day at work, takes out his frustrations on his cat, beating the animal with the poker from his fireplace. Would you want to have a drink with this man? Would you trust him to watch your children? If you wouldn't, then you seem to think that how people treat *animals* has some significant bearing on how they treat *people*. And if that's right, then we have another reason to be thoughtful in the way we relate to animals: who we are with animals affects who we are generally, and in general, we want to be decent people.

Still, we shouldn't pretend that contractualism is nearly as animal-friendly as the other theories we've surveyed. It isn't. So if we conclude that the other theories are implausible for various reasons, and contractualism's problems can be addressed, then we may end up with a moral theory that doesn't lead to such revisionary conclusions. Of course, it's an open question whether that would be a good thing.

Looking Ahead

In this brief introduction, I've tried to give you a sense of the frameworks that have been the most influential in animal ethics in recent years. There are certainly others, and others still are likely to become prominent in the future. For instance, African, Indian, and Asian ethical traditions are only just now being put in dialogue with Western views, and it will be interesting to see how cross-cultural theorizing evolves. By and large, however, the frameworks you have are the ones that have shaped the perspectives of the contributors to this book, either because they endorse one of them or because they're reacting against them.

All that said, it's important to recognize that the frameworks that I just outlined are rejected by most people who work with animals. You won't find many animal researchers who are committed to animal rights; it's the rare farmer who's a utilitarian; there probably aren't any conservation biologists who are committed to ecofeminism. The theories I've sketched are the dominant theories of animal ethics, which bear important relationships to the dominant theories of ethics generally. But most people don't have an ethical framework in the sense that I've described here. This isn't because they don't have ethical views, but rather because they haven't tried to systematize those views in anything like the way we discussed. People who are involved in industries that either use or study animals may indeed be ethically thoughtful in their own way, but it takes some work to figure out how to understand the distinctive features of their moral reasoning. This book is going to give you plenty of opportunities to do that work, and then to put the results in conversation with the ideas that I've set out here. This is, in my view, an enormously important project. Moral theories may be the specialty of philosophers, but moral wisdom almost certainly isn't. And it would be awfully surprising if it were to turn out that smart, sensitive, and morally serious people—the kind willing to contribute to a handbook of animal ethics—wouldn't be able to offer any ideas with which it's worth engaging.

Notes

1. *The Principles of Morals and Legislation*, Chapter XVII, Section 1.
2. Marti Kheel, "From heroic to holistic ethics: The ecofeminist challenge," in Greta Gaard, ed. *Ecofeminism* (Philadelphia: Temple University Press), p. 259.

PART I

Thinking About Animals

Editor's Introduction

According to the standard interpretation, René Descartes believed that nonhuman animals are unfeeling machines. Although they cry out in ways that suggest that they are in pain, the reality is that there's nothing going on inside: it isn't just that the lights are off; on his view, there were never any lights to turn on. If this picture of animals is correct, then it's easy to answer many questions in animal ethics. May we eat animals? Absolutely. May we experiment on them? Of course. May we keep them as pets? Certainly. May we kill some of them—for instance, members of invasive species—to save others? It's hard to see why not. This isn't to say that there aren't any questions remaining. For instance, I can't just do whatever I want to any machine I happen to see. If the machine happens to be yours, then there are some constraints on my behavior. Or if the machine is one that we all share—like a bus or a public computer—then I ought to consider the interests of others in relation to the machine. And we might wonder about how, exactly, to specify the limits of what I may and may not do. But at this point, we're just filling in details. When it comes to the big questions, things are straightforward.

It matters, then, whether this picture of animals is correct. How would we know? That question, among others, is the subject of this section. Our focus in what follows is on how to think about animals, to include questions about how we actually do think about them, as well as how we should.

There are lots of thorny questions to consider. For instance,

- What sorts of biases prevent us from thinking clearly about nonhuman animals? How are we wired—as a result of evolutionary pressures, socialization, and early learning—to think about animals?
- When, if ever, would it be a moral mistake to adopt one standard of evidence over another when it comes to assessing the capacities of animals? In other words, one important task is to determine what counts as "good enough" when we are considering whether the evidence for a hypothesis is good enough to justify believing it. When trying to work that out, is it possible to be too risk-averse? Can we try too hard to avoid error? If so, when?
- Just how much animal behavior can be explained using the simplest sorts of mechanisms? When do we need to appeal to more sophisticated beliefs and desires to make sense of what we see animals do?

- What sorts of emotions can we plausibly attribute to animals? How does our conception of animals change once we think about them as emotional beings?
- To what degree are different animal self-aware, and what exactly does it mean to be self-aware in the first place? What's the moral relevance of being self-aware? To what, if anything, does being self-aware entitle you?
- Forget questions about how we treat animals: are animals themselves moral beings? Can they act well or badly, rightly or wrongly, morally or immorally? If yes, by what standard? And then, back to us again: if animals are moral beings, how does that affect the way we ought to relate to them?
- There are lots of situations where we need to be able to measure well-being. How should we measure animal well-being? How should we quantify it? What are the implications of different ways of quantifying it?
- Assuming that we can quantify animal well-being, how do we make cross-species comparisons? It's one thing to say that we can measure how much pain a dog is experiencing. It's quite another to say that 10 units of dog suffering are worth 50 units of lobster suffering. What are the puzzles that come up when we try to make these kinds of cross-species evaluations?

There are a lot of important connections among these questions. For instance, to the degree that we think that we are biased against appreciating the capacities of animals, we will feel some obligation to correct for that. To the degree that we're worried about being excessively cautious in our judgments about the capacities of animals, we'll be more inclined to attribute sophisticated abilities to them. And to the degree that we attribute to sophisticated capacities, we will interpret subsequent behavior differently. If you are already convinced that animals can contemplate their own futures, you'll probably have a very different view of mourning behavior—such as placing leaves on a dead companion—than you would otherwise. Moreover, insofar as more sophisticated animals have a greater number of interests that can be set back by human activities, you may get very different numbers when you try to compare, say, the cost to human beings of a particular climate change mitigation strategy and the costs to nonhuman animals of our not pursuing that strategy.

This is a handbook of animal ethics, not animal minds. (That handbook exists, however, and is worth your time.) So there are lots of questions worth asking that aren't even touched on in what follows. It should, however, be enough to convince you that there's a lot of work to do even before we get to practical problems in animal ethics. It certainly isn't easy work, but it's fascinating, and as we'll see, the stakes are quite high.

1

PSYCHOLOGICAL MECHANISMS INVOLVED IN HUMAN–ANIMAL INTERACTIONS

How Do Humans Cognize About Animals?

Catherine E. Amiot and Brock Bastian

Back in the 1960s, anthropologist Claude Levi-Strauss (1962, p. 89) wrote that "animals are good to think with." He argued that scholars should pay closer attention to animals and that by investigating how we think about and act toward other species, we might learn more about human nature and understand human societies in new ways. Herein we follow this advice by reviewing research that has investigated how we (humans) think about animals.

Our relations with animals have been ubiquitous in human lives throughout epochs and across cultures (Serpell 1986), and animals are often an integral part of our everyday lives. Nonetheless, our relationship with animals can also be contradictory. Our cognitions, thoughts, attitudes, and beliefs about animals are diverse and complex: some animals we love (e.g., pets, sacred animals), others we hate (e.g., pests, vermin), and others we eat (e.g., farm animals; Herzog 2010).

The goal of the current chapter is to cover the psychological mechanisms that operate in human–animal relations and that explain our different reactions to "them" (Amiot and Bastian 2015). While such mechanisms include individual-level factors such as personality (Mathews and Herzog 1997) and attachment—mostly to pets (Zilcha-Mano et al. 2011a, 2011b, 2012)—societal and group-level processes are also fundamental to understanding the dynamics of human–animal interactions (Plous 1993a, 1993b, 2003). This chapter aims to review these psychological processes; it also aims to highlight areas of contestation that are in need of further research.

Individual-Level Psychological Processes Involved in Human–Animal Relations

Attachment

When we spontaneously think about animals we often also feel a sense of connection and attachment toward them, and particularly toward pets as a proximal subgroup of animals (Mornement 2018). Attachment is not only a classic notion in psychology but also an important factor that has been specifically studied within the human–animal relations literature (McNicholas et al. 2005). Secure attachment, as one type of attachment, refers to the ability of an attachment figure to provide a secure basis, or sense of safety, when the other feels threatened or unsafe. While it is mostly humans who act as caregivers and meet their companion animals' immediate needs (e.g., exercise, food, health), companion animals may also serve as attachment figures for their owners, supplying them with a feeling of comfort and support (Zilcha-Mano et al. 2011a). Hence, humans and their animals can serve as attachment figures for each other. A variety of self-report measures have been developed to assess the attachment construct, for example, the Lexington Attachment to Pets Scale (Johnson et al. 1992) and the Pet Relationship Scale (e.g., Lago et al. 1988). More recently, Zilcha-Mano and colleagues (2011a) have directly applied Bowlby's psychological attachment taxonomy—an established and widely used theoretical framework in psychology—and developed the Pet Attachment Questionnaire, which assesses attachment anxiety and avoidance. A classic procedure in developmental psychology—the Strange Situation, which involves monitoring a child's responses when separated from a parent and then reunited with him or her (Ainsworth 1991)—has been applied to test which type of attachment animals have developed toward their human caregivers (e.g., Marinelli et al. 2007; Palmer and Custance 2008; Topál et al. 2005). Indeed, dogs can display attachment patterns toward their human caregivers that are similar to those patterns observed in human–human relations.

While capturing one's cognitive representation of a relationship, the concept of attachment also has concrete behavioral and well-being consequences: feeling a greater attachment to one's animal is associated with a greater likelihood that the animal will be kept indoors rather than outdoors (Shore et al. 2006), a lower likelihood that the animal will be relinquished (Patronek et al. 1996), a higher satisfaction with one's animal's behavior (Serpell 1996), and higher human well-being (Garrity et al. 1989; Ory and Goldberg 1983). Importantly, which *type* of attachment to animals people have is also associated with the quality of human–animal relations and with (human) adjustment: having an anxious attachment to one's pet has been associated with more psychological distress (Zilcha-Mano et al. 2011a). In contrast, people who have a low avoidant attachment to their pet experienced reduced blood pressure during a stressful event when their pet was either present or recalled to memory (Zilcha-Mano et al. 2012). Together, these findings suggest that attachment is important to understanding the nature of human–animal relations and that—as is the case in human–human relations—a more secure attachment to one's pet per se predicts more beneficial outcomes, for humans at least. The dynamic interplay between the attachment style of humans and their pets, and how these styles may fuel and influence one another, remains an area of further investigation. Research using interactive and systematic behavioral coding systems that apply to both humans and their animals could be particularly useful to capture this interplay (O'Haire 2013).

Personality

Personality is also a central notion in psychology, typically referring to individual differences in characteristic patterns of thinking, feeling, and behaving (Kazdin 2000). When integrating findings from research that has examined the role of personality in understanding and predicting the nature of human–animal relations, the overall trend suggests that individual differences and beliefs that involve a broad, inclusive, and flexible orientation—compared to a conventional, rigid, and hierarchical one—are associated with more positive attitudes and behaviors toward animals. Dispositional empathy, which involves being able to take another's perspective and emotions into account, has been consistently examined within human–animal relations research. Specifically, higher empathic concerns have been associated with more positive attitudes toward animals (Taylor and Signal 2005), greater concern over animal cruelty (Eckardt Erlanger and Tsytsarev 2012; Furnham et al. 2003), and an enhanced capacity to recognize animals' experiences of pain (Ellingsen et al. 2010). In terms of creativity and unconventionality—which are personality traits that should also be associated with more positive attitudes toward animals—individuals categorized as Intuitive-Feeling types (on the basis of their scores on the Myers-Briggs Type Inventory) reported more positive attitudes toward animal welfare issues (Broida et al. 1993). Similarly, being sensitive and imaginative has been associated with more opposition toward animal experimentation (Mathews and Herzog 1997). Another indicator of unconventionality, namely, lower political conservatism, has also been associated with higher concerns toward farm animal welfare (Heleski et al. 2006) and lower support for vivisection (Broida et al. 1993).

Research has also investigated the personality characteristics of animal rights activists as a specific group. Demonstrators at the 1996 March for the Animals were found to report higher dispositional optimism compared to a sample of college students (Galvin and Herzog 1998). Animal rights activists also tend to report lower relativism and higher levels of idealism compared to nonactivist college students (Galvin and Herzog 1992), suggesting that engaging in such actions needs to be fueled by a very determined and optimistic mind-set. Animal activists also report a lower threshold for feeling disgust in general (Herzog and Golden 2009). Possibly as a result of this enhanced sensitivity, activists—compared to nonactivists—consider that animals feel more pain, and they tend to report that all animals (even those who are dissimilar to or distant from humans) possess this capacity (Plous 1991).

An intriguing line of work has furthermore found that people present different personality profiles according to their preferred type of pet (e.g., Podberscek and Gosling 2000). Specifically, people identifying mainly as a "dog person" tend to be more extroverted, more agreeable, more conscientious, less neurotic, and less open to experiences than those who identify themselves as a "cat person" (Gosling et al. 2010). Children who actually own dogs tend to report higher empathy than do those who own cats (Daly and Morton 2003). Compared to dog owners' attachment orientation toward their dogs, cat owners are significantly more avoidant toward their cats (Zilcha-Mano et al. 2011a). Participants also rated a person accompanied by a dog in a picture as more likable than the same person accompanied by a cat (Geries-Johnson and Kennedy 1995). These findings align with the fact that cats are seen as pets who are more independent and solitary while dogs are seen as more interactive and dependent.

Gender

Gender is one of the most stable factors that predicts attitudes and empathy toward animals (see Herzog 2007 for a review), with women generally reporting more positive attitudes toward animals than men. Specifically, relative to men, women have been found to report more empathy toward animals (Hills 1993) and more positive attitudes toward pets (Schenk et al. 1994), and they have been found to be more concerned about protecting the welfare and rights of animals (Herzog et al. 1991; Mathews and Herzog 1997). Among animal health professionals, female veterinary students and female faculty members of an animal science department tend to report greater concerns for animal suffering and animal welfare than do their male colleagues (Heleski et al. 2006).

In a systematic review of gender differences in attitudes toward animals (Herzog 2007), the strength of these gender differences was found to vary according to the extremity of the attitudes and behaviors involved: while men and women are as likely to own pets, women have slightly more positive attitudes toward animals. One area where the gender differences are larger is in the realm of animal activism, with more women than men boycotting circuses, animal experimentation, giving up meat for ethical reasons, and joining animal protection organizations and pressure groups (Galvin and Herzog 1998; Plous 1991). Other important gender differences include women's reduced tendency to act aggressively toward animals (including bestiality) and to hunt, and their increased tendency to hoard animals (Patronek 1999). Further work is needed to disentangle if and when these gender differences are due to socialization (Donovan and Adams 2007; Luke 2007) and/or to evolutionary/biological factors, such as hormonal differences (Handlin et al. 2012; Odendaal and Meintjes 2003).

Ideological Beliefs

Beliefs that support a hierarchy between human social groups and that value obedience to authorities have also been found to play into our attitudes toward animals per se. The more people endorse a social dominance orientation (SDO)—defined as the belief that social groups (of humans) should be organized hierarchically, with those in power being entitled to greater privileges (Sidanius and Pratto 1999)—, the less they tend to believe that humans and animals are similar (Costello and Hodson 2010). Right-wing authoritarianism (RWA; Altemeyer 1998), an ideological belief that specifically involves obeying authorities and adhering to conventional social norms, has also been associated with the tendency to see humans as distinct from and superior to animals (Motyl et al. 2010). Greater endorsement of social hierarchy has been linked to more positive attitudes toward meat-eating (Allen and Ng 2003), a behavior that disregards farm animals' negative treatment.

Recently, a model has been proposed to capture the common causes for dominance of animals and of humans: the social dominance human–animal relations model (SD-HARM). According to the SD-HARM, SDO—more so than RWA—is a central ideological belief that is responsible for the significant association between attitudes toward certain ethnic groups and speciesism (i.e., negative attitudes toward animals; Dhont et al. 2016). The theory suggests that because SDO relates to the endorsement of a direct status difference between social groups, it is the general belief that one's own group is superior that explains prejudice toward both animals and human outgroups. In a series of studies, results confirmed that SDO is a key belief linking prejudice toward human groups and speciesism (Dhont et al. 2016; see also Dhont et al. 2014). Building on Allport's contention that different forms of prejudice are interrelated and share a common cognitive basis, recent research has further revealed that speciesism is associated with higher racism, sexism, and homophobia (Caviola et al. 2018).

Group Processes Involved in Human–Animal Relations

To fully understand the implications of perceiving some social groups as superior to others and the dynamics that are involved in these group processes, adopting an intergroup approach is warranted. Indeed, human–animal relations can involve not only negative, speciesist attitudes toward nonhuman animals as an outgroup (i.e., a group one does not belong to; Singer 2009) but also overt discriminatory behaviors against this outgroup. Such discrimination is observable, for example, in moral dilemmas that pit human (i.e., ingroup) and animal interests against one another and in which most people choose to favor humans over animals (even when the animals are members of endangered species; Petrinovich et al. 1993). As another example, in the family context, people systematically give priority to humans over animals and allocate more resources and more time to humans (Albert and Bulcroft 1987). Understanding human–animal relations as an *intergroup* topic allows for a better understanding of these phenomena. Indeed, animals can be viewed as an outgroup in the same way that members of other cultures, religions, or nationalities are regarded as outgroups. To capture these "us" (humans) versus "them" (nonhuman animals) dynamics (Plous 1993a, 1993b), social psychological theories of intergroup relations, including the social identity approach (e.g., Tajfel and Turner 1986), are particularly useful.

Social Identification, Intergroup Similarity, and Social Status

A central proposition of the social identity approach is that people are motivated to belong to and identify with social groups that are positively distinct compared with outgroups (Tajfel and Turner 1986; Turner et al. 1987). In other words, according to this perspective, humans are motivated to distinguish themselves positively from other social groups. This can also involve nonhuman animals (Plous 1993b, 2003; Serpell 1996). Thus, by extending this theoretical approach to the realm of human–animal relations, we might expect that the more we perceive differences (rather than similarities) between humans and animals, the more this should predict a motivation to affirm humans' superiority relative to animals. These social psychological propositions also sit well with

Kasperbauer's (2017) recent contentions, according to which the intergroup differentiation between animals and humans is reinforced by the idea that humans actively contrast themselves with animals in order to maintain their own unique identity. The phenomenon of infrahumanization, as applied by Kasperbauer (2017), implies that animals are denied uniquely human traits (e.g., love, shame, or hope), while humans and animals are evaluated similarly in terms of more basic emotions (e.g., fear, pleasure, or anger). Even though humans' evaluation of animals is not explicitly negative, animals' status is lowered compared to humans' status due to the denial of these more complex traits. Assigning uniquely human features exclusively to humans allows them to elevate themselves above animals and, consequently, secure their position as the superior species.

Intergroup relations research further suggests that perceiving differences versus similarities between different social groups also has implications for social identification (Amiot et al. 2007; Amiot et al. 2012). Social identity is defined as "that part of the individual's self-concept which derives from his or her knowledge of membership to a social group (or groups) together with the value and the emotional significance attached to it" (Tajfel 1981: 255). To the extent that different social identities share similarities (e.g., being a human implies sharing similar senses, emotions, and physiological characteristics with nonhuman animals) rather than differences, these distinct identities (human and nonhuman animals) will be easier to "connect" and bridge at the cognitive level. Concretely, this will be achieved by creating a superordinate social group that encompasses both subgroups; that is, the superordinate group composed of *all* animals, including human and nonhuman animals. In the realm of human–animal relations, perceiving such similarities between humans and animals indeed leads to beneficial outcomes for animals, including a greater perception of relatedness and empathy, and an increased desire to protect animals' rights (Plous 1993a). Similarly, the tendency to anthropomorphize animals—a psychological mechanism that involves assigning animals humanlike characteristics, such as emotions and cognitions (Waytz et al. 2010), and that, in this sense, implies recognizing our similarities to them—has also been associated with greater concerns for animal welfare (Butterfield et al. 2012). It should be noted that this capacity to perceive similarities between animals and humans occurs despite a general tendency among humans to process and categorize information distinctly about animals and humans, based on different neurological modules (see, e.g., Long et al. 2010; Marinović et al. 2014).

Recent research has investigated the particularly broad and inclusive superordinate identity involving identification with animals and one of its particular dimensions: solidarity with animals (Amiot and Bastian 2017; Amiot et al. 2017). In this work, solidarity with animals is defined as the sense of belonging, psychological attachment, and closeness felt toward other animals. This particular dimension of social identification captures its relational side and the concrete roles that we occupy within a group (see also Leach et al. 2008). It involves investment of the self in coordinated activity with those to whom one feels committed. Because feeling solidarity with animals involves considering them as close to the self and taking their perspective and interests to heart, we expected that this identification dimension would predict more positive attitudes toward animals—such as lower speciesism—and more prosocial behaviors toward them as well as increased intentions to engage in collective actions on their behalf.

In a first series of seven studies, we (Amiot and Bastian 2017) found that solidarity with animals predicted more positive consequences for nonhuman animals, including lower speciesism, more moral concerns for animals, and greater willingness to donate to animal charities (Studies 1 and 7). In terms of personality, higher solidarity with animals was found (in Study 2) to be associated with higher dispositional empathy and openness to experience, as well as a greater tendency to anthropomorphize animals. These findings confirm that solidarity with animals is underpinned by a flexible and inclusive mind-set. Solidarity was also associated with less avoidant but more anxious attachment to one's pet. These associations suggest that people high in solidarity may be more likely to hold close to animals rather than avoid and distance themselves from animals, even if this implies higher feelings of anxiety or dependency toward their pet. Interestingly, solidarity with animals correlated negatively with hierarchy-enhancing ideological beliefs (SDO, RWA) and with different forms of prejudice toward human groups (i.e., racism, ageism; Study 3). These findings imply that solidarity with animals—because it represents a particularly large and inclusive social identity—could lead to a recategorization that also encompasses human outgroups (Bastian, Costello, et al. 2012). Also reflecting the idea that solidarity involves being committed toward animals and feeling close to them (even on a day-to-day basis and when making lifestyles choices), pet owners and vegetarians reported higher solidarity with animals compared to non–pet owners and to meat-eaters (Studies 4a and 4b, respectively).

One experiment directly tested the role of human–animal similarity (vs. difference) as causing greater (vs. lower) solidarity with animals (Study 5). Results confirmed this hypothesis, showing that participants exposed to pictures of animals displaying complex and humanlike emotions reported higher solidarity with animals compared to participants exposed to pictures of animals in stereotypical poses. Study 6 replicated these findings and showed that solidarity helped to explain (mediated) the link between perceiving high levels of similarity with animals and the tendency to hold more positive and supportive attitudes toward animals.

In a second series of three studies (Amiot et al. 2017), we further tested whether the more people perceive human–animal similarities, the more likely they are to feel connected to animals, and whether this, in turn, reduces a person's need to assert humans' superiority relative to animals—that is, the motivation for a positive identity in the social identity tradition. The idea being that if one identifies with all animals, this sense of identification should then be associated with a lower need to consider humans as superior to animals, given that, in this case, nonhuman animals and humans are considered as being part of the same superordinate group. In these studies, identification with animals was assessed with the solidarity with animals measure (Amiot and Bastian 2017) and with a pictorial measure that captures the extent to which people see animals as close to themselves and as included within their own self-concept. In Studies 1 and 2, the more participants tended to perceive a lot of similarities between animals and humans, the more likely they were to also identify with the superordinate group of animals. In turn, this sense of connection to and identification with animals predicted a lower tendency to perceive humans as superior to animals.

In Study 3, we aimed to replicate these findings using a modified method to tap human–animal similarities. Specifically, this study presented people with pictures of animals that are phylogenetically more versus less similar to humans. These were, respectively, pictures of a gorilla, a white rhinoceros, a common crane, an iguana, a flathead catfish, and a May beetle. After seeing each picture, participants completed a set of questions that measured solidarity with each animal and his or her perceived status relative to humans. As expected, we found that participants felt more identified with the more phylogenetically similar animals than with the less similar animals; they also perceived the more similar animals as closer to humans in terms of their social status. Importantly, we found that when participants had viewed the pictures of the more similar to human animals (i.e., gorilla, white rhinoceros, common crane) compared to the less similar animals (i.e., iguana, catfish, beetle), this predicted higher overall solidarity with animals. This increased sense of solidarity, in turn, predicted a reduced overall tendency to perceive humans as superior to animals. These findings have concrete implications for the presentation of effective messages to elicit concerns toward animals. They imply that presenting pictures of animals who are more similar to humans can serve as a springboard to more beneficial perceptions of a *diversity* of animal types, including animals who are similar to humans but also those who are less similar to humans. Future work should seek to identify *how* this sense of identification develops: with exposure to specific animals, such as our pets (Serpell and Paul 1994)? Or possibly also from interactions with important role models in our lives such as parents and teachers (Kidd and Kidd 1997)? More macroscopic variables may also contribute to influencing solidarity with animals. For example, resource scarcity—in line with realistic conflict theory (Sherif 1966)—may impede a sense of connection with animals and promote a zero-sum struggle for these resources. In contrast, framing intergroup relations as cooperative and mutually beneficial for both animals and humans may benefit human–animal relations and humans' sense of belongingness and identification with them.

Distancing Ourselves From Animals: Cognitive Dissonance and Internal Conflict

In contrast to these social identification forces bringing us closer to other animals, intergroup theories also directly account for the ideological forces that legitimize harmful treatment of an outgroup and that are employed to create distance from this group (e.g., Sidanius and Pratto 1999; Tajfel 1981). These factors have been proposed to also operate in human–animal relations.

Maltreatment of meat animals, in particular, can cause considerable psychological conflicts, which creates the need to distance oneself psychologically from this harm (Bastian and Loughnan 2017; Joy 2005). To deal with this unpleasant state, one strategy involves distancing oneself from the harm brought to animals. Concretely, this is done by denying animals human characteristics (consciousness and the capacity to think) and their individuality and sense of morality (Burghardt 2009). For example, when people are reminded of their own meat-eating practices and the harm this brings to animals, they tend to deny mental qualities to the animals they eat (Bastian, Loughnan, et al. 2012; see also Bilewicz et al. 2011). Even just categorizing a novel animal as food reduces our concern for this animal's welfare; this effect is observable both in a situation where the animal was killed for that purpose and in the case where the animal had died naturally but was used for food afterward (Bratanova et al. 2011). Recently, Wegner and Gray (2017) further elaborated on how perceiving higher mind among animals—including their agency or capacity to make decisions and their experience or capacity for feelings and emotions—is associated with more moral concerns toward them. However, they argue that a disconnect exists between real/objective mental capacities and our subjective perception of these capacities. This disconnect could explain why we assign moral obligations to certain animals (e.g., kittens) and not others (e.g., crows, pigs), often regardless of the animals' "real" mental capacities. There are also social and cultural norms guiding which animals are considered edible or not (Herzog 2010), with animals kept as pets (e.g., dogs) being generally considered inappropriate for human consumption (Leach 1964; Serpell 1987, 2009).

Concretely, the words and images we use can also be employed to create psychological distance from animals and legitimize their use by humans (Plous 1993b, 2003; Hyers 2006; Leach 1964; Mitchell 2011). In the medical sciences, animals used in experiments are referred to by numbers rather than names or initials (Lederer 1992), which de-individualizes them and may then facilitate their use (Serpell 1999). This is in contrast to attributing a name to an animal, which in this case highlights their unique individual characteristics and personality (e.g., Sanders 2003). In the context of hunting and trapping, animals are described as "crops," "seed," "surplus," and "renewable resources," and the action of killing animals is referred to as "thinning," "managing," and "controlling" animals (Serpell 1999). Visually, animals we eat are marketed, in most Western countries, at least, without the body parts that remind us of where our meat comes from and that are associated with their personality (Plous 1993a). Logistically, farm animals receive less media coverage than wild animals do (Singer 2009), and intensive farm operations and slaughterhouses are typically located in remote or inaccessible places (Fox 1997). In large-scale commercial farms, executives and managers typically have no direct contact with animals and do not directly see the harm done to them, yet it is their role to decide on the animals' fate. This hierarchy and chain of command are reminiscent of classic work on obedience to an authority (Milgram 1963) and the various ways that people seek to diffuse the burden of individual responsibility for morally troublesome acts (Serpell 1999). Bridging with the notion of social identification discussed earlier, future work could test if identification with animals (vs. identification with humans) predicts whether individuals will be less (vs. more) inclined to accept these distancing and justifying ideologies (see also Reicher et al. [2012] for a discussion of the role of identification in obedience vs. disobedience to authority).

To illustrate and explain why the average North American consumer may feel no qualms about eating animals based on this theoretical and empirical work, let us consider the case of Steven. Steven is a 38-year-old American who eats meat. First, let us zoom out and take the broader social context Steven lives in into account. Indeed, Steven eats meat—as an individual behavior and habit—but many social forces operate in order to promote this individual behavior. For instance, Steven only rarely hears about farm and meat animals when watching television or consuming social media, possibly also because he has never shown a lot of interest in this cause in the past and does not directly associate or have contacts with people who advocate on behalf of farm animals. At the grocery store, there is little to remind Steve of where his food comes from (i.e., on packages, in promotional pictures) and how it is tied to the lives of real, individual animals. Strong social norms also operate in favor of eating meat, and these norms encourage Steve to eat meat during social gatherings. The ubiquitous Thanksgiving turkey is just one example illustrating the strength of these norms. We also know, from social psychological research, that deviating from social norms can be demanding and potentially costly, which could explain why only a minority of individuals currently do.

Second, in terms of internal psychological processes, we know that the feeling of internal conflict experienced when eating meat—that is, a behavior that typically conflicts with seeing oneself as a good and caring person, who also generally likes animals (such as pets)—in itself creates cognitive dissonance. This discomfort should hence motivate Steve to employ psychological strategies specifically aimed at managing and reducing this aversive state. These strategies include denying mind to meat-animals, avoiding any thoughts about how intelligent and sentient they are, and assigning lower moral concerns to them. Such strategies aim to make the harm perpetuated against animals more justifiable and acceptable, even if the harm itself is not directly dealt with.

Conclusion

The motivation to promote positive interspecies relations is gaining momentum. The number of people joining animal rights movements has witnessed a sharp increase over the last three decades (Amiot and Bastian 2015). Specifically, over the last 40 years, there has been a strong social movement toward the recognition of animal rights (Pinker 2011; Regan 1983, 2001; Singer 1981). Greenpeace and the World Wildlife Fund now have memberships that number close to 10 million. PETA (People for the Ethical Treatment of Animals) counted 18 members in 1981, 250,000 in 1990, and more than 3.5 million in 2017. The One Health Initiative (www.onehealthinitiative.com) directly seeks to promote the interconnections that exist between the health of humans and animals and the importance of improving both at once.

As humans become increasingly sensitive to the rights and needs of animals and seek to protect them rather than view them as resources to be exploited, new issues and fields of investigation will emerge. For instance, animals may increasingly become protected by the legal system, with people being punished more harshly for the mistreatment of animals, opening new legal questions about what is considered a "fair" retribution for animal harm-doing. Hunting as a leisure activity is also increasingly questioned, with duck shooting, fox hunting, and game hunting on the decline and denounced by activists. We are currently seeing international conflict over the killing of whales, which are now perceived as deserving of protection, with people risking their own lives to achieve these ends (e.g., Sea Shepherd in Australia). Also, debates are arising over Chinese natural medicines that cause harm to protected/endangered species and that are sometimes extracted in painful ways (e.g., stomach bile from bears). As the value of our relationships with animals grows, animals, in turn, may increasingly be recognized as worthy of rights.

In some domains, placing greater value on our associations with animals will incur direct costs to humans. As farm animals are increasingly valued, pressure will likely be placed on factory farming techniques and inhumane killing practices. The need to ensure that our meat is "ethical" and produced humanely is likely to be beneficial to human–animal relations but will increase production costs and reduce efficiency—a current example of this involves the production of "clean meat." All of these phenomena are couched in broader changes in the way we conceive of our relationships and interactions with animals—those relationships and interactions are becoming more personal and meaningful and will need to be addressed in future work.

While, from an intergroup relations perspective, caring for and assigning increasingly more rights to animals could appear illogical for humans and likely to threaten our resources and erode our well-being, it is also possible that increasing our moral concern for animals could, in certain situations, represent a win–win situation for both humans and animals. This occurs, for example, when people in developing countries take care of their wild animals, which, in turn, helps these animals to thrive, attracting tourism and promoting economic activity. The existence of such mutual benefits also represents the basis of the One Health movement, according to which the positive health statuses of animals and of humans are understood to be interdependent and can be mutually reinforcing (vs. mutually damaging, in the case where animals or humans suffer from diseases). The notion of solidarity with animals covered in this chapter also helps explain why we humans can expand our concerns for other animals beyond our human interests—that is, because we can share a sense of connection and a feeling of belonging with them.

By delving more deeply into the psychological mechanisms involved in human–animal relations, the current chapter not only aligns with this trend toward increasingly recognizing and valuing animals but also questions fundamental assumptions that have guided psychological research from its beginnings. Traditionally, within psychology, animals have been used mainly to inform our understanding of basic human processes (e.g., personality, learning) and assigned a somewhat passive and instrumental role. Instead, and building on an emerging psychology of human–animal relations (Amiot and Bastian 2015), herein we argue that apart from using animals to inform our understanding of human processes, we need to directly account for the interactions and the interdependence that exists between humans and animals.

Further Readings

1. Review of human–animal relations

Amiot, C. E., and Bastian, B. (2015) "Toward a psychology of human—animal relations," *Psychological Bulletin* 141(1): 6. doi: 10.1037/a0038147.

This review paper provides a summary of research conducted on relationships between humans and animals and highlights their importance for various subfields of psychology, from human development to intergroup relations.

2. Distancing self from animals

Bastian, B., Loughnan, S., Haslam, N., and Radke, H. R. (2012) "Don't mind meat? The denial of mind to animals used for human consumption," *Personality and Social Psychology Bulletin* 38(2): 247–256. doi: 10.1177/0146167211424291.

Brock Bastian and colleagues explained the mechanism behind cognitive dissonance related to concurrent moral concerns for animals and the practice of meat eating. In three studies, they showed that such cognitive resonance is resolved by denying mental capacities to meat animals.

3. Moral treatment of animals

Herzog, H. A. (2010) *Some We Love, Some We Hate, Some We Eat: Why It's So Hard to Think Straight About Animals*, New York, NY: Harper.

This book by Hal Herzog synthesizes the multidiciplinary knowledge related to animal ethics and provides important insights into the origin of our controversial treatment of different animal species.

4. Intergroup relations and similarity

Plous, S. (1993a) "Psychological mechanisms in the human use of animals," *Journal of Social Issues* 49(1): 11–52. doi: 10.1111/j.1540-4560.1993.tb00907.x.

In order to better grasp the idea of human–animal interactions as a form of intergroup relations, it is helpful to get familiar with this article by Plous, which highlights the importance of perceived similarities between humans and animals as two different social groups and the intergroup biases operating between them.

5. Ideological beliefs: speciesism

Caviola, L., Everett, J. A., and Faber, N. S. (2018) "The moral standing of animals: Towards a psychology of speciesism," *Journal of Personality and Social Psychology*. Advanced online publication. doi: 10.1037/pspp0000182.

When bridging ideological beliefs with our perceptions of animals, it is useful to apply the concept of speciesism, or the tendency to differentiate animal species in terms of moral treatment. Empirical work conducted by Caviola and colleagues helps to understand the nature of this phenomenon as it relates to other ideological beliefs, such as sexism or social dominance orientation, and predicts our treatment of different animal species.

Bibliography

Ainsworth, M. D. S. (1991) "Attachments and other affectional bonds across the life cycle," in C. M. Parkes, J. Stevenson-Hinde, and P. Marris (eds.) *Attachment Across the Life Cycle*, London: Taylor & Francis.

Albert, A., and Bulcroft, K. (1987) "Pets and urban life," *Anthrozoös* 1: 9–25. doi: 10.2752/089279388787058740.

Allen, M. W., and Ng, S. H. (2003) "Human values, utilitarian benefits and identification: The case of meat," *European Journal of Social Psychology* 33: 37–56. doi: 10.1002/ejsp.128.

Altemeyer, B. (1998) *A 20-Item Version of the RWA Scale*, Winnipeg, MB: University of Manitoba, (n.p.).

Amiot, C. E., and Bastian, B. (2015) "Toward a psychology of human—animal relations," *Psychological Bulletin* 141(1): 6–47. doi: 10.1037/a0038147.

Amiot, C. E., and Bastian, B. (2017) "Feeling solidarity toward animals: Assessing a relevant dimension of social identification with animals," *PLoS One* 12(1): e0168184. doi: 10.1371/journal.pone.0168184.

Amiot, C. E., de la Sablonnière, R., Terry, D. J., and Smith, J. R. (2007) "Integration of social identities in the self: Toward a cognitive-developmental model," *Personality and Social Psychology Review* 11(4): 364–388. doi: 10.1177/1088868307304091.

Amiot, C. E., Sukhanova, K., Greenaway, K. H., and Bastian, B. (2017) "Does human—animal similarity lower the need to affirm humans' superiority relative to animals? A social psychological viewpoint," *Anthrozoös* 30(3): 499–516. doi: 10.1080/08927936.2017.1335117.

Amiot, C. E., Terry, D. J., and McKimmie, B. M. (2012) "Social identity change during an intergroup merger: The role of status, similarity, and identity threat," *Basic and Applied Social Psychology* 34(5): 443–455. doi: 10.1080/01973533.2012.712016.

Bastian, B., Costello, K., Loughnan, S., and Hodson, G. (2012) "When closing the human-animal divide expands moral concern: The importance of framing," *Social Psychological and Personality Science* 3: 421–429. doi: 10.1177/1948550611425106.

Bastian, B., and Loughnan, S. (2017) "Resolving the meat-paradox: A motivational account of morally troublesome behavior and its maintenance," *Personality and Social Psychology Review* 21(3): 278–299. doi: 10.1177/1088868316647562.

Bastian, B., Loughnan, S., Haslam, N., and Radke, H. R. M. (2012) "Don't mind meat? The denial of mind to animals used for human consumption," *Personality and Social Psychology Bulletin* 38(2): 247–256. doi: 10.1177/0146167211424291.

Bilewicz, M., Imhoff, R., and Drogosz, M. (2011) "The humanity of what we eat: Conceptions of human uniqueness among vegetarians and omnivores," *European Journal of Social Psychology* 41(2): 201–209. doi: 10.1002/ejsp.766.

Bratanova, B., Loughnan, S., and Bastian, B. (2011) "The effect of categorization as food on the perceived moral standing of animals," *Appetite* 57(1): 193–196. doi: 10.1016/j.appet.2011.04.020.

Broida, J., Tingley, L., Kimball, R., and Miele, J. (1993) "Personality differences between pro- and anti-vivisectionists," *Society & Animals* 1(2): 129–144. doi: 10.1163/156853093x00037.

Burghardt, G. M. (2009) "Ethics and animal consciousness: How rubber the ethical ruler?" *Journal of Social Issues* 65(3): 499–521. doi: 10.1111/j.1540-4560.2009.01611.x.

Butterfield, M. E., Hill, S. E., and Lord, C. G. (2012) "Mangy mutt or furry friend? Anthropomorphism promotes animal welfare," *Journal of Experimental Social Psychology* 48(4): 957–960. doi: 10.1016/j.jesp.2012.02.010.

Caviola, L., Everett, J. A., and Faber, N. S. (2018) "The moral standing of animals: Towards a psychology of speciesism," *Journal of Personality and Social Psychology* (advanced online publication). doi: 10.1037/pspp0000182.
Costello, K., and Hodson, G. (2010) "Exploring the roots of dehumanization: The role of animal—human similarity in promoting immigrant humanization," *Group Processes & Intergroup Relations* 13: 3–22. doi: 10.1177/1368430209347725.
Daly, B., and Morton, L. L. (2003) "Children with pets do not show higher empathy: A challenge to current views," *Anthrozoös* 16(4): 298–314. doi: 10.2752/089279303786992026.
Dhont, K., Hodson, G., Costello, K., and MacInnis, C. C. (2014) "Social dominance orientation connects prejudicial human—human and human—animal relations," *Personality and Individual Differences* 61–62: 105–108. doi: 10.1016/j.paid.2013.12.020.
Dhont, K., Hodson, G., and Leite, A. C. (2016) "Common ideological roots of speciesism and generalized ethnic prejudice: The social dominance human—animal relations model (SD-HARM)," *European Journal of Personality* 30(6): 507–522. doi: 10.1002/per.2069.
Donovan, J., and Adams, C. J. (2007) *The Feminist Care Tradition on Animal Ethics*, New York, NY: Colombia University Press.
Eckardt Erlanger, A. C., and Tsytsarev, S. V. (2012) "The relationship between empathy and personality in undergraduate students' attitudes toward nonhuman animals," *Society & Animals* 20(1): 21–38. doi: 10.1163/156853012x614341.
Ellingsen, K., Zanella, A. J., Bjerkas, E., and Indrebo, A. (2010) "The relationship between empathy, perception of pain and attitudes toward pets among Norwegian dog owners," *Anthrozoös* 23(3): 231–243. doi: 10.2752/175303710X12750451258931.
Fox, M. W. (1997) *Eating with Conscience: The Bioethics of Food*, Troutdale, OR: NewSage Press.
Furnham, A., McManus, C., and Scott, D. (2003) "Personality, empathy and attitudes to animal welfare," *Anthrozoös* 16(2): 135–146. doi: 10.2752/089279303786992260.
Galvin, S. L., and Herzog, H. A. (1992) "Ethical ideology, animal rights activism, and attitudes toward the treatment of animals," *Ethics & Behavior* 2(3): 141–149. doi: 10.1207/s15327019eb0203_1.
Galvin, S. L., and Herzog, H. A. (1998) "Attitudes and dispositional optimism of animal rights demonstrators," *Society & Animals* 6: 1–11. doi: 10.1163/156853098x00014.
Garrity, T. F., Stallones, L., Marx, M. B., and Johnson, T. P. (1989) "Pet ownership and attachment as supportive factors in the health of the elderly," *Anthrozoös* 3: 35–44. doi: 10.2752/089279390787057829.
Geries-Johnson, B., and Kennedy, J. H. (1995) "Influence of animals on perceived likability of people," *Perceptual and Motor Skills* 80(2): 432–434. doi: 10.2466/pms.1995.80.2.432.
Gosling, S. D., Carson, J. S., and Potter, J. (2010) "Personalities of self-identified 'dog people' and 'cat people," *Anthrozoös* 23(3): 213–222. doi: 10.2752/175303710X12750451258850.
Handlin, L., Nilsson, A., Ejdeback, M., Hydbring-Sandberg, E., and Uvnas-Moberg, K. (2012) "Associations between the psychological characteristics of the human-dog relationship and oxytocin and cortisol levels," *Anthrozoös* 25(2): 215–228. doi: 10.2752/175303712x13316289505468.
Heleski, C. R., Mertig, A. G., and Zanella, A. J. (2006) "Stakeholder attitudes toward farm animal welfare," *Anthrozoös* 19(4): 290–307. doi: 10.2752/089279306785415439.
Herzog, H. A. (2007) "Gender differences in human—animal interactions: A review," *Anthrozoös* 20: 7–21. doi: 10.2752/089279307780216687.
Herzog, H. A. (2010) *Some We Love, Some We Hate, Some We Eat: Why It's So Hard to Think Straight About Animals*, New York, NY: Harper.
Herzog, H. A., Betchart, N. S., and Pittman, R. B. (1991) "Gender, sex role orientation, and attitudes toward animals," *Anthrozoös* 4(3): 184–191. doi: 10.2752/089279391787057170.
Herzog, H. A., and Golden, L. L. (2009) "Moral emotions and social activism: The case of animal rights," *Journal of Social Issues* 65(3): 485–498. doi: 10.1111/j.1540-4560.2009.01610.x.
Hills, A. M. (1993) "The motivational bases of attitudes toward animals," *Society & Animals* 1(2): 111–128. doi: 10.1163/156853093X00028.
Hyers, L. L. (2006) "Myths used to legitimize the exploitation of animals: An application of Social Dominance Theory," *Anthrozoös* 19(3): 194–210. doi: 10.2752/089279306785415538.
Johnson, T. P., Garrity, T. F., and Stallones, L. (1992) "Psychometric evaluation of the Lexington Attachment to Pets Scale (LAPS)," *Anthrozoös* 5(3): 160–175. doi: 10.2752/089279392787011395.
Joy, M. (2005). Humanistic psychology and animal rights: Reconsidering the boundaries of the humanistic ethic. *Journal of Humanistic Psychology* 45: 106–130. http://dx.doi.org/10.1177/0022167804272628
Kasperbauer, T. J. (2017) *Subhuman: The Moral Psychology of Human Attitudes to Animals*, New York, NY: Oxford University Press.
Kazdin, A. E. (2000) *Encyclopedia of Psychology*, New York, NY: Oxford University Press.

Kidd, A. H., and Kidd, R. M. (1997) "Characteristics and motives of adolescent volunteers in wildlife education," *Psychological Reports* 80(3): 747–753. doi: 10.2466/pr0.1997.80.3.747.
Lago, D., Kafer, R., Delaney, M., and Connell, C. (1988) "Assessment of favorable attitudes toward pets: Development and preliminary validation of self-report pet relationship scales," *Anthrozoös* 1(4): 240–254. doi: 10.2752/089279888787058308.
Leach, C. W., Van Zomeren, M., Zebel, S., Vliek, M. L., Pennekamp, S. F., Doosje, B., . . . Spears, R. (2008) "Group-level self-definition and self-investment: A hierarchical (multicomponent) model of ingroup identification," *Journal of Personality and Social Psychology* 95(1): 144. doi: 10.1037/0022-3514.95.1.144.
Leach, E. (1964) "Anthropological aspects of language: Animal categories and verbal abuse," in E. H. Lenneberg (ed.) *New Directions in the Study of Language*, Cambridge, MA: MIT Press.
Lederer, S. E. (1992) "Political animals: The shaping of biomedical research literature in twentieth-century America," *Isis* 83(1): 61–79. doi: 10.2307/233993.
Levi-Strauss, C. (1962) *Totemism*, London: Merlin Press.
Long, C., Liu, Q., Qiu, J., Shen, X., Li, S., and Li, H. (2010) "Neural signs of flexible categorization: Evidence from the flexibility of inclusion of humans in animal/non-animal categorization," *Brain Research* 1337: 64–73. doi: 10.1016/j.brainres.2010.04.023.
Luke, B. (2007) *Brutal: Manhood and the Exploitation of Animals*, Champaign, IL: University of Illinois Press.
Marinelli, L., Adamelli, S., Normando, S., and Bono, G. (2007) "Quality of life of the pet dog: Influence of owner and dog's characteristics," *Applied Animal Behaviour Science* 108(1–2): 143–156. doi: 10.1016/j.applanim.2006.11.018.
Marinović, V., Hoehl, S., and Pauen, S. (2014) "Neural correlates of human—animal distinction: An ERP-study on early categorical differentiation with 4- and 7-month-old infants and adults," *Neuropsychologia* 60: 60–76. doi: 10.1016/j.neuropsychologia.2014.05.013.
Mathews, S., and Herzog, H. A. (1997) "Personality and attitudes toward the treatment of animals," *Society & Animals* 5(2): 169–175. doi: 10.1163/156853097x00060.
McNicholas, J., Gilbey, A., Rennie, A., Ahmedzai, S., Dono, J.-A., and Ormerod, E. (2005) "Pet ownership and human health: A brief review of evidence and issues," *British Medical Journal* 331(7527): 1252–1254. doi: 10.1136/bmj.331.7527.1252.
Milgram, S. (1963) "Behavioral study of obedience," *The Journal of Abnormal and Social Psychology* 67(4): 371–378. doi: 10.1037/h0040525.
Mitchell, L. (2011) "Moral disengagement and support for nonhuman animal farming," *Society & Animals* 19: 38–58. doi: 10.1163/156853011x545529.
Mornement, K. (2018) "Animals as companions," in C. G. Scanes and S. R. Toukhsati (eds.) *Animals and Human Society*, San Diego, CA: Academic Press.
Motyl, M., Hart, J., and Pyszczynski, T. (2010) "When animals attack: The effects of mortality salience, infrahumanization of violence, and authoritarianism on support for war," *Journal of Experimental Social Psychology* 46(1): 200–203. doi: 10.1016/j.jesp.2009.08.012.
Odendaal, J. S. J., and Meintjes, R. A. (2003) "Neurophysiological correlates of affiliative behaviour between humans and dogs," *The Veterinary Journal* 165(3): 296–301. doi: 10.1016/s1090-0233(02)00237-x.
O'Haire, M. E. (2013) "Animal-assisted intervention for autism spectrum disorder: A systematic literature review," *Journal of Autism Development Disorders* 43: 1606–1622. doi: 10.1007/s10803-012-1707-5.
Ory, M. G., and Goldberg, E. L. (1983) "Pet possession and well-being in elderly women," *Research on Aging* 5(3): 389–409. doi: 10.1177/0164027583005003007.
Palmer, R., and Custance, D. (2008) "A counterbalanced version of Ainsworth's Strange Situation Procedure reveals secure-base effects in dog—human relationships," *Applied Animal Behaviour Science* 109(2–4): 306–319. doi: 10.1016/j.applanim.2007.04.002.
Patronek, G. J. (1999) "Hoarding of animals: An under-recognized public health problem in a difficult-to-study population," *Public Health Reports* 114(1): 81–87. doi: 10.1093/phr/114.1.81.
Patronek, G. J., Glickman, L. T., Beck, A. M., McCabe, G. P., and Ecker, C. (1996) "Risk factors for relinquishment of dogs to an animal shelter," *Journal of the American Veterinary Medical Association* 209(3): 572–581.
Petrinovich, L., O'Neill, P., and Jorgensen, M. (1993) "An empirical study of moral intuitions: Toward an evolutionary ethics," *Journal of Personality and Social Psychology* 64(3): 467–478. doi: 10.1037/0022-3514.64.3.467.
Pinker, S. (2011) *The Better Angels of Our Nature: Why Violence Has Declined*, London: Penguin.
Plous, S. (1991) "An attitude survey of animal rights activists," *Psychological Science* 2(3): 194–196. doi: 10.1111/j.1467-9280.1991.tb00131.x.

Plous, S. (1993a) "Psychological mechanisms in the human use of animals," *Journal of Social Issues* 49(1): 11–52. doi: 10.1111/j.1540-4560.1993.tb00907.x.

Plous, S. (1993b) "The role of animals in human society," *Journal of Social Issues* 49(1): 1–9. doi: 10.1111/j.1540-4560.1993.tb00906.x.

Plous, S. (2003) "Is there such a thing as prejudice toward animals?" in S. Plous (ed.) *Understanding Prejudice and Discrimination*, New York, NY: McGraw-Hill.

Podberscek, A. L., and Gosling, S. D. (2000) "Personality research on pets and their owners: Conceptual issues and review," in A. L. Podberscek, E. S. Paul, and J. A. Serpell (eds.) *Companion Animals and Us: Exploring the Relationships Between People and Pets*, Cambridge: Cambridge University Press.

Regan, T. (1983) *The Case for Animal Rights*, Berkeley, CA: University of California Press.

Regan, T. (2001) *Defending Animal Rights*, Champaign, IL: University of Illinois Press.

Reicher, S. D., Haslam, S. A., and Smith, J. R. (2012) "Working toward the experimenter: Reconceptualizing obedience within the Milgram paradigm as identification-based followership," *Perspectives on Psychological Science* 7(4): 315–324. doi: 10.1177/1745691612448482.

Sanders, C. R. (2003). Actions speak louder than words: Close relationships between humans and nonhuman animals. *Symbolic Interaction*, 26: 405–426. http://dx.doi.org/10.1525/si.2003.26.3.405

Schenk, S. A., Templer, D. I., Peters, N. B., and Schmidt, M. (1994) "The genesis and correlates of attitudes toward pets," *Anthrozoös* 7: 60–68. doi: 10.2752/089279394787002041.

Serpell, J. A. (1986) *In the Company of Animals: A Study of Human-animal Relationships*, Oxford: Basil Blackwell.

Serpell, J. A. (1987) "Pet-keeping in non-western societies: Some popular misconceptions," *Anthrozoös* 1(3): 166–174. doi: 10.2752/089279388787058443.

Serpell, J. A. (1996) "Evidence for an association between pet behavior and owner attachment levels," *Applied Animal Behaviour Science* 47(1–2): 49–60. doi: 10.1016/0168-1591(95)01010-6.

Serpell, J. A. (1999) "Sheep in wolves' clothing? Attitudes to animals among farmers and scientists," in F. Dolins (ed.) *Attitudes to Animals: Views in Animal Welfare*, Cambridge: Cambridge University Press.

Serpell, J. A. (2009) "Having our dogs and eating them too: Why animals are a social issue," *Journal of Social Issues* 65(3): 633–644. doi: 10.1111/j.1540-4560.2009.01617.x.

Serpell, J. A., and Paul, E. S. (1994) "Pets and the development of positive attitudes to animals," in A. Manning and J. A. Serpell (eds.) *Animals and Human Society*, London: Routledge.

Sherif, M. (1966) *The Psychology of Social Norms*, Oxford: Harper Torchbooks.

Shore, E. R., Riley, M. L., and Douglas, D. K. (2006) "Pet owner behaviors and attachment to yard versus house dogs," *Anthrozoös* 19(4): 325–334. doi: 10.2752/089279306785415466.

Sidanius, J., and Pratto, F. (1999) *Social Dominance: An Intergroup Theory of Social Hierarchy and Oppression*, Cambridge: Cambridge University Press.

Singer, P. (1981) "The concept of moral standing," in A. Caplan and D. Callahan (eds.) *Ethics in Hard Times*, New York, NY: Plenum Press.

Singer, P. (2009) *Animal Liberation: The Definitive Classic of the Animal Movement*, 4th ed., New York, NY: Harper Collins Publishers.

Tajfel, H. (1981) *Human Groups and Social Categories: Studies in Social Psychology*, Cambridge: Cambridge University Press.

Tajfel, H., and Turner, J. C. (1986) "The social identity theory of intergroup behavior," in S. Worchel and W. G. Austin (eds.) *Psychology of Intergroup Relations*, Chicago, IL: Nelson-Hall.

Taylor, N., and Signal, T. D. (2005) "Empathy and attitudes to animals," *Anthrozoös* 18: 18–27. doi: 10.2752/089279305785594342.

Topál, J., Gácsi, M., Miklósi, Á., Virányi, Z., Kubinyi, E., and Csányi, V. (2005) "Attachment to humans: A comparative study on hand-reared wolves and differently socialized dog puppies," *Animal Behaviour* 70(6): 1367–1375. doi: 10.1016/j.anbehav.2005.03.025.

Turner, J. C., Hogg, M. A., Oakes, P. J., Reicher, S. D., and Wetherell, M. (1987) *Rediscovering the Social Group: A Self-categorization Theory*, Oxford: Blackwell.

Waytz, A., Morewedge, C. K., Epley, N., Monteleone, G., Gao, J.-H., and Cacioppo, J. T. (2010) "Making sense by making sentient: Effectance motivation increases anthropomorphism," *Journal of Personality and Social Psychology* 99(3): 410–435. doi: 10.1037/a0020240.

Wegner, D. M., and Gray, K. (2017) *The Mind Club: Who Thinks, What Feels, and Why It Matters*, New York, NY: Penguin.

Zilcha-Mano, S., Mikulincer, M., and Shaver, P. R. (2011a) "An attachment perspective on human—pet relationships: Conceptualization and assessment of pet attachment orientations," *Journal of Research in Personality* 45(4): 345–357. doi: 10.1016/j.jrp.2011.04.001.

Zilcha-Mano, S., Mikulincer, M., and Shaver, P. R. (2011b) "Pet in the therapy room: An attachment perspective on animal-assisted therapy," *Attachment & Human Development* 13(6): 541–561. doi: 10.1080/14616734.2011.608987.

Zilcha-Mano, S., Mikulincer, M., and Shaver, P. R. (2012) "Pets as safe havens and secure bases: The moderating role of pet attachment orientations," *Journal of Research in Personality* 46(5): 571–580. doi: 10.1016/j.jrp.2012.06.005.

2

UNDERSTANDING THE MORAL IMPLICATIONS OF MORGAN'S CANON

Maria Botero

Introduction

I conducted observations of chimpanzees at Gombe National Park in Tanzania, Africa. Observing chimpanzees in the wild is a unique opportunity, and making these observations in this particular place was even more special. When Jane Goodall arrived in Gombe in the 1960s, it was tradition at the time to identify nonhuman animals by numbers and not names to maintain scientific objectivity. Goodall broke this tradition by giving the chimpanzees who lived at Gombe names instead of numbers (Goodall 2010). Even though this practice is a methodological choice, it had deep ethical consequences. By recognizing chimpanzees with names, Goodall recognized them as individuals capable of having personalities.

This tension between academic rigor and describing nonhuman animal behavior can still be observed to this day when researchers hesitate to describe certain behaviors in terms of friendships or love. In my own research, I have observed this tension in the criticism I received for describing the mother–infant interaction in chimpanzees within a framework of emotions and culture. These methodological decisions, such as whether to describe behavior in terms that may be unique to humans, also have ethical implications; for example, it becomes much more complicated to justify isolation or separation practices with chimpanzees in captivity once we accept that they are capable of complex cognitive capacities such as friendship.

These examples are illustrations from a long history of research wherein focusing on what makes humans similar to and different from nonhuman species has played a fundamental role in animal ethics—in particular, where attributions of cognitive capacities have played a fundamental role in attributing moral status to nonhuman animals. In this chapter, I argue that animal ethicists need to be careful when relying on comparative cognition to justify moral claims. Some comparative psychologists, because of their reliance on Morgan's Canon (1903), engage in a systematic bias that prevents them from attributing complex cognitive characteristics to nonhuman animals. I show that researchers who adopt the canon do not have a clear argument to support this practice. Moreover, I show that even though Morgan's Canon is meant as a methodological principle that recommends epistemic caution, the adoption of this canon by researchers has resulted in an undesirable practical consequence: it stacks the deck against attributing animals complex cognitive capacities. In other words, I show that in practice, following Morgan's Canon is closer to an *a priori* commitment to the impossibility of certain higher-level explanations. I conclude by illustrating this problem in the case of the use of animal models in the research of mental illness.

Animal Ethics and Comparative Psychology

To determine our moral obligations toward nonhuman animals[1] some ethicists rely on what comparative cognition tells us about the mental capacities of animals. For example, moral status is one of the tools used to determine our moral obligations toward different beings in our moral community; if we attribute moral status to a being, others should have an obligation of not interfering with this being in ways that may be detrimental to that being. It is important to notice that the reasons for this obligation cannot be dependent on the welfare of other beings. When we grant a being moral status, the only kind of reasons that justify obligation of not interference (in ways that are detrimental) are ones that are based on the welfare of the being for its own sake. De Grazia (2008) adopts this kind of perspective and argues that: "To say that X has a moral status is to say that (1) moral agents have obligations regarding X, (2) X has interests, and (3) the obligations are based (at least partly) on X's interests" (p. 183). Viewing animals in this way requires that we understand whether animals have interests, and this requires, in turn, that we know whether we can attribute to them mental states that would ground certain interests. For that reason, the knowledge produced by comparative psychologists becomes essential when attempting to attribute moral status to animals.

Other authors have argued that to determine our moral obligations towards animal, we need to determine whether animals can be considered *moral patients* (whether animals can be a legitimate object of moral concern) or *moral agents* (whether they can be evaluated by their motives and actions as a moral agent; Regan, 2004; Rowlands 2012). Traditionally, the answer to this question relies on whether animals have the kinds of cognitive abilities judge necessary to be considered a moral agent or a moral patient—such as intentionality, the capacity to have desires and beliefs, the capacity to deliberate on moral principles, or self-control. Thus, in a way similar to the previous example, animal ethicists have to rely on the findings documented by a comparative psychologist to be able to determine the moral status of animals as *moral patients* or *moral agents*. Finally, even within cases of animals to whom we can attribute complex cognitive capacities, it is important to use studies in comparative psychology to understand the role that these complex mental characteristics will play in the moral obligations towards animals (see Botero 2017 for an example of the moral implications of our understanding whether chimpanzees can assent to participate in experiments).

These examples illustrate how comparative psychology plays an important role in animal ethics, supporting some of the normative claims made in animal ethics. This entails that, to maintain *impartiality*, the knowledge from comparative psychology can only be used if the researchers in that area do not engage in any bias when choosing the direction of their research and when determining the contents of their research. However, as shown in what follows, this is not the case. I argue that through their adoption of Morgan's Canon, comparative cognition researchers are being guided by a series of values that are extraneous to their research and that this has serious consequences for our consideration of moral obligations toward animals. This practice is not unique to comparative psychology. Scientific research, in general, can be influenced by elements extraneous to the scientific research, such as political values (Anderson 2004) or gender and ethnicity (see, e.g., Braun 2014; Schiebinger1999). The aim of describing this bias in comparative cognition is to shed light on the implications that this bias has for normative claims in animal ethics.

Comparative Psychology and Morgan's Canon

Understanding the animal mind entails devoting efforts towards observing animal behavior, as well as documenting the mechanisms that explain how the animal mind works. Understanding the animal mind also requires sensitivity to how knowledge about animal minds is created. Researchers examine how animal behavior is observed and interpreted, and how the resulting hypotheses are verified. Since the beginnings of comparative psychology, researchers have been especially preoccupied with identifying the ideal conditions under which they can provide proof (or, at least, verify) that animals possess cognitive capacities. One of the most influential ways in which comparative psychologists have attempted to maintain rigor when ascribing mental capacities to animals is through what is known as Morgan's Canon. This canon originated with a statement made by Conwy Lloyd Morgan, who argued that

> [i]n no case may we interpret an action as the outcome of the exercise of a higher psychical faculty, if it can be interpreted as the outcome of the exercise of one which stands lower in the psychological scale.
>
> *(Morgan 1894: 53)*

There are several interpretations of this statement (see Fitzpatrick 2018). In the most prevalent, it is argued that following Morgan's Canon means that, when confronted with several explanations for animal behavior, researchers must adopt the explanation that uses "less complex processes" (such as associative learning) as opposed to explanations that involve "more complex cognitive capacities" (such as symbolic or rule-based reasoning and declarative knowledge; see, e.g., Crystal and Foote 2009; Dwyer and Burgess 2011).

Morgan's Canon has been called the most cited canon in comparative psychology (Dewsbury 1984). However, it has also been widely criticized as a valid tool for understanding the animal mind (Andrews and Huss 2014; de Waal 1999; Meketa 2014; Sober 2005; Starzak 2017; Fitzpatrick 2008). In what follows, I join critics of Morgan's Canon but adopt a different perspective: I explore some of the moral implications that this methodological stance has for the moral status of animals.

Morgan's Canon as a Methodological Stance

The way Morgan's Canon has been used since the 1960s and 1970s has been simple. When researchers are presented with two alternative explanations that could potentially explain an observed behavior, they favor the simpler or "lower" form of explanation (e.g., associative learning) instead of more complex cognitive ones.[2] For example, Carruthers (2008), following Morgan's Canon, argues, with regard to recent evidence found in experiments aimed at understanding metacognitive process in animals, that researchers must refrain from ascribing complex cognitive process to these animals and favor explanations using lower cognitive processes instead.

There are two interesting consequences of this use of the canon. First, this use entails a division between higher-order cognitive abilities (such as metacognition or theory of mind) and lower-order processes that require fewer process or representational models (such as associative processes). The difference between higher and lower psychological processes is understood as one of complexity (see Sober 1998; Fitzpatrick 2008; Shettleworth 2010; Meketa 2014). Second, researchers who adopt Morgan's canon, use this canon as a methodological stance that should direct researchers in *all cases* to the simplest lower-process explanation.

These two characteristics make the application of Morgan's Canon problematic. If we take a closer look at the research produced in comparative psychology, it is possible to find that the distinction between lower and higher cognitive skills is not as clear as originally argued. For example, Starzak (2017) argues that the criteria for what constitutes higher and what constitutes lower (psychological processes) are not clear. Buckner (2011) argues that a recent generation of psychological models appears to satisfy existing criteria for both cognition and association, thus blurring the distinction between these two processes.

Moreover, it remains unclear for what reasons must researchers in all cases abstain from attributing complex mental characteristics. Mikhalevich (2018) examines the reasons behind this methodological stance which demands that all animal mental capacities be interpreted as lower ones when two competing explanations (e.g., lower and higher) are available. Mikhalevich argues that instead of a convincing argument, what lies underneath is a kind of bias, a "simplicity heuristic." That is, scientists following this canon are advised, for the sake of simplicity, against developing alternative, more complex explanations of cognitive models. However, as Mikhalevich argues, no clear arguments are given to justify why simplicity is preferable. Moreover, the author warns, always preferring explanations containing lower cognitive processes has the undesirable consequence that a disproportionate amount of intellectual effort and resources are dedicated to designing experiments that focus on "simpler models," shutting down alternative research programs and shaping the setup of future experiments.

If Mikhalevich is right, then using knowledge produced by a comparative psychology research program that follows Morgan's Canon has serious implications for attributing moral status to animals. If ascribing moral status depends on the cognitive characteristics ascribed to the animal by comparative psychologists and if these researchers are not willing to contemplate the possibility that animals may possess higher cognitive abilities, then we can only attribute to animals the kind of moral status that is available to those creatures that possess lower abilities (i.e., associate learning); that is, those creatures can only be seen as moral patients (in the best-case scenario).

It is important to clarify that what is wrong in this case is not the attribution of lower skills and the attributions of limited moral qualities as moral patients to animals (for many species this may be right); there is nothing wrong with this kind of attribution. The problem is in negating the possibility *a priori* (because of Morgan's Canon). This stacks the deck against investigating more complex cognitive abilities in animals that could potentially become a justification for the attribution of higher moral status.

The Origins of This Bias: Fear of Anthropomorphism

When researchers use Morgan's Canon as a methodological principle that directs them in *all cases* to the simplest lower-processes explanation, they are adopting the explanation for a reason not found in the behavior observed. So why begin with this principle? It seems that the appeal of Morgan's Canon stems from the perceived threat of anthropomorphism, and this reason is the one that carries the epistemic load of justifying not choosing an explanation that attributes complex mental capacities to animals.

Several authors (Burghardt 1985; Sober 1998) have argued that Morgan viewed his fellow researchers' interpretations of animal behavior as biased toward anthropomorphism and that this bias needed to be counterbalanced. In other words, Morgan thought of the canon as a way of correcting the human tendency to describe behavior in intentional terms, because such a bias is likely to result in a preference for false explanations or, at least, will undermine our justifications of the ascription of higher psychological processes to nonhuman animals. To this day, the canon continues to be recommended as a bulwark against anthropomorphism (see, e.g., Kennedy 1992).

This fear of anthropomorphism results in researchers adopting Morgan's Canon in an attempt to increase scientific rigor in the study of comparative psychology. This aim of rigor is a desirable value in research enterprises. However, in this particular case, this desire for rigor resulted in an undesirable consequence: the wide adoption of Morgan's Canon has resulted in linking complex cognitive capacities exclusively to cognitive processes that are uniquely human, such as language (see, e.g., Wynne 2007).

This practice, in turn, has consequences for ascriptions of moral considerations toward animals. For example, Bermúdez (2007) argues that, when understanding animal cognition, we must distinguish between animals whose behavior can only be explained through psychological explanations and animals whose behavior can be explained through nonpsychological forms of explanation, that is, explanations that appeal to a mechanism of associative conditioning. He argues that this distinction can be used to "mark" a morally significant line. Moreover, with animals whose behavior can be explained through psychological explanations, it is necessary to distinguish between propositional and nonpropostional thought and the ability to have higher-order thoughts that are language-dependent. As such, Bermudez concludes, any ethical theory must account for these subtle distinctions among different nonlinguistic creatures.

This interpretation of Morgan's Canon, equating complex cognitive capacities with uniquely human capacities, has been criticized by several authors who argue that in assuming that cognitive process is uniquely human, this interpretation begs the question (Andrews 2015; Allen and Bekoff 2007). Buckner (2013) argues that researchers who adopt the canon in this way have engaged in what he calls *anthropofabulations*: unrealistic idealizations of human abilities that set the comparative bar for animals unreasonably high. According to Buckner, researchers fall for these *anthrofabulations* when they assume that the only "true" description of a cognitive capacity is one that entails incremented levels of ability. For example, Buckner shows how some of the current definitions of Theory of Mind (i.e., an agent's ability to attribute others mental states such as desires and believes) used in comparative cognition require that agents must be capable of tracking the mental states of others to a degree that has rarely been observed in the human experimental research.

Moreover, from an evolutionary standpoint, de Waal (1999) and Sober (2005) argue that it is more parsimonious to attribute higher cognitive processes to animals because, in that way, we can argue for a common ancestor rather than two independent evolutionary processes.

Adopting Morgan's Canon as a default position seems to be a biased stance that discriminates based on species—in other words, a return to *speciesism*, because without clear reasons, it forces researchers to attribute complex cognitive capacities only to humans. To clarify, I am not arguing that researchers must attribute higher cognitive mechanisms to every observed animal behavior. Rather, I am arguing that Morgan's Canon is being used as a biased methodological stance that automatically forces researchers to accept or prefer the lowest explanation that is consistent with the behavior observed. In other words, the adoption of Morgan's Canon has become closer to an *a priori* commitment to the impossibility of certain higher-level explanations that could potentially be used to argue for moral status in animals.

For that reason, I am sympathetic to several researchers (Andrews and Huss 2014; de Waal 1999; Sober 2005) who argue that the under-attributions of cognitive skills to non-human animals that result from this kind of application of Morgan's Canon are a problem and that it is as important to correct this under-attributions as to correct over-attribution of cognitive skills.

The Case of Animal Models

One of the main moral consequences of adopting Morgan's Canon is that it allows researchers to treat all animals in ways that are only permissible if those beings have lower-processes mental capacities. As argued, following this canon results in an undesirable practical consequence: animals will always be denied attributions of complex cognitive characteristics. In what follows, I focus on one particular case, animal models in experimentation, and show how the denial of complex characteristics becomes a systematic bias that results in a paradoxical use of animals in research. I show how researchers engage in the paradox of using an animal to test a cognitive ability (i.e., mental illness) and, at the same time, argue that it is morally right to conduct this test because the animal does not possess the ability for which they are testing.

The Case of Animal Models in Mental Illness

Animal models are a common way of studying mental illness (e.g., schizophrenia, depression, bipolar disorder). To understand the complexity of the use of animals in this area, let's consider first how models are scientific tools designed to explain phenomena in the world. According to Craver (2006), scientific models explain when they describe the mechanism they target.[3] Mechanistic explanations are ones where the behavior of a whole is explained in terms of the operation and interaction of the mechanism's parts. For a model to truly be counted as an explanation, Craver argues, "[i]t is insufficient merely to characterize the phenomenon and to describe the behavior of some underlying mechanism. It is required in addition that the components described in the model should correspond to components in the mechanism in [the target system] T" (Craver 2006: 361). Models, he argues, describe the "real components, activities, and organizational features of the mechanism that in fact produce the phenomenon" (p. 361). In other words, for a mechanistic model to explain a feature in the world, such as a cognitive state, it must characterize the *actual* mechanisms of that cognitive state.

When creating an animal model of mental illness, researchers attempt to replicate at least one (ideally several) of the main features of the mental illness they are studying. However, this proves to be challenging because, as Nestler and Hyman (2010) argue, it is not possible to establish whether animals experience the same symptoms used to establish psychiatric diagnoses in humans (e.g., hallucinations, delusions, sadness, guilt). Most researchers, the authors argue, use "reasonable correlates" found in animals (e.g., abnormal social behavior, motivation, working memory, emotion, and executive function), although, as Nestle and Hyman warn us, the correspondence may only be "approximate." However, despite these doubts, animal models are prevalent in the study of mental illness; following Craver's characterization of the scientific model, the animal models are used to characterize actual mental illnesses, since we continue to use all of the pathophysiology and therapeutics research that is produced through animal models.

This also creates a moral contradiction described by Rollin and Rollin (2014) as the *psychologist's dilemma*. The authors argue that research in psychiatric illness is different from other kinds of experimental research in which traditionally differences between animals and humans are used to justify experiments with animals that are not acceptable with human subjects. Employing a valid animal model to study the negative impact of a mental illness in humans implicitly recognizes that the animals used in the experiment have the capability to experience suffering that's importantly like the suffering of as humans who are afflicted by that mental illness. However, despite recognizing this similarity, researchers use animals in experimentation.

I believe that this dilemma can be traced back to the use of Morgan's Canon. This canon creates a series of confusing boundaries between the mental capacities that can be attributed to animals, the mental capacities that researchers need to attribute to animals, and the mental capacities that they end up attributing to animals. As a result we have a paradox: on one hand, when researchers continue to use animal models and continue to apply the resulting knowledge to humans, in practical terms, they seem to disregard Morgan's Canon; that is, researchers seem to choose to believe that these animals do indeed experience something akin to a mental illness, and, therefore, any knowledge about mental illness acquired through these animal models is valid and real. On the other hand, researchers seem to be following Morgan's Canon because they argue that animals do not experience the same kinds of mental illness as humans, and therefore, the researchers are justified in using animals in experiments.

In other words, it seems that researchers are following Morgan's Canon, interpret the behavior of animals in animal models of mental illness (e.g., abnormal social behavior, motivation, working memory, emotion, and executive function) as less complex cognitive capacities to deny that the animals are capable of experiencing a mental illness and therefore, are suited for use in experiments. However, as evident in the paradox, if we examine the practical applications of animal models, this practice of denying a complex cognitive capacity (e.g. mental illness) seems to stem from a form of *speciesism* where human interests are doubly served: mental states are denied animals to preserve the human/animal divide and to justify the research, and at the same time, mental states are attributed to animals, since, otherwise, the research is nonsensical. As Botero and Desforges (forthcoming) argue, only the interests of the human species drive the approval of the use of animals in experimentation.

Why Using Morgan's Canon Is Morally Wrong

The application of Morgan's Canon is morally problematic. It promotes two related wrongs. First, research institutions (e.g. institutions who research animal cognition), without adequate justification, put animals at a disadvantage by denying them the opportunity of having complex cognitive characteristics attributed to them. Moreover, as described earlier, this application of Morgan's Canon also impedes future research opportunities that could potentially provide evidence of those complex cognitive attributes. Second, animals are placed in a position of vulnerability and exploitation because it is assumed that they do not have complex cognitive capacities even in cases in which they are used to investigate those complex cognitive capacities.

Notes

1. For brevity, I refer to nonhuman animals as animals in what follows.
2. There are other interpretations of Morgan's Canon (see Fitzpatrick 2018). The analysis provided is limited to the interpretation described in the previous section.
3. It is important to notice that there are other descriptions of how scientific models explain; for alternative explanations, see Bokulich, 2011.

Bibliography

Allen, C., and Bekoff, M. (2007) "Animal minds, cognitive ethology, and ethics," *Journal of Ethics* 11: 299–317.
Anderson, E. (2004) "Uses of value judgments in science," *Hypatia* 19: 1–24.
Andrews, K. (2015) "A role for folk psychology in animal cognition research," *Experimentation* 66: 1–27.
Andrews, K., and Huss, B. (2014) "Anthropomorphism, anthropectomy, and the null hypothesis," *Biology & Philosophy* 29(5): 711–729.
Bermúdez, J. L. (2007) "Thinking without words: An overview for animal ethics," *The Journal of Ethics* 11(3): 319–335.
Bokulich, A. (2011) "How scientific models can explain," *Synthese* 180(1): 33–45.
Botero, M. (2017) To what kind of research they can dissent? The role that distinguishing between biomedical and behavioral research plays in granting dissent and assent to chimpanzees used in experimentation," *Cambridge Quarterly of Healthcare Ethics* 26(2): 288–291.
Botero, M., and Desforges, D. (forthcoming) "The role of moral values in evaluation of the use of non-human animals in research." *Society & Animals.*
Braun, L. (2014) *Breathing Race into the Machine: The Surprising Career of the Spirometer from Plantation to Genetics,* Minneapolis, MN: University of Minnesota Press.
Buckner, C. (2011) "Two approaches to the distinction between cognition and 'mere association,'" *International Journal of Comparative Psychology* 24(4): 314–348.
Buckner, C. (2013) Morgan's Canon, meet Hume's Dictum: Avoiding anthropofabulation in cross-species comparisons," *Biology Philosophy* 28: 853–871.
Burghardt, G. (1985) "Animal awareness-current perceptions and historical perspective," *American Psychologist* 40: 905–919.
Carruthers, P. (2008) Meta-cognition in animals: A skeptical look. *Mind & Language* 23(1): 58–89.
Craver, C. (2006) "When mechanistic models explain," *Synthese* 153: 355–376.
Crystal, J. D., and Foote, A. L. (2009) "Metacognition in animals," *Comparative Cognition & Behavior Reviews* 4: 1–16.
De Grazia, D. (2008) "Moral status as a matter of degree?," *The Southern Journal of Philosophy* 46: 181–198.
de Waal, F. B. M. (1999) "Anthropomorphism and anthropodenial: Consistency in our thinking about humans and other animals," *Philosophical Topics* 27: 225–280.
Dewsbury, D. A. (1984) *Comparative Psychology in the Twentieth Century,* Stroudsburg, PA: Hutchinson Ross.
Dwyer, D. M., and Burgess, K. V. (2011) "Rational accounts of animal behaviour? Lessons from C. Lloyd Morgan's canon," *International Journal of Comparative Psychology* 24(4): 349–364.
Fitzpatrick, S. (2008) "Doing away with Morgan's Canon," *Mind and Language* 23(2): 224–246.
Fitzpatrick, S. (2018) "Against Morgan's canon," in K. Andrews and J. Beck (eds.) *The Routledge Handbook of Animal Minds* (pp. 437–447), London and New York: Routledge.
Goodall, J. (2010) *Through a Window: My Thirty Years with the Chimpanzees of Gombe,* Boston, MA: Houghton Mifflin Harcourt Publishing Company.

Kennedy, J. S. (1992) *The New Anthropomorphism,* Cambridge: Cambridge University Press.
Meketa, I. (2014) A critique of the principle of cognitive simplicity in comparative cognition. *Biology & Philosophy* 29(5): 731–745.
Mikhalevich, I. (2018) "Simplicity and cognitive models: Avoiding old mistakes in new experimental contexts," in K. Andrews and J. Beck (eds.) *The Routledge Handbook of Animal Minds* (pp. 427–437), London and New York: Routledge.
Morgan, C. L. (1894) *An Introduction to Comparative Psychology,* London: Walter Scott Pub. Co.
Nestler, E. J., and Hyman, S. E. (2010) "Animal models of neuropsychiatric disorders," *Nature Neuroscience* 13(10): 1161–1169. http://doi.org/10.1038/nn.2647.
Regan, T. (2004) *The Case for Animal Rights.* 1983. Updated with a new Preface, Berkeley: University of California.
Rollin, M. D., and Rollin, B. E. (2014) "Crazy like a fox: Validity and ethics of animal models of human psychiatric disease," *Cambridge Quarterly of Healthcare Ethics* 23(2): 140–151.
Rowlands, M. (2012) *Can Animals Be Moral?* Oxford and New York: Oxford University Press.
Schiebinger, L. (1999) *Has Feminism Changed Science?* Cambridge, MA: Harvard University Press.
Shettleworth, S. J. (2010) "Clever animals and killjoy explanations in comparative psychology," *Trends in Cognitive Sciences* 14(11): 477–481.
Sober, E. (1998) "Morgan's cannon," in D. Cummins and C. Allen (eds.) *The Evolution of Mind* (pp. 224–242), Oxford: Oxford University Press.
Sober, E. (2005) "Comparative psychology meets evolutionary biology: Morgan's canon and cladistic parsimony," in G. Mitman and L. Datson (eds.) *Thinking with Animals: New Perspectives on Anthropomorphism* (pp. 85–99), New York: Columbia University Press.
Starzak, T. (2017) "Interpretations without justification: A general argument against Morgan's Canon," *Synthese* 194(5): 1681–1701.
Wynne, C. D. (2007) "Anthropomorphism and its discontents," *Comparative Cognition & Behavior Reviews* 2: 151–154.

3
ANIMAL INTELLIGENCE

John M. Pearce

All animals are able to adapt to changes in their environment. This ability depends on a combination of innate reactions to specific changes (instinct) and acquired reactions that are gained through prior experience (intelligence). For more than a century, experiments with animals have been conducted in order to understand how they are intelligent. The present chapter is based on the premise that intelligence depends on the operation of a number of mental or cognitive processes. Thus, learning, memory, reasoning, and problem solving have all been said to allow an animal to adapt to a changing environment. In order to reveal what is known about animal intelligence, the present chapter presents a summary of what has been learned known about these processes. This knowledge has been acquired through more than 100 years of experimental investigation, generally in the laboratory. The following section provides a summary of how this experimental approach has developed.

Historical Background

The study of animal intelligence was touched on by Darwin (1871) when he raised the possibility that mental abilities, like physical characteristics, were shaped by evolution. Although he did not pursue this idea in detail, it was taken up by a close friend, George Romanes, who was almost 40 years Darwin's junior. In his book titled *Animal Intelligence*, Romanes (1882), adopted an approach to the study of this topic that was laudable in his desire to collect evidence to support his claims concerning animal intelligence but has been heavily criticised for reasons that are as valid today as they were over 100 years ago.

First, Romanes believed there was a progressive increase in intelligence across the animal kingdom. The belief that species can be ranked in this way remains prevalent. Nakajima et al. (2002) asked students to assign a score from 1 to 100 to reflect the intelligence of a large variety of species. A clear pattern emerged. The great apes and dolphins were at the top, with scores around 70; sheep, pigs and penguins were in the middle, with scores around 40; and worms, slugs, and amoeba were at the bottom with scores around 10. To justify such a ranking, Romanes appealed to a phylogenetic scale in which animals are ordered on the basis of whether they are simple or complex, with the assumption that one species evolves from another and, in doing so, becomes more intelligent. Of course, this representation of evolution is unjustifiable. Animals did not evolve in a linear sequence. Instead, it is more realistic to depict evolution as a tree-like structure, where the evolutionary path of each present-day species maps onto a different set of branches to any other present-day species. It is thus wrong to say, for example, that chimpanzees evolved from rats and that chimpanzees are therefore more intelligent than rats. All that can be said is that rats and chimpanzees have different evolutionary histories and that one may, or may not, be more intelligent than the other. To derive any conclusions about the intelligence of different species, it is necessary to test them directly.

To his credit, Romanes (1882) saw the importance of collecting evidence to support his claims, but a second criticism of his contribution was that the evidence was collected in an unsystematic manner by recording anecdotal reports of animals, often cats and dogs, behaving in ways that were deemed to be intelligent. But an isolated incident of a particular behaviour tells us very little about the intelligence of the perpetrator. Did the response occur by accident, was it a result of a deliberate reasoning process, or was it acquired through previous training? To answer these questions, it is necessary to conduct carefully controlled experiments, where the history of the animal is known, and the stimuli that direct its behaviour can be identified. It was for these reasons that Thorndike (1898) and numerous subsequent researchers moved the study of animal intelligence from the natural environment to the laboratory.

A third criticism of the work of Romanes (1882) is that once he had collected his evidence, he was guilty of anthropomorphism when offering an explanation for his observations. For example, if a dog opened a gate in order to escape from a garden, he proposed that the animal acquired this skill in the same way as a human might—through imitation and reasoning about the mechanical properties of the gate. Humans may well learn how to attain a goal in this way, but it does not mean that when an animal behaves in the same way as a human, that it has engaged in the same intellectual processes. To guard against such anthropomorphism, Lloyd Morgan (1894: 53) offered his famous canon:

> In no case is an animal activity to be interpreted in terms of higher psychological processes if it can be fairly interpreted in terms of processes which stand lower in the scale of psychological evolution and development.

In other words, when explaining the behaviour of an animal, always seek the simplest explanation. An alternative explanation for how the dog opened the gate, which is undoubtedly simpler than the one offered by Romanes, is that the dog learned to open it through trial and error. This was certainly the view adopted by Thorndike, who conducted experiments in which a variety of animals were required to escape from a puzzle box in order to gain food. These experiments, which were initiated in order to give the "*coup de grace* to the despised notion that animals reason," led him to propose that animal intelligence was based on nothing more than learning through trial and error.

The impetus to seek anthropomorphic explanations for animal behaviour did not end with Romanes. A hundred years after Romanes conducted his research, Fouts (1997) reports that he had the following experience while working with a chimpanzee called Washoe, who had been trained to communicate with her hands:

> When I looked into Washoe's eyes she caught my gaze and regarded me thoughtfully, just like my own son did. There was a person inside that ape 'costume.' And in those moments of steady eye contact I knew that Washoe was a child."

Of course, according to Morgan's canon, the conclusion drawn by Fouts is valid only if we can be certain that it is not possible to find a simpler explanation for Washoe's behaviour. Perhaps Washoe in the past had received a reward for gazing directly into the eyes of her trainers, and Fouts was the new recipient of this previously acquired habit. Until this explanation, and ones like it, can be rejected, the inference that Washoe had the mind of a child must be treated with caution.

The following discussion summarises what has been revealed by experimental investigations into animal intelligence. In keeping with the spirit of the previous paragraph, this discussion is guided by Morgan's canon. The first three sections outline what are widely regarded as the basic mechanisms of animal intelligence; the remaining sections examine whether the mental abilities of at least some animals extend beyond these simple mechanisms.

Learning

Learning is said to have occurred when an animal undergoes an experience that results in an enduring change in its behaviour. Typically, the experience consists of nothing more than one event being repeatedly followed by another event. An early demonstration of learning in the laboratory was provided by Thorndike, who, as noted earlier, placed a hungry animal in a box where it had to make a certain response, such as press a lever (first event), in order to escape and reach food just outside the box (second event). At first, the subject moved at random until it accidentally made the correct response. With repeated trials, the time taken to make the correct response was reduced until the animal would make it soon after being placed in the box. The animal's response to the box clearly changed as a result of its experience, and thus, learning can be said to have taken place. To explain this change in behaviour, Thorndike proposed that the experience of reward resulted in the formation of an S-R connection between the sight of the lever—the stimulus (S)—and the action of pressing it—the response (R). This connection was assumed to grow in strength with repeated trials and thus account for the more rapid occurrence of the response as training progressed. Thorndike (1911: 244) summarised this view in his Law of Effect:

> Of several responses made to the same situation, those which are accompanied or closely followed by satisfaction to the animal will, other things be equal, be more firmly connected with the situation.

Thus, whenever an animal encounters a novel situation, according to Thorndike (1911), it will behave at random until by chance it makes a response that leads to a reward. The Law of Effect then stipulates that the response will enter into an S–R association that will encourage the animal to repeat the response when it again encounters the same situation. In essence, the Law of Effect was Thorndike's account of animal intelligence.

One enduring influence of Thorndike's theorising is the idea that leaning depends on the formation of connections or associations. However, the associations that can be formed are not restricted to the S–R connections envisaged by Thorndike (1911). Turning to a rather different methodology of training, during the early 1900s, a Russian physiologist, Pavlov (1927), repeatedly presented hungry dogs with an unimportant stimulus, such as a noise, that served as the signal for the delivery of food. At first, the dog showed little reaction to the noise, apart from orienting towards it, but with repeated pairings, the dog would start to salivate as soon as the noise was turned on. Thus, as a result of the experience of the noise being paired with food, there was a change in the reaction to the noise, which provides a demonstration of learning. This learning is generally attributed to the formation of a stimulus–stimulus (S–S★) association, where S represents the noise that serves as the signal for food, S★. By virtue of this association, hearing the noise is assumed to activate a centre in the brain that is responsible for exciting food-related responses, including salivation.

Considerable interest has been directed at elucidating the circumstances under which associative learning will take place. One critically important factor is temporal contiguity. For example, food must be presented shortly after a response, if an S–R association is to develop and shortly after the relevant stimulus if an S–S★ association is to develop. A further ingredient for effective learning is surprise. In a blocking experiment by Kamin (1969), rats first received Pavlovian conditioning with a noise followed by a moderate foot shock, which resulted in rats displaying anxiety during the noise. They then received trials in which the noise was presented simultaneously with a light and followed by the same foot shock. Even though the light was paired with the shock, there was very little evidence of learning about this relationship. The original training with the noise is said to block learning about the light. To explain this outcome, Kamin argued that for learning to take place, not only must two events be paired together, but the second event must also be surprising, or unexpected. During the second stage of the experiment, the presence of the noise will ensure that the occurrence of the shock does not come as a surprise, and little will be learned about the relationship between the light and shock. In a sense, once an important event, such as shock, can be predicted by one stimulus, then there is little to be gained by learning about the relationship between the shock and any other stimulus.

The discovery of the importance of surprise, or prediction error as it is often called, is necessary for associative learning has led to the development of mathematical models of learning, which can predict how effective a particular training episode will be (e.g. Rescorla and Wagner 1972). An intriguing feature of these models is that they have much in common with certain models that account for learning in humans (Gluck and Bower 1988). Such similarity encourages the view that the basic mechanisms of learning in humans and animals are governed by the same fundamental mechanisms.

The significance of these laboratory demonstrations of learning should not be underestimated. The environments inhabited by animals contain many sequential regularities. A particular response might reliably lead to reward, or punishment, while a given stimulus might reliably signal the imminent occurrence of an attractive or aversive event. The foregoing discussion shows that associative learning allows animals to detect these regularities and thereby make use of them in order to behave adaptively.

Discrimination Learning

Thus far, we have been concerned principally with the learning that takes place when one event, a stimulus or a response, is followed by an important outcome such as food. Many tasks in the laboratory, and experiences in the natural environment, do not involve such a simple arrangement. Instead, animals are presented with a discrimination in which the outcome under consideration follows one event but not another. Despite the relative simplicity of this methodology, the study of animal discrimination learning has revealed some remarkable findings concerning animal intelligence.

An early example of discrimination learning can be found in an experiment by Pavlov (1927) in which an illuminated circle, but not an illuminated square, signalled the imminent delivery of food to a hungry dog. Initially, both stimuli were treated similarly, but with continued training the response of salivation was confined largely to trials with the circle. That is, the dog was able to solve a discrimination between the two cues. The theories mentioned above have proved valuable for understanding how discriminations are solved, but the details of this endeavour need not concern us here (see Pearce 2008, chapter 6). What is more important for present purposes is that the ability of the dog to solve the discrimination indicates it can tell the difference between a square and a circle. The study of discrimination learning thus sheds light on the perceptual skills of animals, and the kind of information they use when they learn that an event of importance is signalled by a particular stimulus. Pigeons, for example, are very good at solving discriminations between photographs when one signals food, and the other does not. This ability extends to photographs that are remarkably similar (Vaughan and Greene 1984).

Discrimination learning has also been valuable for investigating the capacity of animal memory. To solve a discrimination, the animal must keep a record of the stimulus and whether it is associated with an event such as food. Fagot and Cook (2006) trained pigeons with a large number of unrelated photographs. In order to gain a reward, the birds had to peck an illuminated disc in the presence of half of the photographs and peck a different disc in the presence of the remaining photographs. The photographs were shown one at a time with a limited number shown in each experimental session. As training progressed, new photographs were introduced to replace those where the birds were making consistently correct responses. The pigeons were eventually able to respond with a high degree of accuracy to more than 800 photographs, which demonstrates a capacity to remember many hundreds of images and how to respond in their presence. This skill is not confined to pigeons. Cook and Fagot gave baboons the same task and found they could remember more than 3,500 photographs.

Dogs, too, have an impressive memory capacity. Pilley and Reid (2011) trained a border collie, Chaser, by saying the name of an object when they gave it to her and repeatedly saying the name as she played with it. She was then told to collect the object and was praised when she did so. By gradually introducing new toys, Chaser was trained to retrieve more than 1,000 toys. Moreover, she was able to retrieve correctly the named toy when it was situated among seven other toys that had been used for training.

The Representation of Knowledge

We have seen that associative learning allows animals to learn about a large number of repeatedly occurring sequences of events, which raises the question of how are the events in a sequence remembered or represented. A simple answer to this question is that a memory for a particular event is no more than a direct copy of the event. The copy of a red circle, for example, might consist of a record of the neurons excited by the circle. The experiments in this section shed light on whether this characterisation of animal memory is accurate, or whether the events encountered by an animal are recorded in a more sophisticated manner.

If a human is shown a set of photographs with some containing a person, and some not, they would have little difficulty in sorting the photographs into two piles based on these categories. The ease with which this task can be performed demonstrates our ability to acquire categories, whereby objects that have something in common—a picture of a person—but differ in other respects, are treated in the same way. The ability to form categories has been said to be unique to humans. For example, Hunt (1982: 48) stated that "[u]nlike any other animal, we have a natural ability to group objects or events into categories."

At first sight, the results from experiments with pigeons appear to challenge Hunt's (1982) conclusion. Aust and Huber (2001; see also Herrnstein et al. 1976), presented pigeons with 80 colour photographs, 40 of which contained an image of a person. Pecks to a Perspex window in front of a photograph resulted in reward but only during the photographs showing a person. The birds soon mastered this task, by pecking rapidly only when a photograph contained a person, which suggests that like humans, they can acquire categories. It is unlikely that this ability depends on retaining a memory of how to respond in the presence of each image, because the birds responded correctly when tested with novel photographs. Other categories that have been used successfully with animals as subjects include trees, bodies of water, a cartoon character, classical music, and paintings by artists.

An insight into how the birds solve categorisation problems can be found in an experimental manipulation by Aust and Huber (2001) in which the colour photographs with and without a person were broken up into small squares that were rearranged randomly. For human participants, this manipulation results in the pictures becoming meaningless and impossible to classify, but the pigeons were still able to perform with a fair degree of accuracy. The explanation for this outcome is that photographs of humans contain areas that are flesh-coloured, and it seems that the birds performed on the basis of whether the picture on display contained an area with the colour of flesh. Once the images are scrambled, provided they contain areas that are flesh-coloured, then the problem can be still solved successfully. Thus, animals appear to solve at least some categorisation problems by identifying a simple physical feature that allows them to differentiate between the two sets of stimuli. Although humans are no doubt capable of acquiring this strategy, they also have at their disposal a more abstract approach that allows them to define categories on the basis of the presence or absence of a person.

For reasons that will be made clear shortly, a considerable amount of effort has been expended on determining whether animals are capable of solving discriminations based on relationships. In order to determine if pigeons are able to appreciate the relationships of sameness and difference, I conducted a variety of experiments in which pigeons were shown patterns on a television screen behind a clear Perspex response panel (Pearce 1988). The patterns were composed of two vertical bars with a gap between them. If the bars were of the same height, then pecking the response panel resulted in food; if the bars were of different heights, then pecking did not produce food. This is a trivially simple problem for humans to solve, but when given to pigeons, they failed miserably by showing no sign of responding more rapidly during the "same" than the "different" patterns. In my hands, at least, pigeons appear to be incapable of identifying whether two objects are the same or different height.

Alternative strategies for teaching animals to solve discriminations based on relationships appear, at first sight, to have been more successful than my attempts. Katz and Wright (2006) showed pigeons pairs of photographs. If they were identical, then making one response led to reward, but if the two photographs were different, then an alternative response was required to gain reward. The birds eventually mastered the task by responding correctly during novel pairs of photographs—either same or different, but the task was not easy. Thousands of training trials were required with a total of 256 different photographs. Monkeys can also master this task. According to Wright and Kelly (2017), the performance of both species reveals they are capable of "higher-order abstraction with a functional manipulation leading to successful concept learning." In other words, the authors believe that pigeons and monkeys are able to appreciate the relationships of sameness and difference.

In contrast to this conclusion, and in keeping with my findings, Premack (1978, 1983; Premack 1983) argued that, with the exception of chimpanzees, animals are insensitive to any sort of relationship, including sameness and difference. To explain results of the kind described by Katz and Wright (2006), he argued that the success of both species depends on nothing more than an ability to differentiate between familiarity and novelty. When presented with two photographs, an animal might look at one and then the other. The sight of the second one will then arouse the sensation of novelty, if it has the same impact on the nervous system as the first one, and the sensation of novelty, if it has a different impact. The task then becomes one of learning to make responses based on the sensations engendered by novelty and familiarity rather than on the perceived relationships of sameness and difference.

Tests such as the one devised by Katz and Wright (2006) therefore offer, at best, ambiguous evidence concerning the ability of animals to apprehend relationships. To provide more convincing evidence, Premack (1983) proposed that we test for the ability of animals to perceive second-order relationships. In such a test, animals are presented with two separate pairs of objects. They then have to judge whether the relationship between the objects in one pair corresponds to the relationship between the objects in the other pair. If an animal can be shown to be capable of making this judgement, only then can it be said to be truly capable of appreciating relationships. Given the complexity of the task, it should not be surprising to learn that it is not easy for animals to solve. Indeed, only one species, the chimpanzee, has been shown to be capable of mastering a second-order relationship, and this was achieved after special training (see Oden et al. 1988, 1990).

Why is it so important to know whether an animal is able to respond to a relationship such as sameness or difference? To answer this question, we must again turn to Premack (1983). He argued there are two ways in which an animal might retain information about a particular event. One is in the form of a *concrete* or *imaginal* code, where a copy of the event is retained. By way of crude example, if a pigeon is exposed to a red circle, then, as noted earlier, it might retain a record of the neurons that were excited by this stimulus. The other is in the form of an abstract code, where an event is represented in an abstract, symbolic manner. By way of a further crude example, I can use the abstract medium of language to represent a red circle with the phrase "I saw a red circle." In order to maintain a memory of an object such as a red circle, Premack was happy to concede that animals possess a concrete code, but he was keen to seek evidence before granting them an abstract code. The evidence he sought was in their ability to appreciate relationships. It is not possible, he argued, to form a record of a relationship in a concrete code. I can form a concrete code, or image, of a man beside a boy, but I can't form an image of the relationship between them. The man might be the boy's stepfather, and to encode this information, it is necessary to use an abstract, symbolic code. This is true for all relationships, including sameness and difference. They do not impinge on the nervous system in the same way as a red circle might. Rather, they are an abstraction by the nervous system, based upon a particular experience. To keep a record of this abstraction, it is necessary to encode it symbolically. Thus, the absence of convincing evidence about whether animals other than chimpanzees can appreciate relationships is profoundly important because it implies that chimpanzees alone have the capacity to make use of abstract information, perhaps by means of a symbolic code.

Problem Solving

When an obstacle is placed between an animal and its goal, it can be said to be confronted by a problem. Animals overcome at least some of the problems they encounter, which raises the question of how they achieve their success. The first researcher to address this question was Thorndike (1898), who suggested there are two possible ways in which a problem can be solved. If the problem is novel then the solution can emerge only through trial and error. Alternatively, if the animal has already solved a similar problem through trial and error, then there may be a stimulus present in the new problem that elicits the correct response because it was present in the original problem. That is, transfer of associative learning from one problem to another is possible, provided they share common features.

Not everyone was happy with this account of how animals solve problems. An early objector was Kohler (1925), who believed that animals are sophisticated problem solvers who reach the correct solution through insight. In one experiment, a group of chimpanzees was released into a cage with a banana suspended out of reach from the ceiling. There was also a box in the cage. When confronted with this problem, one of the chimpanzees paced up and down and then suddenly moved the box nearer to the banana, climbed on the box, and leapt upwards in order to retrieve its goal. There is no suggestion in this example of the animal performing at random to attain the banana. Rather, it seemed to Kohler that the solution occurred quite suddenly to the chimpanzee, as if it had come to him in a flash of inspiration or insight. This interpretation has not gone uncriticised. Prior to this incident, the chimpanzee had experience of playing with boxes, which might have allowed it through trial and error to learn about the benefits of jumping upward from a box in order to gain reward, and his behaviour on the test trial might have reflected nothing more than the transfer of this learning to the novel situation. In support of this interpretation, it has been shown that in the absence of any prior experience with boxes, chimpanzees are unlikely to perform in the manner observed by Kohler (Birch 1945; Schiller 1952). A further criticism is that to claim a behaviour occurs through insight is saying little more than the behaviour occurred suddenly. It offers no explanation for how the problem was solved or why the response occurred when it did. Given these criticisms, the term *insight* is rarely used nowadays, but experiments continue to investigate how animals solve problems.

One line of enquiry has been directed at demonstrating that birds such as rooks and crows solve problems not by trial and error but through an ability to appreciate causal relationships or to engage in analogical and causal reasoning (Bird and Emery 2009). In one task, birds were confronted with a vertical tube partially filled with water and with food floating on the surface of the water. The tube was quite narrow so that birds were unable to reach the food. Their response to this problem was to pick up stones lying nearby and drop them in the water until the level rose to a point where the food could be reached. One interpretation of this finding is that the birds understand that dropping stones into water will raise its level and, thus, make the food accessible, which has been said to reveal a level of causal understanding that matches the ability of a child aged 5 to 7 (Jelbert et al. 2014). This conclusion has not gone unchallenged. Ghirlanda and Lind (2017) were able to explain these and related findings by assuming birds are disposed to pick up functional objects, such as pebbles and then learn through trial and error about the beneficial effects of dropping them into water. The reward that promotes such learning would be the gradual closing of the distance between the bird and food. For a related analysis, see Hennefield et al. (2018).

Results of the sort described by Bird and Emery (2009) show that when confronted with a problem, certain animals behave in a way that some regard as evidence of reasoning. However, as we have just seen, it is possible to find an alternative explanation for this behaviour that is based on nothing more sophisticated than associative learning. There is no doubt that animals can behave in intriguing ways when given problems to solve, but until we can be confident that their behaviour is not a consequence of association formation, it will not be possible to regard their performance as compelling evidence of a sophisticated process such as reasoning. To my knowledge, no study of animal problem solving meets this requirement

Learning From Others

Thus far, animals have been shown to adapt to a changing environment through learning based on their own experiences. We now consider whether they can also adapt to a changing environment through learning based on the experiences of other animals, including humans. An obvious possibility is that once one animal has found a way to gain reward, other animals will copy or imitate its actions in order to gain the same reward.

A variety of tests have been devised to test for imitation in animals. Using the so-called do-as-I-do test, a human trainer performs a novel action, and the animal is expected to copy it. Successful performance on this test has been demonstrated with two chimpanzees who copied such actions as touching the back of their head or clapping their hands (Custance et al. 1995). A parrot has also performed well on this test (Moore 1992). By way of example, the trainer would repeatedly enter the room where the parrot was housed and perform an action while speaking a word. The trainer might nod his head while saying, "Nod." After an incubation period of several months, the parrot was observed to perform the action spontaneously while saying the relevant word. An intriguing feature of these experiments is that the parrot never received a reward for imitating and, although the chimpanzees occasionally received a reward to sustain their interest in the task, it was not dependent on how well the response matched the observed action. Thus, the reasons for the observed imitation are hard to identify.

There are also reports of successful imitation when the demonstrator and the observer belong to the same species. In an experiment by Akins and Zentall (1996), Japanese quail were trained either to step or peck on a pedal in order to gain reward while they were observed by other quail. When placed in the apparatus the observers showed a marked preference for making the same response as the demonstrator. Despite these successes, demonstrations of one animal imitating another are relatively rare. It is also worth noting that when capuchin monkeys watched a demonstrator perform a simple response in order to gain food, despite observing this activity for 50 or more trials, they failed to show any hint of imitating the successful action (Visalberghi 1993; Fragaszy and Visalberghi 2004). It would appear that animals are reluctant imitators, and when they do imitate, it is not necessarily to gain reward.

When imitation does take place, there is the problem of explaining how the correct response is selected. If a chimpanzee sees a human touch the back of his or her head, for example, of all the responses at its disposal, why does it make the correct one? It is possible to appeal to the principles of associative learning to answer this question if the observed response repeatedly leads to reward (see Heyes 2001), but, if it does not lead to reward, a satisfactory answer remains elusive (see Custance et al. 1995).

A second way in which animals might make use of other animals, and humans, in order to gain reward is to draw conclusions about their mental states (e.g. Premack and Woodruff 1978). A simple experiment to test this proposal was conducted by Povinelli and Eddy (1996; see also Povinelli 2000) who first trained chimpanzees to make a begging gesture in front of a trainer in order to receive food. Next, the chimpanzee was confronted with two trainers standing side by side with food between them. One of the trainers, but not the other, wore a blindfold. In order to gain food, it would make sense for the chimpanzee to beg from the trainer without a blindfold as the other trainer would be unable to see this request and therefore be expected to ignore it. While this strategy might seem obvious, it depends on a sophisticated reasoning process. The chimpanzee must infer that she will receive food only if she begs in front of a person who knows she is there. The chimpanzee must also infer that a blindfold will prevent a human from acquiring this knowledge. Thus, to gain reward, the chimpanzee must engage in reasoning based on the assumed mental state of a human or, as it is often described, possess a theory of mind (Woodruff and Premack 1979; Tomasello et al. 2003).

In fact, in this experiment, and many similar ones reported by the authors, there was no indication of the chimpanzee showing a preference of begging towards the person whose eyes were visible. For a similar result with dogs, see Udell and Wynne (2008). The simplicity of the experimental design, and the clear-cut nature of its findings, lends strong support to the view that animals do not make inferences about the mental states of others. Nonetheless, not everyone accepts this conclusion, and additional experiments have been conducted in an attempt to show that animals do indeed possess a theory of mind (e.g. Krupenye et al. 2016). It is important to note, however, that plausible explanations, which do not depend on one animal drawing inferences about the mental state of another animal, have been offered for these findings (e.g. Heyes 2017; Povinelli and Vonk 2004). At best, therefore, the evidence showing that animals have a theory of mind is ambiguous.

Language

Human language consists of sentences, which are composed of words combined through rules of grammar. Language is a powerful tool for expressing and developing our intelligence. Moreover, language has enabled humans to cooperate in extremely sophisticated ways that have transformed dramatically our environment. There is insufficient space to pursue in detail the question of whether animals communicate with each other through language. Instead, we can note that the achievements brought about by the social interactions of any species fall far short of those displayed by humans. One explanation for this gulf between the social achievements of animals and humans is that we possess language and they do not. In this final section, I shall consider if it is possible for animals to acquire language when given suitable training. It hardly needs to be said that if an animal could be trained to communicate through language, the implications for our understanding of animal intelligence would be profound.

The first attempt to teach an animal language that met with any success involved a chimpanzee called Washoe, who was trained to communicate by means of a sign language used among the deaf in North America (Gardner and Gardner 1969). Signs, which were equivalent to words, were created by the hands being placed in a particular position and making a particular action. By combining signs according to rules of grammar it was possible to create sentences. By the age of 5, after four years of training, Washoe had a vocabulary or some 150 words that included nouns, pronouns, and verbs. She was able to create sentences, such as "You tickle me," and she was able to answer questions. When asked, "What's that?" of a swan on a lake, she replied with the novel utterance "water bird" because she did not know the word for *swan*. This reply caused some excitement because it implies that she was using rules of grammar to create a novel utterance, where *water* is an adjective. Additional studies with chimpanzees, but with different methods of training, revealed similar findings (Premack 1976; Rumbaugh 1977). In view of these results, and her related findings with a gorilla, Patterson (1978) drew the exciting conclusion that "language is no longer the exclusive domain of man."

The optimism inspired by these studies, however, was dashed when Terrrace et al. (1979) published an article titled "Can an Ape Create a Sentence?" The article reports the findings of a carefully conducted study with Nim, a chimpanzee trained in much the same way as Washoe. A thorough examination of Nim's utterances revealed little evidence of them being grammatical. Terrace et al. also carefully examined Washoe's communications and again found little evidence of them being grammatical. Even when sentences adhered to grammatical rules, such as "You tickle me," they were explained as being complex responses that the chimpanzee had learned through trial and error in order to gain a particular reward. There has been no convincing rebuttal of the arguments put forward by Terrace et al. and, thus, no good reason to believe animals can create that hallmark of language, a grammatically structured sentence.

Summary and Conclusion

Discussions of the ethical treatment of animals are often informed by assumptions concerning their intellectual skills. If animals have sophisticated intellectual abilities, then they may be viewed in a different light than if their intellectual skills are extremely limited. Unfortunately for such discussions, an examination of the literature reveals two different schools of thought about the nature of animal intelligence. One school of thought is that at least some species possess a reasonably sophisticated intelligence. Animals are assumed to predict the behaviour of humans, and other animals, by means of a theory of mind (Hare et al. 2001), to possess sophisticated problem-solving skills (Bird and Emery 2009), to make use of a mental model of causal structure when imitating (Bates and Byrne 2010), and are capable of mastering language (Patterson 1978). We have seen that many of these claims are not supported convincingly by experimental evidence.

The second school of thought, to which I belong, is firmly guided by Lloyd Morgan's (1894) canon and assumes that animal intelligence is based on nothing more than association formation. Thus, associative explanations have been offered for imitation, for behaviour purported to rely on a theory of mind, for the effects of being trained to communicate through language, and for problem solving. A weakness of this approach is that the explanations offered are not always fully convincing. Thus, by way of example, it is not easy to offer an associative explanation for imitation when it occurs without reward or for problem solving when the successful behaviour is very different to behaviour acquired during prior training. Given the limited experimental evidence relating to these phenomena, it is too early to know if they pose a fundamental challenge to an associative account of animal intelligence. An important goal for future research, therefore, is to determine whether animal intelligence can be understood fully by reference to associative principles or whether some animals are able to behave in ways that involve mental processes that go beyond association formation. Once this has been achieved, we shall be in a better position to use evidence concerning the nature of animal intelligence to inform our discussion about our ethical obligations towards animals.

Further Reading

Boakes, R. A. (1984) *From Darwin to Behaviorism*, Cambridge: Cambridge University Press.
(A wonderfully clear account of the early history of the study of animal intelligence.)

Pearce, J. M. (2008) *Animal Learning and Cognition: An Introduction*, 3rd ed., Hove: Psychology Press.
(An undergraduate text book covering all the topics considered in this chapter.)

Shettleworth, S. J. (2010) *Cognition, Evolution, and Behavior*, Oxford and New York: Oxford University Press.
(Advanced coverage of much of the material considered in this chapter.)

Bibliography

Akins, C. K., and Zentall, T. R. (1996) "Imitative learning in male Japanese quail (*Coturnix japonica*) using the two-action method," *Journal of Comparative Psychology* 110: 316–320.

Aust, U., and Huber, L. (2001) "The role of item- and category-specific information in the discrimination of people versus nonpeople images by pigeons," *Animal Learning and Behavior* 29: 107–119.

Bates, L. A., and Byrne, R. W. (2010) "Imitation: What animal imitation tells us about animal cognition," *Advanced Review* 1: 685–695.

Birch, H. G. (1945) "The relation of previous experience to insightful problem solving," *Journal of Comparative Psychology* 38: 367–383.

Bird, D. C., and Emery, N. J. (2009) "Rooks use stones to raise the water level to reach a floating worm," *Current Biology* 19: 1410–1414.

Custance, D. M., Whiten, A., and Bard, K. A. (1995) "Can young chimpanzees imitate arbitrary actions? Hayes and Hayes (1952) Revisited," *Behaviour* 132: 839–858.

Darwin, C. R. (1871) *The Descent of Man and Selection in Relation to Sex*, London: Murray.

Fagot, J., and Cook, R. G. (2006) "Evidence for large long-term memory capacities in baboons and pigeons and its implications for learning and the evolution of cognition," *Proceedings of the National Academy of Science of the USA* 103: 17564–17567.

Fouts, R. (1997) *Next of Kin*, New York, NY: William Morrow.

Fragaszy, D., and Visalberghi, E. (2004) "Socially biased learning in monkeys," *Learning and Behavior* 32: 24–35.

Gardner, R. A., and Gardner, B. T. (1969) "Teaching sign language to a chimpanzee," *Science* 165: 664–672.

Ghirlanda, S., and Lind, J. (2017) "'Aesop's fable' experiments demonstrate trial-and-error learning in birds, but no causal understanding," *Animal Behaviour* 123: 239–247.

Gluck, M. A., and Bower, G. H. (1988) "From conditioning to category learning: An adaptive network model," *Journal of Experimental Psychology: General* 117: 225–244.

Hare, B., Call, J., and Tomasello, M. (2001) "Do chimpanzees know what conspecifics know?" *Animal Behaviour* 61: 139–151.

Hennefield, L., Hwang, H. G., Weston, S. J., and Povinelli, D. J. (2018) "Meta-analytic techniques reveal that corvid causal reasoning in the Aesop's Fable paradigm is driven by trial-and -error learning," *Animal Cognition* 21: 735–748.

Herrnstein, R. J., Loveland, D. H., and Cable, C. (1976) "Natural concepts in pigeons," *Journal of Experimental Psychology: Animal Behavior Processes* 2: 285–311.

Heyes, C. M. (2001) "Causes and consequences of imitation," *Trends in Cognitive Sciences* 5: 253–261.

Heyes, C. M. (2017) "Apes submentalise," *Trends in Cognitive Science* 21: 1–2.

Hunt, M. (1982) "How the mind works," *New York Times Magazine,* January 24, issue.

Jelbert, S. A., Taylor, A. H., Cheke, L. G., Clayton, N. S., and Gray, R. D. (2014) "Using the Aesop's Fable paradigm to investigate causal understanding of water displacement by New Caledonian crows," *PLoS One* 3: e92895.

Kamin, L. J. (1969) "Predictability, surprise attention, and conditioning," in B. A. Campbell and R. M. Church (eds.) *Punishment and Aversive Behaviour* (pp. 279–296), New York: Appleton-Century-Crofts.

Katz, J. S., and Wright, A. A. (2006) "*Same-different* abstract-concept learning by pigeons," *Journal of Experimental Psychology: Animal Behavior Processes* 32: 80–86.

Kohler, W. (1925) *The Mentality of Apes*, New York, NY: Harcourt.

Krupenye, C., Kano, F., Hirata, S., Call, J., and Tomasello, M. (2016) "Great apes anticipate that other individuals will act according to false beliefs," *Science* 354: 110–114.

Moore, B. R. (1992) "Avian imitation and a new form of mimicry: Tracing the evolution of complex learning," *Behavior* 122: 231–263.

Morgan, C. L. (1894) *An Introduction to Comparative Psychology*, London: Scott.

Nakajima, S., Arimitsu, K., and Lattal, K. M. (2002) "Estimation of animal intelligence by university students in Japan and the United States," *Anthrozoos* 15: 194–205.

Oden, D. L., Thompson, J. K., and Premack, D. (1990) "Infant chimpanzees (*Pan troglodytes*) spontaneously perceive both concrete and abstract same/different relations," *Child Development* 61: 621–631.

Oden, D. L., Thompson, R. K. R., and Premack, D. (1988) "Spontaneous transfer of matching by infant chimpanzees (*Pan troglodytes*)," *Journal of Experimental Psychology: Animal Behavior Processes* 14: 140–145.

Patterson, F. G. P. (1978) "The gestures of a gorilla: Language acquisition in another pongid," *Brain and Language* 5: 72–97.

Pavlov, I. P. (1927) *Conditioned Reflexes (G. V. Anrep, Trans.)*, London: Oxford University Press.

Pearce, J. M. (1988) "Stimulus generalization and the acquisition of categories by pigeons," in L. Weiskrantz (ed.) *Thought Without Language* (pp. 132–155), Oxford: Oxford University Press.

Pearce, J. M. (2008) *Animal Learning and Cognition: An Introduction*, 3rd ed., Hove: Psychology Press.
Pilley, J. W., and Reid, A. K. (2011) "Border collie comprehends object names and verbal referents," *Behavioural Processes* 86: 184–195.
Povinelli, D. J. (2000) *Folk Physics for Apes*, Oxford: Oxford University Press.
Povinelli, D. J., and Eddy, T. J. (1996) "What young chimpanzees know about seeing," *Monographs of the Society for Research in Child Development* 61: 1–190.
Povinelli, D. J., and Vonk, J. (2004) "We don't need a microscope to explore the chimpanzee's mind," *Mind and Language* 19: 1–28.
Premack, D. (1976) *Intelligence in Ape and Man*, Hillsdale, NJ: Lawrence Erlbaum Associates.
Premack, D. (1978) "On the abstractness of human concepts: Why it would be difficult to talk to a pigeon," in S. H. Hulse, H. Fowler, and W. K. Honig (eds.) *Cognitive Processes in Animal Behavior* (pp. 421-451), Hillsdale, NJ: Lawrence Erlbaum Associates.
Premack, D. (1983) "Animal cognition," *Annual Review of Psychology* 34: 351–362.
Premack, D., and Woodruff, G. (1978) "Does the chimpanzee have a theory of mind?" *Behavioral and Brain Sciences* 4: 515–526.
Rescorla, R. A., and Wagner, A. R. (1972) "A theory of Pavlovian conditioning: Variations in the effectiveness of reinforcement and nonreinforcement," in A. H. Black and W. F. Prokasy (eds.) *Classical Conditioning II: Current Research and Theory*, New York, NY: Appleton-Century-Crofts.
Romanes (1882) *Animal Intelligence*, London: Kegan, Paul, Trench & Co.
Rumbaugh, D. M. (1977) *Language Learning by a Chimpanzee: The LANA Project*, New York, NY: Academic Press.
Schiller, P. H. (1952) "Innate constituents of complex responses in primates," *Psychological Review* 59: 177–191.
Terrace, H. S., Petitto, L. A., Sanders, R. J., and Bever, T. G. (1979) "Can an ape create a sentence?" *Science* 206: 891–902.
Thorndike, E. L. (1898) "Animal intelligence: An experimental study of the associative processes in animals," *Psychological Monographs* 2(4, Whole No. 8).
Thorndike, E. L. (1911) *Animal Intelligence: Experimental Studies*, New York, NY: Macmillan.
Tomasello, M., Call, J., and Hare, B. (2003) "Chimpanzees versus humans: It's not that simple," *Trends in Cognitive Science* 7: 239–240.
Udell, M. A. R., and Wynne, C. D. L. (2008) "A review of domestic dogs' (*Canis familiaris*) Human-like behaviors: Or why behavior analysts should stop worrying and love their dogs," *Journal of the Experimental Analysis of Behavior* 89: 247–261.
Vaughan, W., Jr., and Greene, S. L. (1984) "Pigeon visual memory capacity," *Journal of Experimental Psychology: Animal Behavior Processes* 10: 256–271.
Visalberghi, E. (1993) "Capuchin monkeys: A window into tool use activities by apes and humans," in K. Gibson and T. Ingold (eds.) *Tools, Language and Cognition in Human Evolution,* Cambridge: Cambridge University Press.
Woodruff, G., and Premack, D. (1979) "Intentional communication in the chimpanzee: The development of deception," *Cognition* 7: 333–362.
Wright, A. A., and Kelly, D. M. (2017) "Comparative approaches to same/different abstract-concept learning," *Learning and Behavior* 45: 323–324.

4 THE EMOTIONAL LIVES OF ANIMALS

Kristina M. Horback

Why Do We Care About Emotions in Animals?

As Marian Stamps Dawkins (1990) said, "[l]et us not mince words: Animal welfare involves the subjective feelings of animals" (p. 1). Animal welfare science is the study of the biological and psychological state of an animal under human care, such as laboratory, companion, zoo, and farm animals, as it attempts to cope with its environment (Broom 1986). In this definition, welfare includes both pleasurable and unpleasant mental states such as contentment, anxiety, and fear (Fraser and Duncan 1998). At any one point in time, an animal's welfare status lies on the continuum between poor welfare (i.e., suffering; Dawkins 2008) and good welfare (i.e., thriving). The ethical opinions toward ensuring good welfare for captive animals may differ between cultures, regions, time, and individuals of the same culture, region, and time (Cohen et al. 2009). A person may change their opinion on how important it is to prevent a mouse from experiencing negative affective states, like boredom or anxiety, depending on whether the mouse is considered a pet, a laboratory animal, or a pest.

For many years, the predominant method of assessing animal welfare focused on overt physical ailments, such as illness, lameness, and body condition, while ignoring the emotional state of the animal. This is due in large part to the difficulty in validating assumed emotional states in animals. The scarcity of research on emotions in animals may also be due to the reluctance of certain scientific communities from accepting a seemingly controversial topic: that animals experience subjective emotions. This reluctance, however, is in paradox to the decades of considerable experimental work using animals as biomedical models to investigate the mechanism of action for pain relief, drug use (i.e., addiction), and chronic stress (i.e., posttraumatic stress disorder).

Captive animals are exposed to a range of routine management procedures which can trigger negative affective states, such as fear, distress, frustration, or boredom. Acute negative affective states, like fear or hunger, are adaptive for the animal. Proponents of the physiological concept of allostasis, which means stability through change, argue that preventing an animal from experiencing any negative emotion is not the correct path toward enhancing welfare. Instead, allowing animals to demonstrate a capacity to change or adapt to environmental challenges is essential for building positive animal welfare (Korte et al. 2007).

While acute bouts of stress may be beneficial for the animal, chronic negative states, such as distress, can be very detrimental for the animal's physiological and psychological health. *Distress* can be defined as "an aversive and negative state in which coping and adaptation processes in response to stressors fail to return an organism to physiological and/or psychological homeostasis" (NRC, 2008, p. 3). For livestock species, common practices such as shearing, castration, tail docking, dehorning, vaccination, herding, and transportation have all been reported to cause both acute and chronic negative affective states. Animals which experience chronic negative affective states, such as distress, often decrease locomotor activity, feeding behaviors, and display physiological changes, such as altered circadian rhythm, body temperature, or body weight (Boissy et al. 2005).

In order to improve an animal's welfare, we must not only reduce the severity and quantity of adverse situations that can result in a chronic negative affective state; we must also create an environment that allows animals to meet the demands and, thus, experience positive affective states. The absence of negative emotions does not automatically ensure positive emotions, and it is relevant to find and use indicators of positive emotions in animals. The challenge lies in identifying and validating physiological, behavioral, and cognitive indicators of both negative and positive emotional states in captive animals. Advances in modern science, from psychology to neuroscience, will enhance our ability to recognize emotions in animals, which, in turn, may improve the quality of care given and, perhaps, animal welfare.

Defining Emotions in Animals

As verbal human beings, we have the ability to express our subjective experience of an emotion to other members of our species, whether it is a visceral reaction to (i.e., heart rate increase) or cognitive appraisals of (i.e., is this a real threat?) specific stimuli. An emotion can be defined as

> an internal, central (as in central nervous system) state, which is triggered by specific stimuli (extrinsic or intrinsic to the organism) . . . [and] . . . is encoded by the activity of particular neural circuits that give rise . . . to externally observable behaviors, as well as to associated cognitive, somatic, and physiological responses.
>
> *(Anderson and Adolphs 2014: 188)*

Although this definition is debatable, there is agreement in the literature that emotional responses evolved to guide both human and nonhuman animals in achieving survival goals and increase fitness (i.e., seek resources and avoid harm) (Nesse 1990; Oatley and Jenkins 1998).

Through repeated paired associations, organisms learn which elements in their environment (physical or social) may benefit or threaten their fitness. For example, there are similar neuropathways involved in the experience of pleasure leading up to and following consumption of food. It is beneficial for an organism to experience pleasure when eating as it increases the likelihood of the behavior occurring again. In contrast, the anatomy and the connections of neural circuits involved in the experience of fear are also similar across phylogenies. It is also beneficial for organisms to learn to avoid stimuli in their environment that could harm or kill them. The experience of fear will reduce the likelihood of a certain behavior occurring again (e.g., getting close to a predator). The adaptive significance of emotions is the result of three major systems: physiological arousal (e.g., run from a predator), behavioral tendencies and expressions (e.g., social benefits), and cognitive changes (e.g., motivation and planning for future).

For humans, identifying emotions in conspecifics relies on three systems: verbal reports, autonomic response, and physical expression (Oatley and Jenkins 1998). Unfortunately, interpreting the emotional state of a nonhuman animal is not as easy. Emotions in animals can be inferred through physiological measurements (e.g., heart rate and glucocorticoid levels), in combination with behavioral measurements (e.g., body posture, ear posture, pupil dilation) and cognitive assessment (e.g., judgment bias test). With that said, researchers of both human and animal emotions may do their best to quantify the biological, psychological, and cognitive state that the individual is experiencing, but "we cannot know what they feel" (de Waal 2011: 199). A fundamental characteristic of emotions is that the experience is subjective, with varying degrees of consciousness.

The scientific study of emotions in animals often categorizes these subjective experiences into two types: discrete emotions and dimensional affective states. Discrete emotions are rapid, visceral responses to the perception of specific stimuli; when the organism no longer perceives the stimulus, the emotional response fades. For example, the perception through sight, sound, or smell of a predator can elicit a fear response in a prey animal. There are six discrete, or primary, emotions often described in the literature on human emotions: happiness, sadness, fear, surprise, anger, and disgust (Ekman et al. 1969). In the comparative neuroscience field, some argue for a different set of core emotional affects present in both humans and animals: seeking, fear, rage, lust, care, panic, and play (Panksepp 1998).

While some researchers argue for the inclusion or deletion of other core emotions, there is a parallel field of research that defines emotions on a spectrum rather than in mutually exclusive terms. This dimensional approach suggests that there are multiple types of emotions an organism may experience based on a difference in valence (positive to negative) and arousal (high to low; Russell 1980). For example, some emotions may be high arousal but differ in their valence (excited vs. rage), whereas others may be of similar valence but differ in their intensity (content vs. excited; Figure 4.1). While an emotional response is caused by a particular stimulus and will fade once the stimulus is removed, affective states are longer-lasting, diffuse moods that are the result of accumulated experience (e.g., current environmental stimuli and previous experience; Russell 2003; Barrett et al. 2007). The cause of certain chronic affective states, such as anxiety or boredom, is the focus of current efforts in applied animal welfare science (Meagher and Mason 2012).

The predominant method of assessing animal welfare focuses on overt physical ailments, such as skin lesions, lameness, or body condition. While preference, motivation, and aversion testing can be used to infer an animal's subjective experience, there is currently no universal method to gauge the affective state in nonhuman animals objectively. The evaluation of emotions in animals can be deduced from physiological indicators (e.g., heart rate, glucocorticoid levels,

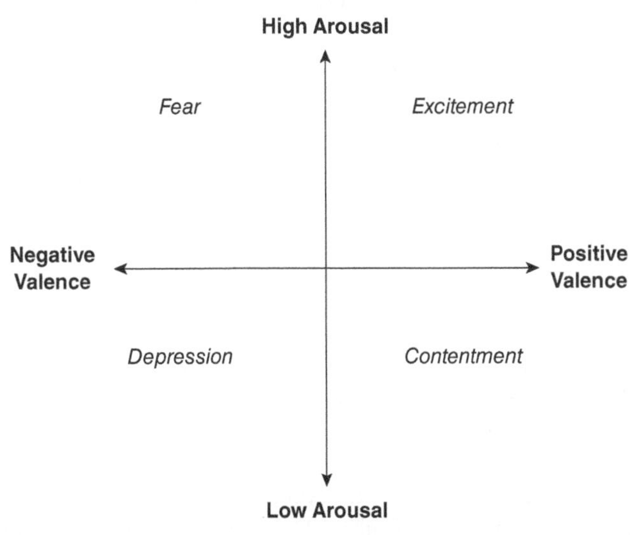

Figure 4.1 The circumplex model of affect, with emotions placed on the axis based on the level of arousal and valence

Source: Adapted from Russell (1980).

Table 4.1 The characteristics of the three types of measurements used to evaluate emotions in animals and examples of each during a high arousal-negative valence state (e.g., animal perceives a threat; fear) and a high arousal–positive valence state (e.g., animal anticipates reward; excited)

Measurement	*Characteristics*	*High Arousal, Negative Valence (ex: Fearful)*	*High Arousal, Positive Valence (ex: Excited)*
Physiological	A visceral state of arousal triggered by the autonomic nervous system.	↑ heart rate ↑ cortisol, adrenaline	↑ heart rate ↑ cortisol, adrenaline
Behavioral	An external expression including facial expressions, body posture, activity, and vocalizations.	↓ Activity (freeze) ↑ Activity (flight) Alert ear posture ↑ vocalization	↑ Activity Alert ear posture ↑ vocalization
Cognitive	Alteration to processing of environmental stimuli, related to appraisal, attention, and memory.	Perceive ambiguous stimuli as threatening; do not approach	Perceive ambiguous stimuli as nonthreatening; quickly approach

hypothalamic–pituitary–adrenal [HPA] and sympathetic–adrenal–medullary [SAM] activity). However, interpretation of these measurements is confounded by the fact that these metrics could reflect an emotion of both positive and negative valence. For example, we can measure an increase in heart rate and a spike in cortisol levels, as well as an increase in locomotor activity (e.g., pacing behavior), in an animal that is experiencing both a high arousal negative state (e.g., anxiety) and a high arousal positive state (e.g., excitement). This is why researchers must take a comprehensive approach including physiological, behavioral, and cognitive measurements when evaluating the emotional experience of nonhuman animals (Table 4.1).

Physiological Assessment of Emotions in Animals

Comparative Neuroanatomy

The majority of vertebrate animals share the same or comparable neurological structures and sensory pathways that are activated during an emotional experience. The occurrence of an emotion is coded by neural structures such as the orbitofrontal cortex, anterior cingulate cortex, insular cortex, hippocampus, amygdala, and nucleus accumbens (Panksepp 2004). Comparative neuroscientists have determined the existence of similar sensory circuits to these major neural structures after such auditory, olfactory, and visual stimulations are processed in their respective sensory cortical areas (Aggleton 1993). Experimental lesions to the amygdala (or its equivalent anatomical structure) in reptiles, birds, and nonprimate mammals have been shown to alter the normal development of emotional responses of the species such as aggressive, sexual, or parental behavior (Kling and Brothers 1992). For example, humans with damaged amygdala don't respond to an expression of fear in other people and cannot associate sounds with shock (Adolphs et al. 1994, 2005), while in rats, damaged amygdala results in a reduced escape responses and can't associate sounds with shock (Blanchard and Blanchard 1972). These amygdala lesions affect not only the avoidance conditioning but also the fear conditioning in mammals and birds (Cohen 1975; Dafter 1976). Fish do not have an amygdala but do have a structure that is considered to be analogous. Similar to mammalian studies, fish with experimental damage to this area are unable to associate a stimulus with shock after repeated pairings (Braithwaite and Boulcott 2007).

In addition to evaluating structure size, placement, and connecting circuitry during an emotional experience, researchers have relied on variation in neurotransmitters, such as serotonin, oxytocin, and dopamine, to infer emotional states in animals. For example, animal models are routinely used to study depression and to test anxiolytics (Willner 1984). In his research of positive emotional states in animals, Berridge (1996) has suggested two distinct stages of pleasure which are mediated by different neurotransmitter systems in the brain: "wanting" (appetitive/motivation) and "liking" (consummatory/experienced). Dopamine reflects the intensity of "wanting" while opioids stimulate the onset of motivated behavior, such as feeding (Levine 2006), social interaction, and play in young animals (Vanderschuren et al. 1995).

Heart Rate Variability

The emotional state of an animal is associated with varying degrees of physiological arousal. This arousal is mediated by the excitatory sympathetic nervous system (SNS) and the inhibitory parasympathetic nervous system (PNS) of the autonomic nervous system (i.e., fight or flight). When an animal is experiencing physical or psychological stress, the SNS becomes active to help the animal meet the challenge. This elevated arousal is characterized by an increased heart rate, increased respiratory rate, and dilation of pupils (to name a few). When an animal is perceiving a safe and stable environment, the PNS takes over to reduce heart rate, reduce respiratory rate, and constrict pupils. Therefore, measures of heart rate are frequently used as an indicator of stress in many vertebrate species. The discrete emotional response of fear can elicit a sharp increase in heart rate, as seen in amphibians when handled (Cabanac and Cabanac 2000) and livestock in response to transport (Fazio and Ferlazzo 2003). In contrast, research has shown that positive social interactions, such as grooming, coincides with decreased heart rate in rhesus macaques (Aureli et al. 1999), horses (Feh and de Mazières 1993), cattle (Schmied et al. 2008), and sheep (Hargreaves and Hutson 1990).

The ease with which an animal can transition between high and low arousal states is dependent on the ability of the autonomic nervous system to switch between the SNS and the PNS. This litmus test of the autonomic nervous system can be evaluated using heart rate variability (HRV). HRV is measured by obtaining a series of inter-beat intervals over a specific period of time and can be used to infer the emotional state of an animal (von Borell et al. 2007). A sustained negative affective state, such as anxiety or stress, has been associated with a lower HRV (Ekman et al. 1983; Porges 1995). For example, there have been many reports of a relationship between HRV and the welfare assessment of horses, from diseases such as grass sickness (Perkins et al. 2000) and laminitis (Rietmann et al. 2004) to the performance of stereotypic crib-biting behavior (Bachmann et al. 2003).

Biomarkers of Stress

A stressor can be defined as any stimulus that provokes the release of the adrenocorticotropic hormone (ACTH) and adrenal glucocorticoids, while "stress" is the emotional state associated with "the biological response elicited when an individual perceives a threat to its homeostasis" (Moberg 2000, p. 1). The experience of stress can be acute (short in duration, lasting minutes or various days) or chronic (lasting weeks, months, or years) and can be caused by social, environmental, metabolic, or immunological changes. While there is no "gold standard" for evaluating stress in animals, researchers tend to evaluate a panel of various biomarkers to assess the activity level of specific systems. For example, activity on the SAM axis is measured via catecholamine concentrations (epinephrine and norepinephrine), while activity on the HPA axis is measured via ACTH and glucocorticoid levels. Elevation of both SAM and HPA are associated with an increase in heart rate and blood pressure following the presentation of threatening stimuli in both humans and animals. Therefore, salivary or blood plasma levels of cortisol or adrenaline are often measured before and after specific experiments to evaluate the relative level of stress in the animal. For example, cortisol levels in dairy calves peak 5 minutes after being restrained in a chute as well as 15 minutes after the dehorning procedure of cauterizing the horn bud (Wohlt et al. 1994).

Salivary alpha–amylase (sAA) is an alternative measurement of the sympathetic nervous system currently used in human research as it is related to physiological and psychological stress (Nater and Rohleder 2009). It is measured in saliva and its activity is correlated with plasma catecholamine concentrations, being a marker of SAM activation. More recently, it has been shown that sAA levels are also increased in pigs after brief periods of stress induced from social isolation and transport (Soler et al. 2013).

Stress-Induced Hyperthermia

Following stimulation of the autonomic nervous system, the heart rate and blood flow change quickly within an organism. This rapid fluctuation in blow flow results in an increase in core body temperature and a corresponding decrease in surface body temperature, called stress-induced hyperthermia (Bouwknecht et al. 2007). This redistribution of blood to internal organs may prevent blood loss caused by potential injuries in the peripheries (Busnardo et al. 2010). For example, Busnardo et al. (2010) reported a decrease in tail temperature in response to acute restraint in laboratory rats. Researchers have been able to monitor this stress-induced hyperthermia in multiple species in order to evaluate whether procedures which are assumed to be stress-inducing cause this temperature change in peripheral regions. A sudden decrease in temperature of the ear pinnae in sheep has been found in response to handling and isolation stress (Lowe et al. 2005), while a sudden decrease in eye temperature has been reported to occur in calves for 5 minutes after disbudding procedure (Stewart et al. 2008) and in chickens after a sudden air-puff blast (Edgar et al. 2011).

Behavioral Assessment of Emotions in Animals

Activity Budgets

Given the current lack of universally accepted physiological indicators of positive emotional states in animals, researchers often rely on behavioral markers. One common method to evaluate the emotional state of an animal is to record changes in activity budget. An activity budget is the proportion of time an animal expends energy in different behavioral states within a time period (e.g., 24 hours). Based on how an animal allocates time toward resting, feeding, being social, or engaging with the environment, researchers can infer that the animal is in a negative emotional state, such as being in pain or anxious, or a positive emotional state, such as being relaxed or excited. For example, it has been reported that barren housing of farmed mink can result in the animals spending more time resting but alert, an indication of boredom (Meagher and Mason 2012). With that said, mink housed in nonenriched cages also displayed a heightened investigation of rewarding stimuli when tested, which suggests that the animals were not experiencing an emotional state of apathy or anhedonia.

Play behavior is most often seen in mammals, fish, amphibians, and reptiles animals whose basic essential needs have been met and whose fitness is not immediately threatened (Bateson 2014; Boissy et al. 2007; Burghardt 2015; Held and Špinka 2011). For example, lambs do not engage in play behavior after castration (Thornton and Waterman-Pearson 2002), and a study of domestic pigs in a seminatural environment reported a significant lack of play behavior during periods of cold weather (Newberry et al. 1988). Given that play occurs only when an animal's primary needs (food, comfort, safety, etc.) have been satisfied, it has been suggested that play may be a sensitive indicator for assessing the welfare of animals (Burghardt 2005; Fagen 1981). For example, Mintline et al. (2013) reported a significant decrease in play behavior in dairy calves significantly decreased up to 27 hours after the painful disbudding procedure.

According to Burghardt (2005), there are five basic criteria for a behavior to be considered play: (1) the behavior is not fully functional; (2) the behavior is spontaneous, voluntary, pleasurable, or autotelic (done for its own sake); (3) the behavior differs from serious behavior in terms of its form, duration, and/or frequency; (4) the behavior is repeatedly performed, but not stereotyped; and (5) the behavior occurs when the animal is healthy and free from stress. From sporadic leaping to combative wrestling, play behavior enhances an animal's rapid response to novel situations, which provides adaptive and competitive advantages. This fine-tuning of reflexive behavior enriches cognitive development in young animals and facilitates the maintenance of social bonds (Bekoff 1984; Byers 1998; Špinka et al. 2001).

Play is classified as a positive affect behavior because it is correlated to high levels of acetylcholine, glutamate, and opioids (i.e., "pleasure" neurotransmitters) and activity of subcortical brain areas that mediate the hedonic properties of reward (Berridge and Kringelbach 2008). Opioid agonists (e.g., morphine) administration increase social play in animals, while opioid antagonists decrease the amount of time animals play (Normansell and Panksepp 1990; Vanderschuren et al. 1995).

Facial Expression

For many diurnal social mammals, from sheep to macaques, visual cues are used to recognize individuals and their emotional state. In addition to body posture, such as bared teeth, arched back, or piloerection of fur/hair, emotional responses in animals can be expressed through alterations to facial muscles. It is these facial "expressions" that Darwin highlighted as evidence of emotional communication across species (Darwin 1872). Researchers have identified specific "action units," or changes to muscles around eyes, ears, mouths, and noses/snouts, which help to determine the affective state of both humans and nonhuman animals (Ekman et al. 1980). For example, species-specific grimace scales can be used to assess the severity of pain an animal is experiencing based on a scale of features, such as orbital tightening, strained jaw muscles, or ear position (flat, backward, or front-facing). Changes facial muscles may also reveal when an animal is experiencing a positive affective state. For example, cattle will display long durations of hanging ears when stroked by a human and during social licking/grooming by conspecifics, which can result in an assumed low-arousal positive state such as content or relaxed (Schmied et al. 2008).

A major objective of emotional research in animals is to enhance our ability to assess the emotional state, and thus welfare, of animals. However, as Darwin noted, it is most advantageous for an animal's fitness to accurately display and identify the emotion of a conspecific using facial expressions. For example, ewes are able to discriminate between calm and stressed or anxious facial displays of familiar conspecifics and they demonstrate a preference for neutral faces of unfamiliar ewes to that of a stressed face of a familiar ewe (Tate et al. 2006). This ability to accurately identify the emotional state of a conspecific using subtle facial cues can allow an observer to avoid potential harm or approach potential reward (i.e., increase individual fitness).

Vocalizations

Auditory signals are another major modality used by animals when communicating emotion. This is because changes in arousal via the autonomic nervous system can influence the constriction of muscles in the vocal tract and respiration rate and, thus, change the quality of sounds produced by the animal. Vocalizations also have a wide transmission range, enabling the sender to communicate with many at one time. For example, White Leghorn chicks will produce distress calls, characterized by their high energy, decreasing frequency, and call duration of about 0.4 seconds, when the animal is isolated (Marx et al. 2001). Animal welfare researchers analyze vocalizations of captive and wild animals in a variety of contexts as a noninvasive method to evaluate an animal's arousal and affective state (Manteuffel et al. 2004). For example, both cows and calves produce an increase in vocalization following sudden separation at weaning (Weary and Chua 2000; Marchant-Forde et al. 2002) and calves branded during restraint produce sounds with a higher maximum frequency and a higher peak sound level as compared to calves that are only restrained and not branded (Watts et al. 2001).

Positive affective states, such as low arousal relaxed or high arousal excitement, can also be inferred from animal vocalizations. For example, purring in cats has been reported to occur exclusively in positive social contexts, such as mother—kitten interactions or during tactile stimulation with inanimate objects as when rolling and rubbing (Kiley-Worthington 1984). Low-pitched bleating in sheep has been associated with some positive-valence situations, as they are produced by males as an estrus female is approaching or by lactating mothers while licking and nursing their lambs (Fisher and Matthews 2001). Lactating sows will produce bouts of low-frequency grunts to attract their piglets at nursing time (Castren et al. 1989). This low-energy sound is also produced in other positive social contexts between sows and piglets, such as reunion after separation (Colonnello et al. 2010).

While investigating reward-seeking behavior in domestic rats, researchers found that rats would approach researchers' hands four times quicker when reinforced with belly tickles, a behavior that imitates the rats' rough-and-tumble play, rather than basic dorsal strokes (Burgdorf and Panksepp 2001). Remarkably, these tickled rats produced an ultrasonic (50 kHz) "laugh-like" repetitive chatter after receiving the belly tickles (Knutson et al. 2002). In the future, animal care staff may monitor this ultrasonic chatter as a noninvasive tool to evaluate the emotional state of laboratory rats following handling procedures.

Cognitive Indicators of Emotions in Animals

Interpretation of biological and behavioral measurements can be difficult because they could reflect an emotion of both positive and negative valence. For example, an increase in heart rate or a surge in cortisol or adrenaline all indicate high arousal but can be associated with escape from predation (negative valence) or with the anticipation of a reward (positive valence). Fish have both the appropriate nerves and pathways to sense and send stress and painful signals, but the motor patterns that express this negative affective state is not expressed the same way as in terrestrial animals. Paradise fish avoid places where they have experienced a single attack by a predator and continue to do so for many months (Csányi and Dóka 1993), and carp learn to avoid bait for up to 3 years after they have been hooked just once (Beukemaj 1970). Given the limitations of physiological and behavioral assessment of animal emotions, novel cognitive methods are carried out to evaluate how the affective state of an animal may alter its cognitive processing (e.g., decision making, attention, appraisal, and memory).

Animal welfare scientists may measure the conditioned response of an animal to infer a subjective pain or pleasure experience. For example, when animals are given a highly valued reward (e.g., food, analgesic drugs) only when placed in a specific environment with particular features, a stimulus–response association can be made called a "conditioned place preference." Conversely, animals may be administered a punishment (e.g., electric shock) only when placed in a specific environment with particular features in order to establish a "conditioned place aversion." If an animal displays conditioned place preference or aversion, then researchers have evidence to support the assessment that the animal experienced the assumed positive (pleasure) or negative (fear, pain) affective state.

Another cognitive test often implemented by animal welfare scientists is the motivation test. Motivation tests measure the strength or willingness with which an animal engages in a behavior (Kirkden and Pajor 2006). For example, injured rats and lame chickens will "work" (i.e., lever press, key peck) to consume more water that contains a pain-relieving drug over pure water (Danbury et al. 2000; Persinger 2003). However, researchers caution that how hard an animal will work to gain access to a resource ("wanting") may not reflect the subjective feeling ("liking") the animal experiences when accessing the resource (Berridge 1996).

Cognitive Bias Testing

Cognitive bias testing allows researchers to infer how an individual's emotional state influences information processing, such as the amount of attention given to an unknown or threatening stimulus, or the evaluation of ambiguous stimuli (is that unknown object a threat or not a threat?). Research on human and nonhuman animals indicates that negative mental states, such as anxiety or depression, can induce pessimistic judgments of ambiguous stimuli (Hallion and Ruscio 2011). For example, barren, crowded, or unpredictable housing conditions can elicit negative cognitive biases, and therefore assumed negative affective states (e.g., anxiety, stress), in rats and piglets (Harding et al. 2004; Mendl et al. 2009; Douglas et al. 2012; Scollo et al. 2014). In addition, researchers are attempting to assess subjective pain experience using cognitive bias testing, such as the Neave et al. (2013) study which reported that dairy calves may display negative cognitive biases following the practice of hot-iron disbudding. The cognitive bias task allows for a priori predictions about the types of biases expected, greater selectivity for the valence of an emotional state as opposed to the level of arousal and the ability to assess positive as well as negative affective states and is a fairly noninvasive method (Paul et al. 2005).

For cognitive bias experiments, operant conditioning methods are used to train animals to distinguish between two stimuli that lie at the ends of a continuous stimulus range. One stimulus may be associated with a highly valued reward while the other is associated with a lower-valued reward or punishment. The conditioned behavioral response can be locomotion-based, such as approach (Figure 4.2), or computer-based method, such as touchscreen responses (i.e., peck, nose, press). Once animals are trained to respond to a positive stimulus ("go") and to not respond when presented the negative stimulus ("no-go"), their cognitive bias can be explored through the presentation of an ambiguous or intermediate stimulus. Animals can be classified as "optimistic" or having a positive affective state if the animal displays behaviors that suggest an increased expectation of reward in the face of ambiguous stimuli. Alternatively, animals can be classified "pessimistic" or having a negative affective state if the animal displays behaviors that suggest an increased expectation of punishment in the face of ambiguous stimuli.

Go/no-go behavioral responses have the advantage of being relatively easy to train and allow for active choice responses, which offer clear data on the choice made (a "no-go" response is classed as a specified time window). The type of reinforcers used may influence the sensitivity of the test and the questions it addresses. Food reinforcers have been used in all animal studies to date but are vulnerable to the effects of individual differences in feeding motivation. Tests using reinforcers that

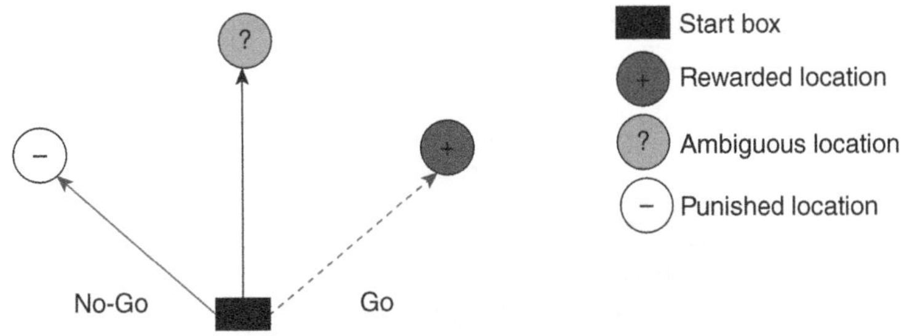

Figure 4.2 Operant conditioning paradigm for judgment bias testing of animals using the "go/no-go" model. Only one conditioned stimulus is present in the arena at a time. Animals learn to quickly approach the rewarded location ("go") and not approach the punished location ("no-go")

are similar in affective terms (e.g., small vs. large quantity of food) may fail to reveal cognitive biases. Tests using positive and neutral reinforcers (e.g., food vs no food) focus on changes in anticipation of positive events (i.e., depression), while the use of negative and neutral reinforcers targets changes in anticipation of negative events (i.e., anxiety). Therefore, tests should use clearly positive and negative reinforcers (e.g., reward and punishment) as they engage a number of affect-related influences on decision making.

Animal Personality and Emotional Experience

While testing of ambiguous cue interpretation allows researchers to infer how an individual's emotional state may influence information processing, it does not take into account the influence of coping styles, or personality traits. An animal's personality can influence the way in which stimuli are processed and evaluated and how the affective state is expressed (e.g., approach/avoid) in experimental psychology tests. Personality is defined by four key elements: (1) it is manifested early in life and is relatively stable throughout development; (2) it is primarily biologically based (genetics and neurobiological mechanisms); (3) it refers to characteristics of behavioral reactions such as intensity, speed, response threshold, latency, and recovery time; and (4) it is most clearly expressed in novel and unpredictable situations (Stamps and Groothuis 2010).

Consistency of individual differences in behavioral response to the same stimuli strongly supports the idea that emotional response in animals is mediated by stable underlying temperamental characteristics. Given that personality is most clearly expressed in novel and unpredictable situations, behavioral tests are constructed to elicit high-arousal responses. For example, the open-field test can be a proxy measurement for activity and/or fear by isolating an animal in a novel arena and recording subsequent behaviors. Freezing behavior and high-pitched vocalizations in an open field test can be used as indicators for the emotion fear (Forkman et al. 2007). Interpretation of this experiment is contingent on the species of interest. Exposure without cover from predators can be fear-inducing for a domestic chicken; however, for a species that evolved for open-ground foraging, such as swine, the open-field test can elicit exploratory behaviors.

Previous research on animal personality suggests a relationship between certain traits and the formation and maintenance of social bonds (Massen and Koski 2014), immunity strength (Segerstrom 2000), the ability to cope with physiological stress (Carere et al. 2010), the performance of abnormal or stereotypic behaviors (Ijichi et al. 2013; Cussen and Mench 2015), and the expression of pain (Ijichi et al. 2014). Personality traits may also influence the affective state of an animal by shaping attentional and informational processing (Figure 4.3). For example, anxious individuals often display

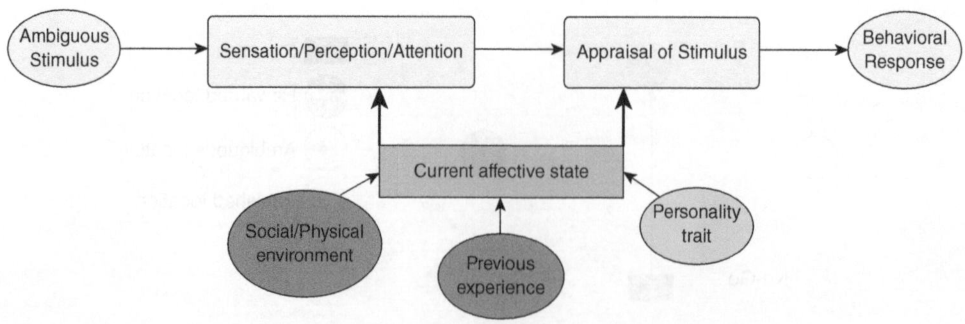

Figure 4.3 The process and influences of an organism appraising and responding to an ambiguous stimulus

Source: Adapted from Mendl et al. (2009).

impaired functioning of the goal-directed attention and an increase in stimulus-driven attention (Eysenck et al. 2007). These cognitive biases can elicit negative affective states, such as the perception of danger during exposure to an actual (fear state) or potential (anxiety state) threat.

While fear is a highly functional and adaptive state, captive animals can be housed in environments that prevent the performance of adaptive behaviors, such as escape or shelter. These environmental constraints can exacerbate fear responses and elicit harmful side effects, including the development of maladaptive behaviors, an increased risk of injury, a decreased immune system, and/or reduced productivity (Boissy 1995; Korte 2001). Therefore, how an animal perceives social and environmental stimuli is influenced by their affective state and how an animal responds to these stimuli are shaped by their personality.

Conclusion

The fundamental approach to studying emotions in animals is rooted in the theory of natural selection, which states that emotions guide animal behavior toward positive experiences (e.g., food, mating, or shelter) and away from negative experiences (e.g., aggressive encounters with conspecifics or predators) to increase individual fitness. As with many theories in science, there are exceptions to the general rule. Sometimes animals behave in manners that do not increase their fitness, like quickly approaching a potentially dangerous, unknown stimulus. It is this behavioral plasticity which allows for a diversity in emotional responses to the same stimulus among animals. As animal personality and comparative cognition research has indicated, individuals may perceive the same situation differently and, thus, experience different affective states in the same context. Conversely, individuals may adopt the same affective state as their social peers even if they themselves did not experience a change in environment.

In the animal emotion literature, researchers often refer to this "spread" of an emotional experience from one animal to other observing animals as an "emotional contagion." The perception of an emotional state in a conspecific, through sight, sound, or smell, can lead to physiological and behavioral changes in the observer which mirrors that of the acting animal. This emotional contagion has been reported in geese (Wascher et al. 2008), ravens (Fraser and Bugnyar 2010), chickens (Edgar et al. 2011), and pigs (Reimert et al. 2015). The strength of an emotional contagion appears to be influenced by previous experience of the observer and the relationship between actor and observer. For example, Goumon and Špinka (2016) reported that when piglets observe another piglet in a stress-induced context (e.g., physical restraint), they will show an increase in behavioral indicators of fear. However, when the observer pigs had previous experience with restraint themselves, they reacted more strongly to observing a conspecific in restraint than the pigs that have never experienced restraint. Female mice will freeze for longer durations when exposed to the pain of a close relative than when exposed to the pain of a more distant relative (Jeon et al. 2010), and hens will display increased alertness, decreased preening behavior, and a decrease in eye temperature when observing their own chicks in a distressed state (caused by air-puff blast) but will not display these behavioral changes when observing the same distressed state in a familiar adult hen (Edgar et al. 2012).

There are various elements in lives of animals which can result in a quick emotional response or a sustained affective state. Both intra- and interspecies assessment of these subjective emotional states is important for animal health, welfare, and genetic fitness. Conspecifics benefit from accurately identifying the emotion in another by learning about the rewarding and risky elements of their environment. Humans benefit from accurately identifying emotions in animals by improving the quality of care given, which, in turn, can improve the animal's biological and psychological health. There are many practical applications for research on emotions in animals which can be applied to those maintained for companionship, as a source of food, as biomedical models for human health, or as educational ambassadors for their wild counterparts. For example, a veterinarian may improve their ability to identify when an animal is experiencing pain (e.g., evaluate the duration of postoperative pain) by incorporating behavioral measurements with clinical assessments. A biomedical researcher may improve their ability to evaluate the efficacy of analgesics in laboratory rodents by recording facial expressions and activity budgets of animals. Animal welfare scientists may combine cognitive measures with traditional assessments of an animal's welfare state in order to determine whether a common husbandry practice causes chronic negative emotional states (e.g., early weaning of dairy calves or piglets) or whether an intervention to alleviate negative emotional states and encourage positive emotional states was successful (e.g., use of environmental enrichment). In the end, while we may never know the true subjective feeling of an animal in a specific context, the methods described in this chapter can be adopted by all disciplines in which humans are responsible for the care of animals in order to reduce the severity and frequency of negative affective states, and promote the experience of positive affective states.

Bibliography

Adolphs, R., Gossein, F., Buchanan, T. W., Tranel, D., Schyns, P., and Damasio, A. R. (2005) "A mechanism for impaired fear recognition after amygdala damage," *Nature* 433: 68–72.

Adolphs, R., Tranel, D., Damasio, H., and Damasio, A. (1994) "Impaired recognition of emotion in facial expressions following bilateral damage to human amygdala," *Nature* 372: 669–672.

Aggleton, J. P. (1993) "The contribution of the amygdala to normal and abnormal emotional states," *Trends in Neurosciences* 16(8): 328–333.

Anderson, D. J., and Adolphs, R. (2014) "A framework for studying emotions across species" *Cell* 157(1): 187-200.

Aureli, F., Preston, S. D., and de Waal, F. B. M. (1999) "Heart rate responses to social interactions in free-moving rhesus macaques (Macaca mulatta): A pilot study," *Journal of Comparative Psychology* 113: 59–65.

Bachmann, I., Bernasconi, P., Herrmann, R., Weishaupt, M. A., and Stauffacher, M. (2003) "Behavioural and physiological responses to an acute stressor in crib-biting and control horses," *Applied Animal Behaviour Science* 82(4): 297–311.

Barrett, L. F., Mesquita, B., Ochsner, K. N., and Gross, J. J. (2007) "The experience of emotion," *Annual Review of Psychology* 58: 373–403.

Bateson, P. (2014) "Play, playfulness, creativity and innovation," *Animl Behavior Cognitive* 1: 99–112.

Bekoff, M. (1984) "Social play behavior," *Bioscience* 34: 228–233.

Berridge, K. C. (1996) "Food reward: Brain substrates of wanting and liking," *Neuroscience & Biobehavioral Reviews* 20: 1–25.

Berridge, K. C., and Kringelbach, M. L. (2008) "Affective neuroscience of pleasure: Reward in humans and animals," *Psychopharmacology* 199: 457–480.

Beukemaj, J. J. (1970) "Acquired hook-avoidance in the pike Esox lucius L. fished with artificial and natural baits," *Journal of Fish Biology* 2(2): 155-160.

Blanchard, D. C., and Blanchard, R. J. (1972) "Innate and conditioned reactions to threat in rats with amygdaloid lesions," *Journal of Comparative and Physiological Psychology* 81: 281–290.

Boissy, A. (1995) "Fear and fearfulness in animals," *The Quarterly Review of Biology* 70: 165–191.

Boissy, A., Fisher, A. D., Bouix, J., Hinch, G. N., and Le Neindre, P. (2005) "Genetics of fear in ruminant livestock," *Livestock Production Science* 93(1): 23–32.

Boissy, A., Manteuffel, G., Jensen, M. B., Moe, R. O., Spruijt, B., Keeling, L. J., . . . Bakken, M. (2007) "Assessment of positive emotions in animals to improve their welfare," *Physiology & Behavior* 92(3): 375–397.

Bouwknecht, J. A., Olivier, B., and Paylor, R. E. (2007) "The stress-induced hyperthermia paradigm as a physiological animal model for anxiety: A review of pharmacological and genetic studies in the mouse," *Neuroscience & Biobehavioral Reviews* 31: 41–59.

Braithwaite, V. A., and Boulcott, P. (2007) "Pain perception, aversion and fear in fish," *Diseases of Aquatic Organisms* 75(2): 131–138.

Broom, D. (1986) "Indicators of poor welfare," *British Veterinary Journal* 142: 524–526.

Burgdorf, J., and Panksepp, J. (2001) "Tickling induces reward in adolescent rates," *Physiology and Behavior* 72: 167–173.

Burghardt, G. M. (2005) *The Genesis of Animal Play: Testing the Limits*. Cambridge, MA: MIT Press.

Burghardt, G. M. (2015) "Play in fishes, frogs and reptiles," *Current Biology* 25: R9–R10.

Busnardo, C., Tavares, R. F., Resstel, L. B. M., Elias, L. L. K., and Correa, F. M. A. (2010) "Paraventricular nucleus modulates autonomic and neuroendocrine responses to acute restraint stress in rats," *Autonomic Neuroscience* 158: 51–57.

Byers, J. A. (ed.) (1998) *Animal Play: Evolutionary, Comparative and Ecological Perspectives,* Cambridge, UK: Cambridge University Press.

Cabanac, A., and Cabanac, M. (2000) "Heart rate response to gentle handling of frog and lizard," *Behavioural Processes* 52: 89–95.

Carere, C., Caramaschi, D., and Fawcett, T. (2010) "Covariation between personalities and individual differences in coping with stress: Converging evidence and hypotheses," *Current Zoology* 56: 728–740.

Castren, H., Algers, B., Jensen, P., and Saloniemi, H. (1989) "Suckling behaviour and milk consumption in newborn piglets as a response to sow grunting," *Applied Animal Behaviour Science* 24(3): 227–238.

Cohen, D. H. (1975) "Involvement of the avian amygdala homologue (archistriatum posterior and mediale) in defensively conditioned heat rate change," *Journal of Comparative Neurology* 160: 13–36.

Cohen, N. E., Brom, F. W., and Stassen, E. N. (2009) "Fundamental moral attitudes to animals and their role in judgment: An empirical model to describe fundamental moral attitudes to animals and their role in judgment on the culling of healthy animals during an animal disease epidemic," *Journal of Agricultural and Environmental Ethics* 22(4): 341–359.

Colonnello, V., Iacobucci, P., and Newberry, R. C. (2010) "Vocal and locomotor responses of piglets to social isolation and reunion," *Developmental Psychobiology* 52(1): 1–12.

Cussen, V., and Mench, J. (2015) "The relationship between personality dimensions and resiliency to environmental stress in orange-winged amazon parrots, as indicated by the development of abnormal behaviors," *PLoS One* 10: e0126170.

Csányi, V., and Dóka, A. (1993) "Learning interactions between prey and predator fish." *Marine and Freshwater Behaviour and Physiology* 23(1-4): 63–78.

Dafters, R. (1976) "Effect of medial archistriatal lesions on the conditioned emotional response and on auditory discrimination performance of the pigeon," *Physiology and Behavior* 17: 659–665.

Danbury, T., Weeks, C., Chambers, J., Waterman-Pearson, A., and Kestin, S. (2000) "Self-selection of the analgesic drug, carprofen, by lame broiler chickens," *Veterinary Record* 146: 307–311.

Darwin, C. (1872) *The Expression of Emotions in Man and Animals,* London: Murray.

Dawkins, M. S. (1990) "From an animal's point of view: Motivation, fitness, and animal welfare," *Behavioral and Brain Sciences* 13: 1–61.

Dawkins, M. S. (2008) "The science of animal suffering," *Ethology* 114: 937–945.

De Waal, F. (2011) "What is an animal emotion?" *Annals of the New York Academy of Sciences* 1224(1): 191–206.

Douglas, C., Bateson, M., Walsh, C., Bédué, A., and Edwards, S. (2012) "Environmental enrichment induces optimistic cognitive biases pigs," *Applied Animal Behaviour Science* 139: 65–73.

Edgar, J. L., Lowe, J. C., Paul, E. S., and Nicol, C. J. (2011) "Avian maternal response to chick distress," *Proceedings of the Royal Society of London B: Biological Sciences* 278(1721): 3129–3134. rspb20102701.

Edgar, J. L., Paul, E. S., Harris, L., Penturn, S., and Nicol, C. J. (2012) "No evidence for emotional empathy in chickens observing familiar adult conspecifics," *PLoS One* 7(2): e31542.

Ekman, P., Freisen, W. V., and Ancoli, S. (1980) "Facial signs of emotional experience," *Journal of Personality and Social Psychology* 39(6): 1125.

Ekman, P., Levenson, R. W., and Friesen, W. V. (1983) "Autonomic nervous system activity distinguishes among emotions," *Science* 221(4616): 1208–1210.

Ekman, P., Sorenson, E. R., and Friesen, W. V. (1969) "Pan-cultural elements in facial displays of emotions," *Science* 164: 86–88.

Eysenck, M. W., Derakshan, N., Santos, R., and Calvo, M. (2007) "Anxiety and cognitive performance: Attentional control theory," *Emotion* 7: 336–353.

Fagen, R. (1981) Animal Play Behavior. Oxford, UK: Oxford University Press

Fazio, E., and Ferlazzo, A. (2003) "Evaluation of stress during transport," *Veterinary Research Communications* 27(1): 519–524.

Feh, C., and de Mazières, J. (1993) "Grooming at a preferred site reduces heart rate in horses," *Animal Behaviour* 46: 1191–1194.

Fisher, A., and Matthews, L. (2001) "The social behaviour of sheep," in L.J. Keeling and H. W. Gonyou (eds.) *Social Behaviour in Farm Animals* (pp. 211–245), Wallingford, UK: CAB International.

Forkman, B., Boissy, A., Meunier-Salaün, M. C., Canali, E., and Jones, R. B. (2007) "A critical review of fear tests used on cattle, pigs, sheep, poultry and horses," *Physiology & Behavior* 92(3): 340-374.

Fraser, D., and Duncan, I. (1998) "Pleasures, 'pains' and animal welfare: Towards a natural history of affect," *Animal Welfare* 7: 383–396.

Fraser, O. N., and Bugnyar, T. (2010) "Do ravens show consolation? Responses to distressed others," *PLoS One* 5: e10605.

Goumon, S., and Špinka, M. (2016) "Emotional contagion of distress in young pigs is potentiated by previous exposure to the same stressor," *Animal Cognition* 19(3): 501–511.

Hallion, L., and Ruscio, A. (2011) "A meta-analysis of the effect of cognitive bias modification on anxiety and depression," *Psychological Bulletin* 137: 940.

Harding, E. J., Paul, E. S., and Mendl, M. (2004) "Animal behaviour: Cognitive bias and affective state," *Nature* 427(6972): 312.

Hargreaves, A. L., and Hutson, G. D. (1990) "The effect of gentling on heart rate, flight distance and aversion of sheep to a handling procedure," *Applied Animal Behaviour Science* 26: 243–252.

Held, S. D. E., and Špinka, M. (2011) "Animal play and animal welfare," *Animal Behaviour* 81: 891–899.

Ijichi, C., Collins, L., and Elwood, R. (2013) "Evidence for the role of personality in stereotypy predisposition," *Animal Behaviour* 85: 1145–1151.

Ijichi, C., Collins, L., and Elwood, R. (2014) "Pain expression is linked to personality in horses," *Applied Animal Behaviour Science* 152: 38–43.

Jeon, D., Kim, S., Chetana, M., Jo, D., Ruley, H. E., Lin, S. Y., . . . Shin, H. S. (2010) "Observational fear learning involves affective pain system and Ca v 1.2 Ca 2+ channels in ACC," *Nature Neuroscience* 13(4): 482.

Kiley-Worthington, M. (1984) "Animal language? Vocal communication of some ungulates, canids and felids," *Acta Zoologica Fennica* 171(83): 8.

Kirkden, R., and Pajor, E. (2006) "Using preference, motivation, and aversion tests to ask scientific questions about animal feelings," *Applied Animal Behaviour Science* 100: 29–47.

Kling, A. S., and Brothers, L. A. (1992) "The amygdala and social behavior," in J. P. Aggleton (ed.) *The Amygdala: Neurobiological Aspects of Emotion, Memory, and Mental Dysfunction* (pp. 353–377), New York: Wiley-Liss.

Knutson, B., Burgdorf, J., and Panksepp, J. (2002) "Ultrasonic vocalizations as indices of affective states in rats," *Psychological Bulletin* 128(6): 961.

Korte, S. (2001) "Corticosteroids in relation to fear, anxiety and psychopathology," *Neuroscience & Biobehavioral Reviews* 25: 117–142.

Korte, S. M., Olivier, B., and Koolhaas, J. M. (2007) "A new animal welfare concept based on allostasis," *Physiology & Behavior* 92(3): 422–428.

Levine, A. S. (2006) "The animal model in food intake regulation: Examples from the opioid literature," *Physiology & Behavior* 89(1): 92–96.

Lowe, T. E., Cook, C. J., Ingram, J. R., and Harris, P. J. (2005) "Changes in ear-pinna temperature as a useful measure of stress in sheep (*Ovis aries*)," *Animal Welfare* 14: 35–42.

Manteuffel, G., Puppe, B., and Schön, P. C. (2004) "Vocalization of farm animals as a measure of welfare," *Applied Animal Behaviour Science* 88(1): 163–182.

Marchant-Forde, J. N., Marchant-Forde, R. M., and Weary, D. M. (2002) "Responses of dairy cows and calves to each other's vocalisations after early separation," *Applied Animal Behaviour Science* 78(1): 19–28.

Marx, G., Leppelt, J., and Ellendorff, F. (2001) "Vocalisation in chicks (*Gallus gallus dom.*) during stepwise social isolation," *Applied Animal Behaviour Science* 75(1): 61–74.

Massen, J., and Koski, S. (2014) "Chimps of a feather sit together: Chimpanzee friendships are based on homophily in personality," *Evolution and Human Behavior* 35: 1–8.

Meagher, R. K., and Mason, G. J. (2012) "Environmental enrichment reduces signs of boredom in caged mink," *PLoS One* 7(11): e49180.

Mendl, M., Burman, O. H. P., Parker, R., and Paul, E. (2009) "Cognitive bias as an indicator of animal emotion and welfare: Emerging evidence and underlying mechanisms," *Applied Animal Behaviour Science* 118: 161–181.

Mintline, E. M., Stewart, M., Rogers, A. R., Cox, N. R., Verkerk, G. A., Stookey, J. M., . . . Tucker, C. B. (2013) "Play behavior as an indicator of animal welfare: Disbudding in dairy calves," *Applied Animal Behaviour Science* 144(1): 22–30.

Moberg, G. P. (2000) "Biological response to stress: Implications for animal welfare," in G. P. Moberg and J. A. Mench (eds) *The Biology of Animal Stress: Basic Principles and Implications for Animal Welfare* (pp. 1–22), Wallingford, OX: CABI Publishing.

Nater, U. M., and Rohleder, N. (2009) "Salivary alpha-amylase as a non-invasive biomarker for the sympathetic nervous system: Current state of research," *Psychoneuroendocrinology* 34(4): 486–496.

National Research Council. (2008). Recognition and alleviation of distress in laboratory animals. Washington, DC: National Academies Press.

Neave, H., Daros, R., Costa, J., von Keyserlingk, M., and Weary, D. (2013) "Pain and pessimism: Dairy calves exhibit negative judgement bias following hot-iron disbudding," *PLoS One* 9: e96135.

Nesse, R. M. (1990) "Evolutionary explanations of emotions," *Human Nature* 1(3): 261–289.

Newberry, R. C., Wood-Gush, D. G. M., and Hall, J. W. (1988) "Playful behaviour of piglets," *Behavioural Processes* 17(3): 205–216.

Normansell, J., and Panksepp, J. (1990) "Effects of morphine and naloxone on play rewarded spatial discrimination in juvenile rats," *Developmental Psychobiology* 23: 75–83.

Oatley, K., and Jenkins, J. M. (1998) *Understanding Emotions,* Oxford: Blackwell.

Panksepp, J. (1998) "The periconscious substrates of consciousness: Affective states and the evolutionary origins of the SELF," *Journal of consciousness studies*, 5(5-6): 566-582.

Panksepp, J. (2004) *Affective Neuroscience: The Foundations of Human and Animal Emotions,* New York, NY: Oxford University Press.

Paul, E., Harding, E., and Mendl, M. (2005) "Measuring emotional processes in animals: The utility of a cognitive approach," *Neuroscience & Biobehavioral Reviews* 29: 469–491.

Perkins, J. D., Bowen, I. M., Else, R. W., Marr, C. M., and Mayhew, I. G. (2000) "Functional and histopathological evidence of cardiac parasympathetic dysautonomia in equine grass sickness," *The Veterinary Record* 146(9): 246–250.

Persinger, M. (2003) "Rats' preferences for an analgesic compared to water: An alternative to 'killing the rat so it does not suffer,'" *Perceptual and Motor Skills* 96: 674–680.

Porges, S. W. (1995) "Cardiac vagal tone: A physiological index of stress," *Neuroscience & Biobehavioral Reviews* 19: 225–233.

Reimert, I., Bolhuis, J. E., Kemp, B., and Rodenburg, T. B. (2015) "Emotions on the loose: emotional contagion and the role of oxytocin in pigs," *Animal cognition* 18(2): 517-532.

Rietmann, T. R., Stauffacher, M., Bernasconi, P., Auer, J. A., and Weishaupt, M. A. (2004) "The association between heart rate, heart rate variability, endocrine and behavioural pain measures in horses suffering from laminitis," *Transboundary and Emerging Diseases* 51(5): 218–225.

Russell, J. A. (1980) "A circumplex model of affect," *Journal of Personality and Social Psychology* 39: 1161.

Russell, J. A. (2003) "Core affect and the psychological construction of emotion," *Psychologival Review* 110: 145–172.

Schmied, C., Waiblinger, S., Scharl, T., Leisch, F., and Boivin, X. (2008) "Stroking of different body regions by a human: Effects on behaviour and heart rate of dairy cows," *Applied Animal Behaviour Science* 109: 25–38.

Scollo, A., Gottardo, F., Contiero, B., and Edwards, S. (2014) "Does stocking density modify affective state in pigs as assessed by cognitive bias, behavioural and physiological parameters?" *Applied Animal Behaviour Science* 153: 26–35.

Segerstrom, S. C. (2000) "Personality and the immune system: Models, methods, and mechanisms," *Annals of Behavioral Medicine* 22(3): 180-190.

Soler, L., Gutiérrez, A., Escribano, D., Fuentes, M., and Cerón, J. J. (2013) "Response of salivary haptoglobin and serum amyloid A to social isolation and short road transport stress in pigs," *Research in Veterinary Science* 95(1): 298–302.

Špinka, M., Newberry, R. C., and Bekoff, M. (2001) "Mammalian play: Training for the unexpected," *The Quarterly Review of Biology* 76(2): 141–168.

Stamps, J., and Groothuis, T. (2010) "The development of animal personality: Relevance, concepts and perspectives," *Biological Reviews* 85: 301–325.

Stewart, M., Stafford, K. J., Dowling, S. K., Schaefer, A. L., and Webster, J. R. (2008) "Eye temperature and heart rate variability of calves disbudded with or without local anaesthetic," *Physiology & Behavior* 93(4–5): 789–797.

Tate, A. J., Fischer, H., Leigh, A. E., and Kendrick, K. M. (2006) "Behavioural and neurophysiological evidence for face identity and face emotion processing in animals," *Philosophical Transactions of the Royal Society of London B: Biological Sciences* 361(1476): 2155–2172.

Thornton, P. D., and Waterman-Pearson, A. E. (2002) "Behavioural responses to castration in lambs," *Animal Welfare* 11(2): 203–212.

Vanderschuren, L. J., Spruijt, B. M., Hol, T., Niesink, R. J., and Van Ree, J. M. (1995) "Sequential analysis of social play behavior in juvenile rats: Effects of morphine," *Behavioural Brain Research* 72(1–2): 89–95.

Von Borell, E., Langbein, J., Després, G., Hansen, S., Leterrier, C., Marchant-Forde, J., . . . Valance, D. (2007) Heart rate variability as a measure of autonomic regulation of cardiac activity for assessing stress and welfare in farm animals—a review," *Physiology & Behavior* 92(3): 293–316.

Wascher, C. A. F., Isabella Scheiber, I. B. R., and Kotrschal, K. (2008) "Heart rate modulation in bystanding geese watching social and non-social events," *Proceedings of the Royal Society B* 275: 1653–1659.

Watts, J. M., Stookey, J. M., Schmutz, S. M., and Waltz, C. S. (2001) "Variability in vocal and behavioural responses to visual isolation between full-sibling families of beef calves," *Applied Animal Behaviour Science* 70(4): 255-273.

Weary, D. M., and Chua, B. (2000) "Effects of early separation on the dairy cow and calf: 1. Separation at 6 h, 1 day and 4 days after birth," *Applied Animal Behaviour Science* 69(3): 177-188.
Willner, P. (1984) "The validity of animal models of depression," *Psychopharmacology* 83(1): 1–16.
Wohlt, J. E., Allyn, M. E., Zajac, P. K., and Katz, L. S. (1994) "Cortisol increases in plasma of Holstein heifer calves from handling and method of electrical dehorning," *Journal of Dairy Science* 77(12): 3725–3729.

5

ANIMAL SELF-AWARENESS

Types, Distribution, and Ethical Significance

David DeGrazia

Self-awareness is often assumed to be a single phenomenon that underlies special moral status. It is also commonly thought to be exclusively human, or nearly so, extending perhaps to a few nonhuman species. These ideas are integrated in Locke's classic investigation of personhood, in which persons are taken to be rational beings who are aware of themselves as persisting over time—making them appropriate subjects of moral accountability (Locke 1694, Bk. II, chap. 27). The idea that self-awareness is a single phenomenon is also reflected in the idea, embraced by many scholars, that a good test for its presence is the mirror self-recognition test (Gallup 1970, 1977).[1] In this chapter, I challenge the two-pronged assumption that self-awareness is a single phenomenon that is more or less exclusively human. I do so in distinguishing and examining four types of self-awareness: narrative identity, introspective awareness, social self-awareness, and bodily agential self-awareness. In examining each type, I address its apparent distribution in the animal kingdom. Finally, I close with brief reflections on the ethical significance of the four types of self-awareness.

Narrative Identity

Narrative identity is a rich sort of self-conception that is characteristic of human beings of sufficient maturity. It involves a sort of biographical self-awareness—an awareness of one's own life as comprising something like a story with most or all of the following elements: a richly detailed set of memories, values and priorities, awareness of important relationships, ongoing endeavors, and intentions or plans for the future. One's narrative identity can help to guide one in making difficult life choices (Glover 1988: 152) and can be threatened in an identity crisis (Schechtman 1996: 74). The idea of narrative identity might also help to flesh out the somewhat vague Lockean concept of a person: perhaps persons are all and only those beings who have a narrative identity.

Yet, even if the idea of narrative identity is more determinate than the concept of a person, it too has blurred boundaries. Still, it seems reasonable to suppose that human beings ordinarily begin to possess such a self-conception when they are three or four years old.[2] Of the four types of self-awareness I distinguish, this is the most likely to be exclusively human. But, in order to determine whether narrative identity is unique to human beings, we cannot recline in the *a priori* armchair, as too many philosophers still do. We must head for the *a posteriori* jungle of available evidence.

Given the conceptual richness of a narrative identity, one might speculate that *only* language users have the requisite cognitive sophistication but also that *any* creature who is a genuine language user is likely to possess a narrative identity.[3] Thus, in considering the possibility of nonhuman narrative identities, let us begin with the few nonhuman animals (hereafter, simply "animals") who have had significant success in learning a language. I consider apes who, as far as I know, have achieved the most extensive linguistic competence.

The bonobo Kanzi learned, by observation without explicit training, to use a keyboard on which his mother was being trained. Although Kanzi's comprehension abilities exceed his productive language abilities, he generates strings of two or three words that have clear meaning in context. Perhaps most impressive, however, is his comprehension of spoken English (even when hearing through headphones without the benefit of visual cues). He can comprehend novel utterances, such as "Take the vacuum cleaner outdoors," and can distinguish strings with the same words but different orders, displaying a type of syntactical mastery—for example, "Pour the coke into the milk" and "Pour the milk into the coke" (PBS 1995; Savage-Rumbaugh 1986; Savage-Rumbaugh and Brakke 1990).

The gorilla Koko, who lived in an environment of American Sign Language (ASL) and spoken English, used a vocabulary of hundreds of signs to produce strings of three or more words. The English vocabulary she understood was larger. Interestingly, she signed to both humans and other language-trained apes. Among her novel definitions are the following: "What's an insult?" "THINK DEVIL DIRTY." "What's a smart gorilla?" "ME." "When do people say darn?" "WORK. OBNOXIOUS." When asked what happened on her birthday, she signed, "OLD GORILLA." Koko also apparently expressed remorse for having bitten a companion on the previous day, signing "SORRY BITE SCRATCH" and "WRONG BITE," explaining that she was mad at her companion (Patterson 1978; Patterson and Gordon 1993).

The late orangutan Chantek mastered more than 150 signs of ASL and learned, without training, how to comprehend much spoken English. He sometimes signed for manipulative purposes, for example signing *dirty* as a pretense to go to the bathroom in order to play with the washing machine. He signed for objects that were not immediately perceivable—for instance, to ask to go to the backyard to look for a favorite cat. Sometimes he created novel strings, such as "DAVE MISSING FINGER" for a person who had lost a finger and "EYE DRINK" for contact lens solution (Miles 1993).

Assuming these descriptions of the three apes' linguistic feats are representative of their acquired language capacities, Kanzi, Koko, and Chantek are fairly promising candidates for animals who have narrative identities. So far, we have considered the possibility of nonhuman narrative identity through the lens of linguistic competence—which young human children have in some significant measure around the time they seem to acquire narrative identities. But we should be open to the possibility that narrative identity can sometimes emerge in cognitively sophisticated creatures who lack linguistic competence. The most promising candidates would seem to be great apes and cetaceans (dolphins, porpoises, and other whales), although elephants may not be far behind. Following are some details about their lives and capacities.

Chimpanzees, bonobos, and gorillas are highly social creatures (in contrast to the semi-solitary orangutans). The social lives of these great apes feature long-term relationships, dominance hierarchies, awareness of kin relationships, non-kin-based alliances, and the tracking of significant interactions, such as fights, groomings, and instances of assistance with particular group members. (For helpful overviews, see Goodall 1986; de Waal 1997; Parker et al. 1999; Russon et al. 1996.) Their penchant for deception, planning, scheming with others, and similar behaviors have led primatologists to characterize some great apes—at least chimpanzees—as "political" (de Waal 2000). With this general understanding of their behavior, it seems plausible to assume that they have fairly extensive episodic memories, substantial social awareness, and perhaps even intentions for the nonimmediate future, such as a plan to topple the current alpha male with the help of a particular group member. On this basis, it might be reasonable to suppose that they have narrative identities. But I leave this open as a possibility, neither confirmed nor disconfirmed.

The case for narrative identity in cetaceans is comparably strong. Their lives feature long-term relationships, dominance hierarchies, intense mother–calf bonding, adult protection of the young, and tracking of significant interactions with other group members. (For background, see Mann et al. 2000). Meanwhile, their natural communicative abilities—or, at least, those of some species—may exceed the complexity found in apes' natural vocalizations and hand gestures. Dolphins recognize each other by their unique "signature whistles" and may call each other by imitating others' whistles; the latter are also modulated, apparently conveying emotional states. Certain bodily postures and gestures seem to convey information about upcoming movements or social information, such as "I want to interact with you." It has been argued, rather plausibly, that because the ocean affords no hiding places from predators, cetaceans evolved under especially intense selective pressure to develop the capacity for cooperation. (For a good overview of dolphins' social lives and communicative abilities, see White 2007, chap. 5; see also Pryor and Norris 1991). On balance, I would say that the possibility of narrative identity in cetaceans is at least as strong as it is in the case of great apes. But, with the information available, we can only speculate. As we will see, there is more rigorous evidence for other types of self-awareness in particular animal species.

Introspective Awareness

Introspective awareness is awareness of one's own mental states, such as believing something, not being sure about something, feeling angry, or being hungry. To head off confusion, we must distinguish *having* a mental state, such as a belief or feeling, from *being aware that* one has that mental state. Although some mental states, such as pain or hunger, may be intrinsically conscious,[4] having such a state is distinguishable from the higher-order awareness, or consciousness, *that one is having the state*. Consider an example. Assuming a human newborn can feel pain prior to possessing even the most rudimentary self-awareness, the infant will at first feel pain without awareness *that* she is feeling pain. She simply feels something that hurts without awareness of herself as a distinct being who is subject to such feelings. On the other hand, as soon as an infant—or any creature—acquires an awareness of himself as a distinct being who is subject to such bodily feelings (see later discussion of bodily agential self-awareness), it makes sense that the experience of intrinsically conscious states such as pain and hunger will ordinarily be accompanied by the introspective awareness of having those states.[5]

Because introspective awareness has a second-order character, it is a type of *metacognition*. Another type of metacognition, the attribution of mental states to other individuals—requiring a "theory of mind"—is sometimes claimed to be necessary for introspective awareness, and vice versa. But I will set aside this thesis and focus on evidence for introspective awareness.

If any animals have a narrative identity, as discussed in the previous section, their self-conception will presumably include not only memories and intentions but also awareness of having these mental states. For example, if a gorilla remembers being thrashed by another gorilla, it seems reasonable to assume that he recognizes the recollection *as a memory*—that is, as indicating what happened to him earlier. Without such a recognition, memories would seem to be useless. Before proceeding, let me clarify that in speaking of memories I mean, more specifically, *episodic memories*, conscious experiences in which one recalls having some earlier experience. These are to be contrasted with *semantic* memories, with which one recalls some fact (without necessarily recalling any associated experience)—for example, the fact that Mexico City is a capital or even a more personal fact such as that *I saw this actor in a movie* (although I can't remember what movie or the experience of seeing it). Perhaps it is possible for a bird to remember (semantically) that there is food in some place where she left it whether or not she has an episodic memory of leaving it there. My present claim is that episodic memories would be useless if a subject were not introspectively aware of now having an experience that represents her past.

Meanwhile, it seems incoherent to posit an intention that its subject does not recognize as an intention. Here, in referring to "intentions" I have in mind a type of conscious experience about what one might do in the future rather than an unconscious disposition to do something. These reflections suggest that some mental states, such as (episodic) memories and (conscious) intentions, have temporal self-awareness—an awareness that one exists over time—built into them. They also suggest that these mental states implicate introspective awareness. Again, memories and intentions would be useless—if they are even possible—without an awareness that one is having these mental states and of what they represent.

Our discussion in the previous section left open whether any animals have narrative identities. Yet, if the reasoning of the previous paragraph is correct, then not only animals with narrative identities, but any animals with memories or intentions would have some introspective awareness. There is considerable evidence that some animals—various mammals including rodents and some types of bird—have episodic memories (see, e.g., de Kort et al. 2005; Babb and Crystal 2005; Schwartz et al. 2005). There is also impressive evidence that certain animals—including primates, rodents, and some birds—plan for the future, implicating intentions, in service of longer-term goals (see, e.g., Raby et al. 2007; Feeney et al. 2011; Roberts 2012). Moreover, if the reasoning presented four paragraphs earlier is correct, then animals with even a rudimentary awareness of themselves as distinct entities subject to certain feelings will typically have introspective awareness of having those feelings when they occur. That may suggest that introspective awareness, despite its second-order character, is very widespread in the animal kingdom—extending beyond mammals and birds to include reptiles and possibly more primitive classes of animals (see later discussion of bodily agential self-awareness). At this point, however, I focus on more direct evidence of introspective awareness in certain animals.

Some of the signings of Koko the gorilla (Patterson and Gordon 1993) may have indicated introspective awareness. Once, when angered, Koko is reported to have signed, "RED MAD GORILLA." On another occasion, she repeatedly asked a companion for juice but was rebuffed. Resorting to drinking water through a straw from a pan on the floor, she allegedly signed, "SAD ELEPHANT." If these anecdotes are accurate, they strongly suggest that Koko had some awareness of her feelings and could express them linguistically.

Let us turn now to experimental evidence, some of which emerged in studies of monkeys by David Smith and colleagues (for summaries, see Smith and Washburn 2005; Phillips 2006). Monkeys were trained to manipulate a joystick to select answers in discrimination tests about visual patterns on a computer screen. Incorrect answers elicited "time-outs" (delays before further trials), which they hated, while correct answers elicited food pellets, which they liked. Later, the monkeys learned the option of choosing an icon for "pass." If they chose this option, they received no pellet but proceeded immediately to the next trial, a result less desirable than immediate food but preferable to a delay without food. Facility with the pass option, which they often used in difficult trials, afforded initial evidence that the monkeys assessed their own level of confidence and perceived that they were *unsure*—an instance of introspective awareness.

Other explanations of the monkeys' behavior are possible. They might have been conflicted about which answer was correct and selected the pass option by default, or perhaps they were simply trying to move faster to a new trial. But further findings cast doubt on such skeptical responses. First, less cognitively complex animals, rats, failed to learn the pass option in one trial (Smith and Schull 1989) while a later trial was ambiguous (Crystal and Foote 2009), suggesting that the monkeys might be performing a higher-level cognitive feat. Second, researchers modified the monkey experiments so that they received food or delays only following a series of trials, rather than after each one. Third, later trials had monkeys demonstrate the ability to *remember* previously shown images rather than discriminate among present images (Hampton 2001, 2005). So monkeys who mastered the task apparently tried to recall an image, compare it with a current image, and decide whether there was a match—suggesting introspective awareness of having a representation of the earlier image. Finally, subsequent research suggested that monkeys who learned to use a pass response in perception tasks could immediately do the same not only in different perception tasks but in memory tasks as well (Kornell et al. 2007).

In view of this experimental evidence for introspective awareness in monkeys, it should not be surprising that similar evidence emerged in trials involving a dolphin, who chose a "pass" option in difficult trials and hesitated before doing so (Smith et al. 1995). Kristin Andrews comments: "This squares well with my own experience working with dolphins, who would respond to difficult tasks by swimming in a tight circle between the two choices before settling on one" (Andrews 2015: 74). If monkeys and dolphins can be aware of their own uncertainty, one might expect the same of great apes—and not just successful language pupils. This hunch was apparently confirmed in a study involving eight chimpanzees, seven gorillas, four bonobos, and seven orangutans, about which the author concludes, "[S]ubjects knew that they could be wrong when choosing" (Call 2010).

Social Self-Awareness

Social self-awareness is awareness of oneself as occupying a particular position within one or more social relationships. It appears to be characteristic of highly social animals including many or most mammals and perhaps some birds.[6] I contend, a bit conservatively, that the members of many mammal species have social self-awareness. The evidence for this capacity consists of individual behavior within groups that seems best explained by the animals' possession of social self-awareness.

Before proceeding, a more explicit characterization of this type of self-awareness might be helpful. Any type of self-awareness is expressible through a sentence with a first-person pronoun, whether or not the being in question actually thinks linguistically. For example, a person's narrative identity might include pieces of self-awareness expressible by the sentences "I am the grandson of a Sicilian immigrant" and "Learning Italian is on my bucket list." Two bits of introspective awareness might be expressed as "My toe hurts" and "Glib talk of social construction irritates me." How about social self-awareness? A chunk of such awareness might be rendered in English as "He is more powerful than I; we have groomed each other; we are allies, so I will help him if he gets in a fight." Exactly how nonlinguistic beings mentally manifest such awareness I don't know. But much behavior seems best explained by the assumed possession of social self-awareness.

Many mammals have complex social lives featuring group living, dominance hierarchies or more equitable relations, a sense of kinship to particular others, shifting alliances, and the like. Individuals often keep track of salient transactions with others, such as fights and episodes of grooming. Each social group member needs to understand her position in the group and her relationship to specific others as well as any expectations that come with these relationships—for example, being an ally to another member of a primate social group might entail coming to his or her assistance in a fight. Such understanding embodies social self-awareness. To the extent that episodic memory is involved—for example, of having been recently groomed by him—such understanding also implicates some temporal self-awareness and, assuming a particular memory is recognized *as a memory* (as discussed earlier), introspective awareness. Following are some observations about particular mammal species.

It is well known that wolf packs feature complex social dynamics. Moreover, domestic dogs—a species that evolved from wolves—appear to engage in pack behavior in human households. If there are other dogs, they will work out a dominance hierarchy. If there are no other dogs, the lone canine will typically work out who is the "alpha" among the human companions and try to forge a strong alliance with that individual. Meanwhile, despite being less actively social than dogs, domestic cats also work out dominance hierarchies among themselves.

There has been extensive ethological study of primate social life. Cheney and Seyfarth's (1990) investigations of vervet monkeys, for example, demonstrate that they know who is a relative, who is dominant, who is a relative of a dominant individual, and how other group members stack up in the hierarchy. The authors conjecture that vervets' innate disposition to sort others in family structures and hierarchies evolved to enhance the ability to predict conspecifics' behavior (Seyfarth and Cheney2003). This conjecture seems equally plausible in the case of other highly social species.

Earlier we discussed some aspects of apes' social lives in discussing the possibility of their having narrative identities. To supplement the earlier discussion, chimpanzees, bonobos, and gorillas recognize individual group members, recall favors bestowed as well as grudges, have enduring relationships, and build and shift alliances (Goodall 1986; Stanford 2001, 2008).[7] The structure of social life within great ape species reveals differences, however. While chimpanzees are very hierarchical and frequently violent, for example, bonobos are more egalitarian and cooperative, communicate with recreational sex, and excel at forging alliances (Stanford 2001, chap. 1).

Despite being more difficult to study due to their aquatic terrain, cetaceans have been found to have exceptionally complex social lives, as noted earlier. The following statement from an eminent dolphin researcher will fill out our sketch:

> Bottlenosed dolphins as well as many other species of toothed whales (odontocetes) live in complexly organized social units (e.g., Connor et al. 1992). To function effectively within these units, the young dolphin must undergo extensive learning about the conventions and rules of the society, about cooperative and collaborative activities, and about the identities and even personalities of group members and associates (Herman 1991). The protracted period of development and dependence of young dolphins on their mothers and other group members allows the time and opportunity for extensive social learning to take place.
>
> *(Herman 2002, at 275)*

The foregoing considerations about the social lives of various mammal species support the attribution of social self-awareness. We have considered wolves, domestic dogs, vervet monkeys, three great ape species, and cetaceans with special attention to dolphins. Without entering into details we may confidently add elephants (see, e.g., Wittemyer and Getz 2007) and probably some other types of mammals. The upshot is that social self-awareness appears to be rather widely distributed in the animal kingdom. As we will find in the next section, another type of self-awareness is distributed far more widely than social self-awareness.

Bodily Agential Self-Awareness

The final type of self-awareness that I consider is what I call *bodily agential self-awareness*: an awareness of one's own body as importantly distinct from the rest of the environment in being directly connected with certain feelings and subject to one's direct control in acting—in short, in being *one's own*. Because of this type of self-awareness, one does not attempt to eat oneself. And one pursues goals. Bodily agential self-awareness, as I understand it, typically includes all of the following: (1) *proprioception*: an awareness of body parts, their position and movement, and overall body position (Bermudez 1998, chap. 6); (2) various *sensations* (e.g., pain, hunger, thirst, sensations of warmth, cold, or tactile pressure) that provide information about what is happening to the body; (3) *spatiotemporal awareness*: an awareness of where one is in relation to nearby objects and of one's persistence through time; and (4) *agential awareness*: awareness that one can do things to control one's body and interact with the environment. These specific forms of awareness, adding up to bodily agential self-awareness, are paramount to creatures that can sense features of their bodies and the environment and respond to this information with flexible behaviors.

For the sake of clarifying this type of self-awareness, consider a creature that lacks it despite being able to sense certain salient conditions and respond in survival-enhancing ways. Suppose the creature can sense the need for food, the need for water, and the presence of noxious stimuli on a body part—and, if sentient, can *feel* hunger, thirst, and pain—and, upon sensing these stimuli, locates food and eats it, locates water and drinks it, and withdraws from the noxious stimulus. Imagine that these three types of action are mechanical, performed automatically in response to relevant sensory stimuli—for example: sensing the need for food + sensing (e.g., by smell) the presence and direction of food → moving toward the food and consuming it. The behavior of this creature has a stimulus-response character (which may or may not be attended by sentience and therefore the experience of feelings). Per our stipulation, the animal lacks a sense of itself as a distinct entity, located within a mental map of space, persisting through time, and capable of acting upon the world in a self-controlling way.[8] In brief, the animal lacks an integrated, internal model of itself within a broader world. This sense of itself, or model, would constitute bodily agential self-awareness.

Now consider a quotidian human example. An infant discovers that every time she touches her foot, she feels something both in her hand and in her foot. When she grabs her plastic block, she feels something in her hand but not in the block. These and similar experiments gradually engender the realization that her hands and feet are special objects—they are part of *her*. The infant also realizes that, if she wants to have that "foot" feeling again, she can reliably produce it by grabbing her foot. The infant has the beginnings of bodily agential self-awareness.

Many animals do as well. Which ones? Although I do not try to answer this question comprehensively, I argue that (1) at least reptiles, birds, and mammals have bodily-agential self-awareness and, perhaps surprisingly, (2) some insects may as well.

My point of departure is a theory that consciousness first evolved in early amniotes, the clade that comprises reptiles, birds, and mammals. Michael Cabanac and colleagues (2009) hypothesize that as land-based lifeforms evolved, "existence required more and more stimulus-response pathways; eventually, a point was reached where it became more efficient, in terms of speed and flexibility, to route all decision making through a single mental space," with a criterion of maximizing expected net pleasure. These newly conscious creatures were capable of pleasure and pain, affording them an experiential basis for selecting behaviors; as the authors put it, "hedonic experience . . . is the common currency that allows motivations to talk to each other" (ibid.: 269). This account focuses on consciousness, but the model of consciousness suggests at least a rudimentary bodily agential self-awareness, as will become clear in a moment.

Cabanac and colleagues (2009) hypothesize that consciousness emerged as an efficient solution to the need to integrate information from multiple sensory modalities and respond flexibly in survival-promoting ways. For example, instead of mechanically retreating from painful stimuli while seeking food, a snake might endure some pain in order to reach the only nearby food source. This sort of behavior is called a "motivational trade-off." (As I understand such choices, they might not be restricted to the hedonic currency of various kinds of pleasant and unpleasant experiences as in Cabanac et al.'s account. For example, trading off between the values of [1] avoiding painful stimuli and [2] finding and consuming food—whether or not doing so is pleasant or is expected to maximize net pleasure over time—would qualify as a motivational trade-off so long as both are health-promoting and innately motivating for the creature.) Consistent with their hypothesis that consciousness and sentience emerged in amniotes, the authors ran trials involving "taste aversion learning"—in which animal subjects could learn to associate a food's pleasant taste with the indigestion that followed, thereafter avoiding the food—and found reptiles but not amphibians to demonstrate this type of learning (Pardis and Cabanac 2004). Moreover, the authors cite literature suggesting that reptiles, when handled, produce physiological responses characteristic of stress (an emotional response)—whereas amphibians do not (Cabanac et al. 2009: 268).

In addition to the authors' arguments that present-day amniotes have the sort of integrated consciousness they describe, there is ample independent evidence that (at least) reptiles, birds, and mammals are capable of performing intentional actions in pursuit of goals. This capability requires bodily agential self-awareness. Rather than review this evidence here, I refer the reader to works in which I do so (DeGrazia 2009: 202–206; DeGrazia 1996, chap. 6) and here offer a single plausible example. A dog runs through the house to the dog door with the intention of exiting the house and entering the backyard in order to get the bone he left there earlier. This dog wants to chew on the bone, remembers having left it outside, has a simple plan for getting it, locates himself in a mental map, and recognizes that he can take action to get what he wants. He has bodily agential self-awareness.

Interestingly, Cabanac and colleagues' assertion that amphibians lack what they call consciousness and I construe as bodily agential self-awareness received some support in an experiment performed almost half a century ago. By way of background, amphibians evolved from fish prior to the emergence of amniotes and include such animals as frogs, newts, and salamanders; they live first in water and then, after a physical metamorphosis, the rest of their lives on land. Things get interesting with a familiar type of amphibian: frogs have one visual stream that allows them to detect and snap at moving objects such as flies and a distinct visual stream that enables them to navigate around barriers. A lack of unified visual perception was demonstrated in an experiment in which surgical rewiring in a frog's brain resulted in a right–left reversal of prey detection without affecting the ability to perceive right and left for purposes of walking around objects (Ingel 1973; discussed in Godfrey-Smith 2016: 89). Although this is just one experiment, it coheres with the hypothesis that frogs—and perhaps amphibians more generally—lack an integrated consciousness of the environment and of themselves acting within it and therefore lack bodily agential self-awareness.

I have argued that at least reptiles, birds, and mammals characteristically possess bodily agential self-awareness. While recognizing that there is some evidence that amphibians lack such integrated awareness, I do not claim with Cabanac and colleagues that *only* amniotes have this endowment. For one thing, fish comprise such an extraordinarily diverse class of animals (Allen 2013: 26) that it seems possible that some presently living fish have a more integrated awareness of themselves than amphibians—or, at least, frogs—do, even though amphibians evolved out of certain fish species. Moreover, there is ample evidence that some cephalopods, especially octopuses, have a high level of cognitive sophistication, and I am confident that the evidence, properly interpreted, would support the attribution of bodily agential self-awareness to these creatures—although I won't defend this claim here (but see Godfrey-Smith 2016). What I argue, building on the work of two scholars, is that there is a significant possibility of bodily agential self-awareness in insects.

Andrew Barron and Colin Klein (2016) have recently argued that insects are conscious. Consciousness, as I conceptualize it, is simply subjective experience or awareness. Although the authors originally define consciousness as I do, their model of what consciousness involves bears some similarity to Cabanac and colleagues' model and embodies what I have called bodily agential self-awareness.

In vertebrates, Barron and Klein argue, the capacity for subjective experience is supported by integrated midbrain structures that create a neural model of the state of a mobile creature in space—a representation of the world from the creature's perspective. Structures in the insect brain function analogously, according to the authors, in relevant respects to the mammalian neocortex—a thin layer of neurons on the outer part of the cerebrum that is thought to be critical to the experience of consciousness in mammals. (It is worth noting that the minuscule brain of the bee, an insect to which Barron and Klein devote much attention, has nearly a million neurons, making it far denser in neurons than the human neocortex [Tye 2017: 152].) In both vertebrates and insects this sort of integrated control system evolved to deal efficiently with (1) the reafference problem (the need to distinguish among the barrage of sensory inputs those that come from one's own actions and those due to the external world),[9] (2) the need to navigate to places beyond one's immediate sensory range, and (3) the need to integrate information from multiple sensory modalities. As Barron and Klein (2016: 4902) state, "[f]or active animals with well-developed spatial senses, it is computationally more effective to resolve the reafference problem once for a unified sensory model than to resolve it in a dispersed and peripheral way for each sense independently." They further contend that the midbrain's integration of different types of information "provides the capacity to resolve competing behavioral priorities or motivations and rank needed resources by both urgency and availability" (ibid.). This claim is consonant with Cabanac et al.'s thesis that consciousness permitted a unified basis for responding flexibly and efficiently to multiple demands on an organism.

I do not know whether Barron and Klein are correct that insects are conscious in a way that would entail bodily agential self-awareness. All I claim is that, given Barron and Klein's theorizing and the evidence they cite, the possibility is worth considering. Many scientists and philosophers (including me) have argued, or simply assumed, that insects cannot possibly have any form of self-awareness. But, as seekers of knowledge, we have to respect the evidence and the world, including its creatures, as they are—not as we might prefer or assume them to be. I consider it an open question whether any or all insects have the rather basic sort of self-awareness we have considered in this section.

On the Ethical Significance of Different Types of Self-Awareness

As noted at the outset, self-awareness is often assumed to ground special moral status and to be exclusively human, or nearly so. The idea might be that, while sentient creatures have moral status, self-aware beings, such as humans and perhaps some great apes, have *full* moral status. The arguments of this paper challenge such thinking by distinguishing several types of self-awareness and demonstrating that some of them extend deeply into the animal kingdom. But these claims are compatible with the idea that self-awareness, in one or more of its varieties, bears moral significance. In this final section, I briefly consider this possibility by noting how the various types of self-awareness are tied to morally significant interests.

As we found in the previous section, a wide class of animals has bodily agential self-awareness. These animals are not only sentient beings, grounding an interest in experiential well-being, but also agents—beings with aims or goals. Their agency grounds some sort of interest in being able to pursue their aims, although different theories will conceptualize this interest in different ways. A libertarian-leaning animal ethics might assert that all such animals have rights to freedom of movement, bodily integrity, and any other conditions vital to agency. A more consequentialist animal ethics might withhold the rights claim while acknowledging the conditions of agency as morally important interests of animal agents—these animals' lives generally go better when they are permitted to function as agents. Whether the conditions of agency are intrinsically important to the animals' well-being, as objective value theories might assert, or important only because instrumental to the animals' experiential welfare, as hedonistic value theories would claim, is a further issue that will divide ethical theories that recognize the moral status of sentient animals and the importance of agency to animal agents.

Earlier we found that members of highly social species exhibit social self-awareness. These animals are not only sentient beings and agents but also highly social creatures. As such, they have interests in being able to socialize with group members and in the maintenance of beneficial (as opposed to antagonistic) relationships. Whether we construe these interests as objects of rights or simply as morally weighty interests, they justify a strong presumption against isolating the animals in question or interfering with the continuation of their beneficial relationships. Thus, for example, highly social animals kept in captivity—whether in zoos, human homes, stalls, or laboratory settings—must have social access to appropriate companions and must not be isolated from such company absent extremely compelling grounds for such separation. This observation condemns primate maternal-deprivation studies and, except in special circumstances (e.g., involving a dangerous infectious disease), the isolation of primates, dogs, and other social animals in separate cages.

Introspective awareness is less straightforwardly connected with important interests than are bodily agential and social self-awareness. However, let me suggest two connections. Introspective awareness seems to facilitate (1) agency and the pursuit of one's goals and (2) social living. For example, in one of the testing paradigms discussed earlier, a monkey who is aware of not being sure which answer is correct can select the "pass" option rather than simply guessing, thereby advancing her interest in getting as much food as possible over a series of trials. Meanwhile, if I face a looming deadline for a manuscript review and notice that I feel ambivalent about working on it now, I can take steps (e.g., a burst of exercise) likely to increase my desire to work on it. Introspective awareness also facilitates social success. For example, if I know that my colleague's comment has made me feel defensive and that I'm prone to say things I later regret when feeling this way, I can switch into mindful-acceptance mode, keep my mouth shut, and decide tomorrow whether any response to the comment is worthwhile. In comparison with reacting impulsively when feeling defensive, the strategy just described is more likely to facilitate a good working relationship with my colleague. Introspective awareness has advantages. So creatures who are introspectively aware have an interest in retaining this capacity. This confirms the obvious thought that we should not damage animals' introspective capacities.

The possession of a narrative identity, meanwhile, generates a very strong interest in remaining alive. Whereas any sentient creature with the prospect of continuing a good life may be harmed by death—insofar as death deprives her of further goods—an individual with the sort of biographical self-conception involved in a narrative identity typically loses much more from premature death. Such an individual loses the opportunity to bring to completion whatever projects make her life meaningful in her own eyes. Such projects, using the term broadly, might include starting a family and engaging in family life for half a century or producing a body of music or scholarship or retiring and giving oneself over to a satisfying mixture of leisure and tutoring in public schools. The biographical shape of life for a being with a narrative identity is perhaps the strongest ground for a right to life—by which I mean an exceptionally stringent moral claim not to have one's life taken involuntarily. If we have reason to judge that some nonhuman animals have narrative identities, then we have reason to ascribe to these animals a right to life in this strong sense of the term.

Acknowledgment and Disclaimer: I would like to thank Bob Fischer for very helpful feedback on a draft. This research was supported, in part, by intramural funds from the National Institutes of Health (NIH) Clinical Center. The views expressed here are the author's own. They do not represent the policy or position of NIH or any other part of the federal government.

Notes

1. For recent examples of scholars who embrace this criterion, see Hyun (2013: 145) and Tye (2017: 42–43).
2. On a biographical note, I was three when I first grasped what people meant when they asked me how old I was. Before that moment, which I remember clearly, I had answered "How old are you?" by supplying the answer I had heard others provide on my behalf. So, when asked my age I simply repeated, "Three." "How old were you before that?" I said, "Two," supplying the answer I remembered had been approved earlier. But when my older sister said I had been one year old before I was two, at first I denied this because I had no memory of this being the correct answer. Then I realized that I must have been one because I understood, finally, that these numbers measured amounts of time in my life. Whether this realization was sufficient for having a narrative identity I am not sure, but it was at least a significant step in the direction of such self-awareness.
3. I use the term *creature* to exclude artificial-intelligence systems, which may possess linguistic capacities while lacking consciousness and self-awareness.
4. By contrast, some mental states, such as beliefs and desires, have a dispositional character and are only conscious when brought to awareness.
5. The qualification "ordinarily" leaves room for some exceptions, such as when one is in such a "flow" experience that one temporarily loses aspects of self-awareness.
6. This point applies only to social species whose members are conscious. There is a sense in which ants and bees are highly social, but if they are not conscious, then they lack awareness in general and therefore lack social self-awareness.
7. This may also be true of orangutans but I am uncertain in the case of this semisolitary ape species.
8. In a similar vein, Godfrey-Smith (2016: 93) argues that pain, thirst, oxygen hunger, and the like do not necessitate having an internal model of the world with oneself in it.
9. This is achieved, in part, through "perceptual constancies," which Godfrey-Smith (2016: 99) helpfully explains in this way: "These are abilities an animal has to re-identify objects despite changes in viewing conditions—distance, lighting, and so on. The animal must factor out the contribution of its own location and perspective to identify the object itself. . . . Perceptual constancies show that an animal is perceiving external objects as external objects—as objects that can stay the same while the animal's vantage point changes."

Further Readings

Allen, C., and Trestman, M. (2016) "Animal consciousness," *Stanford Encyclopedia of Philosophy* (https://plato.stanford.edu/entries/consciousness-animal/), first published 12/23/95, substantive revision 10/24/16.
(A comprehensive overview of the evidence and methodological issues bearing on the attribution of consciousness to nonhuman animals.)

Cavalieri, C., and Singer, P. (1993) *The Great Ape Project: Equality Beyond Humanity*, New York, NY: St. Martin's.
(Contains a wealth of scientific information, accessibly presented, about the evolution, minds, and social lives of the different species of great apes.)

de Waal, F. (2016) *Are We Smart Enough to Know How Smart Animals Are?* New York, NY: Norton.
(An accessible, well-documented volume written by a leading primatologist on the surprising richness of animal cognition.)

Godfrey-Smith, P. (2016) *Other Minds: The Octopus, the Sea, and the Deep Origins of Consciousness*, New York, NY: Farrar, Straus, and Giroux.
(An exploration of the evolution of consciousness and self-awareness with special attention to cephalopod minds.)

Lurz, R. (ed.) (2009) *The Philosophy of Animal Minds*, Cambridge, UK: Cambridge University Press.
(A collection of essays by fourteen philosophers of mind/cognitive sciences on what we can know about animals' mental lives.)

White, T. I. (2007) *In Defense of Dolphins: The New Moral Frontier*, Malden, MA: Blackwell.
(A highly detailed yet accessible exploration of dolphins' intelligence, social lives, and moral importance.)

Bibliography

Allen, C. (2013) "Fish cognition and consciousness," *Journal of Agricultural and Environmental Ethics* 26: 25–39.

Andrews, K. (2015) *The Animal Mind*, Oxford: Routledge.

Babb, S., and Crystal, J. (2005) "Discrimination of what, when, and where: Implications for episodic-like memory in rats," *Learning and Motivation* 36: 177–189.

Barron, A., and Klein, C. (2016) "What insects can tell us about the origins of consciousness," *PNAS* 113: 4900–4908.

Bermudez, J. (1998) *The Paradox of Self-Consciousness*, Cambridge, MA: MIT Press.

Cabanac, M., Cabanac, A., and Parent, A. (2009) "The emergence of consciousness in phylogeny," *Behavioural Brain Research* 198: 267–272.

Call, J. (2010) "Do apes know that they could be wrong?" *Animal Cognition* 13: 689–700.

Cheney, D., and Seyfarth, R. (1990) *How Monkeys See the World*, Chicago, IL: University of Chicago Press.

Connor, R., Smolker, R., and Richards, A. (1992) "Dolphin alliances and coalitions," in A. Harcourt and F. de Waal (eds.) *Coalitions and Alliances in Humans and Other Animals*, Oxford: Oxford University Press.

Crystal, J., and Foote, A. (2009) "Metacognition in animals," *Comparative Cognition and Behavior Reviews* 4: 1–16.

de Kort, S. R., Dickinson, A., and Clayton, N. S. (2005) "Retrospective cognition by food-caching western scrub-jays," *Learning and Motivation* 36(2): 159–176.

de Waal, F. (1997) *Bonobo*, Berkeley, CA: University of California Press.

de Waal, F. (2000) *Chimpanzee Politics*, Baltimore, MD: Johns Hopkins University Press.

DeGrazia, D. (1996) *Taking Animals Seriously: Mental Life and Moral Status*, Cambridge: Cambridge University Press.

DeGrazia, D. (2009) "Self-awareness in animals," in R. Lurz (ed.) *The Philosophy of Animal Minds*, New York, NY: Cambridge University Press.

Feeney, M., Roberts, W., and Sherry, D. (2011) "Black-capped chickadees (*Poecile atricapillus*) anticipate future outcomes of foraging choices," *Journal of Experimental Psychology* 37: 30–40.

Gallup, G. (1970) "Chimpanzees: Self-recognition," *Science* 167, January: 86–87.

Gallup, G. (1977) "Self-recognition in primates: A comparative approach to the bidirectional properties of consciousness," *American Psychologist* 32: 330–338.

Glover, J. (1988) *I: The Philosophy and Psychology of Personal Identity*, London: Penguin.

Godfrey-Smith, P. (2016) *Other Minds*, New York, NY: Farrar, Straus, and Giroux.

Goodall, J. (1986) *The Chimpanzees of Gombe*, Cambridge, MA: Harvard University Press.

Hampton, R. (2001) "Rhesus monkeys know when they remember," *PNAS* 98: 5359–5362.

Hampton, R. (2005) "Can rhesus monkeys discriminate between remembering and forgetting?" in H. Terrace and J. Metcalfe (eds.) *The Missing Link in Cognition*, Oxford: Oxford University Press.
Herman, L. (1991) "What the dolphin knows, or might know, in its natural world," in K. Pryor and K. Norris (eds.) *Dolphin Societies*, Los Angeles, CA: University of California Press.
Herman, L. (2002) "Exploring the cognitive world of the bottlenosed dolphin," in M. Bekoff, C. Allen, and G. Burghardt (eds.) *The Cognitive Animal*, Cambridge, MA: MIT Press.
Hyun, I. (2013) *Bioethics and the Future of Stem-Cell Research*, New York, NY: Cambridge University Press.
Ingel, D. (1973) "Two visual systems in the frog," *Science* 181: 1053–1055.
Kornell, N., Son, L., and Terrace, H. (2007) "Transfer of metacognitive skills and hint seeking in monkeys," *Psychological Science* 18: 64–71.
Locke, J. (1694) *An Essay Concerning Human Understanding*, 2nd ed., London: Awnsham, Churchill, and Manship.
Mann, J., Connor, R., Tyack, P., and Whitehead, H. (eds.) (2000) *Cetacean Societies*, Chicago, IL: University of Chicago Press.
Miles, L. (1993) "Language and the Orang-utan," in P. Cavalieri and P. Singer (eds.) *The Great Ape Project*, New York, NY: St. Martin's.
Pardis, S., and Cabanac, M. (2004) "Flavor aversion learning induced by lithium chloride in reptiles but not in amphibians," *Behavioural Processes* 67: 11–18.
Parker, S., Mitchell, R., and Miles, L. (eds.) (1999) *The Mentalities of Gorillas and Orangutans*, Cambridge: Cambridge University Press.
Patterson, L. (1978) "Linguistic capacities of a lowland gorilla," in F. Peng (ed.) *Sign Language Acquisition in Man and Ape*, Boulder, CO: Westview.
Patterson, L., and Gordon, W. (1993) "The case for the personhood of gorillas," in P. Cavalieri and P. Singer (eds.) *The Great Ape Project*, New York, NY: St. Martin's.
PBS (1995) *Monkey in the Mirror* (documentary about primate cognition).
Phillips, H. (2006) "Metacognition: Known unknowns," *New Scientist*, December: 28–31.
Pryor, K., and Norris, K. (1991) (eds.) *Dolphin Societies*, Berkeley, CA: University of California Press.
Raby, C., Alexis, D., Dickinson, A., and Clayton, N. (2007) "Planning for the future by western scrub-jays," *Nature* 445: 919–921.
Roberts, W. (2012) "Evidence for future cognition in animals," *Learning and Motivation* 43: 169–180.
Russon, A., Bard, K., and Parker, S. T. (1996) *Reaching into Thought: The Minds of the Great Apes*, Cambridge: Cambridge University Press.
Savage-Rumbaugh, S. (1986) *Ape Language*, New York, NY: Columbia University Press.
Savage-Rumbaugh, S., and Brakke, K. (1990) "Animal language: Methodological and interpretive issues," in M. Bekoff and D. Jamieson (eds.) *Interpretation and Explanation in the Study of Animal Behavior*, Vol. 1, Boulder, CO: Westview.
Schechtman, M. (1996) *The Constitution of Selves*, Ithaca, NY: Cornell University Press.
Schwartz, B., Hoffman, M., and Evans, S. (2005) "Episodic-like memory in a gorilla: A review and new findings," *Learning and Motivation* 36: 226–244.
Seyfarth, R. and Cheney, D. (2003) "The Structure of Social Knowledge in Monkeys," in F. de Waal and P. Tyack (eds.) *Animal Social Complexity*, Cambridge, MA: Harvard University Press.
Smith, J. D., and Schull, J. (1989) "A failure of uncertainty monitoring in the rat" (n.p.); cited in Shields, W., Smith, D., Guttmannova, K., and Washburn, D. (2005) "Confidence judgments by humans and rhesus monkeys," *Journal of General Psychology* 13: 165–186.
Smith, J. D., Schull, J., Strote, J., McGee, K., Egnor, R., and Erb, L (1995) "The uncertainty response in the bottlenose dolphin (*Tursiops truncates*)," *Journal of Experimental Philosophy* 124: 391–408.
Smith, J. D., and Washburn, D. (2005) "Uncertainty monitoring and metacognition by animals," *Current Directions in Psychological Science* 14: 19–24.
Stanford, C. (2001) *Significant Others: The Ape-Human Continuum and the Quest for Human Nature*, New York, NY: Basic.
Stanford, C. (2008) *Apes of the Impenetrable Forest: The Behavioral Ecology of Sympatric Chimpanzees and Gorillas*, London: Pearson.
Tye, M. (2017) *Tense Bees and Shell-Shocked Crabs: Are Animals Conscious?* New York, NY: Oxford University Press.
White, T. I. (2007) *In Defense of Dolphins: The New Moral Frontier*, Malden, MA: Blackwell.
Wittemyer, G., and Getz, W. (2007) "Hierarchical dominance structure and social organization in African elephants, *Loxodonta Africana*," *Animal Behaviour* 73: 671–681.

6
THE MORAL ANIMAL

Mark Rowlands

Introduction

Can animals be moral? That is, can they have motivations that are genuinely moral ones, and can they act because of these motivations? Someone who is tempted by a positive answer to this question is likely to find little succor among scientists and philosophers—the possibility of moral behavior in animals has been dismissed by almost all of this demographic. On the other hand, such a person would have at their disposal YouTube: almost certainly the single largest repository of examples of behavior in animals that at least seem to be *candidates* for moral behavior. Check it out any time—although, be warned: it can be addictive. See, for example, a dog lying unconscious on a busy highway. The dog's canine companion, at enormous risk to its own life, weaves in and out of traffic, and eventually manages to drag the unconscious dog to the side of the road. (www.youtube.com/watch?v=-HJTG6RRN4E). Or check out a bear in a zoo, seemingly rescuing a distressed crow from a pond (www.youtube.com/watch?v=TSPgenqMlvQ). Or discover the incredible story of Lilica, the Brazilian junkyard dog, who traveled four miles every night, to get food for her extended family of dogs, cats, mules, and chickens. (www.youtube.com/watch?v=ZnmQ9KWOl28). I could go on—at length—but there's a whole world of YouTube out there just waiting to be discovered, and I would hate to spoil the surprises. Evidence of apparently moral behavior in animals, however, long predates the internet. In 1964, for example, Wechkin and colleagues observed a rhesus monkey refusing to take food when doing so subjected another monkey to an electric shock. The monkey persisted in this refusal for 12 days, nearly starving himself to death. Much of Chapter 4 of Darwin's *The Descent of Man* is a collection of anecdotes of behavior in animals that seem, at least prima facie, to be candidates for moral behavior. Darwin ultimately rejected this possibility, although for reasons that are not, I shall argue, particularly compelling. Marc Bekoff and Jessica Pierce (2009) and de Waal (2006) provide excellent surveys of the empirical evidence for moral behavior in animals.

What should we make of cases such as these? I have twice used the expression "candidates for moral behavior"—together with a liberal sprinkling of the term *seem*. I invoke these cases only to raise the question: Are these examples of moral behavior? I certainly will not assume they answer this question. These cases, like any cases of behavior, are open to interpretation. Two issues, however, should be distinguished. The first is an *empirical* issue about the actual nature of an animal's motivation in any given case. Are these really cases of moral behavior? Or can they be explained in some other way? However, the near blanket dismissal of the possibility of moral behavior in animals, by both philosophers and scientists, is not driven by this sort of case-by-case, empirical consideration but, rather, by *conceptual* considerations. There is, it is commonly thought, something about the nature of moral behavior that renders any animal incapable of engaging in it. We do not need to examine each and every empirical instance of possible moral behavior in animals, and demonstrate that there is an alternative, and preferable, nonnmoral explanation. Rather, the idea is that once we appreciate what moral behavior is and what it requires, we can rule out even the possibility of animals acting morally.

My focus is going to be on this conceptual issue. I shall argue that, on the contrary, there are no insuperable conceptual obstacles to regarding animals as capable of moral behavior. It is true, there are accounts of moral action that would preclude animals acting on the basis of moral motivations. But, equally, there are other accounts that are entirely compatible with this possibility. I shall first identify each type of account, and then present some arguments in favor of the latter.

Moral Motivation: The Kantian Picture

Broadly speaking, accounts of moral motivation can be divided into two sorts. I shall label these *Kantian* and *Humean*. These labels function not so much to pick out specific theories or accounts of moral motivation, but rather, general, underlying pictures of such motivation—they limn the general contours of the sort of thing moral motivation must be. Because of their generality, each label can subsume accounts and figures that would not, in other contexts, be regarded as Kantian or Humean at all.

Christine Korsgaard, an influential defender of the Kantian picture of moral motivation, captures this view very nicely:

> Kant believed that human beings have developed a specific form of self-consciousness, namely, the ability to perceive, and therefore to think about, the grounds of our beliefs and actions as grounds. Here's what I mean: an animal who acts from instinct is conscious of the objects of its fear or desire, and conscious of it as fearful or desirable, and so as to-be-avoided or to-be-sought. That is the ground of its action. But a rational animal is, in addition, conscious that she fears or desires the object, and that she is inclined to act in a certain way as a result. That's what I mean by being conscious of the ground as a ground. So as rational beings we are conscious of the principles on which we are inclined to act. Because of this, we have the ability to ask ourselves whether we should act in the way we are instinctively inclined to. We can say to ourselves: "I am inclined to do act-A for the sake of end-E. But should I?"
>
> *(2004: 148–149)*

The idea of *critical scrutiny* lies at the heart of the Kantian picture. To act morally, I need to be able to critically scrutinize my motivations. That is, with respect to any motivation I currently have, for that motivation to qualify as moral I must be able to (1) ask myself whether they are morally right or wrong, or morally good or bad ones to have, and (2) have some idea of how to go about answering these questions. I satisfy (2) by being able to bring to bear on the motivation relevant moral rules or principles and examining the motivation in their light.

This Kantian picture of moral motivation is a broad one: sufficiently broad to incorporate agents whose preferred moral theories are decidedly not Kantian. This is because condition (2) does not specify which rules or principles should be used to assess the motivations. They might be principles imported from Kant's deontological moral theory. For example, a person might scrutinize her motivations by examining them in the light of the Categorical Imperative: can this emotion be universalized in the way required by the Imperative? However, adherence to Kantian normative theory is not required in order to satisfy (2). What is crucial to the Kantian picture is the idea of critical scrutiny of one's motivation rather than the particular moral theory or principles one employs when engaging in such scrutiny. Thus, a utilitarian might scrutinize his or her motivation by attempting to work out the likelihood of it increasing the overall amount of happiness in the world if it were to be acted upon or increasing the number of satisfied preferences in the world. This satisfies (2) and thus qualifies as an instance of the Kantian picture of motivation. Aristotle also qualifies as an adherent of the Kantian picture. Consider, for example, this famous passage from the *Nichomachean Ethics*:

> But for actions in accord with the virtues to be done transparently or justly it does not suffice that they themselves have the right qualities. Rather, the agent must also be in the right state when he does them. First, he must know that he is doing virtuous actions; second, he must decide on them, and decide on them for themselves; and, third, he must also do them from a firm and unchanging state
>
> *(1999: 1105a27–35)*

For an action to be an expression of a virtue, it must not simply be an example of what would commonly be regarded as a virtuous action (have the "right qualities"). In addition, the agent must (a) know that he is performing a virtuous action and (b) perform the action because it is a virtuous action ("decide on them for themselves"). Thus, for a motivation to be an expression of a virtue, the agent must be able to categorize a motivation as virtuous and perform it precisely because of this categorization. This is a version of the Kantian picture.

The Kantian picture also seems to underlie Darwin's ultimate rejection of the idea that animals are capable of acting morally:

> A moral being is one who is capable of reflecting on his past actions and their motives—of approving of some and disapproving of others; and the fact that man is the one being who certainly deserves this designation, is the greatest of all distinctions between him and the lower animals.
>
> *(1871: 149–150)*

Once again, we find that the ability to critically scrutinize one's motivations—in this case, approving or disapproving of them—is at the heart of the denial to animals of the ability to act morally.

The Kantian picture of moral motivation is an *intellectualist* one. Possession of moral motivations requires significant cognitive sophistication. One must have the metacognitive abilities required to identify one's motivations and the rational capacity to assess these in the light of one's preferred moral principles. It is not unreasonable to suppose this sort of cognitive sophistication is possessed only by humans. Therefore, the Kantian picture seems to preclude the possibility of animals behaving morally. The Kantian picture, however, is not the only game in town.

Moral Motivation: The Humean Picture

The Humean picture provides a very different way of thinking about moral motivation, one where the intellectualist strands of the Kantian picture are largely expunged. Central to any version of the Humean picture is the idea of a *moral emotion*: an emotion that has moral content. But beyond this, the Humean picture, like its Kantian counterpart, is a picture sufficiently broad to incorporate different theoretical articulations. Here, I shall rely on a version of this picture I have developed in earlier work (2012).

Consider an example of a moral emotion—the reactive attitude of indignation. Smith is indignant that Jones snubbed him. That Jones snubbed him is the content of Smith's indignation. There are two ways in which the possession of an emotion might, let us say, *misfire*—which I shall understand as roughly, the analogue of what it is for a belief to be false. The category of misfires is a disjunctive one: an emotion misfires when it is either *misguided* or it is *misplaced*. Smith is indignant because he believes Jones has snubbed him, but he is, in fact, mistaken. Jones didn't snub him at all. Let us say that Smith's indignation is, in this case, *misplaced*. An emotion is misplaced when it is predicated on a factual assertion that is false. If his indignation is not to be misplaced, the factual proposition "Jones snubbed me" must be true.

More important for my purposes, however, is the other category of misfire. Suppose that Jones did indeed snub Smith. However, she has every right to do so—Smith was obnoxious during their previous encounter. Smith, as we might say, deserved no better from Jones in this case. Let us say that Smith's indignation, in this case, is *misguided*. The emotion is misguided, in this case, because it is based on what Smith thinks he deserves—a value judgment rather than a factual one—that is, in fact, erroneous. Thus, generalizing, if an emotion, E, is not misguided, then there is a certain evaluative proposition, *p*, that must be true. It is not necessary that the subject of E be able to entertain p. But the truth of *p* is required if the emotion, E, is not to be misguided: the truth of *p*, we might say, makes sense of E. Smith's indignation, if it is not misguided, guarantees the truth of the evaluative proposition, "Jones was wrong to snub me." An emotion that is not misplaced entails the truth of a factual proposition. An emotion that is not misguided entails the truth of an evaluative proposition. Armed with these ideas, we can define the concept of a *morally laden* emotion as follows:

> An emotion, E, is *morally laden* if: (1) there exists a proposition, *p*, which expresses an evaluative claim, and (2) if E is not misguided, then *p must* be true.

The notion of an evaluative claim, here, is intended to be broad enough to incorporate both straightforwardly moral judgments (e.g., "X is morally wrong") and welfare judgments (e.g., "X is contrary to the welfare of individual P"). In the case of animals, it is welfare judgments that are likely to be to the fore.[1] What is crucial is that to possess a morally laden emotion, an individual needs to be able to entertain neither sort of proposition. The individual, that is, need possess neither the concept of morality nor the concept of welfare. All that is required for the possession of a morally laden emotion is that there exists a moral or welfare proposition and the truth of this proposition is required for the emotion to be a non-misguided one. Given this characterization of a morally laden emotion, we can then define the notion of a *moral subject* as follows:

> An individual, S, is a *moral subject* if and only if S at least sometimes acts on the basis of morally-laden emotions.

I shall talk more about the idea of a moral subject later—and, in particular, its distinctness from the idea of the more traditional idea of a moral *agent*—later. First, however, we must work out what it means to act *on the basis of* a morally laden emotion.

To begin, let us introduce the idea of a *good-* or *bad-making* feature of a situation. Certain features of a situation can make a situation a good one or a bad one. The utilitarian idea of a hedonic calculus provides one obvious way—although, of course, not the only way—of explicating this. Suffering—especially intolerable, immitigable suffering—is generally a bad-making feature of a situation. Happiness and pleasure are good-making features (although their value may be outweighed other malign consequences they bring, as with, for example, a heroin addiction). Claims about the good- and bad-making features of situations are normatively assessable. It is possible to be mistaken about whether a feature is a good- or bad-making one.

Second, suppose now that certain emotions can take these good- or bad-making features as their intentional objects—the emotions are *about* these features. For example, suppose Wechkin's (1964) monkey is distressed *at* the distress of the other monkey, and this is why it declines food.[2]

Being distressed *at* the distress of the other monkey is not the same as being distressed *because* of the other monkey's distress. There are well-known reasons why we should not conflate the cause of an emotion with what that emotion is about. For example, I might be annoyed at someone's irritating behavior, but this behavior would not have bothered me if I were not so tired. Their behavior is the intentional object of my anger, but my tiredness is the cause of my anger.

Third, suppose that the monkey's emotional response to its conspecific's distress is not a one-off, or otherwise unusual, episode. Rather, the monkey is *reliably* distressed at the distress of other monkeys whenever it is aware of it. This reliability will be underwritten by the operations of a (reliable) mechanism whose function is to produce emotions of this sort in these sorts of circumstances. That is, the function of the mechanism is to produce emotional responses to the good- and bad-making features of situations. The mirror neuron system would be the most obvious candidate for this mechanism. Then, we can define what it is to act on the basis of a morally laden emotion as follows:

> A creature, C, acts on the basis of a morally-laden emotion if: (a) C has an emotion that takes a good- or bad-making feature of a situation as its intentional object, (b) this emotion is produced through the operations of a reliable mechanism which functions to produce emotions of this sort to circumstances that contain good- or bad-making features, and (c) this emotion causes C to act in virtue of its intentional object.

For example, suppose the dog who tries to rescue its unconscious companion from the highway has an emotion of a certain sort. We can suppose this emotion will be a form of distress: the dog is distressed at its companion being hit by a car and is now lying prostrate in the road. This is a bad-making feature of the situation in the sense that if the dog's emotion is not misguided, the welfare proposition of the form "Being hit by a car and lying unconscious in the middle of a busy road is a bad thing" must be true. The fact that its companion has been hit by a car and is lying unconscious in the road is what the dog's distress is about—it is the intentional object of this emotion—and it acts on the basis of this. Finally, let us suppose that the dog's having this emotion is not arbitrary but is produced by a mechanism that reliably, or at least reasonably reliably, produces emotions of this sort in the face of these sorts of bad-making features. If these conditions are all met, then we can legitimately regard the dog as acting on the basis of a morally laden emotion. Its motivation is, therefore, a moral one.

I shall assume that the emotion in question is a conscious one. This has one significant consequence. It is common to distinguish acting *on the basis of* a motivation and acting *for* that motivation. In itself, the distinction is a good one. I might be under the impression I am acting for one reason (e.g., someone's irritating behavior) but really am motivated by something else—a reason of which I am entirely unaware (e.g., my tiredness). I act on the basis of the unconscious reason but do not act for it. However, while the distinction is a good one, given the way I shall understand the notion of acting on the basis of a motivation, it essentially amounts to the same thing as acting for that motivation. This is because of the combined requirements that the emotion take a good- or bad-making feature of a situation as its intentional object and that C acts in virtue of this intentional object. In other words, the way I am using the expression "on the basis of" a motivation renders that equivalent to acting "for" that motivation.

This version of the Humean picture requires far less cognitive sophistication than the Kantian model. No metacognitive abilities are required. There is no requirement that moral principles are brought to bear on motivations and used to assess their status. All that is required is the ability to emotionally respond, in a reliable way, to the good- and bad-making features of situations, where the emotional responses take these features as their intentional objects and motivate the individual to act in virtue of these intentional objects.

This account of acting on the basis of a moral motivation is intended as a *sufficient* condition for acting morally not a necessary one. It does not rule out the possibility of nonemotional forms of moral sensitivity to good- and bad-making features. Perhaps a purely rational subject could be sensitive to these features of situations also. I take no stand on this. But it should be noted that the form of sensitivity I appeal to in my development of the Humean picture is specifically *emotional* sensitivity. It does not follow from this willingness to countenance the possibility of other forms of sensitivity that *any* form of sensitivity will do the trick. That the inference from "possibly some other forms of sensitivity will work" to "any form of sensitivity will do" is obviously invalid has not prevented some people drawing it (e.g., see Streiffer 2016).

Problems With the Kantian Model

In earlier work (2011, 2012) I have presented some arguments for preferring the Humean picture to the Kantian alternative. The arguments focused on someone who was very *nice*: Myshkin, who delighted in the happiness of others and tried to promote this happiness wherever possible and, conversely, was horrified by the suffering of others and tried to mitigate this whenever possible. Myshkin was guided by emotion but was unable to scrutinize his motivations in the way required by the Kantian picture. The idea was that it would be implausible to deny that Myshkin acted morally. If so, the Kantian picture does not identify a necessary condition for being moral.

Here, I shall take the converse tack. Instead of focusing on someone very nice, I shall focus on the opposite (see also Rowlands 2017). In 1993, three-year-old Jamie Bulger was abducted, tortured, and murdered by two boys, Robert Thompson and Jon Venables. At the time of the murder, both Venables and Thompson were 10 years old and became the youngest convicted murderers in English history. They kicked Bulger, threw bricks at him, and inserted a battery into his anus, and hit him with an iron bar. Bulger suffered 10 skull fractures and so many injuries that none could be definitively identified as the mortal one. Following the murder, Venables and Thompson placed Bulger's body on railway tracks in the hope that a train would make his death appear an accident. Under questioning, they revealed that they had planned to abduct and murder a child that day and that their initial intention was to take him to a busy road and push him into oncoming traffic—that aspect of the plan later changed.

Venables and Thompson were clearly *motivated*. Their guiding motivation was to abduct and murder a child, and they had given some time and thought to developing this plan for abduction and murder. But suppose it were the case that, because of their young age, or perhaps because of developmental problems, Venables and Thompson were unable to subject their motivations to the kind of critical scrutiny required by the Kantian picture. Even *if* this were true, would we really want to deny that their motivations were morally bad ones? In such circumstances, we *might* want to rescind from morally evaluating Venables and Thompson—they did not understand what they were doing, some might say. But *even if this were true*—even if we cannot morally evaluate Venables and Thompson the individuals—would it really follow that their motivations carry no moral weight? That claim, if one is not in the grip of a peculiarly warped moral psychology, is as counterintuitive as a claim can get. Rescinding from moral evaluation of individuals is one thing; rescinding from moral evaluation of their motivations is quite another.

Note that the argument is stated in hypothetical form. It commits us to no substantive claim about Venables and Thompson other than the claim that their motivation was a morally pernicious one. There is no commitment to the claim that Venables and Thompson were unable to subject their motivations to critical scrutiny. The point is that even *if* this were so—even if they lacked this ability—this would not change the fact that their motivations were morally very, very bad or wrong. The motivations of Venables and Thompson were clearly as morally abhorrent as it is possible to get. Whether they have it or lack it, the ability to scrutinize their motivations is irrelevant to the moral status of those motivations. We must clearly distinguish between the question of the moral status of a motivation (good, bad, right, wrong, etc.) and the question of the moral status of the individual (blameworthy, praiseworthy, etc.) who has that motivation. The inability to critically scrutinize their motivations might be relevant to the moral status of Venables and Thompson—relevant to whether or not they can be morally blamed, or relevant to the extent they can be morally blamed. But it is not relevant to the moral status of their motivations. A motivation can be evaluated as morally good or bad even if the person whose motivation it is cannot.

Morality and Control

I see (following Singer 2011) a small child drowning in a shallow pond. I can easily help and am inclined to do so. Good Kantian that I am, I think to myself: I'm inclined to save this child. Is this a motivation I should endorse or resist? Then, I bring to bear on the motivation my preferred moral principles and decide that I should. This sort of picture, of course, fails as an account of the phenomenology of decision-making. We frequently act—and act morally—without engaging in this sort of reasoning. Indeed, should I reason in this way in this situation, then it might be argued that there is something wrong with me—morally speaking—that I should have to give the situation so much thought. This sort of charge has been raised against the Kantian picture by the virtue-ethical tradition. The Kantian response, of course, is that it is not that we must, or even do, engage in such reasoning whenever we engage in moral action. Rather, we must have the *capacity* to engage in such reasoning. This response, however, merely engenders a further question: Why is it that a capacity that we need not (perhaps ever) exercise is crucial to the moral status of a motivation? This question takes us to the heart of the Kantian picture and at this heart we find the notion of *control*.

It is tempting—at least, many have been tempted—to suppose that a creature that is not able to critically scrutinize its motivations is at the *mercy* of those motivations: they pull her this way or that; she is like a cork bobbing around in a sea of motivations. The ability to critically scrutinize her motivations, however, transforms this picture. Now she can rise above the sea, serenely observe her motivations, and freely decide the extent to which she will let them determine her actions. I don't think this general picture has a hope of working. There is no huge gulf between the (first-order) level of motivations and the (second-order) level of evaluation of those motivations. If issues of control can arise with respect to motivations then they can also arise with respect to our evaluation of those motivations. Various facets of my history or psychology, for example, might make it virtually inevitable that I evaluate a given motivation positively (or negatively). In other words, essentially, the same issues that arise at the first-order level of motivations also arise at the second-order level of evaluations of motivations—for we can be motivated to evaluate our motivations in given ways. (See Rowlands 2011, 2012 for a detailed discussion). Here, however, my focus is on the presuppositions of the picture rather than the picture itself. The guiding presupposition is that a motivation cannot be moral if the subject of that motivation is at its mercy—that is, if the subject has no *control* over that motivation. This is an expression of Kant's famous claim that *ought implies can*.

Imagine someone—we can call him Sigmund—whose motivations are always hidden from him. The motivational component of his mind is akin to a black box: replete with states that successfully guide Sigmund's behavior but to which he has no first-person access. There is an obvious sense in which Sigmund is, as we might put it, "at the mercy" of his motivations. He has no idea what motivates him to act in the way he does and therefore has no control over those motivations. These motivations are, therefore, not the sorts of thing he can embrace or resist. But if he *can* neither embrace nor resist his motivations, then—if ought implies can—it makes no sense to say that he *should* embrace or resist them. If ought implies can, Sigmund's motivations would have no normative dimension. They are not the sorts of things he should endorse or reject. His motivations make no normative claim on Sigmund. However, moral motivations are precisely things that make normative claims on their subjects. Morally good motivations are ones that *should* be embraced by their subject; morally evil motivations are ones that *should* be resisted. Therefore, if ought implies can, it seems that Sigmund's motivations cannot be moral ones.

The widespread acceptance of the idea that *ought implies can* is curious. There are a variety of other forms of "ought," none of which seem to imply "can." The alcoholic ought, *prudentially* speaking, to stop drinking, even though she may be incapable of doing so. The "ought" of prudence does not imply can. Logically, I ought to believe a conclusion that follows validly from true premises, but I may be too dim to see this. The logical "ought" does not imply can. Why, then, should the moral "ought" imply can? Perhaps the moral "ought" is simply idiosyncratic in this way? Kant did regard the moral "ought" as unusual. The prudential ought, for example, is a hypothetical one: if she wants to live a longer and healthier life, the alcoholic ought to stop drinking. But, for Kant, there is nothing hypothetical about the moral ought.

Nevertheless, independently of the question of idiosyncrasy of the moral ought, the idea that a moral motivation requires control over that motivation can be shown to be deeply implausible. Imagine, for example, what we might call *hard determinism world*. In hard determinism world—which may or may not be the actual world—hard determinism is true, and no one is, therefore, ever responsible for what they do. Would we really want to say that, in this world, there is no such thing as moral motivation? When Hitler (or the worldly equivalent thereof) starts a World War and attempts to exterminate various races, would we want to say that his motivations do not count as morally evil? We might, in such a world justifiably rescind from evaluation of Hitler, the person: we might, that is, refuse to blame or hold him responsible for what he does. But refusing to classify his motivations as even falling into the category of the moral is deeply counterintuitive. The principle that ought implies can commits us to this. If Hitler cannot control his motivations, no sense can be made of the idea that he should resist them. His motivations make no normative claim, and therefore cannot qualify as moral. To avoid this, we must reject the idea that ought implies can.

The case of Venables and Thompson, discussed earlier, can be used to make the same claim. Suppose that, due to certain contingencies pertaining to youth or cognitive-emotional development, Venables and Thompson were unable to resist their evil motivations. Would we really want to conclude that, therefore, their motivations are not, in fact, evil? This would, again, be a deeply implausible conclusion. The conclusion rests on conflating the moral evaluation of a subject with the evaluation of his motivations. Rescinding from moral evaluation of individuals is one thing; rescinding from moral evaluation of their motivations is quite another. However, if ought implies can, the former rescindment entails the latter. If "ought implies can" is true, then a person's lack of control over a motivation entails that this motivation makes no normative claim on that person: it would not be the sort of thing he *should* resist because it is not the sort of thing he *can* resist. Therefore, the motivation could not be neither moral nor immoral, because moral motivations are precisely ones that one should endorse and immoral ones are precisely ones that one should resist. If they were not responsible, Venables's and Thompson's motivations would not be morally bad. To avoid this wildly implausible conclusion, we need to reject the idea that "ought" implies "can." The status of a motivation as moral (or immoral) cannot depend on a person having control over that motivation.

Moral Subjects and Moral Agents

If something falls within moral space at all, it is generally thought to belong to one or both of two categories: it can be a moral *patient* or a moral *agent* or both, where:

1. X is a moral *patient* if and only if X is a legitimate object of moral concern.
2. X is a moral *agent* if and only if X is morally responsible for, and so can be morally evaluated for, its motives and actions.

In effect, I have argued for the existence of a third category—an additional, overlooked, region of moral space, occupied by what I have called (2011, 2012, 2017) the moral *subject*:

3. X is a moral *subject* if and only if X is, at least sometimes, motivated to act by moral considerations.

The reason this region of moral space has been overlooked is because of the idea that ought implies can, and the resulting idea that a motivation's qualifying as moral requires that the person whose motivation it is has control over it. Given these assumptions, the category of the moral subject would collapse into that of the moral agent. However, once we reject them, the two categories are clearly distinct. This, in my view, is as it should be. On one hand, there is the question of the moral status of a motivation—whether it is good or bad, right or wrong. On the other, is a quite distinct question—of whether, and the extent to which, a person has control over, is responsible for, or can be praised or blamed for her motivations. We cannot continue to conflate these kinds of questions. Questions of the moral status of a motivation are one thing, and questions of control, responsibility, and praise or blame are quite another.

I doubt any animals are moral agents. And, indeed, I'm not sure humans are. But I think at least some animals are moral subjects.

Notes

1. Conversations with Brad Cokelet helped me clarify this point.
2. Or, following Monso (2017), if you are reluctant to attribute mindreading capacities to the monkey, we can make the same point via mere behavior-reading capacities. The monkey is distressed *that* the other monkey is exhibiting distress behavior.

Bibliography

Aristotle (1999) *Nicomachean Ethics* (T. Irwin, Trans.), Indianapolis, IN: Hackett.
Bekoff, M., and Pierce, J. (2009) *Wild Justice: The Moral Lives of Animals*, Chicago, IL: University of Chicago Press.
Darwin, C. (1871) *The Descent of Man*, London: John Murray.
de Waal, F. (2006) *Primates and Philosophers: How Morality Evolved*, Cambridge, MA: Harvard University Press.
Korsgaard, C. (2004) "Fellow creatures: Kantian ethics and our duties to animals," in G. Peterson (ed.) *Tanner Lectures on Human Values*, Salt Lake City, UT: University of Utah Press.
Monso, S. (2017) "Morality without mindreading," *Mind and Language* 32: 338–357.
Rowlands, M. (2011) "Animals that act for moral reasons," in T. Beauchamp and R. G. Frey (eds.) *Oxford Handbook of Animal Ethics* (pp. 519–546), New York, NY: Oxford University Press.
Rowlands, M. (2012) *Can Animals Be Moral?* New York, NY: Oxford University Press.
Rowlands, M. (2017) "Moral subjects," in K. Andrews and J. Beck (eds.) *The Routledge Handbook of the Philosophy of Animals Minds* (pp. 469–474), New York: Routledge.
Singer, P. (2011) *Practical Ethics*, Cambridge: Cambridge University Press.
Streiffer, R. (2016) "*Can Animals Be Moral?* by Mark Rowlands," *Mind* 125(498): 619–623.
Wechkin, S., Masserman, J., and Terris, W. (1964) "Shock to a conspecific as an aversive stimulus," *Psychonomic Science* 1: 17–18.

7

QUANTIFYING ANIMAL WELL-BEING AND OVERCOMING THE CHALLENGE OF INTERSPECIES COMPARISONS

Mark Budolfson and Dean Spears

Animals,[1] like humans, experience different levels of well-being depending on decisions made by others. As a result, the well-being of animals must be included in any full accounting of the well-being consequences of decisions. However, this is almost never done in large-scale policy analyses and investment analyses, even though it is common to quantify the consequences for human welfare in these decision analyses. This is partly due to prejudice, but increasingly also because we do not currently have good methods for quantifying animal well-being consequences and putting them on the same scale as quantified human well-being consequences. We might call this 'the problem of interspecies comparisons.' This important barrier to including animal well-being in decision-making is the result of an insufficiently developed theory and practice of animal well-being and its relation to human well-being.

This handbook chapter explains the problem of interspecies comparisons, explains recent research that develops methods to overcome this problem, and includes animal welfare in rigorous policy and investment analysis (e.g., in analyses of optimal public policies, analyses of optimal philanthropic investment, and so on). The development of these methods is important: incorporating animal welfare in decision analyses would have an important impact on estimates of what prosocial investments of time and money should be made by individuals,[2] businesses,[3] and charities (including for purposes of 'effective altruism'),[4] and similarly for estimates of optimal public policies for correcting market failures (where the full cost of goods is not reflected in their market price),[5] for sustainable intensification of agriculture that aims to take animal welfare into account (producing more food while reducing the overall impacts of agriculture),[6] for climate change policy (how quickly we should be reducing greenhouse gas emissions),[7] and for wilderness protection policy and other challenges related to natural resource management.[8] In all these cases, if the well-being of animals is taken more fully into account, then decisions by individuals and governments will become better, on utilitarian grounds, and more compassionate toward the plight of animals.

Anthropocentrism and Current Economic Analysis: Animal Welfare as Valuable Only Insofar as It Is Valued by Humans

Anthropocentrism is the view that what makes outcomes better or worse is ultimately entirely a matter of their consequences for humans.[9] On this view, animal welfare is not ultimately valuable per se; instead, animal welfare is only valuable insofar as it is ultimately valued by humans. From this perspective, which is dominant in economic policy analysis, the important problem related to animal well-being is that the marketplace and existing methods of policymaking tend to ignore much of the instrumental value of animal welfare to humans, and thus do not take animal welfare into proper account even from an anthropocentric perspective.

To see how there is a tendency to ignore even the instrumental value of animals, consider an analogy to majestic stands of old-growth forests: until recent decades, the fate of old-growth forests in many nations was entirely determined by a marketplace that did not take into account human society's preference for preserving the most majestic of old-growth forests because that preference was not reflected in market prices for timber. As a consequence, there was a 'market failure' in which the outcome determined by the marketplace was inferior[10] for humans by their own lights than the outcome that would have obtained if human society's preferences for preservation had been incorporated into the price of the most majestic of old-growth forests. This is a classic example of market failure deriving from the existence of 'negative externalities' (costs to society that are not internalized in market prices), which shows how externalities can cause free market transactions to lead to suboptimal outcomes for human society by its own lights. For the purposes of this chapter, the key thing to note is that, in these cases, an outcome is suboptimal from an anthropocentric perspective because market prices do not always reflect everything that is of value to humans. Market failure is a very real and widespread phenomenon, and does not essentially depend on the question of how to value nature, as market failure can arise anytime parties external to a transaction bear costs not internalized by the price mechanism or the parties to the transaction, even when no aspect of nature is in play. The current point is that sometimes externalities do indeed exist for elements of nature, including animal welfare. That is, there are cases where animal welfare is valued by humans in a way that is not reflected in unregulated market prices. In those cases, market failure threatens, and outcomes may be worse for humans by their own lights than other feasible outcomes.[11]

With that background in hand, we now note that there is empirical evidence that the current marketplace and public policies do not adequately reflect the value that humans assign to animal welfare; thus, something is going wrong even by the lights of an anthropocentric view that maintains that animal welfare is valuable only insofar as it is valuable to humans.[12] The relevant empirical evidence here is analogous to the old-growth-forests example: economic analysis indicates that sufficiently many humans are willing to pay to improve animal welfare above current levels in a way that implies that the outcome could be made better by humans' own lights by properly taking that willingness to pay into account. As an analogy, imagine a situation where a single unimportant factory is billowing noise and filth into the air of our community, and many of us are willing to pay to prevent it from doing so. In such a case, the amount we are willing to pay could be more than enough to make the factory owners happy to reduce emissions dramatically if our payments are transferred to them, and we would be better off if we did so because we'd prefer that outcome to continuing to suffer the pollution. So there is an opportunity to make some humans better off without making anyone worse off—clearly a better outcome by the lights of anthropocentrism. The empirical argument of many economists who study these issues is that the same is true regarding animal welfare: we can make all humans better off and none worse off by improving animal welfare in specific ways.[13] A complementary argument for the same conclusion is that some targeted animal welfare improvements would more than pay for themselves by reducing the expected harm to human health from diseases, antimicrobial resistance, and the like, where these harms to human health are not reflected in the market prices of animal products; thus, policies that included targeted animal welfare improvements could yield benefits for everyone in expectation.[14]

The methods of economic analysis that underlie these conclusions are methods of estimating two different categories of anthropocentric value, namely, the (anthropocentric) *use value* of animals (human willingness to pay to use animals) and their *nonuse value*. Use value includes willingness to pay for direct use of animals as well as indirect use, including *ecosystem services* such as the value of pollinators in human agriculture, the value of aquatic mollusks in cleaning water for human use, the value of wildlife to human recreation, and so on. Nonuse value includes the willingness to pay merely for an outcome that includes the existence of animals without their use by humans (*existence value*), as well as the *option value* of keeping animals around for potential future human use.

Substantive methods are needed to estimate these anthropocentric values of animals in many cases, as their values are often not readily reflected in market prices,[15] especially when they have the properties of public rather than private goods. This is almost always the situation in connection with nonuse value, which is why the nonuse value of animals is generally ignored in the marketplace and in policy analysis. At the same time, there are widely known methods for estimating nonuse value, namely, contingent valuation studies and revealed preference methods. *Contingent valuation studies* are generally surveys that elicit self-reported willingness to pay to avoid or bring about particular outcomes. Based on respondents' answers, willingness to pay for nonuse value is estimated and can then be incorporated in decision analyses. However, there are many objections to this method, perhaps the most important of which is the worry that it leads to biased and inflated estimates of willingness to pay, because people do not have to back their answers with real monetary investments.[16] There is some evidence that this leads to a large upward bias in estimates of willingness to pay in comparison to the actual choices that people make when real money is on the line.

This points toward the main alternative method, estimates based on *revealed preferences* (i.e., the valuation implicit in actual choices), often based on the valuation implicit in existing regulation, such as applications of the Endangered Species Act. Using the values that are implicit in existing regulation mitigates a worry that would arise if an analysis were to use the values implicit in individual expenditures instead: namely, a societal collective action problem could explain low expenditure by individuals. This is because the nature of a problem might be such that it would not be mitigated by individual expenditures; it would only be mitigated by widespread collective investment. In such circumstances, the absence of collective investment would make individual investment irrational even if individuals preferred a substantial level of investment by all.[17]

These economic methods allow estimation of different species' value to humans, and are the most widely used methods for doing so. However, most philosophers and many economists would insist that even if perfectly developed, these methods can never be adequate for valuing animal welfare, since by their nature they are incapable of assigning any fundamental value to the well-being of animals. Because it is clear that animals experience well-being that humans are often unwilling to pay to protect, this is unacceptable from the point of view that well-being is the fundamental value that economic analysis aims to promote. This is the point of departure for the next section, which introduces the challenges of both estimating animal well-being and making comparisons between animals and humans.

Animal Welfare as Fundamentally Valuable, and the Problem of Interspecies Comparisons

Many researchers from across disciplines and many citizens believe that an important problem with even the best anthropocentric methodology is that the valuation of animals and other aspects of nature within such a methodology is always merely valuation in terms of what only *humans* value.[18] In other words, valuation is always in terms of the ultimate value of outcomes to humans only, and thus assigns no fundamental value to the well-being of animals. For example, on even the best anthropocentric approach, the deaths of billions of birds due to climate change would have disvalue only insofar as the deaths of those birds have disvalue to humans. But many would object that this way of valuing animal lives is fundamentally incorrect because it ignores the value of the birds' own well-being irrespective of its contribution to human well-being. (Similarly, anthropocentrism assigns no fundamental value to the health of ecosystems as holistic entities, which is a separate criticism that we set aside here.[19])

So, according to critics of anthropocentrism, what is needed is the inclusion of the fundamental value of animals' own well-being, even if it is not valued by humans. In what follows, we side with the critics of anthropocentrism on this issue. We agree that since animals experience well-being and humans often do not value that well-being, any coherent welfarist approach must acknowledge that there is an important question of how to incorporate this ignored fundamental value of animal welfare into decision analysis. Although we frame the discussion in terms of welfarist consequentialism for ease of exposition, the structure of the approach that follows is also compatible with many deontological approaches.[20]

Although most philosophers and an increasing number of practitioners agree that anthropocentrism should be rejected, they also tend to agree that there have not been good methods for quantifying animal well-being consequences and putting them on the same scale as quantified human well-being consequences in a decision analysis. This is 'the problem of interspecies comparisons.'

Recent work by Kevin Wong (Wong 2016) has clarified the most difficult problem that needs to be solved in connection with interspecies comparisons. As Wong notes, the key problem is how to estimate the *well-being capacity* (well-being potential) of members of a nonhuman species relative to the well-being capacity of humans. If we knew how to make those interspecies comparisons of well-being capacity, then we could integrate animal welfare consequences into existing methods of decision analysis. This integration would be made possible by deriving empirically based estimates of animal welfare consequences on the same scale as human consequences that typically underpin welfarist decision-making analyses.

For example, suppose an additional degree of climate change will cause us to lose one million life-years of a particular species of bird, and we want to value this on the same scale as the losses to humans from an additional degree of warming that are (let's assume) already modeled and valued based on an assumption about the value of one average human life-year. Wong's point is that if we had a good estimate of the well-being capacity of that species of bird relative to a human, we could then multiply that estimate by the purely empirical impact estimate of one million life-years lost. This would yield an estimate of the amount of well-being lost by that bird species on the same value scale as the existing estimate of human well-being loss, assuming that one degree of additional climate change does not change the quality of life of those birds. And if one additional degree of warming does diminish the quality of life of the remaining birds of that species, we can simply multiply the number of remaining bird species life-years by a further *quality-of-life adjustment* term that is itself an empirical impact estimate from zoological experts and the like. (We can also use such a term to take into account any antecedent diminishment in the well-being experienced by all of the birds including those that would die before the warming.)

This line of thought leads to the following equation for estimating on a single scale the average well-being experienced by a member of a species s (which we symbolize as $\overline{u}_s$) as a function of the average well-being capacity per unit of time of members of s relative to humans $\left(\overline{\pi}_s\right)$, multiplied by the average duration of a life of a member of s $\left(\overline{\delta}_s\right)$, multiplied by a quality of life adjustment term that estimates how well members of *s* are typically flourishing relative to their species capacity (i.e., a quality of life adjustment term) $\left(\overline{f}_s\right)$[21]:

1. $\overline{u}_s = \overline{\pi}_s * \overline{\delta}_s * \overline{f}_s$

Wong's contribution is to highlight the term $\overline{\pi}_s$ as the key unknown term (as the other terms $\overline{\delta}_s$ and $\overline{f}$ are susceptible to existing empirical methods),[22] where the unsolved problem of how to estimate $\overline{\pi}_s$ is the essence of the challenge of interspecies comparisons.

Wong uses formalisms such as Equation 1 to estimate the valuation of animal well-being that is implicit in some decisions within effective altruism that allocate scarce resources between opportunities to improve animal versus human welfare. However, Wong notes that these implicit valuations are not themselves normatively plausible as the correct way to make trade-offs between humans and animals, and provide no guide to the correct trade-off between animal and human well-being from a non-anthropocentric point of view that sees animal well-being as fundamentally valuable in its own right.

So, the challenging question remains—how can we estimate the well-being capacity of species relative to humans? This question is not about the formal structure of the correct analysis, but is about its content; for example, how do we estimate the well-being level of an average bird versus the well-being level of an average human? This is the difficult problem that needs to be solved, and the theoretical formalism discussed earlier does nothing to help us solve it, although it does help in focusing our attention on the substantive question that needs to be answered.

Overcoming the Problem of Interspecies Comparisons

It is important to see that the problem of interspecies comparisons is not identical to and does not reduce to the fully general and familiar question of how to make interpersonal comparisons between human individuals, as we already have well-accepted methods for making interpersonal comparisons, for example, based on proxies for human well-being such as consumption,[23] human development indicators, and the like.[24] These empirical proxies for human well-being are generally assumed to ground good-enough estimates of individual human well-being levels for use in decision analyses via an assumed-to-be uniform relationship across individuals from the proxy for well-being (e.g., consumption) to the estimated level of well-being for each individual.

In other words, in economics, the challenge of making interpersonal comparisons is familiar. But it is also familiar how this problem is solved in practice, namely, by simply making comparisons based on a method that is believed to involve a good approximation: for example, by estimating a uniform concave function mapping consumption c to utility/well-being, such as

$$2. \quad W^{TU} \approx \sum_{i \in humans} \frac{(c_i)^{1-\theta}}{1-\theta},$$

where θ parameterizes the diminishing marginal utility of consumption.

Typically economic practice is more crude and approximate than this, because economists often use population-level average consumption $\bar{c}$ as the proxy for the consumption of every individual, despite known inequality in individual consumption. For instance, the following equation simply multiplies the utility of per capita average consumption by the size of the human population P_h:

$$3. \quad W^{TU} \approx P_h \frac{(\bar{c})^{1-\theta}}{1-\theta}$$

Mathematical formalism aside, the important point is merely that these formulas take consumption to be a proxy for well-being, and involve a concave transformation of consumption into well-being, which yields diminishing marginal utility of consumption where the rate of diminishment is parameterized by the θ term—which means that different views about the relationship between wealth and well-being can be investigated by varying the θ term. For example, if $\theta > 2$, then additional wealth above some sufficiency threshold generates little increase in well-being.

Our proposal for solving the problem of interspecies comparisons is analogous to the method used in Equations 2 and 3 of taking consumption as a proxy for human well-being: we propose to make interspecies comparisons based on a proxy that is imperfect but yet is as good as is possible in practice. To do this we first need a proxy, call it n, to use as the basis for estimating well-being potentials across species, analogous to the earlier use of consumption (c) as the basis for estimating well-being across humans. As an overly simplistic illustration of this idea (that might nevertheless be useful in practice in some contexts), n might be the number of neurons in the brain of members of a species. Data on number of neurons are readily available and may be a good proxy for well-being potential in some select contexts, such as an enormous global analysis involving billions of individuals where different species are crudely lumped together in small number of bins such as 'mammals' and 'insects;' this is true even if neurons are not the best proxy when fine-grained accuracy between individuals is more important.[25] For example, when greater accuracy is required for specific species or individuals, researchers can set n equal to a more complex metric based on expert analysis of empirical properties that are best correlated with different levels of cognitive capacity and hedonic enjoyment—for example, the number of neocortex-like neurons, cortisol levels, sociality, or other leading factors identified by the scientific community and philosophers as most closely correlated with the capacity to have complex thoughts and feelings, and whatever other empirical properties are found to be necessary for experiencing well-being.[26] At the limit, this sort of metric can reflect the true relationship between empirically measurable facts about individuals and their well-being capacity.

Abstracting for now from those details that are not essential to the core challenge of how to make interspecies comparisons, the first step of the proposal here is to parameterize an empirical proxy n with an exponential weight ψ into comparative well-being capacity for different species. The second step is to multiply this estimate of well-being capacity by a descriptive measure of the degree to which this potential is actually realized, f (i.e., the quality-of-life-adjustment term from Equation 1), to yield the desired well-being estimates:

4. $$W^{TU} \approx \sum_{is} n_{is}^{\psi} f_{is} \delta_{is}$$

(In ordinary language, the total amount of well-being is approximately equal to the sum over all individuals across species of that individual's empirical basis for well-being capacity, raised to the normative exponent [which determines the relationship between the empirical proxy and well-being capacity], multiplied by the flourishing level of that individual relative to its species capacity, multiplied by the duration of that individual's life.)

In practice, it is often more convenient to use species-level averages (where averages are denoted by a bar over the letter) as the proxy for well-being potential, which can then be multiplied by the species population P_s:

5. $$W^{TU} \approx \sum_{s} P_s \bar{n}_s^{\psi} \bar{f}_s \bar{\delta}_s$$

Equations 4 and 5 summarize the basis for our proposed method for making interspecies comparisons. They require an empirical proxy for n (e.g., number of neurons or a more complex empirically based metric), values for ψ grounded in normative and empirical considerations (on analogy with how values for θ in Equations 2 and 3 are grounded in normative and empirical considerations), and empirically determined values for f (based on empirical facts about how well members of the species are actually doing relative to their species potential).

With this method in hand, a decision analysis (e.g., between competing investment possibilities or between alternative public policies) can make use of a sensitivity test that investigates how optimal policy is sensitive to the normative parameter ψ that, in our earlier proposal, can be used to generate different estimates of the comparative well-being capacity of species. In other words, such a sensitivity test can use the preceding equations to capture the range of empirically grounded and principled estimates that represent normative uncertainty over how to estimate the well-being of animals of different species.

Table 7.1 summarizes a standard sensitivity test of this type that illustrates how a sensitivity test that could be incorporated into global policy analyses (which usually only represent animals crudely based on a very small number of classifications). It relies on different principled ways of using the parameter ψ to estimate the potential well-being of a species as a function of the average number of neurons n in a member of that species.

Each estimate is expressed in terms of the well-being capacity of one human life-year ($\overline{n}_h^{\psi}$, for short), and thus, each estimate divides by the estimated well-being capacity of one human life. Estimate $1 = \frac{\overline{n}_s^{\psi}}{\overline{n}_h^{\psi}}$, with ψ set equal to 1 (a higher estimate of the capacity of animals), whereas estimate $2 = \frac{\overline{n}_s^{\psi}}{\overline{n}_h^{\psi}}$, with ψ set equal to 2 (a lower estimate of the capacity of animals). Other estimates are possible, for example, based on different empirical proxies n and other factors, such as different normative parameters and different views about a possible threshold cognitive capacity that arguably is necessary for well-being; see Budolfson and Spears (2019b) for more examples and discussion.

Each estimate can be used to put human life-years (which can be estimated via familiar proxies such as Equations 2 or 3) on the same scale as the life-years of animals of different species, and each estimate does so in a principled way that is empirically grounded. For example, assuming the number of neurons as a basis for well-being estimates, if ψ is set equal to 2 (a principled lower value for animals), then a human life-year is worth almost 120,000 mammal life-years and almost 120,000,000 fish life-years. If instead ψ is set equal to 1 (a principled higher value for animals), then a human life-year is worth about 344 mammal life-years and about 10,700 fish life-years. These alternative estimates appear to represent much of the range of empirically grounded and principled views over the well-being of animals of different species,[27] and can avoid unintuitive implications.[28] It may not even be desirable to attempt to choose between these estimates in policy analysis, if the goal is to take normative uncertainty into account and test the sensitivity of optimal decisions to this range of different reasonable (and empirically and theoretically principled) estimates.

In sum, the method developed in this section allows interspecies comparisons, via Equations 4 and 5, based on both empirically available estimates of species population dynamics and levels of flourishing for members of species[29] and also empirical proxies for well-being capacity n that can be calibrated with the ψ parameter to reflect normative uncertainty about the connection between those empirical proxies and well-being capacity. The key advance is that the term $\overline{n}_s^{\psi}$ provides a tool for making principled and empirically based estimates of the well-being capacity of different species of animals (i.e., $\overline{n}_s^{\psi}$ provides an estimate of the key uncertain term $\overline{\pi}_s$ in Equation 1). Thus, $\overline{n}_s^{\psi}$ provides a framework for articulating principled substantive answers to the question of how to make interspecies comparisons, and allows us to parameterize these comparisons to the range of normative uncertainty about their true value.[30]

Table 7.1 Two alternative estimates of the well-being potential of animal life-years of different species based on the number of neurons in an average member of the species, for illustrative purposes

Wildlife	*n*	*Utility Potential Estimate*	
	Number of neurons (n)	*(ψ = 1)*	*(ψ = 2)*
Mammals	250	0.002907	0.000008450514
Birds	150	0.001744	0.000003042185
Amphibians etc.	15	0.000174	0.000000030422
Fish etc.	8	0.000093	0.000000008653
Insects etc.	0.1	0.000001	0.000000000001
Humans	86,000	1	1

Conclusion

This handbook chapter has explained the problem of interspecies comparisons. It has also explained recent research on developing methods to overcoming this problem, making it possible to include animal welfare in rigorous policy and investment analysis (e.g., in analyses of optimal public policies, analyses of optimal philanthropic investment, and so on). The development of these methods is important: methods of incorporating animal well-being will have an important impact on estimates of optimal prosocial investments of time and money by charities, businesses, or individuals, and similarly for estimates of optimal public policies for correcting market failures that ignore the costs of goods not reflected in their market price, for sustainable intensification of agriculture that aims to take animal welfare into account, for climate change, and for wilderness protection and other challenges related to natural resource usage.

Notes

1. Here and in what follows we use *animals* as shorthand for 'nonhuman animals.'
2. Budolfson 2015.
3. See Berkey forthcoming; Thomas forthcoming; Pacelle 2017.
4. See Wong 2016, GiveWell, Open Philanthropy Project, Animal Charity Evaluators.
5. Cowen 2006; Norwood and Lusk 2011 chapter 10; Jarvis and Donoso 2018.
6. Garnett et al. 2013; Norwood and Lusk 2011.
7. Hsiung and Sunstein 2007; Budolfson and Spears 2019a.
8. Hsiung and Sunstein 2007; Sunstein and Nussbaum 2004; Sunstein 2018 chapter 6.
9. Merely for ease of discussion, we frame all the discussion in this chapter in terms of consequentialism, noting here that deontological views can be represented as consequentialist views in the sort of analyses that are the focus of our discussion here (see Dreier 2011). In saying this, we do not take a stand on whether deontological views are adequately represented at the most fundamental level by 'consequentializing' them, we merely note that consequentialized versions are extensionally equivalent to the fundamental deontological views, and so deontological views can be adequately represented extensionally in the sort of analyses we are interested in here.
10. Technically, the interesting form of suboptimality here is Pareto-inferiority; namely, there is a way of making the outcome better for some that makes the outcome worse for no one.
11. Before concluding from this that therefore government regulations should be enacted to make the outcome better, it must be taken into account that sometimes new regulations would themselves be predictably inefficient, and as a result, the actual consequence of new regulations might in some cases predictably make things worse. See Budolfson 2017 and the references therein for discussion of this important complication for what normative public policy conclusions actually follow or don't follow from the widespread existence of market failure.
12. See Cowen 2006; Norwood and Lusk 2011 chapters 9 and 10 for extended argument for this.
13. See, for example, Norwood and Lusk 2011 chapters 9 and 10.
14. Jarvis and Donoso 2018; Otte and Chilonda 2000.
15. In contrast, when the anthropocentric value of animals is well reflected in market prices—such as, for example, the price of pollination services—market prices are the preferred method of valuation, at least to the extent that the good is a private good traded in a well-functioning marketplace.
16. See Hsiung and Sunstein 2007 and the references therein.
17. Ibid.
18. Schmidtz and Shahar 2018; Sandler 2018; McShane 2018; Palmer et al. 2014; Sarkar 2012; Jamieson 2008; O'Neill et al. 2008; Ng 1995; Singer 1975.
19. See references in previous footnote, and in addition Chan et al. 2016; Frank and Schlenker 2016; Dasgupta 2014; Alcamo 2003; Costanza et al. 1997; Kagan 2019.
20. See the relevant footnote earlier in this chapter on consequentializing moral theories. As an illustration, one can imagine a policy analysis that assumes deontological side constraints and then maximizes welfare subject to those constraints—the methods that follow are equally essential to such calculations as to unconstrained welfare maximization.
21. Compare the term $\bar{f}_s$ to McMahan 2001's concept of *fortune*, a connection Wong 2016 notes.
22. Estimating the flourishing term f can be seen as the focus of existing animal welfare science—see, for example, Fraser 2008; Appleby et al. 2011.

23. Consumption is often understood as income minus savings.
24. For an overview, see Adler and Fleurbaey 2016.
25. See Budolfson and Spears 2019a; Herculano-Houzel 2017; Olkowicz et al. 2016.
26. Herculano-Houzel 2017; Olkowicz et al. 2016; Barron and Klein 2016; Shriver 2014; Dawkins 2012; Appleby et al. 2011; Fraser 2008; see also Tye 2017; Persad 2020; Sebo 2020; Browning 2019; Fischer 2016.
27. Compare Alexander 2019.
28. See Budolfson and Spears 2019b.
29. See for example Fraser 2008; Appleby et al. 2011.
30. See Budolfson and Spears 2019b.

Bibliography

Adler, M., and Fleurbaey, M. (2016) *Oxford Handbook of Well-Being and Public Policy*, New York, NY: Oxford University Press.

Alcamo, J. (2003) *Ecosystems and Human Well-Being: A Framework for Assessment*, Washington, DC: Island Press.

Alexander, S. (2019) *Update to Partial Retraction of Animal Value and Neuron Number*, https://slatestarcodex.com/2019/05/01/update-to-partial-retraction-of-animal-value-and-neuron-number/.

Animal Charity Evaluators, www.animalcharityevaluators.org/.

Appleby, M., Joy Mench, I., Olsson, A., and Galindo, F. (eds.) (2011) *Animal Welfare*, 2nd ed., Cambridge, UK: CABI.

Barnhill, A., Budolfson, M., and Doggett, T. (eds.) (2016) *Food, Ethics, and Society*, Oxford: Oxford University Press.

Barron, A., and Klein, C. (2016) "What insects can tell us about the origins of consciousness," *Proceedings of the National Academy of Sciences* 113(18): 4900–4908.

Berkey, B. (forthcoming) "Prospects for an Animal-Friendly Business Ethics," in N. Thomas (ed.) *Animals and Business Ethics*, Springer.

Browning, H. (2019). *If I Could Talk to the Animals: Measuring Subjective Animal Welfare*, PhD Dissertation, Australian National University.

Budolfson, M. (2015) "Consumer ethics, harm footprints, and the empirical dimension of food choices," in A. Chignell, T. Cuneo, and M. Halteman (eds.) *Philosophy Comes to Dinner*, New York, NY: Routledge.

Budolfson, M. (2017) "Market failure, the tragedy of the commons, and default libertarianism in contemporary economics and policy," in D. Schmidtz and C. Pavel (eds.) *Oxford Handbook of Freedom*, New York, NY: Oxford University Press.

Budolfson, M., and Spears, D. (2018) *Why the Repugnant Conclusion is Inescapable*, Princeton CFI Working Paper.

Budolfson, M., and Spears, D. (2019a) *Optimal Climate Policy Including Animal Welfare*, Princeton CFI Working Paper.

Budolfson, M., and Spears, D. (2019b). *Public policy, consequentialism, the environment, and non-human animals*, in Portmore (ed.) Oxford *Handbook of Consequentialism*, New York, NY: Oxford University Press.

Chan, K. et al. (2016) "Why protect nature? Rethinking values and the environment," *Proceedings of the National Academy of Sciences* 113(6): 1462–1465.

Costanza, R. et al. (1997) "The value of the world's ecosystem services and natural capital," *Nature* 387: 253–260.

Cowen, T. (2006) "Market failure for the treatment of animals," *Society* 43(2): 39–44.

Dasgupta, P. (2014) "Measuring the wealth of nations," *Annual Review of Resource Economics* 6: 17–31.

Dawkins, M. (1980) *Animal Suffering*, New York, NY: Springer.

Dawkins, M. (2012) *Why Animals Matter: Animal Consciousness, Animal Welfare, and Human Well-being*, Oxford: Oxford University Press.

Dreier, J. (2011) "In defense of consequentializing," *Oxford Studies in Normative Ethics*, Vol. 1. New York, NY: Oxford University Press.

Fischer, B. (2016). "What if Klein & Barron are right about insect sentience?", *Animal Sentience* 9(8): 1–6.

Frank, E., and Schlenker, W. (2016) "Balancing economic and ecological goals," *Science* 353: 651–652.

Fraser, D. (2008) *Understanding Animal Welfare: The Science in its Cultural Context*, West Sussex, UK: Wiley.

Garnett, T. et al. (2013) "Sustainable intensification in agriculture: Premises and policies," *Science* 341(6141): 33–34.

GiveWell, www.givewell.org/.

Herculano-Houzel, S. (2017) "Numbers of neurons as biological correlates of cognitive capability," *Current Opinion in Behavioral Sciences* 16: 1–7.

Hsiung, W., and Sunstein, C. (2007) "Climate change and animals," *University of Pennsylvania Law Review* 155: 1695.

Jamieson, D. (2008) *Ethics and the Environment*, Cambridge: Cambridge University Press.

Jarvis, L., and Donoso, P. (2018) "A selective review of the economic analysis of animal health management," *Journal of Agricultural Economics* 69(1): 201–225.

Kagan, S. (2019). *How to Count Animals, More or Less*. New York, NY: Oxford University Press.
Klein, C., and Barron, A. (2016a) "Insect consciousness: Commitments, conflicts and consequences," *Animal Sentience* 9(21): 1–52.
Klein, C., and Barron, A. (2016b) "Reply to Adamo, Key et al., and schilling and cruse: Crawling around the hard problem of consciousness," *Proceedings of the National Academy of Sciences* 113(27): E3814–E3815.
McMahan, J. (2001) *The Ethics of Killing*, New York, NY: Oxford University Press.
McShane, K. (2018) "Why animal welfare is not biodiversity, ecosystem services, or human welfare: Toward a more complete assessment of climate impacts," *Les Ateliers de l'Éthique/The Ethics Forum* 13(1): 43–64.
Mendl, M., and Paul, E. (2016) "Bee happy: Bumblebees show decision-making that reflects emotion-like states," *Science* 353: 1499–1500.
Muehlhauser, L. (2017) *2017 Report on Consciousness and Moral Patienthood*, Open Philanthropy Project.
Ng, Y-K. (1995) "Towards welfare biology: Evolutionary economics of animal consciousness and suffering," *Biology and Philosophy* 10: 255–285.
Ng, Y-K. (2016) "How welfare biology and commonsense may help to reduce animal suffering," *Animal Sentience* 7(1): 1–10. http://animalstudiesrepository.org/cgi/viewcontent.cgi?article=1012&context=animsent
Norwood, F. B., and Lusk, J. L. (2011) *Compassion by the Pound: The Economics of Farm Animal Welfare*, Oxford: Oxford University Press.
Olkowicz, S. et al. (2016) "Birds have primate-like numbers of neurons in the forebrain," *Proceedings of the National Academy of Sciences* 113(26): 7255–7260.
O'Neill, J., Holland, A., and Light, A. (2008) *Environmental Values*, New York: Routledge.
Open Philanthropy Project, Farm Animal Focus, www.openphilanthropy.org/focus/us-policy/farm-animal-welfare.
Otte, M., and Chilonda, P. (2000) *Animal Health Economics: An Introduction*, Livestock Information, Rome, Italy: Sector Analysis and Policy Branch, Animal Production and Health Division (AGA), FAO.
Pacelle, W. (2017) *The Humane Economy: How Innovators and Enlightened Consumers Are Transforming the Lives of Animals*, New York, NY: Harper Collins.
Palmer, C., McShane, K., and Sandler, R. (2014) "Environmental ethics," *Annual Review of Environment and Resources* 39: 419–442.
Persad, G. (2020) Cost-Effectiveness in Animal Health: An Ethical Analysis, in Bob Fischer (ed.) *Routledge Handbook of Animal Ethics*, New York, NY: Routledge.
Rushton, J. et al. (2018) "Initiation of global burden of animal diseases programme," *Lancet* 392(10147): 538–540.
Sandler, R. (2018) *Environmental Ethics: Theory in Practice*, Oxford: Oxford University Press.
Sarkar, S. (2012) *Environmental Philosophy: From Theory to Practice*, West Sussex, UK: Wiley-Blackwell.
Schmidtz, D., and Shahar, D. (eds.) (2018) *Environmental Ethics*, 3rd ed., Oxford: Oxford University Press.
Sebo, J. (forthcoming) "Animals and climate change," in M. Budolfson, T. McPherson, and D. Plunkett (eds.) *Philosophy and Climate Change*, New York, NY: Oxford University Press.
Sebo, J. (2018) "The moral problem of other minds," *The Harvard Review of Philosophy* 25(2018): 51–70.
Sebo, J. (2020) "Effective Animal Advocacy," in Bob Fischer (ed.) *Routledge Handbook of Animal Ethics* (pp. 530–542), New York, NY: Routledge.
Singer, P. (1975) *Animal Liberation*, New York, NY: HarperCollins.
Singer, P. (2016) *Are insects conscious? Project Syndicate*, www.project-syndicate.org/commentary/are-insects-conscious-by-peter-singer-2016-05.
Shriver, A. (2014) "The asymmetrical contributions of pleasure and pain to subjective well-being," *Review of Philosophy and Psychology* 5: 135–153.
Sunstein, C., and Nussbaum, M. (eds.) (2004) *Animal Rights: Current Debates and New Directions*, Oxford: Oxford University Press.
Sunstein, C. (2018) *The Cost-Benefit Revolution*, Cambridge, MA: MIT Press.
Tannsjo, T. (2016) "It's getting better all the time," in A. Barnhill, M. Budolfson, and T. Doggett (eds.) *Food, Ethics, and Society*, Oxford: Oxford University Press.
Thomas N. (ed.) (forthcoming) *Animals and Business Ethics*, Springer.
Tooley, M. (1972) "Abortion and infanticide," *Philosophy & Public Affairs* 2: 37–65.
Tye, M. (2017) *Tense Bees and Shell-Shocked Crabs: Are Animals Conscious?* Oxford: Oxford University Press.
Wong, K. (2016) *Counting Animals: On Effective Altruism and the Prospect of Interspecies Commensurability*, B.A. thesis, Princeton University, Princeton, NJ.

8

COST-EFFECTIVENESS IN ANIMAL HEALTH

An Ethical Analysis

Govind Persad

Introduction

Many national health systems use cost-effectiveness information as part of determining which human health interventions to provide. The two most prominent cost-effectiveness metrics are quality-adjusted and disability-adjusted life-years (QALYs and DALYs, respectively), which typically are based on surveys of individuals' subjective experience of health setbacks. The ethicist Toby Ord has described consideration of cost-effectiveness as a "moral imperative," because replacing less cost-effective treatments with more cost-effective ones can vastly increase the benefits a health system produces (Ord 2013). But other ethicists have criticized the use of cost-effectiveness information in priority-setting for giving no weight to distributional issues or making it more difficult for elderly and disabled individuals to receive health care interventions (Brock 2003). Some have attempted to reconcile these perspectives (John et al. 2017).

There is no similarly systematic discussion of the ethics of using cost-effectiveness information to set animal health priorities.[1] This may once have reflected the difficulty of assessing animals' quality of life. But the increasing availability of quality-of-life information about animals indicates the value of parallel conceptual work on the ethics of using cost-effectiveness methods in animal health. One recent commentator observes that "[s]urprisingly, utility indices are not yet widely used in veterinary decision science or in veterinary practices," even though "quantification of disease burden could be very useful for decision making, priority setting and treatment comparisons in animal health care or simply to get an impression of the [quality of life] of the animal" (Pesie 2012).

This chapter evaluates the ethical issues that using cost-effectiveness considerations to set animal health priorities might present. Ultimately, its conclusions are cautiously optimistic. While using cost-effectiveness calculations in animal health is not without ethical pitfalls, these calculations offer a pathway toward more rigorous priority-setting efforts that allow money spent on animal well-being to do more good. Although assessing quality of life for animals may be more challenging than in humans, implementing prioritization based on cost-effectiveness is less ethically fraught.

In the first part, I review the current state of empirical measures of animal quality of life and consider the conceptual choices that underpin them, including philosophical disagreements about what quality of life amounts to and practical challenges in assessing the quality of life of creatures who cannot self-report their experiences using language. In the second part, I examine how ethical controversies presented by the use of cost-effectiveness information to set priorities for human health translate to the animal health context, such as conflicts between cost-effectiveness and individual rights. I also consider distinctive ethical challenges that the animal health context presents, in particular the problem of making cross-species comparisons of quality of life. In the third part, I consider different ways in which cost-effectiveness assessments could be used to set priorities and argue that doing so is both possible and desirable.

Measuring Cost-Effectiveness in Animal Health

Determining the cost-effectiveness of a health intervention involves assessing two different aspects of an intervention: (1) its cost and (2) its effects on length and quality of life. These two assessments are combined to produce a cost-effectiveness ratio. For instance, a course of cetuximab, a cancer treatment, costs US$80,000 and improves overall survival in humans by 1.2 months, producing a cost-effectiveness ratio of more than $800,000 per life-year (Fojo and Grady 2009).

Assessing the cost of animal health interventions is no more difficult than assessing the cost of similar interventions that improve human health. Assessments of the cost of veterinary medications and other animal health interventions are already frequently conducted as part of cost-effectiveness analyses of zoonotic diseases that affect human health (Schurer et al. 2015). Such assessments are also used in cost–benefit analyses that compare the cost of interventions that improve livestock health to the economic benefits ranchers derive from healthier livestock (Duarte et al. 2015). Likewise, the limitations that exist when assessing costs in human health—including the fact that sticker prices may not reflect costs from a societal perspective and that costs themselves are frequently reflective of societal determinations rather than being entirely independent of those determinations—apply to animal health (Persad 2016).

Cost-effectiveness analyses of zoonotic diseases, however, often entirely ignore how these diseases—and interventions to cure them—affect animals' quality of life. For example, echinococcosis, a zoonotic disease caused by a parasitic worm, can lead to serious illnesses not only in humans but also in host animals like sheep, dogs, and horses (Schurer et al. 2015). Animal health benefits from treating echinococcosis should also be part of the cost-effectiveness calculation. As Cass Sunstein has argued,

> [o]n any plausible view, harm to animals matters, at least to some degree. This judgment is firmly reflected in American law. At the national level, the [Endangered Species Act] is complemented by the Animal Welfare Act, which is designed to protect a wide range of animals against suffering and premature death. Every state attempts to accomplish the same goal through anticruelty laws.
>
> *(Hsiung and Sunstein 2006: 1695)*

It would be at least *prima facie* inconsistent with our willingness to spend money protecting the welfare of laboratory and farm animals to entirely ignore animal quality of life when considering how much to spend treating zoonotic diseases. More generally, if we are willing to give some weight to animal health, it makes sense to include animal health as an outcome in a cost-effectiveness analysis.

In order for cost-effectiveness analysis to incorporate both the costs and the health benefits of animal health interventions, we need an effective way of measuring how these interventions affect animal health. The most common metric for assessing the human health impact of a given intervention is the QALY. Determining QALY impact involves multiplying the number of years of life that the intervention provides by the quality of those years of life. In humans, comparative quality of life is assessed using "time trade-off" or "standard gamble" approaches, where people are asked to say what risk of death they would tolerate, or how much shorter a life they would accept, in order to avoid or eliminate a condition that impairs quality of life.

Because animals typically cannot respond to surveys, other strategies must be used to assess quality of life. A 2000 article in the *Journal of the American Veterinary Medical Association* summarizes research on quality-of-life measurement in animals. The article observes that quality of life in animals is typically assessed by having proxies, such as caregivers or medical professionals, complete surveys that elicit their beliefs about an animal's quality of life. The use of proxy surveys is not unique to animal health proxies are also used to assess quality of life for humans who are incapable of self-reporting their own quality of life, such as infants and those who are severely ill or disabled. Despite the limitation of proxy approaches, the article asserts that "[t]he goal of measuring [quality of life] in animals from the perspective of the animals is not currently attainable" and that development of a proxy survey completed by humans offers the best chance of usefully measuring animal quality of life. Most of the more recent work on quality of life in animals relies on questionnaires completed by proxies (Belshaw et al. 2015; Mullan 2015).

What do these questionnaires ask? The questions asked, and the underlying definition of quality of life at issue, vary widely. A recent review article asserts that few studies of quality of life actually offer a definition of quality of life (Belshaw et al. 2015). Most ask about perceived pain, and many ask about other topics such as function, sociability, and capacity to meet various needs. In general, these assessments have focused on the harms to quality of life that stem from disease, rather than considering how various capabilities might improve animals' quality of life.

Conceptual work on quality-of-life assessments in animals is likewise variable. The three most prominent philosophical accounts of quality of life in humans are the *hedonic* account, on which quality of life is a function of affective responses, the *desire-fulfillment* account, on which quality of life depends on the extent to which one's preferences are being fulfilled, and still others understand it via an *objective list* account, on which quality of life constitutes possessing objectively observable goods or capacities (Crisp 2016). Proxy questionnaires applying each of these approaches have been used for animal health. The hedonic approach is defended by one prominent commentator who contends that

> assessment criteria, like the QOL [quality-of-life] factors they are intended to measure, have value only insofar as they are associated directly or indirectly with affective states. Measurement criteria that are not associated with affective states are not relevant to QOL and, hence, play no role in assessing QOL.
>
> *(McMillan 2000)*

Other discussions propose using a desire-fulfillment account in which those answering questions evaluate how well an animal's preferences are being satisfied (Yeates 2016). One questionnaire proposes an objective list account, where respondents consider whether an animal enjoys various objective goods and capacities (Wojciechowska and Hewson 2005; Wojciechowska et al. 2005).

Is there any alternative to surveying humans about their perceptions of animals' quality of life? Franklin McMillan asserts that "[t]he goal of measuring QOL in animals from the perspective of the animals is not currently attainable" (McMillan 2000: 1908). Despite this claim that there is no alternative to human-completed questionnaires, some quality-of-life assessments for laboratory animals infer quality of life directly from behavior rather than soliciting the perspectives of human raters. Some have proposed that the "Mouse Grimace Scale," based on the idea that examining the facial expressions of mice could serve as an effective way of directly assessing whether mice are in pain (Matsumiya et al. 2012; Leach et al. 2012). Similarly, the ability of mice to perform certain tasks, or their aversive behavior when faced with a stimulus, could be regarded either as indicators of quality of life or as constituents of quality of life (Urban et al. 2011). The strategy of assessing quality of life based on direct observations of behavior could potentially be applied not only to mice but also to companion animals and livestock. As another example, one article proposes looking at hormonal and physiological changes, such as changes in heart rate or hormone concentration (Stockman et al. 2011; Christiansen et al. 2007).

For any of these assessment techniques, it is important to keep in view the fact that whatever is being measured must matter in order to be worth measuring in the first place. Some approaches may provide plentiful and easily accessible quantitative data, but that data may not represent any aspect of animals' lives that is morally significant. Even if perfect consensus cannot be reached on what is morally significant in animal quality of life, however, the strategies presented earlier represent potentially effective ways of assessing quality of life in various animals.

Translating Current Issues in the Ethics of Cost-Effectiveness to Animal Ethics

Quality of life determinations could potentially be used for a variety of purposes, including deciding what sort of clinical care an animal needs, punishing those who cause unjustified pain or injury to animals, or designing zoos, farms, and other types of spaces where animals live. My focus, however, is the use of quality-of-life assessments as part of priority-setting. Quality-of-life assessments are frequently used, in combination with costs, to generate cost-effectiveness figures that are then used to set priorities for human health. For instance, the health care system in the UK uses quality-of-life assessments to determine which interventions fall within a cost-effectiveness threshold (Drummond and Sorenson 2009). They have also been proposed for use in other contexts, such as setting comparative priorities between health care and public health, determining which services are provided in markets or paid for by private insurers, and indicating which interventions should be high priorities for research (Graham et al. 1998; Clement et al. 2009; Persad 2016).

In this section, I consider how several ethical challenges that cost-effectiveness methodologies face when used to set human health priorities might translate to the animal health context. These include the challenges of comparing quality of life across different conditions; of avoiding violations of rights; and of treating people with disabilities or short life expectancies fairly. I also consider a challenge that has special resonance in the animal ethics context: cross-species comparison of quality of life.

Quality of Life Across Conditions

One challenge that is magnified in the animal health context involves the commensurability of quality-of-life measures across conditions. Some argue that it is difficult to compare the quality of life impacts of different medical conditions, such as limited mobility and chronic pain. It may be even harder to compare quality of life across different animals or animal species.

One possibility, if we use the questionnaire approach, is to ask humans these questions on behalf of the animals they are responsible for. This presents a variety of problems. First, humans who are responsible for animals may answer the questions in light of their own moral obligations to animals, obligations that may not track what is good for the animal overall. They may also select outcomes that are better for human caretakers even though worse for animals. For instance, humans may regard outcomes where an animal dies quickly as preferable to outcomes where an animal suffers in pain under their supervision, even if the animal would, in fact, be better off under the second option. Second, moving from the deontic (what decision makers should do) to the axiological (what is valuable), proxy decision makers may, in general, judge outcomes to be different from the way that they seem from the first-person perspective. For instance, people typically prefer pains to be in their own past rather than in their own future but may not exhibit the same time bias when they consider the pains that others experience (Dorsey 2016). Third, human survey respondents may simply lack knowledge of what is actually in the interest of the animals whose lives they are describing.

One alternative would be to ascertain quality-of-life judgments by looking at the risks and tradeoffs animals actually accept—for instance, the risks they are willing to take in order to avoid pathogens or to engage in play activities. We could regard animals' actual behavior as revealing a preference for certain health states over others. However, some might worry that these revealed preferences are hard to interpret. It is difficult to tell whether a given decision reflects a general preference or just a preference in a particular context. It is also challenging to know the extent to which we should try to determine what is good for animals by looking at what they actually do. For instance, the fact that dogs are willing to drink out of toilets might seem to suggest that they are willing to tolerate a high risk of illness in order to pursue their goal. But we might instead conclude that dogs, left to their own devices, are mistaken about what, in fact, is in their own interest. (They are like the person in John Stuart Mill's famous example who is about to cross a rickety bridge he does not realize is rickety.) Similarly, some have argued that the fact that animals often hide when they notice a photographer suggests that they value privacy in the way that humans do and that hidden cameras would set back animal privacy interests just as they set back human ones (Pepper, unpublished). But their decision to hide might reflect a mistake about the photographer's purposes or about what the photographer is holding. This would push us toward an account of animal quality of life that is based on idealized choices (see Heathwood 2006) or instead toward an account that is based on objective experiences or on measurements of hedonic states.

Particularly in the case of dogs and other social or altruistic animals, a further complication of a revealed preference approach is that not all an individual's choices aim to improve that individual's own quality of life. As an example, Amartya Sen argues that we should recognize that individuals sometimes make choices in full knowledge that those choices will worsen their own quality of life in order to honor certain commitments that they have (Sen 1977). The soldier who throws himself on a grenade to save his platoon does not thereby reveal that he judges being pulverized to be an improvement in his quality of life. Rather, he sacrifices his quality of life in order to honor his commitment to his fellow soldiers. This is true for animals as well: various choices animals might make could be understood not to aim at improving their quality of life but, rather, at honoring their commitments. These commitments can be to conspecifics, such as cases where animals sacrifice their lives for fellow pack members, but cross-species commitments are also possible (Casal 2012). Another concern involves what Jon Elster, Sen, and Martha Nussbaum have dubbed "adaptive preferences" (Teschl and Comim 2005). Animals may elect to do various things not because they contribute either to their own quality of life or to their commitments but, rather, because have been led to act in ways that serve the interests of others because of a mistaken sense of self-worth.

These concerns about purely desire-based approaches suggest the attractiveness of either hedonic or objective-list measures of animal quality of life that allow for comparisons across different conditions. Given the complexity of many animals' lives, I would argue that the same considerations that favor objective-list measures over purely hedonic ones—namely, that more matters to people than merely their psychological states—also apply to many animals.

One possibility is to use an externally observed hedonic account of quality of life to assess quality of life in animals that lack sophisticated self-awareness and long-term planning but to use an objective-list account (developed primarily through observation but also potentially through questionnaires) to assess quality of life among more mentally sophisticated animals. This distinction might track David DeGrazia's suggestion that the interests of fish, amphibians, reptiles, and birds be assigned different, and lower, priority than the interests of other sentient animals (DeGrazia 1996). On this view, some animals are capable of having a rich quality of life that goes beyond purely hedonic reactions while others are not.

Quality of Life and Rights

Another challenge involves tensions between the utilitarian bent of most cost-effectiveness approaches and respect for individual rights. The 1991 Oregon Medicaid proposal, which set priorities strictly on the basis of cost-effectiveness, was criticized for providing cost-effective but low-urgency interventions, such as tooth capping, in preference to urgent but costly interventions such as appendectomies (Hadorn 1991). This objection involves what philosophers call the *aggregation* problem—whether many minor burdens can be aggregated to outweigh a major burden. A related objection, termed the *fair chances/best outcomes* problem, questions the exclusion of individuals who are unlikely to benefit, and a third, termed the *separate spheres/indirect benefits* problem, contends that cost-effectiveness approaches give excessive weight to whether individuals who receive interventions will, in turn, produce other benefits for society (Daniels 1993).

Many have suggested that utilitarian reasoning is distinctively appropriate when only animal interests are at issue (McMahan 2002; Nozick 1983).[2] Jeff McMahan effectively summarizes this view when he states that the "morality of respect," which includes nonconsequentialist rights, applies to persons, but the "morality of interests," which does not, applies to sentient animals (McMahan 2002). If this is correct, cost-effectiveness methodologies might face fewer ethical headwinds when used to set priorities in animal health than they do in human health. It might be appropriate to, for instance, prevent many pigs from suffering painful bruises than to prevent a single pig from suffering very severe pain, even if it is inappropriate to make similar calculations where the lives of beings who fall within the morality of respect are at issue. It may also be appropriate to invest no resources at all in curing certain very rare animal diseases if those resources could instead be used to cure much more common diseases. In fact, even more radical utilitarian decisions that are widely agreed to be inappropriate in human health might be appropriate where animals are concerned, such as killing some animals in order to prevent a disease from spreading to many more.

McMahan's view is well argued for and seems normatively plausible. If adopted, it would broaden the scope of permissible use of cost-effectiveness measures in animal health. However, others suggest that animals should be accorded deontological rights rather than being regarded as mere subjects of interests (Regan 1985) or that "it's not in general permissible to cause serious pain and injury to one morally significant entity," including an animal, "in order to benefit others" (Harman 2011). If animals fall within what might be termed the morality of respect, then rights-based concerns about the use of cost-effectiveness criteria to set priorities have more traction. A rights-based view also coheres with some of our practices regarding animals—in particular, the view in animal research that prioritizes refining research over reducing number of animals used on the basis that "the experience of an individual animal is paramount to the number of animals, and that additional suffering for the individual in exchange for a reduced total number of animals is not acceptable" (Boo et al. 2005; see also Ringblom 2016). This echoes the deontological view in human bioethics that recommends saving one person from death over saving many from minor pain.

However, many uses of cost-effectiveness criteria do not conflict with conventionally recognized rights, because they move resources toward more cost-effective uses without violating any rights (Ord 2013). In order to resist these uses of cost-effectiveness criteria, one would have to assign moral weight to much more controversial rights claims, such as the "rule of rescue," which recommends favoring identifiable lives over statistical lives or the view that numbers should not count when aid is being provided (see Persad 2019b).

An intermediate possibility is also interesting. On this view, the appropriateness of utilitarian reasoning in animal health would depend on the relevant animals' capacities and social organization: this suggestion builds on Judith Thomson's speculation that act-utilitarianism is appropriate for eusocial animals but inappropriate for humans (Thomson 1990). Utilitarianism might similarly be inappropriate for animals who are relevantly similar to humans but appropriate for social animals. By this logic, we might use a highly nonconsequentialist ethical theory for balancing the interests of animals that are solitary and typically competitive with conspecifics, such as tigers, in one way; a moderately nonconsequentialist theory for balancing the interests of animals who (like humans) are limited altruists in another way; and a utilitarian theory for balancing the interests of eusocial animals like bees. While this way of developing Thomson's view is interesting, it has two clear limitations. One is that it may base moral obligations and permissions too closely on the natural abilities of specific creatures: that tigers may sometimes attack other tigers in nature does not entail that we should regard their interests as highly separate from a moral point of view. Another is that it does not address the question of how to handle trade-offs between the interests of animals from different species.

Quality of Life and Individual Differences

Cost-effectiveness approaches have been criticized for disadvantaging individuals who, because of comorbidities, can gain less quality or quantity of life from treatment. As an example, some cost-effectiveness approaches to priority setting would assign lower priority for curing the cancer of someone who is blind, because the future life they gain from the treatment will be lower in quality due to their disability. This outcome has been criticized for being unjust and for further disadvantaging the already disadvantaged (Menzel et al. 1999; see also Persad 2019a). Another example is that most cost-effectiveness approaches will also assign lower priority to curing an illness for someone who cannot live for many more years, as opposed to someone who can live for many more years. John Harris and others have argued that this approach unfairly ignores the equal moral worth of each individual (Harris 1987).

The frequently held view that extending animals' lives matters much less from a moral point of view than does improving their quality of life would simplify the latter problem. If additional life-years are assigned little or no value but quality of life has very high value, then animals who can potentially live for a longer time if they receive health benefits will receive little additional priority. However, the view that additional life counts for nothing seems implausible: we are willing to perform painful surgeries on some animals rather than painlessly euthanizing them, and we do so, in part, because we regard quantity, not only quality, of life as valuable for animals. Elizabeth Harman (2011) illustrates this through a case of a young cat undergoing surgery (see also DeGrazia 2016). Nonetheless, even though quantity of life in animals seems to count for something, it is plausible that quantity of life counts for less where animals are concerned. This suggests, for animals, replacing the standard way of calculating the QALY with a different function in which quantity is reduced in importance.

The problem of fairness to individuals with different capacities, in contrast, remains similarly urgent in the animal health context, and perhaps even more urgent given the differences between species. Imagine that wild turkeys have a better quality of life than chickens but the same moral status. Should we be more interested in extending the life of the animal that starts out having a greater quality of life, or should we treat animals of equal moral status equally even if they differ in capacity? (Differences in moral status form a different, and potentially orthogonal, basis for prioritizing some animals over others.) The answer to this question depends on whether there is a way of comparing quality of life across species, a question I take up in the next section. But similar questions can arise even regarding intraspecies decision making—should we assign lower priority to extending the life of a permanently blind horse than a healthy one? Again, the answer to these questions seems to depend in part on whether we regard animals as being within the morality of respect or instead within a morality of interests.

Quality of Life Across Species

Priority setting for animal health can be done with a wider or a narrower focus. Narrow priority setting would determine which interventions are of highest priority for a specific, homogeneous population—for instance, which chicken coop designs are most cost-effective at improving chickens' quality of life. Broader priority setting would attempt to set priorities among wider groups, including multiple species—for instance, to determine which alignment for a train line through a green space would least reduce the quality of life of the animals living in the green space or what treatment for animal disease should be the highest priority for a laboratory that is researching many animal diseases.

Wide priority setting in animal health presents the problem of *cross-species comparisons*. For instance, the green space example might involve considering how various harms caused by pollution or noise affect the quality of life of deer, marmots, bears, and migratory birds. More challenging, they might also involve comparing the relative importance of a given change in quality of life for one type of animal to the importance of that change for other animals. We typically assume that all humans' quality of life interests are given the same moral weight when setting priorities using cost-effectiveness considerations (Drummond and Sorenson 2009). But this conclusion is much less obvious when we are making cross-species comparisons. It might seem mistaken, for instance, to assign the same weight to the loss in quality of life that a salamander would experience as to the loss that a bear or wolf would experience. However, determining what weight to assign to each individual's experiences is challenging. There is no obvious time trade-off or standard gamble—or revealed preference analogue of these—that would compare whether it is better to be a wolf or a salamander. Wolves and salamanders do not face situations where they could become members of the other species. Nor are there "competent judges," in John Stuart Mill's terms (see Brownlee 2016) who have lived both sorts of lives and could recommend which is better.

One response to the challenge of cross-species quality of life comparisons would be to retreat to an approach that offers only a partial ordering of priorities (Ruger 2004): on this approach, cost-effectiveness could be used to set priorities within species but not between species. David DeGrazia seems to favor such a view in his comments on the potential incommensurability of human and animal quality of life:

> [Consider the assumption] that there is a sort of objective super-scale of prudential value, by reference to which one can evaluate the well-being of every actual and possible creature in virtue of its realization of the items enumerated on the massive (infinite?) list. One might find, as I do, the sheer grandiosity of such a conception a reason to doubt it. It seems to require a standpoint of prudential evaluation that is so impartial as to be God-like, or Platonic Form-like. . . . Despite lacking a decisive argument against the idea of an objective super-scale of prudential value, I suspect that this idea is misguided. It seems to me more plausible that assessments of prudential value must be relativized to the sort of creature in question and, in particular, to the native capacities of such creatures.
>
> *(DeGrazia 2016: 514-515)*

However, the frequent skepticism in animal ethics about the fundamental moral weight of biological categories such as species membership (McMahan 2002), and the potential value of an approach that assists decisionmakers who must compare the interests of different types of animals—ranging from veterinarians conducting triage and allocating limited treatments to policymakers setting priorities for animal health investments—indicates the value of a quality-of-life measure that crosses species boundaries.

What are the candidates for such a measure? As discussed earlier, the most natural ones are objective-list measures (either conducted by proxies or external assessment) and hedonic measures. As DeGrazia observes, of course, there are challenges in constructing an objective list measure. To riff on Mill, is it better to be a "salamander satisfied" or a "pig dissatisfied"? Even if we can determine, for example, that it is worse to be a pig in chronic pain than a pig with a hearing impairment, it is hard to identify the list of capacities that would tell us whether it is better to be a pig in chronic pain or a healthy salamander. Notwithstanding these challenges, however, generating broader quality-of-life assessments seems possible. As McMahan observes,

> [i]t might be suggested that individuals should be ranked hierarchically according to their level of psychological capacity or potential. On this view, all individuals with equivalent capacities or potential would count as higher forms of being than individuals with lower capacities and potential. This offers a way of articulating our sense that it is better to be a person than to be a dog. Persons are higher forms of being: their capacities are higher, the range of good accessible to them is higher, and in general their lives are better, or more valuable, or more worth living, than those of dogs
>
> *(McMahan 2002: 160)*

DeGrazia, in discussing similar cases, expresses skepticism about the specific judgment that dogs are better off than human persons, although he does not categorically reject the view. He observes that "[n]o human being has had, for example, the experience of walking through the neighborhood and recognizing the unmistakable scents of dozens of familiar people and dogs. No human being has experienced the magnificent depths of auditory experience that dogs experience every day" (DeGrazia 1996: 4). As a factual matter, DeGrazia is correct. But it seems plausible that the value of complex psychological experiences, such as composing a symphony or designing a fragrance outweighs the value of the sensory experience of smelling many fragrances or hearing many sounds, particularly from an objective-list rather than purely hedonistic perspective. (Think of Beethoven even after his deafness.) DeGrazia himself grants that he lacks a decisive argument against "an objective super-scale of prudential value" (DeGrazia 1996: 4), and the use of such a scale has the advantage that it avoids regarding species membership itself as objectively significant.[3]

In any event, DeGrazia's relativization approach appears too strong—it forecloses the possibility of plausible partial orderings across different types of creatures. Even if one agrees with DeGrazia that the experience of animals like dogs, dolphins, and bats is so different from humans' experience of living their lives as to be incommensurable, there may be some species of animals whose life experiences are "strictly dominated" (to use technical language) by the experiences of other, similar animals—their capacities are close to a proper subset of other animals' capacities. An example might be dolphin species that are less intelligent than the bottlenose dolphin but have the same sensory capacities; more broadly, we might be able to compare very different species of birds and conclude that some have richer lives than others. We might also be able to compare different developmental stages or morphological forms of the same organism: even though tadpoles have some capacities that frogs lack, it seems plausible that frogs overall have a richer set of experiences.

A different way of approaching the problem of cross-species comparisons is to focus on communities of animals or on ecosystems, rather than on the effects of policy decisions or other forces on individual animals. This approach would combine the interests of all animal denizens of the park (as well as potentially other living beings, and even the interests of natural artifacts insofar as they can be said to have interests), and would then assess the impact of the choice on the interests of the ecosystem as a whole. This approach would clearly reject the position taken by thinkers such as Regan and Harman, who regard each individual as separate and worthy of respect, in favor of a more holistic evaluation of the ecosystem's good. However, it would also differ from the position that McMahan defends, which regards the quality of life that different animals enjoy as commensurable. Instead of comparing different animals' quality of life, it would subsume their experiences under a larger unit of evaluation.

Last, reasoning about moral status may pull in a different direction from reasoning about quality of life. We might conclude, based on an objective list of capacities such as complex planning, complexity, and community living, that it is preferable to be a pig than to be a salamander, and even that it is better to be a pig in chronic pain than to be a salamander in good health. (This is what we conclude about the comparison between humans and salamanders.) Yet we may conclude that it is more morally urgent to relieve the pain of a pig than that of a salamander, even though it may be far worse to be a salamander in ill health than a pig in ill health. DeGrazia discusses, although does not explicitly endorse, an "unequal consideration" account of moral status that incorporates not only the capacities an animal has but also what sort of considerability the animal possesses (DeGrazia 2008). Cost-effectiveness approaches could operationalize DeGrazia's unequal consideration account by differentially weighting quality-adjusted life-years or similar metrics according to the moral status of the animal whose health is at issue.

Another factor that could be relevant, and could pull in a different direction from quality of life, is the special obligations that may be owed to certain animals or groups of animals. Recent work by Sue Donaldson and Will Kymlicka makes the case that domesticated animals are owed obligations of co-citizenship, whereas wild animals should instead be afforded sovereignty, and "liminal" animals, which live in human communities but are not domesticated, should be granted a "denizenship" status that differs from citizenship (Donaldson and Kymlicka 2011). Importantly, Donaldson and Kymlicka's approach does not base special obligations on animals' capacities or quality of life—as an example, rabbits could be owed obligations of co-citizenship, sovereignty, or denizenship depending on whether they are pet rabbits, wild rabbits, or rabbits living in proximity to humans. Ultimately, quality of life is unlikely to be the only factor in cross-species priority setting and can be outweighed by considerations of moral status or special obligation.

Operationalizing Cost-Effectiveness to Set Priorities in Animal Health

Perhaps the simplest and least controversial strategy for expanding the use of cost-effectiveness in animal health is to use quality of life measures to create species-specific QALY analogues. One such example could be, for instance, QADYs (quality-adjusted dog years). This approach would not take on the challenge of making cross-species comparisons but could be used to set priorities for research into medical interventions that might benefit dogs or to decide which dogs should receive a scarce medical resource in a triage situation.

A more controversial approach would be to try to find some way of commensurating the species-specific metrics into a unified metric. We might call this the QAALY (quality-adjusted animal life-year).[4] Using a QAALY distributive metric would allow priorities to be set for interventions that affect many different species of animal—for instance, whether to prioritize a given treatment for pigs over another one for chickens.

Even without a QAALY distributive metric in place, it may still be possible to make some priority-setting decisions in light of species-specific quality-of-life measures. This could involve a process of weighing, specifying, and balancing different considerations against one another, just as is done when QALYs are being compared to non-QALY criteria in human health contexts. It could also involve the use of an extended cost-effectiveness analysis that compares costs to both quality-of-life outcomes and other types of outcomes (see Garrison et al. 2017).

One challenging issue for both QAALYs and the species-specific approach involves potential conflicts between human and animal interests. Would we ever be justified in preferring to realize some quantity of animal QALYs rather than a smaller quantity of QALYs for humans? This may seem intolerable where life and death for humans are considered and may favor a "lexical priority" approach in which the lives of humans are incomparably more important than quality improvements in the lives of animals, no matter how many animals benefit (see Zamir 2006: n.3). On the other hand, commonsense practice does allow some trade-offs between human and animal QALYs where human life is not directly at stake, particularly when the amount of human suffering at issue is small—for instance, we might accept a slightly more painful treatment for humans if it can be produced in a way that causes less animal pain (Zamir 2006).

When combined with the common view that some cases of human pain are more morally significant than some risks to human life—which underlies, for instance, the permissibility of risky surgery to cure chronic pain—this could present a problem of intransitivity: a sufficient quantity of human pain can be more important than a human life, but no quantity of animal pain could justify sacrificing a human life, yet some quantity of animal pain could outweigh a quantity of human pain. Similar intransitivities are familiar elsewhere in bioethics (Kamm 1985).

Conclusion

I have argued in favor of greater use of cost-effectiveness criteria to set priorities in animal health. I close with a practical note of caution: a familiar challenge to priority setting discussed by Norman Daniels retains force and relevance in the animal ethics context. Daniels argues that it is difficult to justify comprehensively setting priorities for medical interventions using cost-effectiveness calculations within a health care system, like that in the US, where the savings realized by reducing the provision of interventions with poor cost-effectiveness will not necessarily go toward the provision of more cost-effective interventions, which would lead to net improvements in health but might instead go toward nonhealth programs (Daniels 1986). Some might similarly worry that priority-setting efforts will lead to less money being spent on worthy goals in animal health, and redirected toward other, less important, aims. For priority setting that cuts spending on animal health, rather than reallocating it, to be justified, the alternative spending proposal must be more important.

Notwithstanding Daniels's concerns, incorporating cost-effectiveness considerations into choices regarding animal health could lead to substantial improvements in animal health. Many of the ethical constraints that count against the use of cost-effectiveness to set priorities do not apply, or at least not so clearly, to their use in setting priorities among animals and with regard to animal health issues. Yet there is little current use of cost-effectiveness to set priorities in animal health—much less than in human health. This perhaps stems from the fact that animal health is highly disunified, with care for wild animals, farm animals, pets, and laboratory animals governed by different rules and policies. There is no National Health Service or Medicaid for animals, or National Institutes of Health to organize research into animal health. Yet we spend substantial sums of money on animal health despite this and do so in a disorganized way. As an example, federal law in the US requires that pets be included in emergency planning (Leonard and Scammon 2007). Many millions of dollars are spent on animal health. Even if some of this spending has the goal of promoting human interests, some of it is clearly spent on animals themselves. By considering cost-effectiveness, we can use that money to make animals' lives go better. As Ord observes in human health contexts, "[i]gnoring cost-effectiveness . . . does not mean losing 10 percent or 20 percent of the potential value that a health budget could have achieved, but can easily mean losing 99 percent or more" (Ord 2013: 5). He goes on to state that "[t]he main effect of understanding the moral imperative toward cost-effectiveness is spending our budgets so as to produce greater health benefits, saving many more lives and preventing or treating more disabling conditions" (Ord 2013: 7). This is true for humans, but it is at least as true for animals. It is time to consider cost-effectiveness in animal health.

Notes

1. I use *animal* as shorthand throughout for "nonhuman animal."
2. Elizabeth Harman presents but rejects the view that "[o]ne might try to develop a view on which the kinds of agent-relative constraints that apply to persons do not apply to animals or animal stages. On such a view, it would be permissible to harm one animal or animal stage in order to provide positive benefits to a distinct entity," without discussing McMahan's view on the morality of respect, even though she discusses other parts of McMahan's book (Harman 2011).

3. "But while it seems right that a condition must be evaluated by comparison with what is possible given a certain set of psychological capacities, it seems a mistake to suppose that the relevant capacities are those typical of the members of the individual's own biological species" (McMahan 2002: 325).
4. Pesie similarly suggests that "[t]his research explores the possibilities to develop an Animal Health Utility Index (AHUI), by which reductions in QoL can be quantified to indicate the 'Animal Disease Burden' (ADB) (i.e. the change in QoL during a time period), comparable to the quantification of QALYs in humans" (Pesie 2012: note 4).

Bibliography

Belshaw, Z., Asher, L., Harvey, N., and Dean, R. (2015) "Quality of life assessment in domestic dogs: An evidence-based rapid review," *The Veterinary Journal* 206(2): 203–212.

Boo, J., Rennie, A., Buchanan-Smith, H., and Hendriksen, C. (2005) "The interplay between replacement, reduction and refinement: Considerations where the three Rs interact," *Animal Welfare* 14(4): 327–332.

Brock, D. (2003) "Ethical issues in the use of cost-effectiveness analysis for the prioritization of health care resources," in T. T. T. Edejer, R. Baltussen, T. Adam et al. (eds.) *WHO Guide to Cost-Effectiveness Analysis* (pp. 289–312), Geneva, Switzerland: World Health Organization, www.who.int/choice/publications/p_2003_generalised_cea.pdf (Accessed March 19, 2019).

Brownlee, K. (2016) "The competent judge problem," *Ratio* 29(3): 312–326.

Casal, P. (2012) "Sexual dimorphism and human enhancement," *Journal of Medical Ethics* 39(12): 722–728.

Christiansen, S., Billeschou, S., and Forkman, B. (2007) "Assessment of animal welfare in a veterinary context—a call for ethologists," *Applied Animal Behavior Science* 106(4): 203–220.

Clement, F., Harris, A., Li, J., Yong, K., Lee, K., and Manns, B. (2009) "Using effectiveness and cost-effectiveness to make drug coverage decisions: A comparison of Britain, Australia, and Canada," *JAMA* 302(13): 1437–1443.

Crisp, R. (2016) "Well-being," in E. Zalta (ed.) *The Stanford Encyclopedia of Philosophy*, [online] Summer, https://plato.stanford.edu/entries/well-being/#TheWelBei (Accessed March 20, 2019).

Daniels, N. (1986) "Why saying no to patients in the United States is so hard: Cost containment, justice, and provider autonomy," *New England Journal of Medicine* 314(21): 1380–1383.

Daniels, N. (1993) "Rationing fairly: Programmatic considerations," *Bioethics* 7(2–3): 224–233.

DeGrazia, D. (1996) *Taking Animals Seriously: Mental Life and Moral Status* (p. 280), New York: University of Cambridge.

DeGrazia, D. (2008) "Moral status as a matter of degree?" *The Southern Journal of Philosophy* 46(2): 181–198.

DeGrazia, D. (2016) "Sentient nonpersons and the disvalue of death," *Bioethics* 30(7): 511–519.

Donaldson, S., and Kymlicka, W. (2011) *Zoopolis: A Political Theory of Animal Rights*, New York, NY: Oxford University Press.

Dorsey, D. (2016) "Future-bias: A (Qualified) defense," *Pacific Philosophical Quarterly* 98(S1): 317–344.

Drummond, M., and Sorenson, C. (2009) "Nasty or nice? A perspective on the use of health technology assessment in the United Kingdom," *Value in Health* 12(2): S8–S13.

Duarte, C., Freitas, P., and Bexiga, R. (2015) "Technological advances in bovine mastitis diagnosis: An overview," *Journal of Veterinary Diagnostic Investigation*, [online] 27(6): 665–672. https://doi.org/10.1177/1040638715603087 (Accessed March 19, 2019).

Fojo, T., and Grady, C. (2009) "How much is life worth: Cetuximab, non- small cell Lung Cancer, and the $440 billion question," *JNCI Journal of the National Cancer Institute*, [online] 101(15): 1044–1048. https://doi.org/10.1093/jnci/djp177 (Accessed March 19, 2019).

Garrison, L., Kamal-Bahl, S., and Towse, A. (2017) "Toward a broader concept of value: Identifying and defining elements for an expanded cost-effectiveness analysis," *Value in Health* 20(2): 213–216.

Graham, J., Corso, P., Morris, J., Segui-Gomez, M., and Weinstein, M. (1998) "Evaluating the cost-effectiveness of clinical and public health measures," *Annual Review of Public Health* 19(1): 125–152.

Hadorn, D. (1991) "Setting health care priorities in Oregon: Cost-effectiveness meets the rule of rescue," *JAMA* 265(17): 2218–2225.

Harman, E. (2011) "The moral significance of animal pain and animal death," in T. Beauchamp and R. Frey (eds.) *The Oxford Handbook of Animal Ethics* (pp. 726–737), Oxford: Oxford University Press.

Harris, J. (1987) "QALYfying the value of life," *Journal of Medical Ethics* 13(3): 117–123.

Heathwood, C. (2006) "Desire satisfactionism and hedonism," *Philosophical Studies* 128(3): 539–563.

Hsiung, W., and Sunstein, C. (2006) "Climate change and animals," *University of Pennsylvania Law Review*, [online] 155(6): 1695, https://scholarship.law.upenn.edu/penn_law_review/vol155/iss6/7 (Accessed March 19, 2019).

John, T., Millum, J., and Wasserman, D. (2017) "How to allocate scarce health resources without discriminating against people with disabilities," *Economics and Philosophy* 33(2): 161-186.
Kamm, F. (1985) "Supererogation and obligation," *The Journal of Philosophy* 82(3): 118–138.
Leach, M., Klaus, K., Miller, A., Di Perrotolo, M., Sotocinal, S., and Flecknell, P. (2012) "The assessment of post-vasectomy pain in mice using behaviour and the mouse grimace scale," *PLoS Medicine* 7(4).
Leonard, H., and Scammon, D. (2007) "No pet left behind: Accommodating pets in emergency planning," *Journal of Public Policy & Marketing* 26(1): 49–53.
Matsumiya, L., Sorge, R., Sotocinal, S., Tabaka, J., Wieskopf, J., Zaloum, A., King, O., and Mogil, J. (2012) "Using the mouse grimace scale to reevaluate the efficacy of postoperative analgesics in laboratory mice," *Journal of the American Association for Laboratory Animal Science* 51(1): 42–49.
McMahan, J. (2002) *The Ethics of Killing: Problems at the Margins of Life*, New York: Oxford University Press.
McMillan, F. (2000) "Quality of life in animals," *Journal of the American Veterinary Medical Association* 216(12): 1904–1910.
Menzel, P., Gold, M., Nord, E., Pinto-Prades, J., Richardson, J., and Ubel, P. (1999) "Toward a broader view of values in cost-effectiveness analysis of health," *Hastings Center Report* 29(3): 7–15.
Mullan, S. (2015) "Assessment of quality of life in veterinary practice: Developing tools for companion animal carers and veterinarians," *Veterinary Medicine: Research and Reports* 6: 203–210.
Nozick, R. (1983) "About mammals and people," *New York Times Book Review*, November 27, pp. 11, 29–30.
Ord, T. (2013) *The Moral Imperative toward Cost-Effectiveness in Global Health* [online], Center for Global Development, p. 7, www.cgdev.org/publication/moral-imperative-toward-cost-effectiveness-global-health (Accessed March 19, 2019).
Persad, G. (2016) "Should research ethics encourage the production of cost-effective interventions?" in D. Strech and M. Mertz (eds.) *Ethics and Governance of Biomedical Research: Theory and Practice*, Switzerland: Springer.
Persad, G. (2019a) "Considering quality of life while repudiating disability injustice: The pathways approach to priority-setting," *The Journal of Law, Medicine, and Ethics* 47(2): 294–303.
Persad, G. (2019b) "Will more organs save more lives? Cost-effectiveness and the ethics of expanding organ procurement," *Bioethics* 33(6): 684–690.
Pesie, S. (2012) *Quantification of Animal Disease Burden; Exploring the Possibilities of a Dog-Health Utility Index*, MSc thesis, Wageningen University, http://edepot.wur.nl/244416 (Accessed March 19, 2019).
Regan, T. (1985) "The case for animal rights," in P. Singer (ed.) *In Defense of Animals*, Oxford: Basil Blackwell.
Ringblom, J. (2016) *Worth Weighting For—Studies on Benchmark Dose Analysis in Relation to Animal Ethics in Toxicity Testing*, PhD Karolinska Institutet, pp. 11–14, https://openarchive.ki.se/xmlui/bitstream/handle/10616/45331/Thesis_Joakim_Ringblom.pdf?sequence=1&isAllowed=y (Accessed March 19, 2019).
Ruger, J. (2004) "Health and social justice," *Lancet* 364(9439): 1075.
Schurer, J., Rafferty, E., Farag, M., Zeng, W., and Jenkins, E. (2015) "Echinococcosis: An economic evaluation of a veterinary public health intervention in rural Canada," *PLoS Neglected Tropical Diseases*, [online] 9(7): e0003883., https://doi.org/10.1371/journal.pntd.0003883 (Accessed March 19, 2019).
Sen, A. (1977) "Rational fools: A critique of the behavioral foundations of economic theory," *Philosophy & Public Affairs* 6(4): 317–344.
Stockman, C., Collins, T., Barnes, A., Miller, D., Wickham, S., Beatty, D., Blache, D., Wemelsfelder, F., and Fleming, P. (2011) "Qualitative behavioural assessment and quantitative physiological measurement of cattle naïve and habituated to road transport," *Animal Production Science* 51(3): 240–249.
Teschl, M., and Comim, F. (2005) "Adaptive preferences and capabilities: Some preliminary conceptual explorations," *Review of Social Economy* 63(2): 229–247.
Thomson, J. (1990) *The Realm of Rights*, Cambridge, MA: Harvard University Press.
Urban, R., Scherrer, G., Goulding, E., Tecott, L., and Basbaum, A. (2011) "Behavioral indices of ongoing pain are largely unchanged in male mice with tissue or nerve injury-induced mechanical hypersensitivity," *Pain* 152(5): 990–1000.
Wojciechowska, J., and Hewson, C. (2005) "Quality-of-life assessment in pet dogs," *Journal of the American Veterinary Medical Association* 226(5): 722–728.
Wojciechowska, J., Hewson, C., Stryhn, H., Guy, N., Patronek, G., and Timmons, V. (2005) "Development of a discriminative questionnaire to assess nonphysical aspects of quality of life of dogs," *American Journal of Veterinary Research* 66(8): 1453–1460.
Yeates, J. (2016) "Quality of life and animal behavior," *Applied Animal Behaviour Science* 181: 19–26.
Zamir, T. (2006) "Killing for knowledge," *Journal of Applied Philosophy* 23(1): 17–40.

PART II

Animal Agriculture and Hunting

Editor's Introduction

In the US alone, roughly 9 billion chickens are raised and slaughtered for food each year. This is, on one hand, a staggering achievement of modern agribusinesses. In 1900, chicken and eggs were luxury goods, and most people would only have them for special occasions. In 2020, chicken and eggs are everywhere: chicken is the least expensive meat on every menu, and it's actually difficult to find processed foods without eggs in them. These products are, when consumed in moderation, relatively healthy sources of protein and an array of micronutrients. On the other hand, however, chicken production is very rough on chickens. For instance, most broiler chickens (chickens raised for meat) are raised indoors in crowded, warehouse-like facilities. The high stocking densities increase dust and ammonia concentrations, which adversely affects chickens' respiratory health. Crowding is also correlated with a higher incidence of injury—for example, sores, scabs, and scratches caused by other birds and skin irritation, lesions, or deep ulcers on the backs of the legs and feet (caused by walking and lying in wet, infrequently changed litter). Moreover, selective breeding for larger birds and faster growth create significant welfare problems. The average daily growth rates of chickens have quadrupled in the past 50 years, and this contributes to poor bone health, leg deformities, ruptured tendons, pinching of the spinal cord, accumulation of fluid in the abdominal cavity, enlarged heart, shrunken liver, and sudden death syndrome.

What happened in the last 120 years? How did we get here? Moreover, if we have reservations about the new status quo, who's to blame? Are greedy corporations the bad actors here? Or self-absorbed consumers? Or regulators who were (or are) asleep at the wheel? Or is there no problem at all, and we should simply be grateful for the bountiful and inexpensive food that's now available to us? Moreover, if we *do* think there's a problem, what are the alternatives? Is it any better to eat fish than beef? Are small-scale farms actually any better? What about hunting, which avoids agriculture entirely? This section takes up these and related questions.

Philosophers who write about animal ethics tend to condemn animal agriculture: by and large, they say that it inflicts significant unnecessary suffering on nonhuman animals, and is morally wrong for that reason. However, philosophers who write about animal ethics often endorse claims that are rejected by the public. For instance, unlike these philosophers, most people do not believe that pain is pain, where a certain amount of animal pain is as morally significant as a similar quantity of human pain. Most people do not believe that it is wrong to kill nonhuman animals if it is wrong to kill human beings with comparable cognitive capacities. Most people do not believe that an animal's function in human society is irrelevant to the question of how that animal ought to be treated; they believe, instead, that dogs deserve to be treated very differently than cows, such that it would be wrong to do to a dog what we currently do to cows. Is it possible to defend these parts of common sense? It does seem quite difficult to explain how they could be true. And if they aren't true, then it will be very difficult to defend animal agriculture. (For more on these issues, see the general introduction to the book.)

However, there is more to discuss with respect to animal agriculture than the question of whether it's permissible to engage in it at all. That's obviously a very important question and one that you'll encounter here. But it isn't the only question that we might ask.

Additionally, we might want to think about the best forms of animal agriculture, even if we ultimately say that the best isn't good enough. We might want to think about the future of animal agriculture, given that the economic incentives to continue it are likely to persist, not least because consumer demand is likely to rise. We might want to think about the kinds of impacts that are most important to mitigate—on animals, on the environment, and on public health. We might want to think about the kinds of trade-offs that we will have to accept depending on the remedies that we choose. What's more, we might want to reflect on the way that our thinking about animals affects the way that we understand animal agriculture. In the contributions that follow, for instance, you'll find different views about which sources are trustworthy. Should we doubt the contents of a study because it's funded by industry trade groups? If so, is that skepticism appropriate across the board or only in some limited range of cases? And either way, why? In a similar vein, it's worth thinking about what we are prepared to take for granted in our thinking about animal agriculture. Is it a live possibility that we might *end* animal agriculture? If so, on what time horizon? Is it even possible to substantially *improve* animal agriculture? If so, in what respects, and via what methods? And depending on our answers, what practical conclusions follow?

None of this is to suggest that, from the perspective of the most plausible comprehensive moral theories, we will be able to defend animal agriculture against its critics. It may turn out, as many believe, that this is one of many human practices that ought to go the way of slavery and infanticide—abolished, insofar as possible, to the dustbin of history. Indeed, it would be a bit surprising if we were, in the long run, to reach any other conclusion. If there is a presumption against harming nonhuman animals without sufficient cause, and we can eventually satisfy all our nutritional, economic, and cultural goals without harming animals, then future generations will probably condemn animal agriculture—whatever people say now. In the interim, however, this question isn't the only one that can occupy us, and the contributions that follow invite us to ask many others.

9
THE ORIGINS OF FACTORY FARMING IN THE UNITED STATES
An Overview

James McWilliams

Introduction

In a way, it's a simple story. From the start, the rise of factory farming responded in neoclassical economic fashion to the imperatives of *profit*. Factory farms emerged and developed in the mid-twentieth century to exploit densities of production and economies of scale. They consistently maximized profit and fulfilled global demand for meat specifically and animal products generally. From an economic perspective, the magnitude of agribusiness' success has been explored and acknowledged as a convergence of advanced science, management, and public policy. Profits were the goal; profits were accumulated. Factory farms now produce 99% of our meat. End of story (Daniel 1981; Adams 1988: 453–482; Cronon 1998; Hudson 1994; Whitaker 1975).

But, of course, it's not the end of the story. For all the explanatory power of the profit motive, it's only a partial answer. Most notably, "profit" fails to adequately confront the question of how Americans came to overwhelmingly embrace a system of animal production that confines, abuses, and kills 10 billion animals a year—all to feed consumers excessive amounts of meat. Economically speaking, the profit motive makes perfect sense as an explanatory device. But it ignores critical cultural questions. How did humans emotionally allow such a historical anomaly to happen? Did something fundamental have to change between animals and humans in order for factory farming to become a morally acceptable means of production? And, perhaps most important, how did the necessary changes between humans and animals occur? What, in essence, was the engine of that change?

This chapter explores these related questions to chart the emergence of factory farming. Its focus is on what I see as a pivotal if easily overlooked precondition for the rise of industrial agriculture: the changing relationship between farmers and their animals. I explore this issue in two phases. From 1750 to 1850, agricultural life was overwhelmingly the most common way of life. As a result, Americans interacted intimately with their animals and, in the process, developed a basic respect for them as living beings. The period from 1850 to 1950, however, saw important shifts in the human–animal agrarian relationship. Most notably, it was during this period that Americans achieved what Ralph Waldo Emerson once called a "graceful distance" from the means of meat production. In essence, Americans not only gradually moved into urban settings and pursued urban occupations, but those who stayed on the farm (still a majority) also began to conceptualize their animals as highly functional but ultimately emotionless machines made with interchangeable parts.

Upon achieving this redefinition, American farmers changed too. They established what was perhaps the most critical foundation for factory farming's explosion into the Midwest: general indifference to the welfare of the animals under their care and an exclusive emphasis on mechanized efficiency. This change in perspective, I would argue, joined the profit motive as a necessary condition for the rise of factory farming.

American animal agriculture initially evolved on the East Coast. It was there where the population was the densest and agricultural ideas the most debated. It was also there where progressive/scientific farming initially took root before spreading west through agricultural publications. And thus, it was there where many of the innovations relevant to animal domestication were pioneered. None of this is to say that developments germane to the human–animal agricultural bond did not occur in the trans-Appalachian or trans-Mississippi West, but only that the general thrust of agrarian change migrated east to west. The emotional indifference to animals that emerged on the east coast was, along with rising interest in capitalistic exploitation, precisely the sentiment that allowed factory farming to develop as fully as it did in places west, places as out of sight as they were out of mind.

Guiding Themes

The convergence of culture, science, and ideology influenced the human–animal bond in critical ways and at critical times in American agriculture. Placing the human–animal relationship squarely in the context of agricultural mechanization, quantification, and technological innovation reveals how intimacy yielded to distance, connectedness to alienation, and affection to indifference. This historical process proceeded through distinct phases, all of them tied in one way or another to concrete developments such as the popularization of agricultural communication, consumer preferences for certain food choices over others, and changing ideas of appropriate human–animal relationships. These influences help historicize the changing human perspective on the animal world, a perspective that eventually sanctioned the emergence of factory farming systems that currently produces 99% of the meat we eat today. Four themes make it possible to understand how factory farming came to be:

1. For much of our history, Americans have interacted with a wilderness of seemingly unending biodiversity. European migrants to colonial America were quick to describe their environment as a landscape of profound biological wealth. "If this land be not rich," wrote Thomas Morton of early Massachusetts, "then is the whole world poor?" (Cornon, 33). This rhetoric of environmental exuberance revealed more than a desire to lure potential settlers to a supposed promised land. It reflected sincere wonderment in the unfamiliar cornucopia of animal life—life ranging from the tiniest insect to most hulking ungulate. (Worster 1994: chap 1–2; Greene 1991)

   The popular notion that the American landscape was a treasure trove of novel discoveries had cultural consequences. The most notable was that it sparked naturalistic pride in American exceptionalism. This reverence inspired popular interest in knowing more about North America's indigenous mysteries, if only to portray them as badges of nationalistic honor. When Jefferson bragged to the French naturalist Compte de Buffon that the European reindeer "could walk under the belly of our moose," he was exhibiting this pride (Jefferson). Nothing confirms this characterization more than the colonial and early American reaction to native flora and fauna. (Jefferson 1955; Lewis 2005: 663–696; Peebles 1988)
2. Americans have embraced an unparalleled dedication to mechanization. In an overview of mechanization in the early republic, historian Brooke Hindle titled her article "The Exhilaration of Early American Technology" (McGraw, 40). For most of our history, the United States has been a nation marked by rich natural resources alongside a chronic scarcity of labor. This imbalance of factors has led American producers to eagerly seek ways of maximizing efficiency—whether of goods, knowledge, or services. Of course, every Western nation has experienced a

similar thrust toward increased productivity, but none like the United States. Indeed, the United States has pursued efficiency without hindrance by traditions or customs that might otherwise have moderated our embrace of mechanization.

But efficiency, as we know all too well, has a tendency to eviscerate not only wilderness and biodiversity but also human curiosity in it. When farm animals become conceptualized as machines, as they did after America urbanized and industrialized in the latter part of the nineteenth century, their essence as distinct beings was subsumed under an input/output ethos. As a result, human affection for animal life in general waned and, eventually, diminished altogether. Thus, our relationship with animals has been consistently reshaped by the dominant modes of industrial production. It is no coincidence that Ford's car factory eventually took the slaughterhouse as a model of efficiency (Olmstead and Rhode 1988: 86–112; Rosenbloom 1964: 489–511).

3. It is a critical fact that the United States evolved from a nation where roughly 85% of its citizens farmed in 1776 to one where, in 1900, 64% of its citizens farmed, and, by 1930, it was less than half. At the core of this historical transition is a basic fact that, perhaps because it seems so obvious, historians have ignored: our affection and respect for animals is directly related to our familiarity with them in an agrarian context. The time American farmers spent at work was time they spent around animals—domesticated and wild. And the more time they spent around animals, the more they had a chance to observe the deeper nuances and intricacies of their behavior. Farmers were necessarily observant. Proximity to the animal world would prove to be enormously important in establishing early dominion over it.

   As we moved away from the land, however, humans started conceptualizing animals *strictly* in terms of how they could serve human interests, and they did so without considering their intrinsic worth, not to mention their interest in avoiding unnecessary suffering. When this shift took place, our ability to feel affection for animals as distinct beings worthy of emotional consideration—again, worthy of the benefit of basic welfare—diminished (Olmert 2009; Daston and Mitman 2005; Wilson an Kellert 1993).

4. The American diet has always been meat-centric, so I will not evaluate our changing dietary patterns through the lens of carnivorous consumption. What really matters is growing consumer preference for consistency and uniformity. This desire, perhaps more than any other food-related matter, bears directly on the issue of human regard for animals and the bonds farmers established with them.

   For most of American history, consumers approached their meals with remarkably flexible and forgiving palettes. They ate what was in season, what was on hand, and what had been recently shot or slaughtered. They consumed food with firsthand knowledge of its source, as well as of the human and animal sacrifices required to bring it to the table. They ate to survive (McWilliams 2005).

   By the late nineteenth century, for a variety of reasons, this flexibility yielded to an increasing demand for uniformity. It is quite common to center our ire on the evils of factory farming squarely on producers. But we must not forget about the consumers' role in this historic transition or the impact that it has had on our declining ability to curb the abuses of animal agriculture, something 95 percent of Americans eating an average of 275 pounds of meat a year want to see happen. In many respects, of course, our collective quest for dietary uniformity has made a mockery of our respect for animal welfare (Pillsbury 1998; Levenstein 1988).

These four factors establish the basis for an argument contending that Americans, who were once intimately integrated into, and deeply knowledgeable about, the animal world, became gradually detached from it, increasingly ignorant of animal behavior, indifferent to the suffering of animals as a whole, and open to their mass production and consumption. Once these conditions were met, the more obvious technological and biological catalysts for the industry's growth and consolidation—refrigeration, breeding, synthetic fertilizers and pesticides, and confinement cages—were accepted by producers and consumers alike without much debate.

Chronological Overview

There is a feedlot in Amarillo, Texas, that holds more than 5,000 cattle. They are sent to slaughterhouses that process 2,500 animals a day. How did such places ever come to be? They both emerged in the twentieth century to capture economies of scale and satisfy the growing global demand for meat. Central to achieving this goal has been the economic justification for streamlining and intensifying the production of uniform cuts of animal flesh. The alienation between producer and consumer, as well as between producer and animal, is assumed in the preexisting literature. The question that the Texas concentrated animal feeding operations (CAFOs) and slaughterhouses invite us to ask, however, is one that historians have yet to investigate: What had to happen culturally and emotionally to the human–animal bond in order for humans to foster the development of production systems that demonstrate such disregard for the welfare of the animals being slaughtered to feed the world a meat-based diet? And whatever it was, how did it play out in real time?

1

It is easy to overlook the sincere affection humans had for animals throughout the early American period. Historians have accurately described the foundation and development of colonial America as an extended period of aggressive, often violent, ecological transformation. Rather than work within the comparatively gentle ecological patterns set by Native Americans, Europeans conquered the North American biosphere through brute-force exploitation and resource depletion. In this endeavor, they succeeded beyond anyone's expectations. It's a familiar, if somewhat sad, story. (See Cronon 1983; Merchant 1989; Merchant 1990.)

There is, however, a softer side to the muscular narrative of expansion. Explorers, naturalists, and land conservationists working and writing from 1750 to 1850 left compelling evidence that many Americans nurtured an interest in the mystery and complexity of the animal world irrespective of "progress." This widespread curiosity manifested itself in a powerful undercurrent of affection that farmers showed to domestic animals, whom they often praised for "the humility of their nature" (Anderson 2002: 384).

Many sources from the period reveal an unexpectedly warm relationship between humans and farm animals. The source of that warmth was, above all else, observation. Not unlike animal behaviorists do today, early American farmers took careful, at times obsessive, notes on what conditions upset their animals, what actions soothed them, what different expressions meant, what they preferred to eat, and even what noises relaxed or bothered them. As a result of this intense observation, they naturally came to understand their animals as distinct beings with distinct needs. Killing beasts for food and fiber was a fact of life—few people were questioning the decision to use animals to meet human needs. Americans generally used their animals without apology. But cultural and personal demands ensured that the act was never taken lightly. Affection and attentiveness toward animals were commonplace and culturally sanctioned. Those who refused these values were often publicly shamed.

Much of this early American appreciation of an animal's unique nature had to do with the place of humans and animals in the natural environment. During the first century and a half of British settlement in North America, colonists were in no position to apply strict methods of animal control to their stock. With an abundance of land and a shortage of labor, they had no choice but to allow their stock to roam landscapes with relatively little interference (except, sometimes, from Native Americans). This situation prevailed well into the 1750s, and it predisposed early settlers to think about farm animals as creatures with a sense of their own identity, wants, and needs. When they began to more consistently confine animals in the latter part of the eighteenth century, farmers found themselves prepared to act on that awareness.

The early American diet mattered as well, based as it was on a radical flexibility, acceptance of inconsistency, and the assumption that meals reflected the idiosyncrasies of individual cooks rather than standardized menus adhering to established cookbooks or accepted methods of preparation. The expectation of meat in the American diet was, in effect, an appropriate reflection of the American system of meat production, a system that provided meat when it could and, when it did, served up something slightly different every time.

2

Before 1850 Americans generally nurtured a respectful and at times affectionate relationship with domestic animals, one based on the idea that an animal had unique desires and that a farming system worked best when those desires were respected. Starting in the 1840s, however, this relationship became strained. The force behind this early wedge between humans and farm animals was unlikely but quite critical to future change—commercial fertilizer.

Commercial fertilizer conferred to farmers something they were unaccustomed to having in the world of agriculture: the signature power of precision. The control that mineral-based fertilization gradually gave progressive farmers quickly found a counterpart in the realm of animal husbandry. As "scientific farmers"—an emerging type of farmer who embraced the latest that agricultural scientific research had to offer—began contemplating soil in terms of nutrient balance rather than rest content with making their own compost, they realized that the essence of soil could be broken down, compartmentalized, and reengineered to improve crop yields. Soil, in other words, could be demystified and made more uniformly open to measured inputs. Farmers thus developed a sense that they could dictate "nature" and, in so doing, improve it in predictable ways (Cohen 2010). This control is precisely what Hector St. Jean Crevecoeur, in *Letters from an American Farmer* (1782), had in mind when, discussing the insects in his crops, wrote how the farmer "is obliged to declare war against every ancient inhabitant of this country" (Crevecoeur, 1782: 299–270). Not coincidentally, it was in the midst of this national transition to commercial fertilizer that farmers did something quite important—they began to apply the same logic of manipulation to animal feed.

This transition had subtle but powerful consequences. Prior to the emergence of fertilizer science, farmers were simply focused on whether or not animals got enough calories. The source of those calories mattered relatively little; it was most often a random combination of grass, hay, brewers' waste, and household leftovers. The move toward conceptualizing soil and feed in terms of exact nutrient content and, in turn, measuring the resulting gains in productivity had the effect of turning farm animals from active beings into passive recipients of measured inputs—kind of like they were plants.

Observation did not so much yield to measurement as it did undergo a basic change. Animals no longer ate food; they "converted" it. Farmers thus still carefully observed their animals, but now they were looking for something entirely different than an understanding of the animal's wants and needs. The farmer now searched for the ideal balance of iron, potassium, protein, carbohydrates, vitamins, and minerals. This was the first critical step toward undermining an animal's uniqueness as an animal. The mystery of an animal's nature, in other words, was partially negated when feed became a science, and now it was incumbent on farmers to seek answers to questions regarding the relationship between input and output. The fact that animal manure was being mixed with additives in order to make more nutrient-rich fertilizer only heightened the cyclical connection among fertilizer, feed, and the growing emotional detachment of the enterprising farmers spinning the wheel of agricultural progress.

3

The turn toward micromanaging feed was a small but significant step in the larger process of conceptualizing domestic animals as receptacles for finely calibrated inputs. It was a step that rendered them less animate and more objectified. Writing in *Feeding Animals: A Practical Work upon the Laws of Animal Growth*, Elliot W. Stewart stressed "the mathematical proposition that . . . every additional pound put upon a young animal costs more in food than the previous pound," in order to encourage farmers to fatten animals when they were young (Stewart 1988: 529). While this calculation process began with feed, it continued with even greater force as farmers learned more about—and specialized in—advanced animal breeding, which they did between 1860 and 1900.

Once again, the influence plant science, and its advancements, proved significant. As farmers began to systematically graft, hybridize, and apply selective pressure to commercial crops, they soon realized that, under the right conditions, they could also pursue similarly focused "improvement" with animals. Highly selective breeding had the impact of further compromising the emotional bonds that once held farmers and animals into something approximating mutual dependency (Solomon 1990; Kloppenberg 2005; Kingsbury 2009). None of this is to suggest that scientific progress in agriculture was inherently a bad thing but only to note that the application of science to the animal world—which at the time, was very rarely questioned in ethical terms—led to consequences that, although few could see it happening at the time, reshaped the human–animal bond in ways that helped establish the preconditions for factory farming.

One of the primary reasons that early American farmers treated their animals with a certain level of respect was the fact that their animals were not open books. Their behavior had to be deciphered. Entrusted with this responsibility, farmers, as we noted, sought to understand their animals as animals. Figuring out what pleased or displeased cows and horses, trying to discern what kind of noises sheep preferred to hear, and monitoring the myriad expressions of a pig were ultimately attempts to demystify the animals who served them. This process—this constant observation of what animals desired as animals—brought humans and animals together in ways that consistently reminded the farmer that, despite his ultimate dominion over his beasts, his animals had a distinct set of wants and needs that it was in his interest to nurture.

Human control of animal breeding, however, obviated inquisitive-driven observation and replaced it with a more calculated consideration, one that demanded not so much that farmers understand an animal's behavior but more so its capacity to be manipulated. Even more than animal feed, systematic animal breeding reduced animals from sentient beings to passive objects in constant need of improvement (Derry 2005; Derry 2003; Russell 1984; Carson 2005: 50–73). The discourse of animal breeding is a fascinating historical subject, one that has been largely neglected. This is unfortunate, because it reveals a subtle shift in mentality regarding the human view of the animal world.

The affection and regard for animal welfare that once characterized agrarian writing lost ground to the cold rhetoric of productive efficiency and, almost imperceptibly, the transformation of animals into machines was, by 1900, virtually complete. Writers routinely declared things like "the dairy cow is an intricate piece of machinery," "the pig is the most valuable machine on the farm," and "the pig engine needs to run at a high rate of speed" (Annual Agriculture Reports, Connecticut and Massachusetts).

In a somewhat remarkable parallel, consumer expectations of the place of meat in their diets after the 1850s began to center on consistency and relative abundance. Whereas feed producers and breeders were reducing the human–animal interaction to a series of nutrient measurements, health manuals and dietary guides began to highlight the precise nutrient assay that a diet consistently reliant on meat brought to the human body. It was not long before consumers were demanding the kind of uniformity and abundance that factory farms would eventually excel at meeting.

The powers of observation that once tied farmers so closely to their animals were gradually rendered useless. All that was left to do was ensure that the public shared the farmers' emotional distance from his animals. The proliferation of animal disease, as we'll see, helped do precisely this.

4

The management of feed and breeding habits in the quest to fulfill increasing human demand for meat was something that required farmers to command absolute control over animal mobility. The ability to move with minimal human interference, freedom from undue suffering, and the chance to raise their young are three aspects of animal life critical to their welfare. Radical confinement changed that, and it led to an equally radical change in the human–animal relationship. As already suggested, in the early American period (especially after 1750) farmers loosely confined their animals. Confinement was relatively spacious, attentive to animals' individual needs, and always mixed with some level of pasturage. It was a gentle, sometimes haphazard, manner of confinement, one that fostered the primary goal of observation rather than enhanced productivity while allowing animals to approximate some level of natural behavior (Anderson 2005).

By the 1850s, however, farmers were turning to a method known as "soiling" (Thornton 1986: 189–211). The idea here was to tightly confine animals year-round for the sole purposed of micromanaging diet and reproduction. Comfort and happiness fell from the increasingly streamlined equation of confinement. As the soiling trend gripped agriculture, the logic of industrialism dictated that all inputs—be they material or genetic—become fully maximized to achieve tangible economic results. Cows were now assigned numbers and viewed collectively as "stock." Whatever bonds of affection farmers once nurtured with their animal charges were displaced by a drive to standardize production in order to meet the demands of a changing national diet that craved uniformity as an ideal culinary quality. In this way, rationalization in production and consumption, even as they moved apart, became reinforcing phenomena, ones that helped turn "the meat industry" into a self-organizing system increasingly immune to moral or humanitarian scrutiny.

Highly concentrated confinement of the sort that characterized the soiling movement became linked to a revolution in bacteriological research. Tightly confined animals were obviously more prone to the rapid spread of disease than free-ranged animals. Highly publicized outbreaks of various diseases inspired a growing sense among producers and consumers that domestic animals were necessarily dangerous receptacles of zoonotic illness. Whether it was pleuropneumonia, trichinosis, or bovine tuberculosis, the animal infections that emerged as a result of late nineteenth-century confinement reshaped how an increasingly urbanizing population thought about domestic animals. It was at this juncture, for the first time, that common consumers began to speak about farm animals as dirty, disgusting, stupid, and possibly deadly creatures (Fisher 1988: 215–228; Majewski and Tchakerian 2007: 522–549).

Critical to the process of desensitizing Americans to the inherent worth of farm animals was the mass culling and slaughtering of infected herds. When large numbers of animals were routinely rounded up and incinerated, public estimation for their worth as sentient beings diminished immeasurably. If we assume that any form of mass subjugation requires the subjugator to deem subjects as ultimately disposable, the systematic and highly publicized and visible removal and slaughter of sick animals could be said to have helped pave the way for the massive factory farms that would dominate the twentieth-century midwestern landscape. The fact that the state played a direct role in ordering and enforcing these actions only made the mass culls seem that much more a part of a system of production than a sign of human indifference to animal life.

5

First, there's custom, then the law. The custom of moving animals into isolated regions, treating them as machinery, and maximizing their output in an increasingly confined space (such as a feedlot) laid the foundation for laws that, by the early twentieth century, aggressively mapped the legal landscape of animal production. Since the colonial period, municipal authorities had been passing laws to structure the movement and slaughter of farm animals. However, by the early twentieth century, with feedlots standardized, state authorities took the final step in securing the future of factory farming by legislating where farm animals were and were not allowed to exist. What's most notable about these laws is that they secured the place of factory farms in regions where populations were sparse, environmental laws lax, and visibility very, very low.

Legislating the geography of animal agriculture (which, not incidentally, occurred alongside racial legislation in the Jim Crow South) enabled the state as well as producers to build the final and most imposing barrier between not only farmers and animals but producers and consumers as well. In so doing, the last precondition for factory farming animals fell into place. For a century, American farmers had been—however unconsciously—emotionally distancing themselves from farm animals. This emotional distance allowed for a radical transition toward a method of producing meat that showed callous disregard for animal welfare and environmental sustainability. So callous was this disregard that, should average consumers actually have the chance to see a factory farm at work, they'd be disgusted by it, hence the significance of laws that keep the ugliness of producing meat at a graceful distance. But for the farmers themselves, the attitude that made this transition possible was evident in how one early twentieth-century dairy farmer described his beasts: "milk producing machines."

6

The transition from small-scale, pasture-oriented animal agriculture to confined animal feed operations occurred for more than just economic reasons. There were necessary emotional, cultural, and intellectual changes that had to take place in order for producers and consumers to accept a method of farming that rejected animal welfare concerns and laid the basis for a form of agriculture unprecedented in its cruelty and efficiency. It would be nice to find obvious villains in this narrative. But the matter is not so simple. The origins of factory farming have their roots in situational responses that, in and of themselves, have very little to do with the rise of factory farms. In fact, all the phases outlined earlier—attempting to refine animal feed, systematizing breeding, managing disease, moving animals into closer proximity, seeking culinary uniformity, and passing zoning laws—are, individually speaking, benign examples of Americans attempting to address isolated concerns in rational ways.

Considered collectively, however, these efforts contributed substantially to the emergence of one of the world's most inhumane industries. The idea that an intellectual/emotional transformation was part of the larger move toward factory farming—joining the profit motive—thus is to confront the messiness of historical causation. In essence, we cannot simply place the sole burden of explanation on economic motivation, capitalism, or industrialism, much less Tyson's, McDonald's, or Walmart. Small choices, aimed at dealing with localized problems, have played necessary roles in quietly preparing even the most enlightened members of society to accept what. The rise of factory farming was as much the result of a self-organizing process—one that was largely "unthinking"—as it was the conscious design of those seeking economic gain.

7

With this intellectual and emotional shift completed, and with the preconditions for the rise of farming in place, the path to a 5,000-head twentieth-century feedlot was cleared and, in a way, inevitably achieved. The signature developments underscoring this drive—the rise of synthetic fertilizer, hybrid corn, and continued breeding efficiencies from the 1920s to the 1950s—were natural extensions of developments that took place in the nineteenth century. And because of the developments underscored in this chapter, few Americans in the twentieth century were inclined to question the drive toward raising animals on factory farms. In a sense, the profit motive was released from the deeply engrained cultural and humanitarian barriers to treating farm animals like factory-made objects, so the process proceeded without interference.

The consolidation that followed was indeed impressive—especially for chickens but also for cattle and pigs—as impressive as in any other industry except perhaps cars and microchips. The main challenges that animal producers face today center less on new technological advances as on achieving the ideal balance between levels of confinement and animal health. It is worth noting that, despite the rise in attention to animal welfare among the public and legislators alike, this growing attention to the treatment of animals has been accompanied by a continued rise in factory farms. As recently as between 1997 and 2007, the number of dairy cows kept in factory farms doubled, and their average size in numbers increased by about 25%. During that same period, the number of hogs on factory farms grew by 50% and the farms with more than 2,000 hogs went from 30% to 95% (Factory Farm Nation www.factoryfarmmap.org/wp-content/uploads/2010/11/FactoryFarmNation-web.pdf).

In a way, it's hard to imagine any other course of historical events when it comes to animal agriculture. Once Americans moved off the farm, distanced themselves from the animals they had long cared for in close proximity, and outsourced a food supply they demanded to be cheap, consistent, and readily available, it would have been an anomaly of the utmost historical complexity—a denial of everything we know about human behavior and ingenuity—had anything else but the factory farm emerged as the dominant way to feed America's seemingly limitless desire for meat.

Bibliography

Adams, J. H. (1988) "The decoupling of farm and household: Differential consequences of capitalist development on Southern Illinois and third world family farms," *Comparative Studies in Society and History* 30(3), July: 453–482.

Anderson, V. (2002) "Animals into the wilderness: The development of livestock husbandry in seventeenth-century Chesapeake," *The William and Mary Quarterly* 59(2): 377–408.

Anderson, V. (2005) *Creatures of Empire*, New York, NY: Oxford University Press.

Annual Agriculture Reports, Connecticut and Massachusetts.

Carson, Laurie (Winter 2005) "Forging his own path: William Jasper Spillman and progressive era breeding and genetics," *Agricultural History* 79(1): 50–73.

Cohen, B. R. (2010) *Notes from the Ground: Science, Soil, and Society in the American Countryside*, New Haven, CT: Yale University Press.

Cronon, W. (1983) *Changes in the Land: Indians, Colonists, and the Ecology of New England*, New York, NY: Hill and Wang.

Cronon, W. (1998) *Nature's Metropolis: Brooks Blevins, Cattle in the Cotton Fields: A History of Cattle Raising in Alabama*, Tuscaloosa, AL: University of Alabama Press.

Daston, L., and Mitman, G. (2005) *Thinking with Animals: New Perspectives on Anthropomorphism*, New York, NY: Columbia University Press.

Daniel, C. E. (1981) *Bitter Harvest—A History of California Farm Workers, 1870–1941*, Ithaca, NY: Cornell University Press.

Derry, M. E. (2003) *Bred for Perfection: Shorthorn Cattle, Collies, and Arabian Horses since 1800.*, Baltimore, MD: Johns Hopkins University Press.

Derry, M. E. (2005) *Horses in Society: A Story of Animal Breeding and Marketing Culture, 1800–1920*, Buffalo, NY: University of Toronto Press.

Factory Farm Map, www.factoryfarmmap.org/wp-content/uploads/2010/11/FactoryFarmNation-web.pdf.

Fisher, J. R. (1988) "Cattle plagues past and present: The mystery of mad cow disease," *Journal of Contemporary History* 33(2), April: 215–228.

Greene, J. P. (1991) *The Intellectual Construction of America*, Chapel Hill, NC: University of North Carolina Press.

Hudson, J. C. (1994) *Making the Corn Belt: A Geographical History of Middle Western Agriculture*, Bloomington, IN: Indiana University Press.

Jefferson, T. (1955) *The Papers of Thomas Jefferson, Volumes 11 and 12, January 1 to August 6, 1787*, Princeton, NJ: Princeton University Press.

Kloppenberg, J. (2005) *First the Seed: The Political Economy of Plant Biotechnology*, 2nd ed. Madison, WI: University of Wisconsin Press.

Kingsbury, N. (2009) *Hybrid: The History and Science of Plant Breeding*, Chicago, IL: University of Chicago Press.

Levenstein, H. (1988) *Revolution at the Table: The Transformation of the American Diet*, New York, NY: Oxford University Press.

Lewis, A. J. (2005) "A democracy of facts, an empire of reason: Swallow submersion and natural history in the early American republic," *The William and Mary Quarterly* 62(4), October: 663–696.

Majewski, J., and Tchakerian, V. (2007) "The environmental origins of shifting cultivation: Climate, soils, and disease in nineteenth-century US South," *Agricultural History* 81(4), Fall: 522–549.

McWilliams, J. E. (2005) *A Revolution in Eating: How the Quest for Food Shaped America*, New York, NY: Columbia University Press.

Merchant, C. (1989) *Ecological Revolutions, Nature, Gender, and Science in New England*, Chapel Hill, NC: University of North Carolina Press.

Merchant, C. (1990) *The Death of Nature: Women, Ecology, and the Scientific Revolution*. New York, NY: HarperOne.

Olmert, M. D. (2009) *Made for each Other: The Biology of the Human-Animal* Bond, New York, NY: Da Capo.

Olmstead, A. L., and Rhode, P. (1988) "An overview of California agricultural mechanization, 1870–1930," *Agricultural History* 62(3), Summer: 86–112.

Peebles, R. L. (1988) *Landscape and Written Expression in Revolutionary America*, New York and London: Cambridge University Press.

Pillsbury, R. (1998) *No Foreign Food: The American Diet in Time and* Place, Boulder, CO: Westview Press.

Rosenbloom, R. S. (1964) "Men and machines: Some 19th-century analyses of mechanization," *Technology and Culture* 5(4), Autumn: 489–511.

Russell, N. (1984) *Like Engend'ring Like: Heredity and Animal Breeding in Early Modern England*, New York, NY: Cambridge University Press.

Solomon, D. (1990) *The Business of Breeding: Hybrid Corn in Illinois, 1890–1940*, Ithaca, NY: Cornell University Press.

Stewart, E. W. (1988) *Feeding Animals: A Practical Work upon the Laws of Animal Growth*, New York, NY: Lake View.

Thornton, T. P. (1986) "Between generations: Boston agricultural reform and the aging of new England, 1815–1830," *New England Quarterly* 59(2), June: 189–211.

Whitaker, J. (1975) *Feedlot Empire: Beef cattle Feeding in Illinois and Iowa*, Ames, IA: Iowa State University Press.

Wilson, E. O., and Kellert, S. R. (1993) *The Biophilia Hypothesis*, Washington, DC: Island Press.

Worster, D. (1994) *Nature's Economy: A History of Ecological Ideas*, 2nd ed., chapters 1–2, New York and London: Cambridge University Press.

10

THE ECONOMICS OF INTENSIVE ANIMAL AGRICULTURE

F. Bailey Norwood

If there is one phrase that best describes how market economies evolve it would be: *patterns of specialization and trade*. As long ago as ancient Greece (as revealed in the dialogues of Plato), we have known that when one specializes in the production of a specific good or service, one learns how to produce it at a lower cost and/or higher quality. As people specialize in specific goods, they then must rely on trade with others to acquire the many things they do not themselves produce. This is why human civilization has evolved from small hunter-gathering clans that produced all of their own goods to a modern society where almost everything we consume is produced by someone else. This trend toward specialization and trade is evident in every industry, including agriculture, and has important implications for animal welfare.

This chapter first describes why this specialization has led to the adoption of new technologies that confine animals in cramped spaces and impede their ability to exhibit natural behaviors. It then describes consumer reaction to these new technologies and how consumers and livestock producers interact through both markets and public policy, with a particular emphasis on the story of Proposition 2 in California.

Before modern agriculture, farmers rarely specialized in only one crop or livestock because there were synergies from diversification. Different regions produced different foods according to their climate, of course, but each farm produced a variety of foods. For example, a nineteenth-century farmer who attempted to raise only wheat would find her yields steadily declining each year, as the pests that preyed on the wheat would be able to complete their life cycle each year in the same field, returning each year to inflict greater and greater yield damage. However, if the farmer planted wheat one year and then turnips to feed cattle the next, the wheat-pest cycle is broken, allowing that field to produce lots of wheat one year and lots of beef the next. By rotating what is planted in her different fields, the farmer is more successful than if he or she produced only wheat or only cattle.

Modern fertilizers, pesticides, livestock health care technologies, lower transportation costs, and advances in nutrition science reduced the benefits from diversification. Farmers can now plant the same crop in a field each year, relying on synthetic pesticides to control weeds, insects, and other pests. The quintessential small and diversified family farm then became a corn farm or a swine farm. As they specialize, these farms rely more on trade with others, such as the purchase of pesticides from chemical companies, and as they produce more and more of the same product, they must export the product to more distant regions to find new customers.

In addition to specializing, farms today also produce more intensely, meaning they raise more plants and animals on the same amount of land. More corn seed is planted on each acre. More livestock are grazing in the pastures, and more chickens are perched in the barns. With regard to livestock, this intensification sometimes brings the animals inside barns permanently. A century ago livestock might have only spent the winter in crowded barns, as scientists did not know how to make livestock feed that satisfied all the animals' nutritional needs, so the animals needed to satisfy these other needs on their own when the weather permitted. Today, animal feed has all the nutrients the animal needs for good health, so for many hogs and chickens today, such confinement occurs all four seasons.

The barns used today are very different from their historical counterpart. Parasites were always a problem in livestock production, especially when animals were in winter confinement, where they would sleep on top of their own feces. However, once chickens and hogs were placed in cages or on slatted floors, where manure would drop beneath the floor, parasite infections dropped. The animals became healthier, reproduced faster, and produced more meat for each hour of the farmer's labor. Once it was learned that antibiotics could be added to animal feed, the health of confined livestock

Figure 10.1 Gestation crate used on intensive confinement hog farms

Source: Photo by author.

improved even more. (This does not necessarily imply that animal welfare has improved, for these health improvements came at the expense of denying animals freedom of movement and the ability to exhibit many natural behaviors.)

Of course, if a farm produces only one product, whether it be potatoes, eggs, chicken meat, or pork, it must produce a lot of it to make a living. As these specialized farms grew larger it was learned that they also exhibited economies of scale, whereby the more they produced, the lower their per unit cost of production. Better machinery was invented to make farming more efficient, but these machines were only affordable if used on a large-scale. A farm cannot afford a machine to harvest cotton on just 20 acres (an acre is roughly one football field). However, that machine becomes not only affordable but also highly profitable if used on 1,000 acres. The same can be said for the new buildings used to house livestock. A hog farmer who spends $2 million on a new hog barn incurs that $2 million cost regardless of whether they raise one or 1,000 hogs, so the farmer is financially motivated to place as many hogs as possible inside that barn, thereby minimizing their average fixed costs of production. This gave rise to the gestation crate, shown in Figure 10.1, where pregnant sows have so little room they cannot even turn around.

Whereas the diversified farm was once the more efficient farm, now the efficiency champions are the specialized farms, including intensive animal agriculture. The old-style diversified farm had to either specialize and grow larger or go out of business. Those that did specialize learned that economies of scale force them to grow continually, because if they didn't other farms would and drive them out of business.

Creative Destruction in Agriculture

To see how this plays out in a market economy, consider a stylized story. It is 1970. A small family farm in Iowa has been passed down for four generations since it was first settled, each year growing wheat, rye, buckwheat, corn, five different types of beans, potatoes, four different types of livestock feed, cattle, sheep, hogs, and chickens. As World War II ended chemical fertilizers and pesticides became increasingly effective, as did livestock feed. Although they don't wish to specialize, market forces require it. Around them they see some farmers selling their livestock and producing only crops or selling their plows and using the proceeds to erect livestock barns, perhaps to produce only eggs. These specialized farms produce at a lower cost, and as they flood the market with their abundance, agricultural prices fall. Eventually, these prices fall so low that our diversified farm can no longer turn a profit on anything it produces. Its per unit cost will only be less than the new lower prices if they specialize.

To survive, the family decides to concentrate on hog production. They build large barns that can house up to 500 sows (female hogs for breeding) and other barns to raise the sows' offspring until they reach slaughter weight. Now all their time (both the hogs and the workers) is spent inside the hog barns. The sows with babies are kept in farrowing crates, and pregnant sows are housed in gestation crates, both of which are so narrow that the sow cannot turn around. A sows' offspring have more room than their mother, but they are still kept inside in tight quarters throughout their lives. At all points in time, the animals have only a hard floor to sleep on. Although the animals are denied the ability to exhibit natural behaviors and are likely experiencing lower levels of animal welfare, they are more productive, which means that they grow faster and have higher reproductive rates.

The farm has survived, but the family might not feel like "real" farmers anymore. The livestock market has become increasingly concentrated, meaning they have only one or two buyers of their product. In the past, meat corporations would purchase live animals from auctions, at which point ownership of the animal changed hands from the farmer to the meat processor. However, the corporations want hogs with a more consistent carcass so that they can be easily handled in modern meat processing facilities, so, instead of buying the farmers' hogs, they pay the farmer to raise hogs that are owned by the corporations. The corporation sends the farmer animals, feed, and a variety of other inputs, and the farmer signs a contract stipulating how the animals will be managed. Once the hogs are ready to be slaughtered, the farmer is then paid based on how fast the animals grow. For instance, the farmer might be paid a certain amount of money for each kilogram of weight gained by the animal during its stay at the farm.

Our quintessential diversified family farm is now more like a factory, and the family members feel more like wage laborers than independent farmers. The animals likely exhibit improved health but are now confined to cramped spaces on hard floors, with little ability to express natural behaviors. The family members have lost their independence but they gain income stability. Whereas in the past they were vulnerable to the vacillations of market prices, now they are vulnerable to the greater bargaining power held by the corporation.

The process by which technologies destroy old industries and create new ones is referred to as *creative destruction*, a term coined by the economist Joseph Alois Schumpeter (1883–1950). When the car industry was created it destroyed the horse carriage industry, the internet destroyed much of the newspaper industry, and modern industrialized agriculture destroyed the small diversified farm.

This story seems to make the end result of intensive livestock agriculture unavoidable, but a curious fact accompanies these changes: there is evidence that consumers do not approve of how livestock today are raised. In surveys, Americans state that allowing animals to exhibit normal behaviors and exercise outdoors is more important than low meat prices (Prickett et al. 2010), a majority believe that keeping animals in small cages is inhumane (Lusk et al. 2007), and studies have documented that most people will pay rather large premiums for meat not raised in intensive confinement facilities (Clark et al. 2017; Norwood and Lusk 2011). If this is true, why, then, is the market producing a product of which the consumer disapproves? Answering this requires inquiry into both the supply and demand side of intensive animal agriculture.

Intensive Confinement: The Supply Side

There is little doubt that meat, dairy, and eggs can be produced at a lower cost using intensive production systems, at least in the developed world. Consider the chart in Figure 10.2, where farm size is measured by the total number of pounds gained by pigs from the time they are weaned until they are ready for slaughter. The chart shows that in 2007 it cost small farms around $45 to add a hundred pounds of weight to live pigs, whereas that cost for larger farms was less than $25. Given that large farms can produce at almost half the cost of small farms, it is no wonder that the larger farms are replacing the smaller ones.

The swine industry is just used as an example, but across all types of foods, new technologies in the last century have given the competitive edge to larger, more specialized farms, as many new technologies are not size-neutral but bestow an advantage to larger farms. Size alone does not imply lower welfare standards. A larger farm can provide a better or a worse life for the animal. The relationship between size and animal welfare depends on whether those technologies benefit or harm the animal. Do those technologies enhance animal welfare, or detract from it?

Let us remain with the swine industry as an example to explore this question. Consider the conventional intensive confinement facility that uses gestation crates. The sow stays in a temperature-controlled environment and does not have to worry about other sows bullying her or taking her food. Labor productivity on this farm will be very high, as the farmer will rarely have to move the sow (hogs can be difficult to move) and her food and water are provided using a mechanized system, allowing one person to easily care for hundreds of sows in just a few minutes. Because the crates minimize the amount of room the sow occupies the farmer is able to house many animals under one roof, thereby minimizing the average fixed costs of production.

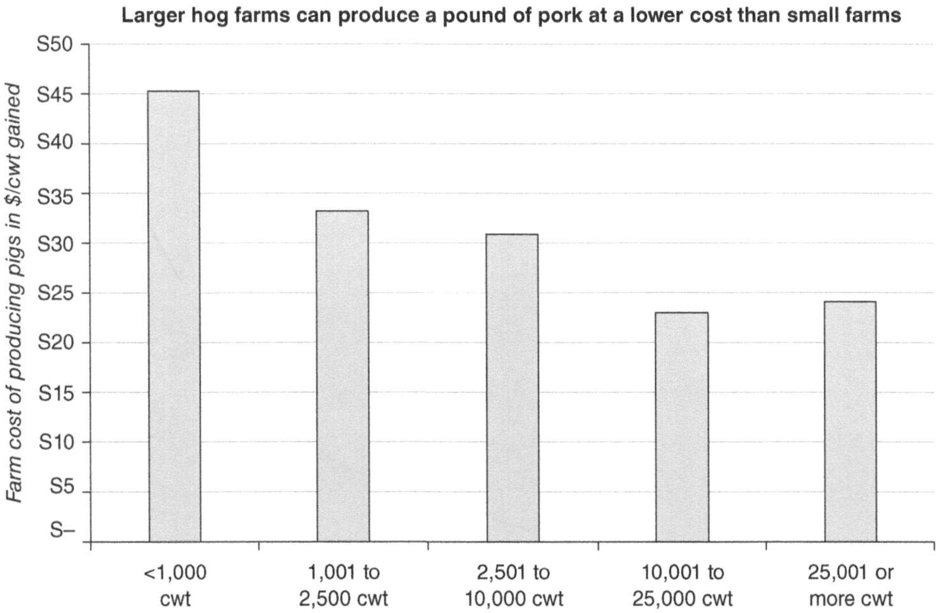

Figure 10.2 Hog production costs across farm size

Source: Key and McBride (2007).

Note: Farm size is measured by cwt (100 lb) gained by weaned pigs on farm each year.

This intensive hog farm is able to produce a lot of pork using few inputs, and as traditional farms are replaced with these intensive hog operations, the price of pork (corrected for inflation) is much lower than in the past. Although wages in the manufacturing sector are 66 times higher in 2003 compared to 1901, as shown in Figure 10.3, bacon and pork chop prices are only about 21% higher (BLS 2006). This means that each hour worked in the US allows consumers to purchase much more pork today than at the beginning of the twentieth century.

However, small farms using less confinement have also improved in productivity since 1901, by adopting more scale-neutral technologies such as scientifically formulated feed and vaccinations. Would inflation-adjusted pork prices had fallen so much if these scale-neutral technologies were adopted but not intensive confinement? That is, what if we never used the gestation crate, allowed hogs outside during nice weather, afforded them freedom of movement, let them socialize with other animals, and gave them comfortable bedding for resting? What if we also employed other technologies, such as vaccinations, that are affordable at a smaller scale? Such a production system would certainly be seen as more humane by most consumers, but would pork still be an inexpensive meat?

Consider my options as I sought to purchase a ham in Oklahoma in 2013. A ham produced in intensive confinement was available in Walmart for about $1.70 per pound. Alternatively, I could have purchased a ham from Whole Foods, which raised the animal without cages or crates—but still indoors on hard concrete floors and little ability to exhibit natural behaviors—for $4.50 per pound. Most consumers would consider this Whole Foods ham to be more humane, but not by much, yet it was 165% more expensive. Does this high premium justify such a small increase in animal welfare?

Research has shown that eliminating gestation crates on conventional farms increases farm costs by only 9%. If all these costs were passed on to the consumer, but the store did not attempt to extract larger profits from the ham, consumers would only pay 2% more for pork (Seibert and

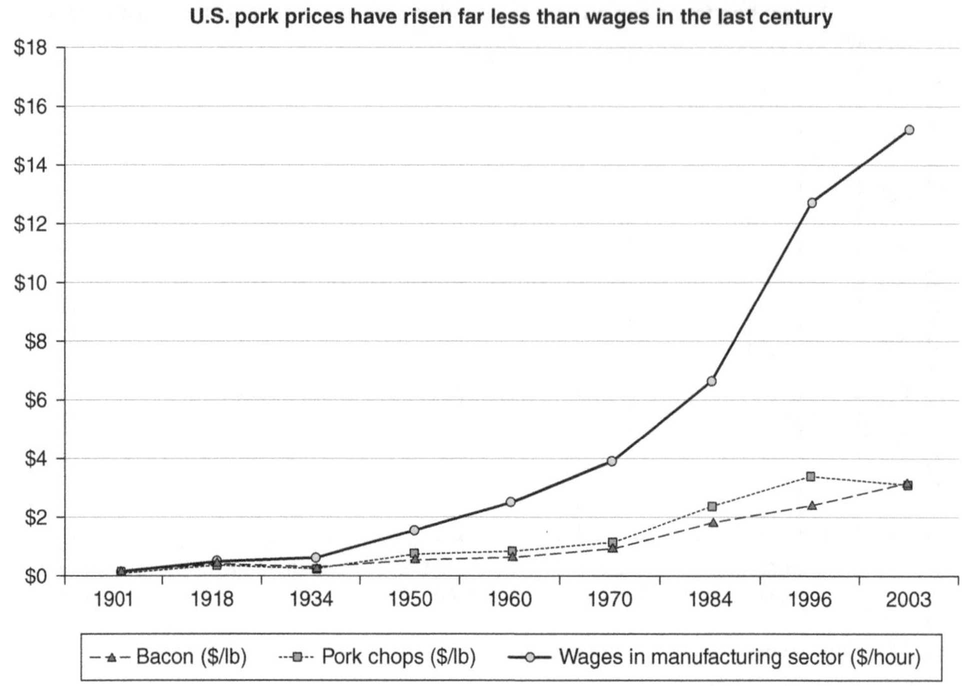

Figure 10.3 Pork prices and wages in the last century

Source: BLS (2006).

Norwood 2011). (The percentage increase at the consumer level is lower than that at the farm level because the cost increase is the same but is being divided by a larger number, as the retail costs reflect all the costs of producing pork, so it includes the cost of processing and distributing the meat in addition to the cost realized at the farm.) If the ham came from a farm with considerably higher levels of animal welfare, where the animal had comfortable bedding for resting, access to outdoors, and the ability to exhibit many of their natural behaviors, the farm costs would increase only by 18% (compared to an intensive farm), which if these costs were passed on to the consumer in the form of higher prices, the consumer would only see a 5% price increase at the grocery store (Seibert and Norwood 2011). If achieving animal welfare only increases the costs of producing retail pork by 2% to 5%, why are consumers seeing prices 165% higher?

Is Whole Foods charging higher prices for their more humane ham than is justified by changes in farm costs, in order to profit more from their consumers? Partly, but we must also consider the food distribution channel. Whole Foods uses a less efficient distribution system than larger stores such as Walmart, and it costs more money to slaughter and process hogs intended for sale at specialty stores compared to pork that will be sold to regular grocery stores. Humane food is currently expensive compared to less humane food not just because it costs more money to raise an animal humanely but also because it is currently a niche market serving affluent customers.

The reader might then ask, "Why don't major food retailers sell more humane pork, using their efficient distribution system to provide humane food more cheaply?" That is a good question, and there are scientific data suggesting the average American would pay more than what is needed to raise hogs more humanely (Norwood and Lusk 2011). It must be that these retailers do not believe these data—and they have good reason not to! To understand why we must explore the demand side of humane foods.

Intensive Confinement: The Demand Side

Estimates of the percent of Americans who are vegetarians vary, but most all find the percentage to be under 10%. For instance, in an internet survey of more than 20,000 Americans, 2% to 6% of Americans identify themselves as vegetarians, vegans, or the like (Lusk and Norwood 2016). It was thus surprising to hear that a survey by the Sentience Institute found that more than 42% of Americans support a ban on slaughterhouses, the facilities where live animals are killed and processed into meat. If the vast majority of Americans eat meat, why do so many of them also wish to ban slaughterhouses?

When I first read these results, I believed they had to be biased somehow, so I replicated the survey myself and found the same results. Maybe the respondents didn't know what a slaughterhouse really was? To check this, I asked a follow-up question to those who indicated they approved of the ban, asking them whether they were aware of what slaughterhouses were and that without them they could not eat meat. More than 70% of them indicated they did indeed know. What this means is that there are many American who eat meat but also say that they wish to ban slaughterhouses, knowing it would prohibit them from eating meat (Norwood 2018). What is going on?

This survey is but one example demonstrating that people's stated preferences for food are often very different than their actual preferences in the grocery store. What people say and what people do are often very different things. Every time voters are asked whether cramped cages in egg production should be banned, the majority vote affirmative, yet still more than 90% of eggs sales are of the cage-egg variety. This contradictory behavior spans the spectrum of many human behaviors (Norwood et al. 2018).

People's actual behavior in experiments is also inconsistent with their actual behaviors. Economic experiments are where human subjects and placed in situations involving trade-offs so that we can learn how they behave and the value they place on various goods. An example would be giving subjects money and allowing them to choose between regular meat and more humane, but also more expensive, meat. The experiments usually occur in an unnatural setting, like a classroom, and the subjects typically know they are being observed. Such experiments usually find that the average American will pay more than what is needed to compensate the farmer for higher production costs (Norwood and Lusk 2011). These high premiums suggest that the food industry could easily improve animal welfare, charge a premium for the products, and make more money than they could raising animals under lower standards of care. However, grocery stores are convinced that if they tried to sell large amounts of animal-friendly foods at high premiums most will go unsold. Why would this be?

Much of it probably has to do with something called social desirability bias, whereby people behave differently when they are watched in order to create a good impression of themselves. For example, we are much more less likely to litter in a public park where we are being watched compared to when we are alone. Social desirability bias can be a conscious choice to alter one's behavior in the presence of others or a form of self-deception (Fisher and Katz 2000).

Contradictory consumer preferences are attributable to more than just social desirability bias, though. The general theory of what drives consumer food behaviors as outlined by Furst et al. (1996) and Connors et al. (2001) explain that food preferences are driven by values, personal factors, resources, social factors, food context, sensory perception, health and nutrition of the food, convenience, cost, and managing relationships. In short, almost anything that could conceivably influence food preferences does influence food preferences.

Consider the context in which a food preference is expressed. Given that our preference for foods depends on trivial factors such as the size of the forks we use (Mishra et al. 2011) or the type of floor on which we stand (Meyers-Levy et al. 2010), then surely our beliefs about how animals should be raised might differ depending on whether we are in the voting booth or the grocery store. There is even a phenomenon called the "mere-measurement" effect, whereby merely asking people whether they are likely to purchase a product increases their likelihood of making the purchase (Morwitz and Fitzsomons 2004). You cannot ask consumers about their preferences without changing their actual preferences!

Decades ago, economists depicted consumer preferences in the form of fixed equations, where one could insert different prices for products and the equations would predict the amount of each good consumers desire. Now, however, most economists agree with psychologists that preferences are anything but fixed and will not protest when psychologist Barry Schwartz's remarks that "studies on the psychology of choice somewhat radically implies that we do not strictly possess preferences and values; instead we construct them in response to the questions the world asks us or the choices it presents" (Schwartz 2007).

People's preferences vary so widely with context that economists sometimes treat the same person as a simultaneous collection of different people. Their differences in the voting booth and grocery store are reflected by saying they act as a "citizen" self when they vote and a "consumer" self as they shop (Alphonce et al. 2014). Other times, they represent the human mind as using different processing units and refer to those different units as multiple selves (Alos-Ferrer 2014). Some amount of deception among these multiple selves exist in the form of self-deception. Self-deception exists partly because it provides the individual with benefits (Johnson and Fowler 2011), such as better performance in sports (Starek and Keating 1991), but studies have found that many consumers actively avoid learning about how farm animals are raised in fear that it will cause them to feel guilty about their eating behaviors (Bell et al. 2017). One version of the self keeps information hidden from another version of the same person.

So, yes, people who learn about the treatment of livestock in the US generally want higher animal welfare standards, but many of those same people do not want to learn about how those animals are raised in the first place. Welcome to the postmodern world of consumer economics, where a person is multiple selves and those multiple selves deceive each other!

While the supply side of intensive animal confinement is relatively straightforward, the demand side is not. The pork, egg, and broiler industry favor intensive confinement because it is cheaper and can arguably produce a better product. However, consumers both favor and dislike intensive confinement at the same time. No wonder, then, that intensive confinement of livestock is so controversial!

The economics of intensive animal agriculture then involves more than just the supply and demand for meat, dairy, and eggs. It includes the supply and demand for animal welfare regulations, and it involves the political economy of animal advocacy activism. To see the dialectic complexity between the livestock industry and the multiple-selves consumer, consider the egg industry in California.

The Legacy of Proposition 2 in California

On November 4, 2008, Californians went to the polls to vote for their presidential candidate of choice, but as usual in California (a state that makes citizen-led initiatives relatively easy), they were asked about a number of other items as well. Twelve propositions were present on the ballot, including Proposition 2, which would require that by 2015, no animal should be confined for a majority of any day such that they could not extend their limbs and turn around freely. California voters approved the proposition with a 63.5% majority.

At the time, most layers (chickens raised for eggs) were housed in a small cage containing a few hens. Most breeds of layers needed 75 square inches of space just to stand comfortably, but egg producers only provided 48 to 67 square inches. For a hen to flap her wings comfortably, she would need 303 square inches. Moreover, the cages were barren, with nothing but a wire floor on which to stand. Although the proposition did not prohibit stores from importing eggs from other states that did not meet the Proposition 2 restrictions, a subsequent law did, so the proposition basically forced anyone who wanted eggs to purchase more humane, but also more expensive, eggs. Note that the proposition did not require cage-free egg production, only more space. As a result only about a third of California egg production is of the cage-free variety, with the remaining produced using larger cages (McGreevy 2017).

It costs about $0.35 more to produce a dozen cage-free eggs than conventional cage eggs. To produce eggs using larger cages than before should cost less than this, so the $0.35 cost increase serves as an upper bound for the farm cost of complying with Proposition 2. If this $0.35 cost is added to the average price of eggs in San Francisco and Oakland of $3.12 (Chang et al. 2010), then retail prices would be 11% higher. That is, if all cage egg production was converted to cage-free production, the same distribution channels were used to deliver the eggs, all the additional costs were passed onto the consumer, and grocery stores did not attempt to charge a premium for the eggs, consumers would face price increases less than 11%. If the same thought experiment were applied to Dallas/Fort Worth, where retail prices are lower, the cost increase would not exceed 21%.

Consider the supply and demand diagram shown in Figure 10.4. This is a model economists use to understand price changes. The supply curve slopes upward, signifying that producers need a higher price in order to produce more eggs. The slope of the supply curve at any quantity equals the marginal cost, or the cost of producing another dozen eggs. Likewise, the demand curve slopes down to reflect the fact that consumers purchase more eggs when the price falls, and fewer eggs when the price rises. Its slope at any specific quantity of eggs equals the marginal value, or the value (in dollars) consumers place in acquiring one dozen. The market is in equilibrium where the supply and demand curves cross.

Before Proposition 2 the market was in equilibrium at P_{BEFORE} and Q_{BEFORE}. After Proposition 2 is passed egg producers must now incur higher costs to produce the same amount of eggs. If the marginal cost of producing each dozen of eggs increases by a fixed amount for each dozen, the supply curve shifts upward by this amount, as shown by the new supply curve below (because the cost increase is fixed, the two supply curves are parallel). The new supply curve and the original demand curve now cross at a higher price, P_{AFTER} and a lower quantity Q_{AFTER}. Consumers pay more for each dozen eggs they purchase, and they consume fewer eggs.

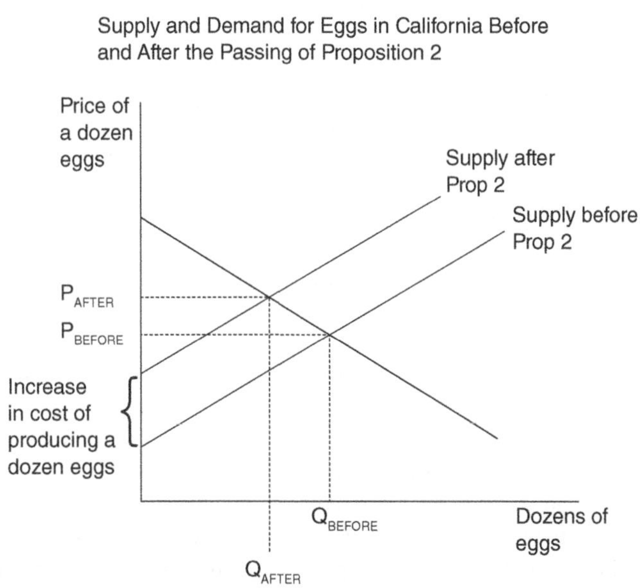

Figure 10.4 Egg price and consumption study by Mullally and Lusk (2017)

Notice that the price increase is smaller than the upward shift in the supply curve, indicating that if it costs X more to produce a dozen eggs, the price of a dozen eggs will rise by less than X. What this means is that some of the extra cost of complying with higher welfare standards is paid for by the consumers, and some is paid for by the producers (not only the egg farmers but all businesses involved in eggs, including grocery stores and food distributors, as well). The percentage of the additional cost paid for by consumers or producers depends on the slopes of the demand curves, which can only be estimated with great difficulty.

A study by Mullally and Lusk (2017) found that California egg prices did rise once Proposition 2 was implemented, and consumption of eggs did fall. Egg prices in California were about 9% to 13% higher in 2016 than they would have been had Proposition 2 never been passed. This means that in 2016 consumers are paying 9% to 13% more for the eggs they purchase due to the stricter animal welfare laws. This is consistent with the 11% maximum increase in costs of producing cage-free eggs for California.

Ideally an agricultural economist would observe the market impacts and calculate whether a policy like Proposition 2 is "good" or "bad," but to do so requires a number of questionable assumptions, such as which version of the consumer (the "citizen" or the "consumer") should be considered. Based on their behavior as consumers, there is little evidence that Californians wanted animal welfare improvements, as few purchased it in the store, in which case Proposition 2 could be said to result in an increase in prices for the same basic good. When combined with the negative financial impacts on the egg industry, this makes Proposition 2 seem disadvantageous for all.

However, given that more than 60% of Californians voted for the proposition and that numerous surveys and experiments show Americans will pay high premiums for animal welfare improvements, paying only 13% for eggs raised under more humane conditions seems like a good deal, even when accounting for the harms to the egg industry.

Given the inconsistent attitude of individuals towards animal welfare, economic theory cannot be used to determine if Proposition 2 was beneficial.

There is one aspect the citizen-consumer dichotomy that might help clear matters, however. Deciding which type of eggs to purchase in a store is different than the decision of how to vote on Proposition 2. Yes, they both concern eggs, but paying higher prices for cage-free eggs at the store only increases animal welfare for that consumers' egg consumption. Passing Proposition 2 increases animal welfare for almost all of Californian's egg consumption.

When people replace their normal eggs with cage-free eggs, not only they, no doubt, feel they are improving the lives of animals, but they also recognize that their purchases have a relatively small impact because they are only one consumer. However, if an election can force all California citizens to purchase more humane eggs, they not only benefit animals from the eggs they purchase but from all the eggs that other Californians purchase also. This presents a reasonable justification for not purchasing cage-free eggs when the decision to do so is voluntary but enthusiastically voting for a measure that forces everyone to do so. The cost of paying more for cage-free eggs is about the same regardless of whether it is voluntary or forced, but the benefits of it being forced can seem much larger because so many others are doing it also.

(This, of course, assumes one is already in the voting booth. Given the minuscule impact of any one vote, one could decide not to vote at all but still decide to purchase cage-free eggs at the store. This presents another reason why "consumers" behave differently than "citizens": they are often different people, as we all eat but we do not all vote.)

In this sense, the regulations begun by Proposition 2 might be attempting to achieve what markets could not: coordination between consumers. If consumers could get together as a group, deliberate, and dictate how animals should be raised, perhaps they would insist that layers be raised in a cage-free environment regardless of the subsequent egg price increase—even if they did not purchase cage-free eggs as a consumer. Such coordination is difficult, but if an animal advocacy group takes leadership and identifies an animal production system that would be acceptable to most and manages to place it on the ballot, then a vote serves as a convenient and effective means of coordination.

What animal welfare regulations have done is to help make animal welfare a public issue instead of a private preference. By voting "yes" on Proposition 2, even vegan Californians were able to help improve the lives of layers. Economist Jayson Lusk (2011) has gone even further, suggesting that a market-based solution to animal welfare could be created where individuals purchase and sell "units" of animal welfare improvements in the same way that people today trade carbon offsets. This would allow a vegan to purchase animal welfare credits, ensuring that the eggs someone else consumed were from chickens raised in a humane setting.

Market Segmentation for Humane Food

Readers who are aware of egg prices might find it curious that cage-free eggs only raise the cost of producing eggs by 10% to 20% and that egg prices in California rose by only 11% after implementation of Proposition 2. After all, in states where cage egg production is allowed, cage-free eggs tend to sell at a premium of about 57% above the conventional egg prices (Chang et al. 2010). Are grocery stores ripping consumers who care about animal treatment off?

When we say that cage-free egg production costs only 10% to 20% more to produce at the level of retail (not farm) prices, we are assuming that the same distribution system is used for cage and cage-free eggs. When both cage-free and cage eggs are sold, it likely costs more to distribute cage-free eggs. The cage-free sector is small, prohibiting them from realizing the economies of scale enjoyed by the cage egg sector.

Also, they are often different types of eggs. Studies have shown that about half of this 57% premium charged for cage-free eggs is due to the fact that cage-free eggs tend to be brown eggs instead of white eggs. Consumers value brown eggs more, and stores have learned that when they bundle brown eggs with a cage-free production system they can charge particularly high prices. Moreover, cage-free eggs are often targeted to more affluent consumers and come in more elaborate packaging. In economics, this is referred to as price discrimination, and grocery stores probably charge a higher premium for cage-free eggs partly because they can.

This introduces a peculiar obstacle to the market provision of animal-friendly foods. For many years, markets have delivered more humane foods to the consumers who value it the most, but it has done so in a way that makes it affordable only to a small percentage of people. Stores often bundle those animal welfare improvements with other attributes, making the good a luxury item. Cage-free eggs are typically brown when they could be white, and the brown color itself raises prices. Pork raised without gestation crates are available in niche stores and would be only slightly more expensive if only the gestation crate were eliminated, but they are typically produced without subtherapeutic antibiotics and are often organic, which leads to quite high prices.

Thus, it seems whenever animal welfare is improved voluntarily to make it more appealing to select consumers, other attributes of the product are also changed to make it a luxury product that only a small minority of affluent consumers would purchase. Perhaps California taught us that if producers and consumers in a region are forced to make animal welfare improvements, then the market finds it harder to segment consumers, and as a result, the price of more humane foods is then affordable to everyone.

Concluding Remarks

The beginning of this chapter remarked that economics studies how societies develop patterns of specialization and trade in the production of goods and services. Economics is also a field that studies how people coordinate with each other. Markets are just one mechanism of coordination. The buying and selling of food in free markets use the profit motive to ensure the right kind and amount of food is produced and distributed across the population. It allows farmers to raise the level of animal welfare on their farms, but only if consumers will compensate them for their extra expense. If animal welfare were a simple matter of farmer costs versus consumer values there would be little controversy regarding the animal welfare topic. Whatever consumers truly wanted, markets would provide.

Yet, people coordinate their activities through means other than just markets. Through activism and politics, they seek to alter the way livestock are treated differently than the market, if left alone, would provide. This chapter shows that the economics of animal welfare involves more than just one group of persons seeking control over what other people eat. In addition to its being a conflict between two different people, it is a conflict of the self. Just as a person might seriously want to exercise more at the same time he or she cannot seem to bring him- or herself to exercise, any one person might want and not want to improve animal welfare at the same time.

Moreover, activism and policy provide a type of coordination that markets cannot. Markets are marvels at aligning the interests of a producer and a consumer, but sometimes they are less adept at aligning the interests of the consumers as a group. The social movements wrought by activism, and the forced behaviors created by regulations, they both allow a person to alter their lifestyle while knowing that many others are doing the same thing. Just as a person might be more likely to exercise if they belong to a group, activism and politics not only helps disseminate information about the treatment of livestock but also helps consumers collectively take a step towards more humane food.

This isn't to say that regulations or activism is always the answer. When done poorly they can perform less well than markets. They point is that they are an alternative means of coordination to markets, and are thus part of the economic story of how the modern world is confronting the changes in how livestock today are raised.

Further Reading

1. *Animal Welfare*, edited by Michael C. Appleby and Barry O. Hughes. A rare book that is both scientifically rigorous and easy to read, covering major aspects of how you assess the welfare of animals from different perspectives and in different systems.

Appleby, M. C., and Hughes, B. O. (eds.) (2011) *Animal Welfare*, Cambridge, MA: CABI Publishing.

2. For an insightful and articulate description of how agriculture changed in the twentieth century, it's hard to beat *The Agricultural Revolution of the 20th Century* by Don and Philip Paarlberg.

Paarlbereg, D., and Paarlberg, P. (2000) *The Agricultural Revolution of the 20th Century*, Ames, IA: Iowa State University Press.

3. A well-written book describing the choices one faces when describing what to eat, *The Way We Eat* is written by perhaps the most well-known philosopher (Peter Singer) and activist (Jim Mason) of the animal rights movement.

Singer, P., and Mason, J. (2006) *The Way We Eat*, Emmaus, PA: Rodale.

4. To better understand how the rearing of livestock has changed relative to Early Modern Period, I suggest *Creatures of Empire*, where a historian describes how the lives of livestock were intertwined with early American colonies.

Anderson, V. D. (2004) *Creatures of Empire*, Oxford, UK: Oxford University Press.

5. What would happen if you raised a pig bred for industrial livestock production as a pet? The answer can be found in the delightful and true story of a couple who raise a pig named Christopher Hogwood. The book is called *The Good, Good Pig*. Montgomery, S. (2006) *The Good, Good Pig*, New York, NY: Ballantine Books.

Bibliography

Alos-Ferrer, C. (2014) "From dual processes to multiple selves: Implications for economic behavior," *Journal of Economic Psychology* 41: 1–11.

Alphonce, R., Alfnes, F., and Sharma, A. (2014) "Consumer vs. Citizen willingness to pay for restaurant food safety," *Food Policy* 49: 160–166.

Bell, E., Norwood, F. B., and Lusk, J. L. (2017) "Are consumers willfully ignorant about animal welfare?" *Animal Welfare* 26: 399–402.

(BLS) US Bureau Labor of Statistics (2006) *100 Years of US Consumer Spending*, US Department of Labor. Report 991.

Carlsson, F., Frykblom, P., and Lagerkvist, C. (2007) "Farm animal welfare—Testing for market failure," *Journal of Agricultural and Applied Economics* 39(1): 61–73.

Chang, J. B., Lusk, J. L., and Norwood, F. B. (2010) "The price of happy hens: A hedonic analysis of retail egg prices," *Journal of Agricultural and Resource Economics* 35(3), December: 406–423.

Clark, B., Stewart, G. B., Panzone, L. A., Kyriazakis, I., and Frewer, L. J. (2017) "Citizens, consumers and farm animal welfare: A meta-analysis of willingness-to-pay studies," *Food Policy* 68: 112–127.

Connors, M., Bisogni, C. A., Sobal, J., and Devine, C. M. (2001) "Managing values in personal food systems," *Appetite* 36: 189–200.

Department of Agricultural Economics (2007) http://asp.okstate.edu/baileynorwood/Survey4/files/InitialReporttoAFB.pdf.

Fisher, R. J., and Katz, J. E. (2000) "Social desirability bias and the validity of self reported values," *Psychology and Marketing* 17: 105–120.

Furst, T., Connors, M., Bisogni, C. A., Sobal, J., and Falk, L. W. (1996) "Food choice: A conceptual model of the process," *Appetite* 26: 247–266.

Heng, Y., Peterson, H. H., and Xianghong, L. (2013) "Consumer attitudes toward farm-animal welfare: The case of laying hens," *Journal of Agricultural and Resource Economics* 38(3): 418–434.

Johnson, D. P., and Fowler, J. H. (2011) "The evolution of overconfidence," *Nature* 477: 317–320.

Key, N., and McBride, W. (2007) "The changing economics of U.S. Hog production," *Economic Research Report Number 52*. Economic Research Service. United States Department of Agriculture.

Lusk, J. L. (2011) "The market for animal welfare," *Agricultural and Human Values* 28: 561–575.

Lusk, J. L., and Norwood, F. B. (2016) "Some vegetarians spend less money on food, others don't," *Ecological Economics* 130: 232–242.

Lusk, J. L., Norwood, F. B., and Prickett, R. W. (2007) "*Consumer Preferences for Farm Animal Welfare: Results of a Nationwide Telephone Survey*, Stillwater, OK: Oklahoma State University.

McGreevy, P. (2017) "Live in California and buy eggs? If voters approve this in 2018, they'll need to be from cage-free hens," *Los Angeles Times*, August 29, www.latimes.com/politics/la-pol-ca-cage-free-eggs-ballot-initiative-20170829-story.html. (Accessed May 29, 2018).

Meyers-Levy, J., Zhu, R., and Jiang, L. (2010) "Context effects from bodily sensations: Examining bodily sensations induced by flooring and the moderating role of product viewing distance," *Journal of Consumer Research* 37(1): 1–14.

Mishra, A., Mishra, H., and Masters, T. M. (2012) "The influence of bite size on quantity of food consumed: A field study," *Journal of Consumer Research* 38(5): 791–795.

Morwitz, V. G., and Fitzsimons, G. J. (2004) "There mere-measurement effect: Why does measuring intentions change actual behavior?" *Journal of Consumer Psychology* 14(1–2): 64–74.

Mullally, C., and Lusk, J. L. (2017) "The impact of farm animal housing restrictions on egg prices, consumer welfare, and production in California," *American Journal of Agricultural Economics* 199(3): 649–669.

Norwood, F. B. (2018) "FooDS: Volume 5, issue 9," January 18, Department of Agricultural Economics, Oklahoma State University.

Norwood, F. B., and Lusk, J. L. (2011) "A calibrated auction-conjoint valuation method: Valuing pork and eggs produced under differing animal welfare conditions," *Journal of Environmental Economics and Management* 62: 80–94.

Norwood, F. B., Lusk, J. L., Norwood, F. B., and Lusk, L. (2011) *Compassion, by the Pound*, New York, NY: Oxford University Press.

Norwood, F. B., Tonsor, G., and Lusk, J. L. (2018) "I will give you my vote but not my money: Preferences for public versus private action in addressing social issues," Applied Economic Perspectives and Policy. doi: 10.1093/aepp/ppy002.

Prickett, R., Norwood, F. B., and Lusk, J. L. (2010) "Consumer preferences for farm animal welfare: Results from a telephone survey of US households," *Animal Welfare* 19(13): 335–347.

Schwartz, B. (2007) "When words decide," *Scientific American Mind,* August/September: 37–42.

Seibert, L., and Norwood, F. B. (2011) "Production costs and animal welfare for four stylized hog production systems," *Journal of Applied Animal Welfare Science* 14(1): 1–17.

Starek, J. E., and Keating, C. F. (1991) "Self-deception and its relationship to success in competition," *Basic and Applied Social Psychology* 12(2): 145–155.

Sumner, D., Rosen-Molina, T., Matthews, W., Mench, J., and Richter, K. (2008) *Economic Effects of Proposed Restrictions on Egg-Laying Hen Housing in California*, University of California Agricultural Issues Center.
Sumner, D. A., Gow, H., Hayes, D., Matthews, W., Norwood, B., Rosen-Molina, J. T., and Thurman, W. (2009) *Economic and Market Issues on the Sustainability of Egg Production in the United States: Analysis of Alternative Production Systems.* Prepared for the Multidisciplinary Project of the Social Sustainability of Egg Production.

11

ANIMAL WELFARE—IS INTENSIFICATION THE PROBLEM?

Joy A. Mench

Animal welfare is one of the most contentious topics affecting contemporary animal agriculture. The decades-long intensification of animal agriculture has led to a variety of concerns about the impacts of production practices on animal welfare, resulting in legislative and/or nonlegislative efforts in many developed countries to regulate or intensify oversight of production practices. In reality, although much of the focus has been on intensive agriculture, animal welfare problems can arise in any production system, whether intensive or extensive—and conversely, both types of systems can have benefits for animal welfare. These topics are explored in this chapter, along with consideration of how the public's views of animals shape the dialogue about agricultural animal welfare and the relationship between animal welfare and the overall sustainability of animal production globally.

What Is Animal Welfare?

Determining what is meant by animal welfare is a crucial first step in discussing this topic. However, defining animal welfare in concrete terms has proved to be a frustrating and elusive goal, largely because people's perceptions of animal welfare are deeply tied to their ethical views about appropriate animal treatment (Fraser 2008). Unfortunately, there is no "consensus" view—instead, it is evident that there are many, often conflicting, opinions about how animals should be housed and managed and indeed whether or not humans should raise animals for consumption at all. This situation has often led to the individuals most involved with animal agriculture—farmers, consumers of animal products, retailers and distributors, veterinarians, and animal scientists—finding it challenging to understand one another's viewpoints and hence to develop successful strategies for addressing animal welfare issues.

In this chapter, I approach this topic from my perspective as a biological scientist rather than as an ethicist or social scientist. From that perspective, it is important to understand that the term *animal welfare* (just like *human welfare*) means something that is inherent to an individual animal (OIE 2018) rather than reflecting human attitudes or behaviors. It is therefore not identical to the human treatment of animals, since an individual's welfare can be affected by a multitude of factors. Human treatment will clearly have a major impact on farm animal welfare, however, particularly for an intensively managed animal, where most aspects of the animal's environment are controlled by humans. Second, welfare exists on a continuum from poor to good and encompasses the experiences of the animal from birth to death, although (also like humans) any animal's welfare can vary from moment to moment depending on its current situation. This makes evaluating animal welfare challenging, because while it is relatively straightforward to assess the impacts on an animal of short-term insults to welfare (for example transport or slaughter), it is much more difficult to understand long-term and lifetime effects of all the factors and practices to which an animal is exposed and how these should be weighed against one another to provide an assessment of the animal's overall quality of life.

Assessment is also complicated by the multidimensional nature of animal welfare. Broadly speaking, welfare can be considered to have three aspects (Fraser 2008): biological functioning (e.g., health, physical condition, normal physical functioning), performance of normal behavior, and affective states (positive and negative feelings such as pain, fear, and pleasure). Sometimes these three aspects intersect. For example, a dairy cow with a serious hoof injury will not only have a health problem but will be in pain and have restricted behavioral activity due to lameness. When the three aspects intersect, it is less likely that people with different ethical attitudes towards animals will disagree with one another—few people would argue that hoof injury/lameness is not an animal welfare problem (although there is no guarantee that those same people will agree about the best solution to that problem!).

More commonly these aspects do not intersect, and this is where serious disagreements arise. An example is the recent controversy about the use of conventional cages for housing egg-laying hens. Globally, most commercial shell eggs (about 90%) are produced from hens housed in conventional cages. These cages severely restrict the behavior of the hens because of their small size and lack of features such as nests and perches. Public concerns about this behavioral restriction have led to conventional cages being banned in the European Union and some US states and to a number of national and global retailers deciding to purchase only noncage eggs (Mench et al. 2011). However, egg producers (and poultry veterinarians) can justifiably claim that conventional cages are generally superior to noncage systems for protecting hen health (AVMA 2019; Mench and Rodenburg 2018)—indeed, hen health considerations were among the reasons conventional cages were adopted by the egg industry in the first place.[1] The controversy here thus arises not because of a disagreement about whether it is important to provide for the welfare of hens but whether behavior should be given priority over health to ensure welfare, or vice versa, or whether an acceptable balance can be found between the two. But more about this later in the chapter.

Farm Animal Production Practices

Animal agriculture production practices can have a wide range of potential effects on animal welfare, and these will vary depending on the type of production system and the species of animal. It is beyond the scope of this chapter to describe the many ways in which farm animals are raised—further details can be found in other sources (sheep: Morris 2017; cattle: Endres and Schwartzkopf-Genswein 2018; poultry: Karcher and Mench 2018; pigs: Pedersen 2018). In general, farm animal rearing systems can be characterized as intensive, extensive, or semi-intensive/extensive. In intensive systems, animals are primarily raised indoors, with automated feeding and watering and environmental controls (temperature, lighting, ventilation). In extensive systems, they are raised outdoors, on pasture or range under natural environmental conditions (although supplemental water/feed or man-made structures for shelter/cooling may be provided). In semi-intensive/extensive systems, animals are mainly raised indoors but given regular access to an outdoor environment (e.g., free-range).

On larger-scale commercial farms in developed countries and emerging economies, poultry raised for meat or eggs, as well as pigs, are typically raised intensively, whereas beef cattle and sheep are raised extensively, and dairy cattle semi-intensively/extensively. However, there are many variants. Beef cattle, for example, are often raised extensively on pasture until shortly before slaughter, at which point they are transported to a feedlot where they are closely confined in outdoor pens and fed a corn-based diet to be fattened for slaughter. On the other hand, there is a growing demand for free-range or pasture-based eggs and thus semi-intensive or extensive raising of laying hens. This demand is overall still relatively small, but the trajectory has been remarkable in some places—in the UK, for example, nearly 50% of hens are now free-range (IEC 2018). In part, this reflects the increasing demand (especially in some European countries) for organic products, with organic programs requiring that animals be given outdoor access.

Rise of Concerns About Animal Welfare

The rise of concerns about farm animal welfare is generally traced to the publication of Ruth Harrison's book *Animal Machines* in 1964 in the UK. In this book, which was serialized in a London newspaper and thus received widespread attention, Harrison described the rise of "industrial" farms and coined the term *factory farming* to characterize the animal production practices that had evolved. The outcry raised by her book led the government to establish a committee of inquiry, referred to as the Brambell Committee after its chair, Bangor University immunology professor F.W.R. Brambell. The committee issued its groundbreaking report in 1965 (Brambell 1967). This report reviewed the current state of animal production in the UK (and Europe), described multiple practices that were of concern to the committee, and made two statements that might now seem obvious to us but that at the time were incredibly forward-thinking—that animals have "innate behavioural urges" that have been "little, if at all, bred out in the process of domestication" and that animals have feelings (Thorpe 1969). The committee accepted that some restriction of behavior was necessary in animal production systems but recommended that, at a minimum, all animals be provided with "five freedoms"—sufficient freedom of movement to turn around, lie down, groom normally, get up, and stretch limbs.

Harrison's book and the Brambell committee report were an outgrowth of the major transformation that agriculture (both plant and animal) underwent in the 20th century. As agricultural ethicist Paul Thompson (2008) discusses, some of the more important changes that characterized agricultural intensification included increased automation and specialization, changes in the nature and quality of inputs (e.g., local versus nonlocal), and confinement of livestock in more restrictive and controlled environments. There was also a consolidation of ownership and vertical integration leading to a steady increase in farm size, along with increased reliance on contract labor. These trends are epitomized in the broiler (meat chicken) industry—in the US, 60% of broilers are produced by only five companies (Alonzo 2016). These companies are all vertically integrated and use contract farms to grow their chickens, although the chick hatcheries and slaughter plants are company-owned.

While the term *factory farming* is often used to condemn current animal production practices, its meaning is ambiguous.[2] In her book, Harrison (p. 1) characterized it as follows:

> farm animals are taken off the fields and the old lichen covered barns are replaced by gawky, industrial type buildings into which animals are put, immobilized through density of stocking and often automatically fed and watered. Mechanical cleaning reduces still further the time the stockman has to spend with them, and the sense of unity with his stock which characterizes the traditional farmer is condemned as being uneconomic and sentimental. Life in the factory farm revolves entirely around profits, and animals are assessed purely for their ability to convert food into flesh ...

I often ask my undergraduate students what they think a "factory farm" is—their answers vary but often incorporate concepts related to operation scale, animal crowding, corporate ownership/operation of farms, profit motivation, and lack of care for individual animals. But one can ask (with all due respect to the wonderful Ruth Harrison and my thoughtful students)—does it really matter to an animal if it is housed in a traditional barn or an industrial-type building or if it is automatically fed and watered instead of hand-fed and watered or if the farm on which it is raised is owned by a family or a corporation? Does it matter to an animal how many other animals are on the farm, or how much time it can spend in contact with a stockperson? What about crowding or lack of regular access to the outdoors?

These are important questions, and it is worth exploring them in light of our definition of *animal welfare* as something inherent to the animal. Many of these criticisms seem on their face to be less related to animal welfare than they are to some people's perceptions of what a farm (and a farmer) "should be"—small, family-run, operated in balance with the environment and the local community, with profit taking a backseat. So how can we tease out the factors that are genuinely important for an individual animal's welfare? This is not a simple matter, because it requires us to try to see the world from the animal's perspective rather than our own.

Assessing the Effects of "Factory Farming" on Animal Welfare

The Brambell Committee not only articulated fundamental concerns about farm animal welfare but set out a path for beginning to evaluate and address those concerns via scientific investigation. In fact, they made it clear that many of their areas of concern were speculative and would need to be proved to be problems (or not) based on subsequent investigation. They particularly highlighted the need for scientific studies of animal behavior and physiology focusing on understanding animals' feelings. In so doing, they launched the field of animal welfare science (Appleby et al. 2018; Fraser 2008). This field has grown exponentially in the last few decades. A recent database search (Walker et al. 2014) found that there were approximately 8,500 published scientific papers focusing on animal welfare. About half of these had been published in the four years preceding the search, and most were about farm animal welfare.

The trajectory of scientific research on animal welfare was strongly influenced by the five freedoms—not the five freedoms of the Brambell Committee but the expansion of those freedoms into a set of guiding principles articulated in the 1990s by the Farm Animal Welfare Council (a UK government advisory body set up in the wake of the Brambell report). These "new" five freedoms (FAWC 2012) were (1) *freedom from pain, injury, or disease* by prevention or rapid diagnosis and treatment; (2) *freedom from discomfort* by providing an appropriate environment including shelter and a comfortable resting area; (3) *freedom from fear and distress* by ensuring conditions and treatment which avoid mental suffering; (4) *freedom from hunger and thirst* by ready access to fresh water and diet to maintain health and vigor; and (5) *freedom to express normal behavior* by providing sufficient space, proper facilities, and company of the animal's own kind. These are clearly aspirational ethical statements—no animal (or human) ever completely achieves these freedoms throughout his or her life. However, they have proven to provide a very useful framework for both scientific inquiry and policy development. I will now describe some general ways in which these freedoms might be infringed by current agricultural practices—further details can be found in Appleby et al. (2018) and in the book chapters cited earlier that describe the different farm animal production systems and their associated welfare challenges.

Freedom from pain, injury, and disease: Injury and disease are obvious welfare problems because they usually also cause pain and distress and a concomitant reduction in the animal's quality of life. Injuries can occur as a result of interactions with other animals (either other animals in the herd or flock or predators), inappropriate handling by humans, or poorly designed or maintained features of the animals' environment (e.g., protruding wire). Diseases can be either infectious (e.g., bacterial, viral, parasitic) or noninfectious (genetic, environmental). This last category of disease has received particular research attention recently because of its relationship to genetic selection and management for high rates of productivity—musculoskeletal disorders are increasingly common and are a significant source of pain and morbidity in farm animals. These include bone fragility and breakage in egg-laying hens and lameness in dairy cattle and meat-type poultry.[3] Painful procedures routinely performed on farm animals, often without anesthesia or analgesia,[4] are also a source of concern—these include castration, dehorning, branding, and trimming/clipping of teeth, tails, beaks, or toes. Pain during slaughter is also an area that has received significant attention.

Freedom from discomfort: Discomfort is not a well-defined term, and there has been very little research specifically addressing this topic. The Farm Animal Welfare Council descriptor for this freedom relates to animals' experiencing discomfort as a result of environmental conditions such as excessive heat or cold in the absence of appropriate shelter or from having to stand or lie for long periods on uncomfortable surfaces (wire, concrete, mud).

Freedom from hunger and thirst: Hunger can be caused by feed that is available in insufficient quantity or that is nutritionally inadequate; similarly, thirst can be caused by an insufficient or contaminated water supply. In addition, there are situations under which farm animals' feed is deliberately restricted to control weight gain. Broiler breeders, the parent stock of broiler (meat) chickens, are an example, but sows are also feed-restricted for similar reasons. Broilers have been genetically selected to consume large quantities of food to achieve rapid weight gain, but if the breeder parents are also allowed to freely consume food, they become obese and have health and reproductive problems. Therefore, their food is restricted to a small quantity per day. This amount of food is sufficient to meet their nutritional needs but insufficient to alleviate their hunger (Karcher and Mench 2018).

Freedom from fear and distress: Like discomfort, distress is a poorly defined term but can be considered to encompass the feelings that can arise in a variety of situations that negatively affect welfare, including those related to the other four freedoms. In terms of fear, any situation that involves novelty (such as transport) can cause fear. Humans presence or contact can also be a source of fear, because farm animals are prey species and humans may be perceived as predators.

Freedom to perform normal behavior: This is, without doubt, the most difficult of the freedoms to address, and there has been much discussion among animal welfare scientists as to what constitutes "normal" behavior (Olsson et al. 2018). Not all normal behaviors are desirable (e.g., predator avoidance), and a hallmark of behavior is that it is flexible and adaptable in different environments. Thus, a great deal of animal welfare research has focused on determining which behaviors are important—either because they are behaviors that the animal is highly motivated to perform regardless of environmental conditions or because they are rewarding to the animal and thus contribute to the animal's quality of life. The freedom to perform normal social and nonsocial behaviors can be infringed because of lack of space, crowding, lack of salient environmental features, or housing in isolation or because the animal is kept in groups of abnormal size or composition (e.g., all-female, all one age). Restriction of normal behavior can also lead to the development of its reverse—abnormal behavior, considered to be an indicator that the welfare of the animal exhibiting the behavior has been compromised at some stage. Abnormal behaviors include repetitive and apparently functionless behaviors like sham chewing or tongue rolling, as well as behaviors that lead to injury of other animals (e.g., feather pecking by poultry, tail biting by pigs).

As may be evident from the lists above, animal welfare problems can occur in any type of production system. Pigs are castrated to reduce boar taint in their meat, and laying hens are often beak trimmed to prevent them injuring one another—whether they are on pasture or in confinement, on a small farm or a large farm, on a corporate farm or a family farm. All animals are killed at the end of production, with the resulting potential for fear and pain. Animal health problems can and do occur in all farming systems, although their type, rate, and severity may vary among systems.

In fact, it appears that many of the criticisms leveled against "factory farms" are only tangentially related to the welfare of the animals on those farms. As just one example, Robbins et al. (2016) recently reviewed 150 papers that evaluated at least one indicator of animal welfare as related to farm size, mainly for dairy farms. They found no clear relationship (either positive or negative) between farm size and animal welfare. Their summary of their own study and the literature they reviewed provide food for thought; some of their conclusions are the following:

- Although smaller farms often raise animals outdoors in extensive conditions, many also use the same controversial practices as their larger counterparts (e.g., painful manipulations, confinement that restricts movement for part of the animal's life).

- Larger farms tend to have fewer workers per animal, but there is little evidence that this has negative effects on the human–animal bond or on animal welfare. The quality of animal care is more dependent on factors other than farm size, such as the technical competence, attitudes toward animals, and job satisfaction levels of the caregivers, and the degree to which human care can be effectively replaced by technology.
- Larger farms are more likely than smaller farms to require the use of standard operating procedures and employ training to reduce human error and improve consistency of care. They are also more likely to consult with veterinarians and nutritionists, and more receptive to implementing science-based recommendations for best management practices to improve animal welfare.[5]

Examining our other assumptions about how "factory farming" affects welfare can lead to similarly surprising results. Take the example of crowding of broilers in intensive production systems. To many people, the combination of tens of thousands of birds housed together and given less than 1 square foot of space per bird seems an obvious welfare problem. However, it may not be as much of a problem for the birds themselves. First, studies show that when large numbers of chickens are housed together, they become socially tolerant of one another and that flock size has relatively little effect on broilers' behavior (Leone et al. 2010). Second, high stocking densities (at least within the limits usually seen in commercial production) seem to have mainly indirect effects on the physical health of broilers by impacting environmental quality (e.g., high bird density leading to excessively soiled bedding and poor air quality). Thus, good environmental control indoors (e.g., ventilation) is more important than stocking density *per se* in terms of protecting broiler health (Dawkins 2018).

A useful way forward is to recognize that all farms—small, large, corporate, family, extensive, intensive—can have practices and conditions that benefit the welfare of the animals on that farm or that pose risks. The approach that animal welfare scientists have taken is to identify these risks and understand the factors that affect them and then to develop risk mitigation strategies where possible (Appleby et al. 2018; Grandin 2015).

In general, risks to welfare in different systems are relative, not absolute. Consider just one example, mortality in different laying hen production systems. Weeks et al. (2016) analyzed data from nearly 4,000 commercial laying flocks in Europe. The mean cumulative mortality in conventional cages was 5.7%, whereas in free-range flocks, it was nearly double that (9.3%), with the difference probably mainly due to predation in the free-range flocks. Thus, one would judge free-range to be worse for the welfare of hens in terms of mortality (and the fear and pain associated with being preyed upon). However, the data also demonstrate the concept of relative risk and the importance of good risk management. The "worst" cage flock had 20% mortality, well above the average for the free-range flocks. The situation in free-range flocks was much more variable than in cage flocks—mortality in the "worst" free-range flock was an astonishing 70%, while in the "best" flocks it was only slightly higher than the average for conventional cages. This indicates that there is a higher relative risk of hen mortality associated with free-range systems but that under good conditions free-range producers can largely manage that risk. Many factors can be important in managing risk, including related to animal health interventions, selection of animals from appropriate genetic backgrounds for particular production situations, modification of the physical or social environment, introduction of technologies that improve animal welfare, and personnel training (Fraser et al. 2013).

There are some situations or systems in which risks to one or more aspects of welfare cannot be managed. Behavioral restriction of egg-laying hens in conventional cages is an example of such a risk. Simply because of the size of conventional cages, hen mobility is limited to basic postural adjustments, and there is no room for structural features that promote normal behaviors. After the EU announced in 1999 that it planned (after further review) to ban conventional cages in 2012, animal welfare scientists, veterinarians and equipment manufacturers worked together to create an alternative system, now called an enriched colony (Mench et al. 2011). This is a much larger cage that allows local movement in addition to postural adjustments and contains features determined by scientific research to be important for hen welfare: nests, perches, and a foraging area.[6] The commercial availability and feasibility of this system, which provided hens with increased behavioral opportunities but also preserved the hen health advantages associated with cages, was a deciding factor in the EU moving forward with the proposed conventional cage ban (Appleby 2003).

Research has resulted in many innovations that can improve animal welfare, although much remains to be done to ensure that these are both practical and cost-effective for farmers to implement. One important key for implementation lies outside of the direct realm of science, although it is often influenced by science. That is animal welfare policy.

Policy Response to Animal Welfare Concerns

In the last few decades, there has been steadily increasing pressure on animal agriculture in many countries to regulate production practices to assure or improve animal welfare. There are two basic types of regulation: involuntary and voluntary. Involuntary regulation involves compliance with laws, regulations, ordinances, or other legal vehicles. Involuntary regulation is the main approach that has been used in the EU (Knierem and Pajor 2018), with the EU banning certain production systems (e.g., battery cages for laying hens, gestation stalls for sows, individual stalls for veal calves), and setting animal welfare standards for rearing, transport and slaughter of pigs, cattle and chickens. These EU regulations are applicable in all member states, although any member state can adopt more stringent regulation.

Voluntary assurance involves a process of identifying and incorporating "best practices" for animal welfare into a set of standards or guidelines that farmers are strongly encouraged, but not required, to follow (CAST 2018). Input into these guidelines can come from a variety of sources, including industry (commodity) groups, farmers, government agencies, nongovernmental organizations (NGOs), and independent scientists/veterinarians. In non-EU countries such as the US, Canada, Australia, and New Zealand, voluntary regulation has been the primary pathway used to assure or improve farm animal welfare. In the US, many animal producers now follow standards developed by their respective commodity groups and conduct first- or second-party audits to ensure continued compliance. Other primary sources of voluntary regulation are the market-driven programs. These can either be animal welfare labeling programs for consumers, such as the Animal Welfare Approved, Certified Humane, Humane Heartland, or Global Animal Partnership programs in the US[7] or, alternatively, programs that are part of the "conditions of doing business" between animal producers and their customers (food retailers and distributors; CAST 2018; Mench 2008). Many of the large retailers and distributors in the US now include animal welfare as part of their corporate social responsibility programs. Requirements vary greatly from one company to another but can include adhering to private or commodity animal welfare standards on farms and/or during transport and slaughter, as well as third-party (independent) auditing to ensure compliance. In many ways, one can consider the retailer/distributor programs to be only quasi-voluntary, since farmers face a real risk of losing a major market if they do not adhere to program standards.

Voluntary and involuntary regulation each have advantages and disadvantages (CAST 2018). Voluntary regulation can stimulate relatively rapid change with minimal market disruption. It can also provide consumers with product choice and provide flexibility in how the underlying animal welfare standards are framed and implemented as science and technology evolve. However, voluntary standards may lack transparency (either in terms of the content of the standards themselves or in terms of the way they are monitored and enforced) and hence confuse consumers. They may also be distrusted if there is a perception of conflict of interest ("fox guarding the henhouse"). In contrast, involuntary regulation is transparent and more likely to be viewed by the public as having "teeth," since it is enforced by a neutral third party. Involuntary regulation can benefit producers by leveling the playing field such that no producer is at an economic disadvantage as a result of improving animal welfare. However, the process of creating involuntary regulation is often protracted, and because of that, laws tend to be relatively inflexible once they are finally passed and an enforcement mechanism established. This raises the possibility that particular involuntary standards generate only short-term or limited benefits for animal welfare.

For these reasons, it is becoming increasingly common to combine voluntary and involuntary regulation for animal welfare (Knierem and Pajor 2018). In the EU, retailer standards are assuming increasing importance—in this case, they go "above and beyond" the EU legal requirements. In addition, the first animal welfare labeling program, Freedom Foods (now called RSPCA Assured), was established in the UK in the 1990s and has served as the inspiration for labeling programs in other countries. Conversely, involuntary regulation is assuming increasing importance in countries with well-developed voluntary programs. In the US, for example, where there is very little federal regulation of farm animal welfare, there has been a recent spate of new state regulation, particularly related to housing standards for egg-laying hens. California passed a particularly sweeping piece of legislation last year regulating the housing of egg-laying hens, swine, and calves.[8] This law applies to animals produced within the state as well as animal products produced in other states and sold in California. Animal welfare NGOs in the US have been particularly effective in utilizing state ballot initiatives to pass farm animal welfare regulations and to exert pressure on retailers via shareholders' resolutions and direct consumer campaigns related to animal welfare.

The Future

In the years since the Brambell Committee report, animal welfare has evolved from being a somewhat "fringe" issue of importance mainly in Northern and Western Europe to one that has the potential to shape the future of animal production globally. The importance that animal welfare concerns will play in that future, however, will be strongly influenced by three, interrelated, factors—the rise in demand for animal products in developing countries and the associated globalization of production and trade in animal products; the importance of minimizing the environmental impact of animal agriculture; and reconciling demands for animal products with the public's evolving expectations about the treatment of animals. All these factors are encompassed within consideration of the long-term sustainability of animal agriculture.

Although the term *sustainability* is often used in an environmental context, sustainability of animal agriculture is multifaceted, involving consideration not only of impacts on the environment but on food safety and quality, food affordability, public acceptability, the health and economic security of rural communities and agricultural workers, and animal welfare. As with the different domains of animal welfare, there are situations in which these aspects of sustainability intersect but others in which they do not.

For example, "animal-friendly" products are often more expensive to produce than conventionally produced products, although exact economic impacts will vary depending on a host of regional factors (Mench and Rodenburg 2018). But to provide just one example, Grethe (2017) outlined the projected farm-level costs associated with implementing German farm animal welfare regulations. Overall, he estimated that meat and milk costs would increase by 13% to 23%. The largest projected increase, 28% to 41%, was for pork, based on providing pigs with more space, reducing confinement of sows and providing material for rooting, as well as anesthesia during castration. Are consumers willing to pay these costs? Most surveys show that, while consumers in developed countries express concerns about the welfare of intensively farmed animals, they often "vote with their pockets" by purchasing lower-cost animal products when given a choice—although as citizens they may vote against their economic interests as consumers when animal welfare measures are on the ballot, believing that farm animals deserve some legal protection (CAST 2018).

Similarly, production systems or practices that address public expectations for animals to have more space or behavioral opportunities, or that reduce the incidence of physical and behavioral problems related to genetic selection, may have greater environmental impacts than more intensive systems or practices. As Place (2018) discusses, increased per-animal productivity has been a strong driver in reducing the environmental emissions from animal agriculture in the US. For example, increased per-hen egg production from 1960 to 2010 was accompanied by a 71% decrease in greenhouse gas emissions per kilogram of eggs produced. This increased production was a result of various factors, including genetic selection and management (e.g., cage housing, nutrition, control of lighting conditions to better manipulate egg production, improved hen health via vaccination and better disease control in indoor environments) of hens to achieve high rates of egg laying, increased egg size, and increased feed efficiency (less feed consumed to produce each kilogram of eggs). In the last 50 years, per-hen egg production has increased 64% and egg mass 83%, while the feed required to produce a kilogram of eggs has decreased by more than 20%. Since feed production is a major contributor to the environmental impact of animal production (Matthews and Sumner 2015), the pressure to increase animal productivity is likely to increase, rather than decrease, because of environmental concerns. And it appears that this pressure will be most intense in developing countries.

In the face of a growing world population, it is projected that the global demand for animal products will also grow. The United Nations Food and Agriculture Organization (FAO) estimates that food production overall will need to increase by 70% by 2050 to feed the projected world population of 10 billion people (FAO 2009) and that meat and egg consumption will increase by 73% and dairy consumption by 58% (McLeod 2011) within that time frame. Little of that increase will occur in North America and Europe—instead, consumption will increase significantly in Latin America and the Caribbean and more than double in Asia and Africa. Given the different economic circumstances, environmental pressures, and attitudes toward animals in different countries and regions, this poses challenges for animal welfare policy and practice broadly. In food-insecure countries and regions, economic and social considerations, such as food affordability and food abundance, are likely to be paramount in driving the adoption of animal production practices. Environmental impact will also assume increasing importance in the wake of projected climate change effects on crop production, as will land-use considerations in the face of a growing human population. All these factors will favor an intensification of animal production similar to that seen historically in the developed countries.

Counterbalancing these factors is an increasing emphasis globally on including animal welfare considerations in farming practice. This is gradually gaining traction as a result of voluntary regulation of the kind referred to previously. The World Organisation for Animal Health (OIE), which historically has had responsibility for setting intergovernmental animal health standards, has recently put forth "General Principles" for animal welfare (Fraser et al. 2013). In many ways these principles mirror and expand upon the Five Freedoms of FAWC, in that they state the importance of selecting breeds that adapt well to different regions and climates, pain minimization, providing appropriate flooring and resting surfaces and conditions that allow for the performance of natural behavior, disease control and prevention, preventing hunger and thirst, maintaining good thermal and air quality conditions, and positive handling.

The OIE has also produced relatively detailed guidelines for animal slaughter, emergency killing, and transport, as well as raising of pigs, dairy cattle, beef cattle, and broilers (Escobar et al. 2018; and see https://oldrpawe.oie.int/index.php?id=280 for full standards). The intent is for these standards to be adopted and implemented globally. Their impact is likely to be most significant in developing countries, where voluntary or involuntary regulation of animal welfare is often weak or absent. In Latin America, for example (Gallo and Tadich 2018), the number of countries with animal welfare legislation, particularly covering animal transport and slaughter, increased in the 10 years following the publication of the initial OIE standards. An OIE Collaborating Center was also established for the region, to assist with dissemination of the standards and capacity building. Overall, the OIE's main approach globally is to assist in strengthening national veterinary services and improving veterinary training in animal welfare to drive enforcement at the local level (Escobar et al. 2018)

Retailers are also increasingly playing a role in the global adoption of animal welfare practices and policies. For example, Unilever announced that, by 2025, it would source only cage-free eggs for its products globally; as of 2017 more than 60% of Unilever's global egg purchases came from cage-free production.[9] A similar example of global reach is McDonald's, which requires an independent annual animal welfare audit of all slaughter plants that provide meat to its worldwide supply chain.[10]

In general, however, facilitating the adoption of good animal welfare practices worldwide will require significant capacity building. A report of the FAO recognized several animal welfare problem areas as being of high priority globally (Fraser et al. 2009): transportation, slaughter, provision of adequate food and water, handling/herding methods, culling and disposition of animals that are sick or of low commercial value, and the keeping of animals under conditions for which they are not genetically suited. These areas, which have already been major areas of focus for voluntary and involuntary regulation in developing countries, provide logical starting points for capacity building efforts. As the FAO report states, human poverty reduction is a significant priority for improving animal welfare globally. Changes are likely to be very slow in certain regions, particularly in situations in which human food sufficiency takes precedence over animal welfare, or where people's concerns about societal treatment of animals (not just of farm animals but also animals in general) are still nascent.

Conclusion

In the title of this chapter, I posed the question as to whether or not intensification is the root of farm animal welfare problems. The answer is—sometimes. Without doubt, behavioral restriction is much more likely to be an issue when animals are kept indoors at high stocking density—although even more extensive systems can be associated with behavioral restriction if they are not optimally configured and managed (e.g., pasture that is denuded and lacking shade/shelter). But as I hope I have shown, many of the costs and benefits to animal welfare can occur in any system, regardless of configuration, farm ownership, or farm size. Day-to-day standards of management often are greater determinants of animals' welfare than the system in which they are housed *per se*. In the US, recent voluntary regulation has resulted in marked improvements in farm animal welfare. However, voluntary programs are by their nature limited in their effects, in that they are designed to represent best management practices that can be applied by the industry as a whole (i.e., commodity group programs), appeal to niche markets (i.e., labeling programs), or cover only products produced for a specific company (i.e., retailer programs).

Although I have approached this topic as a biological scientist, it is clear that little progress will be made without considering the roles of ethics and economics in shaping the future trajectory of animal agriculture. Animal producers are surprisingly constrained in the changes that they can make without a clear mandate. The farm share of each dollar consumers spend on food is only 14.6 cents. The remaining 85.4 cents cover off-farm costs such as processing, wholesaling, distribution, marketing, and retailing (Fatka 2019). In the absence of federal regulation, the market will dictate the changes that producers make (or do not make) in response to concerns about animal welfare, at least if addressing those concerns adds significantly to production costs (as in the German example given earlier). By and large, producers are pragmatic, and if the market (either consumers or customers like retailers) wants and is willing to pay for animal products produced in a different way, then producers will invest in producing those products. This is demonstrated by the marked increase in production of cage-free eggs in the US in the last few years, from 4% in 2010 to 18% in 2018, as more and more retailers announce that they are phasing out their purchases of cage eggs (UEP 2019).

In order for the market to operate effectively, however, the level of communication and awareness needs to increase. The agricultural industries are often not transparent about animal welfare problems in commercial production (Croney and Reynnells 2008). Consumers may have little knowledge about animal production practices, which means that they are susceptible to misinformation and makes a reasonable dialogue about trade-offs impossible. In addition, there is a lack of a coherent ethical framework for this dialogue. Thompson (in press) points out that animal ethicists have mainly approached this topic by simply condemning "industrial agriculture" rather than engaging in the robust discussion that is needed about how animal welfare in such systems (which are the predominant systems for producing animal products in developed countries) could be improved. Animal welfare scientists have now accumulated a large body of evidence about farm animals' behavior, feelings, and biological functioning, as well as risks and risk mitigation, that could serve as the foundation for this discussion.

Notes

1. Although there were certainly economic drivers for moving to cage-rearing, there were also hen health drivers. Because manure in cages drops through the cage floor hens do not ingest feces as they can when they are kept on litter or pasture. This prevents them from becoming infected with the parasite coccidiosis, which causes high rates of morbidity in chicken flocks.
2. In the US, "industrialized" animal farms are often referred to, with a negative connotation, as CAFOs (confined animal feeding operations), although this term does have a "neutral" regulatory definition. The Environmental Protection Agency and US Department of Agriculture define CAFOs as farms with confinement rearing (animals are confined for more than 45 days per year in an area without vegetation) and a scale of production of more than 1,000 animal units (a metric of the number of animals times animal weight that takes account of the different sizes of different species of farm animals) per year.
3. Bone fragility in hens is due, at least in part, to calcium being released from bones to produce the amount of eggshell needed to maintain high rates of egg-laying in hens. Causes of lameness are complex, but the selection for high milk yield in dairy cows and rapid growth in broiler chickens have been major contributors to the problem.
4. There may be a number of reasons for not providing anesthesia or analgesia (CAST 2018), depending upon the species and the particular procedure performed. These include that there are no legally approved or effective anesthetics, cost, traditions, and practicality (time/handling required for administration).
5. I add from my own experience that larger (and more "corporate") farms are also more likely to have the financial capacity to adopt new technologies that benefit animals and also to have the resources needed to become part of third-party auditing programs for animal welfare, mainly because they have enough staff to deal with the record keeping, monitoring, and training requirements of those programs.
6. The fate of the enriched colony is uncertain. Although many hens in some European countries are housed in enriched cages, public sentiment seems to be driving production toward cage-free. At the end of the day, this system may turn out to be the perfect example of an excellent risk mitigation from a scientific perspective that nevertheless fails to address the public's ethical concerns about egg production.
7. Animal Welfare Approved: https://agreenerworld.org/certifications/animal-welfare-approved/; Certified Humane: https://certifiedhumane.org; Humane Heartland: www.humaneheartland.org; Global Animal Partnership: https://globalanimalpartnership.org
8. https://ballotpedia.org/California_Proposition_12,_Farm_Animal_Confinement_Initiative_(2018
9. www.unilever.com/sustainable-living/what-matters-to-you/farm-animal-welfare.html
10. https://corporate.mcdonalds.com/corpmcd/scale-for-good/our-food/animal-health-and-welfare.html

Bibliography

Alonzo, A. (2016) *Top 5 Broiler Producers Dominate US Production*, WattAgNet, www.wattagnet.com/articles/26925-top-5-broiler-producers-dominate-us-production.

Appleby, M. C. (2003) "The European union ban on conventional cages for laying hens: History and prospects," *Journal of Applied Animal Welfare Science* 6: 103–121.

Appleby, M. C., Olsson, I. A. S., and Galindo, F. (eds.) (2018) *Animal Welfare*, 3rd ed., Wallingford, Oxon: CABI.

AVMA (American Veterinary Medical Association). (2019) *AVMA Issues—A Comparison of Cage and Non-Cage Systems for Housing Laying Hens*, www.avma.org/KB/Resources/Reference/AnimalWelfare/Pages/AVMA-issues-A-Comparison-of-Cage-and-Non-Cage-Systems-for-Housing-Laying-Hens.aspx.

Brambell, F. W. R. (1967 reprinted) *Report of the Technical Committee to Enquire into the Welfare of Animals kept under Intensive Livestock Husbandry Systems*, London: Her Majesty's Stationery Office, Command Report 2836, http://edepot.wur.nl/134379.

CAST (Council for Agricultural Science and Technology). (2018) *Scientific, Ethical and Economic Aspects of Farm Animal Welfare*, Task Force Report No. 143, Ames, IA: Council for Agricultural Science and Technology.

Croney, C. C., and Reynnells, R. D. (2008) "The ethics of semantics: Do we clarify or obfuscate reality to influence perceptions of farm animal production?" *Poultry Science* 87: 387–391.

Dawkins, M. S. (2018) "Stocking density: Can we judge how much space poultry need?" in J. A. Mench (ed.) *Advances in Poultry Welfare* (pp. 227–242), London: Woodhead Press.

Endres, M. I., and Schwartzkopf-Genswein, K. (2018) "Overview of cattle production systems," in C. B. Tucker (ed.) *Advances in Cattle Welfare* (pp. 1–26), London: Woodhead Press.

Escobar, L. S., Jara, W. H., Nizam, Q. N. H., and Plavsik, B. (2018) "The perspective of the world organisation for animal health," in J. A. Mench (ed.) *Advances in Agricultural Animal Welfare* (pp. 169–182), London: Woodhead Press.

FAO (Food and Agriculture Organization of the United Nations). (2009) *How to Feed the World in 2050*, Rome: FAO.

Fatka, S. (2019) "Farmer share of food dollar declines," *Feedstuffs*, April 26, 2019, www.feedstuffs.com/news/farmer-share-food-dollar-declines.

FAWC (Farm Animal Welfare Council). (2012 archived) *Five Freedoms*, https://webarchive.nationalarchives.gov.uk/20121010012427/www.fawc.org.uk/freedoms.htm.

Fraser, D. (2008) *Understanding Animal Welfare: The Science in its Cultural Context*, Oxford: Wiley-Blackwell.

Fraser, D., Duncan, I. J. H., Edwards, S. A., Grandin, T., Gregory, N. G., Guyonnet, V., Hemsworth, P. H., Huertas, S. M., Huzzey, J. M., Mellor, D. J., Mench, J. A., Špinka, M., and Whay, H. R. (2013) "General principles for the welfare of animals in production systems: The underlying science and its application," *The Veterinary Jorunal* 198: 19–27.

Fraser, D., Kharb, R. M., McCrindle, C., Mench, J., Paranhos da Costa, M. et al. (2009) *Capacity Building to Implement Good Animal Welfare Practices*, Rome: FAO.

Gallo, C. S., and Tadich, T. G. (2018) "Perspective from Latin America," in J. A. Mench (ed.) *Advances in Agricultural Animal Welfare* (pp. 197–218), London: Woodhead Press.

Grandin, T. (ed.) (2015) *Improving Animal Welfare: A Practical Approach*, 2nd edition, Wallingford, Oxon: CABI.

Grethe, H. (2017) "The economics of farm animal welfare," *Annual Review Resource Economics*: 75–94.

Harrison, R. (1964) *Animal Machines*, London: Vincent Stuart Ltd.

IEC (International Egg Commission). (2018) *Annual Review*, London: IEC.

Karcher, D. M., and Mench, J. A. (2018) "Overview of commercial poultry production systems and their main welfare challenges," in J. A. Mench (ed.) *Advances in Poultry Welfare* (pp. 3–26), London: Woodhead Press.

Knierem, U., and Pajor, E. A. (2018) "Regulation, enforcement and initiatives," in M. C. Appleby, I. A. S. Olsson, and F. Galindo (eds.) *Animal Welfare* (pp. 349–361), Wallingford, Oxon: CABI.

Leone, E. H., Christman, M. C., Douglass, L., and Estevez, I. (2010) "Separating the impact of group size, density and enclosure size on broiler movement and space use at a decreasing perimeter to area ratio," *Behavioural Processes* 83: 16–22.

Matthews, W. A., and Sumner, D. A. (2015) "Effects of housing system on the costs of commercial egg production," *Poultry Science* 94: 552–557.

McLeod, A. (ed.) (2011) *World Livestock 2011: Livestock in Food Security*, Rome: FAO.

Mench, J. A. (2008) "Farm animal welfare in the USA: Farming practices, research, education, regulation and assurance programs," *Applied Animal Behaviour Science* 113: 298–312.

Mench, J. A., and Rodenburg, T. B. (2018) "Sustainability of laying hen housing systems," in J. A. Mench (ed.) *Advances in Poultry Welfare* (pp. 199–225), London: Woodhead Press.

Mench, J. A., Sumner, D. A., and Rosen-Molina, J. T. (2011) "Sustainability of egg production in the United States: The policy and market context," *Poultry Science* 90: 229–240.

Morris, S. T. (2017) "Overview of sheep production systems," in D. M. Ferguson, C. Lee, A. Fisher (eds.) *Advances in Sheep Welfare* (pp. 19–36), London: Woodhead Press.

OIE (2018) "Introduction to the recommendations for animal welfare," *Terrestrial Animal Health Code*, www.oie.int/fileadmin/Home/eng/Health_standards/tahc/current/chapitre_aw_introduction.pdf.

Olsson, I. A. S., Würbel, H., and Mench, J. A. (2018) "Behaviour," in M. C. Appleby, I. A. S. Olsson, F. Galindo (eds.) *Animal Welfare*, 3rd ed. (pp. 160–180), Wallingford, Oxon: CABI.

Pedersen, L. J. (2018) "Overview of commercial pig production systems and their main welfare challenges," in M. Špinka (ed.) *Advances in Pig Welfare* (pp. 3–26), London: Woodhead Press.

Place, S. (2018) "Animal welfare and environmental issues," in J. A. Mench (ed.) *Advances in Agricultural Animal Welfare* (pp. 69–90), London: Woodhead Press.

Robbins, J. A., von Keyserlingk, M. A. G., Fraser, D., and Weary, D. M. (2016) "Farm size and animal welfare," *Journal of Animal Science* 94: 5439–5455.
Thompson, P. B. (2008) *The Ethics of Intensification: Agricultural Development and Cultural Change*, Germany: Springer.
Thompson, P. B. (In press) "The vanishing ethics of husbandry," in B. Bovenkirk and J. Keulartz (eds.) *Animals in our Midst: Animal Ethics in the Anthropocene*, New York: Springer.
Thorpe, W. H. (1969) "Welfare of domestic animals," *Nature* 224: 18–20.
UEP (United Egg Producers). (2019) *Facts and Stats*, https://unitedegg.com/facts-stats/.
Walker, M., Diez-Leon, M., and Mason, G. (2014) "Animal welfare science: Recent publication trends and future research priorities," *The International Journal Comparative Psychology* 27: 80–100.
Weeks, C. A., Lambton, S. L., and Williams, A. G. (2016) "Implications for welfare, productivity and sustainability of the variation in reported levels of mortality for laying hen flocks kept in different housing systems: A meta-analysis," *PLoS One* 11: e0146394.

12

ANYMAL AGRICULTURE AND THE ENVIRONMENT

Lisa Kemmerer

Introduction

If you could erase your worst contributions to environmental degradation—without doing anything more than making different consumer choices—would you do it? To say "yes" to this question confirms a personal moral commitment to the environment—to ecosystems, the world in which we live, humanity, and all living beings. I hope you will not waver from this very sensible moral commitment as you read this chapter.

Who Are We Feeding and at What Cost?

When humans radically disrupt and change landscapes, they destroy ecosystems. When forests and prairies are plowed and planted, local ecosystems are replaced with human crops. When we plant corn and soy, lizards and opossums are killed or driven from their homes, songbirds and black bears lose their safe spaces, including nesting and hibernation sites. For the sake of ecosystems and wildlife, those who care about the environment ought to plow and plant as little land as possible—especially with monoculture crops, which are treated more heavily with pesticides, herbicides, and chemical fertilizers. Like all animals, we must eat to live, but if humans can reduce the damage that we do and yet feed more people, shouldn't we make those changes?

In the United States, 70% of the nation's grain crop is fed to farmed anymals;[1] the European Union feeds 60% of their grains to farmed anymals (Kemmerer, *Eating Earth* 8). Most of these corn and soybean crops are monocultures—planted specifically to feed large herds. Eighty percent of the world's soybeans are grown for and fed to farmed anymals (Reynolds and Nierenberg 13). Poultry are the primary consumers of "crop-based feed" ("Appetite" 10).

Feeding grains to farmed anymals is wasteful, resulting in a net nutritional and caloric loss (Kemmerer, *Eating Earth* 9). Whenever humans cycle grains through farmed anymals they lose much more than they gain: 80% to 90% of the protein, 90% to 96% of calories, and 100% of carbohydrates and fiber (Kaufman and Braun 18). This is because farmed anymals produce hooves, hair, feathers, nails, beaks, and bones and must generate heat, move around, and perpetually engage in the work of being a living organism, including the work of metabolism, breathing, and digesting. All these require calories and nutrients. Additionally, humans can only eat a certain percentage of any given corpse—we cannot eat feathers, skulls, or eyeballs. "A 1000 pound steer will produce a 600 pound **carcass**. 400

pounds are lost in hide, blood, and **inedible** organs" (Whiteheart, 2012). While some means of profitability may be found for portions of the remains, the primary purpose of raising and killing farmed anymals is so that we can eat them. Cycling grains through anymals is extremely wasteful and morally problematic in a world of increasing food stress.

Cows exploited for nursing milk consume a particularly large quantity of grain: A lactating cow eats 56 pounds (24.5 kg) of grain every day (Grant and Kononoff, 2007). How long would it take *you* to eat 56 pounds of grain? In India, a person consumes roughly 440 pounds of grain annually (Brown, "Feeding the World" n.p.): In the course of a year, one adult in India eats less grain than a lactating cow eats in eight days. Calves in feedlots consume more than 1 ton (2,000 pounds) of grain in a handful of months before they are slaughtered (Kemmerer, *Eating Earth* 9). A family of five could live on that amount of grain for a year, with plenty to spare (Kemmerer, *Eating Earth* 11). Factory-farmed fishes (fish farms) also consume grains (as well as other fishes) and like land anymals, consume much more than their little bodies yield ("Do All Farmed Fish" n.p.).

Anymal agriculture requires monumentally more land than any other human enterprise (Steinfeld et al. 133). Consider the following from the United States:

- Corn: 73 million acres (30 million hectares), 80% of which is fed to farmed anymals at home and abroad.
- Soybeans: 73 million acres, of which not quite half are fed to farmed anymals.
- Wheat: 53 million acres (22 million hectares), some 22% of which is used for anymal feed.
- Sorghum: 8 million acres (3.2 million hectares), almost all of which is fed to farmed anymals.
- Hay and alfalfa 60 million acres (24 million hectares)—all for farmed anymals ("Major").
- Grazing: 525 million acres (212 million hectares) all for farmed anymals ("Major Crops"; Kemmerer, *Eating Earth* 32).

In the United States, almost 100 million acres (40 million hectares) are exploited for crops that humans cycle inefficiently through farmed anymals. Consuming the flesh of grass-fed bovines is yet worse because grazing requires more land to yield the same outcomes. A plant-based diet "uses less than half as many hectares as grass-fed dairy and one-tenth as many hectares as grass-fed beef to deliver the same amount of protein" (Matheny, 2003).

When humans consume grains directly, 2.5 acres (1 ha) of land produce 2,200 pounds (1,000 kg) of protein, but when people cycle grains through cattle, 25 acres (10 ha) are required to produce the same "protein" ("protein," in this case, being someone else's body). Humans would need only 37% of current croplands if eating plants directly (Halley, Personal n.p.), returning hundreds of millions of acres to wildlands.

With regard to ecosystems, a

> diet rich in animal products leads to an environmentally devastating chain of events: greater land use, conversion of natural environments to farmland, displacement and death of wild animals living in these regions, and reduced wild populations in remaining wildlands (which leaves these populations more prone to extinction). Thus, the consumption of animal products leads to habitat destruction and species extinction.
>
> *(Haley 159)*

In wealthy, flesh-eating nations, the vast majority of the world's grains are fed to farmed anymals. Given that the nutrient and caloric value are largely (and sometimes completely) wasted when cycled through farmed anymals, and given the environmental effects of farming on ecosystems, sincere and informed environmentalists have a moral obligation to go vegan.

Atmosphere

Those who are disempowered always suffer disproportionately more. Where climate change hurts the most, it hurts the disempowered. The indigenous Alaskan community of Shishmaref recently voted to relocate their entire village because their homes are falling into the ocean (Groc 12). As it turns out, "the biggest climate-driven threats" for wildlife "are likely to come from human communities" relocating to escape the effects of rising seas and unpredictable weather ("Can we Help?"). When humans suffer, anymals suffer yet more—they are the most disempowered.

Climate change ought to be humanity's greatest concern. Already, "the world's least advantaged human populations are unable to escape the impacts of climate change" (Roberts 1). Given the threat of climate change and the fact that climate change is caused largely by the lifestyle choices of wealthier, privileged peoples (such as those likely to be reading this book), it is wealthier and privileged people who must respond—who must change. If you could remove your primary contribution to greenhouse gas emissions (GHGEs)—gas emissions that alter the atmosphere and trap heat against the earth's surface—simply by changing your consumer choices, would you do it?

Anymal agriculture is the leading contributor to anthropogenic (human-caused) GHGEs—carbon dioxide, methane, and nitrous oxide. When humans consume anymal products instead of consuming plants, estimates indicate that we create ten times as many fossil fuel emissions *per calorie* (Oppenlander 18). Globally, anymal agriculture contributes 51% of anthropogenic carbon dioxide (Goodland and Anhang 11), 40% of anthropogenic sources of methane (Oppenlander 6; Steinfeld et al. 82, 95, 112), and close to 100% of human-caused nitrous oxide emissions[2] (Smith and Mukhtar 3; "Emissions of Greenhouse Gases"; Kemmerer, *Eating Earth* 13, 14, 16).

Methane (CH_4) is the second-largest anthropogenic source of GHGE (after carbon dioxide; Oppenlander 6). Methane traps solar radiation 25 times more effectively than carbon dioxide (Goodland and Anhang 13). This lesser-mentioned but extremely potent GHGE holds 72 times more heat against the earth than carbon dioxide when calculated across 20 years (Steinfeld et al. 82; Oppenlander 6; "Methane").

Thawing permafrost is a worrisome source of methane. Permafrost, a frozen layer of soil covering 25% of the Northern Hemisphere, "acts like a giant freezer, keeping microbes, carbon, and soil locked in place" (Resnick n.p.). Permafrost contains "the remains of ancient life," organic matter that has been frozen for 100,000 years—pretty much all of which is made of carbon (Doucleff n.p.). As temperatures rise, permafrost melts, and organic matter rots, releasing methane (Resnick n.p.). Permafrost "contains twice as much carbon as is currently in Earth's atmosphere. . . . That's 1,600 billion metric tons" (Doucleff n.p.). As it turns out, there's more carbon in permafrost than carbon that "humans have spewed into the atmosphere since the Industrial Revolution" (Doucleff n.p.). Even more disconcerting, bacteria that had been frozen for thousands of years, once thawed, is now "converting the carbon that's in dead plants and animals into gases that cause climate change: carbon dioxide and methane" (Doucleff n.p.). As GHGEs builds up, oceans warm, releasing methane that bubbles up from the ocean floor, escaping into the atmosphere (Hickey n.p.). Releasing greenhouse gasses starts a cycle: the more that is released, the more that is trapped, causing more melting of permafrost and ocean ice, and again releasing more methane.

Neither you nor I can plug even one hole leaking methane, but we can make lifestyle changes that diminish our GHGEs. Anymal agriculture is the *greatest* anthropogenic source of GHGEs and therefore the primary reason for the thawing of permafrost and ocean ice, the primary cause of ozone damage. Recall that methane stems from decomposition. As cow, pig, and chicken manure decompose, "large amounts of methane" are released into the environment (Reay n.p.). Additionally, enteric fermentation, the digestive process of the planet's 1.4 billion cattle and 2 billion sheep and goats, creates vast quantities of methane (Robinson et al. n.p.). Ruminants (the scientific name for cud-chewing anymals) belch out almost 8 million metric tons of methane annually—80% of agricultural methane

(Steinfeld et al. 112). Due to methane's potency, this provides the GHGE equivalent of more than 16 million metric tons of carbon dioxide (Cassuto 5; Steinfeld et al. 96). A single cow exploited for dairy emits (mainly burping) between 80 and 120 kilograms of methane annually, which is equivalent in GHGE potency to carbon emissions released by an average family car in the course of an entire year ("The Milky Way" n.p.). Grass-fed flesh produces even more methane because these individuals take longer to fatten (Kemmerer, *Eating Earth* 15) and because grass is more difficult to digest, all of which creates yet more manure and belching. In the course of their lives, grass-fed bovines emit 50% to 60% more methane than grain-fed cattle (Oppenlander 125).

Those who create a demand for farmed anymals are responsible for these extraordinary methane emissions. In the last 15 years, as consumption of anymal products has increased worldwide, methane pollution has increased 145% (Steinfeld et al. 114). If you are concerned about climate change, and if you are aware of the climate change impact of anymal agriculture, you will make different consumer choices.

Water

Water is foundational for life. Would you willingly buy products that damage, deplete, and destroy Earth's essential water reserves?

Intensive anymal agriculture "pollutes more waterways than any other industry" ("How Agriculture" n.p.). The "scale and impact of agricultural pollution is enormous" ("How Agriculture" n.p.). Manure, chemical fertilizers, hormones, and other impurities are flushed from paddocks and fields, slaughterhouses and dairies, into streams and rivers, damaging both freshwater and saltwater.

Most noticeably, manure and chemical fertilizers (nitrogen and phosphorus) create thick algae blooms that block sunlight and then perish, sink, and decompose. As they decompose, they consume available oxygen, creating large anoxic areas, or "dead zones." A "dead zone" is a water body or section of water that has been stripped of oxygen and therefore cannot support life (Rutledge et al. n.p.). While this process is natural (biological), intensive anymal agriculture produces unnatural quantities of nutrient waste, creating massive, unnatural blooms and massive, unnatural "dead zones." For example, "poultry litter" is a "mixture of fecal droppings, feathers, spilled feed, antibiotic residues, heavy metals, cysts, larvae, dead birds, rodents and sawdust" that chickens "sit in for 6 weeks before they are slaughtered" (Davis, "Poultry or Beef" n.p.). According to the Chesapeake Bay Foundation, this

> poultry litter has 4 times the nitrogen and 24 times the phosphorus found in pig and dairy cow operations. Dumped on the environment, [these] mountains of toxic waste burn fragile plant cells, poison the water, and spawn excess algae that consume aquatic nutrients. The algae block sunlight needed by underwater grasses and suffocate fish in the process of decay.
>
> *(Davis, "Poultry or Beef" n.p.)*

The Gulf of Mexico, "polluted by cropland and livestock operations" along the Mississippi River ("How Agriculture" n.p.), holds the world's largest dead zone—nearly 9,000 square miles (Belva n.p.), "the size of New Jersey" (Atkin n.p.). With the growth of anymal agriculture, dead zones have increased exponentially in the last 50 years, moving from about 10 cases in 1960, to at least 169 in 2007, to more than 400 today (Rutledge et al. n.p.; Perlman n.p.).

Based on output per pound, cows exploited for dairy win the big-poop award. Every day, a lactating cow releases well over 10 gallons of waste per 1,000 pounds of cow—adult Holsteins, the primary milking breed in the United States, weigh about 1,400 pounds ("Miscellaneous Facts and Figures").[3] Other bovines produce and release just over 7 gallons of waste per day, per 1,000 pounds (and grow as large as 1,200 pounds), which is similar to pigs, who produce a couple of gallons of waste each day per 300 pounds (roughly what pigs weigh when slaughtered). Chickens exploited for their reproductive eggs, being small, create comparatively little waste (0.026 gallons per day), as do chickens raised only for flesh (0.016)—but there are roughly 20 billion chickens in the world, who collectively create some 400 million gallons of waste annually (20 billion × .02). In Maryland alone, chicken farms produce 300,384 tons of waste (Davis, "Poultry or Beef" n.p.). The massive amount of waste from anymal agriculture is difficult to disperse, so Big Ag has to be creative, pouring poop onto fields as fertilizer, making matters worse when rains inevitably come, washing this thick sludge into Earth's waterways.

While algae blooms and dead zones occasionally create a visual sensation—millions of dead fish and other sea life floating on the surface and washed ashore—humans remain largely (and dangerously) unaware of most of the effects of intensive anymal agriculture. Along with manure, anymal ag leaves our water systems polluted with "antibiotics, pesticides, herbicides, hormones, and cleanings solvents," as well as blood, hair, fat, and other bits of bodies, stemming from anymal agriculture facilities (Kemmerer, *Eating Earth* 23; Hawthorne 36, 38; Hygiene).

How could this not seriously damage water ecosystems? Scientists studying fish in 19 national wildlife refuges in the northeastern United States found an "astonishing 60 to 100 percent "of male smallmouth bass examined "had female egg cells growing in their testes" (Konkel n.p.). "[R]esearchers found the most evidence of intersex fish in areas with a lot of agriculture and wastewater effluent, and large human populations" (Konkel n.p.). The problem is spreading, and "feminized male fish have been discovered in 37 species in lakes and rivers throughout North America, Europe, and other parts of the world" (Konkel n.p.). It turns out that hormones in the water damage fish—who would have guessed?

Big Ag sometimes plants Trenbolone, an anabolic steroid used by bodybuilders, into the ears of cattle or their feed to hasten growth (Renner 196; "Study Finds Traces of Drugs" n.p.). Despite the fact that it is illegal for athletes to use this product, "millions of cattle in U.S. feedlots are legally given these compounds to promote growth. The result could be significant amounts of steroids in runoff from feed lots" (Renner 196). Apparently, that *is* the result. About 10% of steroids pass straight through cattle to persist in the natural environment, and "pharmaceuticals in waterways are damaging wildlife across the nation and around the globe," feminizing male fish, who now create "egg yolk proteins, a process usually restricted to females" and affecting "sentinel species at the foundation of the pyramid of life—such as earth worms" ("Study Finds Traces of Drugs" n.p.). Scientists focused on enhancing Big Ag profits failed to consider the environmental effects of Big Ag steroids slewing into waterways (Renner 197).

Scientists also failed to figure the effects of Big Ag pollutants on human beings. Roughly 70% of antibiotics in the United States—about 29 million pounds (13 kg)—are fed to farmed anymals every year, ending up in the water (Kemmerer, *Eating Earth* 23; Hawthorne 38). Even after running through U.S. water treatment plants, human drinking water contains trace "amounts of pharmaceuticals," including antibiotics and hormones (Doheny n.p.) that have trickled down into US aquifers that are essential because they provide "40 percent of the nation's water supply" ("Study Finds Traces of Drugs"). the problem is widespread, as "pharmaceuticals have been detected in lakes, rivers, reservoirs and streams throughout the world" ("Study Finds Traces of Drugs").

Intensive anymal agriculture also depletes freshwater reserves. Worldwide, anymal agriculture is responsible for 90% of freshwater depletion (Steinfeld et al. 126). The Colorado, Nile, and Yellow Rivers now run dry before they reach the sea, and freshwater fossil (they do not refill) aquifers have been stamped with frighteningly short expiration dates (Dimick n.p.; Brown, *Plan B 3.0*). Freshwater looks to be the next international crisis: "By 2020, up to 250 million people may experience water shortages" (Koneswaran n.p.). Have you ever been *really* thirsty?

Roughly 80% of Earth's freshwater is pumped into agriculture. The vast majority of grain crops are fed to farmed anymals—irrigation for feed crops is the single greatest drain on the planet's freshwater reserves. Wealthy nations waste "1000 tons of water to produce one ton of grain," which they then squander by feeding it to farmed anymals (Brown, *Plan B 3.0*). The Mississippi River basin has one of the world's largest water footprints for crops (Mekonnen 1580). A society that chooses to consume plants directly, rather than cycle plants through anymals, would need to cultivate only 37% of lands now necessary for our anymal based diet (Halley n.p.). A plant-based diet requires only about 60% of crops now raised—crops that disrupt habitat, consume and pollute freshwater, add chemical fertilizers to the land, and burn up petroleum via plowing, planting, and transport.

Farmed anymals—all 24.4 billion of them—also consume freshwater. Thanks to eating habits in wealthier nations, the earth currently supports roughly:

- 20 billion chickens (Robinson et al. n.p.),
- 2 billion sheep and goats,
- 1 billion pigs, and
- 1.4 billion cattle.

How many gallons of water, roughly, does one individual in each category consume in one day and in one year?

Table 12.1 Average water consumed (in gal) per farmed animal

Exploited anymal	*Consumed in one day*	*In one year*
Chicken exploited for eggs	0.13	47
Chicken exploited for flesh	0.09	33
Sheep or goat	1	365
Pig	5	1,825
Bovine, 1,000 pounds, not lactating	9–20 (depending on temperature)	3,240–7,200
Bovine, lactating	20–40 (depending on temperature and gallons of milk produced)	7,200–14,600

Source: Meehan et al. (2015) "Water Intake"

How many gallons of water per year are consumed by all farmed anymals?

Table 12.2 Total water consumed (in gal) by farmed animals globally

Species	*Average Gallons of Water per Year*
All chickens (at 40 gallons per year per hen)	800 billion[4]
Sheep and goats	730 billion
Pigs	1.825 trillion
Cattle (at 7,200 gallons per year per bovine)	10.08 trillion

This accumulates to a whopping 13.4 trillion gallons of freshwater down the gullets of farmed anymals every year. To offer perspective, consider these statistics:

- One trillion people is about 10 times more humans than have ever been born ("Putting Trillions of Dollars in Perspective").
- One trillion dollars piled up, one on top of the other, would reach 63 miles up into the sky.
- One trillion seconds ago, it was 30,000 B.C. ("How much is a Trillion Dollars?").

How much water do cattle drink before slaughter? Bovines are normally slaughtered for their flesh at 12 to 15 months of age. Selecting a middle-of-the-road figure from the preceding estimates of drinking water consumption, water for 10 months (assuming less water when suckling—although their mothers drink for them in order to produce milk) leads to a figure of at least 4000 gallons per bovine. One pound of beef requires an average of about 8 gallons of drinking water. Similarly, 1 pound of pig flesh requires about 6.5 gallons of water,[5] and 1 pound of chicken flesh requires about 2 gallons of water.[6]

These figures[7] reveal heavy water use, yet these figures overlook water required to produce feed crops for these anymals, and *feed crops are, by far, the largest use of water for farmed anymals*: It takes "1000 tons of water to produce one ton of grain" (Brown, *Plan B 3.0*). While it takes about 130 gallons of water to produce one pound of corn, it takes about 500 gallons of water to produce a pound of chicken flesh, 700 gallons of water to produce a pound of pig flesh, and 2,000 gallons of water to produce a pound of bovine flesh ("The Water Content"). It requires 53 gallons of water to produce one egg; the water expended to produce eggs in San Diego County each year would provide "72 years of drinking water for all county residents" (Davis, "Water Used" n.p.). In comparison, a pound of almonds requires only 376 pounds of water[8]—far less than *any* anymal product (Graham n.p.). Have you ever seen anyone eat a pound of almonds in one sitting? What about a pound of flesh? Also worth noting, soybeans have a very low water footprint compared with grains (Mekonnen 1592).

Cattle exploited for their nursing milk consume more water than any other farmed anymal—20 to 40 gallons per day, depending on how hot and dry the weather is and how much nursing milk they are producing. A lactating cow not only drinks 12,000 gallons of water but also chews and swallows 7,500 gallons of water in grains consumed (56 lb per day).[9] That equates to about 2.7 million gallons of water every year for just one lactating cow—and this figure only includes freshwater consumption for feed and drink, not water expended to flush out dairies or slaughterhouses, for example. Moreover, cows exploited to produce dairy products such as yogurt, cheese, and icecream, consume 1,080 more gallons of water than they yield in the form of milk.

Our dependence on anymal agriculture causes "severe environmental degradation" in the form of water pollution and water depletion (Steinfeld et al. 134). Worldwide, anymal agriculture depletes and pollutes Earth's water systems. May I float the idea of a vegan diet?

Deforestation

Billions of generations of South America's unique species evolved in splendid isolation. Some species have ancient connections with anymals in the South Pacific and Africa. South America's "insects, spiders, crabs, centipedes, and millipedes are found nowhere else in the world," including thousands of species "yet to be classified" ("Animal Life" n.p.). South America is home to about 2,700 fish species (including the infamous piranha), 3,000 bird species (including the colorful toucan), a host of unusual amphibians (including a limbless lizard), and a remarkable range of mammals—marsupials, sloths, monkeys, tapirs, and llamas ("Animal Life" n.p.). These splendid beings live in a wide range of dwindling habitat. One-fifth of the world's rainforests were destroyed between 1960 and 1990, and much rainforest lands are now in a "highly fragmented and degraded state" (Reynolds and Nierenberg 11). Pretty much every minute of every day, somewhere in the world, 26 acres of rainforest are destroyed (that's about 20 football fields or 22 European football fields;[10] "How Big is 26 Acres?"). Concurrently, Earth is "losing species at 1,000 to 10,000 times" what might be expected on a vegan planet, "with literally dozens going extinct every day," projecting that "30 to 50%" of species will have forever vanished by mid-century ("The Extinction Crisis").

What does a tree standing in a rainforest in South America have to do with sitting down to dinner in Los Angeles or New York? How do *your* dietary choices affect an endangered black-faced lion tamarin? Most importantly, will you change your diet if doing so protects rainforests and the many beings who are utterly dependent on the rainforests?

The US imports about 80 million pounds of Brazilian beef every year (and 85% of EU beef comes from Brazil). Brazil was home to about 10 million bovines in 1980, but now houses upward of 55 million cattle ("Global Warming Science"):

> Cattle ranching is the leading cause of deforestation in the Amazon rainforest. In Brazil, this has been the case since at least the 1970s: government figures attributed 38 percent of deforestation from 1966–1975 to large-scale cattle ranching. Today the figure in Brazil is closer to 70 percent.
>
> *(Butler n.p.)*

The primary reason for the loss of forests globally is the conversion of lands to agriculture—both for grazing *and for feed crops*. Crops *and* grazing are responsible for about 98% of Brazil's deforestation (Butler n.p.). Most forests have been destroyed to grow soybeans and maize as feed crops (Steinfeld et al. 12). In less than a decade, soybean prices have increased by 70%, providing ample incentive to continue replacing forests with soybean crops (Richards and Hoelle n.p.). Not surprisingly, forest destruction is on the rise in Brazil; "the largest increases in forest loss were recorded in Brazil's leading soybean-producing state, Mato Grosso" (Richards and Hoelle n.p.). This is not the fault of tofu eaters: "*Worldwide, 80 percent of soybean crops are planted, tended, and harvested for farmed animals, implicating those who eat cheese and chicken—not those who eat tofu and tempeh*" (Kemmerer, *Eating Earth* 29; emphasis in original).

For purely selfish reasons, most of us do not wish to see rainforests disappear. Trees hold water and generate rain for drought-plagued landscapes. Trees stabilize weather, maintain soils, and provide wild and beautiful spaces. Rainforests are home to 70% of Earth's land anymals and plants, and "rainforest tree canopies are powerful converters, turning carbon dioxide into oxygen, helping to mitigate some of the effects of climate change" (Kemmerer, "Eating Ecosystems" 189). Rainforests harbor a dense assortment of plants likely "rich with medical and nutritional possibilities," as yet undiscovered by chemists or cooks (Kemmerer, "Eating Ecosystems" 189).

The black-faced lion tamarin is one of many wonderous beings depenedent on South American forests. But I write of just one black-faced lion tamarin, an individual whose photo I found online. This tamarin, whose hair is almost orange (except for a long and thick mane that is rather like a lion's, except pitch black), clings to the bark of strong palo borracho branches with long and delicate hands (Kharvillart n.p.). I assume that she eats and sleeps raises young, and squabbles with friends and relatives, as we do. But unlike you or me, she belongs to a critically endangered species: there are only about 300 black-faced lion tamarins left in the world (Kharvillart n.p.). Does *she* matter to *you*? Do *you* care if she has a place on this planet?

Given that most of Earth's species make their homes in rainforests, and given that the dietary choices of privileged and powerful humans (like you and me) are the leading cause of deforestation, *we* decide *their* fate when we shop. Meat-eating, dairy-consuming, egg-loving humans are chewing black-faced lion tamarins, marsupials, sloths, monkeys, tapirs, llamas, and the world's sleek jaguars out of existence. They need forests to survive. What will it be—a planet with bright-eyed black-faced lion tamarins with shining orange fur clambering through a thick canopy of rainforest, or cheesy chicken's wings?

Soil Degradation

What do overgrazing, agriculture, and deforestation have in common? At this point, you might be inclined to guess—dietary choices that include nursing milk from cows, reproductive eggs from hens, and the flesh of farmed anymals. And you would be right.

Overgrazing (35%), agriculture (28%), and deforestation (30%) are the leading causes (totaling to 93%) of soil degradation ("Land Degradation" n.p.), which leads to desertification. The dietary choices of the privileged—choosing to be omnivores rather than vegans—causes desertification ("Public Lands Ranching"; "Losing Ground"; Kemmerer, *Eating Earth* 31). Giant herds of farmed anymals compact soil—a problem that is not easy to remedy (Clay 2016). Dry areas suffer the most—73% of dry rangeland worldwide is already degraded (Steinfeld et al. 272). When farmed anymals destroy Earth's fragile surface, the result is desertification: plants are destroyed, soil nutrients are depleted, the area turns into a wasteland, and humans pack up and move on (Hawthorne 2012). Because of this process, "every year a new chunk of real estate the size of Rhode Island is being swallowed by sand" (Hawthorne 2012), and that sand has a tendency to travel on the wind.

> Overplowing and overgrazing are converging to create a dust bowl of historic dimensions. With little vegetation remaining in parts of northern and western China, the strong winds . . . remove literally millions of tons of topsoil in a single day, soil that can take centuries to replace.
>
> *(Brown and Tribune n.p.)*

Unfortunately, it is impossible for entire ecosystems to "move on" along with humans, who leave a trail of suffering and death in their wake.

Traditional diets in Asia include little meat and no dairy. Today, dustbowls and dust storms stem from China's quest to satisfy an ever-growing demand for flesh and dairy, inspired by trend-setting nations such as the United States. Gigantic dust storms are rising along with the consumption of anymal products. Deforestation causes desertification. Not just in Asia, but worldwide.

Conclusion

> Because of the enormous mass of humanity currently on the planet, even small changes in lifestyles, such as consuming fewer animal products, can have an enormous environmental impact.
>
> *(Haley 159)*

At the outset, I asked a straightforward question: "If you could erase your worst contributions to environmental degradation without doing anything more than making different consumer choices, would you do it?" Now you *know* how to reduce your personal contributions to the world's most pressing environmental problems by simply changing your consumer choices. You do not need to go hungry even for a moment; you only need to *replace* anymal products (flesh, dairy, and eggs) with a variety of other food options.

The simple decision to change your diet not only mitigates your personal contribution to climate change, degraded ecosystems, freshwater depletion, pollution and dead zones, deforestation, and desertification, but also shares grains with those who do not have a choicein what they eat—the world's poor and hungry. Additionally, changing what you put in your mouth casts a vote against exploitation and premature death for farmed anymals, who do not choose to be exploited or slaughtered to satisfy your taste preferences. Moreover, shifting to a vegan diet is also likely to steer you around the biggest health problems facing humans in wealthy nations (heart attack, stroke, cancers, and obesity). And, if you choose wisely, a vegan diet is also less expensive. In most communities, cooking up a bag of dried beans with a bag of frozen spinach, for example, is *much* cheaper than buying anymal products of any kind and healthier (Kemmerer, *Eating Earth* 122).

The word *choice* is critical. People who have no choice about what they eat—those who only have access to anymal products—are not morally culpable if they consume anymal products, despite causing much misery and destruction. But those who are reading what I have written— *you, that is—are almost surely among those who can choose vegan*. You might prefer to make excuses rather than change your diet, but you are among the informed.

Here are the questions that you now face: If you have the option to make food choices that show solidarity with those who are hungry, will you do so? If you have the option to make food choices that protect vulnerable farmed anymals from exploitation and slaughter, while also protecting your health and reducing food costs, will you do so? If you have the option to make food choices that diminish your environmental footprint in a host of critical ways, will you do so?

Having read what is written here, if you choose *not* to change your diet, it is reasonable to conclude that you care more about your current dietary habits than you do about the world's hungry, farmed anymals and wildlife, plants and water systems, soils and forests and ecosystems—you care more about maintaining your current dietary preferences than you do about the Earth itself. Do you think that is rational?

Notes

1. **Anymal** is a contraction of *any* and *animal*, pronounced as *any* and *mal*. *Anymal* indicates all individuals who are of a species other than that of the speaker/author. In other words, if a human being uses this term, all species except *Homo sapiens* are indicated, but if a chimpanzee signs *anymal*, it references all species (including human beings) except chimpanzees. Using the term *anymal* avoids the use of *animal* as if human beings were not animals; dualistic and alienating references such as *non* and *other*; and cumbersome terms such as nonhuman animals and other-than-human animals. See Kemmerer, Lisa. "Verbal Activism: 'Anymals.'" *Society and Animals* 14.1 (May 2006): 9–14. <http://lisakemmerer.com/Articles/anymal%20article%20Jan%20 2016.pdf>
2. Some authors offer lower GHGE figures for anymal agriculture by leaving out key contributing factors, from piled manure to producing feed crops, from enteric fermentation to transportation of feedstocks, anymals, and anymal body parts. Moreover, those who dispute Goodland and Anhang's numbers are funded by or affiliated with the livestock industry, which makes them a less-than-dependable source of information on this topic.
3. Twelve gallons (45 l) of waste per 1,000-pound (454 kg) bovine, per day.
4. Note how chickens, although small beings, are numerous and therefore have a disconcerting environmental impact.
5. Figuring 250 pounds, about 140 pounds of flesh (57% eatable), 6 months of life drinking about 900 gallons of water.
6. Figuring a 4-pound bird, not quite 3 eatable pounds, 6 weeks of life drinking 6 gallons of water.
7. There are always additional factors that might be considered. For example, anymal byproducts are used in non-food-related products, and the (already-available) alternatives will likely have their own water footprints. Nonetheless, whatever the additional factors might be, they will not lead to water-friendly fried chicken or flank steak.
8. Almonds require only about 1 gallon of water per almond, and 376 almonds weigh about 1 pound.
9. Assuming the estimated 150 gallons of water to produce 1 pound of corn.
10. Or the size of New York's Ellis Island, a little bigger than Alcatraz Island, twice the size of Windsor Castle, and considerably larger than Buckingham Palace.

Bibliography

Akar, T., Avci, M., and Dusunceli, F. (2004) "Barley: Post-harvest operations," *Food and Agriculture Organization of the United Nations: The Central Research Institute for Field Crops*. Last Reviewed June 15, www.fao.org/3/a-au997e.pdf (Accessed December 17, 2017).

"Animal life," *Encyclopedia Britannica: South America: Continent,* www.britannica.com/place/South-America/Animal-life (Accessed January 9, 2018).

"Appetite for destruction: Summary report," *World Wildlife Fund,* www.wwf.org.uk/sites/default/files/2017-10/WWF_AppetiteForDestruction_Summary_Report_SignOff.pdf (Accessed December 12, 2017).

Atkin, E. (2017) "The most overlooked environmental crisis of 2017: Apocalyptic weather dominated the headlines, but meat production also caused severe destruction in 2017," *Republic,* December 13, https://newrepublic.com/article/146195/overlooked-environmental-crisis-2017 (Accessed December 17, 2017).

Baloch, U. K. (1999) "WHEAT: Post-harvest operations," *Food and Agriculture Organization of the United Nations: Pakistan Agricultural Research Council (PARC)*. Last Reviewed October 14, www.fao.org/3/a-ax448e.pdf (Accessed December 17, 2017).

Belva, K. (2017) "Gulf of Mexico 'Dead Zone' is the largest ever measured: June outlook foretold New Jersey-sized area of low oxygen," *National Oceanic and Atmospheric Administration: U.S. Department of Commerce*, August 2, www.noaa.gov/media-release/gulf-of-mexico-dead-zone-is-largest-ever-measured (Accessed December 13, 2017).

Brown, L. R. (2008) *Plan B 3.0: Mobilizing to Save Civilization*. New York, NY: W. W. Norton.

Brown, L. R. (2010) "Feeding the world means hogging less grain," November 10, http://grist.org/article/2010-11-09-improving-food-security-by-strategically-reducing-grain-demand/ (Accessed January 13, 2018).

Brown, L., and the International Herald Tribune. (2003) "The world's biggest dust bowl: China is losing the war on advancing deserts," *New York Times: Opinion*, August 13, www.nytimes.com/2003/08/13/opinion/the-worlds-biggest-dust-bowl-china-is-losing-the-war-on-advancing.html (Accessed December 13, 2017).

Butler, R. (2012) "Soil erosion and its effects," *Mongabay: Rainforests: Disappearing Opportunities*. Updated July 22, https://rainforests.mongabay.com/0903.htm (Accessed December 26, 2017).

Butler, R. (2014) "Deforestation in the Amazon," *MongaBay.com: World Rainforests: Deforestation in Brazil*, Updated July 9, https://data.mongabay.com/brazil.html (Accessed January 6, 2018).

"Can we help wildlife adapt by crowdsourcing human responses to climate change?" *World Wildlife* (Summer 2017): 5.

Cassuto, D. N. (2010) *The CAFO Hothouse: Climate Change, Industrial Agriculture, and the Law*, Ann Arbor, MI: Animals and Society Institute.

Clay, R. (2016) "Does the grazing of cover crops by cattle compact soil?" *No-Till Farmer*, November 21, 2016, www.no-tillfarmer.com/articles/6251-does-the-grazing-of-cover-crops-by-cattle-compact-soil (Accessed January 22, 2018).

Danilo, M. (2003) "Maize: Post-harvest operation," *Food and Agriculture Organization of the United Nations: INPhO Post-harvest Compendium*. Last Reviewed May 15, www.fao.org/3/a-av007e.pdf (Accessed December 17, 2017).

David, P. (2008) "Scientists alarmed by ocean dead zone growth," *San Francisco Chronicle*, August 15, www.sfgate.com/green/article/Scientists-alarmed-by-ocean-dead-zone-growth-3200041.php (Accessed January 8, 2018).

Davis, K. (2015) "Water used in San Diego county," June 29, www.upc-online.org/environment/truth_or_drought/ (Accessed January 22, 2018).

Davis, K. (2016) "Poultry or beef: Which is worse?" *EarthTalk: Questions and Answers about our Environment*, April 30, https://earthtalk.org/poultry-industry/#15167637954921&action=collapse_widget&id=1&data= (Accessed January 22, 2018).

"Definition and classification of commodities (draft): Cereals and cereal products," *FAO (Food and Agriculture Organization of the United Nations)*. 1994. www.fao.org/es/faodef/fdef01e.htm (Accessed December 17, 2017).

"Deforestation: The leading cause of CO2 emissions," *Global Warming Science*. Updated June 13, 2010, www.appinsys.com/globalwarming/deforestation.htm (Accessed January 12, 2018).

Dimick, D. (2014) "If you think the water crisis can't get worse, wait until the aquifers are drained," *National Geographic*, August 21, https://news.nationalgeographic.com/news/2014/08/140819-groundwater-california-drought-aquifers-hidden-crisis/ (Accessed December 17, 2017).

"Do all farmed fish eat the same thing?" *NOAA Fisheries: National Oceanic and Atmospheric Administration*, www.nmfs.noaa.gov/aquaculture/faqs/faq_feeds.html#1what (Accessed January 6, 2018).

Doheny, K. (2018) "Drugs in our drinking water?" *WebMD*, www.webmd.com/a-to-z-guides/features/drugs-in-our-drinking-water#1 (Accessed January 9, 2018).

Doucliff, M. (2018) "Is there a ticking time bomb under the Arctic?" *NPR: Yellowstone Public Radio: History*, January 24, www.npr.org/sections/goatsandsoda/2018/01/24/575220206/is-there-a-ticking-time-bomb-under-the-arctic (Accessed January 24, 2018).

Echolls, T. (2017) "Soil erosion due to rainforest deforestation," *Sciencing: Nature*. Updated April 25, https://sciencing.com/soil-erosion-due-rainforest-deforestation-23042.html.

"Emissions of greenhouse gases methane and nitrous oxide underestimated, research suggests." Delft University of Technology. *ScienceDaily: Science News*, July 1, 2010, www.sciencedaily.com/releases/2010/07/100701090330.htm (Accessed January 13, 2018).

"The extinction crisis." Center for Biological Diversity, www.biologicaldiversity.org/programs/biodiversity/elements_of_biodiversity/extinction_crisis/ (Accessed January 12, 2018).

Goodland, R., and Anhang, J. (2009) "Livestock and Climate Change," *Worldwatch.org*, November/December, www.worldwatch.org/files/pdf/Livestock%20and%20Climate%20Change.pdf (Accessed January 13, 2018).

Graham, B. (2017) "Guess how many gallons of water it takes to produce a single almond," *MRCTV,* www.mrctv.org/blog/guess-how-many-gallons-water-it-takes-produce-single-almond (Accessed January 26, 2017).
Grant, R., and Kononoff, P. J. (2007) University of Nebraska-Lincoln. "Feeding to maximize milk protein and fat yields," February, http://extensionpublications.unl.edu/assets/pdf/g1358.pdf (Accessed December 28, 2018).
Groc, I. (2017) "Open season," *World Wildlife,* Summer: 11–17.
Haley, J. (2014) Personal Communication by e-mail, April 1.
Haley, J. (2015) "So you want to stop devouring ecosystems? Do the math!" in L. Kemmerer (ed.) *Animals and the Environment: Advocacy, Activism, and the Quest for Common Ground* (pp. 151–162), New York: Routledge.
Hawthorne, M. (2012) "Planet in peril," *VegNews,* March–April: 34–41.
Hickey, H. (2014) "Warmer Pacific Ocean could release millions of tons of seafloor methane," *University of Washington: UWNEWS,* December 9, 2014, www.washington.edu/news/2014/12/09/warmer-pacific-ocean-could-release-millions-of-tons-of-seafloor-methane/ (Accessed December 12, 2017).
"How agriculture affects the Mississippi River," *Mississippi River Collaborative,* www.msrivercollab.org/focus-areas/agriculture/ (Accessed December 13, 2017).
"How big is 26 acres?" *The Measure of Things,* www.bluebulbprojects.com/measureofthings/results.php?comp=area&unit=a&amt=26&sort=pr&p=1 (Accessed January 20, 2017).
"How much is a trillion dollars?" *If I had a Trillion Dollars: Youth Film Festival,* https://ihtd.org/festivalguide/resources/how-much-is-a-trillion-dollars/ (Accessed January 20, 2017).
"Hygiene: Liquid and Solid Waste Disposal." *Slaughterhouse Cleaning and Sanitation.* FAO Corporate Document Repository, http://www.fao.org/docrep/003/x6557e/X6557E02.htm (Accessed October 2, 2012).
Kaufman, Stephen R, and Nathan Braun. *Good News for All Creation: Vegetarianism as Christian Stewardship.* Cleveland, OH: Vegetarian Advocates, 2004.
Kemmerer, L. (2014) *Eating Earth: Environmental Ethics and Dietary Choice,* Oxford: Oxford University Press.
Kemmerer, L. (2015) "Eating ecosystems," in L. Kemmerer (ed.) *Animals and the Environment: Advocacy, Activism, and the Quest for Common Ground,* New York, NY: Routledge.
Kharvillart. (2012) "Endangered animals: Black lion tamarin," *YouTube: Arkive: Images of Life on Earth,* June 13, www.youtube.com/watch?v=Hq5f7NCmANs (Accessed January 10, 2018).
Koneswaran, G., and Nierenberg, D. "Global farm animal production and global warming: Impacting and mitigating climate change," *NCB (National Center for Biotechnology Information): U.S. National Library of Medicine.* 2008 May; 116(5): 578–582. https://www.ncbi.nlm.nih.gov/pmc/articles/PMC2367646/ (Accessed June 15, 2018).
Konkel, L. (2016) "Why are these male fish growing eggs?" *National Geographic,* February 3, https://news.nationalgeographic.com/2016/02/160203-feminized-fish-endocrine-disruption-hormones-wildlife-refuges/ (Accessed December 17, 2017).
"Land degradation," *University of Michigan Global Change Program,* January 4, 2010, www.globalchange.umich.edu/globalchange2/current/lectures/land_deg/land_deg.html (Accessed December 26, 2011).
Lantin, R. (1999) "RICE: Post-harvest operations," *Food and Agriculture Organization of the United Nations: International Rice Research Institute, Philippines.* Last Reviewed October 14, 1999, www.fao.org/3/a-ax442e.pdf (Accessed December 17, 2017).
"Major crops grown in the United States," *United States Environmental Protection Agency: Major Crops Grown in the United States | Ag 101 | Agriculture | US EPA.* 2011, www.epa.gov/sites/production/files/2015-07/documents/ag_101_agriculture_us_epa_0.pdf (Accessed January 13, 2018).
Matheny, G. (2003) "Least harm: A defense of vegetarianism from Steven Davis's omnivorous proposal," *Journal of Agricultural and Environmental Ethics,* www.veganoutreach.org/enewsletter/matheny.html (Accessed December 27, 2018).
Meehan, M. A., Stokka, G., and Mostrom, M. (2015) "Livestock water requirements," *NDSU Extension Service,* July, www.ag.ndsu.edu/pubs/ansci/livestoc/as1763.pdf (Accessed December 26, 2017).
Mekonnen, M. M., and Hoekstra, A. Y. (2011) "The green, blue and grey water footprint of crops and derived crop products," *Hydrology and Earth System Sciences,* May 25, www.hydrol-earth-syst-sci.net/15/1577/2011/hess-15-1577-2011.pdf (Accessed December 17, 2017).
"The milky way: Cadbury puts quash on cow gas," *NASA: Global Climate Change: Vital Signs of the Planet,* July 21, 2009, https://climate.nasa.gov/blog/92/ (Accessed January 13, 2018).
"Miscellaneous facts and figures," *Manure Net,* http://agrienvarchive.ca/bioenergy/facts.html (Accessed January 4, 2018).
Oppenlander, R. A. (2011) *Comfortably Unaware: Global Depletion and Food Responsibility . . . What You Choose to Eat is Killing our Planet,* Minneapolis: Langdon Street.
Pimm, S. L., and Raven, P. (2000) "Extinction by numbers," *Nature* 403, February 24: 843–845.

"Public lands ranching," *Western Watersheds Project,* 2010, http://westernwatersheds.org/issues/public-lands-ranching?gclid=CIGqkcarua0CFQ5lhwod V2YG4w (Accessed December 28, 2017).

"Putting trillions of dollars in perspective," *CNBC: Politics* (Associated Press), February 13, 2012, www.cnbc.com/id/46372422 (Accessed January 6, 2018).

Reay, D. (2017) "Methane sources: Waste," *Greenhouse Gas Online: All about Greenhouse Gas Science,* www.ghgonline.org/methanewaste.htm (Accessed December 12, 2017).

Renner, R. (2002) "Do cattle growth hormones pose an environmental risk?" *Environmental Science and Technology,* May 1, http://pubs.acs.org/doi/pdf/10.1021/es022301%2B (Accessed December 17, 2017).

Resnick, B. (2017) "Melting permafrost in the Arctic is unlocking diseases and warping the landscape: The consequences of climate change can be weird and apocalyptic," *Vox Media: Energy and Environment,* October 31, www.vox.com/2017/9/6/16062174/permafrost-melting (Accessed December 12, 2017).

Reynolds, L., and Nierenberg, D. (2012) *Worldwatdch Report 188: Innovations in Sustainable Agriculture: Supporting Climate-Friendly Food Production,* Washington, DC: Worldwatch Institute.

Richards, P., and Hoelle, J. (2016) "Brazil's thriving soy industry threatens its forests and global climate targets," *The Conversation: Academic Rigor, Journalistic Flair: Environment + Energy,* April 17, http://theconversation.com/brazils-thriving-soy-industry-threatens-its-forests-and-global-climate-targets-56973 (Accessed January 13, 2018).

Roberts, C. (2017) "President's letter," *World Wildlife,* Summer: 0–1.

Robinson, T. P., William Wint, G. R., Conchedda, G., Van Boeckel T. P., Ercoli, V., Palamara E., Cinardi G., D'Aietti, L., Hay, S. I., and Gilbert, M. (2014) "Mapping the global distribution of livestock," May 29, http://journals.plos.org/plosone/article?id=10.1371/journal.pone.0096084 (Accessed December 12, 2017).

Rutledge, K., McDaniel, M., Boudreau, D., Ramroop, T., Teng, S., Sprout, E., Costa, Hilary, Hall, H., and Hunt, J. (2011) "Dead zone," *National Geographic: Education,* January 21, www.nationalgeographic.org/encyclopedia/dead-zone/ (Accessed December 12, 2017).

Smith, D. W., and Mukhtar, S. (2017) *The Role of Animal Agriculture on Greenhouse Gas Emissions*, College Station, TX: Texas A&M AgriLife Extension.

Steinfeld, H., Gerber, P., Wassenaar, T., Castel, V., Rosales, M., and de Haan, C. (2006) *Livestock's Long Shadow: Environmental Issues and Options,* Food and Agriculture Organization of the United Nations. Rome: Food and Agriculture Organization.

"Study finds traces of drugs in drinking water in 24 major U.S. regions," *Fox News* (Associated Press), March 10, 2008, www.foxnews.com/story/2008/03/10/study-finds-traces-drugs-in-drinking-water-in-24-major-us-regions.html (Accessed January 3, 2018).

Whiteheart, Rachel. "Yields and Dressing Percentages." CornellCals: College of Agrictulre and Life Sciences. July 10, 2012, https://smallfarms.cornell.edu/2012/07/yields-and-dressing-percentages/ (Accessed October 2, 2019).

"The water content of things: How much water does it take to grow a hamburger?" *USGS: Science for a Changing World,* https://water.usgs.gov/edu/activity-watercontent.php (Accessed December 30, 2017).

"Water intake: A good measure of broiler performance," *The Poultry Site,* December 17, 2003, www.thepoultrysite.com/articles/97/water-intake-a-good-measure-of-broiler-performance/ (Accessed December 10, 2017).

13

INTENSIVE ANIMAL AGRICULTURE AND HUMAN HEALTH

Jonathan Anomaly

This chapter provides an overview of some ethical issues that arise when people use intensive agricultural techniques—often called "factory farming"[1]—to raise livestock for food. I will ignore important issues such as animal suffering and environmental concerns that other authors in this book discuss. Instead, I focus on problems factory farming raises for human health, with special emphasis on antibiotic-resistant bacteria that can arise in animals and infect people. I conclude that there are legitimate disagreements about what we should do about factory farming, but I argue that everyone should agree on two things: first, we should defer to the best available science rather than viewing these issues through the prism of political ideology; second, we should think carefully about the moral trade-offs of different policy proposals rather than pontificating from an armchair.

Introduction

The division between industrial farms and small farms that use traditional techniques is not a clear one. Most animal farms fall somewhere between large and small, and most farmers use varying degrees of confinement to house their animals. More important, the distinction between industrial farming and traditional farming does not track what is morally wrong and right. It is commonly assumed that if there are moral problems with industrial farming, it must be due to the size and scale of the farms. But this view cannot be right, since many companies (like Whole Foods and Chipotle in the United States) source their products from large farms with morally praiseworthy practices (Robbins et al. 2016).

In fact, the high standards and low prices offered by some purveyors of milk and meat and eggs are often made possible by the large scale of the operations from which they buy their food. Moreover, contrary to popular thought, small farms can be worse for the environment than large farms because of the inefficiency of transporting small amounts of food and equipment back and forth (Desrochers and Shimizu 2012) and because sometimes even *small* farms employ techniques such as extreme confinement, which requires them to use antibiotics to prevent infections in their livestock. Finally, small farms in which chickens are left free to roam outside are more likely to expose their animals to pathogens they pick up from wild birds (especially avian flu), which can harm the farm animals themselves and the people who eat them (Greger 2007).

In other words, we should not think of the moral differences between practices as corresponding in any way to the size of the farm or the novelty of technology used in farming. Technology is a tool: it can be used to improve the lives of people and animals, or it can make them worse. What makes intensive animal agriculture such an interesting moral issue is that there are no easy solutions to the problems it poses. Replacing animal protein with plant protein and creating artificial meat from stem cells are not yet viable alternatives for many people around the world. Until everyone has access to cheap sources of protein that doesn't come from animals, we need to consider how to improve current farming techniques that threaten animal welfare and human health.

The Science of Antibiotic Resistance

Antibiotics have probably saved more lives in the twentieth century than any other medicine. It's possible that vaccines against viruses that cause debilitating diseases such as polio and smallpox have *prevented* more diseases than antibiotics have *cured*, but either way, it is worth appreciating the enormous amount of suffering and death antibiotics have helped to prevent. In addition to their ability to cure bacterial infections, antibiotics also make surgery far safer than it would otherwise be. Before the discovery and mass production of antibiotics, a skin laceration or an infected tooth could be lethal because of the opportunity this gave for pathogenic (disease-causing) bacteria to enter the bloodstream and cause organ damage or system failure. The same thing is true for farm animals, although antibiotics are mainly used on factory farms to promote the growth of animals and prevent infections that animals are prone to contract in crowded and stressful conditions (Anomaly 2015).

Most bacteria are not harmful, and in fact, some are helpful to their hosts. Of the approximately 40 trillion bacteria living on and in our body at any given time (Sender et al. 2016), many are *commensal*, meaning that they use us as a source of food and shelter but do not harm us, and some are *mutualistic*, meaning that they provide us with positive benefits in exchange for food and shelter. These benefits can include priming our immune system to distinguish friend from foe, helping us digest food, modulating our hunger, and even altering our mood (Velasquez-Manoff 2013; McAuliffe 2017). Of course, there is no conscious exchange going on. Mutualistic relationships often emerge between organisms when they interact over long periods (Trivers 1971).

Parasitic relationships occur when bacteria thrive at the expense of their hosts. For example, the bacteria that cause tuberculosis can impair our ability to fight off other infections and can kill us when we are especially vulnerable. Some bacteria provide benefits in small concentrations (usually by crowding out pathogenic bacteria) but costs in large numbers, especially when they enter our bloodstream or invade parts of our body where they are not normally found (Blaser 2014). In fact, *mycobacterium tuberculosis* may be an example: some evidence suggests that the complete *absence* of bacteria that cause tuberculosis increases our risk of developing multiple sclerosis, even if the same bacteria can kill us when we have a secondary infection or our immune system is weak (Velasquez-Manoff 2013). This is part of a more general pattern in which eliminating microbes we have coevolved with for millions of years can elevate the risk of developing autoimmune diseases, which occur when our immune system searches for a target that it cannot find and attacks our own cells instead. Our relationship with bacteria is complicated.

Antibiotics are a crucial weapon in fighting off harmful bacterial infections. But the more we use antibiotics, the less effective they tend to become. This is because the creatures that bacteria parasitize have evolved weapons to kill or disable them, and bacteria have evolved mechanisms to resist the effects of these weapons. For example, some antibiotics (produced naturally by living creatures or artificially in the lab) penetrate bacterial cell walls or disrupt the ability of bacteria to replicate. Bacteria often respond by creating thicker cell walls or pumps that eject the antibiotics they absorb. This evolutionary arms race has been going on for billions of years, and all we can do is alter its course by devising new antibiotics and new vaccines and by deciding how to use existing antibiotics responsibly.

According to the most comprehensive analysis yet produced (O'Neill 2016), there is a consensus among researchers that a causal link exists between the use of antibiotics in agriculture and the rise of antibiotic resistance in human populations. It might seem odd that *all* scientists don't agree that agricultural use causes resistance in people. But scientists are cautious in separating correlation and causation. What we can show is that in places where the use of antibiotics in agriculture is high, resistance is also high, and when countries cut back on their use of antibiotics in agriculture (e.g., European Union countries over the last two decades), resistance tends to decline (Spellberg et al. 2016). But it's also true that in places where antibiotic use in animals is high, it tends to be high in people, which means resistance is generally high in those countries in both people and animals. Since resistant strains of bacteria that arise in people can be transferred to animals, it is difficult to determine with certainty what the causal direction of a particular pattern of resistance is: it can spread from people to animals or from animals to people.

Of course, people *are* animals, and we share a bacterial environment with nonhuman animals. Not all bacteria that are dangerous for other animals cause problems in people or vice versa, so some scientists think we should worry more about using antibiotics in agriculture that are medically important for people (e.g., penicillin) rather than antibiotics in general. But since bacteria swap genes with one another and can acquire bits of DNA from bacteria of any species, when we encourage antibiotic resistance in a kind of bacterium that doesn't harm human health, the genes that confer resistance may very well be transferred to other species of bacteria that threaten human health (Marshall and Levy 2011). Thus, the more we use antibiotics in animals, even antibiotics that fight bacteria that aren't harmful to human health, the more we encourage the spread of antibiotic resistance in our general microbial environment. This is an inevitable consequence of evolution by natural selection.

It may be worth mentioning in passing, at least, how devastating the diminished efficacy of antibiotics is for human health. According to a recent analysis by a team of social and natural scientists, at least 700,000 people die of antibiotic-resistant infections worldwide, and many people whose infections are eventually cured have to spend more money on second- and third-generation antibiotics, which are more expensive, because bacteria are becoming increasingly resistant to older antibiotics (O'Neill 2015). Since in most countries the costs of treatment are at least partly borne by taxpayers or members of an insurance pool, they are not fully felt by the individual and are therefore largely ignored. But these costs are growing quickly, and we have good reasons to believe that antibiotics in agriculture play some role in this, even if it is not (yet) a bigger role than the direct use of antibiotics in people.

Ethics and Antibiotic Resistance

Consent

One reason the spread of antibiotic resistance from farm animals to people is a *moral* problem is that it constitutes a harm to which nobody has an opportunity to consent. Adding more resistant bacteria to our shared microbial environment can be thought of as a harmful by-product, or negative externality, of antibiotic use. Resistance rises whether the antibiotics are used to treat an infection, promote growth in farm animals, or are used for no good reason. Bacteria don't care why we use antibiotics. They just tend to respond to antibiotics in their environment by increasing their resistance (more accurately, bacteria without resistance genes die off and make way for their resistant cousins, which harm people because they are more difficult and costly to treat).

In this way, antibiotic resistance is a lot like air and water pollution: even those who don't contribute to pollution bear the costs. Most people who produce pollution don't intend to harm anyone, most victims aren't given a chance to consent to the dangers of pollution, and when individuals or companies are permitted to pollute without compensating victims, we tend to get more pollution than is socially desirable. In the language of economists, we get an "inefficient" amount of pollution when producers and consumers fail to internalize the costs of their transactions.

Collective consequences aside, focus on the following question: Is it ever morally permissible to inflict harms on people without their consent? An obvious case in which this seems justifiable is a parent overriding their children's desires in the short run in order to provide long-run benefits. When I tell my daughter, Loki, that she'll only get cake after she does her school work, I frustrate her desire for instant gratification, and in this sense, I harm her. But in cases like this I'm presumably justified in my *belief* that this will produce a better life for her and justified in my *action* of establishing penalties and incentives for getting her to do the right thing. The hope is that Loki will thank me later. I have good reasons to believe she will because she'll eventually recognize that children have poor impulse control and benefit from developing healthy habits. Even if she thanks me later, however, recognizing why I restricted her liberty doesn't constitute retroactive consent. Nor is it important that, when the child becomes an adult, she explicitly thinks about why her parents' rules benefited her. This suggests that while consent is morally important, it is not the only thing that matters in justifying our actions to other people, even when those actions inflict harm on them.

Of course, farmers adding antibiotics to animal feed are different from parents withholding cake from kids. Antibiotic resistance can be thought of as genetic pollution dumped in the broader microbial environment, which poses a small probability of endangering a large number of people. Unlike parents restricting their children's liberty to promote their welfare, many farmers do not *know* or *care* as much about the people whose welfare they indirectly affect. So even if there are cases in which one person is justified in overriding another's liberty to promote his or her welfare without seeking the other's consent, it is not clear that the use of antibiotics in agriculture is such a case.

Responsibility

Our common moral practices are infused with the assumption that responsibility matters. We praise or punish someone because they *deserve* rewards or penalties, and they deserve their treatment to the extent that they are *responsible* for their behavior. Of course, sometimes people who seem to deserve punishment actually do not, and there are many kinds of exculpatory facts that might lead us to excuse someone for acting in a way they weren't fully responsible for. So, for example, many people think we should punish thieves. But if we find out that a thief was given an ultimatum by a neighbor to either steal a pepperoni pizza or take a bullet to the head, we might excuse the thief because he was under duress.

There are disagreements about the justifications for punishment and praise, but on most accounts, responsibility is crucial. Who is responsible for antibiotic resistance when it arises as a byproduct of voluntary choices by farmers and consumers? Responsibility often involves causal judgments about who produces an outcome that we think of as bad or good. So, for example, we might say farmers who use practices that produce genetic pollution (antibiotic resistance), along with consumers of meat from animals fed antibiotics, are *responsible* for antibiotic resistance. In one sense this is obviously true: both producers and consumers collectively *cause* the outcome we are concerned with, and in most cases, they have the opportunity to behave otherwise.

Notice, however, that any one act of consuming factory-farmed animals, or adding antibiotics to animal feed, does not necessarily cause antibiotic-resistant bacteria to emerge or spread. All we get is a small probability that increases as the quantity of antibiotics involved in these practices grows. Thus, it is hard to say that any one action, or small set of actions, makes a difference (Nefsky 2018). Still, we might say that each person who produces or consumes factory-farmed meat bears some (perhaps probabilistic) responsibility for producing collectively bad consequences. This might not be a decisive reason to change our practices, but it may be a reason to think about how we might change the incentives that predictably lead to the negative consequences of factory farming.

Notice that there is nothing especially unique about antibiotic resistance. Many public health threats, including various kinds of air and water pollution, have a similar structure. Often, chemicals that can threaten human health are benign below a certain threshold. For example, adding a teaspoon of mercury into the municipal water supply or a gallon of arsenic into the ocean will probably not produce any negative health consequences for people. But above some threshold, we may be contributing to a process that leads to adverse health consequences, such as an elevated risk of cancer in adults and neurological disorders in children. Here again, the typical response is not to punish people who are responsible for making a trivial contribution to a social problem. Instead, it is to establish regulations on the amount of pollution that can be produced in a particular industry. Of course, if this is true, it may shift responsibility onto all citizens to lobby governments to establish socially beneficial regulations (Sinnott-Armstrong 2005).

The problem is that most citizens are predictably ignorant about many processes that constitute collective action problems—situations in which what is individually rational is collectively harmful. It may be rational for *each of us* to pollute, even if *all of us* would be better off if each polluted less. Economists use the phrase "rational ignorance" to cover cases in which people have little reason to gather information because the individual costs of doing so exceed the individual benefits (Downs 1957). The idea is that even if all of us would be better off becoming informed about a particular case of pollution and voting for candidates that pledge to cost-effectively regulate it, each of us has an incentive not to be informed. This is because each of us has little power to do much about the problem, even if we understand its precise causes. After all, consuming information is costly in the opportunity sense: we have to give up time and energy and set aside our ideological commitments to confront uncomfortable facts.

The upshot is that responsibility for *causing* the negative effects of intensive animal farming is diffuse, and so is responsibility for *addressing* it. Effectively solving the problem requires many of us to act against our self-interest. To the extent that each of us is responsible for producing the problem, perhaps each of us has a moral reason to gather information about the source of the problem and the set of feasible solutions and to then act on that information at the ballot box, in the market place, and in our ordinary conversations with people. This may seem like a trivial moral obligation. But becoming a responsible voter and consumer, and a sophisticated interlocutor in political conversations, is demanding. As I argue in the next section, it requires us to become aware of our biases, cultivate skepticism, and face up to facts that collide with our prior beliefs.

The Ethics of Belief

In the previous section, I mentioned the concept of rational ignorance from economics and deployed it to try to explain why most consumers and voters don't understand the details of many of the collective action problems they face, including antibiotic resistance. Since each of us barely contributes to the problem, we face a question: why gather any information at all if, once I become informed, I am unlikely to be able to make much difference? Unless I'm a powerful politician or wealthy lobbyist or influential in some other way, it seems difficult to justify the belief that my actions will matter much.

Another way to think about the issue is that there are three separate barriers to solving a collective action problem such as antibiotic resistance. First, enough people have to gather information, even when there are few if any benefits for each of them to do so. Second, each must spend time sifting through the information in order to determine whether the relevant claims are true, whether they're morally important, and which policies or consumption habits are most likely to advance the cause of minimizing antibiotic resistance. Finally, each must *act* on the information by, for example, paying a higher price at the meat counter or convincing people to vote for candidates likely to pass reasonable regulations on antibiotics.

Since antibiotic resistance is only one of many collective action problems, it is understandable, though not necessarily morally excusable, that many people would be ignorant about issues such as the role their own actions play in producing the problem. But the idea of rational ignorance understates the magnitude of the problem. As Michael Huemer (2015) argues, many people are *irrational*, especially in the political realm.

To understand the difference between rational ignorance and irrationality, we need to make a distinction: ***epistemic* rationality** aims at justification, while ***instrumental* rationality** aims at utility. When we seek to match our minds to the world by carefully considering evidence, we are behaving in an epistemically rational way. This sense of the term is captured when we say that a religious fanatic who rejects scientific explanations in favor of magical explanations is "irrational." Epistemically rational people proportion their beliefs to the evidence.

But in economics, and in everyday life, we also use the term *rational* to refer to someone who efficiently satisfies their desires. To take an example, we sometimes say of a brutal dictator that he is behaving in a way that is *wrong* but *rational* if his goal is to stay in power. That is, his actions are morally wrong, but they're effective for achieving his objectives.

Putting these pieces together, as long as we have preferences about how we wish the world were, it can be instrumentally rational to render ourselves epistemically irrational. This is sometimes called rational irrationality, and it is pervasive in politics (Caplan 2001). Beliefs can function like placebos that provide us with meaning (e.g., religion), bond us together in a common cause (e.g., politics), or alter our confidence in ways that affect our performance (e.g., sports).

To take a simple example, since religious and political beliefs often function as tools of social bonding between communities, it can be painful, even devastating, to deviate from the beliefs your classmates, friends, or family members share. And it can be inspiring to believe that whatever hardships we undergo in pursuit of our goals, there is a greater purpose that redeems our effort and suffering. Becoming an atheist or declaring yourself a political heretic can mean losing friends, missing out on social events, and even depriving yourself of a source of transcendent meaning. It is obvious, then, why people not only refrain from forming beliefs when they don't see a point in exerting effort (rational ignorance) but form strong beliefs they cannot possibly justify (rational irrationality). In fact, in political activism and religious practice, it is often a virtue to believe *in spite of* the evidence rather than because of it. True believers are implored to have faith, not use reason.

Why are people especially prone to epistemic irrationality in politics? In addition to political beliefs being tools of social bonding for rival tribes, misguided political beliefs are rarely punished. In markets, consumers get what they pay for, so they are a little more careful to sift through information about the products they buy. When faced with alternative brands of car, consumers have to live with the consequences of what they choose. But in democratic politics, citizens get what *other people* vote for, so they are less inclined to correct their errors of judgment. Voting one way or another doesn't *cause* an outcome to change or a collective action problem to be solved. The fact that we rarely pay any substantial cost for forming irrational political beliefs may explain why otherwise smart people become good rationalizers but poor reasoners when thinking about politics (Haidt 2012).

If political ignorance and irrationality are predictably widespread features of life in a large democracy, it may be that our fundamental moral obligation as citizens is to resist the temptation to believe what we would like to be true, instead deferring to experts on relevant matters. In other words, our main *moral* obligation may be to become *epistemically* rational rather than succumbing to the irrationality that characterizes political discussions.

Committing ourselves to form beliefs on the basis of evidence cuts against the tendency of partisans on all sides. For example, it is common (in modern American politics) for self-described progressives to criticize religious conservatives for rejecting evolutionary explanations about the origin of human beings or for dismissing the idea that human activity is partly responsible for recent changes in the Earth's temperature. But the same people who profess an allegiance to science often reject it when it conflicts with their own political ideology. To take an example, skepticism about the safety of vaccines and genetically modified (GM) food more often comes from progressives than conservatives, despite a consensus that vaccines and GM food are generally safe and socially beneficial (National Academies of Sciences, Engineering, and Medicine 2016).

If these considerations are right, the most fundamental moral obligation consumers and voters arguably have is the duty to proportion their beliefs to the best available evidence and to defer to expert opinion when appropriate. Unless we recognize the obligation to form justified beliefs about politically relevant subjects, it is not clear how we can hope to solve collective action problems such as antibiotic resistance that arises in livestock and spreads to people.

Ethics and Public Policy

I've tried to show that the problem generated by the widespread use of antibiotics in agriculture is one instance of a set of problems that shares a common structure. Like many other global collective action problems, responsibility for producing the problem is diffuse, individuals cannot solve it on their own, and individuals have little incentive to carefully study the nature of the problem or the available solutions. Given this common structure, self-interest pulls us in one direction and morality *should* pull us in another. But we cannot recognize our moral obligations as consumers or voters until we first overcome the strong temptation to form politically motivated rather than epistemically rational beliefs about the nature of the problem.

It is likely that our sense of fairness, or justice, evolved, in part, to solve collective action problems in small groups (Bowles and Gintis 2013). In other words, we likely evolved moral emotions such as guilt and shame because small groups of people with moral emotions can solve collective action problems more effectively than those who need to constantly monitor one another and impose physical costs on people who violate mutually beneficial norms or rules. The problem is that global collective action problems whose consequences spill across generations—such as antibiotic resistance or human-induced global warming—often fail to activate moral emotions that evolved to achieve much more modest goals in small groups. So one obvious question is how to convince individual farmers, consumers, and citizens to care about an outcome they have so little control over but that has potentially devastating effects.

Fortunately, there is some evidence that consumers are willing to pay a tax on products that produce negative side effects, if the tax revenue raised will be used to minimize the problem (Kallbekken et al. 2011). And there is evidence that most American consumers want antibiotic-free meat and are willing to pay at least a little more for it, even if many don't fully understand the problem (Bohne and Halloran 2012). Since ordinary people can't possibly compute the total costs and benefits their consumption choices produce, they cannot be expected to know how much they *should* be willing to pay for products that produce collective harms.

One thought, then, is that the primary moral obligation policy makers have is to try to implement policies that price pollution—understood broadly in this context to include antibiotic resistance—in a way that reflects what people *would* be willing to pay if they knew the relevant information. This is what is known in economics as a Pigovian tax, after the British economist Arthur Pigou, who argued that we should aim at an "optimal" pollution rate. The idea is that some pollution makes productive processes possible. For example, even extremely efficient factories typically emit a little pollution, and preventing or cleaning up the last few particles of pollution becomes increasingly costly relative to the benefits it produces (think of a beach cleanup where it takes many hours to find the last particle of trash: the time and energy could be better spent on other things that produce individual or social value).

In theory, Pigovian taxes reflect the precise social cost of an activity, and in addition to discouraging inefficient levels of pollution, the revenue generated is used to compensate victims of pollution, or it's invested in finding new technologies that produce similar products with less pollution. Since the information required to calculate a Pigovian tax with precision is virtually impossible to gather, many argue for less precise "user fees" on antibiotics that serve a similar function (Battin et al. 2008; Anomaly 2013).

Notice that a Pigovian tax, or a slightly less precise user fee, serves another moral function apart from producing the socially optimal amount of pollution in the form of antibiotic resistance. That is, it doesn't just focus on the collective consequences of our activities, which some philosophers think of as a morally controversial approach to ethics. It also reflects a view of fairness according to which people should pay costs for acting in harmful ways in proportion to the amount of harm their activity can be expected to create. This principle of proportionality goes back at least to Aristotle. The idea is that we should pay for an activity in proportion to the number of social costs we create by engaging in it, especially when we benefit at the expense of others.

Another way of characterizing this idea is to call it a public harm principle. According to a common (although not universally shared) interpretation, John Stuart Mill (1859) argued that people should be free to act as they please provided they don't create morally significant, uncompensated harm to others. What counts as a harm and what counts as a compensating benefit is up for debate. But Mill's framework is meant as a general principle for framing moral debates, not an algorithm that provides us with detailed answers for moral questions. Some extend Mill's harm principle to large number cases (Gaus 1999). On this view, if a set of actions causes an uncompensated harm, and if the harm is serious enough that it merits limiting one person's liberty, then everyone's liberty should be similarly limited.

Moral principles are easier to apply between a small number of people than across a population of millions. One reason for this is that as the number of people involved increases, conflicting judgments about what should be done tends to grow, and solutions are more difficult to implement. When someone steals a car or attacks someone at a pub, it is clear who should pay the costs, and juries don't have too hard a time determining appropriate penalties. But any time we think about political solutions to large collective action problems, we should assume that voters, policymakers, and bureaucrats who implement policies will have imperfect information and conflicting incentives. In order to win election campaigns, policymakers often have to tell people what they want to hear, and we have already discussed why what people want to hear is not always what scientists know to be true.

So quite apart from the problem of politicians having limited knowledge, they do not always have incentives to use their knowledge to promote broader social goals unless voters can be expected to reward them for it. Political feasibility may therefore be a relevant constraint to take seriously in moral theorizing about how we should approach real-world problems like antibiotic resistance in agriculture. When thinking about what to do, we should avoid complacently assuming insuperable voter ignorance and political corruption, on one hand, and perfectly informed and altruistic political actors, on the other.

One of the most important trends in antibiotic resistance is the precipitous rise in antibiotic use in agriculture in the developing world, which is occurring at the same time that antibiotic use in many developed countries is beginning to decline (Van Boeckel et al. 2015). In developing countries, the decline is due, in part, to demand by consumers for antibiotic-free meat and to policy makers' banning the use of antibiotics as growth promoters in the European Union. Nevertheless, as global trade continues to increase in the coming century—a trend that mostly benefits consumers—there is a real danger that many factory farms will migrate to nations with weak regulations on the use of antibiotics in agriculture. Some have therefore called for global agreements to restrict the use of antibiotics in agriculture in order to avoid a situation in which countries with the weakest regulations sell products that produce global harms to people in countries with more responsible production practices (Anomaly 2019).

Conclusion

The widespread use of antibiotics in agriculture has only occurred for about half a century, and the consequences have been anticipated since the beginning. Alexander Fleming, who developed penicillin as a drug to treat serious bacterial infections in people, famously warned that the profligate use of antibiotics would lead to drug resistance, which he described as a *moral* problem. He never anticipated that we might squander the efficacy of antibiotics even faster by feeding them to our farm animals in order to obtain relatively small benefits from cheaper meat.

Banning antibiotics in agriculture would likely benefit people by slowing antibiotic resistance. But it would also prevent farmers from treating animals who get sick even when the farmers engage in otherwise responsible production practices. I leave it up to the reader to decide whether we should opt for a complete ban or instead support some combination of taxes and regulations on the use of antibiotics in agriculture. The trade-offs are difficult, and the benefits and costs of different approaches will be borne by different people.

Note

1. I use *factory farmed* to imply the extreme confinement of animals and the routine use of antibiotics. *Factory farming* is a vague term, but it typically includes these two elements.

Bibliography

Anomaly, J. (2013) "Collective action and individual choice," *Journal of Medical Ethics* 39: 752–756.

Anomaly, J. (2015) "What's wrong with factory farming?" *Public Health Ethics* 8(3): 246–254.

Anomaly, J. (2019) "Antibiotics and animal agriculture: The need for global collective action," in M. Selgelid (ed.) *Ethics and Antimicrobial Resistance* (pp. 1–11), New York: Springer Publishing.

Battin, M. P., Francis, L. P., Jacobson, J. A., and Smith, C. B. (2008) *The Patient as Victim and Vector: Ethics and Infectious Disease*, New York, NY: Oxford University Press.

Blaser, M. (2014) *Missing Microbes: How the Overuse of Antibiotics Is Fueling our Modern Plagues*, New York, NY: Picador-MacMillan.

Bohne, M., and Halloran, J. (2012) "Meat on drugs," *Consumer Reports*, June 2012 issue, www.consumerreports.org/content/dam/cro/news_articles/health/CR%20Meat%20On%20Drugs%20Report%2006-12.pdf.

Bowles, S., and Gintis, H. (2013) *A Cooperative Species: Human Reciprocity and Its Evolution*, Princeton, NJ: Princeton University Press.

Caplan, B. (2001) "Rational ignorance versus rational irrationality," *Kyklos* 54: 3–26.

Desrochers, P., and Shimizu, H. (2012) *The Locavore's Dilemma: In Praise of the 10,000-mile Diet*, New York, NY: Perseus Books.

Downs, A. (1957) "An economic theory of political action in a democracy," *The Journal of Political Economy* 65(2): 135–150.

Gaus, G. (1999) *Social Philosophy*, London: M.E. Sharpe Publishing.

Greger, M. (2007) "The human/animal interface: Emergence and resurgence of zoonotic infectious diseases," *Critical Reviews in Microbiology* 33: 243–299.

Haidt, J. (2012) *The Righteous Mind*, New York, NY: Vintage Books.

Huemer, M. (2015) "Why people are irrational about politics," in J. Anomaly, G. Brennan, M. C. Munger, and G. Sayre-McCord (eds.) *Philosophy, Politics, and Economics: An Anthology*, New York, NY: Oxford University Press.

Kallbekken, S., Kroll, S., and Cherry, T. L. (2011) "Do you not like Pigou, or do you not understand him? Tax aversion and revenue recycling in the lab," *Journal of Environmental Economics and Management* 62(1): 53–64.

Marshall, B. M., and Levy, S. B. (2011) "Food animals and antimicrobials: Impacts on human health," *Clinical Microbiology Reviews* 24(4): 718–733.

McAuliffe, K. (2017) *This Is your Brain on Parasites: How Tiny Creatures Manipulate our Behavior and Shape Society*, Boston, MA: Mariner Books.

Mill, J. S. (1859) *On Liberty*, New York, NY: Henry Holt and Company [1879 ed.].

National Academies of Sciences, Engineering, and Medicine (2016) "Genetically Engineered Crops: Experiences and Prospects," Washington, DC: The National Academies Press, https://doi.org/10.17226/23395

Nefsky, J. (2018) "Consumer choice and collective impact," in A. Barnhill, M. Budolfson, and T. Doggett (eds.) *Oxford Handbook of Food Ethics* (pp. 267–286), New York, NY: Oxford University Press.
O'Neill, J. et al. (2015) "Antimicrobials in agriculture and the environment: Reducing unnecessary use and waste," *The Review on Antimicrobial Resistance*, http://amr-review.org/.
O'Neill, J. et al. (2016) "Tackling drug-resistant infections globally: Final report and recommendations," *The Review on Antimicrobial Resistance*, http://amr-review.org/.
Robbins, J. A., von Keyserlingk, M. A., Fraser, D., and Weary, D. M. (2016) "Farm size and animal welfare," *Journal of Animal Science* 94(12): 5439–5455.
Sender, R., Fuchs, S., and Milo, R. (2016) "Revised estimates for the number of human and bacterial cells in the body," *PLOS Biology* 14(8). https://doi.org/10.1371/journal.pbio.1002533
Sinnott-Armstrong, W. (2005) "It's not my fault: Global warming and individual moral obligations," in W. Sinnott-Armstrong and R. B. Howarth (eds.) *Perspectives on Climate Change: Science, Economics, Politics, Ethics, Advances in the Economics of Environmental Resources*, 5: 221–253, Bingley, UK: Emerald Group.
Spellberg, B., Hansen, G. R., Kar, A., Cordova, C. D., Price, L. B., and Johnson, J. R. (2016) "Antibiotic resistance in humans and animals," *National Academy of Medicine*, discussion paper, www.nam.edu/antibiotic-resistance-in-humans-and-animals.
Trivers, R. (1971) "The evolution of reciprocal altruism," *The Quarterly Review of Biology* 146(1): 35–57.
Van Boeckel, T. P., Brower, C., Gilbert, M., Grenfell, B. T., Levin, S. A., Robinson, T. P., Teillant, A., and Laxminarayan, R. (2015) "Global trends in antimicrobial use in food animals," *Proceedings of the National Academy of Sciences* 112(18): 5649–5654.
Velasquez-Manoff, M. (2013) *An Epidemic of Absence: A New Way of Understanding Allergies and Autoimmune Diseases*, New York, NY: Scribner.

14
SEAFOOD ETHICS
Reconciling Human Well-Being with Fish Welfare

Mimi E. Lam

Introduction

Globally, enough food is produced to feed the world's population (Pogge 2016), yet in 2017, approximately 821 million people (roughly one in nine, but see Pogge 2016, for critique of estimates) suffered from undernourishment or chronic food deprivation, and more than 672 million adults (over one in eight) were obese (FAO et al. 2018). Both undernourishment and obesity are forms of malnutrition that often arise from food insecurity, that is, the lack of access to safe, nutritious, sufficient, and culturally appropriate food (FAO 2001; Cassman 2012). In 2017, an estimated 770 million people—10% of the world's population—experienced severe food insecurity, going without food for a day or more (FAO et al. 2018). Undernourishment and severe food insecurity, respectively, are dominant in Africa (20.4% and 29.4%) and Asia (11.4% and 6.9%), while adult obesity is most prevalent in North America (FAO et al. 2018). The rising trends of world hunger, adult obesity, and food insecurity signal an ethical crisis of distributional inequity in the global food supply, exacerbated by growing demands from a surging, more affluent, yet socioeconomically disparate human population.

Distributional inequity in food provisioning is just one food ethics issue (Mepham 2008). A complex suite of ethical issues arises along the entire food value chain, including food security, human health, social justice, animal welfare, and ecosystem integrity (Lam 2016a, 2016b). Resource access, allocation, and trade decisions can trigger food security or insecurity, ecological sustainability or degradation, peace or conflict, so coordinated, normatively guided actions among diverse institutions are needed for effective yet fair policies and governance (Lam 2016a, 2016b). Food policies are often debated around human well-being, choice, and fairness, but concern is growing over animal welfare and the environmental impacts of food systems. Food ethics issues impact global society vis-à-vis five grand challenges: population growth, climate change, natural resource access, health, and markets (Kaiser and Algers 2016). A concerted and deliberative approach is needed that supports ethical decision-making in food and natural resource policy and governance toward a more sustainable and just global society.

Food ethics issues challenge governance by blurring the distinction between facts and values (Lam 2016a, 2016b). Food governance links three key human institutions that govern political, economic, and social interactions, namely, the government, the market, and civil society (Lam and Pauly 2010; Lam and Pitcher 2012a). Government policies regulate the food sector by trying to minimize risks to public health and the environment, that is, by preventing harm to society, but harm is a normative concept reflecting underlying social judgments of good and bad. The global market is equally incapable of finding equilibria to balance the diverse values brought into the food sector, as social actors act according to their own values, which vary with financial and cultural contexts, as well as constraints from governmental regulations and sectoral norms. Meanwhile, within civil society, non-governmental organizations (NGOs) and scientists often conduct (or fund) policy-relevant research, communicate science to decision-makers, and educate the public, but research findings are often inconclusive. For such value-laden issues in plural democratic societies, a science of deliberation (Dryzek et al. 2019) is needed. This includes structured deliberative processes and robust decision-support tools that can integrate diverse, contested knowledge within ethical governance frameworks (Lam et al. 2019).

In the global seafood sector, diverse social actors harvest, process, trade, and consume fish, often across multiple jurisdictions. Complex seafood value chains (Lam 2016a, 2016b; Drury O'Neill and Crona 2017) thus connect producers, processors, distributors, traders, wholesalers, retailers, and consumers. Actors may interact directly, at a local fish market, or indirectly, through complex global trade networks connecting disparate geographies and cultures. Governing such diverse, complex, and dynamic systems across multiple spatiotemporal scales (Kooiman and Bavinck 2005) is "wicked" (Rittel and Webber 1973). That is, seafood policy and governance problems often have no unique definition, let alone solution (Jentoft and Chuenpagdee 2009; Lam et al. 2019), given the plurality of values guiding multiple actors in their decisions along complex value chains. As the world's seafood production comes primarily from regions with weak governance, improved governance is essential to enhancing the contribution of seafood to food security (Smith et al. 2010). Seafood ethics thus offers a novel, integrated approach to fishery resource sustainability through not only the descriptive and evaluative, but also normative study of values, value-based trade-offs, and ethical dilemmas of multiple stakeholders, including citizens, interacting along diverse seafood value chains (Lam 2016a).

Seafood Production: Fisheries and Aquaculture

Global fish production, including capture fisheries and aquaculture, was a record-high 171 million tonnes in 2016 (Figure 14.1a), reported the United Nations Food and Agriculture Organization (FAO; 2018). FAO (2018) attributes this to a relatively stable capture fisheries production (53% of the total), continued aquaculture growth (47% of the total), and reduced wastage (27% of landed fish and 35% of global catches – if discards prior to landing are added – are lost or wasted). 88% of global production (China produced 19% of this total) is destined for direct human consumption, although the actual amount consumed is less, owing to losses and waste (FAO 2018). Reconstructions of total fish removals estimate marine capture fisheries account for 120 million tonnes (cf. 79 million tonnes reported by FAO; 2018), with declining trends (Pauly and Zeller 2016a; Figure 14.1b). The reconstructed global catches account for all marine fisheries, including illegal, unregulated, and unreported (IUU) fishing, fishing in the (unregulated and largely unreported) high seas, and discarded bycatch, that is, the unintentional catch of fish or marine species when fishing for a target species. The reported marine catches assembled by FAO are from voluntary submissions by its members of national catches in their Exclusive Economic Zones (EEZs). EEZs are the sea zones, stretching 200 nautical miles from the coast, in which states have specified rights and duties of management and use of fishery resources, as prescribed by the third United Nations Convention on the Law of the Sea (UNCLOS III 1982). Pauly and Zeller (2016a) argue that FAO's reported catches underestimate by one third the actual catches for the 1950–2010 period. This controversy (Pauly and Zeller 2017a, 2017b; Ye et al. 2017) of whether fisheries are stable or declining has significant economic, social, and ecological implications for policy and governance.

On behalf of the European Commission, the commissioner for environment, maritime affairs and fisheries recently asked, "How can more food and biomass be obtained from the oceans in a way that does not deprive future generations of their benefits?" In response, the *Food From the Oceans*

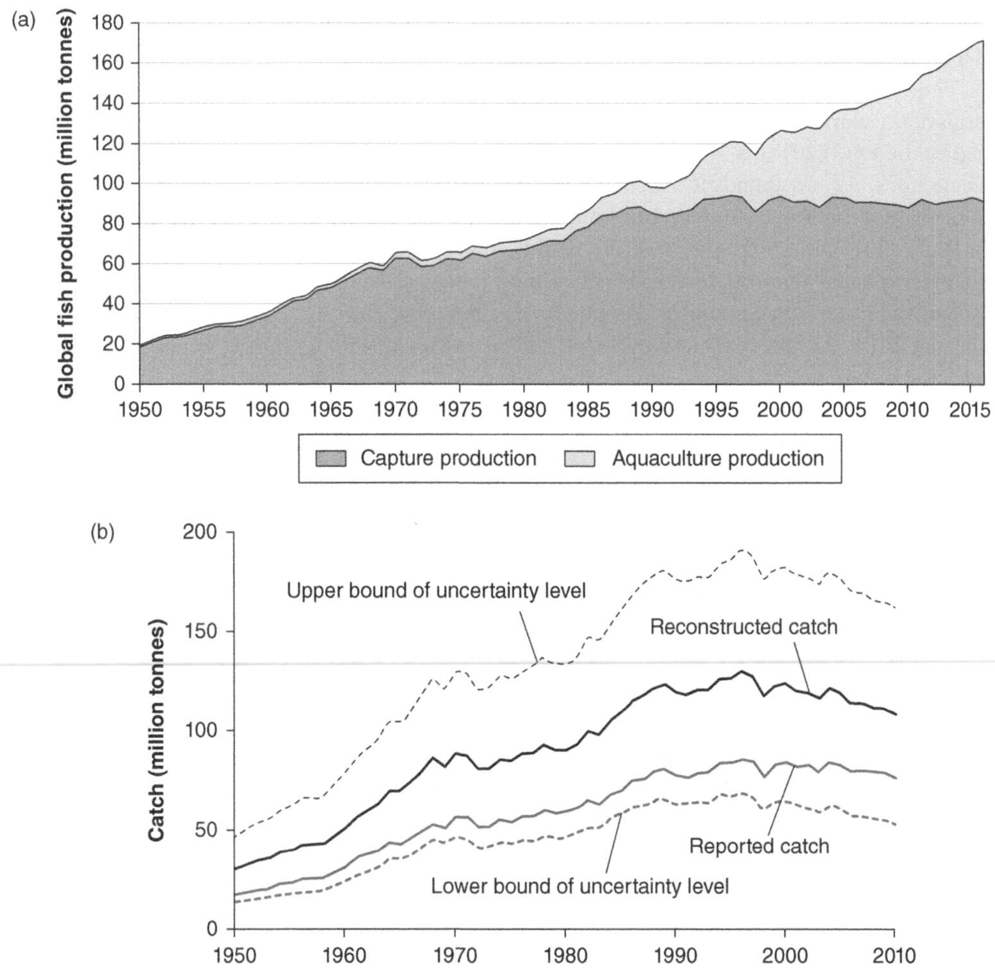

Figure 14.1 (a) World capture fisheries and aquaculture production, reported (for 1950–2016) by the United Nations Food and Agriculture Organization (Redrawn from FAO; 2018), including fish, crustaceans, molluscs and other aquatic animals (henceforth, denoted as 'fish'), but excluding aquatic mammals, reptiles, seaweeds and other aquatic plants, and (b) catch reconstructions of global marine fisheries (for 1950–2015), including illegal, unregulated, and unreported fishing, high-seas fishing, and discarded bycatch, shown with FAO's reported catch and upper and lower uncertainty levels

Source: Redrawn from Pauly and Zeller (2016a).

evidence review report (SAPEA 2017) was written to support the Scientific Advice Mechanism High-Level Group of Scientific Advisors (SAM 2017). The main outcome was to recommend increasing fish capture and culture at lower trophic levels, that is, levels in the ocean food web below the carnivorous levels currently most exploited (SAPEA 2017; SAM 2017). Mariculture, that is, marine aquaculture, notably of herbivorous filter feeders (e.g., molluscs) and omnivorous bottom fish (e.g., mullet), was identified to have the greatest potential for direct human consumption. Mariculture and cultivated algae were promoted as more ecologically efficient sources of feed for farmed marine carnivores (e.g., most finfish and shrimp). Pilot fishing of unexploited lower trophic-level species and vast but diffuse populations of mesopelagic fish was also promoted (SAPEA 2017; SAM 2017). These recommendations resonate with the recent finding that farming molluscs and pelagic fish species were the most environmentally friendly among a variety of food production systems (Hilborn et al. 2018).

These recommendations to increase global fish production, however, come at a time when the status of the world's fish stocks shows alarming trends (Figure 14.2). One third of fish stocks are classified by the FAO (2018) as overfished, that is, fished at biologically unsustainable levels (Figure 14.2a). These stocks are less abundant than that needed to produce maximum sustainable yield (MSY), that is, the largest catch that can be taken from a fish population sustained at its maximum growth rate. In Pauly (2016), more than 40% of the world's stocks (nearly 60% in Pitcher and Cheung 2013) are overexploited (biomass less than half of the biomass at MSY) or collapsed (biomass less than 10% of the historic maximum of unfished levels), and more than 80% (almost 90% in Pitcher and Cheung 2013) are fully exploited, overexploited, or collapsed, with increasing trends (Figure 14.2b).

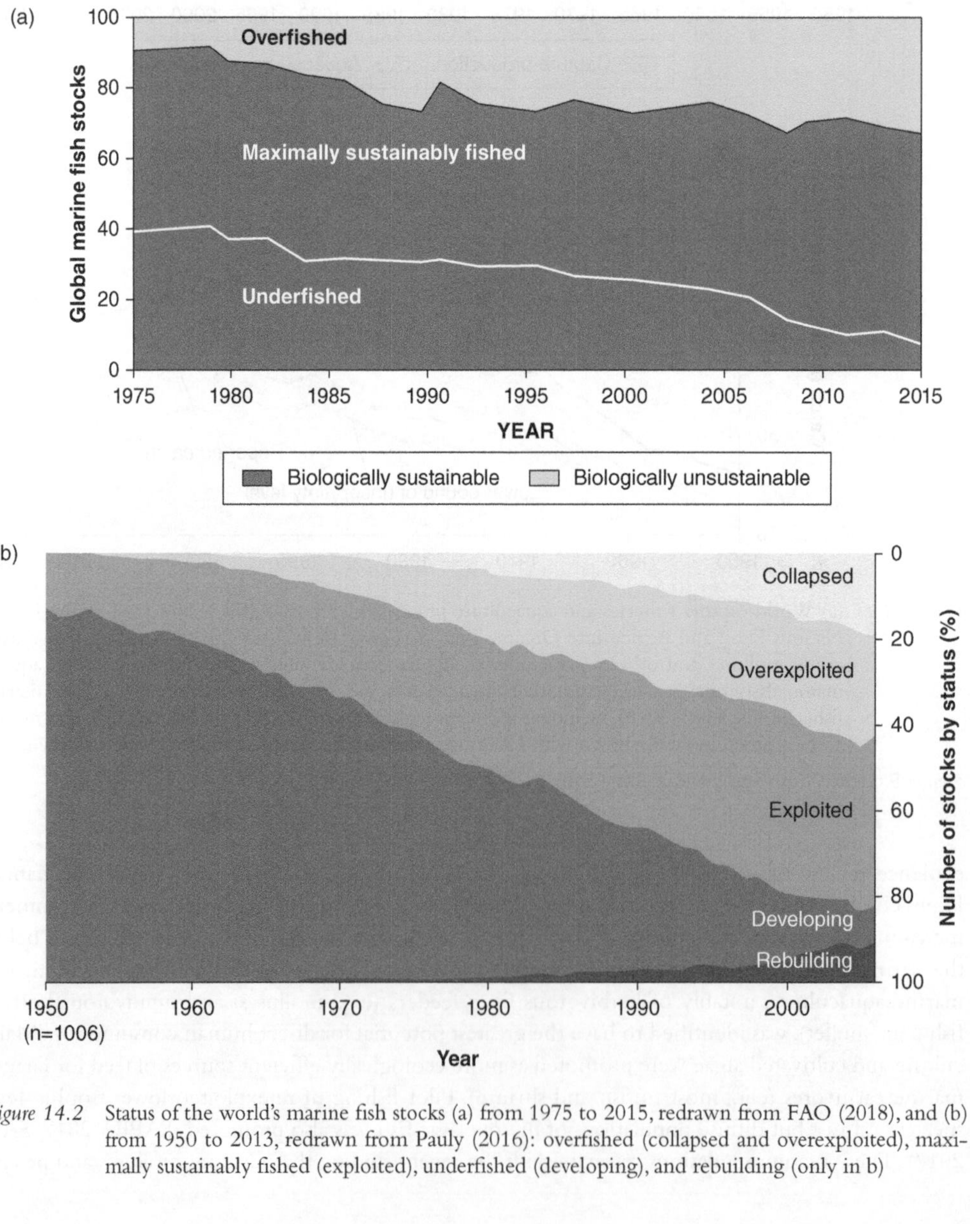

Figure 14.2 Status of the world's marine fish stocks (a) from 1975 to 2015, redrawn from FAO (2018), and (b) from 1950 to 2013, redrawn from Pauly (2016): overfished (collapsed and overexploited), maximally sustainably fished (exploited), underfished (developing), and rebuilding (only in b)

Such discrepancies in global stock status, as in global fish production, arise from different definitions, assessments, and methodologies used to interpret catch statistics and biomasses. Fisheries science, consequently, is subject to opposing views of "doom and gloom" (Pauly 2007) and "management successes" (Hilborn 2007). These scientific uncertainties need to be acknowledged, along with methodological, epistemological, and societal dimensions of uncertainty, when communicating science for policy, and shift instead to knowledge quality assessments to inform decision-making (van der Sluijs 2017).

Fishing is the capture of aquatic organisms in marine, coastal, and inland areas. Capture fisheries are typically categorized as either large- or small-scale, though the distinctions, often by vessel size or engine power, can vary by country and fishing sector. Large-scale, industrial fisheries use large specialized vessels, such as trawlers, purse seiners, long liners, and gill netters, with salaried crews. Small-scale, artisanal fisheries tend to rely on small owner- or family-operated crafts, some non-motorized, operated mostly inshore with more selective, multiple fishing gear technologies. Fishing causes ecological impacts through fishing mortality and collateral impacts (to species and habitats), leading to incidental mortality and decline in mean trophic level and, ultimately, altered ecosystem structure and function (Figure 14.3; Morgan and Chuenpagdee 2003).

Decline in mean trophic level is also known as "fishing down marine food webs," as high-trophic-level piscivorous bottom fish are caught, decreasing in numbers and size, followed by serial depletions

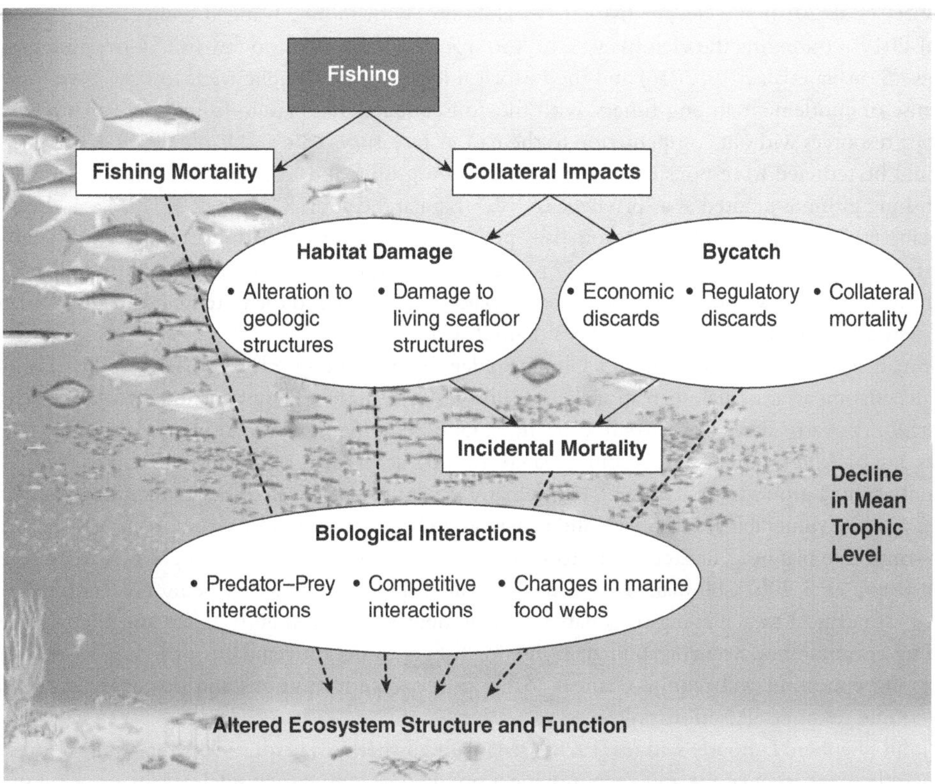

Figure 14.3 Ecological impacts of fishing: fishing mortality and collateral impacts, such as habitat damage and bycatch, leading also to incidental mortality. Fishing often leads to a decline in the mean trophic level, as large fish are removed from the food web, and altered ecosystem structure and function

Source: Redrawn from Morgan and Chuenpagdee (2003).

of smaller species, leaving lower-trophic-level planktivorous pelagic fish (Pauly et al. 1998). Fishing gear types, however, have vastly different impacts on fish and ecosystems: some (high-impact) gears can destroy critical habitats, while others (with low selectivity) result in high bycatch and discard levels of (both target and nontarget) species. The most severe ecological impacts are associated with bottom trawl gear, bottom gillnets, and dredges, while pots and traps and purse seines have intermediate impacts, and hook and line (e.g., rod and reel, troll, and hand lining) have the lowest impacts (Morgan and Chuenpagdee 2003; Fuller et al. 2008). Large-scale fisheries contribute almost twice as much fish globally for human consumption as small-scale fisheries, but their fuel use, government subsidies (Schuhbauer et al. 2017), catch for industrial reduction (as feed for aquaculture; Cashion et al. 2017a, 2017b), and discards at sea (Zeller et al. 2018) are much greater, while the number of fishers employed (FAO 2015) is much less (see summary in Figure 14.4, from Pauly and Zeller 2016b; Lam 2016b).

Of all people directly dependent on capture fisheries, 90% work in the small-scale fisheries sector, often supplying fish for direct consumption within their own households and local communities (FAO 2015).

Large-scale fisheries, comprising industrial commercial fishing fleets that are generally of high volume and profit, have contributed to global overfishing and declining fish stocks (Pauly et al. 2002; Worm et al. 2009; Pitcher 2012; Pitcher and Cheung 2013; Pauly and Zeller 2016a; Costello et al. 2016). After World War II, national fishing fleets were heavily subsidized to expand, with environmental impacts largely unrecognized. As global fish stocks have declined, industrial fleets fish farther (Swartz et al. 2010) and deeper (Morato et al. 2006), with more advanced technology (Rousseau et al. 2019), continuing the vicious cycle of "too many boats chasing too few fish." Government subsidies (Sumaila et al. 2010, 2016) and the historical legacy of the "public right to fish" have fostered a sense of entitlement among fishers, with the consequence that private fishing enterprises exploit public resources without compensation to the national treasury (Lam and Pauly 2010). Fishing rights should be tethered to responsibilities of resource stewardship and marine conservation, with access to fishing grounds granted as a "privilege to fish" (Lam and Calcari-Campbell 2012). To address this missing ethical dimension, "pay-as-you-fish" policies have been proposed that charge extraction fees for fish landed (Bromley 2008, 2009). This could be incorporated by adopting a harm principle in fisheries that charges a progressive tax or restricts access according to the potential damage to fish and ecosystems caused by vessel or gear types (Lam 2012).

Small-scale fisheries (Jentoft and Chuenpagdee 2015; Chuenpagdee and Jentoft 2016), meanwhile, encompass artisanal, recreational, and subsistence fisheries. They contribute to food security, nutrition, poverty alleviation, livelihoods, and rural development, vitally in many developing countries, but are often unreported or unregulated in open-access regimes or in poorly managed community fishing grounds (Lam 2016b). Problematic aspects of small-scale fisheries include income and asset poverty, vulnerability to climate and other variability, exclusion from decision-making, a lack of recognition in planning, limited access to social services and infrastructure, and political marginalization (Béné et al. 2007). Human rights (Allison et al. 2011, 2012) and well-being (Weeratunge et al. 2014) are critical to achieving sustainable development for small-scale fisheries and the supported fishing communities. Securing human rights for fishers to decent standards of living, work, health care, and education will require changes to the governance institutions and power structures that determine resource allocation and access.

John Shepherd famously said (ca. 1978): "Managing fisheries is hard: it's like managing a forest, in which the trees are invisible and keep moving around." The voluntary guidelines for securing small-scale fisheries in the context of food security and poverty eradication address the vital importance of small-scale fisheries (FAO 2015). It complements the FAO's Code of Conduct for Responsible Fisheries (1995), which sets out principles and international standards of behaviour for responsible practices, but has poor compliance (Pitcher et al. 2008, 2009). The only legally binding international

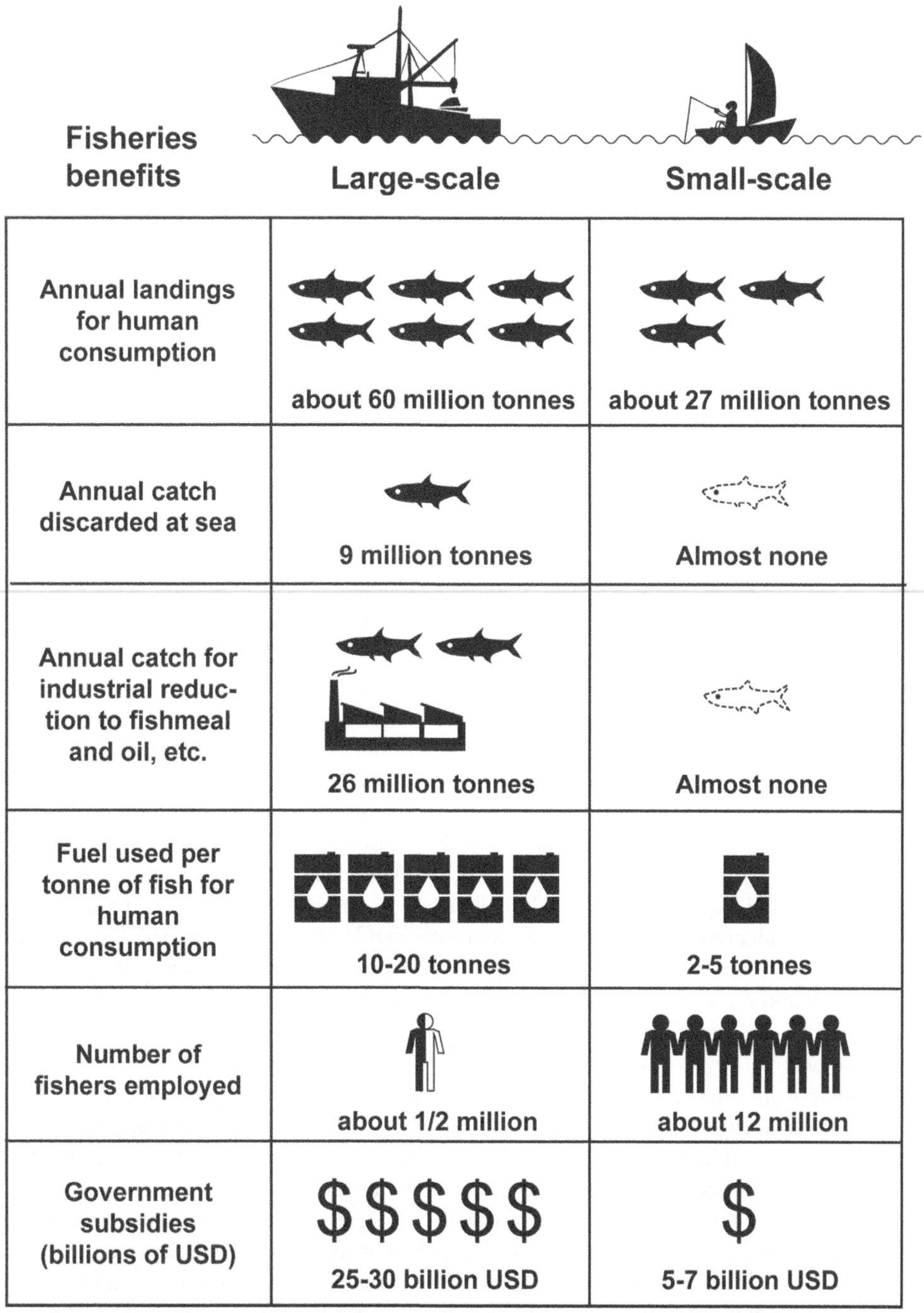

Figure 14.4 Comparison of benefits for large- and small-scale fisheries, with data for the 2000–2010 period reconstructed from Pauly and Zeller 2016a (Figure 14.4, Pauly and Zeller (2016b)

Source: Copyright © 2016 by the authors. Reproduced by permission of Island Press, Washington, DC.

framework for the oceans and responsible fisheries management is UNCLOS III (1982), which specified national jurisdiction of coastal states to within their EEZs. Few judicial mechanisms exist to enforce compliance by coastal states within their EEZs, whereas outside their EEZs, in the High Seas, freedom to fish currently prevails. Calls for a complete ban of fishing in the international open waters of the high seas (White and Costello 2014), however, is gaining traction. Such a ban would create a huge marine protected area that would allow wild fish to recover from overfishing, leading to more fish overall without reducing global catches, as well as creating more equitable economic benefits for smaller fishing nations to fish in their EEZs (Sumaila et al. 2015). The current inequity is because the high-seas fishing fleet is heavily subsidized and dominated by six large fishing nations (Sala et al. 2018).

Meanwhile, the phenomenal growth in aquaculture, that is, the cultivation or farming of aquatic organisms, such as fish, shellfish, and aquatic plants, promises enhanced resilience through diversification of global food production systems, but policies must incentivize resource efficiency, equity, and environmental protection (Troell et al. 2014). Aquaculture systems vary greatly in their intensification, integration, technology, and farmed species (Lam 2016b). Commercial farms use intensive methods to produce commodities for global and regional markets, while family and cooperative farms rely on less intensive practices and low-value species for household subsistence or local markets. Marine fish and crustaceans are typically reared near shore and in coastal ponds, respectively, modifying habitats and degrading ecosystem services, while farmed freshwater finfish are often integrated within agricultural and polyculture systems, causing less environmental damage. Culture system improvements, alternative feed strategies, and species selection offer some remedies (Klinger and Naylor 2012), but less harmful, more expensive technologies are less competitive in the marketplace, as aquaculture systems are not charged for discharges of wastes and effluents or other adverse environmental impacts (Lam 2016b). Farmed carnivorous species rely on wild fish inputs, making them fish consumers, while omnivorous and herbivorous species are often net fish producers, consuming primary producers (e.g., aquatic plants and plankton), organic detritus, and/or low-trophic-level forage fish. More than half of aquaculture production is internationally traded, with a net flow from developing to developed countries, generating issues of distributional equity, energy use, and food miles. As well, global aquaculture production does not substantially displace (to lower resource consumption), but rather, supplements fisheries capture (Longo et al. 2019), despite it often being touted as a solution to food security. This suggests that aquaculture is creating more pressure overall on marine resources through its dependence on feeds, as well as freshwater and land resources for facilities, while exacerbating food inequity. Some of these ecological and social issues may be addressed by enhancing ethical frameworks for aquaculture standards (Haugen et al. 2017) and practices (Lam 2016b).

Farming high-value, high-trophic-level, carnivorous species (e.g., salmon and shrimp) requires wild-caught fish for feed (Naylor et al. 2000, 2009; Naylor and Burke 2005). Atlantic salmon, for example, are fed fishmeal and fish oils from the reduction of forage fish (e.g., herring, sardines, and anchovies), with (wet biomass) ratios of wild fish input to total farmed-fish output (fish in–fish out, FIFO) that were as high as 4:1 (Naylor et al. 2000). Farmed salmon are exported predominantly to affluent countries in the global North, but feed predominantly on forage fish often consumed as food by the poor in the global South and that provide essential ecosystem services. Aquaculture feed efficiency is improving (Naylor et al. 2009), however, with inputs from plant-based sources (e.g., soybeans and grains), single-cell protein and oil, insect and krill protein, and animal and fish processing by-products that were previously waste (fish by-products are now 25%–35% by volume; FAO 2018). The Marine Ingredients Organisation (2017) reported the FIFO for the aquaculture sector overall to be 0.22 in 2015, down from 0.63 in 2000. And for the first time, the FIFO for farmed salmon and trout dropped below 1 to 0.82, making them net fish producers, compared with 2.57 in 2000 (cf., Naylor et al. 2009, Bendiksen et al. 2011). Given their mixed feed, farmed salmon are now "more like pigs" (Cressey 2009), but with reduced health benefits due to their lower omega-3 unsaturated fatty acid contents.

Aquaculture facilities that confine large numbers of carnivorous fish or crustaceans in ponds, tanks or cages in coastal waters, especially mangroves and wetlands, can cause unmitigated environmental damage (Naylor et al. 2000, 2001a, 2001b) and fish welfare concerns. Environmental concerns include nursery habitat destruction; chemical pollution and eutrophication via excessive nutrients, wastes, and antibiotics, causing algal blooms and oxygen dead zones; threats to wild species by farmed-fish escapes (interbreeding and competing with wild populations), parasites (e.g., sea lice), and disease outbreaks (Naylor et al. 2000, 2001a, 2001b; Klinger and Naylor 2012; Ahmed and Thompson 2019). Selective breeding and gene transfer technology are introducing fast maturation and disease resistance traits in farmed species, but raise ecological (Li et al. 2014) and human health concerns (Bremer et al. 2013, 2015). Closed-pen recirculation systems treat and reuse wastewater, which avoids potential hazards of open cages in the ocean, but this eco-friendly technology has costs not borne by open systems unregulated for environmental damage (Lam 2016b). Offshore farms have higher water quality and fewer conflicts with recreational water users than do near-shore farms, but have more complex engineering and licensing requirements (Cressey 2009).

Omnivorous (e.g., Nile tilapia) and herbivorous (e.g., carp) species and filter feeders (e.g., oysters, clams, and mussels) are lower in the food web, with fewer adverse impacts on ecosystems. The Nile tilapia, an omnivorous freshwater fish named the "aquatic chicken" for its speedy and efficient growth, is a net fish producer, with a FIFO ratio of 1:3 (Cressey 2009). Tilapia are less likely to build up toxins such as mercury in their flesh and have a sweet and inoffensive flavour, but are viewed as bland and not favoured in the West, being a third less valuable than Atlantic salmon. They also have a tangible danger of escaping and becoming invasive. Industrial-scale farming of omnivorous and herbivorous species may still divert food away from human consumption, livestock production, and other industrial uses, as they use roughly 15% fishmeal and fish oil for compound feeds, with lower omega-3 contents than in carnivorous fish. Ethical issues in aquaculture (and capture fisheries) production are thus highly complex, with differential benefits and impacts on diverse human populations, fish species, and ecosystems.

Fish Welfare

Beyond the values that humans associate with fish, the ethical aspects of fish welfare also need to be considered in the context of the global seafood industry. Fish have neuro-architecture analogous to mammals: they can process and integrate complex information and organize behaviour in response to environmental stimulation (Huntingford et al. 2006; Andersen et al. 2016). Cognitive and emotional capacities and behavioural patterns of fish vary, both between and within species (Braithwaite et al. 2013). Individual fish can even exhibit distinct "personalities" (Kalueff et al. 2012), behaviours (Martins et al. 2012), and stress-coping styles (Braithwaite et al. 2013), suggesting fish are sentient beings capable of pain, fear, stress, and suffering. Animal welfare is defined as the absence of suffering and is predicated on animal health and wants (Dawkins 2008). Ethically, it has been argued that fish should be given the benefit of the doubt and treated humanely (Sneddon 2006, 2011). Expanding the "moral circle" (Singer 2011) to include fish so that they receive moral consideration in their own right (Meijboom and Bovenkerk 2013) would ensure their basic welfare (Lund et al. 2007). Fish welfare could adopt the five freedoms of animal welfare, namely, freedom from hunger and thirst; freedom from discomfort; freedom from pain, injury or disease; freedom from fear and distress; and freedom to express normal behaviour (Farm Animal Welfare Council 2009). However, normative theories differ in how fish should be treated, depending on how fish physiology is assessed (Broom 2016; Rose et al. 2014) and whether fish are given moral consideration or moral significance (Bovenkerk and Meijboom 2012, 2013).

For seafood producers, fish welfare and other ethical considerations are often tightly interwoven with pragmatic decisions of market choices and technological constraints. Within the aquaculture industry, it is recognized that promoting fish welfare is sound business practice, for both product quality and social acceptability. Thus, knowledge of fish behaviour and ecology are being incorporated into aquaculture practices, driven largely by economic considerations and state regulations of fish welfare requirements in the industry. There is active research and development into improved feeds and feeding methods, treatment for and reduction of conditions leading to sea lice infestations, other parasites, and diseases, and enhanced habitats through reduced densities, recirculating aquaculture systems, and integrated multi-trophic aquaculture. From an ethical perspective, aquaculture could be designed so that animals are treated respectfully and are killed without suffering, i.e., painlessly and unknowingly (Lund and Olsson 2006).

In capture fisheries, however, little to no recognition and regulations of fish welfare exist. Caught wild fish typically asphyxiate or are gutted while still alive on board fishing vessels before being transported for processing or distribution (Metcalfe 2009). Ethical issues in capture fisheries (FAO 2005; Sandøe et al. 2009; Lam 2016b) vary with the individual fish, population, species, and fishery. Target and nontarget species can either be captured (dead or alive) or escape (damaged or undamaged), with differential survival rates (Metcalfe 2009). How fish are treated in the harvesting process (capture, handling, and killing) varies with gear type and selectivity, the type and value of the fish, and whether the fish is killed upon capture or upon landing or processing (Metcalfe 2009). Welfare issues identified in current fishing practices include the specificity of gear and capture method and the experience of pain suffered by the target and nontarget fish, whether the captured target species are killed humanely, and damage to escapes and discards (Metcalfe 2009). Steps towards increasing fish welfare in capture fisheries include bans by the European Union (EU) on "ghost" drift nets lost at sea, in which fish, birds, and cetaceans often become entangled, and live-finning of sharks, as well as bans on the discard of non-target fish species within Norway and the EU Common Fisheries Policy.

Fish welfare in transport and handling through the processing, distribution, and trade of fish is particularly important for live fish and local niche markets and community-supported fisheries. The stress and suffering of fish under various conditions of human treatment have been investigated biologically (Braithwaite and Huntingford 2004; Huntingford et al. 2006; Dawkins 2008; Braithwaite et al. 2013) and ethically (Bergqvist and Gunnarsson 2013). Again, the ethical treatment of farmed fish is more advanced than that of wild fish, owing to differential welfare regulations and the paucity of scientific attention devoted to fish welfare in general. Research into fish cognition and behaviour is growing, however, and new regulations are emerging that, while not explicitly citing fish welfare, do protect aspects of fish welfare, and arguably, instil nascent notions of seafood ethics in fisheries management (Lam 2012; Lam and Pitcher 2012a).

Despite the attention being devoted by global NGOs and prominence within civil society of animal welfare issues, consumer awareness of fish welfare issues is low (Röcklinsberg 2015). Ethical consumerism, however, is rising, particularly in Western developed nations, owing to environmental and animal welfare concerns and recent highly publicized human rights violations in seafood production. For consumers, making sound ethical decisions, such as whether to buy local, organic, or fair-trade seafood, depends on a complex suite of factors specific to the seafood system and opaque interactions with other value chains (Lam 2016b). Consumer decisions, particularly with regard to health and safety concerns, require adequate information. Seafood certification and tracing schemes and awareness campaigns (Ward and Phillips 2010), such as through eco-labels, buying guides, and rating systems, notably the Marine Stewardship Council (Jacquet et al. 2010) and Aquaculture Stewardship Council (Bush et al. 2013), aid consumers to choose sustainably harvested or ethically sourced seafood (Lam and Pitcher 2012a). Meanwhile, ethical consumers have been advised to avoid eating farmed carnivorous, "smart" octopus and instead, to opt for filter-feeding bivalves, such as clams, oysters, and mussels (Jacquet 2017).

Human Values of Fish

The global seafood industry encompasses both wild capture fisheries and aquaculture. Seafood demand is rising along with global human population, affluence, and per capita consumption, while seafood supply is rising with advances in aquaculture, despite declining wild fish stocks. How governments regulate these two seafood production systems has ethical implications for global food security, local resource access, income distribution, and ecological sustainability. Meanwhile, the global seafood market is being influenced by consumer demand for sustainable and ethical fish products and heightened corporate social responsibility. Human rights violations in the global fishing industry, notably forced labour of migrant crews in Thailand (Chantavanich et al. 2016) and on South Korean foreign charter vessels in the New Zealand deep-sea fishery (Simmons and Stringer 2014), often co-occur with other illegal fishing practices, such as discharging bilge oil, high-grading quota species, and dumping rubbish and nonquota species. The sustainability and ethics of the global seafood industry are intricately linked (Lam 2016b): the profits, livelihoods, and choices of social actors along diverse seafood commodity, labour, and value chains directly impact the resilience and well-being of human fishing and coastal communities, as well as wild fish stocks and marine ecosystems. Understanding the complex suite of values that humans associate with fish can foster human well-being and ethical decision-making in seafood policy and governance. These fish values include food and nutritional, socioeconomic (livelihoods and income), cultural, and ecological values (Figure 14.5).

Fish, crustaceans, molluscs, and other aquatic animals provide about 10% of global caloric intake (UNEP 2012). They account for approximately 17% of global animal protein and 7% of all proteins consumed, and over half of the animal protein and minerals for some of the world's poorest (FAO 2018). From 1961 to 2016, global food fish consumption increased annually on average by 3.2%,

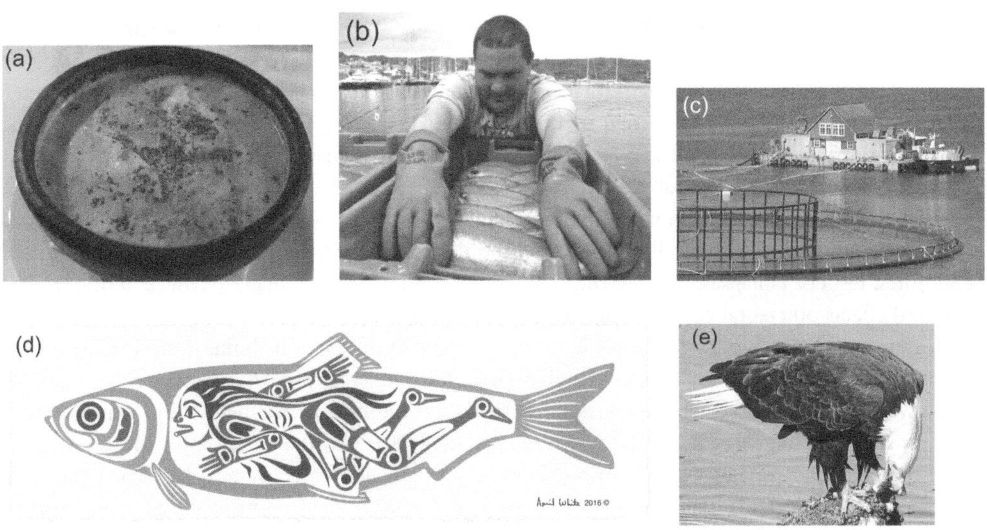

Figure 14.5 Diverse values of fish: (a) food and nutritional value, depicted by congrio chowder, a Chilean national dish (photo: Mimi E. Lam); (b) livelihood value, represented by a small-scale fisher in the Azores, Portugal (photo: Mimi E. Lam); (c) income value, shown by a salmon aquaculture facility in Hardangerfjord, Norway (photo: Mimi E. Lam); (d) cultural value, as depicted in *Herring People* art series showing different predators of herring, here, a female human, inside its silvery outline (© Haida artist, April S.-J. White); and (e) ecological value, captured as a bald eagle feeds upon a spawning herring, Hornby Island, Canada (video still: Barb Biagi)

twice the population growth, and in 2017, was 20.5 kilogram per capita (FAO 2018). Nutritionally, seafood products are important sources of proteins, lipids, vitamins, and micronutrients, with regular fish consumption related to positive health benefits, such as decreased risks of coronary heart and cardiovascular diseases, inflammatory diseases, and cancer (Tilami and Sampels 2017). Analysis of the micronutrient content of 367 marine finfish species worldwide suggests that fish-based food strategies have the potential to substantially contribute to global food and nutrition security, particularly for coastal populations suffering from micronutrient deficiencies (Hicks et al. 2019, Pauly 2019).

Fish have important socioeconomic value as a source of income and livelihoods. Fish products are the most traded food commodities in the world (Lam and Pitcher 2012), accounting for over USD 152 billion in global export trade value in 2017, an increase of 7% from 2016 (FAO 2018). A single bluefin tuna sold at a Japanese fish market for a record USD 1.74 million in 2013, purchased by a restaurateur for its auspicious entry into the New Year. Of the total first sale value of seafood production in 2016, estimated at USD 362 billion, almost two thirds came from aquaculture, the faster-growing food sector. The fish sector supports the livelihoods of 10% to 12% of the global population, with 59.6 million people in 2016 engaged in the primary sector of capture fisheries (40.3 million) and aquaculture (19.3 million) (FAO 2018). Fish workers are concentrated in Asia (85%), Africa (10%), and Latin America and the Caribbean (4%), with 96% of all aquaculture workers in Asia (FAO 2018). Historically, commercial herring fisheries in the North Sea contributed to not only the local food and economy, but also the social fabric, sustaining community relationships and trade in the Hanseatic League for centuries (Pitcher and Lam 2015).

Fish also embody cultural identities, relations of reciprocity, and shared knowledge (Lam and Pitcher 2012b), notably traditional ecological knowledge of indigenous communities (Berkes 2012) and local ecological knowledge of fishers (Haggan et al. 2007). Subsistence and informal exchange economies around fish have preserved its relational value in indigenous cultures, such as for the Saami in Norway (Lam and Borch 2011), the Māori in New Zealand (Lam 2015), and the Haida in Canada (Lam et al. 2019; Pitcher et al. 2017; Jones et al. 2017). In the Pacific Northwest, salmon, herring, and other fish species are believed by indigenous people to be kindred spirits that must be treated respectfully to ensure their return. Cultural taboos and indigenous fishing protocols often proscribe to take only what is needed to meet immediate dietary needs, to gift fish for food, and to not be wasteful. Formalized as Haida ethics (Jones and Williams-Davidson 2000; Jones et al. 2010) or Māori rāhui (McDowall 2011), for example, this may have promoted sustainable practices (Johnsen 2010; Lam 2015). For maritime nations, such as Norway (Kolle et al. 2017), indigenous and local people have relied historically on fish, shaping their values, beliefs, identities, communities, and economies. Fishery collapses, notably the northern Atlantic cod in Canada (Rose 2006), have had profound effects on coastal communities reliant on fishing as a way of life.

Fish are important ecological nodes in complex webs of interactions linking marine organisms. This is particularly evident for forage fish, small pelagic fish that eat planktonic invertebrates and are themselves prey to larger vertebrate consumers, such as predatory fish, marine mammals, and seabirds (Cury et al. 2011; Pikitch et al. 2012, 2014; Sydeman et al. 2017; Surma et al. 2018). The ecosystem-provisioning role of forage fish has inspired international calls for ecosystem-based management (Pikitch et al. 2012, 2014) over the single-species-based management traditionally practised around the world. Not managing fisheries for the ecological interactions of target fish species can lead to unintended consequences with ethical implications. The collapse of Atlantic Canada's cod fishery led to a regime shift in the ecosystem, with cod's invertebrate prey species, for example, lobster, crab, and shrimps, and their lucrative fisheries, now predominating, displacing not only the cod but also the fishers that had fished it for centuries (Levin and Möllmann 2015). Sustaining fisheries and fishery resources thus can be seen as an ethical dilemma of how to reconcile trade-offs among conflicting human values, particularly given the differential costs and benefits impacting diverse social actors along seafood value chains and across sectors.

Seafood Ethics

Seafood ethics adds a value dimension to marine resource management that has the potential to weigh diverse stakeholders' values in public decision-making and shift it towards more ethical governance. These diverse human values along seafood value chains, however, need to be balanced against the intrinsic value of fish and ecosystems. Seafood policies and governance typically focus on promoting ecological sustainability and human well-being, but integrated, comprehensive frameworks that can evaluate the sustainability and ethics of seafood systems are emerging (Lam 2016a, 2016b). Understanding the relationship between sustainability and ethics along integrated seafood value chains is critical to meeting food security, social justice, and environmental challenges. Recent calls for socially responsible seafood have focused on three pillars: to protect human rights and dignity and respect access to resources, to ensure equality and equitable opportunities to benefit, and to improve food and livelihood security (Kittinger et al. 2017; Teh et al. 2019).

To rationally analyse and balance the diverse global seafood values and interests, practical ethics tools (Kaiser 2006) are needed that offer robust ethical deliberation and decision support. One tool, the ethical matrix (Mepham 2000), has been adapted to evaluate the ethics of fisheries (Kaiser and Forsberg 2001; FAO 2005; Lam and Pitcher 2012b), aquaculture (Kaiser and Stead 2002; Kaiser et al. 2007), and seafood production systems generally (Lam 2016b). The (Western) ethical principles of well-being, freedom and justice are evaluated for their impacts on seafood interest groups among the natural system (individual fish, fish populations, and the ecosystem) and the human system (consumers, producers, government agents, other stakeholders, society, and future generations). A specific in-depth ethical analysis will depend on developing quantitative indicators for the seafood system and incorporating the cultural values and ethical perspectives of the stakeholders within the affected community or society. However, to spur on ethical reflection and deliberation in relation to seafood systems now, the ethics of large- and small-scale fisheries (Figure 14.6) and farming carnivorous versus omnivorous species (Figure 14.7) are qualitatively compared (after Lam 2016b).

Another practical ethics approach is to identify empirically stakeholder values and preferences in relation to resource management. For instance, aquaculture stakeholders were asked to rank values, to map value landscapes for development scenarios, and to produce sustainability indicators (Bremer et al. 2016). Stakeholder values and principles were elicited to guide fishery governance (Song and Chuenpagdee 2015) and to examine how people prioritize outcomes of marine management (Loring and Hinzman 2018). Practical ethics also was incorporated into a value- and ecosystem-based management approach (VEBMA) developed to help resolve the Pacific herring fishery conflict in western Canada (Lam et al. 2019). VEBMA combined a participatory approach to reveal diverse stakeholders' values, knowledge, and preferences for herring fishery management with ecological modelling to reveal ecological and socio-economic impacts and risks associated with different management scenarios. It culminated in a science-policy table, a deliberation and decision-support tool that made explicit the societal and ecological implications of alternative fishery policy choices. VEBMA is thus both descriptive and evaluative and prepares the ground for normative judgments in ethical decision-making and governance in resource management (Lam et al. 2019).

Ethical considerations of humans, of fish, and of the environment should be considered along the entire seafood value chain. Narrowly focusing on growing seafood production systems to address global food security often misses distributional inequity and other food ethics issues. Regulations restrict the treatment of farmed, but not wild fish, with largely economic drivers of fish welfare. Meanwhile, the legal landscape vis-à-vis nature is changing, as New Zealand's Whanganui River is now a person under domestic law, India's Ganges River has been granted human rights, and in Ecuador and Bolivia, nature has a "right to integral respect" (Tanasescu 2017). Progress in seafood ethics can be made via governance that invokes not only government regulations, but also the participation, and indeed, cooperation, of the seafood industry with civil society to establish acceptable ethical

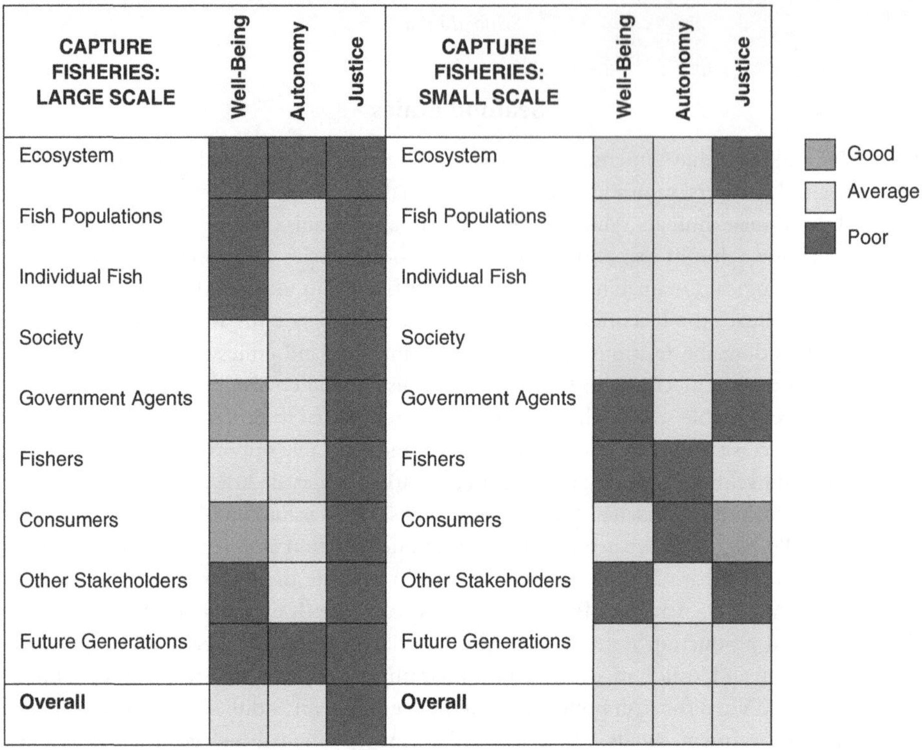

Figure 14.6 Comparing the ethics of global large- and small-scale fisheries

Source: Redrawn from Lam (2016b).

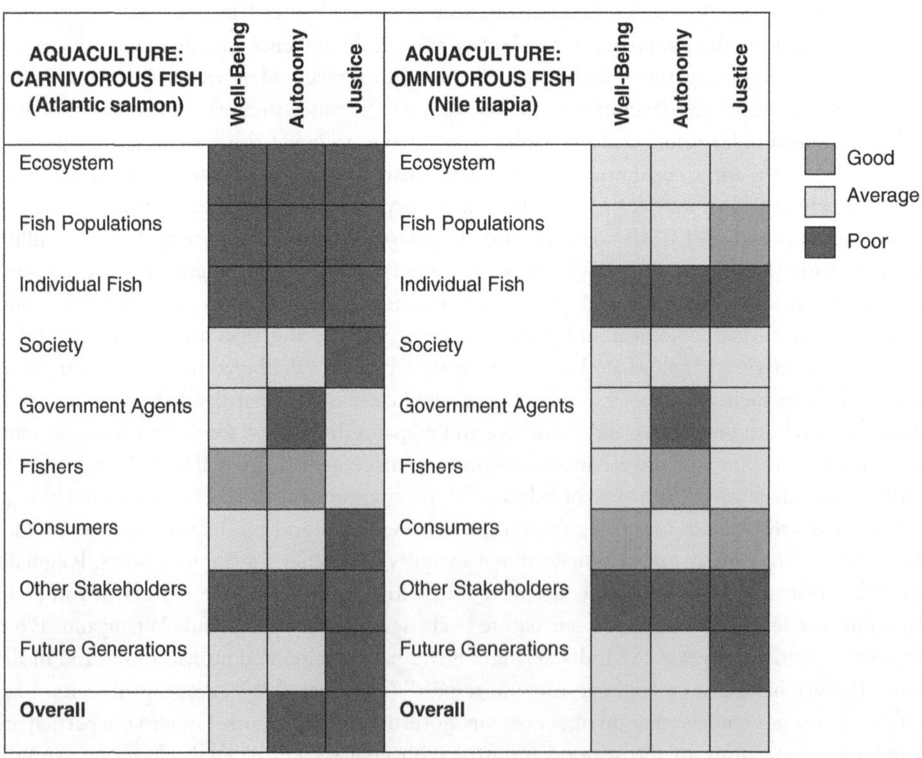

Figure 14.7 Comparing the ethics of farmed carnivorous and omnivorous species

Source: Redrawn from (Lam 2016b).

standards and performance metrics. In the Seafood Business for Ocean Stewardship (SeaBOS) Initiative, chief executive officers of the major global seafood companies, in collaboration with scientists, have made a declaration in connection with the United Nations Sustainable Development Goals to actively work to increase sustainable seafood production, corporate social responsibility, and healthy oceans (Österblom et al. 2017). Incorporating ethical considerations into the management of fishery resources might achieve more sustainable seafood systems. This can be accomplished through ethical governance, that is, participatory, deliberative, transparent, and accountable decision-making, designed to synthesize diverse sources of knowledge and reconcile a plurality of values (Lam et al. 2019).

Conclusion

Resolving ethical issues in seafood policy and governance is a "wicked" problem, as there is no one solution and even the problem definition is complex and varies with one's values and needs. Fish can mean a source of food, income, livelihood, or culture, depending on the fish species or the individual person. Fish also have non-instrumental, intrinsic value and can represent a moral commitment of all humanity to care for and respect non-human animals and the environment. Reconciling these diverse values of fish necessitates explicit answers to questions not typically asked when managing fishery resources, which tend to focus on how much fish to catch and by whom, where, and when. Instead, the conversation needs to expand to include not just fishery stakeholders and managers but all of society, both to ask and to answer questions, such as why and how are we fishing, who is benefitting and who is being harmed, and is it really necessary to fish at all, at least, here and now? That is, instead of racing to grow or catch more, we should learn from the biblical parable of the "Feeding of the 5000" to concentrate our energies on less fish production and waste and more reflection and sharing (Figure 14.8).

Figure 14.8 Logo for Marie Skłodowska-Curie Individual Fellowship project, *eSEAS: Enhancing Seafood Ethics and Sustainability* (funded by the European Union), designed to promote greater reflection in understanding humanity's relationships with, and impacts on fish (© Mimi E. Lam)

Acknowledgments

I am indebted to Bob Fischer for his invitation to contribute this chapter and for his encouragement throughout the writing process. I also would like to thank Tony J. Pitcher and Matthias Kaiser for their constructive comments on my manuscript draft. Finally, I gratefully acknowledge funding for my *eSEAS* project from the European Union's Horizon 2020 research and innovation programme under the Marie Skłodowska-Curie grant agreement No 753937.

Bibliography

Ahmed, N., and Thompson, S. (2019) "The blue dimensions of aquaculture: A global synthesis," *Science of the Total Environment* 652: 851–861.

Allison, E. H., Åsgård, B., and Willmann, R. (2011) "Editorial: Human rights approaches to governing fisheries," *MAST: Maritime Studies* 10(2): 5–13.

Allison, E. H., Ratner, B. D., Åsgård, B., Willmann, R., Pomeroy, R., and Kurien, J. (2012) "Rights-based fisheries governance: From fishing rights to human rights," *Fish and Fisheries* 13(1): 14–29.

Andersen, B. S., Jørgensen, C., Eliassen, S., and Giske, J. (2016) "The proximate architecture for decision-making in fish," *Fish and Fisheries* 17: 680–695.

Béné, C., Macfadayen, G., and Allison, E. H. (2007) "Increasing the contribution of small-scale fisheries to poverty alleviation and food security," *Fisheries and Aquaculture Technical Papers 481*, Rome: FAO.

Bendiksen, E. Å, Johnsen, C. A., Olsen, H. J., and Jobling, M. (2011) "Sustainable aquafeeds: Progress towards reduced reliance upon marine ingredients in diets for farmed atlantic Salmon (*Salmo salar*)," *Aquaculture* 314(1): 132–139.

Bergqvist, J., and Gunnarsson, S. (2013) "Finfish aquaculture: Animal welfare, the environment, and ethical implications," *Journal of Agricultural and Environmental Ethics* 26: 75–99.

Berkes, F. (2012) *Sacred Ecology*, 3rd ed., New York, NY: Routledge.

Bovenkerk, B., and Meijboom, F. L. B. (2012) "The moral status of fish: The importance and limitations of a fundamental discussion for practical ethical questions in fish farming," *Journal of Agricultural and Environmental Ethics* 25: 843–860.

Bovenkerk, B., and Meijboom, F. L. B. (2013) "Fish welfare in aquaculture: Explicating the chain of interactions between science and ethics," *Journal of Agricultural and Environmental Ethics* 26: 41–61.

Braithwaite, V. A., and Huntingford, F. A. (2004) "Fish and welfare: Do fish have the capacity for pain perception and suffering," *Animal Welfare* 13: S87–S92.

Braithwaite, V. A., Huntingford, F., and van den Bos, R. (2013) "Variation in emotion and cognition among fishes," *Journal of Agricultural and Environmental Ethics* 26: 7–23.

Bremer, S., Haque, M. M., Haugen, A. S., and Kaiser, M. (2016) "Inclusive governance of aquaculture value-chains: Co-producing sustainability standards for Bangladeshi Shrimp and Prawns," *Ocean Coastal Management* 131: 13–24.

Bremer, S., Millar, K., Wright, N., and Kaiser, M. (2015) "Responsible techno-innovation in aquaculture: Employing ethical engagement to explore attitudes to GM Salmon in Northern Europe," *Aquaculture* 437: 370–381.

Bremer, S., Reddy, G. P., and Millar, K. (2013) "Using engagement tools to encourage ethical reflection in bioscience and innovation research: Examining the GM Animals case with Indian researchers using the ethical matrix," *Asian Biotechnology & Development Review* 15(2): 21–39.

Bromley, D. W. (2008) "The crisis in ocean governance: Conceptual confusion, spurious economics, political indifference," *MAST: Maritime Studies* 6(2): 7–22.

Bromley, D. W. (2009) "Perspective: Socioeconomics. Abdicating responsibility: The deceits of fisheries policy," *Fisheries* 34(6): 280–290.

Broom, D. M. (2016) "Fish brains and behaviour indicate capacity for feeling pain," *Animal Sentience: An Interdisciplinary Journal on Animal Feeling* 1(3): 4.

Bush, S. R., Belton, B., Hall, D., Vandergeest, P., Murray, F. J., Ponte, S. et al. (2013) "Certify sustainable aquaculture?" *Science* 341: 1067–1068.

Cashion, T., Le Manach, F., Zeller, D., and Pauly, D. (2017a) "Most fish destined for fishmeal production are food-grade fish," *Fish and Fisheries* 18: 1–8.

Cashion, T., Tyedmers, P., and Parker, R. W. R. (2017b) "Global reduction fisheries and their products in the context of sustainable limits," *Fish and Fisheries* 18: 1026–1037.

Cassman, K. G. (2012) "What do we need to know about global food security?" *Global Food Security* 1: 81–82.
Chantavanich, S., Laodumrongchai, S., and Stringer, C. (2016) "Under the shadow: Forced labour among sea fishers in Thailand," *Marine Policy* 68: 1–7. doi: 10.1016/j.marpol.2015.12.015.
Chuenpagdee, R., and Jentoft, S. (eds.) (2016) *Transdisciplinarity for Small-scale Fisheries Governance: Analysis and Practice*. Mare Publication Series Volume 21, Cham, Switzerland: Springer.
Costello, C., Ovando, D., Clavelle, T., Strauss, C. K., Hilborn, R., Melnychuk, M. C. et al. (2016) "Global fishery prospects under contrasting management regimes," *Proceedings of the Royal Academy of the Sciences USA* 113: 5125–5129.
Cressey, D. (2009) "Aquaculture: Future fish," *Nature* 458: 398–400.
Cury, P. M., Boyd, I. L., Bonhommeau, S., Anker-Nilssen, T. et al. (2011) "Global seabird response to forage fish depletion—one-third for the birds," *Science* 334: 1703–1706.
Dawkins, M. S. (2008) "The science of animal suffering," *Ethology* 114: 937–945.
Drury O'Neill, E., and Crona, B. (2017) "Assistance networks in seafood trade—A means to assess benefit distribution in small-scale fisheries," *Marine Policy* 78: 196–205.
Dryzek, J. S., Bächtiger, A., Chambers, S., Cohen, J., Druckman, J. N., Felicetti, A., Fishkin, J. S., Farrell, D. M., Fung, A., Gutmann, A., Landemore, H., Mansbridge, J., Marien, S., Neblo, M. A., Niemeyer, S., Setälä, M., Slothuus, R., Suiter, J., Thompson, D., and Warren, M. E. (2019) "Policy forum: The crisis of democracy and the science of deliberation," *Science* 363(6432): 1144–1146.
Farm Animal Welfare Council (FAWC) (2009) *Farm Animal Welfare in Great Britain: Past, Present, and Future*, London: FAWC.
Food and Agriculture Organization of the United Nations (FAO) (1995) *Code of Conduct for Responsible Fisheries*, Rome: FAO.
FAO (2001) *Ethical Issues in Food and Agriculture*, FAO Ethics Series 1, Rome: FAO.
FAO (2005) *Ethical Issues in Fisheries*, FAO Ethics Series 4, Rome: FAO.
FAO (2015) *Voluntary Guidelines for Securing Sustainable Small-Scale Fisheries in the Context of Food Security and Poverty Eradication*, Rome: FAO.
FAO (2018) *The State of World Fisheries and Aquaculture 2018 — Meeting the Sustainable Development Goals*, Rome: FAO. Licence: CC BY-NC-SA 3.0 IGO.
FAO, IFAD, UNICEF, WFP, and WHO (2018) *The State of Food Security and Nutrition in the World 2018. Building Climate Resilience for Food Security and Nutrition*, Rome: FAO. Licence: CC BY-NC-SA 3.0 IGO.
Fuller, S. D., Picco, C., Ford, J., Tsao, C.-F., Morgan, L. E., Hangaard, D., and Chuenpagdee, R. (2008) *How We Fish Matters: Addressing the Ecological Impacts of Canadian Fishing Gear*, Delta, Canada: Ecology Action Centre, Living Oceans Society, and Marine Conservation Biology Institute.
Haggan, N., Neis, B., and Baird, I. G. (eds.) (2007) *Fishers' Knowledge in Fisheries Science and Management*, Paris: Coastal Management Sourcebooks 4, UNESCO Publishing.
Haugen, A. S., Bremer, S., and Kaiser, M. (2017) "Weaknesses in the ethical framework of aquaculture related standards," *Marine Policy* 75: 11–18.
Hilborn, R. (2007) "Moving to sustainability by learning from successful fisheries," *Ambio A Journal of the Human Environment* 36(4): 296–303. doi: 10.1579/0044-7447(2007)36[296:MTSBLF]2.0.CO;2.
Hilborn, R., Banobi, J., Hall, S. J., Pucylowski, T., and Walsworth, T. E. (2018) "The Environmental Cost of Animal Source Foods," *Frontiers in Ecology and the Environment* 16(6): 329–335. doi: 10.1002/fee.1822.
Huntingford, F. A., Adams, C., Braithwaite, V. A., Kadri, S., Pottinger, T. G., Sandøe, P., and Turnbull, J. F. (2006) "Current issues in fish welfare," *Journal of Fish Biology* 68: 332–372.
Jacquet, J. (2017) "Why oysters, mussels and clams could hold the key to more ethical fish farming," *The Guardian*, www.theguardian.com/sustainable-business/2017/jan/23/aquaculture-bivalves-oysters-factory-farming-environment.
Jacquet, J., Pauly, D., Ainley, D., Holt, S., Dayton, P., and Jackson, J. (2010) "Seafood stewardship in crisis," *Nature* 467: 28–29.
Jentoft, S., and Chuenpagdee, R. (2009) "Fisheries and coastal governance as a wicked problem," *Marine Policy* 33: 553–560.
Jentoft, S., and Chuenpagdee, R. (eds.) (2015) *Interactive Governance for Small-Scale Fisheries: Global Reflections*, MARE Publication Series Volume 20, Cham, Switzerland: Springer.
Johnsen, D. B. (2010) "Salmon, science, and reciprocity on the northwest coast," *Ecology and Society* 14(2): 43.
Jones, R., Rigg, C., and Lee, L. (2010) "Haida marine planning: First nations as a partner in marine conservation," *Ecology and Society* 15(1): 12.
Jones, R., Rigg, C., and Pinkerton, E. (2017) "Strategies for assertion of conservation and local management rights: A haida gwaii herring story," *Marine Policy* 80: 154–167.

Jones, R., and Williams-Davidson, T.-L. (2000) "Applying haida ethics in today's fishery," in H. Coward, R. Ommer, and T. Pitcher (eds.) *Just Fish: Ethics and Canadian Marine Fisheries* (pp. 100–115), Social and Economic Papers No. 23, St. John's, NF: Institute of Social and Economic Research.

Kaiser, M. (2006) "Practical ethics in search of a toolbox: Ethics of science and technology at the crossroads," in J. Gunning and H. Søren (eds.) *Ethics, Law and Society,* Vol. 2 (pp. 35–44), Cardiff, UK: Ashgate Publishing.

Kaiser, M., and Algers, A. (2016) "Food ethics: A wide field in need of dialogue," *Food Ethics* 1: 1–7.

Kaiser, M., and Forsberg, E.-M. (2001) "Assessing fisheries—using an ethical matrix in a participatory process," *Journal of Agricultural and Environmental Ethics* 14: 191–200.

Kaiser, M., Millar, K., Thorstensen, E., and Tomkins, S. (2007) "Developing the ethical matrix as a decision support framework: GM fish as a case study," *Journal of Agricultural and Environmental Ethics* 20: 65–80. doi: 10.1007/s10806-006-9023-8.

Kaiser, M., and Stead, S. M. (2002) "Uncertainties and values in European aquaculture: Communication, management and policy issues in times of changing public perceptions," *Aquaculture International* 10: 469–490.

Kalueff, A. V., Stewart, A. M., Kyzar, E. J., Cachat, J., Gebhardt, M., Landsman, S. et al. (2012) "Time to recognize zebrafish 'Affective' behaviour," *Behaviour* 149: 1019–1036.

Kittinger, J. N., Teh, L. C. L., Allison, E. H., Bennett, N., Crowder, L. B. et al. (2017) "Committing to socially responsible seafood," *Science* 356(6341): 912–913.

Klinger, D., and Naylor, R. (2012) "Searching for solutions in aquaculture: Charting a sustainable course," *Annual Review of Environment and Resources* 37: 247–276. doi: 10.1146/annurev-environ-021111-161531.

Kolle, N., Nielsen, A. R., Døssland, A., and Christensen, P. (2017) *Fish, Coast and Communities: A History of Norway*, Bergen, Norway: Fagbokforlaget.

Kooiman, J., and Bavinck, M. (2005) "The governance perspective," in J. Kooiman, M. Bavinck, S. Jentoft, and R. Pullin (eds.) *Fish for Life: Interactive Governance for Fisheries,* 1st ed. (pp. 11–24), Amsterdam: Amsterdam University Press.

Lam, M. E. (2012) "Of fish and fishermen: Shifting societal baselines to reduce environmental harm in fisheries," *Ecology and Society* 17(4): 18. doi: 10.5751/ES-05113-170418.

Lam, M. E. (2015) "Aboriginal freshwater fisheries as resilient social-ecological systems," in J. Craig (ed.) *Freshwater Fisheries Ecology* (pp. 422–437), Oxford: Wiley-Blackwell.

Lam, M. E. (2016a) "Enhancing Seafood Ethics and Sustainability (eSEAS): A values and ecosystem-based management approach," *Marie Curie Individual Fellowship (MSCA-IFEF-ST) Project 753937*, European Commission.

Lam, M. E. (2016b) "The ethics and sustainability of capture fisheries and aquaculture," *Journal of Agricultural Environmental Ethics* 29: 35–65. doi: 10.1007/s10806-015-9587-2.

Lam, M. E., and Borch, T. (2011) "Cultural valuing of fishery resources by the Norwegian Saami," in L. Westra, K. Bosselmann, and C. Soskolne (eds.) *Globalisation and Ecological Integrity in Science and International Law* (pp. 361–376), Cambridge: Cambridge Scholars Publishing.

Lam, M. E., and Calcari Campbell, M. E. (2012) "The privilege to fish," *Ecology and Society* 17(4): 19. doi: 10.5751/ES-04156-170419.

Lam, M. E., and Pauly, D. (2010) "Who is right to fish? Evolving a social contract for ethical fisheries," *Ecology and Society* 15(3): 16, www.ecologyandsociety.org/vol15/iss3/art16/.

Lam, M. E., and Pitcher, T. J. (2012a) "The ethical dimensions of fisheries," *Current Opinion in Environmental Sustainability* 4: 364–373.

Lam, M. E., and Pitcher, T. J. (2012b) "Fish commoditization: Sustainability strategies to protect living fish," *Bulletin of Science, Technology & Society* 32(1): 31–40.

Lam, M. E., Pitcher, T. J., Surma, S., Scott, J., Kaiser, M., White, A. S. J., Pakhomov, E. A., and Ward, L. (2019) "Value- and ecosystem-based management approach: The pacific herring fishery conflict," *Marine Ecology Progress Series* 617: 341–364. https://doi.org/10.3354/meps12972.

Levin, P. S., and Möllmann, C. (2015) "Marine ecosystem regime shifts: Challenges and opportunities for ecosystem-based management," *Philosophical Transactions of the Royal Society B* 370: 20130275. http://dx.doi.org/10.1098/rstb.2013.0275

Li, L., Pitcher, T. J., and Devlin, R. (2014) "Potential risks of trophic impacts by escaped transgenic Salmon in marine environments," *Environmental Conservation* 42(2): 152–161.

Longo, S. B., Clark, B., York, R., and Jorgenson, A. K. (2019) "Aquaculture and the displacement of fisheries captures," *Conservation Biology* 33(4): 832–841.

Loring, P., and Hinzman, M. S. (2018) "'They're All Really Important, But . . .': Unpacking how people prioritize values for the marine environment in Haida Gwaii, British Columbia," *Ecological Economics* 152: 367–377.

Lund, V., Mejdell, C. M., Röcklinsberg, H., Anthony, R., and Håstein, T. (2007) "Expanding the moral circle: Farmed fish as objects of moral concern," *Diseases of Aquatic Organisms* 75: 109–118.

Lund, V., and Olsson, I. A. S. (2006) "Animal agriculture: Symbiosis, culture or ethical conflict?" *Journal of Agricultural and Environmental Ethics* 19: 47–56.

Marine Ingredients Organisation (2017) *Fish In: Fish Out (FIFO) Ratios for the Conversion of Wild Feed to Farmed Fish, Including Salmon*, www.iffo.net/fish-fish-out-fifo-ratios-conversion-wild-feed.

Martins, C. I. M., Galhardo, L., Noble, C., Damsgård, B., Spedicato, M. T., Zupa, W., Beauchaud, M., Kulczykowska, E., Massabuau, J.-C., Carter, T., Rey Planellas, S., and Kristiansen, T. (2012) "Behavioural indicators of welfare in farmed fish," *Fish Physiology Biochemistry* 38(1): 17–41.

McDowall, R. M. (2011) *IKAWAI: Freshwater Fishes in Māori Culture and Economy*, Christchurch: Canterbury University Press.

Meijboom, F. L. B., and Bovenkerk, B. (2013) "Fish welfare: Challenge for science and ethics—Why fish makes the difference," *Journal of Agricultural and Environmental Ethics* 26: 1–6.

Mepham, B. (2000) "A framework for the ethical analysis of novel foods: the ethical matrix," *Journal of Agricultural and Environmental Ethics* 23(2): 165–176.

Mepham, B. (ed.) (2008) *Food Ethics*, London: Routledge.

Metcalfe, J. D. (2009) "Welfare in wild-capture marine fisheries," *Journal of Fish Biology* 75: 2855–2861.

Morato, T., Watson, R., Pitcher, T. J., and Pauly, D. (2006) "Fishing down the deep," *Fish and Fisheries* 7: 24–34.

Morgan, L. E., and Chuenpagdee, R. (2003) *Shifting Gears: Addressing the Collateral Impacts of Fishing Methods in U.S. Waters*, Pew Science Series, Washington, DC: Island Press.

Naylor, R. L., and Burke, M. (2005) "Aquaculture and ocean resources: Raising tigers of the seas," *Annual Review of Environment Resources* 30: 185–218.

Naylor, R. L., Goldburg, R. J., Primavera, J., Kautsky, N., Beveridge, M. C. M., Clay, J. et al. (2000) "Effect of aquaculture on world fish supplies," *Nature* 405: 1017-1024.

Naylor, R. L., Goldburg, R. J., Primavera, J., Kautsky, N., Beveridge, M. C. M., Clay, J. et al. (2001) "Effects of aquaculture on world fish supplies," *Issues in Ecology* 8: 12.

Naylor, R. L., Hardy, R. W., Bureau, D. P., Chiu, A., Elliott, M., Farrell, A. P., Forster, I., Gatlin, D. M., Goldburg, R. J., Hua, K., and Nichols, P. D. (2009) "Feeding aquaculture in an era of finite resources," *Proceedings of the National Academy of Sciences of the USA* 106(36): 15103–15110.

Naylor, R. L., Williams, S. L., and Strong, D. R. (2001) "Aquaculture—A gateway for exotic species," *Science* 294(5547): 1655–1656. doi: 10.1126/science.1064875.

Österblom, H., Jouffray, J. B., Folke, C., and Rockström, J. (2017) "Emergence of a Global Science—business Initiative for Ocean Stewardship," *Proceedings of the National Academy of Sciences of the United States of America* 114(34): 9038–9043.

Pauly, D. (2007) *Ocean Fisheries: Doom and Gloom*, www.ecoshock.org/2007/03/ocean-fisheries-gloom-doom-daniel-pauly.html.

Pauly, D. (2016) "On the importance of fisheries catches, with a rationale for their reconstruction," in D. Pauly and D. Zeller (eds.) *Global Atlas of Marine Fisheries: A Critical Appraisal of Catches and Ecosystem Impacts* (pp. 1–9), Washington, DC: Island Press.

Pauly, D., Christensen, V., Dalsgaard, J., Froese, R., and Torres, Jr. F. (1998) "Fishing down marine food webs," *Science* 279(5352): 860–863. doi: 10.1126/science.279.5352.860.

Pauly, D., Christensen, V., Guénette, S., Pitcher, T. J., Sumaila, U. R., Walters, C. J. et al. (2002) "Towards sustainability in world fisheries," *Nature* 418: 689–695.

Pauly, D., and Zeller, D. (2016a) "Catch reconstructions reveal that global marine fishery catches are higher than reported and declining," *Nature Communications* 7: 1–9. doi: 10.1038/ncomms10244.

Pauly, D., and Zeller, D. (2016b) "Toward a comprehensive estimate of global marine fisheries catches," in D. Pauly and D. Zeller (eds.) *Global Atlas of Marine Fisheries: A Critical Appraisal of Catches and Ecosystem Impacts* (pp. 171–182). Washington, DC: Island Press.

Pauly, D., and Zeller, D. (2017a) "Comments on FAOs *State of World Fisheries and Aquaculture* (SOFIA) 2016," *Marine Policy* 77: 176–181.

Pauly, D., and Zeller, D. (2017b) "The best catch data that can possibly be? Rejoinder to Ye et al. FAO's statistic data and sustainability of fisheries and aquaculture," *Marine Policy* 81: 406–410.

Pikitch, E., Boersma, P. D., Boyd, I. L., Conover, D. O. et al. (2012) *Little Fish, Big Impact: Managing a Crucial Link in Ocean Food Webs*, A Report from the Lenfest Forage Fish Task Force, Washington, DC: Lenfest Ocean Program.

Pikitch, E. K., Konstantine, J. R., Essington, T. E., Santora, C. et al. (2014) "The global contribution of forage fish to marine fisheries and ecosystems," *Fish and Fisheries* 15: 43–64.

Pitcher, T. J. (2012) "Fisheries," in R. C. Anderson et al. (eds.) *The Berkshire Encyclopedia of Sustainability. Volume 10: The Future of Sustainability* (pp. 108–115), Great Barrington, MA: Berkshire Publishing Group.

Pitcher, T. J., and Cheung, W. W. L. (2013) "Fisheries: Hope or despair?" *Marine Pollution Bulletin*, http://dx.doi.org/10.1016/j.marpolbul.2013.05.045

Pitcher, T. J., Kalikoski, D., Pramod, G., and Short, K. (2008) *Safe Conduct? Twelve Years Fishing Under the UN Code*, Gland: World Wildlife Fund.

Pitcher, T. J., Kalikoski, D., Pramod, G., and Short, K. (2009) "Not honouring the code," *Nature* 457: 658–659.

Pitcher, T. J., and Lam, M. E. (2015) "Fish commoditization and the historical origins of catching fish for profit," *Maritime Studies (MAST)* 14(1): 1–19.

Pitcher, T. J., Lam, M. E., Kaiser, M., White, A. S. J., and Pakhomov, E. (2017) "Hard of herring," in P. Tortell, M. Young, and P. Nemetz (eds.) *Reflections of Canada: Illuminating our Opportunities and Challenges at 150+ Years* (pp. 112–119), Vancouver, Canada: Peter Wall Institute of Advanced Studies.

Pogge, T. (2016) "The hunger games," *Food Ethics* 1: 9–27.

Rittel, H. W. J., and Webber, M. M. (1973) "Dilemmas in a general theory of planning," *Policy Sciences* 4: 155–169.

Röcklinsberg, H. (2015) "Fish consumption: Choices in the intersection of public concern, fish welfare, food security, human health and climate change," *Journal of Agricultural and Environmental Ethics* 28: 533–551.

Rose, G. A. (2006) *Cod: The Ecological History of the North Atlantic Fisheries*, St. John's, NL: Breakwater Books, Ltd.

Rose, J. D., Arlinghaus, R., Cooke, S. J., Diggles, B. K., Sawynok, W., Stevens, E. D., and Wynne, C. D. (2014) "Can fish really feel pain?" *Fish and Fisheries* 15(1): 97–133.

Rousseau, Y., Watson, R. A., Blanchard, J. L., and Fulton, E. A. (2019) "Evolution of global marine fishing fleets and the response of fished resources," *Proceedings of the National Academy of Sciences*, www.pnas.org/cgi/doi/10.1073/pnas.1820344116.

Sala, E., Mayorga, J., Costello, C., Kroodsma, D., Palomares, M. L. D., Pauly, D., Sumaila, U. R., and Zeller, D. (2018) "The economics of fishing the high seas," *Science Advances* 4: eaat2504.

Sandøe, P., Gamborg, C., Kadri, S., and Millar, K. (2009) "Balancing the needs and preferences of humans against concerns for fishes: How to handle the emerging ethical discussions regarding capture fisheries," *Journal of Fish Biology* 75: 2868–2871.

Schuhbauer, A., Chuenpagdee, R., Cheung, W. W. L., Greer, K., and Sumaila, U. R. (2017) "How subsidies affect the economic viability of small-scale fisheries," *Marine Policy* 82: 114–121.

Scientific Advice Mechanism (SAM). (2017) *Food from the Oceans Report*, High Level Group of Scientific Advisors Scientific Opinion No. 3/2017, Brussels: European Commission.

Science Advice for Policy by European Academies (SAPEA). (2017) *Food from the Oceans: How Can More Food and Biomass Be Obtained from the Oceans in a Way That Does Not Deprive Future Generations of their Benefits?* Berlin: SAPEA.

Shepherd, J. (ca. 1978), quote from unpublished lecture, Princeton University, www.jgshepherd.com/thoughts.

Simmons, G., and Stringer, C. (2014) "New Zealand's fisheries management system: Forced labour an ignored or overlooked dimension?" *Marine Policy* 50: 74–80. doi: 10.1016/j.marpol.2014.05.013.

Singer, P. (2011) *The Expanding Circle: Ethics, Evolution, and Moral Progress*, Princeton, NJ: Princeton University Press.

Smith, M. D., Roheim, C. A., Crowder, L. B., Halpern, B. S., Turnipseed, M., Anderson, J. L. et al. (2010) "Sustainability and global seafood," *Science* 327: 784–786.

Sneddon, L. U. (2006) "Ethics and welfare: Pain perception in fish," *Bulletin of the European Association of Fish Pathologists* 26(1): 6–10.

Sneddon, L. U. (2011) "Cognition and welfare," in C. Brown, K. Laland, and J. Krause (eds.) *Fish Cognition and Behavior*, 2nd ed. (pp. 405–434), Oxford: Wiley-Blackwell.

Song, A. M., and Chuenpagdee, R. (2015) "Eliciting values and principles of fishery stakeholders in South Korea: A methodological exploration," *Society and Natural Resources* 28: 1075–1091.

Sumaila, U. R., Khan, A. S., Dyck, A. J., Watson, R., Munro, G., Tyedmers, P., and Pauly, D. (2010) "A bottom-up re-estimation of global fisheries subsidies," *Journal of Bioeconomics* 12: 201–225.

Sumaila, U. R., Lam, V., Le Manach, F., Swartz, W., and Pauly, D. (2016) "Global fisheries subsidies: An updated estimate," *Marine Policy* 69: 189–193.

Sumaila, U. R., Lam, V. W. Y., Miller, D. D., Teh, L., Watson, R. A., Zeller, D., Cheung, W. W. L., Côté, I. M., Rogers, A. D., Roberts, C., Sala, E., and Pauly, D. (2015) "Winners and losers in a world where the high seas is closed to fishing," *Scientific Reports* 5, Article Number 8481.

Surma, S., Pitcher, T. J., Kumar, R., Varkey, D., Pakhomov, E. A., and Lam, M. E. (2018) "Herring supports northeast pacific predators and fisheries: Insights from ecosystem modelling and management strategy evaluation," *PLoS One* 13: e0196307.

Swartz, W., Sala, E., Tracey, S., Watson, R., and Pauly, D. (2010) "The spatial expansion and ecological footprint of fisheries (1950 to Present)," *PLoS One* 5(12): e15143.

Sydeman, W. J., Thompson, S. A., Anker-Nilssen, T., Arimitsu, M., Bennison, A., Bertrand, S. et al. (2017) "Best practices for assessing forage fish fisheries-seabird resource competition," *Fisheries Research* 194: 209–221.

Tanasescu, M. (2017) *When a River Is a Person: From Ecuador to New Zealand, Nature Gets Its Day in Court*, https://theconversation.com/when-a-river-is-a-person-from-ecuador-to-new-zealand-nature-gets-its-day-in-court-79278.

Teh, L. C. L., Caddell, R., Allison, E. H., Finkbeiner, E. M., Kittinger, J. N. et al. (2019) "The role of human rights in implementing socially responsible seafood," *PLoS One* 14(1): e0210241. https://doi.org/10.1371/journal.pone.0210241

Tilami, S. K., and Sampels, S. (2017) "Nutritional value of fish: Lipids, proteins, vitamins, and minerals," *Reviews in Fisheries Science & Aquaculture* 26(2): 243–253. doi: 10.1080/23308249.2017.1399104.

Troell, M., Naylor, R. L., Metian, M., Beveridge, M., Tyedmers, P. H., Folke, C. et al. (2014) "Does aquaculture add resilience to the global food system?" *Proceedings of the National Academy of Science of the United States* 111(37): 13257–13263.

United Nations Convention on the Law of the Sea (UNCLOS III, concluded December 10, 1982, entered into force November 16, 1994), 1833 United Nations Treaty Series 397, UnitedNationsDocumentA/Conf.61/122, www.un.org/Depts/los/convention_agreements/texts/unclos/unclos_e.pdf.

United Nations Environment Programme (UNEP). (2012) *Avoiding Future Famines: Strengthening the Ecological Foundation of Food Security through Sustainable Food Systems*, Nairobi: UNEP.

van der Sluijs, J. P. (2017) "The NUSAP approach to uncertainty appraisal and communication," in C. L. Spash (ed.) *Routledge Handbook of Ecological Economics: Nature and Society* (pp. 301–310), London: Routledge.

Ward, T., and Phillips, B. (2010) "Seafood ecolabeling," in R. Q. Grafton, R. Hilborn, D. Squires, M. Tait, and M. Williams (eds.) *Handbook of Marine Fisheries Conservation and Management* (pp. 608–617), Oxford: Oxford University Press.

Weeratunge, N., Béné, C., Siriwardane, R., Charles, A., Johnson, D., Allison, E. H. et al. (2014) "Small-scale fisheries through the wellbeing lens," *Fish and Fisheries* 15(2): 255–279. doi: 10.1111/faf.12016.

White, C., and Costello, C. (2014) "Close the high seas to fishing?" *PLoS Biology* 12(3): e1001826. doi: 10.1371/journal.pbio.1001826.

Worm, B., Hilborn, R., Baum, J. K., Branch, T. A., Collie, J. S., Costello, C. et al. (2009) "Rebuilding global fisheries," *Science* 325: 578–585.

Ye, Y., Barange, M., Beveridge, M., Garibaldi, L., Gutierrez, N., Anganuzzi, A., and Taconet, M. (2017) "FAO's statistic data and sustainability of fisheries and aquaculture: Comments on Pauly and Zeller," *Marine Policy* 81: 401–405.

Zeller, D., Cashion, T., Palomares, M., and Pauly, D. (2018) "Global marine fisheries discards: A synthesis of reconstructed data," *Fish and Fisheries* 19: 30–39.

15

SMALL-SCALE ANIMAL AGRICULTURE

Donald W. Bruckner

Introduction

This chapter addresses the ethics of producing and consuming the products of small-scale animal agriculture. "Small-scale animal agriculture" is meant as a contrast to conventional ("intensive," "industrialized," or "factory") farming of animals. The farming operations under discussion are those that aim to distinguish themselves from modern, conventional, industrialized farms, especially with regard to animal welfare and environmental impacts. For example, such operations do not keep laying hens in battery cages but often in portable shelters, where they can go outside during the day to scratch and peck for insects and eat vegetation. Meat chickens are raised similarly. Pigs are kept outdoors, never in gestation crates, and engage in natural behaviors of wallowing, rooting, and foraging for dropped fruit and nuts. Cattle and other ruminants (sheep and goats) are not confined to feedlots but graze in managed pasture rotations. In such conditions, hens' beaks do not need to be trimmed, nor pigs' tails docked. While animals receive necessary medicine and veterinary care, they are not given antibiotics preventively or to induce growth, nor do they receive growth-inducing substances such as hormones and steroids. The farmer grows or sources forage and grain raised with sound environmental goals such as maximizing the physical and biological integrity of the soil, minimizing pesticide and synthetic fertilizer inputs, and minimizing fossil fuel use. The feed, and animals, might even be Certified Organic. Animals are either slaughtered on the farm or handled, transported, and slaughtered in such a way as to minimize stress.

Operations engaging in such practices abound and are becoming more widespread. The American Pastured Poultry Producers Association has 651 member farms. Humane Farm Animal Care has approved 196 farms as Certified Humane.[1] LocalHarvest (2018) lists more than 4,400 small farms in the US that sell some sort of meat. The vast majority of these farms describe their practices as humane, free-range, pastured, sustainable, organic, biodynamic, low-till, or pesticide-, antibiotic-, hormone-, or genetically modified (GM)–free. In sum, it is easy to find products of animals raised on nonconventional farms.[2]

The aim of this chapter is to outline some of the philosophical questions that need to be answered and empirical data that need to be considered when discussing the morality of producing and consuming products of small-scale animal agriculture. I conclude with a suggestion for how to think about the morality of small-scale animal agriculture in the context of thinking about the morality of the practices of modern life more broadly.

Most philosophical arguments concerning agricultural production and consumption practices focus on the harm that is caused by one set of practices in comparison to other practices. The harms canvassed usually are to animals, the environment, and humans. So the discussion here will be divided into those three categories. In each case, the discussion will be subdivided further into theoretical questions and empirical questions.

Animals

Some actions are claimed to harm farm animals. Just what a *harm* is has been debated by philosophers. On one intuitively appealing account of harm—the counterfactual comparative account—an animal is harmed by an action if it does less well when that action is performed than it would have done had that action not been performed.[3] Asserting that the animal does better or worse under one scenario than under another is a claim about the animal's *well-being* or *welfare*. While much has been written in the history of philosophy about human well-being, or what makes a human's life go well (see Bradley 2015; Fletcher 2016 for introductions), very little has been written about animal well-being. To be sure, many philosophers and others make claims about specific actions or practices that contribute to or detract from animal welfare, but philosophers have not paused sufficiently to develop a theoretical account of what animal well-being is in order to ground such claims.[4] Scientists researching animal welfare have done better in thinking about what animal welfare is, but the development of robust philosophical theories has not been their focus.[5]

Answering this philosophical question about what animal welfare is matters to claims about the moral status of certain husbandry practices. A hedonist theory of animal welfare would say that what is good for animals is pleasure and what is bad for them is pain. It seems reasonable to think that the squeal a pig makes when it is castrated without anesthesia indicates that it is in pain and that the enthusiastic move toward feed dumped in a trough and subsequent contented grunting indicates pleasure. So, on a hedonic theory, pigs can be doing quite well when kept in common industrialized farming conditions: confined in indoor pens with good bedding, ventilation, food and water, medical care, and safety from painful predation. However, people who advocate for the sort of husbandry under discussion often claim that animals kept in this way have poor lives even if they have little pain and much pleasure, because they do not get to engage in their natural behaviors, such as rooting and wallowing in the case of pigs, grazing for ruminants, and scratching for insects, eating vegetation, and roosting at night for chickens. So this welfare claim supposes that hedonism about animals is false, and there is more to a good animal life than pleasure and the absence of pain. To support such a claim, the proponent might appeal to a theory of animal well-being according to which the good life for an animal is one that allows it to develop and live according to its specific nature. Of course, this theory will also need accounts of porcine nature, poultry nature, bovine nature, and so on.

Supposing we had a well-developed and defended account of animal well-being and we had a husbandry practice that gave animals the highest possible level of well-being, there would still be questions about the morality of the husbandry practice. In particular, the sort of husbandry practices under consideration call for killing animals, which seems obviously to harm them. Philosophers have found it surprisingly difficult to justify the common belief that death is a harm even to humans. To apply the counterfactual account of harm to death, we would have to compare how well the animal would be doing if it were not killed to how well it would be doing if it were killed. If it is not killed, it will be doing quite well, rooting and wallowing as a pig, say. If it is killed, how well is it doing after its death? Apparently, there is no such thing as the pig's welfare after its death because it is not living at all, let alone well or badly. We cannot compare its positive welfare in the first scenario with its nonexistent welfare in the second.

In brief, the harm of death is puzzling.[6] If it is a harm, it appears that we will need an account of harm other than the otherwise intuitively appealing counterfactual comparative account. However, we should not from the start rule out the possibility that, due to the kinds of beings that livestock animals are, painless death does not harm them. One account of the harm of death for humans is that it frustrates desires they have for their own long-term futures. Since animals do not have such desires, a painless death may not harm them.[7]

A related question that is at least as difficult to answer is whether it can harm or benefit an animal to create it. If it does benefit an animal to create it (if it lives a life that is high in well-being), then it might turn out to be a net benefit to the animal to raise it for food even if its death is a harm to it. Some of the issues involved in thinking about the benefit or harm of creation are the same as the questions about the harm or benefit of death. We cannot just compare, for example, the animal's welfare if we create it to its welfare if we do not.[8]

Another important theoretical question is whether and how we should weigh the well-being of animals against each other. Some people (e.g., Warren [1986] in response to Regan [1983]) think that animals with higher degrees of cognitive sophistication and sentience are due more moral consideration than those lower on those two scales, presumably because the greater those capacities are, the greater well-being the animal is capable of. Suppose we agree, perhaps for reasons of cognitive sophistication, that pigs have a much higher moral status than chickens (as Budolfson [2016: 171] claims). Does it make sense to say that the moral weight of one pig life is, say, 10 times the moral weight of one chicken life? If so, and if it takes 30 chickens to produce the same nutritional value as one pig, and if it is wrong to kill chickens and pigs, is it then three times as wrong to kill 30 chickens as it is to kill one pig? Or are the moral values of animal lives not comparable in this way? We seem to make such comparative judgments all the time: we kill rabies vectors such as rats that could spread disease to humans or pets and we kill a dog whose dangerousness to humans is established by repeated attacks, but we never kill either dogfight organizers or toddlers whose dangerousness to dogs is established by repeated abuse. Maybe making these judgments as we do is either a mistake or relevantly different from comparing chicken and pig lives. Again, much will depend on what animal well-being is and which animals are harmed by death and to what extent.

The moral status of different production and consumption practices will depend on our answers to these theoretical questions. The sort of husbandry practices mentioned in the introduction will be morally preferable to conventional practices on most answers to most of the theoretical questions adumbrated earlier. The issue becomes more difficult when comparing agricultural practices needed for a vegan diet with the husbandry practices under consideration. This leads us to important empirical questions.

The sort of agriculture needed to support a plant-based diet causes much incidental harm and death to animals such as mice, rabbits, and other mammals, ground-nesting birds, reptiles, amphibians, and insects through plowing, discing, planting, and harvesting. Davis (2003) provides a calculation based on empirical studies that shows that an exclusively plant-based diet would take more animal lives than a diet that included meat from some large pasture-raised herbivores, because grazing them would produce far less incidental harm to field animals than the multiple intense field operations needed for producing food from plants.[9] Our answers to the theoretical questions about well-being, death, and comparability now have significant bearing. Is one painless bovine death equally bad (if it is bad) as one painless mouse or bird death? If so, then a diet including some pasture-raised herbivore meat is morally preferable to a plant-based diet on the basis of comparing numbers of animal lives lost. If a bovine death is many times worse than a mouse or bird death, then the omnivorous diet is worse and the vegan diet is better than Davis's original calculation represents.

Another relevant empirical question about the agriculture supporting a vegan diet is how many field animals are killed painlessly in field operations in comparison to how may suffer slow, painful deaths from injuries or are painfully injured and then recover. If it is good, on balance, for a cow to lead a natural albeit short life ended by a painless death, and if a significant number of field animals die ghastly deaths in cropping for vegan food, then this may tip the balance back in favor of omnivorism.

Suppose we have an agricultural practice that gives animals very good lives and that we have answers to the theoretical questions about the benefit of coming into existence and the harm of death that make it good for those animals overall to be brought into existence, even to die for our pleasure. Even with that, animals do not go from living to dying automatically. One might claim that there are well-known welfare abuses at slaughterhouses that are standard practice, as documented by horrific internet videos. There are also harms associated with transportation to distant slaughterhouses, such that animals sometimes arrive stressed, dehydrated, or injured. So even if it is good for animals to live according to the sort of husbandry practices under discussion, the horror of the circumstances surrounding their deaths might tip the balance again back toward the moral requirement to abstain from raising and consuming them.

An advocate of small-scale husbandry has responses to these practical challenges. First, it is not clear how much we *know* about the abuses that occur in slaughterhouses. Surely images and accounts of abuses abound, but so do images and accounts of very bad professorial practices. Such images and accounts are at best anecdotal and at worst highly selective propaganda.

Second, many of the recorded abuses are due to the sheer volume of animals processed at an industrial slaughter facility, which often does not allow pausing the disassembly line if, say, a cow is not properly rendered unconscious before the disassembly continues. These giant industrial slaughter facilities are not the sort where small-scale farmers are likely to have their animals slaughtered, since such facilities most often serve large suppliers who wholesale large numbers of animals to the facilities, which then prepare the meat for wholesale to distributors. Smaller producers have opted out of the commodity meat industry and are therefore more likely to use a small, local, custom or federally inspected processor. These processors are small enough to allow the farmer, and often the farmer's individual customers, to place orders for how the animal will be portioned into specific cuts of meat and then pick up the finished product directly from the slaughter facility. Smaller producers often have the option of using such facilities, which are not rare.[10] In 2016, for example, there were 650 federally inspected plants that slaughtered cattle in the US. Of these 650 plants, 471 slaughtered between 1 and 999 head of cattle that year, while 13 plants each slaughtered more than a million cattle that year (United States Department of Agriculture 2017: 60). Arithmetic shows that the *largest* small facility slaughters about 3.8 cattle per day while the *smallest* of the mega-plants slaughters about 3,800 cattle daily. It is reasonable to think that far more abuses occur at the latter than the former, again, due to the sheer volume.

The Environment

Another consideration frequently offered against animal agriculture is that it is bad for the environment. A first theoretical question is whether morality requires not harming the environment. One answer is yes, because harming humans is wrong and harming the environment harms humans. But not every action that harms the environment also harms currently existing humans. Although we may be obligated not to make each other's lives go less well, it is unclear whether we are obligated not to make the lives of people who do not yet exist go less well. The so-called Non-Identity Problem makes the issue even more complicated. To illustrate: suppose that a society has a choice between (A) providing incentives to parents to produce fewer children and (B) not interfering but letting population grow and the associated environmental destruction continue unabated. Under (A), many parents will make different choices about when to conceive and fewer people will be created than under (B). The lives of those who come to exist under (A) will be better than those under (B) because of the better condition of the environment under (A), but the sets of people who exist under the two scenarios will be largely disjoint. Some of the same problems now arise as when we discussed creating farm animals. If we choose (B), do we harm those people who come to exist? It is not clear we do, because they would not have existed had we not also created the worse environmental conditions under which they exist. Perhaps we even benefit them by making it so that they come to exist, at least as long as they have worthwhile lives. If that is right, then perhaps it is better to pursue (B), because although people are not doing as well on average, the total amount of well-being is greater under (B) than under (A).

For these reasons, basing a moral obligation to protect the environment on obligations to humans is controversial. An alternative position is that we have an obligation to preserve the environment and its resources independent of any obligations to humans or animals. Biocentric ethicists (e.g., Taylor 1986) argue that living things, in general, have morally relevant intrinsic value, so we have an obligation not to take life unnecessarily. So we should not kill trees unless the value to humans (here the Non-Identity Problem may arise again: Which humans?) is sufficiently weighty. Maybe instead of basing environmental obligations on the value of individual living things, we should base them instead on the value of collections of living things like a forest, a whole ecosystem, or most broadly "the land," as Leopold (1949) calls it, and say that such collections deserve moral consideration. These questions cannot be resolved here. This suffices to indicate some of the difficult theoretical issues involved in grounding moral reasons not to harm the environment.

Supposing that we settle the theoretical questions in favor of the existence of moral reasons not to harm the environment, interesting empirical questions arise. One motivation for the sort of husbandry under discussion is the avoidance of environmental and animal harms. There are possible tensions between these two goals, again, depending on our answers to theoretical questions about animal welfare. If we suppose that it is better for an animal's well-being not to be raised in confinement, then we have an animal-welfare-based reason in favor of pasturing livestock, as small-scale operations typically do. There are environmental reasons both for and against pasturing livestock.

According to one study (Capper 2012), because a conventional system for raising beef (including growth-enhancing technologies such as steroids and hormones) is more efficient, it requires fewer animals and less water, land, and fossil fuel energy to produce a unit of beef than a pasture system.[11] Another study finds that energy consumption, soil erosion, fertilizer runoff, and pesticide contamination are higher in a conventional system than in grazing systems (Modernel et al. 2013). While these studies agree that the carbon footprint of pastured beef is larger than that of conventionally raised beef, CleanMetrics (2018), whose Food Carbon Emissions Calculator is based on a variety of models and many data sources, claims that the greenhouse gas (GHG) emissions of pastured beef are slightly lower than those of conventional beef. The issue is complicated and there is scholarly disagreement.

There are additional environmental impacts to consider. Tilling soil annually for crops to feed to livestock rather than allowing soil to remain intact as pasture is harmful to the structure of the soil and the support of microbes that benefit plant nutrition. Moreover, keeping animals in confinement concentrates manure and may require bedding (made from plant products) in order to keep the animals dry and healthy. When manure and bedding are spread—using fossil fuel—at low rates over crop fields, this benefits the soil for the nutrients, organic matter, and increased microbial activity. In some confinement systems, however, more manure is produced than can be absorbed locally. When manure is left concentrated in feedlots or piled outdoors after cleaning livestock pens, it can run off and introduce pathogens such as *E. coli* into underground aquifers used for human drinking water (Mekonnen and Hoekstra 2012: 413). Manure deposited by animals in a well-managed pasture (rotated to maximize biomass production and prevent topsoil runoff and parasite concentration from overgrazing), however, improves the soil.

These considerations largely pertain to beef, sometimes claimed to be the worst environmental offenders among livestock due mostly to their comparatively low feed-conversion efficiency and high GHG emissions from enteric fermentation. Nevertheless, this case illustrates the general point that the moral status of producing and consuming meat from livestock raised according to different methods will depend (at least) on the answers to (a) the theoretical question about what animal well-being is, (b) whether and to what extent we have moral reasons not to harm the environment, and (c) how these considerations trade off against each other. Our judgment will also depend on answers to empirical questions about (d) the environmental impacts of different agricultural practices as well as the specific (e) animal welfare and (f) environmental practices on a given farm.

What is perhaps more interesting and more complicated is comparing the environmental impact of a diet including some animal products to that of a plant-based diet. On many measures, many diets including animal products from small-scale animal agriculture are worse for the environment than most vegan diets.

Most studies conclude, for example, that a diet including meat is associated with higher GHG emissions than a diet without meat (Scarborough et al. 2014). One reason for the higher emissions is that obtaining nutrition from meat, however it is raised, is relatively inefficient. Instead of growing crops for human consumption, crops are grown for livestock consumption, and more nutritional value has to go into the livestock than comes out in consumable meat.

One must take great care to avoid overgeneralization or overstatement about GHG emissions of a diet including meat. Almost any diet will be responsible for the emission of some GHGs from production and transportation. For example, rice, tomatoes, potatoes, and apples all have large carbon footprints that rival or exceed chicken and pork per unit of nutrition.[12] That is, we don't simply compare the GHG emissions per pound of product, since the nutritional profiles of a pound of cabbage, a pound of corn, and pound of pork, for example, are quite different. Instead, we compare kilograms of carbon dioxide (CO_2)–equivalent emissions per 10,000 kilocalories of energy or per kilogram of protein. When we do this, some fruits and vegetables are found to be higher emitters than some meats.

Another environmental consideration is the quantity of water used in livestock production. A leading scientific source puts the water footprint of beef at 1847 gallons per pound (Mekonnen and Hoekstra 2012: Table 15.1) and claims that, generally, "[m]eat-based diets have a larger water footprint compared to a vegetarian diet" (2012: 410). Philosophers have mostly accepted uncritically the claim that the large water footprint of livestock production is a reason not to produce and consume meat.

Again, however, we should take care to avoid overgeneralization or overstatement. The water footprints of some vegan protein staples such as almonds, lentils, and beans often rival and sometimes exceed the water footprints of some meats per unit of nutrition.[13] For example, it takes more water to produce a kilogram of protein from almonds than from beef, lamb, pork, or chicken.

Pastured livestock, however, has a much larger water footprint than confined livestock. This is surprising because pastured animals get a lot of their hydration from the water content of the plants they graze. But there is room to doubt the relevance of water footprints as commonly used. To see why, understand that a water footprint has three components. *Blue water* is water drawn from surface water or underground aquifers for drinking or for irrigating crops grown for livestock. *Gray water* is the water needed to dilute pollutants to acceptable levels. *Green water* is water evaporated from the soil or transpired by crops or pasture forage (Mekonnen and Hoekstra 2012: 402). Pastured livestock has a larger water footprint overall than confined livestock because it has a much larger green water footprint. The reason is that even though industrially raised livestock requires feed, which requires a lot of green water to grow, the animals convert these cereal grains to meat much more efficiently than they convert pasture forage to feed. So much more pasture forage and, hence, land and green water, is needed for pastured livestock (Mekonnen and Hoekstra 2012: 407).

But is green water a relevant metric? It accounts for 87.2% of water used for livestock production globally (Mekonnen and Hoekstra 2012: 408), but it is not at all clear that its consumption by crops or pasture forage imposes any environmental cost. Anything that grows in the soil will consume water. Indeed, some natural vegetation, such as conifers or fruit trees, consumes more water than crops or pasture forage (Allen et al. 1998: Table 12), so in comparison to many other uses, agriculture is a net negative user of green water. The agricultural use of blue and gray water is more clearly environmentally costly, at least in places where water is not abundant.

Speaking for the moment just about ruminants and not chickens and pigs, conventionally raised ruminants have much larger blue- and gray-water footprints than ruminants raised exclusively on pasture, due to irrigation and fertilization of feed crops and water pollution from manure (Mekonnen and Hoekstra 2012: 408). As some leading scientists acknowledge, freshwater problems relate to blue-water scarcity and water pollution as opposed to competition over green water (Mekonnen and Hoekstra 2012: 408). Because green-water consumption is of questionable relevance and because pastured ruminant meat has much smaller blue- and gray-water footprints (although chicken and pork do not—see Mekonnen and Hoekstra 2012: Table 14.1), this is a point in favor of small-scale production of pastured ruminants over their industrial production.

Finally, to return to a point discussed above in considering confinement versus grazing, there are environmental benefits that come with pastured livestock that are absent in crop production, whether the crops are for human or livestock consumption, including better soil health and prevention of soil, fertilizer, and pesticide runoff.

It is probably true that taking into account *all* the environmental harms and benefits of omnivorous diets that include products of small-scale animal operations in comparison to vegan diets, most of the former are worse for the environment than most of the latter, although whether and to what extent they are will depend considerably on the details of the farms being compared. This should not close the case on such omnivorous diets. Single-minded focus on environmental considerations ignores animal welfare that, as outlined earlier, may argue in favor of extensive livestock operations over production of plants for human consumption. The harms and benefits to humans from agricultural practices must also be considered.

Humans

Just as we asked about animal well-being, we need to ask about the nature of human well-being. Does the good life consist in pleasure and the absence of pain, the satisfaction of one's desires, or something more objective, such as the development and expression of one's human nature? Is human well-being such that it can be traded off against animal well-being? Some philosophers think so. Others think that weighing human and animal well-being against each other is a mistake, because animals have rights that cannot be infringed for the sake of good consequences for humans (Regan 1983), just as enslaving a group of humans would be wrong even if total human well-being would be higher with slaves than without.

Assuming for the moment, however, that human and animal well-being are fungible, there may be important values associated with small-scale animal agriculture that contribute to human well-being. Agrarian philosophy is a neglected line of thinking about nature, in general, and agriculture, in particular (see Thompson 2010; Lipscomb 2016). One guiding agrarian thought is that producing food and fiber helps us connect our humanity with nature and see our place in nature. Other things (hiking, studying environmental science) connect us with nature but not in the way that farming connects us to the very source of our life through the elemental experience of producing our own food. Even if one does not get one's hands dirty, one can still participate in agriculture vicariously by supporting small-scale producers rather than eating nameless, placeless, mass-produced food.

While small-scale husbandry is an important source of agrarian value, it may not be unique. Hunting is another source. Hunting is not agriculture, but much agrarian thinking is rooted in a connection to nature in much the same way that hunting (and angling) can be (see Cerulli 2012). If it is important to animal well-being to live as naturally as possible, then hunting for food will come out ahead of husbandry on that measure. Additionally, there is no reason to think that small-scale plant agriculture cannot also realize the agrarian values at issue. Therefore, small-scale animal husbandry may not uniquely embody the important sources of human value recognized in agrarian thinking. Even if human and animal well-being are fungible, considering the value of agrarian activities for humans does not seem to add to the case for small-scale animal agriculture.

Turning now to empirical questions, if human and animal well-being can be compared and traded off, then producing and consuming animal products may nevertheless be justifiable on consequentialist grounds. Lomasky (2013: 192) argues that even with industrialized farming conditions as they are, "the case for eating meat is overwhelmingly positive" due to the extremely high social, cultural, gustatory, and aesthetic value of food involving meat. Lomasky thinks that industrially raised animals have lives that are, at worst, neutral. Although that is dubious, there is a plausible case, outlined earlier, that the overall welfare impact on animals from small-scale animal agriculture is better than from plant-based agriculture. If that is right, then producing and consuming such animal products is even more strongly justified when we consider the benefits to humans. And even if plant agriculture is better for animals, if the benefit to humans from consuming animal products is sufficiently high, then we would still have a justification for eating meat that considers both human and animal well-being. Crisp (1988) argues on similar consequentialist grounds that the benefits to humans and animals together imply a requirement to eat the flesh of nonintensively reared animals.

Moreover, if we are concerned with the overall harm footprint of our food choices, we would do well to consider the spectrum of human harm involved in its production. Work in slaughter mega-plants is dangerous (Schlosser 2001). One expects that conditions are considerably better in the smaller facilities more common with small producers. There is much human harm associated with some vegan foods. Demand for quinoa allegedly priced the staple out of reach for locals who depended on it. The avocado trade has been rife with extortion by criminal gangs. Farmworkers on large vegetable and fruit plantations are often migrants who lead generally poor lives performing back-breaking work in harassing and degrading conditions with no escape (see Budolfson 2016: 164, 171–172).

Implications

The morality of producing and consuming products of small-scale animal agriculture can only be assessed after answering myriad complicated theoretical and empirical questions about animal, environmental, and human harm, including those adumbrated above. Reasonable people can disagree on the answers to these questions due to the thorny philosophical issues and the empirical complexities involved. On defensible answers to those theoretical and empirical questions, a typical diet including products of small-scale animal agriculture is less harmful than a typical plant-based diet. On other defensible answers to those questions, even a conscientious diet including such animal products is more harmful than a conscientious plant-based diet.

I now argue that even if a conscientious omnivorous diet is more harmful, one is not obligated to adopt a less harmful plant-based diet. The reason is that if the suboptimal harm footprint of a conscientious omnivorous diet makes it morally impermissible, then a sweeping array of lifestyle practices that enrich modern life is also impermissible. Since these practices are not impermissible, neither is an omnivorous diet.

The sort of practices I have in mind are suboptimal in some of the very ways that producing and consuming products of small-scale animal husbandry might be suboptimal. Take, as a first example, keeping dogs and cats as pets. In the US, dogs and cats (hereafter "pets") consume 33% as much animal-derived energy as humans do. Pets are therefore responsible for 25% to 30% of the total environmental impacts of livestock production, in terms of land, water, fossil fuel, phosphate, and biocides (Okin 2017). In brief, pets are responsible for about 25% of the total harm to the environment and to animals that comes from raising animals for food. Recognizing these harms of feeding pets animal-derived foodstuffs, Milburn argues that dogs "should be converted to vegan diets" (2017: 199). While a vegan diet for a dog may be less suboptimal than one including animal products, a vegan diet still has nonnegligible environmental and animal-harm footprints.

Keeping pets is unnecessary. The opportunity cost of keeping pets is high because the financial resources used to keep pets could be put, instead, toward benefiting humans in serious need, if only by diverting the nutrition (whether plant- or animal-derived) currently used for pets to humans. Keeping even vegan pets, therefore, is suboptimal. Yet it is hard to believe that keeping even vegan pets is morally impermissible on these grounds. So the impermissibility of a practice does not follow from its suboptimality, at least if it is true that keeping vegan pets is morally permissible.

Some may be willing to bite the bullet on grounds of consistency and declare that keeping even vegan pets is morally impermissible. If that is right, then it is not difficult to keep the ball rolling. Many other things are morally impermissible if small-scale animal agriculture is, all on the grounds of the suboptimality for the environment, humans, or animals. Lamb is the most GHG-intensive meat, releasing about one-and-a-half times the GHG per pound as beef and about six times that of chicken. In comparison, consider flying round trip from Chicago to London for an academic conference (or a similar distance for domestic conferences)—standard practice for many academics. Flying such a distance in one year produces as much additional GHG emissions as eating an additional 699.6 quarter-pound lamb burgers in a year (or 1.92 lamb burgers every day of the year) on top of one's normal diet.[14] If eating that many extra lamb burgers is impermissible just in light of its egregious GHG suboptimality, so is flying such a distance. On the same grounds, it is also impermissible to travel long distances for family vacations or to have a habit of making short leisure trips to see a movie, go out to dinner, visit a friend across town, or drive one's dog to the dog park (a double offense!).

It is easy to continue the list of unnecessary, suboptimal practices. According to one well-researched source (Klimek 2014), 54 million acres are required globally to grow the grain and grapes needed for alcohol production. Enormous numbers of field animals are displaced, injured, or killed on this cropland. Massive amounts of water and energy are used in the production and transportation of alcoholic beverages. In the US alone, about 107,000 people die annually due to alcohol consumption—their own or someone else's (Centers for Disease Control and Prevention 2016). So the production and consumption of alcoholic beverages have immense environmental, animal, and human harm footprints. If suboptimality implies impermissibility, drinking alcohol is impermissible.

We are concerned with anthropogenic harms to animals, the environment, and humans. One way that humans produce harm is by producing humans. An individual who does not procreate produces one human's worth of harm—his own. An individual who does procreate thereby produces, *additionally*, somewhere between one human's worth of harm (the one child) and a boundless sum of harm from generation after generation of reproduction. That additional harm due to procreation is far greater than the additional harm that an omnivore produces over a vegan. So if omnivory is impermissible due to its harm footprint, reproduction is also impermissible.

If suboptimality entails impermissibility, and if the considerations above show that small-scale animal husbandry is suboptimal, then not only is small-scale animal husbandry impermissible, but so is a wide array of suboptimal modern practices. It seems absurd that these practices are impermissible, so there must be something wrong with the standard vegan argument that infers impermissibility from suboptimality.

One thing that could be wrong with the vegan's argument is the assumption that we should minimize our harm footprint.[15] It is this assumption that implies that we should jettison our way of life, for we cannot minimize our harm footprint while we keep pets, travel unnecessarily, drink alcohol, or procreate. Perhaps that very strong assumption should be replaced by the assumption that each individual should keep her harm footprint within some budget, as suggested by Budolfson (2016:

167). I won't hazard a guess as to what a reasonable harm budget would be, except to note again that the harm of producing one child is greater than the harm of an omnivorous diet. From this, we can infer that a reasonable harm budget must be higher than the harm of an omnivorous diet, at least if producing one child is morally permissible. In any case, if harm minimization is not required by morality and only keeping harm within some budget is required, then we are both freed of the implication that jettisoning our modern way of life is required and also free to find many ways to reduce harm other than by changing our food production and consumption practices.

So let's entertain the assumption that it is not harm minimization, but staying under some budget, that is required. With that assumption, and remembering the many harm-producing elements of our lifestyles noted earlier, the strength of the case for abstaining from producing and consuming products of small-scale animal husbandry is considerably weakened. There are so many ways of producing or reducing harm that abstinence from production and consumption of animal products loses the spotlight. Again, we are assuming that, holding all else constant, a careful vegan diet produces less harm than a conscientious omnivorous one. Consider, however, some of the many ways in which not all else may be constant.

A nontraveling, pet-owning conscientious omnivore with children who works for a social services agency providing family planning and contraception probably has a lower harm footprint than a childless non-pet-owning vegan working in academia who travels regularly to academic conferences.

A childless, SUV-driving conscientious omnivore who heats her modest home to 60 degrees in winter, buys her clothes secondhand or through free-trade channels, and works as a lawyer for victims of domestic violence probably has a lower harm footprint than a hybrid-driving stay-at-home vegan parent of two who heats his larger home to 68 degrees in winter and sometimes buys retail clothing manufactured in sweatshops.

Depending on the details, all these individuals may be living permissibly under the maximum harm footprint allowed by morality. If one's goal is to keep one's harm footprint under budget, one can do better than to focus on food production and consumption decisions.

Despite the current preoccupation with the morality of agricultural and dietary choices, there may be nothing morally special about them generally, much less about choices involving small-scale animal agriculture, in particular, in comparison to other life choices. The harms and benefits of dietary choices and such agriculture are among all the harms and benefits that should be considered in connection with the overall impact of one's activities on animals, the environment, and humans. There may be good reasons to abstain from producing and consuming animal products. These reasons are no stronger than the parallel reasons to abstain, for instance, from procreation, keeping pets, inessential travel, drinking alcohol, or occupations not aimed at reducing animal, environmental, and human harm.

Acknowledgments

Thanks to Bob Fischer for constructive feedback on a draft of this chapter.

Notes

1. See American Pastured Poultry Producers Association (2018) and Humane Farm Animal Care (2018).
2. This may not have been the case two decades ago when Fox claimed that "the opportunities for obtaining 'environmentally friendly' meat are extremely rare in practice" (2000: 166) and Engel complained of "[t]he 'free range' fantasy" (2000: 880) and claimed that "there are no humane livestock transportation companies and no humane slaughterhouses" (2000: 881).
3. See Klocksiem (2012) for an entry into the literature on the counterfactual comparative account.
4. For two rare exceptions, see Rollin (2005, 2006, 2014) and Moore (2017).

5. See Fraser (2014) for a helpful overview of conceptualizations of animal welfare by scientists.
6. For an entry into the literature on death, see Purves (2017).
7. Scruton (2000: 142) and Belshaw (2016) hold views like this, while Harman (2011) and Bower and Fischer (2018) argue against it.
8. Gardner (2016) is one entry into relevant literature on the possibility of benefiting a person by creating her.
9. Matheny (2003) and Lamey (2007) challenge the calculation.
10. Fischer (2018), McWilliams (2015), and DeGrazia (2009) all seem to assume that all (off-farm) slaughter happens at the same plants and that abuses of handling, transportation, and slaughter are therefore "standard even in small-scale agriculture" (Fischer 2018: 240). DeGrazia, for example, claims that "farm animals [from traditional family farms] typically experience harms associated with transport to the slaughterhouse, rough handling from farm to truck to slaughterhouse, and the process of slaughter itself (see earlier description of factory farms)" (2009: 160).
11. Some questionable assumptions of this study are that 50% of pasture for beef is irrigated and that grass-fed cattle raised on a small scale are transported, on average, the same distance (100 mi) to slaughter as conventionally raised cattle are (Capper 2012: 137, 132).
12. See Budolfson (2016, 2018) and the references therein for the carbon footprints of common meats and vegan foods.
13. Again, see Budolfson (2016, 2018).
14. I used 2.63 kilograms of CO_2-equivalent emissions per quarter pound of lamb meat, the average of the figures given by Ledgard et al. (2010: 6), Environmental Working Group (2018), and CleanMetrics (2018). This emissions estimate is considerably higher than most found in the scientific literature—see Nijdam et al. (2012) and the references therein. For the roundtrip flights, I used 1.84 metric tons, the average of the outputs given by Carbon Footprint Ltd. (2018), The Climate Protection Partnership (2018), and Offsetters (2018).
15. Budolfson (2016: 163–167) also casts doubt on this assumption, although for different reasons than presented here.

Bibliography

Allen, R. G., Pereira, L. S., Raes, D., and Smith, M. (1998) "Crop evapotranspiration—Guidelines for computing crop water requirements," *FAO Irrigation and Drainage Paper 56*, Food and Agriculture Organization of the United Nations, Rome, www.fao.org/docrep/X0490E/X0490E00.htm.

American Pastured Poultry Producers Association (2018) *Pastured Poultry Farmer Directory*, www.apppa.org/Directory.

Belshaw, C. (2016) "Meat," in B. Bramble and B. Fischer (eds.) *The Moral Complexities of Eating Meat*, Oxford: Oxford University Press.

Bower, M., and Fischer, B. (2018) "Categorical desires and the badness of animal death," *Journal of Value Inquiry* 52(1): 97–111.

Bradley, B. (2015) *Well-Being*, Cambridge, UK: Polity Press.

Budolfson, M. (2016) "Consumer ethics, harm footprints, and the empirical dimensions of food choices," in A. Chignell, T. Cuneo, and M. C. Halteman (eds.) *Philosophy Comes to Dinner: Arguments about the Ethics of Eating* (pp. 163–181), New York, NY: Routledge.

Budolfson, M. (2018) "Food, the environment, and global justice," in A. Barnhill, T. Doggett, and M. Budolfson (eds.) *The Oxford Handbook of Food Ethics* (pp. 67–94), Oxford: Oxford University Press.

Capper, J. L. (2012) "Is the grass always greener? Comparing the environmental impact of conventional, natural, and grass-fed beef production systems," *Animals* 2(2): 127–143.

Carbon Footprint Ltd (2018) *Carbon Footprint Calculator, Flights*, https://calculator.carbonfootprint.com/calculator.aspx?tab=3.

Centers for Disease Control and Prevention (2016) "Alcohol and public health: Alcohol-Related Disease Impact (ARDI) application," *National Center for Chronic Disease Prevention and Health Promotion*, https://nccd.cdc.gov/DPH_ARDI/default/default.aspx.

Cerulli, T. (2012) *The Mindful Carnivore: A Vegetarian's Hunt for Subsistence*, New York, NY: Pegasus Books.

CleanMetrics (2018) *Food Emissions Calculator*, www.foodemissions.com/foodemissions/Calculator.aspx.

Climate Protection Partnership (2018) *Flight Calculator*, https://co2.myclimate.org/en/flight_calculators/new.

Crisp, R. (1988) "Utilitarianism and vegetarianism," *International Journal of Applied Philosophy* 4(1): 41–49.

Davis, S. (2003) "The least harm principle may require that humans consume a diet containing large herbivores, not a vegan diet," *Journal of Agricultural and Environmental Ethics* 16(4): 387–394.

DeGrazia, D. (2009) "Moral vegetarianism from a very broad basis," *Journal of Moral Philosophy* 6(2): 143–165.
Engel, Jr., M. (2000) "The immorality of eating meat," in L. P. Pojman (ed.) *The Moral Life: An Introductory Reader in Ethics and Literature*, New York, NY: Oxford University Press.
Environmental Working Group (2018) *Meat Eater's Guide: Report; Climate and Environmental Impacts*, www.ewg.org/meateatersguide/a-meat-eaters-guide-to-climate-change-health-what-you-eat-matters/climate-and-environmental-impacts/.
Fischer, B. (2018) "Arguments for consuming animal products," in A. Barnhill, T. Doggett, and M. Budolfson (eds.) *The Oxford Handbook of Food Ethics* (pp. 241–266), Oxford: Oxford University Press.
Fletcher, G. (2016) *The Philosophy of Well-Being: An Introduction*, London: Routledge.
Fox, M. A. (2000) "Vegetarianism and planetary health," *Ethics and the Environment* 5(2): 163–174.
Fraser, D. (2014) "Animal welfare in the context of animal production," in P. Thompson and D. Kaplan (eds.) *Encyclopedia of Food and Agricultural Ethics*, Dordrecht: Springer.
Gardner, M. (2016) "Beneficence and procreation," *Philosophical Studies* 173(2): 321–336.
Harman, E. (2011) "The moral significance of animal pain and animal death," in T. L. Beauchamp and R. G. Frey (eds.) *The Oxford Handbook of Animal Ethics* (pp. 726–737), Oxford: Oxford University Press.
Humane Farm Animal Care (2018) *Producers Who Are Certified Humane*, https://certifiedhumane.org/whos-certified-2/.
Klimek, O. (2014) "The environmental impact of alcoholic drinks production," *Dramming: Everything Whiskey*, www.dramming.com/2014/08/22/the-environmental-impact-of-alcoholic-drinks-production/.
Klocksiem, J. (2012) "A defense of the counterfactual comparative account of harm," *American Philosophical Quarterly* 49(4): 285–300.
Lamey, A. (2007) "Food fight! Davis versus regan on the ethics of eating beef," *Journal of Social Philosophy* 38(2): 331–348.
Ledgard, S. F., Lieffering, M., McDevitt, J., Boyes, M., and Kemp, R. (2010) *A Greenhouse Gas Footprint Study for Exported New Zealand Lamb*, New Zealand: AgResearch.
Leopold, A. (1949) *A Sand County Almanac*, Oxford: Oxford University Press.
Lipscomb, B. J. B. (2016) "'Eat Responsibly': Agrarianism and meat," in A. Chignell, T. Cuneo, and M. C. Halteman (eds.) *Philosophy Comes to Dinner: Arguments about the Ethics of Eating* (pp. 56–72), New York: Routledge.
LocalHarvest (2018) "Farms," [search results for 'meat'], www.localharvest.org/search.jsp?&ty=0&nm=meat.
Lomasky, L. (2013) "Is it wrong to eat animals?" *Social Philosophy and Policy* 30(1–2): 177–200.
Matheny, G. (2003) "Least harm: A defense of vegetarianism from Steven Davis's omnivorous proposal," *Journal of Agricultural and Environmental Ethics* 16(5): 505–511.
McWilliams, J. (2015) *The Modern Savage: Our Unthinking Decision to Eat Animals*, New York, NY: Thomas Dunne Books.
Mekonnen, M. M., and Hoekstra, A. Y. (2012) "A global assessment of the water footprint of farm animal products," *Ecosystems* 15(3): 401–415.
Milburn, J. (2017) "The animal lovers' paradox? On the ethics of 'Pet Food,'" in C. Overall (ed.) *Pets and People: The Ethics of Our Relationships with Companion Animals*, Oxford: Oxford University Press.
Modernel, P., Astigarraga, L., and Picasso, V. (2013) "Global versus local environmental impacts of grazing and confined beef production systems," *Environmental Research Letters* 8(3): 1–10.
Moore, A. (2017) "Objectivism about animal and alien well-being," *Analysis* 77(2): 328–336.
Nijdam, D., Rood, T., and Westhoek, H. (2012) "The price of protein: Review of land use and carbon footprints from life cycle assessments of animal food products and their substitutes," *Food Policy* 37(6): 760–770.
Offsetters (2018) *Flight Emissions Calculator*, www.offsetters.ca/education/calculators/flight-emissions-calculator.
Okin, G. S. (2017) "Environmental impact of food consumption by dogs and cats," *PLoS One* 12(8): 1–14.
Purves, D. (2017) "Desire satisfaction, death, and time," *Canadian Journal of Philosophy* 47(6): 799–819.
Regan, T. (1983) *The Case for Animal Rights*, Berkeley, CA: University of California Press.
Rollin, B. E. (2005) "Animal happiness: A philosophical view," in F. D. McMillan (ed.) *Mental Health and Well-Being in Animals* (pp. 235–242), Oxford: Blackwell.
Rollin, B. E. (2006) *Animal Rights and Human Morality*, 3rd ed., Amherst, NY: Prometheus.
Rollin, B. E. (2014) "*Telos* and farm animal welfare," in P. Thompson and D. Kaplan (eds.) *Encyclopedia of Food and Agricultural Ethics*, Dordrecht: Springer.
Scarborough, P., Appleby, P. N., Mizdrak, A., Briggs, A. D. M., Travis, R. C., Bradbury, K. E., and Key, T. J. (2014) "Dietary greenhouse gas emissions of meat-eaters, fish-eaters, vegetarians and vegans in the U.K.," *Climatic Change* 125(2): 179–192.
Schlosser, E. (2001) *Fast Food Nation: The Dark Side of the All-American Meal*, New York, NY: Houghton Mifflin.
Scruton, R. (2000) *Animal Rights and Wrongs*, 3rd ed., London: Metro Books.

Taylor, P. (1986) *Respect for Nature: A Theory of Environmental Ethics*, Princeton, NJ: Princeton University Press.

Thompson, P. B. (2010) *The Agrarian Vision: Sustainability and Environmental Ethics*, Lexington, MA: University Press of Kentucky.

United States Department of Agriculture (2017) "Livestock slaughter: 2016 summary," *National Agricultural Statistics Service*, usda.mannlib.cornell.edu/usda/current/LiveSlauSu/LiveSlauSu-04–19–2017.pdf.

Warren, M. A. (1986) "Difficulties with the strong animal rights position," *Between the Species* 2(4): 163–173.

16
SUBSISTENCE HUNTING

Raymond Anthony and Gary Varner

Introduction

There is no widely agreed-on definition of *subsistence hunting*. In this chapter, we use a discussion of subsistence hunting as a way to compare and contrast Euro-American and Indigenous ways of thinking about human–animal relationships in general. In this introductory section, we briefly discuss the variety of conceptions of *subsistence* and subsistence *hunting* before turning to separate discussions of (1) subsistence hunting in the literature of Euro-American philosophy and (2) the broader, more general notion of subsistence *cultures* and the place of hunting within those cultures. Our focus there is on subsistence cultures of the Circumpolar North (i.e., above the Arctic Circle) and some other indigenous groups (namely, the Yup'iit) in western, southwestern, and southcentral Alaska and the Russian Far East.[1]

One definition of subsistence hunting is "hunt[ing] to provide nutrition that cannot be grown" (Woods et al. 2009: 500). Aboriginal Inuit peoples (the generic name for culturally related human groups that inhabit the Arctic Circle regions of Alaska, Greenland, and Canada), could not have survived on plant-based foods alone. For them, hunting was clearly necessary for adequate nutrition and most people think that humans are justified in killing animals when it is *necessary* for human survival. We will refer to this as "subsistence hunting *narrowly construed*."

The term *subsistence* is itself used in different ways, however, as George Wenzel observed in his study of controversy surrounding seal hunting in the Canadian Arctic circa 1990. In one sense, it refers to when "human life is sustained at the barest possible margin." In another sense, it describes "a self-contained system," in which people do not take inputs from or send outputs to a larger socio/economic system. In the latter sense, as soon as Inuit peoples began using imported technology, such as snowmobiles and guns, and began exporting pelts and other by-products of hunting, they were no longer practicing "subsistence" hunting (Wenzel 1991: 57–58). Wenzel argues, however (and we concur), that consumption and use of the products of traditional hunting and fishing often play an essential role in the maintenance of aboriginal communities even after they are economically and technologically intertwined with modern European and North American economies. Thus, while hunting is no longer necessary for physical survival, subsistence hunting is still necessary for the continuation of time-honored practices (despite the use of some modern devices) and to the survival of aboriginal, subsistence *cultures*.

Relatedly, anthropologists increasingly see the term *subsistence* as a contested concept. Early anthropologists such as Lewis Henry Morgan saw subsistence hunters as "fully dependent on nature, while farmers had cultural means—technical, social, mental, and moral—that placed them in a superior position to nature" (Moss 2010: 123–124). Some contemporary anthropologists are emphasizing the need to rethink the concept of subsistence and the ways anthropological research contributes to challenges faced by indigenous peoples. For instance, Polly Wheeler and Tom Thornton note that "Euro-American conceptions tend to be static, restrictive and minimalist, often defining subsistence as 'the minimum resources necessary to support life,'" whereas "Alaska Natives typically define subsistence in dynamic, broad, and holistic ways, as 'our culture,' 'our way of being,' or 'our life'" (Wheeler and Thornton 2005: 70). And Madonna Moss (2010) calls for research that more accurately depicts how prehistoric subsistence hunters in Alaska managed their takes of various animals and actively adapted to changing environmental conditions.[2]

In this chapter, we do not defend any specific definition of *subsistence hunting*, but the cases that we discuss all involve what we would describe as *hunting in the service of vital human needs or interests*. In the second section, we discuss what we are calling subsistence hunting *narrowly construed* (where hunting is necessary for physical human survival), and we note some reasons that have been offered for relying on hunting for food by people in industrialized societies for whom food insecurity is not an issue. In the third section, we turn to the notion of subsistence cultures. For some or many indigenous peoples, subsistence is not just a matter of getting "nutrition that can't be grown," nor does it refer to just hunting, fishing, and gathering. As we emphasize, practices associated with hunting, foraging, fishing, and gathering in subsistence cultures are connected to and shape a people's metaphysics, epistemology, ethics, and day-to-day living.

Subsistence Hunting in Euro-American Philosophy

Although much has been published by academic philosophers on hunting in general, very little has been published on subsistence hunting in particular. For instance, in late 2018, a keyword search for "subsistence" and "hunting" in *The Philosopher's Index* returned only one entry (Comstock 2004), and the entry on "Hunting and Fishing" in the *Encyclopedia of Environmental Ethics and Philosophy* (Callicott and Frodeman 2009) includes sections on "recreational hunting" and "ecological hunting" but not on subsistence hunting. The reason for the lack of attention may be that where hunting is necessary for physical human survival—what we are calling subsistence hunting narrowly construed—almost everyone agrees that hunting is justified. To illustrate why, in this section we briefly describe both *utilitarian* and *rights-based* justifications for subsistence hunting in such cases.

"Utilitarianism" refers to a family of theories that hold one thing in common: that the right thing to do is arrange things so that aggregate happiness is maximized. One way of justifying subsistence hunting where it is necessary for physical human survival turns on a common theme among utilitarian authors. Utilitarians from John Stuart Mill in the 19th century (1957 [1861]) through the 20th century's Peter Singer (1990, 2011) have held that the cognitive capacities of human beings (at least typical adults) can enhance their conscious experiences in ways that make a human life add more happiness to the world than does the life of a nonhuman animal. This belief would support an argument for subsistence hunting from a utilitarian perspective insofar as it supports the conclusion that a human dying from lack of food would remove more happiness from the world than the death of an animal.

As understood in terms of underlying philosophical commitments, however, the distinction between *animal rights* and *animal welfare* is related to the difference between such aggregative, utilitarian thinking and rights-based thinking in ethics. In this spirit, philosophers sometimes characterize individual rights as "trump cards against utilitarian thinking in ethics." The idea is that if an individual has moral rights, then the aggregated good that could come to others cannot justify harming that individual (at least without their consent). On an animal rights view, then, the fact that a human's continued life would contribute more happiness to the aggregate than would an animal's does not suffice to justify the human in killing the animal, even if that would be the only way for the human to survive.

Nevertheless, that subsistence hunting would be justified—at least in "the paradigm case"—is acknowledged even by Tom Regan, whose book *The Case for Animal Rights* (2004 [1983]) is the most widely discussed rights-based view in opposition to Singer's utilitarianism. It is difficult to do justice to the arguments of Regan's 400-page book in a short section of this chapter. Here, we focus only on two principles that are fundamental to his position, without describing various detailed arguments he gives for these principles and for expanding the scope of rights in the animal kingdom.

Regan argues that an animal rights view ultimately depends on the kind of respect that is due to individuals whose lives have what he calls "inherent value." He compares the value that utilitarian thinking ascribes to individuals (both humans and animals) to thinking of them as "utility receptacles" so that the value accorded an individual is actually a kind of *instrumental* value, a question of how much happiness an individual's "cup" is capable of holding. Regan contrasts this kind of the value of an individual with what he calls *inherent* value, or the value that individuals have "in themselves" rather than as means to some end, including the utilitarian end of maximizing aggregate happiness. Noting that we commonly think of humans as having such inherent value, Regan gives detailed arguments for extending inherent value and individual rights to a range of nonhuman animals, including at least mammals and birds.[3]

Regan then gives detailed arguments for two principles that express the appropriate attitude of respect toward all those with individual rights. The first is this:

> *Regan's worse-off principle:* When it is necessary to choose between violating the rights of *n* individuals and the rights of *m* individuals, and the harms that one or more of the *n* individuals will suffer via those rights violations will be *non-comparably worse* than any harm that any of the *m* individuals will suffer, then we ought to choose to avoid overriding the rights of that *worse-off individual or individuals.*
>
> *(This is a paraphrase of Regan 2004 [1983]: 308)*

We unpack the notion of a "non-comparable harm" and a "worse-off individual" in the following discussion, but first, here is Regan's other principle:

> *Regan's miniride principle:* When it is necessary to choose between violating the rights of the many and the rights of the few, and each of the individuals will suffer a *comparable* harm via those rights violations, then we ought to choose to override the rights of the few.
>
> *(This is a paraphrase of Regan 2004 [1983]: 308)*

Regarding these principles, two important points of clarification are needed.

The first is that Regan admits that, *where it applies*, the second principle leads to the same conclusions as utilitarian reasoning. For *where it applies*, the miniride principle holds that we ought to be minimizing the number of "comparable harms" that we cause to individuals, and this would seem to require doing the same thing as if our goal were to maximize aggregate happiness, in this case by minimizing aggregate harm. Although his miniride principle implies the same conclusions as utilitarian thinking, Regan argues that the reasoning behind it is fundamentally different: when the harms involved are comparable, showing equal respect to individuals with inherent value means not counting anyone's right against being harmed to count for more than anyone else's (paraphrasing 2004 [1983]: 305).

The second is that, as we noted earlier, the concept of "harm" needs unpacking. Regan conceives of harms as (roughly) lost opportunities for desire formation and satisfaction. On this conception of harm, a painless, unanticipated death is still a harm, because it forecloses all future opportunities for desire formation and satisfaction, and thus, for any given individual, death is the most serious harm that the individual can suffer. At the same time, however, Regan thinks it is clear that death to a human being (at least a typical adult) is a worse harm than death is to any nonhuman animal. This is because human beings have the ability to form and satisfy a larger range of desires.

To illustrate the implications of his worse-off principle, Regan introduces a fanciful "lifeboat case":

> Imagine five survivors are on a lifeboat. Because of limits of size, the boat can only support four. All weigh approximately the same and would take up approximately the same amount of space. Four of the five are normal adult human beings. The fifth is a dog. One must be thrown overboard or else all will perish. Whom should it be?
>
> *(2004 [1983]: 285)*

Given his conception of harm, Regan thinks it clear that the dog should be thrown overboard:

> Now, the harm that death is, is a function of the opportunities for satisfaction it forecloses, and no reasonable person would deny that the death of any of the four humans would be a greater prima facie loss, and thus a greater prima facie harm, than would be true in the case of the dog. Death for the dog, in short, though a harm, is not comparable to the harm that death would be for any of the humans.
>
> *(2004 [1983]: 324)*

Of course, a precontact Inuit would have to kill indefinitely many nonhuman mammals to survive over the course of a lifetime, but, as Regan points out regarding his lifeboat case, the worse-off principle is not sensitive to the numbers involved, and thus, a human should be saved even if it required throwing "any number of dogs" overboard. For to think otherwise, he says, "will inevitably involve one in aggregative considerations—the sum of the losses of the million dogs over and against the losses for one of the humans—an approach that cannot be sanctioned" if we think of individuals as having non-utilitarian *inherent* value (2004 [1983]: 325).

Although Regan's case refers to the lifeboat's weight-carrying capacity, the same reasoning would apply if the only way to save any of the five human survivors were to kill one (or "any number" of) dogs so that a human could avoid starvation. This is why even Regan's animal rights view implies that subsistence hunting is justified, at least where it is necessary for physical human survival. With regard to humans in modern conditions, however, Regan—and ethical vegans generally—argue that we can thrive on plant-based diets, and thus, we are not in a situation "[w]hen it is necessary to choose between violating the rights" of one group of individuals or those of another.

In the next section, we discuss subsistence hunting among indigenous peoples after they are economically and technologically intertwined with modern Euro-American economies. We close this section, however, by briefly considering two complications.

The first is that, from a utilitarian perspective, killing a nonhuman animal removes some happiness from the world. So even in situations where humans need to hunt in order to survive, killing larger animals would be better, insofar as fewer animals would have to be killed (MacClellan 2013). This observation complicates a utilitarian analysis of subsistence hunting narrowly construed, because it suggests that whaling might be better than sealing, hunting moose, or fishing. Of course, popular Euro-American culture tends to assume special moral status for cetaceans, and various utilitarian philosophers have argued that certain cognitive capacities give individuals' lives special moral significance. For instance, in the first edition of *Practical Ethics*, Singer stipulatively defined *persons* as "rational and self-conscious" beings (1979: 76), and there and in subsequent work he has argued (1) that persons are "not replaceable" and (2) that this would, for various reasons, block the preceding kind of reasoning where cetaceans are involved. Singer's definition of *persons* and his argument against "replaceability" for persons have varied over the years. While we cannot but provide this a very brief overview here, clearly, attributing varying moral significance on the basis of the cognitive capacities of animals could complicate a utilitarian analysis of subsistence hunting still further (see generally Varner 2012, with details on Singer on replaceability in §9.4).

The second complication is related to Regan's rights view and its implications for the dietary choices of people living in affluent, developed nations. Dan Demetriou and Bob Fischer—two authors who are sympathetic to ethical veganism—argue that what they call "dignitarian hunting" is a respectful alternative to a vegan diet that is based on contemporary, intensive plant agriculture. Their argument takes off from a discussion of the number of "field animals" that are killed annually using the intensive, mechanized processes of current, conventional plant agriculture. Animal scientist Steven Davis (2003) estimated that in the US, 1.8 billion wild animals are killed annually producing plant-based foods, and Demetriou and Fischer (2018) argue that vegans could save animal lives by converting part of their diet to meat from animals that they have hunted and killed themselves.[4] They also argue that animals killed during row-crop production suffer a kind of harm to their dignity that animals hunted in an "adversarial" context do not. Demetriou and Fischer argue that all of this is consistent with Regan's two principles and with respecting the inherent value of animals. The "dignitarian hunting" that Demetriou and Fischer defend is not subsistence hunting narrowly construed, because dignitarian hunters could get all their nutrition without *directly* killing any animals. Dignitarian hunters recognize that by eating commercially available, plant-based foods they kill animals *indirectly*, however, and that by substituting some hunted animals, they can lower the total number of animal deaths for which they are responsible, either directly or indirectly.

In this section, we have described how each of the two most-discussed approaches to thinking about animal ethics in Euro-American philosophy could support subsistence hunting when it is narrowly construed as necessary for physical human survival. As we noted in the introduction, however, there is no widely agreed-on definition of the term *subsistence hunting*, and a wide range of practices have been described as "subsistence" hunting. For instance, should ethical vegans deciding to replace part of their diets with meat from respectfully hunted animals as Demetriou and Fischer describe be described as "subsistence" hunting? What about the killing of wildlife to prevent damage to crops or future attacks on human beings? What about when trophy hunters pay to cull dangerous or overpopulated elephants and the meat is distributed among local villagers for food?

In the following section, instead of pursuing such nuances about how exactly "subsistence hunting" should be conceptualized, we turn to a discussion of what we referred to in the introduction as "subsistence *cultures*." This term emphasizes how indigenous peoples often think of hunting, fishing, and gathering as more than ways to get food and other needed supplies, and—as we emphasize in the following section—indigenous peoples often experience humans' ethical relationships with animals very differently than Euro-American philosophers have framed them.

Subsistence Cultures and Animals

For many indigenous peoples (including the Inuit peoples and other Alaska Native and Russian Far East groups on whom we focus here) hunting is not just a matter of practical necessity for personal and community food security. Rather, subsistence hunting carries tremendous symbolic weight of sovereignty, respect, caring, and gratitude to subsistence animal species, and conveys a sense of connectedness to place and tradition (Kawagley 1995). As a Yup'ik elder put it, "[s]ubsistence is directly related to and affected by everything that is happening . . . in the way of education, land use, economic development, wildlife management and other areas of public policy. Subsistence really is an entire way of life" (Harold Napoleon, quoted in Yupiktak Bista 1974: 2).

Practicing such culturally based subsistence hunting is vital to the well-being of the adherent and should be understood within the context of its strong connectedness to specific customs, cultural beliefs, foundational narratives, and myths; focal practices and social, economic, and governance schemes regarding self-determination; and rights to land, waterways, territories, and natural resources (United Nations 2007). For indigenous peoples, losing grip on subsistence hunting and foraging activities can lead to cultural loss and mourning (Barnhardt and Kawagley 2005; Cunsolo-Willox 2012). A recent case in point: the Gwich'in people, a tribe in the northern part of Alaska and Canada, have relied on the Porcupine Caribou Herd for thousands of years, for food, cultural, and spiritual needs. The Porcupine Caribou Herd migrates to the Arctic National Wildlife Refuge coastal plain every year to produce their offspring. The coastal plain known as "Iizhik Gwats'an Gwandaii Goodlit" (The Sacred Place Where Life Begins) is hallowed ground to the Gwich'in. The herd and Gwich'in way of life will be under threat if oil and gas exploration and development are permitted in Section 1002 of the Arctic National Wildlife Refuge-coastal plain.

For indigenous peoples, traditional knowledge about how to adapt is a central part of their resilience in what is sometimes perceived to be inhospitable environments. Indigenous subsistence lifestyles, in general, and subsistence hunting, in particular, have been a natural part and/or customary norm of many indigenous communities for centuries and are connected to human survival and/or well-being. Indigenous subsistence activities have been endorsed by the United Nations and many member states as a basic human right. The right of indigenous peoples to subsistence hunt, gather, and forage is constitutive of the constellation of their collective rights and "are vested in indigenous individuals that organize themselves as peoples" (United Nations 2013: 7). The normative significance of subsistence hunts that are conducted by indigenous peoples may be contested from within and without, when it is not done in accordance with traditional standards, where the hunts lead to waste, and when the animals are not respected.

The subsistence constructs that are found in narratives and foundational myths of indigenous peoples have played an essential role in helping communities understand the regularities of the world in which they inhabit. They convey detailed observations of animal behavior in conjunction with ways to decipher patterns of order in nature like seasonal cycles (Barnhardt and Kawagley 2005). They also contain descriptions of particular prowess and extraordinary capacities of animals, and they have symbolic meaning and pedagogical value for members of a particular society. For subsistence hunters steeped in these myths and traditions, animals are clearly not passive subjects as urban dwellers might encounter with domesticated animals. Hunting, in indigenous cosmology, is an activity tied to ritual, moral responsibility, and spirituality. Each animal possesses its own way of being (agency or spirit), meaning and connection with humans, for example, a specific willingness among prey species to come to the hunter (Stuckenberger 2007; Peter et al. 2002). Certain animals, for example, the bear, wolf, raven, eagle, and beaver represent power, and each species possesses certain distinguishing characteristics (e.g., the bear represents strength, the wolf social organization, the raven the ability to remain airborne for great lengths of time, and the eagle visual acuity).

The relationship between hunter and animals is shaped by the following overlapping elements: ecosophy, epistemology, ethics, engagement with nature and tradition, and a view of animals as agents in themselves.

Ecosophy

Indigenous ecological philosophy, or "ecosophy" (Kawagley 1995), is based on the view that all environmental entities are endowed with consciousness. This contrasts with the Euro-American impulse to compartmentalize natural entities as either "conscious" or "unconscious," to place animals at arm's length from humans as things that we consume, and to think of them as in need of our protection or care. Instead, indigenous ecosophy emphasizes the nearness of human–animal relationships, the roles animals play in the mythological crossover between the profane and sacred, and the richness of the contributions of animals to a shared human/animal/natural community. For subsistence communities, animals have broader significance or cultural value than merely being objects of moral concern.

They are also beings to be venerated or honored, they serve as role models with admirable capacities and beings to be feared and respected. Animals can also be our teachers (e.g., Raven in the Iñupiat creation story), and through careful observation of them and how they interact with conspecifics, other species, and the world, we can learn how to cope and deal with various challenges, thus providing instruction on how humans ought to live and a check on human hubris (Inuit Circumpolar Council 2015; Barnhardt and Kawagley 2005; Peter et al. 2002).

Epistemology

Indigenous ecosophy is a form of reliabilism and animistic pragmatism. It is reliabilism in the sense that knowledge and justification for activities and norms are based on lived experiences that reinforce current and future processes. Knowledge of the relationship between environmental elements is formed by a reliable process of observation that has been shaped over time (Kawagley 1999). It is pragmatic in the sense that the knowledge that is passed down through narratives is naturalistic, developmental, and fallible and aimed at problem solving (McKenna 2004). Abstract principles or overarching systems of ethics are not what is fundamentally action guiding, and they do not take precedence over bottom-up solutions to moral and practical dilemmas. Instead, contextual wisdom and concrete norms are formed from lived experiences that reflect day-to-day intricacies with human–environment–animal relationships. Through narratives that ground focal practices wedded to tradition, subsistence cultures have a way to orient themselves regarding their relationship to nature and to refine and test solutions as environmental challenges emerge (Anthony 2013).

Ethics

For indigenous communities that are committed to the subsistence lifestyle, ethical norms are drawn from the lived, shared experiences of ancestors and elders, and appropriate ecological knowledge is based on a keen sense of the elders' engagement with an integrated ecological-cultural community.

The ethical norms are learned by the young from their elders, generally during participation in real-world activities (Kawagley 1995; Nee-Benham and Cooper 2000). The elders are specialists in understanding the interconnected natural forces at work, including being able to predict weather events based on subtle signs (Barnhardt and Kawagley 2005). They also provide moral leadership by recognizing essential interdependencies, balances, and diversity that help both humans and nature thrive (Barnhardt and Kawagley 1999). The elders embody or know the appropriate narratives to employ in a given situation with the intimacy of personal discovery, for they themselves have had direct participation in real-world subsistence activities. This intimacy gives them a particular understanding of the moral reasons behind specific normative truths to convey and apply (Anthony 2013). Their intimate connection to the past and to natural phenomenon will include moral reasons that shape the narratives or are revealed in them (Cajete 2000; Eglash 2002).

In terms of the norms' content, the "bush consciousness" of indigenous Alaskans stresses the greater good of oneness: community, harmony, balance, reciprocity, and gratitude; cooperation and sharing; self-reliance; and the integration of useful knowledge into a holistic and internally consistent worldview (Scollon and Scollon 1981). The narratives often convey "embodied truth," meaning that they portray focal habits or activities that promote a kind of moral ecology of virtue to care for a place in ways that protect future hunts and harvests (Anthony 2013; Kawagley 1995).

Engagement

Humans' active participation in the natural world is emphasized in a network of spiritual beliefs and norms, and focal activities that are tied to an understanding of a people and place. This stress on human participation in nature has adaptive value insofar as it reminds us that humans must live sustainably and adapt to the changing circumstances.

Spiritually, by emphasizing the participation of human beings to their natural environments, human beings are seen as contributors to and partners in the structure and sustainability of ecosystems (Barnhardt and Kawagley 2005). Overfishing, overhunting, and overforaging can have detrimental implications for a subsistence community, and subsistence hunters are instructed to take care of what they have harvested and to harvest only what is needed (Peter et al. 2002; Inuit Circumpolar Council 2015). Such norms of mutual respect between the species promote a culture of care about community and environment and an attitude of humility, gratitude, and reverence for environmental entities.

Participation in subsistence practices is among the most highly valued parts of an indigenous culture. In Alaska and throughout the circumpolar north, for example, the inherited traditions associated with subsistence hunting preserve not only a people's sense of cultural identity and connectedness to each other but also their connection to natural spaces and animal beings (Inuit Circumpolar Council 2015). The remains of animals are incorporated into housing, clothing, and other cultural symbols, such as drums. In the Native Youth Olympics, each event represents some skill needed for a successful hunt: the stick pull represents the strength needed to pull a seal from water, the two-foot high kick is a signal to communicate a successful hunt among groups across ice flows, and the seal hop reflects the endurance and stealth needed to sneak up on animals like seals across the ice (adapted from www.ankn.uaf.edu/curriculum/NativeGames/games.html).[5]

Animal Agency

Many indigenous narratives focus on a profound and immediate connection between humans, other animals, and the environment. With no stark duality between human and nonhuman, indigenous animism acknowledges a rich view of animal agency or soul. Animals play a central role in indigenous cosmology, ecosophy, and religious-spiritual beliefs, which determines how animals should be viewed and treated and which animals can be hunted for food. The role of animals is an important ingredient in the adaptive capacity and flourishing of a community and is central in aboriginal storytelling and forensics.

The foundational myths contain "accrued knowledge" associated with hunting different species of animals and describe an intimate and symbiotic relationship between hunter and the hunted (Barnhardt and Kawagley 2005). They encourage humility during the hunt so that animals continue to return in future seasons. Every environmental entity is seen as a part of the living community and as contributing to the health of every other member. The view is both biocentric and egalitarian in nature, and each spirit or "agency" has a defined role that contributes to overall ecological harmony and prosperity of place (Peter et al. 2002). Insistence on the agency of animals can be seen as a result of direct contact with them. The agencies of animals could be influenced—especially by elders or shaman who are thought to have mystical powers—through rituals, charms, chants, and dance. For hunters who experience the unrestricted capacities of animals in the wild, they are autonomous beings who have spiritual agency making moral determinations on their own, and charges of anthropomorphism are dismissed as going against one's survival interests and underlying normative truth (Stuckenberger 2007; Hanson 2002; Linderman 2003 [1932]). The agency or spirit of human and animals often intertwine in indigenous beliefs and inform beliefs about reincarnation and the cycle of regeneration and influence which and how animals should be hunted. Animals are believed to have power over natural elements such as water, the weather, and the food supply. For the Yup'iit or Yup'ik people (aboriginal peoples of western, southwestern, and southcentral Alaska and the Russian Far East, who do not reside in the Arctic Circle per se), all things of the environment are "endowed with consciousness," or *Ellam Yua* (Creative Force). The Yup'iit live among spirits of their ancestors, animals, plants, and other environmental entities (Kawagley 1999). The Yup'iit are cautioned to never harm or make fun of animals, since animals and everything else will always know if misuse, abuse, or disrespect has occurred. Such behaviors would result in the spirits not returning to Earth to be born and renew their kind again, resulting in extinction for the Yup'iit (Kawagley 1999).

Accordingly, subsistence hunting activities are carefully timed and carried out in a manner that ensures sustainability and respects the dignity of the animals. Hunters are informed by their elders and by foundational narratives of the innate capacities of animals and what respect for animals should look like. In some traditions such as the Iñupiat of Seward Peninsula, Alaska, there are a set number of animals that exist between the real and spirit worlds. There is a dynamic "responsive interaction" (Haraway 2008) or communication between hunter and animal, and mutual learning and co-shaping of behavior occur during a hunt. Hunters and animals are "invited" to participate in activities governed by the "ministry of nature," that is, rhymes and rhythms of nature that sustain the ecological health of a particular place (Brower Sr., an Inuit hunter as cited in Wohlforth 2004). An animal that has allowed a hunter to kill it is thought, in this tradition, to have willingly given itself to the hunter. This willingness is a central aspect of the relationship between human and animal agencies. The hunter who has been chosen to receive this sacrifice has to be open to communicating and listening to the animal's spirit directly (Peter et al. 2002). After a hunt, the agencies of animals are celebrated in the feast promoting communal solidarity with human and animal agents, and gratitude is offered to the self-sacrificing animals. These ceremonial rituals include bidding the spirits of killed animals back to the spiritual realm as a way to ensure blessing for the next season's hunt (Stuckenberger 2007). Respectful hunts are rewarded by future availability of animals to hunters, while animals that are disrespected or treated with callousness are trapped in spirit form in the Spirit world. Practically, disrespect or abuse of animals threatens food security. Thus, rituals relating to cultivating respect for animals and interspecies interdependence are ubiquitous and permeate day-to-day living (Bodenhorn 1990).

Subsistence *Hunting* and Lessons from Subsistence *Cultures*

As we saw earlier, Euro-American philosophers have paid little attention to subsistence hunting. We suggested that this is because "subsistence *hunting*" tends to be construed narrowly, as covering only situations where hunting is nutritionally necessary for human survival, and we described how both utilitarians and rights theorists can justify hunting in such cases. Precontact Inuit and some other Alaska Native and Russian Far East indigenous peoples fit this description of subsistence hunting narrowly construed. As we noted in the introduction, however, the term "subsistence" is itself used in multiple ways, including to describe sustaining life "at the barest possible margin" or in a community that is cut off from trade with larger (or at least modern) societies. Today, these previously mentioned indigenous groups are not cut off from trade with outsiders, but there is another sense of "subsistence" that accurately describes their modern situation in contact with Euro-Americans.

As we described earlier, for the Inuit (and other indigenous peoples) subsistence in the sense of "living off the land"—through activities such as hunting, fishing, and foraging—is a way to simultaneously nourish one's people and one's community identity. These activities are core features of the practitioners' heritage and cultural or traditional identity. Thus, hunting in a subsistence culture serves a significant moral interest beyond providing nutrition (Sandler 2017), a spiritual act (Kawagley 1999), even when imported technologies are used today in place of traditional ones and where there is gradually more intermingling between indigenous and Euro-American-based economies. In such a culture, subsistence involves adapting to a harsh environment via careful utilization of natural resources governed by the lived ecological knowledge of elders regarding what to eat (e.g., including meat, fish, herbs, berries, and other wild plants) and when and where each is available (Nilsson 2015).

Among the beliefs passed on by elders in subsistence cultures are beliefs about animals that sound counterintuitive to many Euro-Americans. As much as thinking about animal cognition has gained scientific credence in recent decades, the idea that a hunted animal will "give itself to" a virtuous hunter sounds patently false to many (if not most) Euro-Americans. For now, we have nothing to say about how to reconcile such conflicting aspects of Euro-American and Indigenous ethics and metaphysics. We do believe, however, that beliefs characteristic of subsistence cultures provide an important counterpoint to the types of Euro- American-based moral philosophy that have generally dominated discussions of animal ethics.

For instance, both utilitarians and rights theorists tend to cast questions about humans' reliance on animals for subsistence in terms of conflicts, and they tend to speak of humans as separate from "nature." In contrast, people in subsistence cultures often see humans and animals as cooperating or in harmony in ways that "Westerners" dismiss as mythical, and they often see humans as essentially enmeshed in nature. The orientation here is one of antidominance, and there is a general recognition that the boundaries between human and nature, between "wild" and "culture," are necessarily ambiguous (Barnhardt and Kawagley 2005). Subsistence practices such as hunting are part of cosmology, linked to rituals involving reincarnation and rebirth, and they reflect a dynamic embodying of human–animal interfaces. These practices influence ideas about animals and their symbolic value for a people, and the role of humans in the natural world. Animals are seen as meaning producers in their own right who inspire and challenge the way humans conceive of what there is (Kawagley 1995).

In addition to serving as a counterpoint to Euro-American assumptions about animal natures and about the human/natural divide, we believe that indigenous cultures' pragmatic epistemology and engagement with nonhuman nature provide an aspirational model for humanity in the age of global climate change. People in subsistence cultures around the world feel as if their lives and heritage are in peril due to climate change. Climate change is inhibiting access to and availability of traditional food sources and diminishing access to wild areas due to coastal erosion, reduction in sea ice, and loss of habitat for various species. Traditional knowledge and narratives, and the capacity of elders to sustain their community and culture are becoming less and less relevant due to competing Euro-American knowledge sources that identify self-worth with productive function and rely on technological fixes that tend to focus on domination of the natural world (Anthony 2013; Wohlforth 2004; Barnhardt and Kawagley 2005).

As the whole world faces challenges due to climate change, subsistence cultures' time-tested emphases on pragmatism, adaptation, "flexibility, innovation and change" (Schneider 1982: 169), and experiential knowledge provide us with an opportunity to revisit our current unsustainable practices and with instructive suggestions about how we can relate to the nonhuman world and confront and adapt to climate and weather challenges or environmental hazards that are unpredictable, uncommon, or extreme. As a start, Euro-American and Indigenous scholars should rethink the role academia continues to play in "the bureaucratization of subsistence" (Moss 2010: 132) and reexamine the concept of "subsistence" and its practical implications in contemporary struggles over conservation and habitat protection. In the process, this cross-cultural endeavor may yield fruitful opportunities to nurture traditional ecological knowledge and integrate it into Euro-American wise management strategies. We can also draw inspiration from Indigenous peoples as we work to scale up environmental virtues that promote interdependence, sharing and cooperation, gratitude, balance, and mindfulness. Subsistence relies on hunting animals as a normal part of an intimate ethical and social economy with the land, sea, and animals that goes back thousands of years and that helps to ensure sustainable access and use. Subsistence cultures' belief systems also offer a model of how to soften the typical Euro-American oppositional binary between humans and animals/nature and how to recognize the symbolism or iconic value of animals in ways that promote respectful treatment and appreciation of nature.

Acknowledgment

The authors would like to thank Maria Williams and Beth Leonard of the Alaska Native Studies Program at the University of Alaska–Anchorage for highlighting some of the relevant literature and for clarifying cultural aspects of categorizing Alaska Native groups. Any shortcomings in the text of this chapter are due to the authors.

Notes

1. There is not a neat overlap between the traditional geographical homes of Indigenous peoples and the languages spoken in these regions. Many other Indigenous peoples also call these regions home. Hence, there is a great diversity of languages spoken and cultural groups here. In using language designations to describe groups of Indigenous peoples, we have tried to be sensitive to the diversity in this Eskimo-Aleut language family (www.uaf.edu/anlc/languages/).
2. The contested nature of the concept of "subsistence" is reflected in the Alaska National Interest Lands Conservation ACT (ANILCA). While not privileging ethnicity as a special claim to subsistence rights, the law recognizes that there are some relevant differences between indigenous and nonindigenous subsistence rights: "[Subsistence] is essential to Native physical, economic, traditional and cultural existence" but only "to non-Native physical, economic, traditional, and social existence" (94 STAT 2372, Section 801).
3. Regan argues that there is conclusive evidence in the case of mammals and birds, but, in order to avoid controversies about "line-drawing," he does not discuss just how much further individual rights should be extended into the animal kingdom.
4. For the calculation of deaths caused per acre in contemporary row-crop agriculture, see Davis 2003. For skeptical reaction to Davis' reasoning, see Matheny 2003; Lamey 2007; and Cahoone 2009.
5. See also https://citci.org/partnerships-events/nyo-games/competitive-events/.

Bibliography

Anthony, R. (2013) "Animistic pragmatism and native ways of knowing: Adaptive strategies for overcoming the struggle for food in the sub-arctic," *International Journal of Circumpolar Health* 72(1). doi: 10.340/ijch.v72i0.21224.

Barnhardt, R., and Kawagley, A. O. (1999) "Education indigenous to place: Western science meets indigenous reality," in G. Smith and D. Williams (eds.) *Ecological Education in Action* (pp. 117–142), New York, NY: State University of New York Press.

Barnhardt, R., and Kawagley, A. O. (2005) "Indigenous knowledge systems and Alaska native ways of knowing," *Anthropology and Education Quarterly* 36(1): 8–23.

Bodenhorn, B. (1990) "I'm not the great hunter, my wife is: Iñupiat and anthropological models of gender," *Ê'tudes/Inuit/Studies* 14(1–2): 55–74.

Cahoone, L. (2009) "Hunting as a moral good," *Environmental Values* 18: 67–80.

Cajete, G. (2000) *Native Science: Natural Laws of Interdependence*, Sante Fe, NM: Clear Light Publishers.

Callicott, J. B., and Frodeman, R. (eds.) (2009) *Encyclopedia of Environmental Ethics and Philosophy*, Vol. 1, Farmington Hills, MI: Gale Cengage Learning/Macmillan Reference.

Comstock, G. L. (2004) "Subsistence hunting," in S. F. Sapontzis (ed.) *Food for Thought: The Debate Over Eating Meat* (pp. 359–370), Amherst, NY: Prometheus Books.

Cunsolo-Willox, A. (2012) "Climate change as the work of mourning," *Ethics & the Environment* 17(2): 137–164.

Davis, S. (2003) "The least harm principle may require that humans consume a diet containing large herbivores, not a vegan diet," *Journal of Agricultural and Environmental Ethics* 16: 387–394.

Demetriou, D., and Fischer, B. (2018) "Dignitarian hunting: A rights-based defense," *Social Theory and Practice* 44(1): 49–73.

Eglash, R. (2002) "Computation, complexity and coding in native American knowledge systems," in J. E. Hankes and G. R. Fast (eds.) *Changing the Faces of Mathematics: Perspectives on Indigenous People of North America* (pp. 251–262), Reston, VA: National Council of Teachers of Mathematics.

Hanson, K. H. (2002) *Resilient Thesis, Wordless Synthesis: Women's Perceptions on Iñupiaq Expression and Transformation of Protestant Belief and Practice*, unpublished dissertation, Emory University, Atlanta, GA.

Haraway, D. J. (2008) *When Species Meet*, Minneapolis, MN: University of Minnesota Press.

Inuit Circumpolar Council-Alaska (2015) *Alaskan Inuit Food Security Conceptual Framework: How to Assess the Arctic From an Inuit Perspective: Summary Report and Recommendations Report*, Anchorage, AK, https://iccalaska.

org/wp-icc/wp-content/uploads/2016/03/Food-Security-Summary-and-Recommendations-Report.pdf (Accessed December 2, 2018).

Kawagley, A. O. (1995) *A Yupiaq World View: A Pathway to Ecology and Spirit*, Prospect Heights, IL: Waveland Press.

Kawagley, A. O. (1999) "Alaska Native education: History and adaptation in the new millenium," *Journal of American Indian Education* 39(1): 31–51.

Lamey, A. (2007) "Food fight! Steven Davis vs. Tom Regan on the ethics of eating meat," *Journal of Social Philosophy* 38: 331–348.

Linderman, F. (2003 [1932]) *Pretty-shield: Medicine Woman of the Crows*, Lincoln: University of Nebraska Press.

MacClellan, J. (2013) "Size matters: Animal size, contributory causation, and ethical vegetarianism," *Journal of Animal Ethics* 3(1): 57–68.

Matheny, G. (2003) "Least harm: A defense of vegetarianism from Steven Davis's omnivorous proposal," *Journal of Agricultural and Environmental Ethics* 16: 505–511.

McKenna, E. (2004) "Pragmatism and the production of Livestock," in E. McKenna and A. Light (eds.) *Animal Pragmatism: Rethinking Human-nonhuman Relationships* (pp. 160–178), Bloomington, IN: Indiana University Press.

Mill, J. S. (1957 [1861]) *Utilitarianism*, Indianapolis, IN: Bobbs-Merrill.

Moss, M. L. (2010) "Rethinking subsistence in Southeast Alaska: The potential of zooarchaeology," *Alaska Journal of Anthropology* 8(1): 121–135.

Nee-Benham, M. K., and Cooper, J. E. (eds.) (2000) *Indigenous Educational Models for Contemporary Practice: In Our Mother's Voice*, Hillsdale, NJ: Lawrence Erlbaum Associates.

Nilsson, L. M. (2015) "Aboriginal food security in the Arctic—a Utopia or a matter of attitude?" *EURSAFE Newsletter* 17(3): 2–3.

Peter, A., Ishulutak, M., Shaimaiyuk, J., Kisa, N., Kootoo, B., and Enuaraq, S. (2002) "The seal: An integral part of our culture," *Études/Inuit/Studies* 26(1): 167–174.

Regan, T. (2004 [1983]) *The Case for Animal Rights*, 2nd ed., Berkeley, CA: University of California Press.

Sandler, R. (2017) *Environmental Ethics: Theory in Practice*, New York, NY: Oxford University Press.

Schneider, W. S. (1982) "Subsistence in Alaska: A look at the issue over time," in P. Cornwall and G. McBeath (eds.) *Alaska's Rural Development* (pp. 169–180), Boulder, CO: Westview Press.

Scollon, R., and Scollon, S. (1981) *Narrative Literacy and Face in Interethnic Communication*, Norwood, NJ: Ablex Publishing Corporation.

Singer, P. (1979) *Animal Liberation*, 1st ed., New York, NY: Avon Books.

Singer, P. (1990) *Animal Liberation*, 2nd ed., New York, NY: Avon Books.

Singer, P. (2011) *Practical Ethics*, 3rd ed., New York: Cambridge University Press.

Stuckenberger, N. A. (2007) "Thin ice: Inuit life and climate change," in N. Nadeau (ed.) *Thin Ice: Inuit Traditions with a Changing Environment* (pp. 29–44), Hanover, NH: Hood Museum of Art Dartmouth College.

United Nations (2007) *United Nations Declaration on the Rights of Indigenous Peoples*, New York and Geneva: United Nations.

United Nations Human Rights Office of the High Commissioner (2013) *Indigenous Peoples and the United Nations Human Rights System Fact Sheet No. 9/Rev. 2*, New York and Geneva: United Nations.

Varner, G. E. (2012) *Personhood, Ethics, and Animal Cognition: Situating Animals in Hare's Two-Level Utilitarianism*, New York, NY: Oxford University Press.

Wenzel, G. (1991) *Animal Rights, Human Rights: Ecology, Economy and Ideology in the Canadian Arctic*, Toronto: University of Toronto Press.

Wheeler, P., and Thornton, T. (2005) "Subsistence research in Alaska: A thirty year retrospective," *Alaska Journal of Anthropology* 3(1): 69–103.

Wohlforth, C. P. (2004) *The Whale and the Supercomputer: On the Northern Front of Climate Change*, New York, NY: North Point Press.

Woods, M., Gunn, A. S., Varner, G., Evans, J. C., and Preston, C. (2009) "Hunting and fishing," in J. B. Callicott and R. Frodeman (eds.) *Encyclopedia of Environmental Ethics and Philosophy*, Vol. 1, Farmington Hills, MI: Gale Cengage Learning/Macmillan Reference.

Yupiktak Bista (1974) "Does one way of life have to die so that another can live?" in A. Davidson (ed.) *A Report on Subsistence and the Conservation of the Yup'ik Lifestyle* (p. 6), Bethel, AK: Yupiktak Bista.

PART III

Animal Research and Genetic Engineering

Editor's Introduction

Suppose you discover a new chemical compound. You want to know what its effects are going to be on human beings. What should you do?

One option is to inject the compound into several human beings and see how they respond. Unsurprisingly, this may not be a way of becoming friends with those human beings. What do you do if no one wants to take the risk? If money were no object, then you could keep offering people higher and higher payments; eventually, you would probably get your research subjects. But first, that isn't guaranteed: if there is a serious risk of dying, perhaps no one will take you up on the offer. Second, money almost always *is* an object. Third, many people would have moral objections to paying people large amounts of money to be experimental subjects. They would worry that the poor would end up selling their bodies to escape their poverty and that this would constitute an objectionable form of exploitation. (People have qualms about organ sales for the same reason.) Finally, there are some subjects who simply *can't* consent, even in principle. If you want to study the impact of the compound on infants or on people with a certain severe cognitive disability, then you can't get their consent by offering more money, as you can't get their consent at all.

At this juncture, another option presents itself: you could administer the compound to a group of nonhuman animals. You can't get their consent either, but you might think that you don't need it: they are "mere" animals. And depending on how the experiment goes, you will either be able to tell that the compound is dangerous, in which case it certainly shouldn't be administered to human beings, or safe, in which case further testing may reveal potential benefits.

That, at a certain level of abstraction, is the way we might think about the motivations of scientists who use animal models—that is, those who use animals to make predictions about how, say, the human endocrine system works, or how cancer cells in humans will be affected by a new radiation therapy, or what have you. At each point along the way, however, it's clear that there are assumptions we can question.

For instance, why does it matter whether we get consent from a human being but not whether we get it from animals? Why is it permissible to use them, when, presumably, it wouldn't be permissible to use comatose human beings? How reliable are animal models, and how reliable do they need to be to justify various interventions? And crucially, how should we think about these questions in institutional contexts, where scientists need approval from their peers for their research protocols?

Moreover, while we often experiment on animals for anthropocentric reasons—for the sake of medical knowledge or promoting human safety—we sometimes experiment on them for non-anthropocentric reasons. We cause considerable pain to animals when we raise them for food. Is it permissible to genetically engineer farmed animals so that they suffer less? May we turn them into beings who won't be harmed as badly by the environments in which we want to raise them? Or consider the incredible impact that human beings have had on the environment, such that many now refer to our age as the Anthropocene. One of its characteristic features is dramatic species loss—estimated to be occurring at more than 100 times the normal rate of species loss—resulting in major changes to ecosystems the world over. We now have technologies available to us, however, that might help us reverse this trend: it's possible to bring back the passenger pigeon and, perhaps, even the woolly mammoth. Should we do it? Either way, why?

There are several important connections between all these issues, but I'll just mention three. First, in every case, we are proposing to experiment on some to benefit others. The others differ, of course. In many cases, the others are human beings. But they can also be other nonhuman animals of the same species, or other nonhuman animals of different species, or something more complex, such as an ecosystem. In the vast majority of cases, the proposed research is nontherapeutic: it doesn't benefit the individuals on whom the experiments are performed. Are nontherapeutic research programs ever permissible? If so, when? Are some nontherapeutic research programs easier to justify, simply based on who they are supposed to benefit? If so, which ones and why?

Second, in every case, there is considerable uncertainty about outcomes. We *might* be able to make some plausible predictions about what would happen if, say, a certain kind of bird were to be produced via genetic engineering and released into the wild. But ecosystems are massively complex, and it's very difficult to know, or even to state with any high degree of confidence, that a certain outcome will occur. Likewise, of course, for cancer research and attempts to produce more heat-tolerant pigs. How confident do we need to be about the results before an experimental program is warranted? What burdens can we impose on nonhuman animals when our degree of confidence is relatively low?

Third, it's important to remember that research on animals occurs in a radically different context than does animal agriculture. Researchers are professionals, not in the sense that they belong to some vaulted class of individuals, but in the sense that they are members of specialized groups that see themselves as being bound by specific ethical norms. Doctors take the Hippocratic oath; lawyers take oaths that commit them to following the ethical guidelines of the jurisdictions in which they serve. Likewise, researchers are bound by ethics codes, often established by either their university or by professional societies of which they are members. How well does animal research live up to the ethical guidelines that researchers publicly endorse? Even if no member of the public complains about what animal researchers are doing, can the researchers themselves do the work with a clean conscience? If, for instance, researchers agree that they ought to abide by the 3Rs—a framework that calls for researchers to reduce the number of animals on which they experiment, to refine their methods so that they are less harmful, and to replace animals with nonanimal models where possible—then which research programs should be ruled out? Obviously, there are no easy answers here, but it's important to recognize that the ethical discussion isn't just about independent moral critique: it's also about whether researchers meet their own standards.

17
INSTITUTIONALIZED ETHICAL ASSESSMENTS OF ANIMAL EXPERIMENTS

Bernice Bovenkerk and Lonneke Poort

Introduction

The use of animal experiments has long been a societally contested issue. On the one hand, we find those who argue that even though it would be better not to use animals, for the foreseeable future we cannot do without them if we want to fight human and animal diseases: It is a necessary evil. On the other hand, we find those who argue that the harm done to animals through experimentation is either categorically wrong or simply does not outweigh the achieved benefits. In most countries, animal experiments are allowed for medical purposes, under the provision that there is a reasonable chance of the experiment's success, the results cannot be obtained in any other way, and the animals suffer as little as possible.[1] In order to establish whether these conditions are met, in many countries researchers have to apply for a permit, and this application is then assessed by an animal experimentation committee (AEC). In this chapter, we zoom in on the assessment made by such committees. How do they weigh the experiments' goal's justification against the harm done to the animals? What sort of moral dilemmas and institutional challenges do they face?

There are many different kinds of committees in different regions that monitor the way animals are used, and they vary in their constitution, powers, and aims. Given their many differences, and the limited space available here, it would be impossible to discuss them all. In this chapter, we can merely offer a glimpse into this world by discussing Dutch AECs. However, many of the issues facing Dutch AECs will be familiar to those in other countries, and we are certain that some common challenges exist. By way of illustration, we therefore point out some differences and similarities between the Dutch and US systems.

In the Netherlands, the creation of genetically modified animals used to be assessed separately from their use. This assessment was carried out by the Committee for Animal Biotechnology (CAB), which advised the Minister of Agriculture on whether to grant a license. While this is now a 'sleeping committee' that is only woken when someone applies for a potentially contentious biotechnological procedure, it is still interesting to examine its functioning. There are two reasons for this. First, this committee has been analysed quite extensively, and we can draw lessons from these analyses to inform similar processes in other parts of the world. Second, we think that the use of new technologies, such as CRISPR-Cas9, will bring new applications and more intensive use of modified animals into view. This means that we may have to re-open the moral debate about animal biotechnology and wake up the CAB. While we only discuss the CAB here, in previous work we have also looked at the Swiss and Australian animal biotechnology committees and at the Danish system (see Box 17.1).

First, we briefly discuss the rationale for carrying out an ethical assessment of animal experiments in the second section; after this, we discuss the set-up and functioning of AECs in the third section, and some of the ethical and institutional challenges they face in the fourth. In the fifth section, we focus on the Committee for Animal Biotechnology and, in the sixth section, we focus on some of the lessons we can learn of its past functioning. We close off with a brief conclusion in the seventh section.

The Rationale for an Ethical Assessment

In 2017, 530,568 animal experiments were performed in the Netherlands. The number of animals registered is lower, as some animals are used in multiple experiments. Moreover, in contrast to the Netherlands, according to European regulation, surplus animals that are only killed but that were not part of a test (and instead, say, were used for their organs) do not need to be registered. The official number of registered test animals was 477,550 (NVWA 2019). This indicates an increase of almost 18% of registered test animals with respect to 2016.[2] The most commonly used animals are mice (43.1 %), rats (19.2 %), chickens (11.6 %), and recent years have witnessed a particular increase in the use of zebrafish (10.9 %) due to the ease of studying and manipulating their ontogenetic development (Bradford et al. 2017). Moreover, higher numbers of cats and dogs were used for experiments. The use of the latter were a special cause for public concern, even though they each made up only 0.2% of the total number of animals used (NVWA 2019). In comparison, in Great Britain in 2017, 3.79 million biomedical procedures were carried out involving living animals, accounting for a 4% decrease compared to 2016.[3] It is difficult to make a comparison with the United States, as most mice, rats, and fish are excluded from registration under the US Animal Welfare Act.[4] However, research by Goodman et al. (2015) shows that if these animals, in particular mice, are taken into account, the number of animals used in the United States has increased significantly over the last 15 years. It is estimated that 115 million animals are used anually in experiments worldwide.[5]

Researchers only need to apply for a licence for carrying out experiments on vertebrates, and are exempt from assessment in the case of invertebrates. There are a number of notable exceptions to those exemptions, such as experimenting on octopus, which due to their supposed intelligence are deemed able to suffer. Moreover, animals bred for experimental purposes are not registered as experimental animals. This has been criticized: It leads to the perception that fewer animals are used than really are, and these animals can still suffer from the condition in which they are held. Also, it is argued that they are in a sense 'commodified,' or turned into objects for sale. As Ferrari (2019: 195) observes,

> in experiments that make use of genetically altered animals, many individuals are used for the realization and maintenance of a, so-called, transgenic animal line, and are not counted in the statistics. Furthermore, many transgenic lines are bred in commercial facilities to be ready for use, so that scientists can order them from a catalogue.

Animals are used for different purposes in biomedical research: fundamental (knowledge-driven) research; education, veterinary medicine; as disease models for human diseases, "in behavioural and cognitive ethological studies; as bioreactors to produce fluids or bodily parts which contain therapeutic substances for human beings (i.e., "gene-pharming"); and as sources of cells, tissues, and organs for human transplantation" (Ferrari 2019: 194–195).

Many animal ethicists oppose the use of animals in experimentation, either categorically because it violates the animals' inherent value or rights (in Regan's deontological view), or on balance, as experiments do not yield enough benefits to justify harming and killing animals (in Singer's utilitarian view) (Regan 2004; Singer 1975).[6] Other arguments in favour of prohibition are, firstly, that animals should not be used to solve human problems. According to this view, a conflict is constructed between two magnitudes that are, in fact, not comparable, namely potential future human benefits and the harm and killing of animals (Ferrari 2019). Second, it is argued that only when researchers are forbidden from using animal models will they have strong incentives to look for alternative strategies. Necessity is the mother of invention. Currently, relatively little government funding is spent on searching for alternatives to animal experiments.[7]

Yet, a common line of reasoning—which, in fact, appears to be accepted by most people (Crettaz von Roten 2012)—is that animal experiments are allowable as long as explicit care for the animals' welfare is taken.[8] If the use of a small number of animals leads to medical treatment for a large number of people, this would be warranted, as long as the discomfort to the animals is minimized as much as possible. A number of assumptions underlie this view: First, attributing moral considerability to animals does not mean that their interests are as significant as those of humans.[9] This view could be based on an idea that, by definition, human interests matter more than animal interests (a view that animal ethicists claim is speciecist), or on the idea that humans, because of their cognitive complexity, have more interests and stand to suffer more from similar harms (see Bovenkerk and Kaldewaij 2014). Second, on this view, killing animals is generally not considered a moral harm. In animal experimentation committees, therefore, the actual killing of animals is not assessed, except insofar as the method of killing inflicts discomfort.[10] Generally, animals are 'euthanized' after experiments. This is often necessary because of the research set-up (e.g., when the animals have to be dissected afterwards) or because otherwise the animals would suffer too much. So-called humane endpoints can be established prior to the experiments, to determine under what condition the animals have to be killed, even if the experiment is not finished yet (Stokes 2002). However, as Franco and Olsson (2016) argue, sometimes animals are killed unnecessarily when they could, in fact, be rehabilitated and re-homed or taken to a sanctuary.

Consistency

A commonly heard complaint from researchers is that when they want to use an animal in an experiment, they have to go through a lengthy and expensive assessment procedure, and follow many guidelines in order to minimize the animal's suffering. Yet, anyone can harm and kill a mouse they come across in their pantry in whatever way they like. At first sight, there indeed appears to be an asymmetry here; can this be explained with the help of ethical theory? Both utilitarian and deontological animal ethicists have argued that treating an animal differently in different contexts is inconsistent. A wild rabbit is treated differently from a lab rabbit, a pet rabbit, or a rabbit that is kept for consumption, yet these rabbits all have similar capacities for suffering and enjoyment, or have the same intrinsic value. Pointing out the inconsistency in the treatment of different yet similar animals is, in fact, a central strategy for utilitarians and deontologists who also argue that it is inconsistent to pamper our pets while at the same time subjecting lab animals to tests (or eating production animals). Relational ethicists, on the other hand, can give us an argument for imposing harsher rules on animal experiments, although they would not condone inflicting unnecessary harm on the mouse in our pantry either. They argue that the different relationships we have with animals give rise to different moral commitments. Palmer (2010), for example, argues that we have stronger moral obligations towards domesticated animals than towards wild animals. We only have negative duties towards the latter—in other words, the duty to leave them alone as much as possible—but positive duties to care for the welfare of the former. This is because a lab animal is in a cage because the researcher put him or her there, thereby making the animal completely dependent on the researcher. This gives rise to a moral duty on the part of the researcher. In fact, the animal's whole identity and constitution are determined by the researcher and breeder. If the animal is in a study about cancer, chances are that he or she has been bred to develop cancer. This animal's situation is quite different from that of a mouse one encounters in the pantry. We did not invite the mouse in, and the mouse is not dependent on us for his or her sustenance or well-being. It is therefore part of researchers' professional responsibility to care for the animals in their experiments as best as they can; and the assessment by the AEC is there to support researchers and ensure that they take this responsibility seriously.

AECs

In the Netherlands, the use of animals for biomedical research is regulated under the Animals Experiment Act, and researchers need to apply for a license for each experiment. The application is assessed by an animal experimentation committee (AEC), which is usually tied to the research institution (the licence holder), and which gives advice to the Central Committee on Animal Experiments (CCD). The CCD assesses whether the AECs carry out their work properly, and has the ultimate decision on licences. Moreover, the CCD can signal dilemmas that arise for AECs and inform the Secretary of state for Economic Affairs, Agriculture and Innovation, who, in turn, can ask for the advice of the Dutch Council for Animal Affairs. An example of such a dilemma was whether to grant a licence for research that enables intensive livestock farming.[11] Furthermore, in the Netherlands, there is the National Committee for policy advice regarding animal experiments (NCAD), whose particular focus is on improving experimental animals' welfare and on the use of alternatives.[12]

AECs are made up of at least seven members, more than half of whom should be independent from the licence holder and who represent the following expertises: alternatives to testing, ethics, methodology and statistics, care of experimental animals, and (animal experimentation) medicine. AECs follow several steps in their assessment: They ask whether the aim of the research itself is morally acceptable, whether the aim of the research outweighs the discomfort caused to the animals, whether there are alternatives to the experiment in terms of replacement, reduction, or refinement (Russel and Burch 1959), and whether methodological improvements are possible. The use of non-animal methods are replacement alternatives. Examples of these alternatives include the use of human or animal tissue, computer models, cells in culture, chemical-based analytical tests, or fake animals used for educational purposes. Reduction can be achieved, for example, by re-using animals in several tests or by using better statistical methods. Refinement refers to all measures taken to reduce the animals' discomfort, such as anaesthesia, better housing conditions, or housing with conspecifics (for more on alternatives, see Vieira de Castro 2017).

The AECs usually meet once a month to discuss all pending applications, and their aim is to have a recommendation within 40 working days after the application. In practice, their decision can take longer, for example when they have to ask the applicants additional questions, or because meetings may be rescheduled. For researchers, the application to an AEC may form an unwelcome delay and an additional research expense—a frustration which is likely to be heightened by the fact that the AEC's assessment is generally at the end of the pipeline, when the experimental set-up has already been designed and funding applied for. Prior to the instalment of the CCD, the process went faster, as only local AECs had to give a recommendation. Because of the time and expenses involved, some researchers now prefer to carry out their research in countries with less regulation.[13]

By comparison, since 1985, the US Department of Agriculture (USDA) animal welfare regulations and the Public Health Service Policy on the Humane Care and Use of Laboratory Animals (PHS Policy) in the United States require research institutions to install local Institutional Animal Care and Use Committees (IACUCs) to oversee animal experimentation when they receive federal grants from the National Institutes of Health (Hansen 2013). An obvious difference with the Dutch situation is that in the United States, privately funded research is not overseen by an IACUC. However, most research in the United States is federally funded. Each IACUC must be composed of at least five members, including a chief executive officer appointed by the institution, a veterinarian with training or experience with experimental animals, a researcher who has experience with animal experimentation, a non-scientist, and someone who is not affiliated to the institution.[14] IACUCs are mandated to represent "society's concerns regarding the welfare of animal subjects" (Silverman et al. 2014: 23). Specific criteria need to be met before an application is approved, similar to the criteria of Dutch AECs. These criteria relate to the minimization of animal stress and discomfort, care and housing conditions, the three Rs, humane end points, the expertise of the researchers, and the assurance that experiments are not duplicated.[15] This latter point is distinct as compared to the Dutch system, where avoidance of duplication is not a specific criterion.

In the United States, there is no overarching committee, like the Dutch CCD. There can be marked differences in the review of different committees, as each institution has its own local culture and system. This means that the ethical assessment of animal experiments may not be carried out consistently between different committees, as a 2001 study found (Plous and Herzog 2001). Moreover, there is no overarching committee like the Dutch CCD that discusses more fundamental or contentious issues, nor is there explicit assessment of research with genetically modified animals.

Challenges for AECs

In order to give insight into what kind of dilemmas members of AECs face, we now discuss a number of challenges that have led to discussion, without having the pretension of completeness.

Paradoxically, the very use of 'vivisection' in the past has led to the start of the anti-animal cruelty movement. Because of the experiments, researchers gained more insight into animals' bodies and, in particular, into their similarities to human bodies. We can discern a tension here: Animal experiments are deemed to work because animals have a similar physiology to humans and respond similarly to treatments. But one could also argue that these similarities form a reason not to use the animals, as they also ground moral status (see Bovenkerk and Kaldewaij 2014). Yet, the extrapolation between the results of animal experiments and the human situation is increasingly called into question. There is an epistemic discussion about the usefulness of animal experiments; it is argued that the animal model is often not predictive of the human condition. In fact, some even argue that the use of animal experiments is dangerous. For example, if the translation from the animal model to human disease cannot be properly made, we may erroneously think that certain substances that are not toxic for animals are not toxic for humans either. As Ferrari (2019: 200) argues,

> [h]istorically, some animal experiments were consistent with hypothetical-deductive methods, in that they were useful to gain knowledge. However, in the present day, with scientific and technical advancements in alternative methods, the potential of molecular biology (together with proteomics and genomics, among others), as well as computer models, animal models have become obsolete and poor scientific practice. As a result, scientists who promote animal models are not adhering to good scientific practice, and continued use of animal models may prevent the attainment of human-relevant results. . . . The epistemic critique of animal experiments is supported by considerable literature from retrospective studies, which have established the poor clinical value of animal models.

One of the claims of this epistemic critique is that animal models are wrongly taken as the gold standard, because better alternatives exist.

What Counts as an Alternative?

Alternatives are another widely-discussed area. Each committee should have a member whose expertise is alternatives in the sense of replacement, reduction, and refinement. But this member is in the first place a technical expert, while many of the questions surrounding alternatives actually have a moral nature. For example, should the stated alternative be an alternative to the direct aim of the research proposal or to the ultimate aim of the overall research? In the second case, there is often more room for alternative courses of action than in the first. For example, should prevention of diseases by campaigning for healthier lifestyles be regarded as an alternative? And which of the three Rs should be prioritized? Russel and Burch (1959) proposed that replacement should be the first priority, followed by reduction, and then refinement. However, as Franco et al. (2018) have found, researchers reverse this prioritization in practice. They asked researchers to respond to several dilemmas, such as 'if you had to choose to refine your experiment by pair-housing instead of single housing mice (as companions to the mice in your experiment), would you choose using fewer mice (with more suffering)?[16] Do you give post-operative analgesia even if this could slightly interfere with your research results?' They found that "Refinement was considered more feasible than Replacement, as well as more urgent, and was also favored over Reduction" (Franco et al. 2018: 1). In other words, most researchers are more concerned about the animals' level of discomfort than about the number of animals used, which could be explained by the unpleasantness of seeing animals suffer.

A related problem appears to be that, in practice, smaller animals, such as mice, rats, and fish are sometimes regarded as an alternative for bigger animals, such as horses, dogs, and cats, while the bigger animals may have better predictive value. As De Vries et al. (2012: 427) argue, "by skipping the small models and using only one large preclinical model, it is indeed possible to restrict the number of animal models, thereby reducing the number of laboratory animals used." From an animal ethics perspective, only the amount of suffering and enjoyment experienced should matter, not the size of the animal. As there are no self-evident reasons to expect that, say, a rat suffers less than a cat, favouring the cat over the rat could be considered speciecist. However, this is socially contentious, as was illustrated by the recent outcry in the Netherlands when it turned out that more cats and dogs had been used in experiements than in previous years.[17]

Animal Experiments Are the Gold Standard

Another challenge to AECs is to establish what type of research warrants the use of animals. For example, does fundamental (purely curiosity-driven) research justify the use of animals, or should a clear medical benefit always be in sight? Of course, assessing experimental protocols requires making predictions about their likely success or failure. This is not always easy for an AEC to do. Moreover, the methodological quality of experiments turns out to be "an important factor hampering the translation of results from animal studies to a clinical setting" (Van Luijk et al. 2014: 1). For both reasons, it is important to carry out systematic reviews of experiments that have taken place in the past. This could provide input both for making more informed assessments in the future and for improving methodological quality. However, research by Van Luijk et al. (2014) shows that the methodological quality of systematic reviews is often quite poor. Moreover, publication bias is a problem. First, if only those studies that had a positive outcome are published, and research that did not confirm the hypothesis or is methodologically flawed is not published, then there is a very real risk that bad research is duplicated. Second, publication bias confirms animal models as the gold standard, as researchers tend to think that "results from animal studies make it easier to publish in high-ranking journals" (Franco et al. 2018: 10).

Another problem hampering the use of alternatives is that many animal tests are required by European law, and therefore, no replacement alternatives are accepted. For example, the REACH initiative requires manufacturers of chemical substances to perform animal toxicity tests before taking them to market.[18] Therefore, even if alternatives exist, animal experiments are still often the gold standard.

Ethical Relativism

An often-heard criticism of AECs is that they only make a marginal assessment: Most applications are approved, and the committees only make slight technical recommendations. A similar argument is made against IACUCs. As Hansen (2013: 188) argues, "[r]ather than making ethical judgments, IACUCs have restricted themselves to technical or advisory roles focused on reworking the details of some animal-use protocols, but ultimately approving almost all of them." Three causes for this can be pointed out. Firstly, committees tend to fulfill a 'gatekeeper function,' deterring researchers from applying for experiments that are not likely to be approved because they do not serve a very important goal or are very harmful to animals, for example. Secondly, the composition of committees could cause a bias. Hansen (2013) argues that because 67% of IACUC members and 93% of the chairpersons of IACUCs are themselves animal researchers, the committee members have a vested interest in the approval of applications.

Thirdly, while AECs tend to look very critically at the research methodology and use of alternatives of license applications, committee members tend to be reluctant to make an actual ethical judgment—weighing the benefit of the potential outcome of the experiment against the discomfort caused to the animals. In our view, this is because they implicitly espouse a morally relativist position that asks whether ethics is really more than a collection of subjective opinions. The view is that, in the end, the assessment comes down to fundamental personal values, and 'who are we to make this decision for other people?' However, moral relativism is a normative position that most ethicists reject. After all, if moral positions are simply personal preferences—like a preference for vanilla over chocolate ice cream—no moral discussion would be possible in the first place. Ethicists generally hold that morality is something that can be argued over; moral positions can be tested for their tenability (do they conform to generally accepted facts?) and logical consistency (are argumentative fallacies used?). Even if different—even opposing—moral positions can both be valid, this does not mean that all positions are valid. In other words, through argumentation, it is at least possible to point out wrong solutions to moral dilemmas. In our view, therefore, members of AECs should be encouraged to hold a real moral discussion.

Legitimacy and Transparency

Yet, we could still ask how legitimate the outcome of such a discussion is if the AEC is composed solely of (scientific) experts—who tend to, for example, attach more than average value to scientific knowledge. At least in the Netherlands, neither members of the public nor representatives of animal protection organizations are included in AECs. The latter often do not want to be part of an AEC because their membership would be difficult to explain to their support base. Moreover, a problem of including lay members of the general public is that, after a number of months, they can no longer be considered laypersons, and might have become 'desensitized' to the experiments. Moreover, a potential problem could be that the layperson would be overshadowed by the committee experts.

A related point is that of transparency. Animal protection groups often argue that the decision-making by AECs should be made more transparent, and information about exactly what experiments are performed by whom should be publicized. Besides the obvious privacy and safety issues for researchers, there is another pitfall to transparency. Does the public even want to know about animal experiments? Or do they, in fact, have a right not to know? McGlacken (2019) argues that members of the public are often strategically ignorant when it comes to animal experimentation, as they feel uncomfortable about them, and, at the same time, have no power to change the system. She argues: "Recognising the potential gains to be made by greater involvement in the public debate and the costs of not contributing, furthering public suspicions around the secrecy of the sector, the bioscience community in the UK recently embraced an agenda of greater openness." (McGlacken 2019: 129) But this might, in fact, have the opposite effect. Many people use a coping strategy because they are ambivalent about animal experimentation, in that they reject information about animal experimentation and use, and do not engage in discussions about these practices. According to McGlacken (2019), "highlighting the basis of medicine production in animal research at a point in which there is little choice to consume medical products arising from non-human animal alternatives may leave some publics feeling unable to rectify their moral conflict." This raises the question of what kind of communication strategy is necessary and efficacious to inform the public on animal experiments.

Committee for Animal Biotechnology

With the advent of ever more sophisticated biotechnological techniques, an increasing number of lab animals have been genetically modified, leading to discussion not only about suffering caused by scientific experiments, but also about the actual creation of experimental animal lines. How are biotechnological procedures with animals dealt with in the Netherlands? Under Dutch law (art. 23.2 Animals Act), gene editing with animals is prohibited, unless a licence is granted. Biomedical procedures are excluded from licensing under the Animals Act. Furthermore, gene editing with animals is prohibited for sports and pleasure (Article 23.1 Animals Act). Section 4 of the act prescribes that the minister will only grant a licence if the biotechnological procedure has no unacceptable consequences for the health and well-being of the animals. Furthermore, a license can only be granted if no (other) ethical objections exist against those procedures. The legal framework, thus, requires an ethical review of biotechnological procedures with animals. Paradoxically, in regulatory practice, there is no official institution or licensing procedure in which ethical review can take place.

From 1997 until 2014, all biotechnological procedures carried out on animals were subject to licencing by the Dutch minister for agriculture after the minister had received advice from the Dutch Committee for Animal Biotechnology (CAB). This committee weighed the importance of the goal of the creation of genetically modified animals against the damage done to the health, welfare, and integrity of the animals in question. Their ethical review system was based on the following questions:

1. Is the objective of the research of substantial interest?
2. Do realistic alternatives to the use of animal biotechnology exist?
3. Does the research pose unacceptable risks of harm to health and welfare of the animals?
4. Does the research involve an unacceptable violation of animal integrity?

In 2014, the work of this committee was discontinued. Reasons for discontinuation were that the ministry decided that all biotechnological research carried out with animals in a biomedical context was allowed, and in practice, only licences had been applied for in such a context. Also, it was noted that in the years of the committee and the years prior to its instalment, all the relevant arguments regarding animal biotechnology had been discussed, and that animal biotechnology was no longer the contentious public issue it was at the time of instalment of the CAB.[19] Biomedical researchers who create genetically engineered animals for their research are now automatically licensed, and their research is only subject to ethical review by an AEC. The difference between the reviews of the former CAB and the AEC (and the CCD) is that the CAB's advice pertained to the creation of genetically modified animals, while the AEC deals primarily with the discomfort caused as a result of the experiment that is carried out with genetically modified animals.

Besides the AEC and the CCD there is another committee in the Netherlands that deals with genetically modified organisms in general: the Netherlands Commission on Genetic Modification (COGEM). This independent advisory committee is composed of scientists, and advises the Dutch

Minister for the Environment. It provides information on the health and environmental risks associated with laboratory experiments and genetically modified organisms (GMOs). It advises particularly on risks and safety measures and on the adaptation or interpretation of the regulations in case of new technological developments. It is thereby part of the application process for permits for the use and release of GMOs. Moreover, it can inform the Dutch government of ethical and societal issues linked to genetic modification.[20] In this capacity, it also fulfils an agenda-setting function.[21] The COGEM has not taken over the tasks of the discontinued CAB, as it does not have the explicit task of weighing the consequences of genetic modification on animals' health, welfare, and integrity. It can signal ethical issues, but this is not part of a licensing procedure; so, the status of this signalling function is unclear.

Lessons for Biotechnology Ethics Committees

The rapid developments in gene editing that we are now witnessing as a result of technologies such as CRISPR-Cas9 can be both scientific and ethical game-changers. According to Schultz-Bergin (2018: 235), CRISPR is an ethical game-changer because it changes the debate landscape. He argues,

> In making possible new scientific endeavors, it opens new avenues for genuine ethical consideration. While we could (and did) contemplate the ethics of de-extinction previously, CRISPR has made the debates more salient. Rather than simply considering the ethics of various thought experiments, we are now forced to confront real possibilities.

As we are now confronted with real possibilities, some ethical concerns are reduced and other concerns become more pressing. For example, one of the main intrinsic objections against genetic engineering involves crossing species boundaries. CRISPR, however, does not necessarily involve the use of foreign DNA, as it can be used to induce a natural modification that is already there, such as polled (dehorned) cattle. Instead of creating polled cattle through selective breeding systems, CRISPR offers a more precise alternative. Consequently, genetic engineering does not necessarily involve crossing species boundaries, and therefore, this can no longer frame the general debate, but only debate about certain applications (Schultz-Bergin 2018: 224). On the other hand, objections to gene editing related to animal welfare might increase. CRISPR may or may not create new welfare problems, but in any case, due to the versatility of CRISPR technology, a greater number of animals will be used for gene editing experiments (Schultz-Bergin 2018: 232). Moreover, CRISPR may raise new objections beyond welfare when applications such as the resurrection of extinct species come into view. These developments bring in regulatory challenges for dealing with these new technologies.

The Dutch legal framework requires ethical review for those issues, but currently has no operational institution to do so. Consequently, in our view, the CAB should be awakened. We think that when the CAB is awakened, some lessons from the past should be taken seriously.

First of all, the CAB has been criticized for aiming to combine three different roles, which can be in tension with each other: participant of public debate (communicative role), licensing (legal role), and deepening debate (analysing role). In previous work, we showed that these roles are in tension with each other, as the legal or advisory role may easily overshadow the communicative role. After all, when a committee has to defend itself against potential legal complaints, it is less free to consider different viewpoints publicly. At the same time, if the terms of reference are solely focused on public debate, ethics committees can become toothless (Bovenkerk and Poort 2008). They are then considered just voice among many in the debate, and their specific combination of expertise is not utilized. We argued that ethics committees' primary task involves deepening debate, by stocktaking and analysing relevant moral considerations. To avoid making ethics committees toothless, ethics committees could still have an advisory role in legal decision-making. In 2016, we sharpened our analysis of ethics committees' role; we argued that some of the problems regarding biotechnology ethics committees and their mandate were caused by regulators', politicians', and stakeholders' erroneous expectations that ethics committees be arbiters of the moral truth. We argued that these expectations should be changed, and that committees' ethical expertise lies in their communicative function of deepening debate and social interaction, not in their moral decision-making (Poort and Bovenkerk 2016).

This communicative role or function involves both making explicit the implicit value assumptions in moral discussions, and analysing specific arguments put forward in the debate. To that extent, ethics committees can provide insights into specific moral dilemmas, thereby contributing to deepening debate (Poort and Bovenkerk 2016). These insights can also feed into the broader public debate. At the same time, ethics committees can bridge public and political debates by feeding political debate with their insights about the values expressed in public debate. Here, ethics committees' advisory role and their role in public debate seem to intertwine. Their role becomes one of moderating debate.

This role change has consequences for these committees' ethical review. The outcomes of ethical review should not contain concrete advice on the moral acceptability of an application, as is often expected. Ethical experts do not provide the right answer. As Verweij et al. (2000) argue, ethical experts should not have the last say in moral reasoning, as it takes away scientists' own moral responsibility. Instead, the ultimate goal of ethics committees lies in their communicative and expressive functions, enabling scientists and policymakers to make, and share, their reflective judgments.

A second problem that the CAB faced was that its ethical review obtained the character of a 'ritual dance' after a number of years. The committee drafted predictable recommendations, and in the public consultation rounds, the same concerns were voiced time and again, to which the committee provided the same predictable responses. The public did not feel that it was taken seriously in the consultation process, and the ethical debate seemed exhausted. As moderator of debate, an awoken CAB would need to find ways to break through this stalemate. In this role, it must be clear that ethics committees do not provide the morally right answer, and neither do they replace public debate. Rather, they can be a driver and informer of public debate. In this capacity, they should not have a formal role in the legal structure. In order to give the CAB the opportunity to function as driver and informer of debate, public debate should be facilitated via a different 'track,' separate from the formal legal structure.

Box 17.1 International Comparison of Committees

Many of the issues within the Dutch practice will be familiar to those in other countries and we have revealed in earlier work that some common challenges are faced: challenges regarding the inclusion of ethical considerations and regarding public input into the formal structures of licensing. In Australia, the Gene Technology Ethics Committee (GTEC) is assigned to stimulate public debate. Its terms of reference leave no formal role in licensing. On one hand, that leaves room for (public) debates on the more fundamental matters. On the other hand, it made the GTEC rather toothless, as ethical review solely found a place outside the legal framework lacking mechanisms to include the outcomes of (public) debate in licensing (Bovenkerk and Poort 2008; Bovenkerk 2012). While in Switzerland the Ethics Committee on Non-Human Gene Technology (ECNH) is not obliged to advice on the moral status of an experiment carried out with genetically modified animals, it can respond to those license applications they consider morally controversial. To that extent, the ECNH can focus on new trends and think through the consequences and moral impact of those trends. At the same time, when the ECNH does not respond, the official institutions can assume the experiment is not morally controversial. There is no further official mechanism to ensure ethical review or public input into licensing (Bovenkerk and Poort 2008). The assessment carried out within the CCD is rather close to the Danish one. In Denmark, the Animal Experiment Directorate deals with the discomfort caused as a result of the experiment carried out with genetically modified animals. It is reasoned that an ethical review of the moral status of the creation of genetically modified animals is covered within the legislative process that resulted in the current licensing practice. Consequently, in reality, there is no room within this practice for debates on the more fundamental matters (Poort 2013).

Conclusion

In this chapter, we have analysed the ethical assessment made of animal experiments. We did so by focusing on the assessment in the Dutch context. In our view, the Dutch case study provides an interesting and informative example of the challenges and dilemmas that animal ethics committees faced in general, and we expect that similar challenges are faced elsewhere.

Regarding the AECs, we have pointed out several questions that have given rise to discussion: To what extent can we extrapolate the results of animal experiments to the human context? What counts as an alternative to the experiment under application? What are the prospects for an animal-free-research future if animal experiments are regarded as the gold standard, by researchers themselves, by publishers, but also legally? Do AECs make a full ethical assessment if their members do not always regard themselves as morally competent to do so, due to an inherent morally relativistic stance? To what extent are committees that are composed solely of experts legitimate decision-makers? On the other hand, to what extent does the public have a 'right not to know' the specifics of animal experimentation? What would constitute a better way of informing the general public about animal experiments?

Furthermore, we have updated the results of our earlier research in light of new developments in biotechnology, such as CRISPR-CAS9, that are believed to be both scientific and ethical game-changers. These new developments and the need for renewing the ethical review gives room to learn from the past, and to prevent history from repeating itself. Following these lessons, we have argued that ethics committees' role should be one of debate moderation. We have emphasized that ethics committees should not have the last say and neither do they provide the right answer. Ethics committees can drive public debate and enable decision-makers to make their own reflected opinions. AECs play a different role, as they have a strictly advisory role regarding specific applications. The CCD, however, can play a role as a driver of public debate, because aside from its role in licensing, it has the mandate to discuss more overarching ethical questions and to set specific issues on the political agenda. In this context, we should bear our discussion about AECs' legitimacy and transparency challenges in mind. Perhaps these could be ameliorated, albeit partially, if the CCD would play a clearer role in public debate.

Notes

1. We will leave testing for commercial purposes, such as cosmetics, out of this discussion. This practice has been banned in many parts of the world, among which are the EU, New Zealand, India and Israel, and is in the process of being banned in many other parts of the world, such as the US.
2. Several reasons for this increase can be pointed out: numbers fluctuate from year to year; a number of large experiments were started in 2016 but finished in 2017, and the animals are registered the year the experiment finishes; more zebrafish were used and cancer research was intensified. See NVWA 2019.

3. https://assets.publishing.service.gov.uk/government/uploads/system/uploads/attachment_data/file/724611/annual-statistics-scientific-procedures-living-animals-2017.pdf (Accessed April 19, 2019).
4. https://speakingofresearch.com/2017/06/19/usda-publishes-2016-animal-research-statistics-7-rise-in-animal-use/ (Accessed April 19, 2019). See also Goodman et al. 2015.
5. www.npofocus.nl/artikel/7905/waarom-schaffen-we-dierproeven-niet-af (Accessed April 24, 2019).
6. For a discussion of the main ethical theories in relation to practical cases of animal experimentation, see Meijboom 2017.
7. In the Netherlands, 1 million euros per year is spent on the search for alternatives to animal testing, while a lot more funding is applied for. See www.rijksoverheid.nl/onderwerpen/dierproeven/alternatieven-voor-dierproeven (Accessed April 24, 2019). Of course, it is a normative claim that this is little funding, but it should be kept in mind that a lot more funding is spent on actual animal experiments.
8. Although, some polls suggest that, in the US in particular, public support for animal experiments is decreasing and, among women and young people in particular, the majority opposes animal experimentation. (See Hansen 2013).
9. But see Bovenkerk and Meijboom 2012 for a critical discussion of the view that moral status comes in degrees.
10. However, there is a lively debate in animal ethics about the question of whether and why killing animals constitutes a moral harm. See for example Višak and Garner 2015.
11. Annual Report Central Committee for Animal Experiments 2017. www.centralecommissiedierproeven.nl/documenten/jaarverslagen/18/3/8/jaarverslag-ccd-2017 (Accessed July 20, 2018).
12. www.ncadierproevenbeleid.nl/ (Accessed April 24, 2019). For a historical overview of animal experimentation legislation in a European context, as well as the current situation, see Cvek 2017.
13. Personal communication with researchers.
14. See www.aaalac.org/about/index.cfm (Accessed May 9, 2019).
15. See www.nal.usda.gov/awic/overview (Accessed May 9, 2019).
16. Of course, one way to lessen this dilemma would be to reuse the companion mice in other tests, although sometimes the pair housing would last throughout the mice's lifetimes.
17. For example, after the publication of the NVWA report about the increase in animal experiments in 2017, nearly all newspaper, radio, and television reports focused on the increased use of cats and dogs and an animal rights organization is organizing a protest in front of the institute that performs a lot of experiments on cats and dogs. See www.animalrights.nl/duizenden-honden-en-katten-dodelijke-dierproeven (Accessed April 23, 2019).
18. See http://ec.europa.eu/environment/chemicals/reach/animal_en.htm (Accessed April 19, 2019).
19. See annual report of the CAB 2010. https://zoek.officielebekendmakingen.nl/blg-75099.pdf (Accessed July 20, 2018).
20. The tasks and structure of COGEM are laid down in the Environmental Protection Act.
21. See https:// www.cogem.net/index.cfm/nl/over-ons/ (Accessed July 20, 2018).

Bibliography

Bovenkerk, B., and Kaldewaij, F. (2014) "The use of animal models in behavioural neuroscience research," in G. Lee, J. Illes, and F. Ohl (eds.) *Current Topics in Behavioural Neuroscience*, Berlin: Springer.

Bovenkerk, B., and Meijboom, F. L. B. (2012) "The moral status of fish: The importance and limitations of a fundamental discussion for practical ethical questions in fish farming," *Journal of Agricultural and Environmental Ethics* 25: 843–860.

Bovenkerk, B. (2012). *The Biotechnology Debate. Democracy in the Face of Intractable Disagreement* (pp. 17–46), Dordrecht: Springer.

Bovenkerk, B., and Poort, L. M. (2008) "The Role of Ethics Committees in Public Debate," *International Journal of Applied Philosophy* 22(1): 19–35.

Bradford, Y. M., Toro, S., Ramachandran, S., Ruzicka, L., Howe, D. G., Eagle, A., Kalita, P., Martin, R., Taylor Moxon, S. A., Schaper, K., and Westerfield, M. (2017) "Zebrafish models of human disease: Gaining insight into human disease at ZFIN," *ILAR Journal* 58(1): 4–16. doi: 10.1093/ilar/ilw040.

Crettaz von Roten, F. (2012) "Public perceptions of animal experimentation across Europe," *Public Understanding of Science* 22 (6): 691–703.

Cvek, K., Varga, O. E., Schuppli, C., Ormandy, E., Guillén, J. Gruber, F., and Fernandes, J. (2017) "Regulation and Legislation: Overview and Background," in H. Röcklinsberg, M. Gjerris, and A. Olsson (eds.) *Animal Ethics in Animal Research* (pp. 68–90), Cambridge: Cambridge University Press.

De Vries, R. B. M., Buma, P., Leenaars, M., Ritskes-Hoitinga, M., and Gordijn, B. (2012) "Reducing the number of laboratory animals used in tissue engineering research by restricting the variety of animal models: Articular cartilage tissue engineering as a case study," *Tissue Engineering: Part B* 18(6): 427–435.
Ferrari, A. (2019) "Contesting animal experiments through ethics and epistemology: In defense of a political critique of animal experimentation," in K. Herrmann, and K. Jayne (eds.) *Animal Experimentation: Working Towards a Paradigm Change* (pp. 194–208), Leiden, Boston: Brill.
Franco, N. H., and Olsson, A. (2016) "Killing animals as a necessary evil? The case of animal research," in F. Meijboom, and E. Stassen (eds.) *The End of Animal Life: A Start for Ethical Debate: Ethical and Societal Considerations on Killing Animals* (pp. 187–202), Wageningen: Wageningen Academic Publishers.
Franco, N. H., Sandøe, P., and Olsson, A. S. (2018) "Researchers' attitudes to the 3Rs—An upturned hierarchy?" *PLoS One* 13(8): 1–22.
Goodman, J., Chandna, A., and Roe, K. (2015) "Trends in animal use at US research facilities," *Journal of Medical Ethics* 41(7): 567–569.
Hansen, L. A. (2013) "Institution animal care and use committees need greater ethical diversity," *Journal of Medical Ethics* 39: 188–190.
McGlacken. R. (2019) "Animal research and publics: Knowing and not knowing in the UK debate," in E. Vinnari and M. Vinnari (eds.) *Sustainable Governance and Management of Food Systems. Ethical Perspectives* (pp. 129–134), Wageningen: Wageningen Academic Publishers.
Meijboom, F. L. B. (2017) "Applying ethical thinking and social relevance," in H. Röcklinsberg, M. Gjerris, and A. Olsson (eds.) *Animal Ethics in Animal Research* (pp. 41–67), Cambridge: Cambridge University Press.
NVWA (The Netherlands food and consumer product safety authority) (2019) "Zo doende 2017. Jaaroverzicht dierproeven en proefdieren van de Nederlandse Voedsel- en Warenautoriteit," www.nvwa.nl/documenten/dier/dierenwelzijn/zo-doende/publicaties/zo-doende-2017-jaaroverzicht-dierproeven-en-proefdieren-van-de-nvwa (Accessed April 18, 2019).
Palmer, C. (2010) *Animal Ethics in Context*, New York, NY: Columbia University Press.
Plous, S., and Herzog, H. A. Jr. (2001) "Reliability of protocol reviews for animal research," *Science* 293: 608–609.
Poort, L. M. (2013) *Consensus and Controversies in Animal Biotechnology: An Interactive Legislative Approach to Animal Biotechnology in Denmark, Switzerland, and the Netherlands*, The Hague: Eleven International Publishing.
Poort, L. M., and Bovenkerk, B. (2016) "Changing expectations of experts: The symbolic role of ethics committees," in B. C. van Beers, B. M. J. van Klink, and L. M. Poort (eds.) *The Symbolic Dimensions of Biolaw* (pp. 269–287), Switzerland: Springer's Legisprudence Series.
Regan, T. (2004) *The Case for Animal Rights: Updated with a New Preface*, Berkeley, CA: University of California Press.
Russel, W. M. S., and Burch, R. L. (1959) *The Principles of Humane Experimental Technique*, London: Methuen and Co Ltd.
Schultz-Bergin, M. (2018) "Is CRISPR an ethical game changer?" *Journal of Agricultural and Environmental Ethics* 31: 219–238.
Silverman, J., Sukow, M., and Murthy, S. (2014) *The IACUC Handbook*, 3rd ed., Boca Raton, FL: CRC Press.
Singer, P. (1975) *Animal Liberation*. New York, NY: Harper Collins.
Stokes, W. (2002) "Humane endpoints for laboratory animals used in regulatory testing," *Ilar Journal* 43: S31–S38.
Van Luijk, J., Bakker, B., Rovers, M. M., Ritskes-Hoitinga, M., De Vries, R. B. M., and Leenaars, M. (2014) "Systematic reviews of animal studies: Missing link in translational research?" *PLoS One* 9(3): 1–5.
Verweij, M., Brom, F. W. A., and Huibers, A. (2000) "Do's and don'ts for ethics committees: Practical lessons learned in the Netherlands," *International Perspectives* 12(4): 344–357.
Vieira de Castro, C. (2017) "The 3Rs and good scientific practice," in H. Röcklinsberg, M. Gjerris, and A. Olsson (eds.) *Animal Ethics in Animal Research* (pp. 15–40), Cambridge: Cambridge University Press.
Višak, T., and Garner, R. (2015) *The Ethics of Killing Animals*, Oxford: Oxford University Press.

Websites

https://assets.publishing.service.gov.uk/government/uploads/system/uploads/attachment_data/file/724611/annual-statistics-scientific-procedures-living-animals-2017.pdf (Accessed April 19, 2019).
https://speakingofresearch.com/2017/06/19/usda-publishes-2016-animal-research-statistics-7-rise-in-animal-use/ (Accessed April 19, 2019).
wwwss.centralecommissiedierproeven.nl/documenten/jaarverslagen/18/3/8/jaarverslag-ccd-2017 (Accessed July 20, 2018).
www.animalrights.nl/duizenden-honden-en-katten-dodelijke-dierproeven (Accessed April 23, 2019).

http://ec.europa.eu/environment/chemicals/reach/animal_en.htm (Accessed April 19, 2019).
https://zoek.officielebekendmakingen.nl/blg-75099.pdf (Accessed July 20, 2018).
https:// www.cogem.net/index.cfm/nl/over-ons/ (Accessed July 20, 2018).
www.rijksoverheid.nl/onderwerpen/dierproeven/alternatieven-voor-dierproeven (Accessed April 24, 2019).
www.ncadierproevenbeleid.nl/ (Accessed April 24, 2019).
www.npofocus.nl/artikel/7905/waarom-schaffen-we-dierproeven-niet-af (Accessed April 24, 2019).
www.aaalac.org/about/index.cfm (Accessed May 9, 2019).

18

ANIMAL MODELS

Problems and Prospects

Pandora Pound

Introduction

This century is witnessing huge changes in the field of animal research. The evidence-based medicine movement that has for the past few decades been revolutionizing clinical research and human medicine (Cochrane 1972) is at last slowly infiltrating the field of preclinical animal research. As it does so, weaknesses in the design, conduct, and reporting of animal studies are being revealed, weaknesses that raise significant doubts about the validity of findings derived from animal research and about their translation from animals to humans. While these doubts are increasingly reflected in the works of philosophers and bioethicists working in the field of animal research (e.g., Garrett 2012; Bass 2012; DeGrazia and Sebo 2015), there is perhaps a need to grapple to a greater extent with the evidence if ethical theory is not to stagnate or become too abstract. The empirical bioethics approach seeks to answer bioethical questions by drawing on empirical data alongside philosophical theory and analysis, with the aim of firmly grounding the ethical analysis in evidence. As such, empirical data can be used to inform or refine theory, and the resulting integration of empirical data and normative analysis can be used to draw conclusions that are evidence-based (Davies et al. 2015). This chapter explores recent evidence relating to the quality, validity, and value of preclinical animal research (i.e., animal research that is expected to have relevance for humans), in the hope that it will provide a firm grounding for an updated ethics of animal research. The discussion is limited to the use of research animals as "models" of human conditions, where animals are used as proxies for humans in the hope that the understanding of biological phenomena gained in animals will be generalizable to humans.

The evidence considered relates to (1) the quality of animal research, that is, the extent to which animal researchers design valid experiments, for example, by taking measures to reduce the risk of bias or by controlling for factors that might confound the results; (2) the ways in which animal research is reported and published; (3) the validity of animal models; and (4) the extent to which findings from animal research translate to humans. The chapter concludes with a reflection on prospects for the future. We begin, however, with a brief consideration of the ethical framework that governs the use of animals in research.

Applied Ethics in Animal Research

The concept of "animal rights" does not apply within animal research; animals do not consent to being research subjects and their vital interests are sacrificed for human interests. In addition to being subject to harm, animals used as models for humans will not benefit in any way from the research they are used in (Garrett 2012). Researchers are, however, expected to follow guidance relating to the use of laboratory animals; they are asked to consider *replacing* animals with alternative research methods, to *reduce* as far as possible the number of animals used in experiments, and to *refine* their research procedures so that harms to animals are minimized. This approach, known as the 3Rs, was developed 60 years ago (Russell and Burch 1959). In practice, opportunities to implement the 3Rs are frequently missed (van Luijk et al. 2013; Balcombe et al. 2013); the uptake of nonanimal techniques remains low (Gruber and Hartung 2004), the number of animals used in research continues to increase (Home Office 2017), and thousands of animals continue to suffer severe harms (Pound and Nicol 2018).

Animal research is situated within a utilitarian (Gregory 2000), instrumentalist, goals-based, or consequentialist (Foster 2001) approach to ethics, whereby the end is seen to justify the means. In a wider context, animal research sits within an ethical approach in which humans are deemed to count morally "more" than animals, an approach that allows animals to be harmed if the outcome is deemed to benefit humans. This is the ethical context within which the harm–benefit analysis (HBA) sits. The HBA is a prospective evaluation conducted by a competent authority as part of the approval or licensing of research projects involving animals. The evaluation involves weighing the anticipated benefits of the research against its predicted harms to animals. It has been a legal requirement in the UK since the Animals (Scientific Procedures) Act 1986 and in EU member states since 2013 (European Union 2010), where it is a cornerstone of animal research regulation.[1] The aim of the HBA is to assess "whether the harm to the animals in terms of suffering, pain and distress is justified by the expected outcome" and to consider whether the research "may ultimately benefit human beings, animals or the environment" (European Union 2010: Article 38). Consequently, the issue of human benefit is central to defenses of animal research, and as Garrett (2012) notes, few people would be willing to defend animal research if it offered no benefits at all.

All too often, the benefits of animal research to humans are simply assumed or asserted without any supporting evidence (e.g., Royal Society 2015). Here we use the best available sources, such as systematic reviews and meta-analyses, to explore whether the evidence supports these assumptions and assertions.

The Quality of Animal Research

Students undertaking courses in research methods learn about threats to the validity of experiments, that is, factors that may undermine the research and make it difficult to trust the findings. When designing research studies, it is vital to take steps to minimize the risk of any subjective bias so that the study's conclusions are as robust and trustworthy as possible. The only difference between experimental groups should be the different treatments they receive. When it comes to animal studies, then, each animal should have an equal chance of entering any of the experimental groups. Steps to reduce bias include a random method for allocating animals to treatment or control groups (random allocation), a way of concealing this allocation from those assigning animals to treatment or control groups (allocation concealment), and a way of concealing from those assessing the outcome of experiments that animals are in the treatment or control groups (blinded outcome assessment).

Unfortunately, awareness of the risk of bias appears low amongst researchers using animals (Reichlin et al. 2016), and only a small proportion take steps to minimize bias. Bebarta et al. (2003) found that only 32% of the studies they reviewed reported using random allocation of animals to experimental groups. In 2009, Kilkenny et al. found an even lower proportion: only 12% of the studies reviewed. Since then various low rates of random allocation use have been found, including 29% (Hirst et al. 2014), 25% (Macleod et al. 2015), and 37% (Henderson et al. 2015). The problem is that failure to use random allocation is likely to result in the benefits of treatments being overstated; that is, studies that do not use randomization are more likely to be positive than studies that do (Bebarta et al. 2003). Or, to put it another way, randomized trials reduce treatment effect sizes (Hirst et al. 2014). Similarly, few animal studies are found to report adequate allocation concealment. None of the studies reviewed by Henderson et al. (2015) reported using allocation concealment, while Perel et al. (2007) found a rate of only 12% and Hirst et al. (2014), only 15%. In the field of stroke, animal studies that were unblinded overestimated the effect of the intervention by 13.1% compared with studies that included blinding (Crossley et al. 2008).

None of the studies reviewed by Henderson et al. (2015) reported using blinded outcome assessment. Others have found variously low rates of reported blinded outcome assessment use, including 11% (Bebarta et al. 2003), 21% (Perel et al. 2007), 14% (Kilkenny et al. 2009), 35% (Hirst et al. 2014), and 29.5% (Macleod et al. 2015). Studies that did not report using blinded outcome assessment were more likely to be positive than studies that did (Bebarta et al. 2003). Or, put another way, studies reporting blinding found significantly smaller treatment effect sizes than those that did not report blinding (Vesterinen et al. 2010).

There is also a risk of bias if researchers do not report performing formal power calculations in advance to determine the necessary sample size for their study. It is important to state the necessary sample size up front; otherwise, readers will not know whether the researchers simply continued accumulating data until a significant result was found. Few animal studies report performing formal power calculations in advance. Kilkenny et al. (2009) found that none of the 48 studies they assessed discussed how the sample size was chosen, while Macleod et al. (2015) found that only 0.7% of the studies in their sample reported a sample size calculation.

The validity of research findings may also be affected by confounding variables. These are factors that may alter the experiment unless researchers control or exclude them. Everitt (2015) outlines many of the potential confounding factors in animal research, including factors inherent to the animal (e.g., stock/strain, source, sex, age, weight, pathogen status), factors relating to the animal facility (e.g., diet, bedding, housing, water delivery, lighting, noise, vibration, temperature, humidity), factors relating to animal handling, such as the sex of the researcher, and factors relating to dose administration. Additionally, both pain and the drugs that manage pain (e.g., surgical anesthesia and postoperative analgesics) can affect research outcomes and skew data (Carbone and Austin 2016), as can animal stress and suffering (Friese 2013).

Within animal research, samples have typically been small. This leads to underpowered studies (Sena et al. 2014) that cannot give reliable results (Button et al. 2013). Furthermore, simplistic statistical analyses are often used that do not account for potential confounding variables. For example, in the field of motor neurone disease, Scott et al. (2008) observed that despite at least 50 publications describing drugs that extended the life span of mice, only one of them (Riluzole) had shown any efficacy in humans. The authors identified factors that they felt might be confounding animal research into motor neurone disease (e.g., the sex of the animal) and then retested several of the drugs, this time correctly controlling for the confounding variables. They found that none of the agents (including Riluzole) now extended the life span of mice and concluded that the majority of the published treatment effects had most likely been due to confounding variables as opposed to any actual drug effect.

The concept of reproducibility relates to the ability of researchers to independently reproduce a study under the same conditions as the original and obtain the same outcome. As such, it provides an indication of the scientific rigor of the original study. Pharmaceutical companies conducting in-house validation of data coming from academia find that many studies are not reproducible (Prinz et al. 2011; Begley and Ellis 2012; Mobley et al. 2013). Because many factors contribute to the lack of reproducibility of animal studies, including poor experimental design, poor use of statistics, poor scientific conduct, and a lack of measures to reduce the risk of bias, the problem of reproducibility is often used as a shorthand for the current lack of scientific rigor in animal research.

The Reporting and Publication of Animal Research

Another important issue within the scientific enterprise is the reporting and publication of research and the biases associated with this; these issues can have a major impact on the way animal research findings are perceived. Ideally, research is reported as fully as possible so that the reader has a clear idea of exactly how the research was conducted, how many animals were involved, and how many (if any) were lost from the study and why. The reporting of findings should also be complete and unbiased; that is, it should include *all* the findings, not just those with the "best" results. Finally, all studies should be published, not just those with positive findings.

However, many publications describing animal research suffer from incomplete reporting, which renders them of limited value to inform policy or practice (Kilkenny et al. 2010). Even basic information, such as the number of animals used in experiments, is often very poorly reported (Pound and Nicol 2018), as is reporting about the loss of animals from a study through death or exclusion (known as attrition). Holman et al. (2016) note that, particularly given the small sample sizes in animal research, losses of animals can dramatically alter the results of a study. Furthermore, attrition may be indicative of side effects or toxicity, all vital information that should be reported.

In the field of stroke, for example, animals might be excluded from a study because the experiment has not sufficiently reduced blood flow to the brain or because they die. Removing these animals will affect the results of the study, but if this is not reported its effect cannot be accounted for. Holman et al. (2016) found that attrition was not adequately reported in the fields of stroke and cancer and concluded that, as a result, treatment effect sizes in these fields were likely to be overestimated. Using a simulation technique, they found that the biased removal of animals, such as removal of outliers, dramatically increased the probability of false-positive results.

Some researchers have been found to select, from among the many outcomes studied and analyses performed, only those with the "best" results. This practice, known as selective analysis and outcome reporting, leads to an overestimate of beneficial treatment effects, leading ultimately to a body of evidence with an inflated proportion of studies with statistically significant (positive) results. Tsilidis et al. (2013) used a statistical technique to estimate whether the number of published animal studies with positive results was "too large to be true." They assessed 4,445 animal studies for 160 potential treatments of neurological disorders and concluded that whereas only 919 studies would be expected to have significant results, in fact, 1,719 reported positive results. The authors suggest that this was indicative of selective analysis and outcome reporting. A recent example of selective outcome reporting is provided by the British Medical Journal, which claims that an Oxford University research group was selective in the reporting of their animal study results in order to gain funding and approval for human trials of a tuberculosis vaccine. The group gained funding for the human trials, which ultimately failed. An independent systematic review of the animal data concluded that insufficient evidence had existed to support claims about the efficacy of the vaccine and that these claims had been overstated (Cohen 2018).

Bias can also result from funding sources. So that readers can judge whether bias has been introduced for this reason, researchers are encouraged to be transparent about any conflicts of interest relating to funding. However, Henderson et al. (2015) found that only 26% of the studies they reviewed declared a conflict of interest, while Macleod et al. (2015) found an even lower rate of 11.5%.

Publication bias arises as a result of studies being more likely to be published if they present "positive" findings, that is, findings that a treatment has been effective. If "negative" or inconclusive findings are rarely published, then the effectiveness of whole bodies of research can be overestimated. Publication bias is a significant problem in animal research (Korevaar et al. 2011; Mueller et al. 2014). In the field of stroke, for example, it has been estimated that the benefits of animal studies are overstated, possibly by as much as one third, as a result of negative studies not being published

(Sena et al. 2010). Similarly, Henderson et al. (2015) estimated that publication bias could account for a 45% overestimation of the effect size of the cancer drug Sunitinib.

The Use of Animals as Models for Humans

The assumption underpinning preclinical animal research is that animals are good models of humans, in other words, that they mimic the relevant human disease accurately, so that the results obtained from studies using the animal model can be generalized to humans. In addition to mimicking the human disease in question as closely as possible, in a "good" animal model the conditions under which a drug is tested will be as close as possible to the conditions in which humans are given the drug, and the animal sample will reflect as closely as possible the human population that suffers from that disease.

However, a lack of attention to key important details may render animal models unrepresentative of the human condition. Almost 30 years ago, Wiebers et al. (1990) noted that the experimental techniques used in animal models of stroke introduce factors that are not found in the human stroke situation, including skull trauma, external blood vessel injury, intracranial pressure, the effects of general anesthesia if this is used, or stress if no general anesthesia is used. On the other hand, they observed that while animals used in stroke models were usually young and fit, in humans, an ischemic stroke might result from years of atherosclerosis and that a human with atherosclerosis might also smoke, be taking a range of medications, and have chronic hypertension and perhaps diabetes mellitus. Many potential treatments for stroke have been found to be less effective in the presence of hypertension (Howells and Macleod 2013) and experimental studies of stroke that use healthy rather than comorbid animals have been found to overestimate the effects of the intervention (Crossley et al. 2008). Similarly, in models of osteoarthritis (OA) animals tend to be young (e.g., 10–12-week-old mice) and of normal weight and are given drugs prophylactically or in the early stages of OA, whereas clinical trials focus mainly on older people with obesity in the late stages of OA (Malfait and Little 2015). In the field of multiple sclerosis (MS) Vesterinen et al. (2010: 1053) highlight the challenges involved in creating an animal model that reflects the situation in humans, where MS "reflects a complex interplay between inflammation, demyelination, remyelination and neurodegeneration with a temporal shift in pathological emphasis from inflammation to neurodegeneration."

The evidence suggests that animal models also fail because drugs are not tested under clinically relevant conditions. Tirilazad, for example, successfully treated animals that had a stroke induced experimentally; however, the animals were given treatment within 10 minutes of the stroke, a situation that would be highly unlikely for humans unless they were already in hospital and had access to emergency care with acute stroke expertise. In clinical trials, humans were given Tirilazad within a more realistic five hours, and the trials were unsuccessful (Howells and Macleod 2013). In the field of MS, experimental drugs are most commonly administered to animals some days before the development of neurological impairment. As these drugs may work by blocking the induction of the disease, they are not relevant to the human condition as there is no way of identifying human patients prior to the onset of their MS; animal models of MS will only be relevant to the human condition if treatment can be successfully started *after* the onset of symptoms, not before (Vesterinen et al. 2010). A similar situation is found with animal models of Parkinson's disease (Zeiss et al. 2017).

A more intractable issue is the possibility that differences in the underlying biology of humans and animals mean that animals cannot be reliable predictors of human conditions. Species differences tend not to be acknowledged within animal research, argues Preuss (2006), and if they are, they "tend to be soft-peddled" and treated simply as confounding variables. Similarly, Langley (2014) notes that the animal model paradigm tends to discourage a critical appraisal of the differences between species and encourages a view that animal-based findings are generally applicable to humans. Yet increasingly scientists are highlighting important differences between the species that need to be acknowledged

(Wall and Shani 2008; Lynch 2009; Geerts 2009; Greek and Hansen 2013; Seok et al. 2013; Bailey 2014; Bailey and Taylor 2016). Bailey (2014), for example, argues that the frequently quoted 90% to 93% genetic similarity between nonhuman primates (NHPs) and humans is only superficial and belies key differences between humans and NHPs in all aspects of gene expression and protein function. Similarly, Bailey and Taylor (2016) observe that in complex living systems, even ostensibly minor differences can cause significant disparities in biological processes and outcomes. Greek and Hansen (2013) argue that it is challenging to predict the outcome from perturbations to evolved complex adaptive systems (CASs), especially when this occurs at higher levels of organization, as in responses to drugs and diseases. They note that it is difficult to predict outcomes for perturbations to a single evolved CAS, let alone using one evolved CAS to predict outcomes for a second of a different species. In the field of cholestatic liver disease, Noor (2015) argues that significant species differences exist in relation to liver immunology and biliary physiology, which lead to differences in pathogenesis and progression of liver diseases in humans when compared with other animals. Langley (2014) notes that evolutionary biology means that the species barrier cannot be overcome and that significant differences between animal models and human diseases will continue to frustrate progress. She suggests that it is "hard to envisage how animal models, limited by inter- as well as intra-species variations, could expedite the development of more personalised medicine" (Langley 2014: 1115). The development of transgenic mouse models, once touted as a means of enhancing the validity of animal models, has not improved matters (Geerts 2009).

Increasingly and importantly, the argument is being made that generalizability from animals to humans needs to be demonstrated empirically rather than simply presumed. Lynch (2009), for example, points to evidence that gene functions and gene networks can diverge through evolution and argues that equivalence between the animal and human gene with respect to the particular function under study should be demonstrated rather than assumed, to avoid spurious conclusions. Similarly, noting that virtually every drug and drug candidate functions at the molecular level, Seok et al. (2013) argue that researchers should clearly specify the extent to which the animal model mimics the molecular behavior of the key genes and key pathways thought to be important for the human disease under investigation.

Some have suggested that the use of animals as models of humans involves flaws in reasoning because it involves making crude inferences about the properties of one group based on observations from another group, simply because both groups have some other property in common (Sjoberg 2017). Sjoberg notes, for example, that if Jack is clumsy then it might be inferred that his sibling Jill is also clumsy; however, there is no evidence that Jill is clumsy, and the argument is based solely on the observation that Jack and Jill have genetic properties in common. Consequently, the use of animal models involves applying knowledge gained from animals to humans because of superficial similarities, but without any evidence to support this application. As an example, while the SOD1 transgenic mouse appears to mimic humans in terms of some of the characteristics of motor neurone disease, this does not guarantee that the same mechanisms are involved (Greek and Hansen 2013). However— and here another problem of reasoning appears (the "extrapolator's circle")—if we want to determine whether a mechanism in humans is sufficiently similar to the mechanism in animals to justify extrapolation, one must know how the relevant mechanism in humans behaves. But, if we already have knowledge about the mechanism in humans, then the initial study in animals may be redundant (Howick et al. 2013), depending on the purpose of the animal study. Or, as Greek and Hansen (2013) put it, if validity can only be established in retrospect, then the animal model is not predictive.

The Translation of Findings From Animals to Humans

If preclinical animal studies were conducted according to agreed scientific standards, if the animal models used in those studies were generalizable to humans, and if the studies were reported fairly and objectively, then we would expect them to "translate" into the clinical setting—that is, to deliver benefits for humans. However, given the problems outlined so far, it may not come as a surprise that few animal studies do translate into the clinical context. Drugs with little or no therapeutic potential are frequently taken forward into large, expensive clinical trials where, for the most part, they fail (Lindner 2007). The documents that present the findings from animal studies to ethical review boards and that are intended to persuade the board that a proposed clinical trial is worth the risk (investigator brochures) are overly optimistic and often lack the information necessary for reviewers to evaluate how good the animal evidence is (Wieschowski et al. 2018). In aggregate, it is estimated that between only 5% (Contopoulos-Ioannidis et al. 2003) and 10% of animal study interventions result in approved use in patients (Kola and Landis 2004; van der Worp et al. 2010). However, the rate varies widely by disease. Looking at the leading causes of death, for example, cancer is estimated to have a 5% success rate (Kola and Landis 2004), Alzheimer's disease a 0.4% success rate (Cummings et al. 2014), and stroke a dismal success rate of 0.1% (Howells et al. 2012). The highest rate—18%—appears to be for cardiovascular disease (Vatner 2016). Miscellaneous other fields show similarly poor rates of translation, including spinal cord injury, where, of the 22 drugs shown to provide benefit in animal models, none worked in the clinical situation (Geerts 2009); central nervous system (CNS) diseases, where, despite the extensive use of rodent models for CNS drug discovery, the success rate of new CNS drugs entering phase I clinical development is only about 8% (Geerts 2009); and inflammation, where correlation between human and mouse data is virtually absent for burns, trauma, endotoxemia, infection, and sepsis (Seok et al. 2013). Every single approach to treating sepsis that had been successful in animals has failed in humans (Leist and Hartung 2013). In relation to motor neurone disease, all but one of the experimental treatments that had ameliorated disease in animals failed in clinical trials, and the survival benefits of that one successful treatment are marginal (Perrin 2014).

Translation rates are a good indicator of whether animal models are able to benefit humans, but no systematic evaluations of the clinical relevance of animal research exist. This was highlighted in a 2016 lecture to animal researchers by Sir Mark Walport, who at the time was the UK government's chief scientific advisor. He asked,

> To what extent have we as a community, ever subjected our claims about how vital animal research has been to human health to the same level of scrutiny we'd apply to those claiming to have discovered a new cure? And I think if not, we must.
>
> *(Walport 2016)*

Although systematic reviews of animal studies are increasing in number, these do not tend to explore clinical relevance. An exception is provided by Perel et al. (2007), who identified six interventions for which there was unambiguous systematic review evidence of a treatment effect (either good or bad) in humans, then searched for all controlled animal studies for the same six interventions and systematically reviewed them. Comparing the results from the systematic reviews of animal studies with the systematic reviews of clinical studies, they found that two interventions were concordant (i.e., the findings from the animal studies agreed with the findings from the human studies), one was partially concordant, and three were discordant. These findings indicate that the majority of studies did not represent or model the human condition adequately and thus were not clinically relevant.

It is not unusual to switch on the news and hear an optimistic report about a medical breakthrough. The journalist might report that excellent results using a certain drug for a certain condition have been found in mice, and there will usually be a statement about the relevance of these findings for human health. However, there will be few caveats about the difficulties involved in extrapolating such findings to humans (Woloshin et al. 2009). This "promissory discourse," as Hobson-West (2012) puts it, holds out the promise of animal research without actually having to deliver anything at the time. However, it does serve to secure support and funding for such research (Brown 2003). It is unlikely that anyone will note down that news item and follow it up in 10, 15, or 20 years' time to see if the study has indeed delivered its promised benefits for humans. However, when such news items are reported, it is worthwhile considering how likely it is that these benefits will be delivered. Among other things, the possibility of that study resulting in benefits for humans will depend on how well the animal study was conducted (i.e., whether potential sources of bias were accounted for, whether the researchers controlled for confounding variables, whether the sample size was adequate, how impartial the analysis was, etc.), whether the animal model accurately mimicked the relevant human condition, whether the animals used adequately represented the relevant human population, and how likely is it that findings can be generalized from one species to another.

Prospects for the Future

The shortcomings of animal research would not have been revealed so irrefutably had researchers not started at the beginning of this century to use systematic review methodology to examine and synthesize the evidence from animal studies. The revelation of these shortcomings has made it more acceptable to challenge the science of animal research and has brought acknowledgment of the need for change. The necessity for training in experimental design has been acknowledged (Festing 2013; Collins and Tabak 2014; Dukes 2016), and a variety of guidelines and checklists now exist to help scientists plan, design, and conduct their studies (Henderson et al. 2013; Smith et al. 2018; Percie du Sert et al. 2017); identify risk of bias (Hooijmans et al. 2014); and improve reporting (Hooijmans et al. 2010; Kilkenny et al. 2010; Martins and Franco 2015). There have been calls to register animal study protocols in advance to reduce selective outcome reporting bias and publication bias (Howells and Macleod 2013; Muhlhausler et al. 2013) and to increase readers' confidence that the hypothesis tested has not shifted (Wieschowski et al. 2016). It has also been suggested that the results of animal studies, including negative results, should be deposited in publicly accessible databases to enable knowledge to be shared and ensure that the burdens endured by animals retain their moral justification (Kimmelman and Anderson 2012). Other suggestions include ensuring that sufficient good-quality evidence exists before proceeding to clinical trials by conducting systematic reviews of animal studies in the relevant field (Pound et al. 2004; Perel et al. 2007; van Luijk et al. 2012; Ritskes-Hoitinga and Wever 2018), investigating the validity of the relevant animal model, and testing proof of mechanism (Wendler and Wehling 2012).

Nevertheless, underlying all these suggestions is an assumption that improved scientific rigor will ultimately lead to better rates of translation. Dirnagl and Endres (2014) in the field of stroke, for example, are typical in arguing that as long as scientific standards are improved, animal models will eventually bear fruit. Sutherland et al. (2012), however, strike a note of caution. They note that the 1999 Stroke Treatment Academic Industry Roundtable (STAIR) recommendations were intended to improve the quality of animal studies in stroke, yet "more than 10 years after the first STAIR recommendations were published, the ultimate proof that plain standardization of procedures in fact increases the rate of successful translation from bench to bedside in stroke research is still missing" (p. 415). Sutherland et al. wrote this in 2012, and it remains the case seven years on.

Given this situation, it is not surprising that there are increasing calls for a complete change of paradigm. These calls are gaining momentum due to promising new *in silico* (i.e. mathematical, statistical, modelling, and computer science tools) and human *in vivo* methods; human induced pluripotent stem cells, for example, can generate unique disease-relevant tissue and human-specific cell models for liver disease (Noor 2015) and even for genetically complex diseases, such as Alzheimer's disease (Langley 2014), which along with micro-physiological systems, such as "organs on chips," can be used for studying disease pathways. Langley (2014) argues that new technologies such as these create the potential for a new human biology and systems-based understanding of disease.

Industry is increasingly backing the call for a paradigm change. In 2015, "[a] non-animal technologies roadmap" was drawn up by the UK's Innovation Agency. It identified nonanimal technologies as one of a series of emerging technologies with the potential to drive future UK economic growth, noting that the market potential was huge (Innovate UK 2015). In 2018, a major report from the UK BioIndustry Association and the Medicines Discovery Catapult stated that "humanising" the process of drug discovery and testing would ease the productivity crisis in the pharmaceutical industry. It identified a range of emerging nonanimal technologies with potential to make the early stages of research more predictive of how drugs actually work in humans (BioIndustry Association and the Medicines Discovery Catapult 2018). The US Food and Drug Administration (FDA) is encouraging the development, validation, and integration of the most promising new technologies into the product pipeline, with the aim of reducing animal use (FDA 2017), whilst 16 US federal government agencies, including the FDA, have stated their aim to "expedite the use of 21st-century science to protect and improve public health" and ensure the use of new approaches that reduce or eliminate the need for testing in animals and that are more relevant to human health (ICCVAM 2018). Finally, in a bold move, the Dutch government has declared that it will phase out animal testing for regulatory purposes by 2025 (NCad 2016), while the US Environmental Protection Agency has recently announced that it aims to eliminate all requests and funding for studies using mammals by 2035 (Environmental Protection Agency 2019)

Conclusion

The evidence has shown that scientists conducting animal research commonly fail to take measures to prevent bias and are frequently selective in the reporting of their studies. Together with publication bias (failure to publish "negative" findings), this has led to the benefits of animal studies being overestimated and a body of evidence containing a falsely inflated proportion of studies with statistically significant (positive) results. Furthermore, the evidence suggests that animal models frequently fail to represent the complexity of human disease and that the timing of interventions in animal models may not be relevant to the human situation. It has also been suggested that generalizability from animals to humans is assumed rather than demonstrated empirically and that the use of animal models involves flaws in reasoning. As a result of simplistic animal models and because methodologically flawed animal studies produce over-optimistic conclusions about efficacy, drugs with little therapeutic potential are tested in clinical trials and, for the most part, fail. Consequently, we can conclude that (1) most animal research is at present of such poor quality that no reliable conclusions may be drawn from it, (2) rates of translation from animals to humans indicate that much animal research fails to benefit humans, and (3) there is no systematic evidence to support the assertion that animal research benefits humans.

However, as DeGrazia and Sebo (2015) argue, an expectation of sufficient net benefit to humans must be a condition of morally responsible animal research. So where does this leave the ethics of animal research? If we accept the conclusions presented earlier, then the traditional utilitarian argument is gravely undermined and a revised ethical approach is required. Bass (2012: 94–95) has begun to explore the implications for the utilitarian argument of this doubt about human benefit:

> Since the best working hypothesis is that the human benefits of animal research are either small or unclear, we are not in a position to claim justification. If the benefits are small, they cannot outweigh large harms to animals; if they are unclear, even as to probabilities, then we do not know there to be any outweighing benefit.

The task here is to combine such arguments with the available evidence in order to challenge and update the existing ethical frameworks used within animal research regulation, namely, the HBA.

However, the evidence points to another factor that ethicists need to seriously grapple with, namely, the poor quality of animal research. At present, an unethical situation exists because poorly conducted animal studies produce unreliable and inconclusive findings. This means that any suffering endured by animals loses its moral justification because their use cannot possibly contribute toward clinical benefit. The same applies in cases where negative findings from animal studies are not published; the animals used cannot possibly contribute toward scientific knowledge because the findings are not made public. In this case, there is an additional risk that the animal experiments may be repeated unnecessarily because findings from the original unsuccessful experiments are not publicized. Consequently, an updated ethics of animal research needs to add research quality and integrity into the harm/benefit equation (Pound and Nicol 2018).

The ethics of animal research has to date been too abstract, and this has resulted in stagnation. However, if bioethicists begin to engage to a greater extent with the evidence, there is a chance that the field can be renewed and reinvigorated in order to become a real force for change.

Note

1. The United States Department of Agriculture (USDA) Animal Welfare Act does not require an HBA to be performed; however, the US Institutional Animal Care and use Committee (IACUC) is obliged to weigh the objectives of each study against its potential harms to animals.

Recommended Readings

Cohen, D. (2018) "Oxford TB vaccine study calls into question selective use of animal data," *BMJ* 360: j5845.
(A very interesting and recent investigation into the selective reporting of animal data by Oxford academics).

Ioannidis, J. P. A. (2012) "Extrapolating from animals to humans," *Science Translational Medicine* 4(151): 1–3.
(A classic paper that explores some of the reasons why it is difficult to extrapolate findings from animal studies to humans.)

Leist, M., and Hartung, T. (2013) "Inflammatory findings on species extrapolations: Humans are definitely no 70-kg mice," *Archives of Toxicology* 87: 563–567.
(An article that sums up some of the findings suggesting species differences make animal studies poor predictors of human response.)

Tsilidis, K. K., Panagiotou, O. A., Sena, E. S., Aretouli, E., Evangelou, E., Howells, D. W., Salman, R. A. S., Macleod, M. R., and Ioannidis, J. P. (2013) "Evaluation of excess significance bias in animal studies of neurological diseases," *PLoS Biology* 11(7): e1001609.
(A study that found evidence of strong biases in the field of animal studies of neurological disorders and that illustrates how these biases make that research field appear more successful than it actually is.)

Woloshin, S., Schwartz, L. M., Casella, S. L., Kennedy, A. T., and Larson, R. J. (2009) "Press releases by academic medical centers: Not so academic?" *Annals of Internal Medicine* 150(9): 613–618.
(This article illustrates how academics frequently use press releases to promote research of uncertain relevance to human health without acknowledging its limitations.)

Bibliography

Bailey, J. (2014) "Monkey-based research on human disease: The implications of genetic differences," *Alternatives to Laboratory Animals* 42: 287–317.

Bailey, J., and Taylor, K. (2016) "Non-human primates in neuroscience research: The case against its scientific necessity," *Alternatives to Laboratory Animals* 43(1): 43–69.

Balcombe, J., Ferdowsian, H., and Briese, L. (2013) "Prolonged pain research in mice: Trends in reference to the 3Rs," *Journal of Applied Animal Welfare Science* 16(1): 77–95.

Bass, R. (2012) "Lives in the Balance: Utilitarianism and Animal Research," in J. R. Garrett (ed.) *The Ethics of Animal Research: Exploring the Controversy* (pp. 81–106), London: MIT Press.

Bebarta, V., Luyten, D., and Heard, K. (2003) "Emergency medicine animal research: Does use of randomization and blinding affect the results?" *Academic Emergency Medicine* 10(6): 684–687.
Begley, C. G., and Ellis, L. M. (2012) "Raise standards for preclinical cancer research," *Nature* 483: 531–533.
BioIndustry Association and the Medicines Discovery Catapult (2018) *State of the Discovery Nation 2018 and the Role of the Medicines Discovery Catapult*, https://s3-eu-west-1.amazonaws.com/media.newmd.catapult/wp-content/uploads/2018/01/16220811/ MDC10529-Thought-Leader_v10_Interactive_v1.pdf
Brown, N. (2003) "Hope against hype-accountability in biopasts, presents and futures," *Science Studies* 16(2): 3–21.
Button, K. S., Ioannidis, J. P., Mokrysz, C., Nosek, B. A., Flint, J., Robinson, E. S., and Munafò, M. R. (2013) "Power failure: Why small sample size undermines the reliability of neuroscience," *Nature Reviews Neuroscience* 14(5): 365–376.
Carbone, L., and Austin, J. (2016) "Pain and laboratory animals: Publication practices for better data reproducibility and better animal welfare," *PLoS One* 11(5): e0155001.
Cochrane, A. L. (1972) *Effectiveness and Efficiency: Random Reflections on Health Services*, London: Nuffield Provincial Hospitals Trust.
Cohen, D. (2018) "Oxford TB vaccine study calls into question selective use of animal data," *BMJ* 360: j5845.
Collins, F. S., and Tabak, L. A. (2014) "NIH plans to enhance reproducibility," *Nature* 505(7485): 612.
Contopoulos-Ioannidis, D. G., Ntzani, E. E., and Ioannidis, J. P. A. (2003) "Translation of highly promising basic science research into clinical applications," *American Journal of Medicine* 114: 477–484.
Crossley, N. A., Sena, E., Goehler, J., Horn, J., van der Worp, B., Bath, P. M., Macleod, M., and Dirnagl, U. (2008) "Empirical evidence of bias in the design of experimental stroke studies: A metaepidemiologic approach," *Stroke* 39(3): 929–934.
Cummings, J. L., Morstorf, T., and Zhong, K. (2014) "Alzheimer's disease drug-development pipeline: Few candidates, frequent failures," *Alzheimer's Research & Therapy* 6(4): 37.
Davies, R., Ives, J., and Dunn, M. (2015) "A systematic review of empirical bioethics methodologies," *BMC Medical Ethics* 16(15). doi:10.1186/s12910-015-0010-3.
DeGrazia, D., and Sebo, J. (2015) "Necessary conditions for morally responsible animal research," *Cambridge Quarterly of Healthcare Ethics* 24(4): 420–430.
Dirnagl, U., and Endres, M. (2014) "Found in translation," *Stroke* 45: 1510–1518.
Dukes, J. (2016) "We are getting animal research wrong—only education can fix it," *Guardian,* December 10.
Environmental Protection Agency (2019) *Efforts to reduce animal testing at EPA*, https://www.epa.gov/environmental-topics/efforts-reduce-animal-testing-epa
European Union (2010) "European union directive 2010/63/EU of the European parliament and of the council of 22 September 2010 on the protection of animals used for scientific purposes," *Official Journal of the European Union* 276: 33–79.
Everitt, J. I. (2015) "The future of preclinical animal models in pharmaceutical discovery and development: A need to bring in cerebro to the in vivo discussions," *Toxicologic Pathology* 43(1): 70–77.
[FDA], Food and Drug Administration (2017) "FDA'S predictive toxicology roadmap," www.fda.gov/downloads/ScienceResearch/SpecialTopics/RegulatoryScience/UCM587831.pdf.
Festing, M. F. W. (2013) "We are not born knowing how to design and analyse scientific experiments," *Alternatives to Laboratory Animals* 41: 1–3.
Foster, C. (2001) *The Ethics of Medical Research on Humans*, Cambridge: Cambridge University Press.
Friese, C. (2013) "Realizing potential in translational medicine: The uncanny emergence of care as science," *Current Anthropology* 54(S7): S129–S138.
Garrett, J. R. (2012) "The ethics of animal research: An overview of the debate," in J. R. Garrett (ed.) *The Ethics of Animal Research: Exploring the Controversy* (pp. 1–16), London: MIT Press.
Geerts, H. (2009) "Of mice and men: Bridging the translational disconnect in CNS drug discovery," *CNS Drugs* 23(1): 915–926.
Greek, R., and Hansen, L. A. (2013) "Questions regarding the predictive value of one evolved complex adaptive system for a second: Exemplified by the SOD1 mouse," *Progress in Biophysics and Molecular Biology* 113: 231–253.
Gregory, R. T. (2000) "The failure of traditional arguments in the vivisection debate," *Public Affairs Quarterly* 14(2): 159–182.
Gruber, F. P., and Hartung, T. (2004) "Alternatives to animal experimentation in basic research," *ALTEX* 21(suppl. 1/04): 3–31.
Henderson, V. C., Demko, N., Hakala, A., MacKinnon, N., Federico, C. A., Fergusson, D., and Kimmelman, J. (2015) "A meta-analysis of threats to valid clinical inference in preclinical research of sunitinib," *Elife* 4: e08351.
Henderson, V. C., Kimmelman, J., Fergusson, D., Grimshaw, J. M., and Hackam, D. G. (2013) "Threats to validity in the design and conduct of preclinical efficacy studies: A systematic review of guidelines for in vivo animal experiments," *PLoS Medicine* 10(7): e1001489.

Hirst, J. A., Howick, J., Aronson, J. K., Roberts, N., Perera, R., Koshiaris, C., and Heneghan, C. (2014) "The need for randomization in animal trials: An overview of systematic reviews," *PLoS One* 9(6): e98856.

Hobson-West, P. (2012) "Ethical boundary-work in the animal research laboratory," *Sociology* 46(4): 649–663.

Holman, C., Piper, S. K., Grittner, U., Diamantaras, A. A., Kimmelman, J., Siegerink, B., and Dirnagl, U. (2016) "Where have all the rodents gone? The effects of attrition in experimental research on cancer and stroke," *PLoS Biology* 14(1): e1002331.

Home Office (2017) *Statistics of Scientific Procedures on Living Animals, Great Britain 2016*, www.gov.uk/government/statistics/statistics-of-scientific-procedures-on-living-animals-great-britain-2016

Hooijmans, C. R., Leenaars, M., and Ritskes-Hoitinga, M. (2010) "A gold standard publication checklist to improve the quality of animal studies, to fully integrate the three rs, and to make systematic reviews more feasible," *Alternatives to Laboratory Animals* 38(2): 167–182.

Hooijmans, C. R., Rovers, M. M., de Vries, R. B., Leenaars, M., Ritskes-Hoitinga, M., and Langendam, M. W. (2014) "SYRCLE's risk of bias tool for animal studies," *BMC Medical Research Methodology* 14(1): 43.

Howells, D. W., and Macleod, M. R. (2013) "Evidence-based translational medicine," *Stroke* 44: 1466–1471.

Howells, D. W., Sena, E. S., O'Collins, V., and Macleod, M. R. (2012) "Improving the efficiency of the development of drugs for stroke," *International Journal of Stroke* 7(5): 371–377.

Howick, J., Glasziou, P., and Aronson, J. K. (2013) "Problems with using mechanisms to solve the problem of extrapolation," *Theoretical Medicine and Bioethics* 34(4): 275–291.

[ICCVAM], Interagency Coordinating Committee on the Validation of Alternative Methods (2018) *A Strategic Roadmap for Establishing New Approaches to Evaluate the Safety of Chemicals and Medical Products in the United States*, https://ntp.niehs.nih.gov/iccvam/docs/roadmap/2018_01_04_roadmap-fd-508.pdf

Innovate UK (2015) *A Non-Animal Technologies Roadmap for the UK: Advancing Predictive Biology*, https://nc3rs.org.uk/sites/default/files/documents/NonAnimalTechCO082_ RYE_4_nrfinal2.pdf

Kilkenny, C., Browne, W. J., Cuthill, I. C., Emerson, M., and Altman, D. (2010) "Improving bioscience research reporting: The ARRIVE guidelines for reporting animal research," *PLoS Biology* 8(6): e1000412.

Kilkenny, C., Parsons, N., Kadyszewski, E., Festing, M. F., Cuthill, I. C., Fry, D., Hutton, J., and Altman, D. G. (2009) "Survey of the quality of experimental design, statistical analysis and reporting of research using animals," *PLoS One* 4(11): e7824.

Kimmelman, J., and Anderson, J. A. (2012) "Should preclinical studies be registered?" *Nature Biotechnology* 30(6): 488–489.

Kola, I., and Landis, J. (2004) "Can the pharmaceutical industry reduce attrition rates?" *Nature Reviews Drug Discovery* 3: 711–715.

Korevaar, D. A., Hooft, L., and Ter Riet, G. (2011) "Systematic reviews and meta-analyses of preclinical studies: Publication bias in laboratory animal experiments, *Laboratory Animals* 45(4): 225–230.

Langley, G. R. (2014) "Considering a New paradigm for alzheimer's disease research," *Drug Discovery Today* 19(8): 1114–1124.

Leist, M., and Hartung, T. (2013) "Inflammatory findings on species extrapolations: Humans are definitely no 70-kg mice," *Archives of Toxicology* 87: 563–567.

Lindner, M. D. (2007) "Clinical attrition due to biased preclinical assessments of potential efficacy, *Pharmacology and Therapeutics* 115: 148–175.

Lynch, V. J. (2009) "Use with caution: Developmental systems divergence and potential pitfalls of animal models," *The Yale Journal of Biology and Medicine* 82(2): 53–66.

Macleod, M. R., McLean, A. L., Kyriakopoulou, A., Serghiou, S., de Wilde, A., Sherratt, N., Hirst, T., Hemblade, R., Bahor, Z., Nunes-Fonseca, C., and Potluru, A. (2015) "Risk of bias in reports of in vivo research: A focus for improvement," *PLoS Biology* 13(10): e1002273.

Malfait, A. M., and Little, C. B. (2015) "On the predictive utility of animal models of osteoarthritis," *Arthritis Research and Therapy* 17(1): 225.

Martins, A. R., and Franco, N. H. (2015) "A critical look at biomedical journals' policies on animal research by use of a novel tool: The EXEMPLAR scale," *Animals* 5(2): 315–331.

Mobley, A., Linder, S. K., Braeuer, R., Ellis, L. M., and Zwelling, L. (2013) "A survey on data reproducibility in cancer research provides insights into our limited ability to translate findings from the laboratory to the clinic," *PLoS One* 8(5): e63221.

Mueller, K. F., Briel, M., Strech, D., Meerpohl, J. J., Lang, B., Motschall, E., Gloy, V., Lamontagne, F., and Bassler, D. (2014) "Dissemination bias in systematic reviews of animal research: A systematic review," *PLoS One* 9(12): e116016.

Muhlhausler, B. S., Bloomfield, F. H., and Gillman, M. W. (2013) "Whole animal experiments should be more like human randomized controlled trials," *PLoS Biology* 11(2): e1001481.

[NCad], Netherlands National Committee for the Protection of Animals Used for Scientific Purposes (2016) *Transition to Non-Animal Research*, file:///C:/Users/User/Documents/Documents/Animals/Safer%20medicines%20trust/NCad+Opinion+Transition+to+non-animal+research.pdf

Noor, F. (2015) "A shift in paradigm towards human biology-based systems for cholestatic-liver diseases," *The Journal of Physiology* 593(23): 5043–5055.

Percie du Sert, N., Alfieri, A., Allan, S. M., Carswell, H. V., Deuchar, G. A., Farr, T. D., Flecknell, P., Gallagher, L., Gibson, C. L., Haley, M. J., and Macleod, M. R. (2017) "The IMPROVE guidelines (Ischaemia models: Procedural refinements of in vivo experiments), *Journal of Cerebral Blood Flow & Metabolism* 37(11): 3488–3517.

Perel, P., Roberts, I., Sena, E., Wheble, P., Briscoe, C., Sandercock, P., Macleod, M., Mignini, L. E., Jayaram, P., and Khan, K. S. (2007) "Comparison of treatment effects between animal experiments and clinical trials: Systematic review," *BMJ* 334: 197.

Perrin, S. (2014) "Make mouse studies work," *Nature* 507: 423–425.

Pound, P., Ebrahim, S., Sandercock, P., Bracken, M. B., and Roberts, I. (2004) "Where is the evidence that animal research benefits humans?" *BMJ* 328: 514.

Pound, P., and Nicol, C. (2018) "Retrospective harm benefit analysis of pre-clinical animal research for six treatment interventions," *PLoS One* 13(3): e0193758.

Preuss, T. M. (2006) "Who's afraid of homo sapiens?" *Journal of Biomedical Discovery and Collaboration* 1(1): 17.

Prinz, F., Schlange, T., and Asadullah, K. (2011) "Believe it or not: How much can we rely on published data on potential drug targets?" *Nature Reviews Drug Discovery* 10: 328–329.

Reichlin, T. S., Vogt, L., and Würbel, H. (2016) "The researchers' view of scientific rigor—survey on the conduct and reporting of in vivo research," *PLoS One* 11(12): e0165999.

Ritskes-Hoitinga, M., and Wever, K. (2018) "Improving the conduct, reporting, and appraisal of animal research," *BMJ* 360: j4935.

Royal Society (2015) "Statement of the royal society's position on the use of animals in research," https://royalsociety.org/topics-policy/publications/2015/animals-in-research/

Russell, W., and Burch, R. (1959) *The Principles of Humane Experimental Technique*, London: Methuen.

Scott, S., Kranz, J. E., Cole, J., Lincecum, J. M., Thompson, K., Kelly, N., Bostrom, A., Theodoss, J., Al-Nakhala, B. M., Vieira, F. G., and Ramasubbu, J. (2008) "Design, power, and interpretation of studies in the standard murine model of ALS," *Amyotrophic Lateral Sclerosis* 9(1): 4–15.

Sena, E. S., Currie, G. L., McCann, S. K., Macleod, M. R., and Howells, D. W. (2014) "Systematic reviews and meta-analysis of preclinical studies: Why perform them and how to appraise them critically," *Journal of Cerebral Blood Flow & Metabolism* 34(5): 737–742.

Sena, E. S., van der Worp, H. B., Bath, P. M., Howells, D. W., and Macleod, M. R. (2010) "Publication bias in reports of animal stroke studies leads to major overstatement of efficacy," *PLoS Biology* 8(3): e1000344.

Seok, J., Warren, S., Cuenca, A., Mindrinos, M., Baker, H., Xu, W., Richards, D., . . . and the Inflammation and Host Response to Injury Large Scale Collaborative Research Program (2013) "Genomic responses in mouse models poorly mimic human inflammatory diseases," *Proceedings of the National Academy of Sciences* 110(9): 3507–3512.

Sjoberg, E. A. (2017) "Logical fallacies in animal model research," *Behavioral and Brain Functions* 13(1): 3.

Smith, A. J., Clutton, R. E., Lilley, E., Hansen, K. E. A., and Brattelid, T. (2018) "PREPARE: Guidelines for planning animal research and testing," *Laboratory Animals* 52(2): 135–141.

Sutherland, B. A., Minnerup, J., Balami, J. S., Arba, F., Buchan, A. M., and Kleinschnitz, C. (2012) "Neuroprotection for ischaemic stroke: Translation from the bench to the bedside," *International Journal of Stroke* 7: 407–418.

Tsilidis, K. K., Panagiotou, O. A., Sena, E. S., Aretouli, E., Evangelou, E., Howells, D. W., Salman, R. A. S., Macleod, M. R., and Ioannidis, J. P. (2013) "Evaluation of excess significance bias in animal studies of neurological diseases," *PLoS Biology* 11(7): e1001609.

van der Worp, H., Howells, D. W., Sena, E. S., Porritt, M. J., Rewell, S., O'Collins, V., and Macleod, M. R. (2010) "Can animal models of disease reliably inform human studies?" *PLoS Medicine* 7(3): e1000245.

van Luijk, J., Cuijpers, Y., van der Vaart, L., de Roo, T., Leenaars, M., and Ritskes-Hoitinga, M. (2013) "Assessing the application of the 3Rs: A survey among animal welfare officers in the Netherlands," *Laboratory Animals* 47(3): 210–219.

van Luijk, J., Leenaars, M., Hooijmans, C., Wever, K., de Vries, R., and Ritskes-Hoitinga, M. (2012) "Towards evidence-based translational research: The pros and cons of conducting systematic reviews of animal studies," *Alternatives to Animal Experimentation* 30(2–13): 256–257.

Vatner, S. F. (2016) "Why so few new cardiovascular drugs translate to the clinics," *Circulation Research* 119(6): 714–717.

Vesterinen, H. M., Sena, E. S., French-Constant, C., Williams, A., Chandran, S., and Macleod, M. R. (2010) "Improving the translational hit of experimental treatments in multiple sclerosis," *Multiple Sclerosis Journal* 16(9): 1044–1055.

Wall, R. J., and Shani, M. (2008) "Are animal models as good as we think?" *Theriogenology* 69: 2–9.

Walport, M. (2016) *Animal Research: Then and Now*, 80th Stephen Paget Memorial Lecture, www.understanding animalresearch.org.uk/resources/video-library/80th-paget-lecture-2016/

Wendler, A., and Wehling, M. (2012) "Translatability scoring in drug development: Eight case studies," *Journal of Translational Medicine* 10(1): 39.

Wiebers, D. O., Adams, H. P., and Whisnant, J. P. (1990) "Animal models of stroke: Are they relevant to human disease?" *Stroke* 21(1): 1–3.

Wieschowski, S., Chin, W. W. L., Federico, C., Sievers, S., Kimmelman, J., and Strech, D., (2018) "Preclinical efficacy studies in investigator brochures: Do they enable risk—benefit assessment?" *PLoS Biology* 16(4): e2004879.

Wieschowski, S., Silva, D. S., and Strech, D. (2016) "Animal study registries: Results from a stakeholder analysis on potential strengths, weaknesses, facilitators, and barriers," *PLoS Biology* 14(11): e2000391.

Woloshin, S., Schwartz, L. M., Casella, S. L., Kennedy, A. T., and Larson, R. J. (2009) "Press releases by academic medical centers: Not so academic?" *Annals of Internal Medicine* 150(9): 613–618.

Zeiss, C. J., Allore, H. G., and Beck, A. P. (2017) "Established patterns of animal study design undermine translation of disease-modifying therapies for Parkinson's disease," *PLoS One* 12(2): e0171790.

19

APPLIED ETHICS IN ANIMAL EXPERIMENTATION

Larry Carbone

In the research laboratory, animals face the prospect of pain, disease, and early death so that humans will gain valuable knowledge, mostly to benefit humans. Many millions of animals live and die in American laboratories every year. The financial costs of animal-based research are high; how about the ethical costs?

Research animal ethics shares much with general issues in animal ethics, most notably, the question of when, if ever, we can morally justify harming animals for human benefit. And it raises questions about medical research more generally. For example, how much money, time, or animal harm can we justify spending in pursuit of new and high-tech medical knowledge when so much of the world is currently not receiving or not able to afford the benefits of the knowledge we already have? Can we justify harming animals to study medical conditions we largely bring on ourselves, such as smoking-related diseases or weapons-related injuries?

I come to this project as a historian of laboratory animal welfare policy[1] and as a practicing laboratory animal veterinarian. In my veterinary work, animals in laboratories are my patients, scientists that use those animals are my clients. I work for my campus Institutional Animal Care and Use Committee (IACUC). (See Latham's chapter in this volume.) My perspective here includes a focus on our need for clear empirical facts—what we know about animals and how we know it, what we know about the knowledge produced in animal experimentation—that we mesh with our values to develop norms and policy for animal care and use.

Scientists study laboratory animals in many ways and for many reasons. Zoologists and ecologists study animals to learn about animals, how they move and live in the world, sometimes with an eye to applied conservation biology. Animal scientists and veterinarians study companion and farm animals to find better ways of producing meat and milk or better treatments for household animals. Toxicology laboratories study animals to learn how substances can affect the body's health or, more particularly, how safe a specific drug or a batch of vaccine might be for humans. And, of course, medical researchers use millions of animals in laboratories throughout the world as "models" of human health and biology, with an eye to preventing and curing human diseases. (See Pound's chapter in this volume.)

For animal research to be justifiable, as a large-scale enterprise, as well as for individual projects, two general conditions must both be met: (1) it must be ethically defensible to use animals in harmful ways for humans ends, and (2) animal research must produce truly useful knowledge.[2–4]

Case Study: Guinea Pigs in Arthritis Research

With animals used in so very many ways in research, we can start with an example as a case study that will illustrate the ethical issues. Dr. Cavy is working to develop new drugs to treat arthritis and wants to make sure they are both safe and effective in animals before testing them in human arthritis patients. She decides that she will model human arthritis by causing arthritis in Guinea pigs, with a chemical she injects into their knees.[5] She will then give her new drugs to half of the animals, her experimental group, and a placebo drug to the other half, her control group. At two weeks, and again at four weeks after the injections, she will test how well they move on a treadmill and do a CT scan of their knees. If her experimental group walks significantly better than her placebo group, she will have evidence that her new drug is an effective arthritis drug. Before she buys her Guinea pigs and starts her experiments, she will need approval from her university's IACUC and funding from the National Institutes of Health (NIH), and they will want to know how much pain the animals will be in, whether the placebo group can receive pain medications if they are in pain, how she will know if they are in pain, and how likely it is that she will produce valuable data that apply to humans as well as Guinea pigs. They must balance the potential animal suffering against the potential human benefits.

The laws that require IACUCs and ethics committees presume that human interests can sometimes outweigh animal interests but not without limits. In the research laboratory, animal interests are challenged by their confinement and living conditions, but most important, experiments that cause pain and distress, intentionally or contingently, to animals must be accounted for when balancing against human interests. This cannot be allowed if humans' moral worth is not considerably greater than that of animals (see this book's general introduction). This chapter cannot resolve this question and so starts with the reality that in our current world, our laws and practices reflect a widely shared moral belief that (most) humans do have higher moral worth and receive greater privileges and protections than (most) nonhuman animals.[6]

In animal research, as in animal farming, hunting and other uses of animals, people rarely intend to cause animal suffering. But farmers and scientists actively do things to animals that they should know will cause suffering as a side effect of producing meat or producing research data. In the Guinea pig arthritis lab, Dr. Cavy is producing a month of pain in her animals, even though she only tests them twice. If she did not think it would disrupt her experiments, she would treat them with known, effective medicines for all but the two testing days so that they avoid a full month of suffering. Were she simply studying animals who developed arthritis in the course of their natural lives, trying her best candidate treatments for the animals' benefit, the ethical questions would be different and likely less intense. Instead of these alternative approaches, she stays with her plan to induce a month of pain in her animals.

There are compelling arguments for abolishing the use of animals in laboratories outright or, at least, the use of sentient animals (see Pound's chapter in this volume). Among those who accept some use of sentient animals in research, consequentialist analysis of harms (mostly to the animals) and benefits (mostly to humans) is the common language. Deontologically, some argue that we do these analyses precisely because sentient animals have a right to this consideration, though a weak right that we frequently override for human interests.

To the extent that regulations reflect ethical norms and drive ethical deliberations, some sort of explicit weighing of harms and benefits is becoming the norm in animal research. Since the 1980s, American law and policy have obliquely touched on this. The *United States Government Principles for the Utilization and Care of Vertebrate Animals Used in Testing, Research, and Training* are the basis for the law that requires animal welfare programs for government-funded research. Principle II states that "[p]rocedures involving animals should be designed and performed with due consideration of their relevance to human or animal health, the advancement of knowledge, or the good of society" but without quite saying that these must be weighed against the harms to animals or raising the possibility that the cost-to-benefit could render a project indefensible and unapproveable. The Animal Welfare Act Regulations' references to "justification" and justification "for scientific reasons," mostly formulated in the late 1980s, are even more amorphous.

Through several editions, the *Guide for the Care and Use of Laboratory Animals* referenced the US government *Principles* without fleshing out guidance on who should consider proposed studies' relevance or how. In its eighth edition in 2010, the *Guide* took the step of calling on IACUCs to "weigh the objectives of the study against potential animal welfare concerns." Failure to do this could jeopardize an institution's access to NIH research funds or the accreditation status of its animal care and use program.

And so, how do we evaluate the benefits of animal research? The harms? How do we balance the one against the other?

Evaluating the Benefits of Animal Experimentation

This chapter would be simpler and very different if a scientist could say with certainty that 1,000 mice will experience *x* units of harm in her laboratory to increase five-year survival of glioblastoma brain tumors in humans. Or that there was an 80% chance of that outcome. It would be different if the mice had some voice in whether those odds of helping humans were somehow compelling to the mice themselves and could justify their sacrifice. Unfortunately, in animal research, the harms and benefits are often unknown and probabilistic. If scientists only studied animals with naturally occurring conditions—arthritic pet Guinea pigs in a neighborhood veterinary hospital, for example—the ethical questions would be different. But in most biomedical research, most harms to the animals are at humans' hands.

Modern medical advances are a compelling testament to the success of animal research. Survival rates for childhood cancers have increased dramatically over recent years, as have survival rates for some other cancers.[7] HIV infection is a manageable chronic illness, not a death sentence, in societies in which patients have access to antiviral medications. Twenty-first-century vaccines are available for Hepatitis A, Hepatitis B, Human Papillomavirus/Cervical cancer, and Ebola, with potential malaria vaccines on the horizon. All these advances have involved animal experiments in their development. Now are those advances, and even others in progress, a compelling case for the continued value of animal experimentation?

There are counterarguments and complications before we proclaim the benefits of animal research too boldly. Just as scientists can point to important discoveries and advances, there are likewise many instances in which animal studies have failed to yield fruit or have yielded (or would have yielded) incorrect information. A commonly noted example is penicillin—had it been studied in Guinea pigs before its usefulness in people was already well known, it would have been discarded as a toxic and dangerous chemical.[8] Add to this the many unknown potentially beneficial compounds and medicines, unknown because any initial examination in the animal lab suggested, possibly incorrectly, that they had no place in human medicine.

Another brake on our certainty of the benefits of animal studies is the reality that in most modern countries, law requires that some basic level of demonstrated efficacy and safety in animals is necessary before something can be studied in humans. One of the earliest statements in the ethics of human experimentation, the Declaration of Helsinki, requires this.[9] But a requirement that we test medicines in animals before we test them in humans is not equivalent to a scientific conclusion that the animal studies associated with any list of discoveries is *causal*.

Scientists are constantly developing new research methods, and some animal experimental methods that were once necessary (i.e., necessary to get the data that resulted) are now obsolete or soon will be. When Russell and Burch wrote their 1959 book *The Principles of Humane Experimental Technique*, Guinea pigs were in wide use for the diagnosis of tuberculosis.[10] Scientists had no successful way to grow tuberculosis bacteria to confirm a diagnosis other than in live animals. That is no longer true. Monoclonal antibodies are an important tool in research labs and in cancer medicine. Throughout the 1990s, animal rights groups pressured the NIH to prohibit using mice in their production; now, monoclonal antibodies are mostly grown in flasks and bioreactors and very rarely in mice.[11] What other currently necessary animal methodologies are on their way out?

At best then, weighing the potential benefits to humans is a challenge. Past successes convince this author that animal research does produce valuable knowledge, but prior necessity of animal studies does not mean they are as valuable going forward. One metric of success is the "translation success rate," of drugs that have fared well enough in animal labs to progress to human clinical trials. How to evaluate a concern that only about 10% of psychiatric drugs succeed in translation?[11] That represents a lot of money and animals ultimately resulting in failure, unless the 10% success rate is actually odds we can justify. But that requires evaluating how much animal harm goes into that level of benefit.

Evaluating the Harms to Animals in Laboratories

Animals are at risk of myriad harms in the animal laboratory. Some experiments may cause pain. Others may cause forms of distress that differ from pain, whether primarily physical (hunger, thirst, some diseases) or psychological (loneliness, fear, anxiety, aggression). In some situations, the pain or distress is the scientist's intent, such as causing arthritis pain in Dr. Cavy's Guinea pig experiments. More often, the pain or distress is incidental or contingent, such as when the scientist must house an animal in a solitary cage for experimental observations or conduct a surgery to create a condition (e.g., surgery to create a rat version of a heart attack, or myocardial infarction) to later potential study treatments for it. Dr. Cavy's Guinea pigs may move less, eat less, and sleep less well if their pain is severe enough—unnecessary suffering that is not only a challenge for the animals' welfare but could affect their response to the test drugs too.

In addition to pain, distress, and suffering that the experiment causes, simply housing animals for use in experiments can also harm animals. Animals live in cages far smaller than their natural range, in groups not of their choosing or in solitary cages. Their busy lives in the natural state are frequently quite impoverished in a cage, with no need to forage and strategize how to get food and no social interactions to occupy them.

Not all pain and not all stressors rise to the level of ethically significant suffering. Every year, millions of people voluntarily subject themselves to the pain of a flu vaccine and their dog to a parvo vaccine or endure small injuries in their chosen sport. In Dr. Cavy's lab, she is aiming for a level of pain that will be detectable when her Guinea pigs are forced to walk on a treadmill. At other times, like arthritic humans, the animals may be fine if they have soft bedding, easily accessible food, and are not required to walk around if their knees are feeling sore; they may be in mild pain that is not ethically significant suffering. In the American legal system, the Animal Welfare Act sets a threshold of concern of "more than slight or momentary" pain or distress. Procedures that exceed this require special consideration, and in most cases, special efforts, including pain medications and possibly euthanasia, to minimize the pain and distress. In other countries, scientists must make a "severity assessment" or their planned animal procedures, and animal-use proposals are assigned a graded category of potential harm to animals.

Animal Welfare Science and Critical Anthropomorphism in Quantifying Animal Harms

It is not always straightforward evaluating what might cause an animal significant pain or distress.

One approach, embedded in various laws and policies, is *critical anthropomorphism*. American public policy and many others state that "unless the contrary is established, investigators should consider that procedures that cause pain or distress in human beings may cause pain or distress in animals."[12] Signs of pain in animals can be subtle, and the less humanlike the animal—zebrafish, frogs, fruit flies—the greater the challenge in recognizing whether they are suffering. This principle challenges us to err on the side of taking their potential pain and distress seriously, even if we do not see it.

Critical anthropomorphism is an excellent precautionary principle, but it lacks the precision required for day-to-day treatment of animals in laboratories. Reasoning from "If I were a frog or a Guinea pig . . ." can sometimes mislead us, and it is not precise enough for every question. For example, Dr. Cavy works on the assumption that experimentally induced arthritis is painful to Guinea pigs in a roughly similar way to knee arthritis in humans. This drives her ethical commitment to minimize their pain to the extent that she can. And, of course, this human–nonhuman similarity convinces her that her work in Guinea pigs will be relevant and translatable to human medicine. But she and her veterinarian also need to know what medicines at what doses may safely alleviate that pain, and a general extrapolation from human to Guinea pig is unlikely to answer those questions. And we want to get this right, given the power we have over the animals. It's one thing for a person to decide to risk the pain of a flu vaccine for her own body's health. By contrast, her dog does not have a voice in getting his parvo shot, and so his human guardian must shoulder that responsibility.

Animal welfare science works to supplement the information and the details that critical anthropomorphism alone cannot provide. In the case of housing for laboratory animals, scientists may use observational data, preference testing, and other research tools to ask the animals what they want and how strongly they want it. In the case of laboratory mice, various shelters and nesting material are important, as an empathetic human could predict. Mice will make nests in their cage if they have the right materials, and it is simple observation to score how complete the nest that is possible with different types of nesting material. Scientists can then set up two choice-testing apparatuses to see if mice will spend more time in a chamber with their preferred nesting material or if they will work, by pressing a lever or going into a chamber that is uncomfortably brightly lit, to get access to some nesting material. With these techniques for measuring animals' preferences, careful science can elucidate what types of shelters and nests best fit the animals. And paradoxically, careful animal welfare science can also uncover seeming improvements in animal environments that are intuitively beneficial but demonstrably harmful. For example, shelters and houses result in more aggression when male mice are housed together in a cage, possibly because they now have something to fight over, the various territories in the cage that these structures create that they would not have in the barren cage.[13]

In our concern for animal pain, scientists must study the signs of pain in various species so that they can quantify not only just how severe a particular experiment may be but also whether or not the candidate pain medications to alleviate that pain are actually going to help the animals. Even this sort of science can include animal welfare costs: to quantify the facial grimaces, the twitches, the vocalizations, and other behaviors that signify the animals' pain requires having animals in pain available for examination, and in most cases, that means actively inducing pain in present animals to develop pain management medications for future animals.

Minimizing Animal Harms: The Three Rs Alternatives Framework

Many regulations and policies employ the "Three Rs Alternatives" framework for evaluating and mitigating harms to laboratory animals: Replacement, Reduction, and Refinement. The Three Rs framework serves to standardize attempts to evaluate and minimize animal harms in laboratories, whether those harms result from experiments or simply from housing the animals, and whether the harms in experiments are intentionally, directly caused or are incidental, contingent side effects of the experiment.

Replacement is a straightforward concept: replace sentient animals who might suffer in experiments with cells in culture or computer models or bacterial cultures or data gleaned from human patients' medical records. With technological advances, such as noninvasive ways to image different parts of a person's brain, scientists may even sometimes replace nonhuman animals with human volunteers. Before she causes pain in Guinea pigs, Dr. Cavy might be able to do some work in cell culture, making sure that the drug she is developing binds to cartilage cells grown *in vitro* in a dish.

More complex is the question of replacing sentient animals with animals we believe to be less sentient: Are frogs significantly less sentient than mice to ethically drive a search to swap out mice for frogs in the laboratory? Or fish? Or fruit flies? Among mammals, is there an ethical mandate to replace monkeys with mice?

Reduction likewise is a simple concept: scientists should do their best to use a smaller number of animals, especially in harmful experiments, than they would otherwise. Can better statistical tests allow comparisons of smaller groups? Can the scientists reduce the amount of genetic or environmental variability in their animal study population, so that experimental results are clearer with smaller animal numbers? Can they eliminate control groups from their experiments, and work with historical published data as the comparison in experiments?

Refinement is the most complex form of alternatives in animal research, comprising all the myriad ways to mitigate pain and distress in animal experiments. Refinement, in my experience, is the primary agenda of an IACUC review, looking at maximizing the use of analgesic pain medications, enriched animal housing, training animals to cooperate with procedures rather than forcing them, using x-rays and magnetic resonance imaging scans to noninvasively monitor disease progression, training researchers so that every animal manipulation is expertly performed, and so much more. Refinement also encompasses setting criteria for humane intervention in experiments, such as painlessly killing (euthanizing) animals if an experimental disease progresses to the point of causing significant, untreatable suffering.

The Three Rs may work in harmony. For example, Dr. Cavy may decide that instead of studying whole, live animals in all her arthritis experiments, she can study Guinea pig cartilage cells *in vitro*. Even if that requires killing some animals to obtain fresh cells, she has replaced sentient animals in arthritis pain with nonsentient isolated cells and reduced her animal use to the small number required to produce the cells for study. An experimental refinement that keeps the animals healthier and in less pain may mean that fewer animals need to be removed from study midway through the experiment, and so fewer animals are initially enrolled in the experiment from the start.

And the Three Rs can compete with each other. A laboratory may reduce its overall number of animals by using the same animals in more than one experiment. If that reuse leads to greater individual harm, the drive for reduction is competing with the goals of refinement. The attempt to replace highly sentient animals (such as monkeys) with possibly fewer sentient species (such as mice) may end up requiring larger numbers of animals.

Unalleviated Pain and Distress

Scientists face an ethical and regulatory imperative to minimize animal pain and distress in their experiments, but the imperative in current practice is not absolute. Current regulations allow for situations in which scientists such as Dr. Cavy may intentionally opt to leave pain untreated when they have reason to fear that pain medications or other refinements will somehow interfere with the conduct of the experiment, or the interpretation of the data. The standards for justification are higher, but many such experiments are approved and published. Under the US Animal Welfare Act, for example, the scientist must tell the IACUC how she knows—literature searches and other methods—that alternatives are not appropriate, must describe her rationale for allowing unalleviated pain or distress, and then the institution must annually report the number of such animals in Column E of their annual report to the US Department of Agriculture.

Justification for experiments with unalleviated suffering include careful assessments of the amount of pain, the ability of pain medicines or other refinements to reduce the pain, the effects that pain medicines might have on the body and on the data being generated, the very real effects that untreated pain may have on the body and on data. Commonly, full information on all of these questions is lacking, and scientists and IACUCs must decide in the face of partial information on how to proceed. Institutions vary in the strength of the justification they require for allowing unalleviated pain.[14]

IACUC Review of the Guinea Pig Arthritis Experiments

Dr. Cavy, her university veterinarian, and the IACUC can use the Three Rs framework to evaluate the animal harms and their mitigation and to consider whether untreated pain is an acceptable part of this experiment.

As part of the evaluation, a funding agency such as the NIH will evaluate the benefits of the proposed work at two levels. First, in a world with dozens of pain medicines already on the market, is there a pressing need for better pain management in arthritis? Second, if there is a need, does this proposed experiment have a good chance to meet it? Are the methods sound to find improved pain management in the treated animals if the medicine is effective? If yes, do Guinea pig arthritis data translate reliably to predicting human outcomes?

Separately, assuming the benefits evaluations are favorable, the scientist, the veterinarian, and the IACUC negotiate a pain management strategy for the animals' welfare. Is it possible to completely replace the animals? An early step might be seeing how the novel test medicine works in tissue culture with the types of cartilage cells that the drug is supposed to target.

If Guinea pigs are still deemed necessary, the scientists and the IACUC must consider reducing animal numbers. They must search for refinements to mitigate animal suffering. Historically, arthritis studies have required euthanizing animals for tissue analysis at several time points during an experiment; modern imaging technologies such as microtomography may allow Dr. Cavy to study a smaller cohort of animals over the entire course of the condition. What about historical data instead of a concurrent untreated cohort of animals? Are historical control data precise enough, or is the method of causing arthritis variable enough that improvement with treatment is only measurable if compared with current control animals?

Next, they must work with the veterinarian to review when the pain is likely to occur, how great it is likely to be, and how the scientists can recognize it. A Guinea pig with painful arthritis will not want to move around much, so Dr. Cavy will likely force the animals to move or press and palpate the swollen joints to see what vocalization, lameness, or other pain indicators she can elicit. With a plan in mind for measuring painfulness, she must consider options to minimize the pain but to still see the effects of the painkiller, if it works. Can she compare her novel therapy not against a placebo but against the best current drug? In a human clinical trial, scientists would avoid comparing their new arthritis treatment to an inactive placebo without some sort of "rescue" plan—for example, allowing patients on placebo (or on the test drug, if it's not as highly effective as hoped) to take a known effective drug. Can a parallel strategy be used with animals? If drugs must be avoided, what are ways to refine the animals housing—with soft bedding, easy access to food, minimal need to move around in the cage—such that the occurrence of pain is minimized?

It is possible that the use of pain medicines will, in fact, impact the results of the experiment. Early use of non-steroidal anti-inflammatory drugs (NSAIDs, such as ibuprofen, meloxicam, or aspirin) could affect the course of the arthritis and confound interpretation of pain and relief from pain. But the scientist must also consider what untreated pain might do to the data. The animals may move around less. They may eat less. Pain may disrupt their sleep patterns or their social interactions. Both critical anthropomorphism and animal welfare science suggest that these effects on the body can affect the animals' perception of pain and the response to test pain medicines.

Only once it has scrutinized all these various factors can the IACUC ethically approve the experiment.

Who Weighs the Benefits and Harms in Animal Studies?

If animal experiments' ethical justification requires somehow weighing the potential benefits (mostly to humans) against the likely harms to animals, surely this calls for high levels of expert evaluation. In current practice, the various levels of review are often fragmented. The people who decide the level of investment in a certain area of science (whether government funding agencies, donors to medical research charities, or corporate executives in pharmaceutical companies) are typically not the group that evaluates the merit and scientific validity of the proposed experiments. And these groups are not typically the people who seriously weigh the harms to animals and the ways to mitigate them; that task belongs to veterinarians and IACUC members who may not have the discipline-specific knowledge to evaluate the experimental methods and may not know the details of the relevant peer experts' review. Absent a robust communication among the bodies that conduct these three levels of review, a true weighing of benefits and harms is truly challenging to achieve.

This theoretically produces some sort of score, a three-part accounting of worthiness of the research, quality of the research, and animal welfare in the research. Challenging though it is to make this calculus, it still does not answer the ethical question of how strong the human benefit, how great the animal suffering must be to approve or negate a research proposal.

Conclusion

Animal experiments may be ethically justified when the harms to animals are clearly outweighed by the potential benefits, mostly to humans. I have focused on the many empirical questions of fact that are necessary to truly quantify the benefits and the harms. Whether accurately predicting the potential value of a research program or accurately measuring animal pain distress and suffering, the assessments are quite complicated. Sound ethical decisions require these expert evaluations, but the facts alone do not determine the ethical norms or make the decision for when it is justified, if ever, to harm animals in experiments conducted to advance human interests.

Acknowledgment

David Takacs provided very helpful commentary on this chapter.

Notes

1. Carbone, L. (2004) *What Animals Want: Expertise and Advocacy in Laboratory Animal Welfare Policy*, New York: Oxford, 291p.
2. Beauchamp, T., and DeGrazia, D. (2019) *Principles of Animal Research Ethics*, New York: Oxford University Press.
3. Carbone, L."The utility of basic animal research (in: Animal research ethics: Evolving views and practices, hastings center report)," *Hastings Center Report 2012, Special Report* 42(6): S12–S15.
4. DeGrazia, D., and Sebo, J. (2015) "Necessary conditions for morally responsible animal research," *The Cambridge Quarterly of Healthcare Ethics* 24(4): 420–430. Epub 2015/09/15. doi: 10.1017/S0963180115000080. PubMed PMID: 26364777.
5. Kim, J. E., Song, D. H., Kim, S. H., Jung, Y., and Kim, S. J. (2018) "Development and characterization of various osteoarthritis models for tissue engineering," *PLoS One* 13(3): e0194288.

Epub 2018/03/14. doi: 10.1371/journal.pone.0194288. PubMed PMID: 29534084; PubMed Central PMCID: PMCPMC5849317.

6. Russell, W. M. S., and Burch, R. L. (1959) *The Principles of Humane Experimental Technique*, London: Methuen & Co. Ltd., 238p.
7. Cancer CfCs. 5-Year Survival Rate 2018, https://curesearch.org/5-Year-Survival-Rate (Accessed November 16, 2018).
8. Greek, C. R., and Greek, J. S. (2000) *Sacred Cows and Golden Geese: The Human Cost of Experiments on Animals*, New York, NY: Continuum.
9. World Medical Association (1997) "Declaration of Helsinki," *Journal of the American Medical Association* 277(11): 925–926.
10. Malakoff, D. (1999) "Alternatives to animals urged for producing antibodies," *Science* 284(5412): 230. Epub science.sciencemag.org/content/284/5412/230.1.full. PubMed PMID: 10232965.
11. Garner, J. P. (2014) "The significance of meaning: why do over 90% of behavioral neuroscience results fail to translate to humans, and what can we do to fix it?" *ILAR journal/National Research Council, Institute of Laboratory Animal Resources* 55(3): 438–456. doi: 10.1093/ilar/ilu047. PubMed PMID: 25541546; PubMed Central PMCID: PMC4342719.
12. IRAC [Interagency Research Animal Committee]. 1985. U.S. Government Principles for Utilization and Care of Vertebrate Animals Used in Testing, Research, and Training. Federal Register, May 20, 1985. Washington: Office of Science and Technology Policy. Available at https://olaw.nih.gov/policies-laws/phs-policy.htm#USGovPrinciples; accessed October 3, 2019.
13. Howerton, C. L., Garner, J. P., and Mench, J. A. (2008) "Effects of a running wheel-igloo enrichment on aggression, hierarchy linearity, and stereotypy in group-housed male CD-1 (ICR) mice," *Applied Animal Behaviour Science* 115(1–2): 90–103.
14. Carbone, L. (2014) "Justification for the use of animals," in J. Silverman, M. A. Suckow, and S. Murthy (eds.) *The IACUC Handbook*, 3rd ed. (pp. 211–251), Boca Raton: CRC Press.

20
GENETIC ENGINEERING OF NONHUMAN ANIMALS

Adam Shriver

Although related ideas had previously been raised in discussions about the selective breeding of animals,[1] philosophical debate about the ethics of genetically engineering animals was kicked off in earnest by Bernard Rollin's publication of "That Frankenstein Thing" (Rollin 1986). From the outset, Rollin recognized not only that there was something off-putting about the idea of dramatically modifying animals but also that this discomfort could not be neatly captured within many existing philosophical frameworks. Since his initial article, using thought experiments designed to test and clarify moral intuitions, philosophical debates about genetic engineering have led to discussions of headless chickens, pain-free cows, and animals genetically programmed to die early.

Recent years, however, have seen dramatic improvement in genetic technology, such that gene-editing experiments on nonhuman animals are now vastly cheaper, easier, and more precise (Schultz-Bergin 2018; Shriver and McConnachie 2017). This precision, rather than leading to the proliferation of even more dramatic thought experiments, will push philosophers to grapple with the very real practical applications that are currently being used or are likely to be used in the near future. Gene editing animals is occurring at present, and it is no longer clear that discussing the abstract implications of the technology is more important than simply developing a framework that can coherently address current practices.

As examples of current applications of gene editing on nonhuman animals that raise interesting questions, consider the following. Research scientists are using genetic modification in laboratory animals in attempts to better understand the underlying biology behind negative mental states such as pain, anxiety, and depression, as well as adverse health conditions such as Parkinson's disease, Alzheimer's disease, and stroke. In light of shortages in organ availability, scientists are actively using gene editing to attempt to facilitate xenotransplantation, a technique where human organs are grown in other animals (e.g., pigs) for the purpose of supplying humans in need. At the same time, the agricultural industry, which was once resistant to the suggestion that there were any problems that needed solving via gene editing, is now repeatedly making public statements in favour of gene editing livestock for the sake of "dramatic animal health gains ... reduced financial risk for farmers ... less need to use antibiotics to care for livestock and reduced environmental impact from more efficient farm operations" (National Pork Producer's Council 2019). And scientists are even using genetic engineering to explore hypotheses about evolution such as predictions about how chicken beaks evolved (Bhullar et al. 2015).

Given the vast and growing number of applications of genetically engineered nonhuman animals, the number of potential critiques is also quite broad. Some arguments against the technology follow straightforwardly from moral criticisms of other human uses of animals. Other arguments have developed specifically to explain why genetic modification of nonhuman animals is wrong even in cases where it seems compatible with previous accounts of our obligations to nonhuman animals. In order to provide an overview of the debate, discussion is organized as follows: I first describe the potential benefits of genetically engineering animals to answer the question, "Why would anyone want to do this in the first place?" I then describe ethical objections to certain applications of genetic engineering that can be justified only by dramatically discounting the interests of nonhuman animals. These applications are vulnerable to objections of speciesism and, in particular, a minimalist version of speciesism that most people would be willing to accept. Finally, I discuss a different set of ethical objections to applications of gene-editing technology that would presumably be permissible according to standard harm-benefit analyses of the type utilitarian philosophers might be inclined to use. These include both objections from more traditional animal rights approaches as well as those based on concepts which have been recently introduced into this debate specifically to deal with new ethical issues raised by biotechnology.

Benefits of Genetic Engineering

From the moment of fertilization, the genetic code contained in DNA plays a central role in directing animals' development, determining which cells are created and how they are organized. Virtually every aspect of human existence is determined at least in part by our genetic code, and consequently, the potential to exercise direct control over this code has vast implications for the future of our species including, no doubt, ideas that have yet to be imagined. As such, genetic engineering in theory has a mind-boggling potential for positively transforming life, first through improving our understanding of biology and later by allowing us to use this understanding to directly shape the biology of humans and of other species.

Our ability to use genetic engineering to investigate biological mechanisms has been greatly improved since the time genetic engineering was first developed in the 1970s. Initial genetic engineering techniques made use of viruses that were capable of introducing DNA to other cells. By tinkering with the DNA targets of the viruses, researchers were able to examine the role of DNA in shaping future developments. However, these methods generally lacked precision as they involved a fairly haphazard integration of foreign genetic material into the genome with a high risk of unintentional changes, and they also could be performed only on particular species depending upon the virus being used.

Recent years have seen the development of techniques relying on the use of designer nucleases, including most prominently the CRISPR-Cas systems. These techniques can be used on any species and allow for much more precise changes with decreased risks of off-target effects. They provide researchers with the ability to "knock out" and "knock in" particular genes in order to observe the effects on an organism and this allows researchers to study the effects that the presence or absence of particular genes can have on biological function. For example, using "the subtraction method," if researchers remove a section of DNA and then observe changes in biological function, they can reasonably conclude that that section of DNA played a role in the function, and can take steps to try to better understand what that role may be. Similarly, if a gene is "knocked in" to an organism and a particular change is observed, this implies that the gene is playing a role in the change. Of course, things are much more complicated than this in practice, but the basic underlying logic should be clear.

This technique can be used to investigate a startling array of biological processes. Using nonhuman animal models, researchers routinely knock-out genes to see if doing so eliminates evidence of states such as pain, anxiety, or depression. They also "knock in" genes to see if doing so can make the animals resistant to various diseases. And, of course, similar techniques can be used to study virtually any aspect of gene expression in living organisms. In fact, interventions used in research are becoming even more precise than this. A technique known as "optogenetics" involves genetically modifying cells so that certain properties are activated only when the cells are exposed to light. Because researchers can use extremely precise lasers to target particular cells with light, this technique allows certain properties to be "turned on" and "turned off" in exact groups of cells with a precision of milliseconds. Needless to say, this can allow for incredibly fine-grained interventions that can be used to test hypotheses about many different levels of organization in animals, from overt behaviour down to the level of the operation of individual cells.

So it is no exaggeration to say that genetic engineering has radically changed scientific practice and has allowed for the investigation of aspects of biology at a level of detail previously unattainable. This understanding of underlying biological mechanisms undoubtedly has been and will continue to be important for medical breakthroughs related to physical and mental health, and will thereby contribute importantly to human and animal well-being. There are some who have questioned whether animal models have contributed anything to science, but it is genuinely hard to imagine having anything close to the detailed understanding of the basic workings of the nervous system or biology that we currently possess in the absence of some use of genetic engineering.[2]

More questionable, however, is how effective gene editing has been in directly leading to cures to human maladies using animal models. There have been numerous instances of scientists discovering that gene modifications in nonhuman animals "cure" negative affective states such as chronic pain, anxiety, or depression, yet these models thus far have failed to lead to any "silver bullet" breakthroughs for the most serious afflictions of humans. As previous animal models using animals such as mice or rats have failed, researchers have also been keener to use genetic engineering to make nonhuman animals even more closely resemble humans with various diseases, which arguably adds new ethically problematic dimensions. There is a complicated debate about the efficacy of various animal models of human conditions that is beyond the scope of this chapter, but needless to say, the debate is highly relevant for evaluating the extent to which genetic engineering technology will impact future human health. But it is reasonably safe to say that animal models are generally valuable in helping to lead to understanding maladies in humans and in leading to their improvements.

Thus far, the discussion has centered on how genetic engineering can lead to knowledge about biology, which, in turn, will likely have important positive consequences. But the far more discussed aspect of genetic engineering involves its potential to be directly utilized to change organisms in ways that will be beneficial for humanity. Here is a brief list of some of these potential benefits.

In the case of livestock, animals might be engineered to grow faster or larger, to be resistant to diseases; to require less food, water, or space; to have less of an impact on the environment; or to lead to more nutritious products. Gene drives might be introduced into populations of disease-carrying organisms, such as mosquitos or rats, that ultimately lead to their population ceasing to reproduce. Animals have historically been used for "labour" for humans, from agricultural to military contexts, and genetic engineering presumably could lead to improvements in the traits relevant for these activities. And we can imagine any number of changes to animals that might result in increased "entertainment" or "aesthetic appeal" for humans, such as bioluminescent pets. These are all, to greater or lesser extent, benefits that could accrue to humans via genetic engineering of nonhuman animals.

Moreover, if we assume that animal models do successfully inform gene editing in humans, then this research on nonhuman animals could also potentially lead to other direct interventions on humans (although using the techniques on humans come with their own set of ethical concerns). Perhaps least controversially, gene editing could be used to prevent common genetic illnesses that lead to conditions that cause suffering. More controversial would be using gene editing to prevent conditions such as blindness and deafness, among others, as there currently is vigorous debate as to whether we should accept a medical or social model of disability and whether these conditions truly represent diminished opportunities for well-being or other aspects of a good life (Barnes 2016; Campbell and

Stramondo 2017). And finally, and quite controversially, gene editing could presumably be used in the future to enhance humans in many different ways. This could involve making humans bigger, stronger, faster, more intelligent, more empathic or even more faithful (Savulescu 2012).

In other words, to the question, "Why even engage in gene editing in the first place?" there are a number of potential answers. Thus, there is genuine positive motivation for pursuing genetic engineering. But whether any of these potential applications are ultimately morally justified will, of course, also depend on the strength of the objections to them. I turn next to these objections.

Speciesism Critiques of Human–Nonhuman Interactions

One important set of critiques of some of the applications of gene editing mentioned earlier follows rather straightforwardly from other criticisms of human treatment of nonhuman animals. According to these criticisms, many current practices unjustifiably discount or completely ignore the interests of nonhuman animals, even interests that are similar in all of the relevant ways to human interests. These types of critiques are often summarized by suggesting that current human practices and policies are "speciesist" (Horta 2010).

The term *speciesism* was initially coined by Richard Ryder but was popularized by Peter Singer. In *Animal Liberation* (1975) and *Practical Ethics* (1993), Singer develops the concept of speciesism by starting from the assumption that enlightened readers could agree that all humans are deserving of equal moral consideration. He examines what could justify this belief. It is not, Singer says, because individual humans are actually equal in all of their abilities, since that is clearly false. Nor is it because there are never any differences between groups of people, since we also wouldn't want to say that "people with an IQ above 90" should be given preference over others despite the fact that there is some relevant group difference. Rather, Singer suggests, when we say that all humans are equal, we are committing to equality as a *moral* ideal. In particular, he suggests, we are committing to what he calls the Principle of Equal Consideration (PEC) of interests, which tells us to "give equal weight in our moral deliberations to the like interests of [every being] affected by our actions" (Singer 1993, p. 21)

The PEC can be justified via a closer examination of a key aspect of morality known as universalizability (Hare 1981; Varner 2012). Moral judgments should be universalizable; that is, if I think that I ought to behave in a particular way in a particular situation, it would be irrational or incoherent to think that someone else in an identical situation should behave differently. Identical situations are the easy cases, but we should also be able to agree that *relevantly similar* situations should also be treated the same. For example, if I think it's wrong to steal money from the church till in a church with mahogany benches, I should also agree that it would be wrong to steal money in a church with cedar benches, since the type of wood used is clearly irrelevant for the evaluation of the action.

In cases of racism and sexism, Singer argues, irrelevant considerations are used in order to discount the interests of humans. Treating people's interests as less valuable because of their gender or their skin pigmentation is wrong because it uses irrelevant criteria to discount the interests of people. However, once we have agreed to this, there's nothing preventing us from also extending the principle of equal consideration of interests to cover nonhuman animals. After all, if attributes such as skin color and gender and even intelligence are not relevant for whether one's interests are treated the same, species membership seems like an equally arbitrary criterion on which to base consideration in many cases. For example, if we think that the unpleasantness of the feeling of pain is what makes it the case that it is wrong to inflict pain on other humans, all things being equal, and if we have good reason to believe that chimpanzees feel pain in an identical manner to humans, then appealing to species membership to discount the interests a chimpanzee has in avoiding pain relative to humans is clearly utilizing irrelevant criteria to justify differential treatment and thus in violation of the PEC.

Singer is a utilitarian and believes that the PEC ultimately leads to the utilitarian recommendation that we ought to maximize the happiness and minimize the suffering of all sentient being affected by our actions. However, it's worth noting that the PEC as stated and the definition of speciesism based upon it are compatible with other types of views. For example, a moral rights theorist who believes that it's wrong to cause serious harm to an individual human for the sake of benefitting five others might also extend this principle to say that treating the interests of animals equivalently would entail also refraining from causing harm to an animal for the sake of benefitting five humans. And a theorist who claims that pleasure and pain are not relevant to evaluating a situation but, rather, thinks that some other set of criteria, X, is important, should thereby be committed to thinking that X is also relevant insofar as it also applies to situations involving nonhuman animals.[3] So speciesism is not exclusively a utilitarian principle, but it is a principle that, at minimum, involves a commitment to the view that using species membership in the absence of any morally significant differences to justify discounting interests of different species is wrong.

So how does the speciesism criticism work in regards to current practices? Consider the most common target: modern industrial farming. Many humans like the taste of meat better than the taste of meat alternatives such as, say, flavored tofu. But they would not be willing to undergo extended periods in captivity, be deprived of the ability to engage in natural behaviors, and be subjected to numerous painful episodes in exchange for the opportunity to eat meat rather than an equally nutrient-rich but slightly less tasty alternative to meat. Yet modern industrial agriculture and particularly confined animal feeding operations subject animals such as pigs and chickens to all the conditions mentioned in the previous sentence. Since humans are ignoring the similar interests of animals for the sake of relatively trivial benefits for themselves, humans are engaging in speciesist approaches to these animals and violating the PEC. In fact, actual laws protecting welfare in the agricultural context are so minimal in some countries that one might argue that animals' interests are not taken into consideration at all. Such practices are speciesist even on a relatively minimal interpretation of speciesism that simply holds that relevantly similar interests of nonhuman animals should be given some weight in our moral deliberation, something which most people would likely accept.

On the other hand, consider an argument that we shouldn't genetically edit humans because doing so might lead to ostracism and social stigmas between different groups of humans or to further inequality. Since this idea relies on features that are unique to human social structures, it seems clear that these moral reasons wouldn't apply to gene editing nonhuman animals. Some of the reasons we have for worrying about genetically engineering humans do not apply to nonhuman animals.

As such, commitment to the PEC doesn't necessarily commit one to strong claims about how equivalently humans and nonhumans should be treated in practice; not all uses of genetic technology on nonhuman animals need be considered vulnerable to speciesist critiques. But nevertheless, on the minimalist conception of speciesism, which most reasonable people would accept, there are certain practices involving genetic technology, including those currently being practiced on animals, that are clear violations of the principle of equal consideration of interests. In what follows I detail several of these types of practices.

Consider the use of gene editing for the sake of improving the cost-effectiveness of farming. This could include creating animals that grow faster, or naturally carry more meat, or require less food or water. These would essentially be done for the sake of improving the profit margins of farming. In many ways, they would be similar to current selective breeding practices, which have created lines of chickens which grow much faster than previously and have bone and joint problems as a result, or the US pork industry's insistence on using ractopamine to make pigs grow faster despite clear evidence of negative effects on welfare (Marchant-Forde et al. 2003).

The similarity with current practices also suggests how such practices could be vulnerable to charges of speciesism. If such technology progresses the way farming practices have for much of the 20th century, then the interests of the animals would be given almost no weight whatsoever in determining regulations. In the United States, for example, while there may be environmental or health regulations that place some minimal restrictions on how animals can be raised, there are virtually no pure welfare requirements for raising animals save laws governing transportation and slaughter practices. The result of this has essentially been a system where animals are treated as objects rather than as sentient beings, and their interests are discounted entirely, as revealed by an almost never-ending string of undercover videos showing conditions on modern factory farms.

And even if one held, implausibly, that those undercover videos only showed completely atypical situations that do not occur relatively frequently in the industry, there are numerous reasons to think that livestock in confined animal feeding operations (CAFOs) suffer. First, many of them have been bred to grow as fast as possible, which has an effect of causing them to be in a near-constant state of hunger, which presumably is unpleasant. Moreover, because they often grow faster than their bodies were initially designed for, many animals develop bone and joint problems. And finally, even if one thought that slaughter regulations were being followed perfectly despite limited enforcement capacities, the laws still allow for animals to be slaughtered after being improperly stunned, provided that this number doesn't cross a certain threshold.

As such, if the ethos of doing anything to maximize profit was followed alongside permissive use of genetic engineering, and regulations were focused only on ensuring that genetic technologies did not pose any new risks to human consumers, it seems likely that many animals would suffer. Evidence suggests that many surprises occur in the process of testing genetic technology, so during the research phase, many animals would be born who are unable to grow into adulthood. Moreover, if animals were bred simply to grow faster or bigger, presumably many of the same welfare problems from selective breeding would be exacerbated. In short, if genetic changes led to decreases in welfare that were not reflected in losses of productivity or profit or increases in risks to humans, the current paradigm seems unlikely to be willing to change them. As such, animals' interests would be discounted and likely ignored. This would provide strong moral reasons against adopting these practices in the absence of dramatic changes in regulations or industry practices regarding welfare.

Turning to animal research, we can also envision examples of practices involving genetic engineering that are clear violations of minimal conceptions of speciesism. Consider hypothetical research that inflicts negative states on animals merely for the sake of scientific curiosity, with little or no chance of leading to future benefits for humans or other animals. The connection between exploratory research and clinically relevant research is complicated, and sometimes research can lead to unintended discoveries of significance, but nevertheless, clearly a line should be drawn somewhere to prevent experimenting merely for the sake of seeing what is possible.

Moreover, many critiques of current research suggest that animals' interest is not given sufficient weight in assessing the value of particular research protocols. To the extent that such critiques are accurate, genetic engineering is likely to exacerbate them, since it dramatically increases the number of possible interventions on nonhuman animals and the ways in which they can be used to arguably mimic human conditions. Since such techniques are tinkering with the basic building blocks of organisms, they have the potential to cause extremely serious unanticipated problems and perhaps even problems that are difficult to detect.

Finally, there's a risk that in the absence of proper regulations, genetic engineering will be used on nonhuman animals for trivial purposes by private companies. Think of cats or dogs that glow in the dark or use camouflage similar to squid or chameleons or anything else that some small group of consumers might deem "entertaining." The track record of humanity suggests that private companies would be willing to cater to such groups if they are legally allowed to do so. So, on a minimalist conception of speciesism, where the term simply refers to similar interests of nonhuman animals being given virtually no weight in our moral deliberation, there are many practices involving nonhuman genetic engineering that are likely to be unethical unless substantive protections are devised.

However, there are many applications of genetic engineering that would seem to be permissible on this minimalist conception. Making livestock disease-resistant would presumably increase productivity for farmers and likely increase the welfare of the livestock themselves. Causing some harms to animals in biological research could be outweighed if this research resulted in substantial improvements to the health and/or welfare of future humans or other animals. In fact, even genetic changes that resulted in fairly cosmetic improvements but which did not cause any decreases in welfare could theoretically be justified on this minimal notion of speciesism. As such, many have suggested that stronger or additional criteria provide reasons for criticizing a wider range of practices.

Beyond Minimal Speciesism

As noted earlier, the concept of speciesism and the PEC can be interpreted in Singer's way, but can also be compatible with more right-based approaches (Regan 1983). Turning to a more demanding rights-based notion of speciesism, additional applications of genetic technology would be prohibited. For example, rights-based approaches generally hold that it is not permissible to cause serious harm to one for the sake of benefitting others. Thus, for example, using gene editing to study the mechanisms of pain and suffering in mice would not be permissible on these views, even if such practices could on balance reduced the total amount of suffering by increasing the probability of finding better analgesics or treatments in the future.

However, rights-based approaches face an interesting question in relation to some practices. According to the non-identity problem, if you cause a new animal to exist who is slightly worse off than another animal who would have existed, but this new animal would not have existed unless you had made the changes, then it seems inaccurate to say that you have "harmed" the new animal, since it would not have existed otherwise (Palmer 2011). So, for example, if one genetically alters cows to grow faster and the new cows have worse welfare than the old cows, they nevertheless would not have existed otherwise, so it would be strange to say that they had been harmed. Many would argue that this problem only truly applies if both of the potential animals have lives that are, overall, worth living, since it seems like we are definitively harming an animal by bringing it into a miserable life even if it would not have lived otherwise. If this is correct, a right-based approach could more consistently condemn *all* uses of genetic engineering that bring animals into the world that have lives that are not worth living . . . more utilitarian approaches, on the other hand, have to be open to the possibility that it may sometimes be permissible to bring unhappy animals into existence if doing so results in an overall improvement of consequences for all affected.

Rights-based versions of speciesism and the PEC seem to rule out additional practices involving genetic engineering of nonhuman animals but not all. But many people seem to think that *all* such applications are morally objectionable. For this reason, various authors have developed new concepts designed to capture the wrongness of genetically modifying animals.

To see what motivates such views, consider the following hypothetical example, developed by Gary Comstock responding to Bernie Rollin:

> Picture yourself fifty years from now standing in the middle of a huge antiseptic warehouse staring at rows of tan colored objects that look something like footballs. Shiny stainless steel pipes descend from the ceiling and disappear into mouth-like orifices on top of each object. Black rubber tubes are attached by suction cups to the bottoms. The only attendant in the building tells you that the pipes bring water and rations to what he calls "the birds," while the rubber tubes carry excrement and urine to a sewer beneath the floor. Every twelve hours each bird drops a no cholesterol egg onto a conveyor belt . . . You are staring at thousands of living egg machines, transgenic animals genetically engineered to

convert feed and water into eggs more efficiently than any of their evolutionary ancestors, layer hens.

(Comstock 2000: 152)

Would there be anything wrong with this system? Note that for both minimal and rights-based conceptions of speciesism, there would not be, since according to both approaches, sentience is required for organisms to have interests that can be taken into consideration. Nevertheless, according to several authors, this example strikes most people as intuitively wrong.

It is clear that at least some people believe that there is something wrong with genetically engineering completely insentient animals for the sake of human consumption. One such approach appeals to the *telos* or the nature of the animals being engineered. Bernard Rollin introduced the notion of *telos* to debates about animal ethics as part of a criticism of approaches to animal welfare that focused only on the prevention of pain. Rollin defined an animal's *telos* as "the set of needs and interests which are genetically based and environmentally expressed, and which collectively constitute or define the 'form of life' or way of living exhibited by that animal, and whose fulfilment or thwarting matter to the animal" (1998: 162). Rollin claimed that in addition to avoiding pain, we need also to respect the *telos* of animals and to allow the animals to engage in the behaviour expressed by their natures.

Rollin himself has suggested that genetically modifying animals does not necessarily run afoul of his emphasis on *telos*, since changing genetics gives animals a new *telos* by changing their underlying biology rather than violating their old *telos*. But some authors have suggested that we should instead appeal to a species-typical notion of *telos* or related ideas that capture what it is like to be an average member of the species. According to these views, by changing the nature of a chicken to be a brainless, headless, egg-laying machine, you have thereby prevented the resulting chickens from realizing the full potential of their species-typical *telos*.

Others have argued elsewhere that this perspective seems callous insofar as it suggests that it would be preferable to cause sentient animals suffering rather than violate the "dignity" defined in relation to species-typical functioning (Thompson 2008; Shriver and McConnachie 2018). But for the purposes of this chapter, it's worth noting that this "species-typical" conception of *telos* (as opposed to Rollin's) doesn't merely add an additional moral constraint beyond the principle of equal consideration of interests but, in fact, seems to directly contradict the principle. The PEC is explicitly formulated such that the interests of individuals are taken into account rather than a notion of interests abstracted from what a "typical" species member looks like. That is, it centrally involves the idea that we base our moral commitments to individuals on the properties that those individuals actually possess rather than features that are generally associated with certain groups to which they belong.

One famous argument that cut against the PEC was Carl Cohen's suggestion that cognitively impaired humans should not be regarded as equal to similarly intelligent animals because the impaired humans are "of a kind" with humans, who on average are more cognitively sophisticated than other animals (Cohen 1986). Most people who support consideration of animal interests reject arguments such as Cohen's that appeal to "types" or "kinds" in order to establish the moral standing of individuals because doing so ignores the properties of the individual beings. Yet the preceding argument about *telos* seems to diverge from this line of reasoning and leave open arguments discounting the interests of animals based on the claim that they aren't "of the right kind" to be comparable to humans. If we accept that appealing to "species-typical" features can be invoked to claim harms to nonhuman animals who are not typical, it is hard to see what principled objection can be used against authors appealing to "species-typical" traits to claim that humans inevitably possess stronger interests than nonhuman animals.

Other arguments against headless chickens avoid this complication, at least at first glance. Bovenkerk et al. (2001), for example, introduced a particularly influential critique of headless chickens that focused on the notion of bodily integrity, a concept often considered by many to be important in human bioethics. According to these views, we might be "harming" a chicken by violating its bodily integrity, which they define, citing Bart Rutgers, as the "wholeness and intactness of the animal and its species-specific balance, as well as the capacity to sustain itself in an environment suitable to the species" (p. 17). These harms should not be thought of as necessarily more important than the harm of suffering, but nevertheless might be sufficient to suggest that the overall practice of genetic engineering animals is morally impermissible. And the view doesn't run afoul of the PEC since it merely states that bodily integrity is one consideration relevant to an individual's interests in addition to other interests such as avoiding pains.

But it is not entirely clear how it works out in practice without appealing to ideas that once again appeal to species-typical notions of naturalness. Taking away a capacity, such as sight, or a body part, such as a brain, seems to be violating bodily integrity on normal usage, but not all genetic modifications fit easily into this model. What about a modification that, for example, causes an animal to grow faster? This doesn't seem to be violating bodily integrity as stated. Moreover, we can produce extremely contradictory results simply by altering whether we are causing a particular change by "taking away" or by "adding" from the organism. For example, if we engineered an animal to be incapable of feeling pain by removing nociceptors, this would appear to violate the integrity clause. But what if we eliminated pain not by taking away nociceptors but, rather, by causing an animal to be constantly high on endogenous opioids (and presumably pain-free as a result)? Assume further that the resulting animals in these two cases behaved identically. It would be quite bizarre to say that one of these is morally impermissible while the other is not simply because one involved a "lack of intactness" while the other involved adding something further, and this divergence suggests that integrity may not fully capture our intuitions.

Could we then instead lean more heavily on the clause emphasizing, "the capacity to sustain itself in an environment suitable to the species" to capture what's wrong with certain genetic interventions? But this seems to go back, once again, to notions of what a typical member of the species is like, and thus encounters the same objections mentioned above.

And perhaps, especially given the difficulty philosophers have in finding a coherent framework that fits with all the purported intuitions about genetic engineering, this should serve as a reminder about how important it is to be clear about just what those intuitions actually are. Although most philosophers writing about gene editing animals have suggested that people find it distasteful, many of these claims have been based primarily on the philosophers' own intuitions, on anecdotes accumulated through conversations and presentations, or on extrapolations from surveys looking at related topics. But there's a whole field of philosophy, known as experimental philosophy, that has set about empirically investigating whether or not the public truly has the beliefs attributed to them by philosophers (Doris and Moral Psychology Research Group 2012). And though the exact extent to which this field is relevant to traditional philosophy is still being debated, there have been some unquestionably clear examples of studies where the public's intuitions about particular cases have radically diverged from what philosophers have predicted or what philosophers have claimed to be "obvious to all." So it may very well be that public intuitions about, say, headless chickens are not as straightforward as some have claimed, and there is at least some evidence that public views towards gene editing for welfare improvements are not as negative as some might expect (McConnachie et al. 2019).

Conclusion

Genetic technology is progressing rapidly, and those in various industries that rely on the use of animals are now thinking seriously about how to use this technology to achieve various goals. The technology has already led to tremendous advancements in our understanding of biology and is likely to continue to do so. And many in the agriculture industry are now pushing for the power of genetic technology to be used to improve efficiency as well as address various moral concerns related to modern agriculture. Finally, in the absence of serious regulation, it also seems likely that genetic modification of animals could be used for human entertainment.

While public sentiment has generally pushed politicians to take seriously concerns about potential risks to humans, it is not clear that consideration of animal interests is yet being taken seriously by policymakers in relation to this issue. Many potential uses of genetic technology on animals seem to violate the principle of equal consideration of interests, and can only be justified by discounting the interests of animals. Other potential uses would involve causing harm to individual animals for the sake of some greater good, and these may also be viewed as wrong by many ethicists and members of the public. And finally, many people seem to object to gene editing even if it does not obviously cause harms to animals in the traditional sense.

A wide variety of proposals have been put forward by ethicists that attempt to clarify which practices involving animals are problematic, but thus far, there's been little clarity about how these views are related to public sentiments. Much work is needed to better understand how the public truly feels about these issues and, more important, what we *should* feel once we fully understand the facts and values at play.

Notes

1. Some examples include the fictionalized cow bred to want to be eaten in Douglas Adams's *The Restaurant at the End of the Universe* (Adams 1980), Ali and Cheng's (1985) and later Sandøe *et al.*'s (1999) discussion of the welfare of genetically blind chickens, and Beilharz and Zeeb's (1981) discussion of "fitting the horse to the harness."
2. Whether such research, which involved invasive procedures on nonhuman animals, was thereby morally justified is, of course, a further question.
3. It is also important to note that a proper understanding of the interests at stake with regards to speciesism is deeply dependent upon knowledge about empirical facts about the world. If one takes sentience to be a morally relevant feature, then neuroscientific, behavioural, and genetic facts about the likelihood of sentience in other animals are relevant. If, on the other hand, moral notions such as autonomy or a plan for how one's life overall will go are thought to be relevant, then empirical evidence for more sophisticated cognitive capacities than mere sentience is required.

Bibliography

Adams, D. (1980) *The Restaurant at the End of the Universe: Hitchhiker's Guide 2*,Vol. 2. London, UK: Tor UK.

Ali, A., and Cheng, K. M. (1985) "Early egg production in genetically blind (rc/rc) chickens in comparison with sighted (Rc?/rc) controls," *Poultry Science* 64(5): 789–794.

Barnes, E. (2016) *The Minority Body: A Theory of Disability*, New York, NY: Oxford University Press.

Beilharz, R. G., and Zeeb, K. (1981) "Applied ethology and animal welfare," *Applied Animal Ethology* 7(1): 3–10.

Bhullar, B. A. S., Morris, Z. S., Sefton, E. M., Tok, A., Tokita, M., Namkoong, B., . . . Abzhanov, A. (2015) "A molecular mechanism for the origin of a key evolutionary innovation, the bird beak and palate, revealed by an integrative approach to major transitions in vertebrate history," *Evolution* 69(7): 1665–1677.

Bovenkerk, B., Brom, F. W. A., and van den Bergh, B. J. (2001) "Brave new birds: The use of integrity in animal ethics," *Hastings Center Report* 32(1): 16–22. https://doi.org/10.2307/3528292

Campbell, S. M., and Stramondo, J. A. (2017) "The complicated relationship of disability and well-being," *Kennedy Institute of Ethics Journal* 27(2): 151–184.

Cohen, C. (1986) "The case for the use of animals in biomedical research," *New England Journal of Medicine* 315: 865–870.

Comstock, G. (2000) *Vexing Nature? On the Ethical Case Against Agricultural Biotechnology*, Norwell, MA: Kluwer Academic Publishers.

Doris, J. M., and Moral Psychology Research Group. (2010). *The Moral Psychology Handbook*. Oxford, UK: Oxford University Press.

Hare, R. M. (1981) *Moral Thinking: Its Levels, Method, and Point*, Oxford: Oxford University Press.

Horta, O. (2010) "what is speciesism?" *Journal Agricicultural and Environ Ethics* 23(2010): 243–266. doi: 10.1007/s10806-009-9205-2.

Marchant-Forde, J. N., Lay Jr, D. C., Pajor, E. A., Richert, B. T., and Schinckel, A. P. (2003) "The effects of ractopamine on the behavior and physiology of finishing pigs," *Journal of Animal Science* 81(2): 416–422.

McConnachie, E., Hötzel, M. J., Robbins, J. A., Shriver, A., Weary, D. M., and von Keyserlingk, M. (2019) "Public attitudes towards genetically modified polled cattle," *PLoS One* 14(5): e0216542. doi: 10.1371/journal.pone.0216542.

National Pork Producer's Council (2019) "Gene editing development stalled: NPPC renews call for USDA oversight," *Press Release*. Online: http://nppc.org/gene-editing-development-stalled-nppc-renews-call-for-usda-oversight/ (Accessed on August 19, 2019).

Palmer, C. (2011) "Animal disenhancement and the non-identity problem: A response to Thompson," *NanoEthics* 5(1): 43–48.

Rana & Cramer (2018) "Big tongues and extra vertebrae: The unintended consequences of animal gene editing," *Wall St Journal*, updates December 14, www.wsj.com/articles/deformities-alarm-scientists-racing-to-rewrite-animal-dna-11544808779?mod=e2tw.

Regan, T. (1983) *The Case for Animal Rights*, Berkeley, CA: University of California Press.

Rollin, B. E. (1986) The frankenstein thing," in J. W. Evans and A. Hollaender (eds.) *Genetic Engineering of Animals: An Agricultural Perspective*, New York: Plenum Press.

Rollin, B. E. (1998). On telos and genetic engineering. In *Animal Biotechnology and Ethics*. Boston, MA: Springer.

Sandøe, P. B., Nielsen, L., Christensen, L. G., and Sørensen, P. (1999) "Staying good while playing God—the ethics of breeding farm animals," *Animal Welfare* 8(4): 313–328.

Schultz-Bergin, M. (2018) "Is CRISPR an ethical game changer?" *Journal of Agricultural and Environmental Ethics* 31(2): 219–238.

Savulescu, J. (2012) *Unfit for the Future: The Need for Moral Enhancement*, Oxford: Oxford University Press.

Shriver, A., and McConnachie, E. (2017) "Genetic engineering and animal welfare," in P. B. Thompson and D. M. Kaplan (eds.) *Encyclopedia of Food and Agricultural Ethics*, 2nd ed. Dordrecht: Springer.

Singer, P. (1975) *Animal Liberation*, New York: Random House.

Singer, P. (1993) *Practical Ethics*, New York: Cambridge University Press.

Thompson, P. B. (2008) "The opposite of human enhancement: Nanotechnology and the blind chicken problem," *Nanoethics* 2(3): 305–316.

Varner, G. (2012) *Personhood, Ethics, and Animal Cognition*, New York, NY: Oxford University Press.

21
BUILDING ETHICAL DE-EXTINCTION PROGRAMS
Considerations of Animal Welfare in Genetic Rescue

Ben J. Novak

Genetic Rescue to De-extinction

In recent centuries, extinction rates have accelerated dramatically due to human activities, leading scientists to recognize the modern era as the "sixth mass extinction" (Ceballos et al. 2015). While this biodiversity crisis is typically quantified by recent species loss, human activities have coincided with catastrophic extinction events for the past fifty thousand years (Burney and Flannery 2005). Perhaps more problematic than species loss are dramatic reductions in population abundance, which leaves species in precarious states of vulnerability, and often crippling ecological functions (Ceballos et al. 2017). The goal of conservation is to halt species loss and reverse population reductions to avoid the long-term consequences of mass extinctions.

Accomplishing conservation goals has proved difficult, demanding constant innovation. Strategies have evolved over time to confront varying challenges imposed upon wildlife by humans. The most well-known and understood drivers of species loss/reductions are overharvesting, habitat alteration, and pollutants, all of which have been addressed through what are considered "traditional conservation methods," including regulated hunting and habitat protection. Isolated successes worldwide have proven that, given a chance to recover, ecosystems are more resilient than commonly perceived. However, the resilience of many species is hindered by the genetic impacts of population reductions (i.e., inbreeding, compromised adaptability), necessitating interventionist approaches. Increasingly, conservation has been implementing "genetic rescue," whereby individuals from healthy populations are translocated to struggling populations to increase genetic diversity. While the exact causal relationship between genetic diversity and population viability is unknown, increasing genetic diversity in genetically impoverished populations has produced population gains in more than 90% of genetic rescue applications (Frankham 2015).

Resilience at the ecosystem level can also be restored using wildlife translocations, typically for locally extinct populations, a process known as reintroduction (Seddon et al. 2014). The most famous example is the reintroduction of wolves to Yellowstone National Park in 1995. Seventy years after local extirpation, gray wolves from Alberta, Canada, were introduced. A cascade of beneficial trophic and habitat changes ensued, among which was an increase in tree growth, allowing the unassisted recolonization of another locally extinct species, the beaver. In turn, this led to wetland habitat creation and improved stream hydrology (Beschta and Ripple 2016).

Table 21.1 De-extinction terminology

Term	*Definition*
Precise hybridization	The process of de-extinction using GE.
Extinct species	A species with no living individuals of any ontogenic state—active or torpid—in other words, a species from which no biological material exists to produce sexually reproductive individuals by any means (naturally or assisted). In the context of de-extinction, this is the species being replaced in a de-extinction program.
Template species	The closest living relative of the extinct species being replaced in an ecosystem. The template species is the species whose genome will be edited to produce the hybrid proxy.
Reproductive surrogate	The organism that generates the sex cells and/or gestates the hybrid proxy offspring. This species can be the template species, another species, a combination of multiple species, or in some cases artificial alternatives may be used in place of living organisms.
Hybrid proxy	The de-extinct organism, the result of precise integration of alleles (i.e., GE) from the extinct species into the genome of the template species, producing selectively (i.e., precisely) bred hybrid organisms displaying phenotypes predominantly from the extinct species, thereby reproducing the extinct species' ecotype in the wild.
The "wild"	Any open habitat or environment in which an organism is free-ranging. In this sense, the "wild" referred to by *in situ* conservation pertains to urban and synanthropic habitats, not just traditional perceptions of wilderness.

The reintroduction of wolves to Yellowstone National Park addressed the local extinction of a still-extant species, but other ecosystems are compromised due to the global extinction of a species. To address problems stemming from global extinction conservationists have looked to replacement and rewilding strategies (Seddon et al. 2014), both of which require suitable living proxies, that is, compatible ecotypes, to restore the role of the extinct species and communities in question. This practice has similarly produced beneficial conservation gains but is severely limited when confronting the loss of completely unique species for which no living compatible ecotype exists. Viable solutions to overcome problems resulting from the extinction of irreplaceable species have been beyond reach, until now. With rapid recent advancements in paleogenomics, reproductive technologies, and genome editing (GE), new opportunities now exist to produce ecological proxies of extinct species, a practice dubbed "de-extinction" by the media but more properly known as "precise hybridization" (Novak et al. 2018). Collectively, efforts to reintroduce and replace extinct populations via translocation or precise hybridization can all be viewed as methods of de-extinction (Novak 2018), sharing the same conservation goal but simply differing in the source of the animals introduced to the ecosystem. As de-extinction and genetic rescue both stem from the translocation of individuals, they can be seen as a continuum of conservation practice, whereby genetic rescue represents an intervention to benefit organisms at the population level and de-extinction extends benefits to the ecosystem level. Translocating animals to restore or rescue populations can trace its origins to the 1830s (Jorgensen 2013), making it the oldest of modern conservation practices (Novak 2018).

Animal Welfare in Conservation Interventions

De-extinction projects via precise hybridization, notably efforts to re-create the woolly mammoth and passenger pigeon, have garnered a great deal of attention (Novak 2018), particularly debates and criticisms regarding the welfare of animals in current and proposed programs (Kasperbauer 2017; Kohl 2017; Smith 2017; Turner 2017). Many of the arguments often neglect to consider precise hybridization as an extension of reintroduction practices. Given the longevity of de-extinction practice via reintroductions, most welfare concerns over *ex situ* and *in situ* animal handling and care have been well addressed, nullifying some points of criticism and debate; however, precise hybridization, with its use of GE, presents new issues to address because it unites what were previously separate worlds of animal welfare consideration: (1) animals in laboratory research (*in silico*, *in vitro*, *in vivo*), (2) animals in conservation breeding programs (*ex situ* conservation), and (3) the management of populations in the wild (*in situ* conservation). The ethical parameters of all three areas of animal welfare have been thoroughly debated, resulting in mature laboratory and *ex situ* welfare frameworks, although *in situ* welfare still has not yielded consensus. For *in situ* intervention, the question is whether or not welfare should be considered from the perspective of the individual organism, the priority of animal welfarists, or by the perspective of the welfare of the population, the priority of conservationists (Paquet and Darimont 2010).

These perspectives—individualism versus populism—are mistakenly viewed as mutually exclusive. For instance, laboratory animal welfare prioritizes the individual, but extends to the welfare of future individuals, embracing the "replace, reduce, and refine" ethic. In order to identify methods that replace or reduce the use of animals or refine procedures to eliminate discomfort or distress, it is necessary to conduct research on a pilot group of individuals. Alternate procedures must be validated, requiring a control group of individuals. In this case the driver of the research is to ensure individual welfare; however, given that "all future individuals" can benefit, the research can be seen as beneficial at a population level as well. There are other instances in which laboratory research is blatantly beneficial to populations, such as vaccine development. In these studies, a test group of animals is used in research to benefit all future untested individuals far beyond the lab. This ethical paradigm extends to *ex situ* conservation. For example, zoos have developed noninvasive means to monitor hormone levels crucial to timing breeding. Noninvasive hormone screening methods include fecal and urine collection, as opposed to screening hormones from blood, which requires invasive procedures. But, confirming that noninvasive methods provide equal reliability to blood samples requires a control sample of blood. A blood draw is no minor procedure; for some animals, it may require sedation. For wildlife seldom researched medically, this carries unknown risks. But to eliminate such risks requires a test group of individuals. Unreliable data leads to mistakes in animal husbandry, which may lead to negative welfare. For example, in territorial species it can be risky to introduce a male to a female that isn't receptive. Either individual may act aggressively, leading to injury. Alternatives to natural mating are no less risky, as they once again require invasive procedures. It is easy to deduce why noninvasive procedures are preferred, both for the safety of the animals as well as their handlers.

The *ex situ* to *in situ* release of individuals is when populationist viewpoints become pertinent, but once again can be deduced to the level of the individual. For example, the survival of individuals in the wild is what ensures the survival of the population. Survival in the wild hinges on adaptation to selection pressures, often missing from *ex situ* environments. The wild imposes inevitable negative factors to which individuals must respond, demanding *ex situ* conditioning before wild release—which leads to counterintuitive welfare measures that induce temporary fear, distress, and possibly discomfort. A premiere example of this is the power-line-aversion conditioning necessary to ensure the survival of California condors. When condors perch on power lines in the wild they are electrocuted to death. Prerelease condors are housed with low-voltage power lines that give the birds a nonlethal shock, imprinting power-line avoidance in the wild.[1] These brief moments of pain and distress prevent future suffering in the wild. Conditioning individuals for wild release (where de-extinction ends) exemplifies some of the nuances of welfare inherent to conservation that are absent for domestic and laboratory environments (where de-extinction begins).

Table 21.2 Aspects of animal use and livelihoods throughout the five stages of de-extinction

		In Silico	*In Vitro**	*In Vivo*	*Ex Situ*	*In Situ*
Definition		Bioinformatics research (e.g. genomics, transcriptomics, proteomics)	Manipulations in cell, tissue, and embryonic culture systems	Manipulations in live animals, captive breeding for research purposes	Captive breeding for propagation purposes	Introduction (release), monitoring, and ongoing management in the wild
Purpose/Parameters		To identify the alleles of the extinct species needed to recreate the extinct species phenotypes, thereby conferring the needed ecotype (i.e., hybrid proxy) to be reintroduced to the former range of the extinct species.	To validate phenotypes of candidate alleles and produce the cellular origins of gametes or embryos of the hybrid proxy	To validate the phenotypes of candidate alleles (functional genomics) and establish the founder individuals of the hybrid proxy population	Propagation of the hybrid proxy population from the founders to sufficient numbers for release to the wild and conduct research pertinent to successful wild release and *in situ* activities	To establish a viable, self-sustaining hybrid proxy population that fills the ecological role of the extinct species
Animal uses		Tissue sampling	Tissue sampling, embryo production	Breeding; phenotypic research	Breeding, behavioral & physiological research	Observational research; ecological function; ecosystem services
Living environments of Animals involved from birth to death (birth/death)	Wild/Wild Animals	Yes	Possibly			Yes
	Wild/Captive Animals	Possibly	Possibly	Yes	Yes	
	Captive/Captive Animals	Possibly	Possibly	Yes	Yes	
	Captive/Wild Animals				Yes	Yes
Invasive Procedures**		Yes	Possibly	Yes	Yes	Possibly

* Examples include *in ovo* and *ex ovo* research for egg-laying species.

** Examples include blood draws, biopsies, oocyte and semen collection, artificial insemination, embryo implantation, radio-collar fitting, sedation, general health checkup, and so on. In general, any procedure that involves handling the animal physically can be considered an invasive procedure.

De-extinction is unique in that it brings the use of laboratory research to conservation and it brings conservation-oriented research into the laboratory through feedback loops created by the five stages of de-extinction (Table 21.2). The de-extinction research process isn't linear. Each stage provides information pertinent to answering research questions in other stages. For example, *in silico* genomics reconstructs population dynamics, giving scientists insight for how species responded to historic environmental changes, thereby informing predictions of how a de-extinct population may respond to present environmental conditions. *In situ* ecology and *ex situ* research provides feedback for *in silico* work, particularly aiding bioinformatic strategies to identify candidate genes correlated to specific phenotypes.

Answering the questions necessary for building a successful de-extinction program involves using different individual animals for different purposes, in different environments. Wild individuals may undergo invasive *in situ* or temporary *ex situ* procedures or be transferred to *ex situ* propagation permanently, while some individuals will spend their entire lives *ex situ*, and others born *ex situ* will be released to the wild (Table 21.2). No matter the use or environment of an animal in scientific research, there are three tenets that shape welfare considerations in science: (1) that individuals should lead natural lives befitting their eco-evolutionary proclivities, (2) that individuals should be free from prolonged negative states, and (3) that individuals should function in normal and good health both physiologically and behaviorally over the full span of their lives (Fraser et al. 1997). With careful adaptations to protocols, it is possible to satisfy these tenets for every individual animal in a de-extinction program and still achieve the *in situ* end goal.

De-extinction in the Laboratory

The technology that makes de-extinction of species like the passenger pigeon and woolly mammoth possible is GE, particularly CRISPR-Cas9 GE (Novak et al. 2018). GE can be seen as a "precise-breeding tool," allowing scientists to change genetics in ways that produce organisms that could arise through selective breeding, but in a manner that is much faster and avoids inheritance of deleterious alleles that would otherwise be linked to the trait under selection (Novak et al. 2018). An example is the production of hornless dairy cattle via GE (Carlson et al. 2016). Dairy cattle possess horns. Angus cattle, a meat breed, don't possess horns, thanks to a mutant version of the *POLLED* allele. One can cross the two breeds, but the result will be an individual that is suboptimal for meat and dairy production. Rescuing quality dairy production while selecting for the *POLLED* allele will take many generations and may never yield the original quality attained in horned Holsteins or reach fixation (Bastiaansen et al. 2018). GE allows the *POLLED* allele to be inherited specifically, without loss of painstakingly optimized dairy genetics. The genomes of these GE cattle aren't altered or modified; they are entirely bovine. The horned allele has simply been overwritten with the *POLLED* Angus allele, a process that would happen via recombination during meiosis when interbreeding Angus/Holstein F2 generations (Figure 21.1). The intermixing of alleles in this case is between two variants of the same species. De-extinction via GE takes this to a higher taxonomic level, hence "precise hybridization." But the end result is still the same as in the case of the hornless Holsteins: a de-extinct passenger pigeon is a pigeon generated entirely from pigeon genetics and is an organism that in theory could be generated through the selective interbreeding and subsequent back breeding of passenger and band-tailed pigeons, if passenger pigeons were alive.

While GE is the tool that enables precise hybridization, the first step is comparative genomics: sequencing the genome of the extinct species and the template species to identify the alleles needed for creating the hybrid proxy via GE. This step presents the first use of animals in de-extinction: tissue sampling. The deteriorated state of ancient DNA makes it implausible to assemble an extinct species' genome from the sequenced fragments, so an extinct species genome is analyzed by aligning matching sequences to their position on a living species' fully assembled reference genome (Shapiro

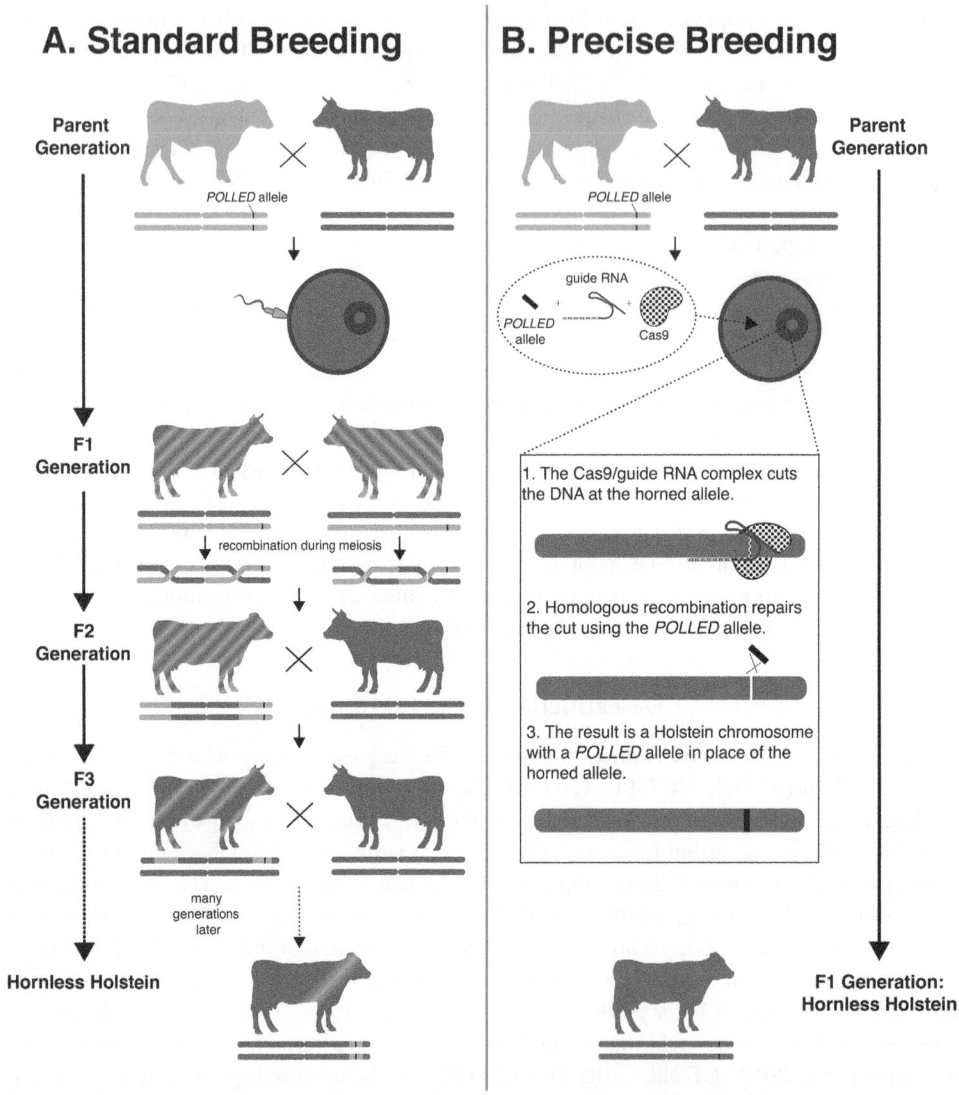

Figure 21.1 Comparison of methods to produce hornless Holstein cows by the combination of Angus (light gray) and Holstein (dark gray) genetics by (A) natural fertilization (standard breeding), represented by the union of sperm and egg, versus (B) genome-editing (precise breeding). In B, the inset box shows the genome-editing processes going on inside a Holstein cell used to generate an individual. The inheritance of the *POLLED* allele, responsible for hornless phenotypes, is tracked on simple representations of chromosomes

2016). Producing a reference genome requires tissue samples, which can be obtained from nonlethal sampling (skin biopsy, plucked feathers or hairs, blood draws, etc.) to sequence DNA. However, to accurately annotate a genome (i.e. identify genes, regulatory factors, etc.), scientists need to sequence all the transcribed RNA elements in the genome (dubbed the transcriptome). Almost all an organism's phenotypes are derived from the protein and nonprotein coding elements of the transcriptome, making it the most important part of the genome to understand for de-extinction. Annotating a genome presents the first issue for animal welfare in genomics, as it is difficult to acquire tissue samples for comprehensive transcriptome annotation in nonlethal ways. The issue is that every tissue type only transcribes a subset of the transcriptome—for example, heart tissue only uses the genes necessary to create the heart and doesn't express the same genes that create lungs. So, to discover all a genome's transcribed elements, a multitude of diverse tissue samples are required. One means of avoiding animal suffering for transcriptomics is to sacrifice an embryonic individual unable to cognitively synthesize consciousness. However, this does present limitations, as individuals express different sets of genes at different life stages: embryo, neonate, juvenile, subadult, to adult. To account for genes that may not be present in RNA sequence data, genomicists employ predictive bioinformatic pathways to avoid the use of further animal tissues.

Tissue sampling for DNA, RNA, and proteomic sequencing represent the only animal use during the *in silico* stage. Acquiring tissues for wild species often involves the capture of wild individuals. Capturing wild animals inevitably introduces distress to individuals, especially small animals that can't be sedated using darts shot from a distance. When possible, scientists obtain tissues from individuals raised in captivity that are accustomed to human handling and veterinary procedures. When nonlethal sampling is infeasible, scientists attempt to use subsamples of cryopreserved tissues previously collected for various natural history archives (i.e., museum collections and frozen zoos). If possible, researchers will also attempt to obtain samples from naturally deceased individuals from captive settings or randomly encountered in the wild, although RNA degrades quickly after death rendering most randomly found carcasses unusable. Another alternative pathway to obtaining tissues is sampling legally harvested individuals during regulated hunting seasons, preventing excess take from populations for scientific purposes. All these methods eliminate the need to sacrifice individuals expressly for the purpose of genomics.

Identifying candidate alleles for specific phenotypes doesn't verify the genotype-to-phenotype relationship. To discover the phenotype of a candidate allele requires functional genomics analyses, representing the *in vitro* and *in vivo* stages of de-extinction. Studying phenotypes can be done in cell cultures, which can be derived from the same tissue samples used for *in silico* research. This is especially true for mammalian species, in which there has been a great deal of advancement in the generation of induced pluripotent stem cells (iPSC) from other cell types and their necessary culture conditions. iPSC technologies offer the most powerful means of replacing animal subjects in de-extinction research. iPSCs can be acquired nonlethally from biopsies. They are immortal cell lines—meaning a single culture can be used indefinitely, preventing the need to perform biopsy procedures more than once for any individual and reducing the number of invasive procedures to the absolute possible minimum. From iPSCs, multiple tissue types can be generated. In combination with three-dimensional printing and other techniques, the functional genomics of genome edits at the tissue and organ levels can be analyzed (Takebe et al. 2017), enabling scientists to observe phenotypes without generating individual organisms, reducing the number of individuals used for *in vivo* functional genomics.

The power of iPSCs to reduce animal welfare issues in de-extinction is most substantial when generating founder individuals for a de-extinct population. Stem cell embryogenesis, the process by which embryos are derived from stem cells rather than the fertilization of oocytes with spermatozoa, has been steadily progressing through research with mice. To date, mice pups have been born from stem-cell-generated sperm-like cells (Zhou et al. 2016) and also from stem-cell-derived oocyte-like cells (Hayashi et al. 2012; Hikabe et al. 2016). These methods mean that it may be possible in the short term to develop mammalian embryos for implantation using sperm and eggs derived from skin biopsies, rather than conducting invasive procedures to collect oocytes for somatic cell nuclear transfer (SCNT; i.e., cloning) or spermatozoa for artificial insemination.

Gestating hybrid proxies represents the most intensive animal use demands for de-extinction. Once a hybrid-proxy cell line has been established, it must become an embryo. While stem cell embryogenesis may afford ways of bypassing animal resources for that step, the embryo must subsequently develop to term. For placental mammals, this occurs completely *in utero*. The reproductive surrogate to bare hybrid proxy offspring is often assumed to be the template species; however, for some species, it may be feasible to use domesticated relatives to bring hybrid offspring to term. Interspecies cloning of wildlife, using domestic surrogates for both oocyte harvest and embryo implantation, has produced viable offspring for several wild mammalian taxa (Wisely et al. 2015). The use of domestic varieties as reproductive surrogates could be applied for de-extinction of felid, canid, porcine, ovine, caprine, bovine, and equine species—all taxa widely cloned commercially. Using domestic reproductive surrogates is hugely important for adequate animal welfare in de-extinction practice, as most reproductive surrogates will spend their entire lives in captivity and undergo multiple invasive procedures. It is therefore paramount that reproductive surrogates be acclimated to such a lifestyle, for which most template species are not. However, with proper innovation, it may be possible to conduct most mammalian de-extinction programs independently of reproductive surrogates, as scientists work towards developing artificial uterus systems. Recent attempts were able to gestate late-stage sheep fetuses normally for a period of one week (Usuda et al. 2017). In the long term, it will be possible to generate live mammalian offspring from a skin biopsy with no use of reproductive surrogates.

Developing synthetic embryogenesis in egg-laying species is more difficult to conceive, and isn't possible in the foreseeable future. This means live individuals are necessary to propagate de-extinct birds. SCNT methods have produced viable offspring of insects, fish, and amphibians but not yet in avian or reptilian species (Kjelland et al. 2017). The current means of generating genetically manipulated birds is a process known as germ-line transmission (van de Lavoir et al. 2006). In SCNT, the genetically manipulated cell is fused with an oocyte to produce an embryo. Multiple types of cells can be used to fuse with the oocyte, allowing a degree of flexibility when conducting *in vitro* stages of work. However, for germ-line transmission in birds, scientists must work with primordial germ cells (PGCs). This means that the first step for *in vitro* and *in vivo* avian de-extinction is establishing PGC cultures from the template species. This task requires obtaining consistent quantities of fertile eggs for testing PGCs culture conditions. Once a cell culture recipe is developed, the need to sample embryos for culture greatly reduces, but at least one embryo will be required to develop every new cell line. PGC culture conditions currently exist only for domestic chickens, meaning that a large degree of replicative work will be necessary to establish PGC cultures for wild birds, potentially requiring many embryos for each species. While embryos may not experience negative affective states, the sacrifice of many embryos presents a significant physiological demand on breeding adults. Many avian template species have not been conditioned for extensive reproductive cycling; therefore, refining PGC isolation and culture methods to reduce the number of embryos necessary for PGC culture refinement is a significant welfare factor for avian de-extinction. Preliminary research with diverse poultry and domestic species may yield universal culture conditions or refined procedures applicable to PGC technology developments in wild species.

Once cultured, PGCs can be edited to produce a hybrid proxy cell line. This cell-line can then be implanted into the gonads of a specialized reproductive surrogate, known as a germ-line chimera. At this stage, it may be possible to use domestic varieties to replace the template species as the reproductive surrogate, a process known as interspecies germ-line transmission (van de Lavoir et al. 2012). For example, a pair of band-tailed pigeon germ-line chimeras may be able to produce one or two hybrid proxy passenger pigeon offspring per year under nonmanipulated conditions and possibly 36 offspring if breeding is manipulated (Marini and Novak 2015). In comparison, domestic pigeon germ-line chimeras may produce approximately 14 offspring naturally and more than 60 offspring per year under manipulated conditions. Chicken germ-line chimeras could theoretically produce nearly 300 hybrid proxy offspring per year. Domestic chickens and pigeons have been selectively bred for captive lifestyles and increased reproductive output for more than 8,000 years. The selection for domestication is apparent for pigeons: feral populations almost exclusively live in synanthropic habitats. Domestic pigeons offer plausible surrogates for de-extinction programs for the passenger pigeon, the dodo, and other extinct pigeons, although the ultimate goal for avian de-extinction is the ambitious development of domestic chicken strains capable of being the universal germ-line chimera.

De-extinction at the Zoo: *Ex Situ* Propagation

Once a founder population of hybrid proxies is produced, the process of propagating a viable population begins *ex situ* by private animal breeders and zoos. Welfare concerns in zoos and other animal breeding institutions are very similar to those in laboratory research, although the purpose of *ex situ* conservation is to support the recovery of species *in situ*. While it is acknowledged that a species' survival ultimately trumps animal welfare concerns, the Association of Zoos and Aquariums (AZA) and the Society for Conservation Biology have formalized obligations for the humane treatment of animals (Minteer and Collins 2013). The AZA in particular urge facilities to provide quality of life exceeding legal minimums (Whitham and Wielebnowski 2013). However, there will always be precarious periods involved when developing optimal welfare conditions for species novel to *ex situ* environments, which is the case for most template species. While the hybrid proxy is expected to display physiological and behavioral phenotypes of its extinct genetic parent, it is crucial to understand the husbandry and welfare needs of the template species to develop an effective propagation program accounting for the hybrid proxy's differing traits, given knowledge of an extinct species' husbandry and welfare needs are entirely absent for most extinct species.

It is important to expand the founding population quickly. Prolonged generations at low numbers erode genetic diversity, leading to inbreeding depression, reducing fertility and increasing frequencies of negatively affective (even lethal) genetic disorders (Charlesworth and Willis 2009; Räikkönen et al. 2009). Prolonged generations in captivity may inadvertently select for individuals adapted to *ex situ* environments, potentially compromising population viability in the wild. Rapid expansion of the population not only preserves the founding genetic diversity, but a genotype-guided pedigree may make it possible to increase a hybrid proxy population's effective genetic size. The benefit that a hybrid proxy population possesses over bottlenecked endangered species is that the founders for de-extinction can be purposefully selected from the diversity of the template species for maximum breeding viability, while the genetic diversity preserved after species' bottlenecks is beyond control. In other words, de-extinction practitioners begin breeding from a custom-tailored genetic fabric, while conservationists recovering bottlenecked species must work with what threads remain after a population has deteriorated.

Rapid expansion of a founder population is achieved in the following fashion: the founders' reproductive cycles are manipulated for maximum output to produce excess offspring that are raised via fostering strategies. Certain welfare concerns pertaining to fostering and manipulating reproduction cycles are nullified by emerging biotechnologies. As described earlier, domestic species can be used as reproductive surrogates for de-extinct hybrid proxies, precluding the manipulation of founder reproduction cycles. The major welfare concerns will be fostering oriented. Fostering strategies vary from using excess foster parents of the same species (for de-extinction this would be the template species), to using foster parents of a different species, puppet rearing, or hand rearing. All these strategies have been employed for endangered species with differing success rates, which arguably stem from the amount of knowledge pertaining to the life histories of the species being fostered. The major concern with fostering is false imprinting, when the offspring bonds to the foster species inhibiting integration to their own kind. This is one reason that puppet rearing is employed over hand rearing, as it is assumed that the visual image of the proper parent species imposes proper imprinting. However, wild pigeons in zoos and in private aviculture have been cross-fostered with domestic pigeons or doves without negative imprinting (pers. communication, David Oehler). The most extreme example is the use of Nicobar pigeons to raise crowned pigeons, a species six times larger with extremely different plumage. Despite the tremendous differences, crowned pigeons raised by Nicobar pigeons don't exhibit issues integrating into crowned pigeon flocks. Cross-foster research should be conducted with the template species to identify which strategies will provide the best results for eventual hybrid proxies. With the appropriate system in place, cross-fostering should produce a hybrid proxy population with their own natural behaviors.

Founders, fosters, and many individuals in a growing *ex situ* population will spend their entire lives in captivity, and as such will need to be provided optimal *ex situ* conditions. Human–animal relationships can serve as a vital tool for enriching the lives of captive-bound individuals (Claxton 2011). Symptoms of negative affective states can be subtle, and this presents another benefit to human–animal relationships: the observations of animal caretakers have shown to be qualitatively reliable when assessing animal welfare needs (Meagher 2009). In short, if an animal appears in a negative affective state to the caretaker that has a bond to that animal, then it is very likely the animal is experiencing a negative affective state that needs correcting. Animals restricted to captivity will undergo multiple invasive procedures, which present yet another major reason to foster human–animal interactions. Animals accustomed to human handling will experience far less distress, if any, when various procedures are performed. For de-extinction programs, it is important to treat individuals restricted to captivity more in alignment with the treatment of domesticated individuals with rich human interaction and to treat individuals intended for release as wild animals with limited human interaction.

Eventually, the population will be large enough to consider release to the wild. Prior to this threshold, a group of animals in the *ex situ* program must be raised under natural conditions, even if they spend their entire lives in captivity, in order to prevent problems of *ex situ* adaptation. Individuals raised by naturally conditioned parents will then undergo a preconditioning period prior to release. The preconditioning period is the beginning of a divergence from intuitive welfare concerns, as exemplified by the case of the California condors mentioned earlier. Preconditioning individuals for wild release involves exposing the individuals to conditions they will encounter in the wild, which have associated mortality risks. Although exposing individuals to elements that will cause distress and discomfort is necessary to ensure the welfare of those individuals when released to the wild, it doesn't need to result in mortality in the process. During the preconditioning phase, it is possible to identify individuals that struggle to cope with wild conditions and remove those individuals for use in further propagation or as educational individuals for the public.

Once a group of individuals has been preconditioned for release to the wild, the *in situ* stage of de-extinction begins. However, preparations for *in situ* release begin *ex situ*. While optimizing animal husbandry and welfare protocols represents the majority of zoo-based animal research, there has been a progressive emphasis towards research that can improve the welfare of *in situ* management from initial release to monitoring and handling practices (Minteer and Collins 2013). *Ex situ* animal research allows scientists to learn things about species and populations that would be difficult to achieve from *in situ* observations, although for de-extinction a combination of *in situ* observational research of the template species and *ex situ* experimentation with both template species and hybrid proxies can be combined to synergistically produce the knowledge needed for *in situ* success.

Reintroducing Extinct Ecotypes: The *In Situ* Stage

Going from initial release to a self-sustaining population presents the most precarious stage for conservation programs establishing wild populations from *ex situ* individuals. The scientific and ethical process of *in situ* reintroductions has formulated to an informed discipline in recent decades, as *ex situ* to *in situ* interventions have increased (Seddon et al. 2007; Seddon et al. 2014). The most prevalent invasive procedures occurring during *in situ* release involve the long-term monitoring and capture/recapture handling of released and subsequently wild-born individuals. Difficulty in this task has driven innovation in remote sensing, which has advanced from short-distance transmitting radio collars to long-range solar-powered global positioning transmitters, making it possible today for conservationists to download live movements of animals via satellite at their desk (Thomas et al. 2011). The size of remote-sensing monitors is decreasing, meaning that de-extinction programs will impose far fewer invasive procedures on wildlife than historic reintroductions and translocations.

Monitoring is essential to track the success of *in situ* management strategies, allowing practitioners to enact changes to the program as necessary as well as to intervene for individual welfare purposes (e.g., an individual at human–animal conflict risk). However, gauging welfare in the wild requires different considerations than in *ex situ* situations. Ecosystems aren't farms or laboratories, and thus, the welfare issues at play are quite different. The ecological process is intrinsically incompatible with Barrel's five freedoms: (1) freedom from hunger and thirst; (2) freedom from discomfort; (3) freedom from pain, injury, or disease; (4) freedom to express normal behavior; and (5) freedom from fear and distress. Barrel's five freedoms, developed for the keeping of livestock, become naively inapplicable in the wild.

In nature, individuals suffer distress and hardship for many reasons ranging from aberrant seasonal weather and threat of predation to natural cataclysms. Constant factors that cause suffering on individuals shape the evolution of populations, selecting adaptations to cope with those factors. The go-to example of this evolutionary process is the arms race between predators and prey, and in this eco-evolutionary dynamic, we encounter a fundamental inability to satisfy Barrel's fifth freedom. The fear and distress from predation can only be provided in predator-free environments, which may provide improved welfare to some individuals but will inevitably lead to the suffering of other individuals, which can only be mitigated by further intervention. *In situ* environments globally are altered by human activities and are vulnerable to uncontrollable factors, which imposes the need for wildlife, land, and natural resource management to maintain conservation goals. Wildlife management is a primary issue in public animal welfare debates owing to widely held misconceptions of ecology.

An exemplary case of clashes of animal welfare concerns in wildlife management is Oostvaardersplassen, a human-created habitat on reclaimed lands in the Netherlands. The grazing herds at Oostvaardersplassen are an example of rewilding. Managers released small cohorts of konik horses (proxies for extinct tarpans), heck cattle (proxies for extinct aurochs), and red deer to mirror historic European ecosystems. The environment is a closed park system, surrounded by urbanization and agriculture, devoid of apex predators. The small size of the park makes predator introductions infeasible and public objections have excluded human hunting to manage the grazing herds. As an alternative to predation, managers have opted to allow natural processes to play out—meaning that as populations rise above carrying capacity, they die back due to starvation. Overall the rewilding experiment of Oostvaardersplassen has been an immense success; the park supports higher densities of diverse species than many wilderness areas throughout the world owing to the dynamic processes taking place on the landscape. However, this form of management has generated immense backlash from animal welfarists, who see the starvation of the animals in the park as unnecessary suffering. Pressure from welfare groups has required managers to intervene and cull individuals before they succumb to starvation. Some years require greater culling than others, leading to heightened social tensions as herds that had grown greatly thanks to several mild winters had to be culled by nearly 90% due to a prolonged and harsh winter in early 2018. The magnitude of this cull incited public outcry, leading government officials to end the practice of "natural process" management and set caps to herd sizes (Barkham 2018), which will require continuous culling in the future.

Oostvaardersplassen highlights a key impasse for animal welfarists: natural processes or hunting? Natural processes induce suffering via starvation. Human hunting, although able to provide painless death in some instances and less painful deaths than nonhuman predators, most often induces a brief period of fear, distress, and pain, and during problematic kills can lead to prolonged suffering. Culling doesn't present a higher degree of welfare over hunting, as truly painless culling would require the capture and handling of individuals for chemical euthanasia, which will require darting for sedation or trapping; no terminal intervention is absolutely free of distress. In the end, welfarists must reconcile this absolute fact: death is part of ecology.

Hunting and culling are necessary parts of conservation management, and to believe otherwise reveals a fundamentally flawed grasp of ecology and evolution. Predation isn't only a natural element it is a universal component of ecosystems. The absence of predation inevitably leads to major ecosystem changes, including community collapse (Estes et al. 2011). When considering conservation *in situ* it is important to remember that hominids have been an apex predator and an ecosystem engineer globally for tens of thousands of years (hundreds of thousands of years throughout Africa and Eurasia). Where apex predators are absent, scientifically and ethically regulated hunting by humans maintains important ecological processes. Although seemingly counterintuitive to animal welfare, it can be argued that the true success of a de-extinction program finally occurs when the introduced hybrid proxy is prolific enough to allow regulated hunting.

Woolly Mammoth Welfare: A De-extinction Case Study

The primary purpose of mammoth de-extinction is to convert tundra habitats to their Pleistocene grassland state, generating expansive ecosystems beneficial to long-term climate stability. The major evidence supporting this motivation comes from decades of rewilding in Siberia, at a site known as "Pleistocene Park," that shows the reintroduction of grazing animals restores grasslands (Zimov et al. 2012). Independent analyses of reindeer grazing in Norway corroborate that intensified grazing favors grasses over other forms of vegetation (Väisänen et al. 2014; Ylänne et al. 2018). Furthermore, research in northern Norway has shown that grasslands store more carbon in soils than other northern habitats (Vedel-Sørensen et al. 2018), and in general, globally, grasslands are one of the highest carbon-sequestering habitat types, much more so than forests (Amthor et al. 1998). Not only does grass sequester carbon, herding grazers expose ground to winter air when foraging, keeping permafrost frozen and its store of millennia-old sequestered carbon (the largest carbon store in the world) locked away. It's not a short-term fix to climate change, but it is a long-term solution to sustainable climate maintenance, one that is gaining support among scientists, as reflected by two recent comprehensive review papers that both show the value of animal management for combatting climate change (Olofsson and Post 2018; Schmitz et al. 2018).

The end goal of the mammoth de-extinction is admiral, but does this justify the means?

Even de-extinction scientists have expressed concerns over the value of mammoth de-extinction and the welfare of animals in the program (Shapiro 2015). Living elephants are threatened in the wild. Providing sufficient *ex situ* habitats is difficult. But de-extinction practitioner George Church has committed to developing a program that not only avoids conflicts with Asian elephant conservation but will also help save elephants. The *in vitro/in vivo* stages are developing synthetic embryogenesis/gestation for elephants (pers. comm.). The program's plan is to bring the first hybrid proxies to term with no reproductive surrogates. The science to do this is paving the way for GE solutions to combat ivory trade as well as deadly viruses (Mack 2018). Elephant herpes virus is fatal to elephant calves. *In vitro* viral culture has failed, preventing the development of vaccines/treatments. Incentivized by de-extinction, the Church lab is synthesizing a culturable strain of the virus from its genomic sequence and will employ *in vitro* cell assays in pursuing solutions for immunity/treatment.

The largest benefits for elephants come directly from de-extinction, not peripherally enabled projects. The adaptation of elephants to cooler climates is demographically beneficial: increasing global range while simultaneously providing refuge from illegal poaching in inhospitable environments. Mammoth de-extinction is beneficial on an evolutionary scale as well. New genomic research has revealed that adaptability of elephants through time was facilitated by extensive hybridization (Palkopoulou et al. 2018), a process now impossible due to extinctions/range reductions. Precise hybridization using alleles from diverse extinct proboscideans may be key to the long-term adaptability of the three surviving species.

The road to colonizing new environments is a difficult one. Mammoth hybrid proxy founders will possess thermal optimums much cooler than elephants, presenting a welfare challenge in raising the offspring. However, Asian elephants in several cold climate zoos are observed to enjoy short periods playing in winter snow—even rolling giant snowballs. The first de-extinct mammoths will find foster families at such institutions and upon maturity may be transferred to more open facilities in northern latitudes. Initial *in situ* habitat already exists at tundra sites of converted grassland. Progress will be slow and gradual, but there is no reason to suspect that hybrid proxy mammoths should suffer negative welfare along the way.

Mammoths were highly social. This is perhaps the greatest welfare concern to the program. The hybrid proxies must not only be able to form social bonds *ex situ*; they must be introduced to the wild in cohorts large enough to prevent density-dependent problems ensuing from improperly expressed behavioral and even physiological social adaptations. Contrary to common perceptions of de-extinction, the process doesn't generate a "single resurrected" founder. The surrogate reproductive strategies in place will generate initial cohorts. It should be possible to establish small herds of mammoths *in situ* initially at sites such as Pleistocene Park and, with adequate augmentation from *ex situ* herds, propagate viable populations capable of expanding their range over several decades, similar to most conservation recovery periods for rare species (Suckling et al. 2012; Suckling et al. 2016), not centuries as some critics have mused.

Pleistocene Park? Why not Jurassic Park?

The most common popular-culture association with de-extinction is the Jurassic Park film franchise, prompting many to wonder, now that scientists are recreating mammoths, when might we see Mesozoic dinosaurs? The short answer is never, at least not by means of precise hybridization. DNA doesn't survive much more than 1 million years (Allentoft et al. 2012), precluding the methods being employed to re-create woolly mammoths. However, research into altering the development of birds (living dinosaurs), including studies of tail development (Rashid et al. 2014), ankle formation (Botelho et al. 2016), snout formation (Bhullar et al. 2015), and even mutations giving chickens teeth (Harris et al. 2006), is paving the way for potentially creating dinosaur-like birds in the form of "chickenosaurus" (Miller 2015).

There is no ecological place for a Mesozoic dinosaur ecotype in today's world and therefore no place in wildlife conservation. However, the alteration of species to produce companion animals has been continually performed for thousands of years (e.g., cats and dogs). Chickens are sometimes raised and kept as companion animals, so there could be an ethical place in the world for chickenosaurus as domestic pets—but that ethical debate is far afield from that centered around woolly mammoths and other de-extinction conservation efforts.

It is hard to imagine engineering sauropod or ceratopsian morphotypes from chickenosaurs may ever be possible, but they would all be facsimiles with no genetic contributions from their extinct templates. Ultimately, the vision of Jurassic Park is unattainable, rendering its relevance to de-extinction debates null and void. But the return of Pleistocene megafauna, like the woolly mammoth, is no longer science fiction.

De-extinction to Next-Generation Genetic Rescue

There are many ways de-extinction programs are working to improve animal welfare internally. In this chapter, I have highlighted just a few of the ways in which de-extinction is advancing methods to replace, reduce, and refine animal uses in conservation interventions. The broader application of de-extinction technologies to extant wildlife has expanded the scope of genetic rescue to include all conservation strategies rooted in genetic knowledge, ranging from traditional approaches enhanced via genetic insight (e.g., habitats/populations protection, translocation, or captive breeding pedigree decisions informed by population genetics) to facilitated adaptation (i.e., using GE to introduce alleles necessary to adapt to challenges outpacing natural selection processes, such as disease or climate change, Thomas 2013), to suppression or extirpation of damaging invasive species through gene drives (Novak et al. 2018). Gene drives are an immense development for invasive animal welfare, offering the first truly humane, nonlethal alternative to pesticides. Many "next generation" genetic rescue programs have formed since GE-based de-extinction began, including facilitated adaptation for the endangered black-footed ferret (Novak et al. 2018), stem cell embryogenesis to save the Northern White Rhinoceros (Saragusty et al. 2016), and genomic insight to guide kakapo recovery (Iorns 2017). With enough genetic rescue innovation and strategic implementation, a sustainable future for global wildlife is possible, depending on far fewer major interventions.

Note

1. The history of California Condor *ex situ* recovery can be read in greater detail in John Nielson's book *The Condor: To the Brink and Back—the Life and Times of One Giant Bird*, published by Harper Perennial 2007.

Bibliography

Allentoft, M. E. et al. (2012) "The half-life of DNA in bone: Measuring decay kinetics in 158 dated fossils," *Proceedings of the Royal Society B: Biological Sciences* 279(1748): 4724–4733.

Amthor, J. et al. (1998) *Terrestrial Ecosystem Responses to Global Change: A Research Strategy*. Oak Ridge National Laboratory, Oak Ridge, Tennessee, Report No.: ORNL/TM--1998/27; DOI: 10.2172/666209.

Barkham, P. (2018) "Dutch rewilding experiment sparks backlash as thousands of animals starve," *The Guardian*, www.theguardian.com/environment/2018/apr/27/dutch-rewilding-experiment-backfires-as-thousands-of-animals-starve (Accessed May 21, 2018).

Bastiaansen, J. W. M. et al. (2018) "The impact of genome editing on the introduction of monogenic traits in livestock," *Genetics Selection Evolution* 50(1): 18, https://gsejournal.biomedcentral.com/articles/10.1186/s12711-018-0389-7 (Accessed May 21, 2018).

Beschta, R. L., and Ripple, W. J. (2016) "Riparian vegetation recovery in Yellowstone: The first two decades after wolf reintroduction," *Biological Conservation* 198: 93–103, http://dx.doi.org/10.1016/j.biocon.2016.03.031.

Bhullar, B-A. S. et al. (2015) "A molecular mechanism for the origin of a key evolutionary innovation, the bird beak and palate, revealed by an integrative approach to major transitions in vertebrate history," *Evolution* 69(7): 1665–1677, http://doi.wiley.com/10.1111/evo.12684 (Accessed December 28, 2018).

Botelho, J. F. et al. (2016) "Molecular development of fibular reduction in birds and its evolution from dinosaurs," *Evolution* 70(3): 543–554.

Burney, D. A., and Flannery, T. F. (2005) "Fifty millennia of catastrophic extinctions after human contact," *Trends in Ecology and Evolution* 20(7): 395–401.

Carlson, D. F. et al. (2016) "Production of hornless dairy cattle from genome-edited cell lines," *Nature Biotechnology* 34(5): 479–481.

Ceballos, G. et al. (2015) "Accelerated modern human—induced species losses: Entering the sixth mass extinction," *Sciences Advances* 1(5): e1400253.

Ceballos, G., Ehrlich, P. R., and Dirzo, R. (2017) "Biological annihilation via the ongoing sixth mass extinction signaled by vertebrate population losses and declines," *Proceedings of the National Academy of Sciences*: 201704949, www.pnas.org/lookup/doi/10.1073/pnas.1704949114

Charlesworth, D., and Willis, J. H. (2009) "Fundamental concepts in genetics: The genetics of inbreeding depression," *Nature Reviews Genetics* 10: 783–796, ftp://193.134.202.10/pub/gugerli/ConservationGenetics/CharlesworthWillis_NatRevGenet2009.pdf (Accessed May 30, 2018).

Claxton, A. M. (2011) "The potential of the human—animal relationship as an environmental enrichment for the welfare of zoo-housed animals," *Applied Animal Behaviour Science* 133(1–2): 1–10, http://linkinghub.elsevier.com/retrieve/pii/S0168159111000827 (Accessed May 21, 2018).

Estes, J. A. et al. (2011) "Trophic downgrading of planet earth," *Science* 333(6040): 301–306.

Frankham, R. (2015) "Genetic rescue of small inbred populations: Meta-analysis reveals large and consistent benefits of gene flow," *Molecular Ecology* 24(11): 2610–2618.

Fraser, D. et al. (1997) "A scientific concept of animal welfare that reflects ethical concerns," *Animal Welfare* 6(2): 187–205.

Harris, M. P. et al. (2006) "Report the development of archosaurian first-generation teeth in a chicken mutant," *Current Biology* 16: 371–377, https://ac.els-cdn.com/S0960982206000649/1-s2.0-S0960982206000649-main.pdf?_tid=ec66f735-80bb-466a-b14a-82ae321201bc&acdnat=1545997267_86ef91d9efea4853c8fa58a59b120193 (Accessed December 28, 2018).

Hayashi, K. et al. (2012) "Offspring from oocytes derived from in vitro primordial germ cell-like cells in mice," *Science (New York, N.Y.)* 338(6109): 971–975, www.ncbi.nlm.nih.gov/pubmed/23042295 (Accessed May 21, 2018).

Hikabe, O. et al. (2016) "Reconstitution in vitro of the entire cycle of the mouse female germ line," *Nature* 539(7628): 299–303, www.nature.com/doifinder/10.1038/nature20104 (Accessed May 21, 2018).

Iorns, D. (2017) "Science exchange to complete Kakapo 125 Project," *Science Exchange Blog*, https://blog.scienceexchange.com/2017/07/science-exchange-enables-completion-of-the-kakapo-125-project/ (Accessed May 30, 2018).

Jorgensen, D. (2013) "Reintroduction and De-extinction," *BioScience* 63(9): 719–720, http://bioscience.oxfordjournals.org/content/63/9/719.full (Accessed October 28, 2014).

Kasperbauer, T. J. (2017) "Should we bring back the passenger pigeon? The ethics of de-extinction," *Ethics, Policy & Environment* 20(1): 1–14. http://dx.doi.org/10.1080/21550085.2017.1291831.

Kjelland, M. E. et al. (2017) "Manipulating the avian egg: Applications for embryo transfer, transgenics, and cloning," *Avian Biology Research* 10(3): 146–155.

Kohl, P. (2017) "Using de-extinction to create extinct species proxies: Natural history not included," *Ethics, Policy & Environment* 20(1): 15–17. http://dx.doi.org/10.1080/21550085.2017.1291832

Mack, E. (2018) "Ancient mammoth DNA could save threatened elephants," CNET. https://www.cnet.com/news/woolly-mammoth-genes-study-traces-path-to-elephant-hybrid/

Marini, P., and Novak, B. J. (2015) "Effect of controlled lighting on band-tailed pigeon (Patagioenas fasciata) breeding," p. 5, http://reviverestore.org/wp-content/uploads/2015/09/Effect_of_controlled_lighting_on_BTP_breeding_6.18.15.pdf

Meagher, R. K. (2009) "Observer ratings: Validity and value as a tool for animal welfare research," *Applied Animal Behaviour Science* 119(1–2): 1–14.

Miller, M. (2015) "Jurassic world: How dinosaurs are making a real comeback, thanks to science | PEOPLE.com," *People Magazine*, https://people.com/movies/jurassic-world-how-dinosaurs-are-making-a-real-comeback-thanks-to-science/ (Accessed December 29, 2018).

Minteer, B. A., and Collins, J. P. (2013) "Ecological ethics in captivity: Balancing values and responsibilities in zoo and aquarium research under rapid global change," *ILAR Journal* 54(1): 41–51.

Novak, B. (2018) "De-extinction," *Genes* 9(11): 548, www.mdpi.com/2073-4425/9/11/548

Novak, B. J., Maloney, T., and Phelan, R. (2018) "Advancing a new toolkit for conservation: From science to policy," *The CRISPR Journal* 1(1): 11–15. http://online.liebertpub.com/doi/pdfplus/10.1089/crispr.2017.0019

Olofsson, J., and Post, E. (2018) "Effects of large herbivores on tundra vegetation in a changing climate, and implications for rewilding," *Philosophical Transactions of the Royal Society of London: Series B, Biological Sciences* 373(1761): 20170437, www.ncbi.nlm.nih.gov/pubmed/30348880 (Accessed December 28, 2018).

Palkopoulou, E. et al. (2018) "A comprehensive genomic history of extinct and living elephants," *Proceedings of the National Academy of Sciences*: 201720554, www.pnas.org/lookup/doi/10.1073/pnas.1720554115

Paquet, P. C. D., and Darimont, C. T. (2010) "Wildlife conservation and animal welfare: Two sides of the same coin?" *Animal Welfare* 19(2, SI): 177–190.

Räikkönen, J. et al. (2009) "Congenital bone deformities and the inbred wolves (Canis lupus) of Isle Royale," *Biological Conservation* 142(5): 1025–1031, http://allthingscanid.org/Congenital bone deformities and the inbred wolves.pdf (Accessed May 30, 2018).

Rashid, D. J. et al. (2014) *From Dinosaurs to Birds: A Tail of Evolution*, www.evodevojournal.com/content/5/1/25 (Accessed December 28, 2018).

Saragusty, J. et al. (2016) "Rewinding the process of mammalian extinction," *Zoo biology* 35(4): 280–292.

Schmitz, O. J. et al. (2018) "Animals and the zoogeochemistry of the carbon cycle," *Science* 362(6419): eaar3213, http://science.sciencemag.org/content/362/6419/eaar3213 (Accessed December 28, 2018).

Seddon, P. J., Armstrong, D. P., and Maloney, R. F. (2007) "Developing the science of reintroduction biology," *Conservation Biology: The Journal of the Society for Conservation Biology* 21(2): 303–312, www.ncbi.nlm.nih.gov/pubmed/17391180 (Accessed July 10, 2014).

Seddon, P. J. et al. (2014) "Reversing defaunation: Restoring species in a changing world," *Science* 345(6195): 406–412.

Shapiro, B. (2015) *How to Clone a Mammoth: The Science of De-extinction*, 1st ed., Princeton, NJ: Princeton University Press.

Shapiro, B. (2016) "Pathways to de-extinction: How close can we get to resurrection of an extinct species?" *Functional Ecology* 31(5): 996–1002.

Smith, I. A. (2017) "De-extinction and the flourishing of species," *Ethics, Policy & Environment* 20(1): 38–40. http://dx.doi.org/10.1080/21550085.2017.1291834

Suckling, K., Greenwald, N., and Curry, T. (2012) *On Time, On Target: How the Endangered Species Act is Saving America's Wildlife*, www.esasuccess.org/report_2012.html.

Suckling, K. et al. (2016) *A Wild Success: A Systematic Review of Bird Recovery Under the Endangered Species Act*, www.esasuccess.org/2016/report.html.

Takebe, T., Zhang, B., and Radisic, M. (2017) "Synergistic engineering: Organoids meet organs-on-a-chip," *Cell Stem Cell* 21: 297–300. http://dx.doi.org/10.1016/j.stem.2017.08.016 (Accessed May 21, 2018).

Thomas, B., Holland, J. D., and Minot, E. O. (2011) "Wildlife tracking technology options and cost considerations," *Wildlife Research* 38(8): 653–663.

Thomas, M. A. (2013) "Gene tweaking for conservation," *Nature* 501(7468): 485–486.

Turner, D. D. (2017) "Biases in the selection of candidate species for de-extinction," *Ethics, Policy & Environment* 20(1): 21–24. http://dx.doi.org/10.1080/21550085.2017.1291835.

Usuda, H. et al. (2017) "Successful maintenance of key physiological parameters in preterm lambs treated with ex vivo uterine environment therapy for a period of 1 week," *American Journal of Obstetrics and Gynecology* 217(4): 457.e1–457.e13, www.ncbi.nlm.nih.gov/pubmed/28646647 (Accessed May 21, 2018).

Väisänen, M. et al. (2014) "Consequences of warming on tundra carbon balance determined by reindeer grazing history," www.nature.com/natureclimatechange (Accessed December 28, 2018).

van de Lavoir, M. C. et al. (2006) "Germline transmission of genetically modified primordial germ cells," *Nature* 441(7094): 766–769, www.nature.com/doifinder/10.1038/nature04831 (Accessed June 28, 2016).

van de Lavoir, M. C. et al. (2012) "Interspecific germline transmission of cultured primordial germ cells," *PLoS One* 7(5): 1–6.

Vedel-Sørensen, M. et al. (2018) "Draining the pool? Carbon storage and fluxes in three alpine plant communities," *Ecosystems* 21(2): 316–330, www.researchgate.net/publication/317752755 (Accessed December 28, 2018).

Whitham, J. C., and Wielebnowski, N. (2013) "New directions for zoo animal welfare science," *Applied Animal Behaviour Science* 147(3–4): 247–260.

Wisely, S. M. et al. (2015) "A road map for 21st century genetic restoration: Gene pool enrichment of the black-footed ferret," *Journal of Heredity* 106(5): 581–592. https://academic.oup.com/jhered/article-lookup/doi/10.1093/jhered/esv041 (Accessed May 21, 2018).

Ylänne, H. et al. (2018) "Consequences of grazer-induced vegetation transitions on ecosystem carbon storage in the tundra E. Dorrepaal, ed.," *Functional Ecology* 32(4): 1091–1102. http://doi.wiley.com/10.1111/1365-2435.13029 (Accessed December 28, 2018).

Zhou, Q. et al. (2016) "Complete meiosis from embryonic stem cell-derived germ cells in vitro," *Cell Stem Cell* 18(3): 330–340, www.ncbi.nlm.nih.gov/pubmed/26923202 (Accessed May 21, 2018).

Zimov, S. A. et al. (2012) "Mammoth steppe: A high-productivity phenomenon," *Quaternary Science Reviews* 57: 26–45.

PART IV

Companion Animals

Editor's Introduction

Many people have very strong emotional attachments to the animals with whom they live. They are not simply guard dogs or mouse chasers; they are family members (or something near enough). It wasn't always this way, so how did we get here? And what moral questions come up when we reflect on the practice of living with animals?

There are different ways we might approach the latter question, but three deserve special mention.

First, we might start thinking about the ethical issues here from the perspective of one of the two most prominent moral frameworks in the animal ethics literature, namely, utilitarianism and the rights view (see the general introduction for an overview). If we do, then we need to ask questions like these:

- What are the costs and benefits for animals when they are kept as pets? What are the costs and benefits for human beings? On net, is pet keeping a utility-maximizing practice? Are particular dimensions of pet keeping utility-maximizing? For instance, does spaying and neutering promote more happiness overall? What about the practice of creating and continuing specific breeds? What about the practice of feeding nonhuman animals in whom we have little emotional investment (e.g., chickens) to other nonhuman animals in whom we have great emotional investment (e.g., cats)?
- Are the rights of nonhuman animals violated by any part of the practice of pet-keeping? For instance, is it morally permissible to confine animals against their will? Do nonhuman animals have a right to sexual autonomy, and does spaying and neutering pets violate that right? Is it morally permissible to prevent animals from engaging in natural behaviors, such as preventing cats from killing songbirds?

When we approach the ethics of keeping pets in this way, then we see pet keeping as a new context in which to apply familiar principles.

Second, we might think about these issues from an historical and relational perspective. If we do, then we will end up asking questions such as these:

- What, exactly, have we done in domesticating wild animals? How should we think about the new kinds of beings we've produced through selective breeding programs?

- What sort of interests do companion animals have? How are they different from the interests of wild animals, or the animals we raise for food, or the animals in our labs?
- What special obligations do we have to companion animals in virtue of our historical relationship with them? We can ask this both at the individual and at the group level. What special obligations do *you* have, as someone with a companion animal, to that animal in virtue of her history? Likewise, what special obligations do *we* have, collectively, to companion animals in virtue of their history with human beings?
- What special obligations do we have to companion animals in virtue of the particular relationships we have with them? For instance, how should we care for them as their lives come to an end? Should we ever end those lives? If so, when?

Of course, we can ask these questions as utilitarians or rights theorists as well, and we can ask the preceding questions from a historical and relational perspective. The point here is just that different starting points make different questions salient, and it's worth noticing that our starting point may have an outsized influence on the kind of moral inquiry in which we engage.

Third, however, our relationships to companion animals may be an opportunity for us to reflect not just on what we ought to do but also on who we are. For instance, what does the impulse to domesticate—and to keep pets specifically—say about us? Why do we see that as something that we are entitled to do? Now that we don't need animals for their labor—most people don't have cats to address a mouse problem—why are we so interested in living with domesticated animals? What do our relationships to our pets reveal about us? What does violence toward animals indicate about a person? Is there such a thing as excessive kindness toward companion animals? If so, what does that reveal about us? Finally, how can companion animals serve as a way of understanding ourselves in relation to animals more generally? In what follows, you'll find discussions of these and many other questions.

22
PETS

Katherine C. Grier

Judging from the millions posted online, making and sharing videos of pets, and watching other people's cats, dogs and parrots in these short films, constitutes an unquantifiable yet significant source of pleasure for millions of Americans. Both the making and the viewing suggest the growing place of pet ownership in the private lives of Americans. This is further confirmed by the market research of the American Pet Products Association, which estimated that more than 84 million households contained one or more animals in 2016 (American Pet Products Association 2018). The number of television advertisements for products intended to keep pets healthy and happy, the explosion of popular books on our relationships with animals, the vibrant media coverage of such topics as the rise of the "emotional support animal": these suggest that understanding the practice of keeping pet animals in the home is one avenue toward greater understanding of American social life in the twenty-first century and that one thing it demonstrates is the fundamental good nature of pet owners. At the same time, pet keeping has unintended consequences for both people and animals; lots of things can, and do, go wrong. It has a dark side, from the transmission of zoonotic diseases to humans to the dire effects of poor care, neglect, or cruelty on the animals themselves. Good statistics on these poor outcomes are difficult to collect and interpret, but Shelter Animals Count, a national nonprofit that manages a national database of animals in shelters reports that, in 2017, 2,801 reporting shelters (of an estimated 3,900 shelters and rescue groups) took in over 2.1 million cats and dogs and euthanized around 411,000 of these animals (Shelter Animals Count 2018; Rowan 2009).

Contrary to common perceptions, pet keeping is not confined to wealthy Western societies, nor is it solely an artifact of modern life. Some form of the practice appears repeatedly in cultures and societies throughout history. It doesn't appear in all places, nor does pet keeping in long-gone communities share all the characteristics of pet keeping today (Serpell 1996). Researchers have recorded pet animals as a feature of daily life in non-Western, small-scale societies, and the practice turns up repeatedly in the historical record. Historians have demonstrated that, in the Atlantic world, pet keeping became increasingly complex over the course of the nineteenth century. What distinguishes contemporary pet keeping in prosperous industrial and postindustrial societies from its earlier manifestations, however, is its deep penetration into the institutions that shape daily life—government, work, commerce, mass media—while, at the same time, it is regarded as part of private life, family-centered, and often deeply individual, even quirky.

This essay considers modern pet keeping, but its focus is not on formal debates in the animal studies community considering the rightness of keeping animals in private homes. Rather, it focuses on unpacking the dynamic, if not necessarily coherent, ethical discourse embedded within the practices, daily routines, and popular ideas about pets. After presenting the complexities of defining the "pet," it considers the historical period when pet keeping, a long-standing traditional practice, was incorporated into a humane discourse linking the treatment of animals to the treatment of dependent human beings. Its case study is the history of pet keeping in the United States from the late 1700s to the present, emphasizing Euro-American culture and its evolving ideas about both the natures of animals and what people owed to the animals in their care. This relatively narrow focus makes it possible to hone in on the changes in pet keeping during this crucial period. In fact, it is clear that, even by the late 1800s, the idea of the "pet" animal had already become the most complicated kind of relationship people had with animals and that the word itself encompassed ideas and practices that were sometimes in conflict with one another. Furthermore, this new self-consciousness was challenged by the often harsh realities of life when animals were workers more often than they were companions. The chapter concludes with a consideration of further changes in American pet ownership since the mid-twentieth century. These built upon the nineteenth century's conception of kindness and the role of the pet, but the evolving discourse associated with pets reveals powerful new metaphors for the relationship between pet and caregiver, and new tensions as the traditional social boundaries between human beings and at least some pet animals is challenged.

Defining the Pet

The problem of defining what a pet is also suggests the complexity of the "everyday" or practical ethics associated with the practice, so it's a good place to begin. First, the pet animal is chosen by one or more people who take responsibility (however imperfect or inadequate their efforts may be) for its care and survival. The owner, a designation that is itself under debate in animal welfare/animal rights circles, intervenes routinely in the pet's life course. Second, the pet animal is kept in close proximity to its caregivers, almost always residing in or around the dwelling. Third, the pet animal is not required to "earn" its living except insofar as it contributes to the emotional economy of private life; it is a source of delight and the focus of moments of leisure. Often pets are regarded as a necessary part of the experience of childhood. Fourth, many pets are recognized as distinctive individuals by their owners; these pets are described as companions and are often the object of deep emotional ties. Exceptions here might include tanks of tropical fish or flocks of cage birds. In these cases (fifth), pet animals may be kept as a hobby, and their care may require considerable expertise that is an avenue for social life such as clubs. Sixth, some pet animals are kept, partly or wholly, for their beauty. Beauty is found in their appearance or in specific behaviors, such as the singing of canaries. Seventh, the animal may be the source or a signal of social status. Such pets may be rare or "purebred" (created from a closed genetic pool of similar individuals) or have some other special characteristic, as when their high status is associated with actual or implied potential danger associated with ownership, as with snakes. Eighth, the status of the pet is contingent. An animal can be raised into the status of the "pet" or lose that status through abandonment or other failures of care. Shelters are full of animals demonstrating just how imperfect human stewardship can be, and their staffs work hard to restore the special status of the "pet," with its implied promise of care, to abandoned animals.

Most pets only meet some of these characteristics or conditions, and while being a companion is one of the most important attributes of some species such as dogs, many pet animals play other roles. Aquarium keepers, for example, often view their tanks both in relation to a wider interest in natural history, and the hobby becomes an avenue for social life. Human intervention in the life course of the pet is so fundamental to the practice, however, that it is invisible, and pet owners seem rarely to consider it except in moments of crisis. Setting aside how competent the care in any one instance is, all pets are dependent on humans for nourishment, a safe environment, and care that enables and perhaps extends their lives. The pet's fertility is controlled, whether by surgical interventions or controlling its access to potential mates. And caregivers may control the time and circumstances of the pet's death. Sometimes this death is the result of incompetent care, abandonment, or active cruelty. But often the pet owner has determined that the animal is suffering from illness or declines from advanced age, or that the owner is unable to provide the required care. Note, too, that this set of characteristics and conditions is not a legal definition of the pet animal. While a lively legal debate on the legal status of animals, including pets, is ongoing, daily life with pets is shaped by a set of widely agreed-on practices that reflect both longstanding custom and evolving ideas.

The Domestic Ethic of Kindness

In the United States, changing popular attitudes about human obligations to animals, along with evolving standards for their welfare, went hand in hand with the expansion of pet keeping in the nineteenth century. What can be called a "domestic ethic of kindness" made life with animals at home the starting point for defining standards of animal welfare. Its appearance was associated with the coalescence of a distinctive middle class in the first half of the nineteenth century (Grier 2006). Being "middle class" was a new socioeconomic status that reflected changes in the American economy and ways of earning a living; in religious belief and practice and associated thought; and in patterns of daily living that focused on women's emotional authority in the domestic sphere and more intensive child-rearing practices. Middle-class culture in the United States was shaped in distinctive ways by three interlocking sets of ideas: gentility, liberal evangelical Protestantism, and domesticity. Gentility and liberal Protestantism stressed self-development and personal excellence based on self-control; domesticity redefined the nature of home life and made it the model of the world as it should be, where traditions of hierarchy and power were softened by self-restraint, love, and care. All three sets of ideas stressed the importance of active stewardship toward "dependent" others: children, the aged and infirm, the poor, the enslaved, the mentally ill—and nonhuman animals. Domesticated animals around the household became the starting point and medium for teaching children what advice author Lydia Sigourney called "kindness to all around" (Grier 2006: 127).

A crucial precondition for the rise of the organized animal welfare movement in the United States following the Civil War, the domestic ethic of kindness recast pet keeping as a positive social good as well as a source of pleasure and private leisure. Incorporating elements from English Enlightenment philosophy, along with the influential writings of liberal clergy such as New England Congregational minister Horace Bushnell, the ethic emphasized both the negative effect that poor treatment of animals had on the development of children and the entitlement of animals themselves to lives free from physical suffering. The new ethic stigmatized some traditional interactions with animals, especially cruel sports such as dogfighting and public physical abuse of working animals such as cart horses. These behaviors not only led to suffering by animals, but they also brutalized their human agents, who in turn would brutalize other people. This reasoning underlay the effort, only partly successful, to stigmatize and eliminate blood sports such as cock- and dogfighting.

Pet keeping was already present in many households when this new ethic of care for animals was promulgated in Sunday school lessons, schoolbooks and children's books, and popular prints by the 1820s. European settlers had carried to the American colonies the practice of keeping caged songbirds and companion dogs such as small spaniels, along with working dogs and cats that sometimes became prized companions and were called "favorites." Now, however, pets became more visible and more important to a happy family life. The domestic ethic of kindness argued that children needed pets to become responsible, loving adults. Parents not only were encouraged to allow pets but also were urged to supervise children in their routines of care so that no taint of cruelty or neglect entered the family circle. Some types of pet keeping—keeping an aquarium (by the late 1850s) or a flock of ornamental pigeons, for example—were regarded as "rational amusements," forms of leisure that improved their practitioners.

Animals were not simply vehicles for enacting and promoting kind behavior, however. Popular thought, and what may be considered folk tradition, had long regarded animals as intelligent, feeling beings who, at least in the case of some animals such as dogs, had the capacity to be moral actors (Thomas 1983). Popular media published an increasing number of news stories and fiction that discussed the admirable character of pet animals, including the model domestic lives of songbirds and chickens, the gentleness of rabbits, the loyalty and intelligence of family dogs, and the home-loving nature of cats. In the domestic ethic of kindness, some household pets became middle-class individuals, presaging the way they are regarded by some pet owners today. However, their individuality had particular limits since pets could be sagacious and loving but were "voiceless" to press their own interests. The domestic ethic of kindness never rejected the concept of social hierarchy; rather, hierarchy was embodied in its conception of stewardship, where humans were still firmly in charge, but it was limited by a duty of care toward dependents. Failure to recognize and act on that duty was, at the least, a moral failing and, at worst, a sin. By the late nineteenth century, the concept of the pet as part of family life was deeply ingrained and widely supported as a social good by groups such as the American Humane Education Society, which was successful in introducing to public school classrooms textbooks on kindness to animals; the lessons centered mainly on household pets.

While advice literature and other printed media offer abundant clues to the public conversation around keeping pets in the past, reconstituting the relationships of ordinary Americans to their cats, dogs, and canaries requires some ingenuity including examining visual sources and surviving artifacts. Family papers for ordinary people are rarely found in public archives, but collections for famous Americans such as Samuel Clemens (known as Mark Twain) or presidential families often contain correspondence, memoirs, and portraits of beloved pets. Portraiture was mainly available to well-to-do Americans until 1840, when the technology of photography arrived from France. As soon as photography studios appeared, ordinary people began to bring dogs, and even some patient cats and cages of canaries, to have their portraits made, sometimes with their owner and sometimes as individuals alone. The relationship between the human and animal sitter is evident in poses, which often include the animal sitting on a chair or a tabletop, in the sitter's lap, or posed at his or her feet. Until the 1880s, almost all photography took place in professional studios, but the rapid rise of amateur photography, culminating in the introduction of the Brownie snapshot camera by Eastman Kodak in 1900, meant that pet owners could begin to document the presence of their pets at home and their activities, from a dog chasing a ball to a cat inspecting the family's caged canary. The development of the wildly popular photographic "snapshot" postcard around 1900 not only provides telling images of daily life with pets, but the written messages often also include details such as names, ages, information about behaviors, and evidence of feelings. Artifacts are also revealing (Grier 2005). The design of surviving cages reveals one aspect of the proxemics of pet keeping: they show that caged birds were carried around the house and either placed on tabletops or hung from walls or window frames. Surviving dog collars engraved with names seem to have been kept as posthumous mementos, and gravestones in pet cemeteries (Hartsdale Canine Cemetery, the largest in the United States, was founded in 1896) reveal that deep attachment of pet owners extended beyond dogs and cats to parrots, monkeys, and rabbits (Martin 1997).

Ethical Dilemmas

While the domestic ethic of kindness insisted that animals were entitled to gentle treatment, it was less clear what "kindness" meant outside the household, in the context of a society wholly dependent on animals for power, raw materials, and food. Even city folk had contact in the streets with animals living their lives completely outside the domestic spaces that were intended to perform the ideal of kindness in daily life. (Early humane efforts on behalf of working horses were sometimes mocked as domestic sentiments inappropriate to the rough world of men's work.) Furthermore, tensions in the ethic in relation to pet keeping became especially apparent in the context of two household dilemmas—the status of the livestock pet and the circumstances surrounding euthanizing.

Often, town-dwelling and suburban families that kept pets also kept livestock: a small flock of chickens, a family cow or pig, and perhaps a horse. These seem to have been the families where livestock pets appeared, although there is evidence of farm children having pet calves and poultry. In the case of chickens, children might have one or two as named pets, exempt from slaughter because of their special identity, while the rest of the flock was available for both meat and eggs. The fate of calves and pigs, often named but inevitably doomed, was sometimes the source of emotional trauma for sensitive children. It's rarely clear how conversations about these boundaries took place, but children may have imbibed the lesson simply by observing the routines of care and consumption. The domestic ethic of kindness insisted that such animals were entitled to kind treatment while in the family's direct care, but it never challenged the idea that their meat was necessary for the human family itself to thrive (Grier 2006).

The debate around euthanizing cats and dogs also reveals the ways that practical realities shaped ideas about kindness. Pet owners often faced the dilemma of animal suffering, and the domestic ethic of kindness asked that they take responsibility for either ameliorating or ending it. The primitive nature of medical care (for both animals and people) meant that most pets died from infectious diseases now prevented by vaccines or from injuries that they would now recover from thanks to modern veterinary medicine. Controlling the numbers of some species was also a problem. Cats bred freely, leaving owners to cope with multiple litters; the fertility of female dogs was controlled to some extent by shutting them away from suitors. The domestic ethic's concept of stewardship taught that abandonment of a dependent was cruel. Thus, conscientious owners were left to face the task of euthanizing but without the support of veterinary medicine, which largely neglected small animals until the 1920s. Praise for owners who offered an animal a good death, and even instructions on the proper methods appeared in some surprising places, including children's literature. For example, Harriet Beecher Stowe published one children's story explicitly supporting the common practice of drowning unwanted kittens as a kindness. A paperback edition of Anna Sewell's groundbreaking anticruelty appeal *Black Beauty*, distributed by the American Humane Education Society in the 1890s, included a diagram showing the correct spot on a dog's skull for a bullet to cause a quick and painless death.

Pet Keeping Enters the Consumer Economy

Beginning in the mid-nineteenth century, pet keeping became a tiny but telling segment of the general expansion of the United States' growing markets for consumer goods. What differentiates the pet trade from other consumer activities, however, is that the animal itself was (and is) often the initial object of an economic transaction that then set off a lifetime of expenditure. The goods that supported its life included both purchases directly intended for the pet and the redirection of preexisting family resources, such as shelter and food, already present in the household. Well into the twentieth century, the most popular pure pets were not the dog or cat but cage birds, which were also the beneficiaries of the first array of products for their care. Caged songbirds were popular because they were, in fact, the only ambient music in households until the radio came along in the 1920s. American songbirds—mockingbirds, goldfinches, and cardinals, among others—were captured and sold in city markets and even sent overseas to wealthy European bird enthusiasts in the eighteenth century. Imported canaries occasionally appear in newspaper advertisements as early as the 1750s. However, a well-organized international trade in cage birds, featuring canaries raised for sale in Germany, appeared in the United States by the 1840s when steamship travel reduced the length of the sea journey. It was important enough to support a new set of businesses that provided not only animals themselves but also an expanding array of products (Grier 2016). Cage birds were the first pets to be served by packaged foods and remedies, which appeared in "bird stores" (the forerunner of the pet store) as early as the 1840s. By the early 1900s, almost a quarter-million canaries arrived in the United States from Europe each year; domestic production of these little birds and others, such as parakeets, was also widespread. Cage birds were so inexpensive and so popular that the early five-and-ten stores, such as F. W. Woolworth, featured pet departments by the 1920s.

Dogs were the next to benefit from a full array of products for their care, including foods, over-the-counter medicines, and, by the turn of the twentieth century, extras like toys, coats, and special beds. The evolving status of dogs, moving them further into family life as pure companions, was shaped in part by the development of the modern "purebred" dog in the mid-nineteenth century. Individual dogs such as small spaniels were imported to the United States both as pets and to establish breeding stock for hunting kennels as early as the late eighteenth century. Organized activity to establish pure lines, closing the gene pool through the use of registries and clubs, was associated with the rise of the British dog fancy. At first, hunting dogs were the most popular purebred dogs, but by the late 1870s, the first fads for purebred companion dogs appeared for pugs, fox terriers, and collies. The first large dog show in the United States was at the Centennial Exposition in Philadelphia in 1876; the American Kennel Club (AKC) was founded in 1884. Americans soon created their own distinctive breeds such as the Boston terrier, one of the most popular dogs through the 1930s. Most Americans did not own pedigreed dogs, but the number of AKC registrations, less than 2,000 in 1885, increased steadily even through the 1920s and 1930s, and jumped from 77,400 in 1944 to 206,978 in 1946, when buying a purebred dog became another purchase in support of the postwar "good life" for middle-class families (O'Neill 1985). The first commercial dog food, a British import from a company called Spratts Patent Ltd., appeared for sale in the United States in the 1870s, but it was so expensive that only well-to-do dog owners, typically breeders of purebred dogs and elite hunting packs, used it. The breakthrough decade for commercial pet food was the 1930s, which seems counterintuitive since so many Americans were impoverished during the Great Depression but which reflects the relative comfort of many, the efforts of the meat industry to use its by-products, and the work of these companies to market their products as both healthy and convenient. Still, many dog owners followed the old practice of cooking for their cats and dogs, who dined on table scraps and starches made into "dog stews." How well the family dog ate was directly correlated to how well his human family ate, and the dogs of the poor, no matter how well loved, often scavenged for their meals (Grier 2008). Dogs were also the first beneficiaries of small-animal veterinary medicine; the University of Pennsylvania established the first American dog clinic in 1885. However, in the absence of effective vaccines, most dog owners relied on over-the-counter remedies, both made in their local drugstores and packaged for sale by individuals, such as Polk Miller, a Richmond, Virginia, pharmacist who began to sell his own dog remedies under the name Sergeant's Pet Products in 1886. All these concrete changes in dog care demonstrate that Americans were increasingly willing and able to spend money on behalf of their pets, and believed that dogs deserved to be healthy and comfortable. And some new products made it easier for owners to be close to their pets—literally. Beginning around 1870, companies introduced carbolic disinfectant soaps, a boon for household and institutional sanitation, as "dog soaps" guaranteed to kill fleas and treat skin diseases such as mange. This was the first step toward modern flea control.

By the 1890s, modern full-service pet shops proliferated in cities, selling millions of animals along with goods to house and care for them. A general trend toward more intensive care of pet animals is apparent, although it was uneven, reflecting both class and regional differences. The lives of dogs, cats, and other household animals differed substantially in relation to their access to resources, as did the lives of their human owners. Historically, it is impossible to know how these differences correlated with attitudes; was the cat owned by a poor family loved less than the cat of a wealthy household? Modern research into the demographics of pet keeping suggests that this is not the case (Humane Society of the United States 2014). What was different, however, was the ability of poor pet owners to afford supplies, such as medicines, or dog licenses, which cities required as early as the 1820s in the case of New York City. Then, as now, the pets of the poor were more vulnerable, more likely to suffer from poor nutrition, parasites, and infectious disease or incursions by the community dogcatcher.

Pet Keeping Grows After World War II

By the mid-twentieth century, pet keeping was the primary mode through which most Americans had contact with nonhuman animals. Livestock almost completely disappeared from urban places by the 1930s, thanks to public health laws and motorized transportation. As the industrialization of animal agriculture and slaughter was perfected and animal experimentation disappeared into academic research facilities, two of the most troublesome arenas of animal use virtually disappeared from public view. Pet keeping enjoyed a bump in popularity following World War II, when it became one marker of happy child life and family life after decades of austerity. But it has experienced another marked jump in the past two decades according to the American Pet Products Association, which began a survey on pet ownership in 1988. At the same time, despite the wide interest in pet keeping, it remains a relatively small part of the consumer economy. Direct expenditure of products and services for pet animals was close to $67 billion; to date, no one seems to be measuring such indirect expenses as lawn care, household damage, and visits to the doctor for treatment of allergies or bites.

The evolving status of cats is one of the most striking changes in pet keeping. From their first arrival in America in the seventeenth century, cats were important, if sometimes ill-treated, household and community workers. Wherever people and livestock lived, so too did rodents, and cats, often unclaimed by people, were the only effective way to control them. Some cats, however, also enjoyed the status of prized pets, and efforts to create a parallel set of registries and clubs for cats appeared around the same time as the rise of a purebred dog establishment. By the 1930s, the market was strong enough that canned foods and toys, including scratching posts, were offered for sale. However, practicalities—including their uncontrolled fertility and the challenge of housekeeping—worked against keeping cats as indoor pets. Some owners kept boxes of sand or torn paper, but among ordinary families, the rise of "indoor cat" correlates with the appearance of effective clay litter-box fillers. Especially since the 1960s, cat ownership has become more intensive as spay–neuter has become routine veterinary practice and animal welfare groups have made the case for keeping cats indoors on the grounds of both safety and the negative impact of outdoor cats on bird populations. While a smaller percentage of American households own cats (30.4% compared to 36.5% for dogs), cat-owning households average two animals (American Pet Products Association 2017). As keeping cats indoors has become more common, the debate over the millions of unowned cats, both feral animals and abandoned pets, has grown more heated. Should these cats be rounded up, sterilized, and allowed to live in stable colonies ("trap–neuter–release")? Or should they be trapped and euthanized? The conflict here reflects two profoundly different perceptions of cats: first, as an invasive species; second, as individuals with the right to live out their lives with human supervision, though perhaps as "community cats" with relatively little direct intervention (AnimalSheltering.org 2018).

It seems clear that most American pet owners make an effort to adhere to what veterinarian and animal advocate Michael W. Fox describes as humane stewardship, a position that, with its emphasis on human responsibility and belief in the priority of human decision-making about animal welfare, recalls the earlier domestic ethic of kindness (Fox 1983). However, pet keeping has expanded in ways that nineteenth-century advocates would never have anticipated. Modern small-animal veterinary medicine now offers complex services—including effective care for chronic conditions, cancer treatment, and organ transplant—that have moved the goalposts for discussions about the end of life. Especially in the case of chronic disease, deciding when suffering should end is more difficult and upsetting. However, bereft owners are now able to draw on a variety of services to help them process and interpret grief, including a sizable literature, support groups, and websites (see, e.g., Wolfeldt 2004). The rate of surgical sterilization for cat and dogs has increased dramatically since the 1960s (90% of owned cats are now sterilized), thanks, in large measure, to the efforts of animal welfare organizations publicizing the grim situation of unwanted pets beginning in the 1970s.

Not all trends are good news, however. While cats, dogs, birds, and an array of domestically raised small animals constitute the pet population in most homes, a market for exotics, including reptiles, developed by the mid-twentieth century and now includes an international black market for wild animals, including endangered and dangerous species. The statistics on animal numbers, sheltering, abandonment, and euthanizing rates collected by the Humane Society of the United States as part of its Animal Sheltering initiative (www.animalsheltering.org/) and the nonprofit Humane Research Organization (https://faunalytics.org/) suggest that, even as overpopulation has declined significantly, pet keeping still has substantial spillover effects that impact community life, including the cost of animal "control" by municipal and county governments ($2.4 billion in 2007) (Rowan 2009). Tens of thousands of cats and dogs are still abandoned or experience poor care, which may be an unintended outcome of family crisis or poverty, but may also reflect ignorance or carelessness on the part of owners. More than ever before, pet care requires disposable income. Aside from initial costs of acquiring an animal, the American Society for the Prevention of Cruelty to Animals estimates that the annual cost of ownership for a medium dog or a cat is about $700 (after surgical sterilization); rabbit care costs around $475; a small bird, $317 (ASPCA; "Pet Care Costs," accessed 1 June 2018). Issues of appropriate stewardship seem especially thorny when nontraditional pets, such as reptiles and exotic, possibly endangered animals, are the subject. While a handful of American pet owners kept "exotic" animals such as parrots and monkeys in the eighteenth and nineteenth centuries, ownership of captive wild animals was limited by such basic considerations as the general absence of central heat in houses. By the 1960s, a handful of businesses sold and promoted exotic animals, and a few were common features in neighborhood pet stores. Today, the exotic pet business in the United States relies on auctions, the internet, and networks of enthusiasts that hold shows. The quest for market novelty has led to the arrival on the scene of small animals such as African pygmy hedgehogs, but the expansion of exotic pet keeping is most striking in the case of reptiles. More research is needed into the motivations of owners to keep exotic animals such as tropical snakes, but it is clear that these pets now present extraordinary challenges to the shelter community. Furthermore, the release of pythons and iguanas once purchased as pets has had harmful consequences on ecosystems in Florida. Current efforts to limit the numbers of exotic pets include proposals for more education for prospective owners through a uniform labeling system (Warwick et al. 2018).

The development of the internet into the "critternet" provides abundant material for evaluating attitudes and ethics associated with pet keeping in the United States. "Home movie" videos often feature animals at play, animals demonstrating intelligence in a way that surprises human observers, and occasions when the pet faces a situation where it may require human help. Thus, pet videos share characteristics with videos of small children. The language used by today's owners and animal welfare groups to describe their pets—"best friend," "fur baby," and "pet parents," among others—suggests that attitudes and feelings on the part of some owners have grown more intense since the domestic ethic of kindness began to shape pet keeping in the nineteenth century. Describing animals, especially dogs, as "companions" appears as early as the 1600s, but the appearance of the term *companion animal* which, according to the *Oxford English Dictionary*, began to appear in print in the late 1970s, now suggests something more than simply keeping company. Rather, the modern usage is intended to suggest a relationship where the traditional hierarchy between owner and animal is blurred and where the animal is entitled to make claims on the owner. Further complicating things, a 2011 Harris Poll of 2,164 American adults reported that more than 90% of respondents regarded their pet as "member of the family" (Harris Poll 2011). It is unclear what this means in terms of actual behavior or to the everyday ethics of pet keeping. However, public debates over animal care (as in discussions of ear and tail docking for some dog breeds) and the varied adoption requirements associated with privately run shelters, which have embraced the moniker "rescues," demonstrate that there is still considerable variation in ideas about what is appropriate practice.

With its focus on relationships between people and animals in the idealized household, the nineteenth-century domestic ethic of kindness contained an undeveloped critique of animals as commodities. This critique finally received full articulation within the sheltering community, where it is visible in the language associated with its work as well as its efforts to regulate or eliminate commercial dog breeders, labeled "puppy mills." The application of the concept of "adoption," a word originally used to describe taking into the household and accepting responsibility for a dependent human being, to animals, suggests how concepts of stewardship toward pets included a critique of the workings of an unfeeling commercial market for at least some species of pet animals. Applying *adoption* to animals seems to have appeared for the first time in the 1870s and was in common use in the animal welfare community in the 1930s, according to the *Oxford English Dictionary*. The appearance of the more recent descriptor *rescue*, used by many all-volunteer organizations, also suggests heroic action on the part of the humans who intervene and care for abandoned or "failed" pets.

Conclusion

The challenges of interpreting recent trends in pet keeping are considerable, but it is clear that the practice is evolving in ways that are synchronous with other changes in American social life. For example, prospective dog owners can now shop for purebred and "designer" mixed-breed puppies online, and commercial dog breeders now routinely make sales through the internet. Furthermore, as with other fashionable consumer goods, the market for purebred puppies relies on a network of institutions and businesses that together use a full range of communication media and sophisticated publicity to mystify and obscure the fact that pet animals are themselves a multibillion-dollar industry (Kavin 2016). While owning animals in the household is still regarded as an individual, private act, it has an impact on community well-being, including public health and ecology, as the debate over the impact of domestic cats on bird populations and released pythons on the native species of the Florida Everglades reveals. Most institutions still regard pet animals, for legal purposes, as a form of sentient private property. However, the growing field of animal law is working to alter definitions of what constitutes acceptable care, and standards for government intervention when cruelty or neglect are identified are evolving more rapidly than in the past (Favre 2011). A cohort of scholars and policy makers also study the public impacts along with the more private benefits and costs associated with keeping pet animals. Beginning in the 1980s, psychologists, biologists and other scholars undertook research on our relationships to pets, resulting in studies suggesting that pet ownership improves physical and mental health. Work on the benefits of pet keeping is now mainstream enough to be included on the website of the Centers for Disease Control and Prevention (Centers for Disease Control and Prevention 2018).

Pet owner behavior suggests that there is broad consensus about some issues such as the desirability of sterilization and vaccinations for cats and dogs, for example, although there is considerable difference in the ability of pet owners to pay for these services. Definitions of inappropriate behavior vary among pet owners, however. Is it cruel to hit a dog for the purposes of training? Is docking a young puppy's ears acceptable? Is keeping a nondomesticated animal in a cage as a pet cruel? Must cats really spend their lives indoors? What about declawing? Unfortunately, no one has yet done a study that fully examines pet owners' attitudes toward these and other questions in a systematic way, although debates on these matters are lively. Furthermore, front-line animal welfare activists, both within the animal welfare establishment and in the grassroots community of rescue, harbor profound disagreements about such matters as "no-kill" policies. Thanks to the relative ease of establishing nonprofit organizations, the growth of private rescues reflects a particularly American approach to solving the problem of unwanted pet animals. By the 1970s, private rescues, which began with breed-specific groups dedicated to saving purebreds turned in to shelters and exotics such as parrots, appeared. Private rescues vary considerably in their quality and policies. Many are excellent, but they are also unregulated, with policies toward adoption that sometimes reflect the particular concerns of their founders. Furthermore, the ongoing tension between pet ownership as a fundamentally private and consumer-driven practice, and legal and philosophical debates about the personhood of animals become especially apparent in relation to activities like cloning, which received fresh attention with the news that singer Barbra Streisand had cloned her favorite dog.

Because pet keeping is integrated into the daily lives of Americans, its successes and failures reflect the messiness of domestic life in both heartwarming and distressing ways. Millennials are now the biggest demographic cohort among pet owners (35%), and a recent survey reported that 65% felt that they "would find it more stressful to be separated for a week from their pet . . . than their cell phone" (Pet Product News 2018). Yet the private nature of pet keeping means that animals can be the unwitting victims of the tragedies of family life. Because perpetrators of family violence often hurt or kill family pets, and because victims are often unwilling to leave behind animals that they know will suffer, 32 states, the District of Columbia, and Puerto Rico now provide for pets in orders of protection. "Safe haven" programs with shelters keep pets safe while abuse victims are under protection or in transition (Animal Welfare Institute n.d.). There is also some evidence that pet owners themselves may want to understand their leisure pursuit better. Returning to pet videos, a study that asked watchers to voluntarily fill out a questionnaire about their cat-video viewing habits received more than 7,000 responses. The study's results, more suggestive than definitive, offered the interpretation that watching cat videos provides viewers with a "pick-me-up" in daily life (Myrick 2015 rev. 2018).

In sum, pet keeping is a complicated set of behaviors with a rich history, undergirded by an equally complex set of ideas held by pet owners, shaped and interpreted for public consumption by a wide array of popular media, supported by government institutions and nonprofit organizations, and sustained by billions of dollars' worth of goods and services each year. No matter how complex pet keeping is, it is still shaped fundamentally by the deep, unresolvable tension between the animal as an object, a consumer good, even an expendable, and the animal as a subject, a declared family member, a being with needs, intelligence, and a wholeness that will always remain slightly inaccessible to even the most dedicated pet owner.

Bibliography

American Pet Products Association (2018) *The 2017–2018 APPA Pet Owners Survey Debut: Trusted Data for Smart Business Decisions*, https://americanpetproducts.org/Uploads/MemServices/GPE2017_NPOS_Seminar.pdf.

American Society for the Prevention of Cruelty to Animals (n.d.) "Pet care costs," https://www.aspca.org/sites/default/files/pet_care_costs.pdf (Accessed June 1, 2018).

AnimalSheltering.org. "Community cats: Scientific studies and data," www.animalsheltering.org/page/community-cats-scientific-studies-and-data (Accessed July 1, 2018).

Centers for Disease Control and Prevention (n.d.) "Healthy pets, healthy people," www.cdc.gov/healthypets/index.html (Accessed June 30, 2018).

Davis, J. M. (2016) *The Gospel of Kindness: Animal Welfare and the Making of Modern America,* New York: Oxford University Press.

Favre, D. F. (2011) *Animal Law: Welfare Interests and Rights*, 2nd ed., New York, NY: Aspen Elective.

Fox, M. W. DVM. (1983) "Humane ethics and animal rights," *International Journal for the Study of Animal Problems* 4(4): 286–289.

Grier, K. C. (2005) *At Home with Animals: People and Pets in America*, Columbia, SC: McKissick Museum, University of South Carolina.

Grier, K. C. (2006) *Pets in America: A History,* Chapel Hill, NC: University of North Carolina Press.

Grier, K. C. (2008) "Provisioning man's best friend: The early years of the American pet food industry, 1870–1942," in W. Belasco and R. Horowitz (eds.) *Food Chains: From Farmyard to Shopping Cart* (pp. 126–141), Philadelphia, PA: University of Pennsylvania Press.

Grier, K. C. (2016) "'To the admirers of the feathered creation': The birth of the pet industry and the trade in songbirds in nineteenth-century America," in D. Ohrem (ed.) *American Beasts: Perspectives on Animals, Animality and U.S. Culture, 1776–1920* (pp. 51–74), Berlin: Neofelis Verlag.

Humane Society of the United States (2014) *2014/Pets for Life: An In-Depth Community Understanding,* Washington, D.C.: Humane Society of the United States.

Kim, K. (2016) *The Dog Merchants: Inside the Big Business of Breeders, Pet Stores, and Rescuers,* New York, NY: Pegasus Books.

Martin, Jr. E. C. (1997) *Dr. Johnson's Apple Orchard: The Story of American's First Pet Cemetery*, Hartsdale, NY: Hartsdale Canine Cemetery, Inc.

Myrick, J. G. (2015, rev. 2018) "Study shows the paw-sitive effects of watching cat videos," *The Conversation*, Indiana University, http://theconversation.com/study-shows-the-paw-sitive-effects-of-watching-cat-videos-43454 (Accessed June 30, 2018).

O'Neill, C. A. T. (ed.) (1985) *The American Kennel Club 1884–1984.* New York, NY: American Kennel Club.

Pet Product News (2018) "Survey examines how millennials shop for their pets," February 21, 2018, www.petproductnews.com/News/Survey-Examinies-How-Millennials-Shop-For-Their-Pets/ (Accessed August 26, 2018).

Poll, H. (2011) "Pets really are members of the family," https://theharrispoll.com/americans-have-always-had-interesting-relationships-with-their-pets-whether-that-pet-is-a-cat-dog-parakeet-or-something-else-the-pet-industry-is-thriving-and-for-good-reason-more-than-three-in-f/ (Accessed June 1, 2018).

Rowan, A. N. (2009) "Animal Sheltering Trends in the U.S.: A Historical Lesson from—and for—U.S. Shelters," www.humanesociety.org/animal_community/resources/timelines/animal_sheltering_trends.html.

Serpell, J. (1996) *In the Company of Animals: A Study of Animal-Human Relationships*, Cambridge: Cambridge University Press, rev. ed. Shelter Animals Count, www.shelteranimalscount.org/ (Accessed June 15, 2018).

Thomas, K. (1983) *Man and the Natural World: A History of the Modern Sensibility,* New York: Pantheon Books.

Warwick, C. et al. (2018) "Exotic pet suitability: Understanding some problems and using a labeling system to aid animal welfare, environment, and consumer protection," *Journal of Veterinary Behavior* 26, July–August: 17–26.

Wolfeldt, A. D. (2004) *When Your Pet Dies: A Guide to Mourning, Remembering and Healing*, Ft. Collins, CO: Companion Press.

23

THE ETHICS OF DOMESTICATION

Jessica du Toit

Approximately 12,000 years ago, at the end of the last global Ice Age, an estimated 10 million human beings inhabited the earth. From that point until about 500 years ago, the human population grew (relatively) slowly but steadily from 10 million to 1 billion. Today, just a few centuries later, there are at least 7.6 billion of us. While the recent exponential growth of the human population has had a devastating effect on the population sizes of most other animal species, it has resulted in the skyrocketing of domesticated animal[1] populations. Indeed, while the current population sizes of animals such as dogs, cats, sheep, chickens, and cows range from approximately 500 million (cats) to 50 billion (chickens), the wild ancestors of these animals are either extinct or dangerously close to extinction.

Domesticated animals have fared so well in this evolutionary sense because of the central role that they have come to play in human life. While human society in general uses domesticated animals (and their body parts) for a startling array of purposes,[2] most individual humans utilize these animals for purposes ranging from food to clothing to companionship.

Given our tremendous reliance on domesticated animals, we have needed to keep them close at hand. Thus, it is unsurprising that today's domesticated animals differ in a number of ways from the wild animals from whom they descended. Most notably, they are much more docile and trusting and much less aggressive. These characteristics confer a clear benefit on domesticated animals themselves: they ensure that living with, or in close proximity to, humans is less stressful than it would otherwise be. But these characteristics also render domesticated animals more manageable and malleable and, thus, better suited to serving humans' ends.

On these grounds, many have concluded that domestication was a process deliberately initiated by our ancestors in order to improve human life.[3] Given, then, that many domesticated animals live relatively short lives of utter hell and that even the best domesticated animal lives[4] include significant frustration, anxiety and boredom,[5] it is tempting to think that humans ought not to have domesticated animals and, furthermore, that we ought not to perpetuate the existence of domesticated animals. In what follows, I shall consider and evaluate both of these (possible) conclusions. In doing so, I shall focus primarily on (the domestication of) those animals we tend to keep as companions. The reason for this is that, generally speaking, these animals will lead the best possible domesticated animal lives. Thus, if it turns out that humans ought not to have domesticated, or ought not to perpetuate the existence of, *these* animals, then the implications for other domesticated animals—farmed animals, for example—are obvious.

Dogs and cats are paradigmatic examples of animals we tend to keep as companions. Thus, although much of my argument makes reference to these species in particular, most of what I shall say can be applied to other species of companion animals.

I

Although some domesticated animal species—dogs and cats, for example—are many thousands of years old, the vast majority of today's domesticated animal species are only a few centuries old. Interestingly, however, irrespective of when they were domesticated, and in spite of the evolutionary distance between their wild ancestors, all domesticated animal species have a number of traits in common. These traits—known collectively as the "domestication syndrome"—distinguish domesticated animal species from the wild animals from whom they descended. They include not only the increased docility and trust and reduced aggression that I mentioned earlier, but also reduced brain size, greater variation in coat colour and texture, and alterations in skull shape and tooth crowding.

The ubiquity of the domestication syndrome—together with the lack of an obvious connection between the relevant common traits—has long been thought to suggest a single domestication process. More specifically, it has long been thought to suggest that domestication is a process deliberately initiated and directed by humans. However, the latest archaeological and genetic research on the subject posits that there are in fact three major pathways to domestication, namely the "commensal pathway," the "prey pathway," and the "directed pathway."[6] While I shall say a bit more about each of these in due course, the important point at this stage is that only the directed pathway is a (clearly) human-initiated process. By contrast, the commensal pathway is an animal-initiated process: the relevant animal begins its journey into domestication by establishing a "commensal" relationship with humans, that is, a relationship in which the animal derives some benefit, but the relevant humans derive little, if any, benefit (or harm).[7] (With time, the relevant animals start to become socially or economically valuable to the relevant humans, who thus begin to derive some tangible benefit from the association.)[8]

The wild ancestors of today's dogs and cats are widely believed to have followed the commensal pathway into domestication. Indeed, while there is much disagreement regarding the time and place that dog domestication,[9] for example, occurred, there is widespread consensus that the process was initiated by wolves. More specifically, it is widely agreed that dog domestication was initiated by wolves who were both attracted to certain elements of the prehistoric human niche—especially to human food waste—and able to tolerate greater proximity to humans.[10] Of these wolves, those who were the least fearful and aggressive would have been the most successful scavengers. They would therefore have produced more offspring than their "wilder" packmates. Thus, a closer association with humans resulted in natural selection for (greater) tameness, which was the first step towards "dogness," or a differentiated genotype.[11]

Natural selection for tameness resulted in a host of other, seemingly unrelated, changes in the wolf-dog over time.[12] These changes included drooping ears, shorter, occasionally upturned tails, shortened snouts, and shifts in the developmental timing of various other characteristics. Thus, natural selection for tameness played a surprisingly large role in the emergence of a new phenotype. And, as wolf-dogs gradually became more numerous and "at home" around the human niche, there was a corresponding gradual relaxation of the selective pressures that operate on truly wild wolves.[13] That is to say, as wolf-dog populations became more established in the human environment, humans began to play an increasingly important—even if non-conscious—role in providing for and protecting these animals. As a result, various characteristics that would previously have been selected against would have started to find expression in some wolf-dogs. For example, greater variation in coat colour would have become possible, as spotted and other non-camouflage coat colours would not have been disadvantageous within the human niche.[14]

In fact, given the human tendency to place a premium on novelty, the emergence of non-camouflage coat colours very likely conferred a fitness advantage within the human niche.[15] That is to say, our ancestors very likely bestowed preferential treatment on those wolf-dogs with novel coat colours. Soon enough, however, our ancestors likely started to play an even more active role in ensuring the appearance of non-camouflage coat colours, by encouraging the breeding of certain wolf-dog individuals with one another.[16]

But, while our ancestors' interests in selective breeding might initially have been confined to achieving various aesthetic ends, they soon expanded to achieving various functional ends. Indeed, depending on the particular human environment in which wolf-dogs lived, some would have been selectively bred for their speed and endurance, thereby rendering them very effective hunters in rocky terrains, for example. By contrast, other wolf-dogs would have been selectively bred for their size and strength, thereby allowing them to successfully bring down large game. Still others would have been selectively bred for their superior herding skills. This, then, marked the beginning of the transition to proto-dog breeds.[17]

Although there was undoubtedly a long period of transition between the sort of selective breeding described earlier and the sort that gave rise to most modern dog breeds,[18] it is clear that humans have long been a conscious force in the evolution of the dog.[19] That said, however, it should also be clear that by the time humans started engaging in selective breeding, the domestication process was already well underway. Thus, even without any (deliberate) human intervention, wolves would have become at least partly domesticated. More specifically, although they would likely have been much more like today's squirrels and raccoons, for example, wolf-dogs would have become liminal members of our communities even if humans had never (deliberately) inserted themselves into the wolf-dog breeding process.

It follows from this, then, that, at least in the case of animals such as dogs and cats, it cannot sensibly be said that humans ought not to have domesticated their wild predecessors. But, while domestication is not necessarily a process initiated by deliberate human action, deliberate human action—specifically, the selective breeding of certain animals so as to achieve certain human ends—features in all domestication processes.

Now, this sort of selective breeding—as opposed to, for example, that which aims to serve the relevant animals' interests[20,21]—certainly raises some moral questions. For example, one might wonder whether it is permissible for humans to exert some control over the reproduction of dogs and cats for the purposes of achieving certain human ends, but where this does not harm the relevant animals or their offspring, all things considered. Indeed, assuming it were possible to selectively breed dogs to detect various human cancers in their earliest stages,[22] would it be wrong to engage in this kind of selective breeding?

Some might well think that it would be wrong. They might think that even this kind of selective breeding treats the relevant dogs as means to human ends, and that treating dogs in this way is morally impermissible. I have my doubts about this particular argument. But I also have no clear intuitions about the moral permissibility of the particular kind of selective breeding in question here. Nonetheless, the evolutionary trajectory of most, if not all, modern dog and cat breeds includes a kind of selective breeding that involves myriad wrongs: most modern dog and cat breeds are the result of a process of selective breeding that sought to achieve certain (trivial) human ends (e.g., a dog's having a very short snout or flat face) at the expense of fundamental animal interests (in well-being).[23,24] Thus, it can—and must—be said that humans ought not to have engaged in *this* kind of selective breeding. What is more, humans ought to desist from engaging in this sort of selective breeding in the future.

The clear implication of this is that humans ought to desist from breeding *some* companion animals, and even from allowing *some* such animals to reproduce with one another. Here I am thinking of so-called purebred or thoroughbred companion animals, and the relevant obligation applies even when the reproduction of these animals involves no confinement or forced breeding.[25] This is because "thoroughbred" companion animals have been, and continue to be, bred to conform to specific human preferences, despite the fact that they are, as a direct consequence, usually plagued by various diseases and disorders.[26]

For example, virtually all "purebred" dogs suffer from some or other genetic ailment, the most common being cancer. In fact, cancer occurs in these animals at frequencies that in humans would be considered an epidemic.[27] Large and giant breeds of dog are especially prone to cancer and, in particular, to bone cancer. This is largely a result of their accelerated growth rate: cancer is more likely to occur where growth, and thus cell division, is more rapid.[28] On the other end of the spectrum, small breeds of dog—especially those with short legs—are predisposed to dysplasias and patellar dislocations.[29]

The problem of the very high incidence of heritable diseases and disorders among "thoroughbred" companion animals is exacerbated in those breeds where the quest to uphold, or even amplify, breed standards lead to the overuse of popular sires.[30] This overuse results in greatly attenuated gene pools and, thus, a very high degree of inbreeding. The inevitable consequence of this, of course, is the accumulation of additional deleterious mutations.[31]

But the woes of "purebred" companion animals do not end with genetic diseases and disorders. Many of the physical features characteristic of particular breeds of companion animal carry a heavy cost in and of themselves. Consider, for example, that dog breeds with protruding eyes are prone to ulceration and irritation of the eye.[32] Dogs bred to have unnaturally short noses and flat faces are highly susceptible to varying degrees of upper airway obstruction.[33] And, since dogs rely primarily on panting to regulate their body temperature, dog breeds with "squashed" faces are especially vulnerable to overheating.[34] There are also many disorders associated with wrinkled skin or excessive skin folds. These include skin fold dermatitis, which is caused by friction between the skin surfaces. This, in turn, leads to ulcers, infections, and the discharge of pus and fluid.[35] One final example of many possible others is that the skulls of breeds such as the Cavalier King Charles Spaniel and the Chihuahua are often too small to accommodate the entire brain (which may also be abnormally large).[36] While the consequences of this condition vary, they often include the relevant animals' suffering immense pain and gradually becoming paralyzed.[37]

All of this explains why "purebred" companion animals have much shorter life spans (on average) than "non-thoroughbred" companion animals living in the same conditions.[38] This, taken together with the fact that the diseases and disorders mentioned above usually have a profoundly negative impact on the quality of life of those who suffer from them, provides an excellent reason for thinking that we should desist from breeding, or allowing the reproduction of, "thoroughbred" companion animals.[39]

But what are our obligations when it comes to the reproduction of "*non-thoroughbred*" companion animals? Let us assume, once again, that where humans have some measure of control over the reproductive activities of these animals, there is no confinement or forced breeding. Unlike "purebred" companion animals, "non-thoroughbreds" are not bred to conform to specific human ends, and as a result, they do not stand a significantly elevated chance of suffering immensely[40] and dying prematurely. Thus, our breeding "non-thoroughbreds" and allowing them to reproduce with one another seems, at least on the face of it, to be innocuous. But is it, in fact, permissible for us to bring or allow more of these animals to be brought into existence? It is to this question that I shall turn in the next part of this chapter.

II

There are a number of arguments that can be made to the effect that while humans ought to continue to care for those companion animals who already exist, they ought not to breed *even* "non-thoroughbred" companion animals.[41] The most controversial of these arguments—which I shall consider only very briefly—are the anti-natalist arguments. According to this set of arguments, coming into existence is either so risky or so harmful that we ought not to bring any sentient being, whether human or animal, into existence.[42] One straightforward implication of this is that breeding even "non-thoroughbred" companion animals is wrong.

Of course, all things being equal, most "non-thoroughbred" companion animals suffer much less harm than do most "purebred" companion animals who, in turn, suffer much less harm than do the vast majority of domesticated animals reared for human consumption, for example. The harms that "non-thoroughbred" companion animals suffer are nonetheless more substantial than most humans typically recognize. To see this, consider that even the best companion animal lives include pain, suffering, frustration, anxiety, boredom, and death. And while it might strike some as melodramatic to worry about the frustration, anxiety, and boredom experienced by dogs who are well fed and well loved, for example, this is because most humans fail to appreciate just how frequently even lucky dogs find themselves in these emotional states.

Although lucky dogs are seldom, if ever, caged or chained,[43] their movement is nonetheless constrained by closed doors, house walls, garden fences, and leashes. These restrictions hinder—and often entirely thwart—their pursuit of various scents and other (sensory) pleasures. These restrictions must therefore be immensely frustrating in and of themselves. But they also often lead to separation anxiety, especially when the relevant dogs are left at home alone. And, if they are left alone for long periods, boredom can also result. Humans also restrict lucky dogs' freedom by controlling what, when, and how much they eat.

To be sure, these restrictions on the freedom of lucky dogs are (almost always) justified by the welfare of these animals: many lucky dogs would stand a very high chance of being seriously injured or killed if they were allowed to roam freely. The point, however, is that animals are typically unable to understand this. Therefore, (immense) frustration, anxiety, and boredom are part of the *daily* experience of even lucky dogs.

In spite of all of this, some might think that at least some lucky dogs (can) lead good lives, and thus that it cannot always be wrong to bring "non-thoroughbred" companion animals into existence. But even if the anti-natalist granted, for the sake of argument, that some sentient beings (can) lead good lives, he would argue that it remains true that every sentient being brought into existence runs the risk of suffering *significant* harm at some stage in its life. This is important as it highlights one very serious problem with all breeding, including human procreation: Who are we to decide that the risks associated with bringing another sentient being into existence are acceptable, when we will not be the (primary) victims of whatever harms that new sentient being might have to endure?[44]

While I have not presented a comprehensive argument for the anti-natalist conclusion, I hope that I have said enough for the reader to get a sense of why the anti-natalist thinks that we ought not to breed existing companion animals, including "non-thoroughbred" companion animals. But the anti-natalist also thinks that we ought not to allow existing ("non-thoroughbred") companion animals to reproduce with one another. This purported obligation requires some additional justification as, in the absence of long-term animal contraception, preventing these animals from breeding with one another requires that we either restrict these animals' access to mates or sterilize them. Both these options involve the infliction of some harm on existing "non-thoroughbred" companion animals.

On the one hand, thwarting these animals' drive to engage in sexual intercourse—and thus also denying them the pleasure than can be derived from such activity[45]—would certainly cause them (additional) frustration.[46] This would, in turn, have a negative effect on the quality of their lives. And while sterilizing these animals would avoid that problem, sterilization requires surgery, which usually requires general anaesthesia. Thus, even if the procedure goes as well as it could possibly go, the newly sterilized animal will have to endure the adverse effects of general anaesthesia as well as at least some post-operative pain and discomfort. This, too, will have a negative effect on the relevant animal's quality of life, even if for only a relatively short time.

The anti-natalist will not dispute the fact that sterilization necessarily entails the infliction of some non-trivial harms on the animal sterilized. He will simply point out that in failing to sterilize existing companion animals, we will be allowing the creation of a great many more such beings, (at least) the vast majority of whom will suffer much greater harms. But while some might be satisfied with the anti-natalist's response, others may have concerns. They might worry, for example, that if we take animal rights seriously, then the imposition of non-trivial harms on the animal sterilized simply cannot be justified by the fact that the imposition will result in a great many others being spared tremendous suffering.[47]

To see why, let us briefly consider the best-known theory of animal rights, that proposed by Tom Regan in the early 1980s.[48] According to Professor Regan's account, since all sentient animals are beings with beliefs, desires, a sense of themselves over time, and interests in their own fate, all sentient animals are, what he calls, "subjects-of-a-life." This is important as, according to Professor Regan, all subjects-of-a-life possess inherent value and, as a result, the right to respectful treatment. In essence, this is the right to be treated as a self-governing agent and, thus, not to be used merely as a means to secure the best overall consequences. And, if all sentient animals possess this right, then it seems clear that harming some companion animal—by sterilizing it—cannot be justified by the fact that doing so will spare many others much greater suffering, and so will produce the best consequences overall.

There are at least two ways to respond to this kind of argument. The first is to challenge Professor Regan's theory of animal rights. Alasdair Cochrane provides a very compelling critique of this theory in his *Animal Rights Without Liberation: Applied Ethics and Human Obligations*.[49] The gist of his critique is as follows: it is possible for a being to have beliefs, desires, a sense of itself over time, interests in its fate, and so on, without also having the capacities of moral and autonomous agency. The vast majority of sentient animals are a case in point: they do not have the capacities to reason, reflect, and act on self-chosen moral principles or life goals. Thus, while all sentient animals are subjects-of-a-life who possess inherent value, most sentient animals are *not* self-governing agents. But if this is true, then there is no reason to think that these beings must (always) be treated as self-governing agents.[50]

A second response takes the form of a thought experiment. Imagine the case of a human adult who is sufficiently cognitively impaired that she would be unable to care adequately for any offspring that she might produce, but not so cognitively impaired that she would be unable to consent to sexual intercourse. In this case, it would seem that we are justified in sterilizing the relevant human adult, even though our doing so would impose harms on her. This is partly because bringing new beings into existence is wrong if you cannot ensure that they will be well cared for, partly because the harms imposed on the relevant human adult are minor in comparison to the harms that we thereby prevent, and partly because sterilizing her may be the only way in which we can effectively prevent her from imposing serious harm on others, albeit unwittingly.

What is more, to think that we are justified in sterilizing the relevant human in this case is not in tension with thinking that she has the moral right to be treated as a self-governing agent. Rather, it is in tension with thinking that she has an absolute right to reproductive freedom. And this is by no means a strike against the notion that sterilization is justified in the relevant case.[51]

Bearing in mind the two responses outlined earlier, it does seem that the anti-natalist is justified in his assertion that we ought to sterilize existing companion animals. To concede this does not commit one to accepting the anti-natalist's main thesis, however. And for those who remain unconvinced by anti-natalism, there is a second set of arguments against reproducing even "non-thoroughbred" companion animals that makes no assumptions about the inherent risks and/or harms of coming into existence. Indeed, although these "abolitionist" arguments also call for the sterilization of all existing companion animals, this is on the grounds that there is something morally problematic about the category "companion animal." Thus, the institution of keeping companion animals must be brought to an end, and systematic sterilization is the only ethically defensible means of achieving this.

In what follows, I shall consider two abolitionist arguments, namely what I shall call the "property argument" and the "dependency argument."

The Property Argument

As far as the law is concerned, companion animals are the personal property of the humans who keep them. Another way of putting this is that humans *own* the companion animals they keep. However, humans, as well as our legal systems, distinguish companion animals from *in*animate items of property in that they recognize that humans ought not to treat companion animals in certain ways and that they owe at least some duties *directly* to these animals.[52]

While the *status quo* may be acceptable to some, others find it objectionable. Gary Francione falls into the latter group. He acknowledges that existing legal systems offer *some* protection for companion animals' interests but insists that they are unable to offer *sufficient* protection for the interests of these animals.[53] Thus, he thinks that while humans ought to continue to care for those companion animals who *already* exist, they ought not to breed these animals or even allow them to reproduce with one another.[54] I shall refer to this view as the "property argument."

While one might agree with Professor Francione that the legal status of companion animals raises some serious moral questions, one might question whether one is therefore committed to bringing about the (eventual) extinction of these animals. One might, for example, contend that having the legal status of property need not harm companion animals. To see why, consider that those humans who reject the law's view of animals and regard their companion animals as sentient beings whose interests must be safeguarded and promoted to the greatest extent possible would not harm the animals in their care. Thus, perhaps these humans, and these humans alone, are morally permitted to breed "non-thoroughbred" companion animals and allow them to breed with one another, on the condition that they ensure that the resultant offspring will be kept by humans with the same attitude.

But, while companion animals need not suffer any harm as a result of their being owned, it might be thought that they are nonetheless necessarily *wronged*. To see why, consider that acknowledging that companion animals possess rights might be thought to commit one to thinking that we cannot class these beings as the property of humans, irrespective of whether they suffer any harm as a result of their being categorized in this way. This is because to place companion animals in the category of property is to impugn the moral status or inherent value of these animals. And this, of course, is to wrong companion animals.

However, on an interest-based account of moral rights such as the one proposed by Alasdair Cochrane,[55] recognizing that companion animals possess interest-based rights does not by itself tell us *which* rights these animals possess. Thus, we need not be committed to thinking that these animals have the right not to be owned. Whether we are committed in this way depends on the precise set of interests that the relevant animals possess. And, according to Professor Cochrane, most sentient animals do not have an intrinsic interest in not being owned as they are not self-governing agents: they do not have the capacity to frame, revise, and pursue their own conceptions of the good life.[56] This being the case, they are not wronged when they are prevented from leading their own freely chosen lives—as is the case when they are owned—provided that they suffer no harm as a result.[57]

While Professor Cochrane's argument is a powerful one, it does have some potentially very unpalatable implications. For example, consider that with regard to being owned, there is no morally relevant difference between very young children or severely cognitively impaired humans and (most of) those animals who tend nowadays to be kept as companions. Young children, severely cognitively impaired humans, and companion animals are all sentient, and none is a self-governing agent. Thus, if we think that Professor Cochrane has provided good grounds for thinking that companion animals are not necessarily wronged by being categorized as property, then we are committed to thinking that it is sometimes permissible to own at least some other humans. However, most of us abhor this idea: most of us think that even if they suffer no harm as a result of their being categorized as property, "property" is simply not an appropriate category in which to place *any* human being. What is more, my guess is that our feelings about this would remain equally strong even if the ownership in question was limited and highly regulated such that interests of the relevant human beings were always protected. Thus, most of us seem to be committed to thinking that it is wrong to place a companion animal in the category of property, even if that animal suffers no harm as a result.

In response to the foregoing argument, it might be said that it is necessary to distinguish between a companion animal's (a) being wronged by the law and (b) being wronged by the humans in whose home it is kept. With regard to (b), it is necessary to distinguish furthermore between a human's (1) accepting or endorsing the law's categorization of companion animals as property and (2) rejecting the law's categorization of companion animals as property.

Given that the law categorizes companion animals as property, these animals are clearly wronged by the law. But it is my view that companion animals are wronged by the humans in whose homes they are kept *only if* those humans accept or endorse the law's categorization of companion animals as property. Those humans who reject the law's categorization might still own their companion animals, but they do so merely in a technical, legal sense. Therefore, *they* do not wrong their companion animals by "owning" them.

Let us suppose that I am right about this. That is, let us suppose that those humans who reject the law's categorization of companion animals as property, do not themselves wrong their companion animals. These humans are best thought of as the "keepers," "custodians," or "caretakers" of their companion animals. To be the keeper, custodian, or caretaker of a companion animal is much like being the (legal) guardian of a child, for example. In the same way that the guardian of a child is charged with caring for and safeguarding the interests of that child, the keeper, custodian, or caretaker of a companion animal is charged with caring for and safeguarding the interests of his companion animal. The question, then, is whether our continuing to bring even "non-thoroughbred" companion animals into existence would be morally permissible if we merely kept and did not own them (or owned them only in a technical legal sense).[58] This question brings us to the second abolitionist argument.

The Dependency Argument

According to Gary Francione,[59] even if humans merely kept companion animals—and did not own them, even if only in a technical legal sense—we would nonetheless be morally obligated to ensure that no more of these animals came into existence. This is because he thinks that there is something inherently wrong with keeping companion animals and not merely with owning them.

In Professor Francione's view, the inherent wrongfulness of keeping these animals has to do with the fact that they are domesticated animals, and domesticated animals are—as I have mentioned before—much more docile and trusting than the wild animals from whom they descended. While this renders them better suited to living with, or in close proximity to, humans, it also renders them less effective predators and too trusting to be successful in evading (other) predators.[60] The upshot of this is that domesticated animals are almost entirely dependent on (the goodwill of) humans for the satisfaction of their fundamental needs and desires.

To be so dependent on another is, of course, to be in a position of extreme vulnerability. And, to find oneself in this position is to run a very high risk of leading a short and miserable life. For this reason, Professor Francione deems it morally indefensible to place or allow another creature to be placed in such a position. Given, then, that our continuing to bring or allow domesticated animals to be brought into existence is to place or allow these animals to be placed in such a position, he thinks that we ought not to do these things. Thus, he thinks that we ought to desist from reproducing even "non-thoroughbred" companion animals and, therefore, ultimately, to bring an end to the practice of keeping companion animals.

In responding to the dependency argument, I shall restrict my attention and arguments to kind and caring companion animal keepers alone. Now, although it is *strictly speaking* true that even these keepers' animals are at risk of leading short and miserable lives, they are precisely the kind of keepers who can be relied upon to satisfy their companion animals' needs and desires (to the greatest extent possible). Thus, if the dependency objection is to make any sense in these cases, it must be because *even if* the relevant animals' needs and desires are met and their interests safeguarded, it is morally unconscionable that companion animals are so dependent on (the goodwill of) their respective keepers.

Before I go any further, I should clarify that dependency itself is not necessarily a bad thing. Both humans and animals depend on other humans or animals for a variety of things throughout their respective lifetimes. For example, while many of us might rely on other humans to grow food for us or to look after us when we become ill, those other humans might rely on us to teach their children or to invest their savings. This is not to say that we would not prefer to be (more) independent of one another. But, in general, we do not think it deeply troubling that we depend on others for a variety of things.

With this in mind, it might be suggested that it is only in a very small minority of cases—such as that of a blind human and his guide dog—that humans and companion animals can really be said to rely on one another. In the overwhelming majority of cases, so the argument runs, while the relevant humans depend on their companion animals for very little, the relevant companion animals are almost entirely dependent on their respective humans for the satisfaction of even their most basic needs and desires. And this kind of dependency relationship is allegedly much more perturbing, both because it is extremely asymmetrical and because it involves fundamental needs and desires.

I am not entirely convinced by the foregoing argument. While it is true that humans are seldom, if ever, dependent on their companion animals for their continued existence, many humans derive something of immense value from their relationships with their companion animals. That is, the quality of these humans' lives would be significantly adversely affected if these relationships were to come to an end. Thus, it is at least arguable that humans are often emotionally dependent on their companion animals and, thus, furthermore, that their dependency involves something of import. This being the case, it is at least arguable that humans and their companion animals are often *inter*dependent in an important sense.

Some[61] think that once we recognize this interdependence—as well as all the other, innumerable interdependencies that pervade our existence—we will be (more) motivated to rearrange society in such a way that dependency and its associated vulnerability are minimally oppressive and disabling. This will enable us to (start to) bring an end to our fear and loathing of dependency and vulnerability. According to these thinkers, then, there is nothing *inherently* objectionable about dependency, even that involving fundamental needs and desires. Thus, according to these thinkers, if companion animals' needs and desires are satisfied and their interests safeguarded, it is not morally unconscionable that they are so dependent on (the goodwill of) their respective keepers.

While I cannot provide a thoroughgoing argument against this position, my strong sense is that the extreme dependency of companion animals is morally problematic, *even if* the relevant human carers can be relied on to satisfy these animals' needs and desires to the greatest extent possible. Indeed, my strong sense is that even if some kind of dependency is inevitable, being dependent is always regrettable, and more so the more basic the need(s) and desire(s) the relevant dependency involves. It is always preferable not to have to rely on another, even if that other can be trusted.

In an attempt to "flesh out" and understand why this might be so, let us imagine a hypothetical situation in which someone began to breed genetically altered chimpanzees who are ill suited to living in the wild but well suited to living with, or in close proximity to, humans. While these domesticated chimpanzees would have mental capacities equivalent to those of wild chimpanzees, they would be much less aggressive and more docile, for example, than their wild relatives. Since they would therefore be unable to live independently of humans in the human world, the domesticated chimpanzees would be perpetually dependent on humans for the satisfaction of their fundamental needs and desires. Is there something morally problematic about creating a breed of such chimpanzees?[62]

My prediction is that people's intuitions about this matter will differ. That is to say, I suspect that while some will see nothing wrong with creating a breed of domesticated chimpanzees so as long as humans satisfy their needs and desires, others will feel that it is morally unacceptable to choose to bring into existence a creature who will, perpetually, be so dependent on another, even if the relevant creature will be well cared for. There will also likely be others who are sympathetic to both these intuitions.

What can be said in the face of these conflicting intuitions and the absence of any clear way of choosing between them? Perhaps it is the case that proponents of the dependency argument must bear the burden of proof? They are, after all, placing a demand on others to change their current practices. Since these proponents have not been able to provide a compelling case, however, it might be thought that we are entitled to dismiss the dependency argument.

But it is just not clear that proponents of the dependency argument should, in fact, bear the burden of proof. This being the case, the clash of intuitions that we are faced with implies that the soundness of the dependency argument simply cannot be evaluated; at least for now, it is unclear whether, on the basis of the dependency argument, it is wrong to continue to breed "non-thoroughbred" companion animals and to allow them to reproduce with one another.

Fortunately, even if it is not clear that we have a duty grounded in anti-natalist or abolitionist arguments to desist from breeding "non-thoroughbred" companion animals and from allowing them to breed with one another, there are excellent practical reasons for thinking that these practices are morally impermissible. One such reason is rooted in the fact that there are far too many companion animals who are brought into existence but never homed, or homed for a short while before being abandoned.

The Humane Society of the United States (HSUS), for example, estimates that there were approximately 6 to 8 million dogs and cats who entered animal shelters between 2012 and 2013.[63] The United Kingdom's Royal Society for the Prevention of Cruelty to Animals (RSPCA) estimates that during 2017, 114,584 animals entered its shelters.[64] Australia's RSPCA estimates that between during the 2016–2017 financial year, more than 100,000 cats and dogs were taken into their various facilities.[65] And in Canada, shelters took in an estimated 120,000 cats and dogs during 2016.[66] Only some of these animals are rehomed. A great many are eventually killed.[67]

Thus, it should be clear that by continuing to breed companion animals and allowing them to reproduce with one another, we are adding to the large number of such animals who are condemned to lead short and miserable lives. Sometimes this effect will be direct because the particular animals who come into existence end up being abandoned or killed. On most occasions, however, the effect will be indirect: although the particular companion animals who come into existence are themselves homed, their existence reduces the "demand" for other, existing animals who have been abandoned or will be killed.

This is not mere supposition on my part. While the number of cats and dogs adopted from shelters or rescue groups seems to be increasing each year, according to the latest figures, only about 40% of all cats and dogs in American households come from these sorts of organizations.[68] The rest come from pet stores, puppy mills,[69] and small-scale breeders, for example.

Bearing all of the aforementioned figures in mind, there is a very powerful argument that existing companion animals should be adopted before more are bred. The practical effect of this, as things currently stand, is that no one should breed companion animals and everyone should sterilize animals in his or her care in order to avoid the birth of more animals who will themselves be discarded or killed, or who will doom other animals in need of adoption to this fate.

Of course, if the companion animal population eventually decreased such that the demand for these animals exceeded the number of them *already* in existence, then there would no longer be *this* reason for thinking that we ought to desist from breeding such animals and to prevent them from reproducing with one another. Whether breeding even "non-thoroughbred" companion animals would then be permissible would depend on the cogency of the anti-natalist and abolitionist arguments.

III—Conclusion

Contrary to what many have assumed (and to what some continue to assume), domestication is not necessarily a process deliberately initiated and directed by humans. Although it is true that all recent domesticates are the result of such a process, the wild forebears of dogs and cats, for example, are widely thought to have been the instigators of their respective journeys into domestication. Thus, although humans *became* an increasingly conscious force in the evolutionary processes of these animals, even without any (deliberate) human intervention, the wild ancestors of today's dogs and cats would have become at least partly domesticated. The implication of this is that, at least in the case of animals such as cats and dogs, it cannot be said that humans ought not to have domesticated their wild predecessors. But it can—and must—be said that humans ought not to have engaged in the kind of selective breeding that formed part of the evolutionary trajectory of these animals. Furthermore, humans ought not to engage in this kind of selective breeding in the future.

The clear implication of this, as I have argued, is that humans ought not to breed "purebred" companion animals or even allow these animals to reproduce with one another. I have also argued that we are, *at least at present*, morally obligated to sterilize all "non-thoroughbred" companion animals in our care: until such time as the demand for companion animals exceeds the number of these animals already in existence, we should not allow the birth of any more such animals, even though we should continue to care for those companion animals who already exist. However, given just how many millions of unwanted animals there currently are, we are nowhere near that point. Thus, as things stand, we cannot avoid the conclusion that reproducing (even "non-thoroughbred") companion animals is wrong.[70]

Notes

1. Although many people (seem to) have great difficulty accepting this simple truth, humans—like lions, warthogs, and brown spider monkeys—are a species of animal. Thus, rather than using the terms *human(s)* and *animal(s)*, I ought really use the terms *human animal(s)* and *non-human animal(s)*. However, to do this would likely render my writing unduly clumsy. I shall therefore continue to use *human(s)* as shorthand for *human animal(s)* and *animal(s)* as shorthand for *non-human animal(s)*.
2. In this regard, see Christien Meindertsma's TED Talk titled "How pig parts make the world turn" (July 2010), www.ted.com/talks/christien_meindertsma_on_pig_05049 (Accessed April 29, 2018).
3. One needs only to consider various classic definitions of the concept of domestication to appreciate this point. See D. Morey's *Dogs: Domestication and the Development of a Social Bond* (New York: Cambridge University Press, 2010), and particularly the chapter called "Domestication of Dogs and Other Organisms," for some examples of classic definitions.
4. Here I am thinking of companion animal lives.
5. I shall say more about this frustration, anxiety, and boredom in due course. For now, however, it should be noted that (much of) the significant frustration, anxiety, and boredom experienced by even the luckiest domesticated animals is a direct consequence of their close association with humans.
6. M. Zeder, "The Domestication of Animals," in *Journal of Anthropological Research*, Vol. 68(2), 2012, pp. 161–190; G. Larson & D.Q. Fuller, "The Evolution of Animal Domestication," in *Annual Review of Ecology, Evolution, and Systematics*, Vol. 45, 2014, pp. 115–136.

7. Ibid.
8. M. Zeder, "The Domestication of Animals," in *Journal of Anthropological Research*, Vol. 68(2), 2012, p. 171.
9. According to Larson et al ("Rethinking Dog Domestication by Integrating Genetics, Archaeology, and Biogeography," in *Proceedings of the National Academy of Sciences USA*, Vol. 109(23), 2012, pp. 8878–8883), dogs may well have multiple origins. That is to say, dog domestication may well have occurred more than once.
10. M. Zeder, "The Domestication of Animals," in *Journal of Anthropological Research*, Vol. 68(2), 2012, pp. 171–172.
11. R. Francis, *Domesticated: Evolution in a Man-Made World* (New York: W.W. Norton & Company, Inc, 2015), p. 28.
12. Dmitry Belyaev was the first to demonstrate that the traits that distinguish domesticated animals from their wild forebears—known collectively as the domestication syndrome—were not individually selected for but, rather, emerged as a result of natural selection for a single behavioural trait, namely tameness.
13. M. Zeder, "The Domestication of Animals," in *Journal of Anthropological Research*, Vol. 68(2), 2012, p. 164; G. Larson & D.Q. Fuller, "The Evolution of Animal Domestication," in *Annual Review of Ecology, Evolution, and Systematics*, Vol. 45, 2014, pp. 120–130.
14. M. Zeder, "The Domestication of Animals," in *Journal of Anthropological Research*, Vol. 68(2), 2012, p. 164; G. Larson & D.Q. Fuller, "The Evolution of Animal Domestication," in *Annual Review of Ecology, Evolution, and Systematics*, Vol. 45, 2014, pp. 128–130.
15. G. Larson & D.Q. Fuller, "The Evolution of Animal Domestication," in *Annual Review of Ecology, Evolution, and Systematics*, Vol. 45, 2014, pp. 129–130.
16. Ibid.
17. Francis, *Domesticated: Evolution in a Man-Made World*, 2015, p. 38.
18. More specifically, the sort characterized by the coercive confinement and forced breeding of animals, without any concern for the animals' own interests.
19. Francis, *Domesticated: Evolution in a Man-Made World*, 2015, p. 2.
20. Indeed, one might, for example, breed animals in such a way that their offspring are spared some congenital defect.
21. In due course, I shall consider a position known as anti-natalism, according to which coming into existence is either so risky or so harmful that we ought not to bring any sentient being, whether human or animal, into existence. If the anti-natalist is correct, then there can be no variety of selective breeding that in fact serves an animal's interests, all things considered.
22. This would be on the basis of breath, tissue, or urine odour, for example. This is because the waste products released by cancerous cells are different from those released by healthy cells in the human body.
23. I shall say more about this in due course.
24. It should be noted at this stage that I am obviously assuming that the interests of animals are morally significant and indeed sufficiently significant that only quite weighty human interests can outweigh them. While there remains *some* debate about this in the literature, I am not going to argue for this assumption here.
25. Although the reproduction of "thoroughbred" companion animals is often under human control, not all forms of controlled reproduction involve harms to the relevant animals. Humans might, for example, simply facilitate the meeting of specific animals, allowing them to mate if and when they chose.
26. L. Asher et al., "Inherited Defects in Pedigree Dogs. Part 1: Disorders Related to Breed Standards," in *The Veterinary Journal*, Vol. 182, 2009, p. 402.
27. Francis, *Domesticated: Evolution in a Man-Made World*, 2015, p. 49.
28. L. Asher et al., "Inherited Defects in Pedigree Dogs. Part 1: Disorders Related to Breed Standards," in *The Veterinary Journal*, Vol. 182, 2009, p. 406; Advocates for Animals, "The Price of a Pedigree: Dog Breed Standards and Breed-Related Illness," p. 14; R. Francis, *Domesticated: Evolution in a Man-Made World*, 2015, p. 50.
29. L. Asher et al., "Inherited Defects in Pedigree Dogs," p. 406; Advocates for Animals, "The Price of a Pedigree: Dog Breed Standards and Breed-Related Illness," p. 13.
30. E.A. Ostrander, "Both Ends of the Leash—The Human Links to Good Dogs with Bad Genes," in *The New England Journal of Medicine,* Vol. 367, 2012, pp. 637–638.
31. J. F. Summers et al., "Inherited Defects in Pedigree Dogs. Part 2: Disorders That Are Not Related to Breed Standards," in *The Veterinary Journal*, Vol. 183, 2010, p. 40; W. Pacelle, *The Bond: Our Kinship with Animals, Our Call to Defend Them* (New York: William Morrow, 2011), p. 218; Francis, *Domesticated: Evolution in a Man-Made World*, 2015, p. 49.
32. L. Asher et al., "Inherited Defects in Pedigree Dogs," p. 406; Advocates for Animals, "The Price of a Pedigree: Dog Breed Standards and Breed-Related Illness," 2006, p. 9, www.onekind.org/uploads/publications/price-of-a-pedigree.pdf.
33. Advocates for Animals, "The Price of a Pedigree: Dog Breed Standards and Breed-Related Illness," p. 10.
34. R. Francis, *Domesticated: Evolution in a Man-Made World*, 2015, p. 50.

35. L. Asher et al., "Inherited Defects in Pedigree Dogs," p. 407; Advocates for Animals, "The Price of a Pedigree: Dog Breed Standards and Breed-Related Illness," p. 12.
36. CavalierHEALTH.org, "Syringomyelia (SM) and the Cavalier King Charles Spaniel," https://cavalierhealth.org/syringomyelia.htm.
37. Ibid.
38. Francis, *Domesticated: Evolution in a Man-Made World*, 2015, p. 49.
39. Even those who, because of the non-identity problem (see Derek Parfit, *Reasons and Persons* [Oxford: Clarendon Press, 1984], Part IV), would deny that the animals brought into existence are harmed, could still recognize that worse (and often unacceptable) outcomes result from breeding "thoroughbreds" than from "non-thoroughbreds."
40. While it is true that "non-thoroughbred" companion animals are likely to experience no more suffering than the average sentient being, David Benatar has argued (see, e.g., *Better Never to Have Been: The Harm of Coming into Existence* [New York: Oxford University Press, 2006]) that the harms of existence are (nonetheless) considerable. I shall say a bit more about Professor Benatar's anti-natalist argument(s) in due course.
41. Many of the ideas presented in this section emanate from my and David Benatar's paper, "Reproducing Companion Animals," published in *Pets and People: The Ethics of Our Relationships with Companion Animals* (New York: Oxford University Press, 2017).
42. There are a number of ways of reaching the anti-natalist conclusion. While I am not going to elaborate on any of these (in very much depth) in this chapter, David Benatar provides a full account of them, with a focus on human reproduction, in his *Better Never to Have Been: The Harm of Coming into Existence* (New York: Oxford University Press, 2006).
43. Crate-training may form part of a lucky dog's potty-training program, for example.
44. The implications of the anti-natalist conclusion for human procreation are so drastic and so unpalatable that many assume that there must therefore be a flaw in one of the premises.
45. Most pets must derive (some) pleasure from sexual intercourse. It would otherwise be very difficult to explain why so many of them engage in it when they have the opportunity to do so.
46. Being prevented from having sex (when one has the relevant desire) is frustrating in and of itself. But preventing an animal from having sex may require some sort of confinement of the relevant animal, which would likely be a source of additional frustration.
47. Although he is ultimately unconvinced that the sterilization of individual animals cannot be justified by appealing to the easily preventable suffering of a great many others, David Boonin considers this challenge seriously in his "Robbing PETA to Spay Paul: Do Animal Rights Include Reproductive Rights?" in *Between the Species*, Issue III, August 2003, https://digitalcommons.calpoly.edu/cgi/viewcontent.cgi?referer=www.google.co.za/&httpsredir=1&article=1050&context=bts.
48. See his *The Case for Animal Rights*.
49. New York: Columbia University Press, 2012.
50. It is worth clarifying that while Professor Cochrane is critical of Professor Regan's account of animal rights, and while he does not believe that animals have the right to be treated as self-governing agents, he nonetheless thinks that all sentient animals do have rights. According to Professor Cochrane, however, animal rights are grounded in animal interests, and because most animals are not capable of moral and autonomous agency, most animals do not have interests in such agency.
51. The preceding thought experiment and argument were proposed by David Benatar during a discussion about this section of my chapter.
52. Among other things, various animal welfare laws prohibit the unnecessary killing, torturing, maiming, starving, and underfeeding of companion animals, and require that these animals receive adequate veterinary or other medical attention whenever necessary.
53. See, for example, his *Animals, Property and the Law* (Philadelphia: Temple University Press, 1995); and *Animals as Persons: Essays on the Abolition of Animal Exploitation* (New York: Columbia University Press, 2009).
54. Francione, *Animals as Persons*, 1995, p. 13; "'Pets': The Inherent Problems of Domestication" in *Animal Rights: The Abolitionist Approach*, 31 July 2012, www.abolitionistapproach.com/pets-the-inherent-problems-of-domestication/ (Accessed October 21, 2012).
55. In *Animal Rights Without Liberation: Applied Ethics and Human Obligations* (New York: Columbia University Press, 2012).
56. Cochrane, *Animal Rights Without Liberation*, 2012, p. 11.
57. Confining companion animals in cages almost always harms them. However, since this is not an essential part of owning companion animals, Professor Cochrane argues that it is the confinement as opposed to the ownership of companion animals that is morally problematic. Given then that companion animals have only

an instrumental—and not an intrinsic—interest in liberty, they possess no fundamental right to be free and, thus, no fundamental right not to be owned.

58. This question remains even if it is true that those humans who reject the law's categorization of companion animals as property, nonetheless wrong their companion animals. To see this, consider that if companion animals are wronged by being owned in a technical legal sense, then we should either (a) ensure that no more companion animals are brought into existence or (b) ensure that companion animals no longer have the legal status of property. If we managed to change the legal status of companion animals, then those humans who have companion animals would no longer be the owners of those animals. Rather, they would be the "keepers," "custodians," or "caretakers" of those companion animals. Thus, we must ask whether our continuing to bring "non-thoroughbred" companion animals into existence would be morally permissible if we merely kept and did not own them.
59. "'Pets': The Inherent Problems of Domestication" in *Animal Rights: The Abolitionist Approach*, 31 July 2012, www.abolitionistapproach.com/pets-the-inherent-problems-of-domestication/ (Accessed October 21, 2012).
60. There are packs of feral dogs, for example, but their lives tend to be "nasty, brutish and short."
61. Here I am thinking of people like Sunaura Taylor (see, e.g., "Beasts of Burden: Disability Studies and Animal Rights," in *Qui Parle*, Spring/Summer 2011, Vol. 19(2), pp. 191–222). Eva Feder Kittay takes a similar position, although she focuses on human dependency alone (see, e.g., "The Ethics of Care, Dependence and Disability," in *Ratio Juris*, Vol. 24(1), 2011, pp. 49–58).
62. It might be suggested that breeding permanently dependent humans is a better analogy and that our intuitions are clearly against such a practice. However, one confounding variable with this case is that the cognitive capacities of the bred humans would be stunted relative to the species norm, and this might explain our opposition to it.
63. W. Pacelle, *The Bond: Our Kinship with Animals, Our Call to Defend Them* (New York: William Morrow [an imprint of Harper Collins Publishers], 2011), p. 197. Also, see the Humane Society of the United States' fact sheet titled "Pets by Numbers," 30 January 2014, www.humanesociety.org/issues/pet_overpopulation/facts/pet_ownership_statistics.html (Accessed September 2, 2018).
64. See the UK RSPCA's fact sheet titled "Facts and Figures," https://media.rspca.org.uk/media/facts (Accessed September 2, 2018).
65. See RSPCA Australia's "National Statistics 2016–2017" report, www.rspca.org.au/sites/default/files/RSPCA%20Australia%20Annual%20Statistics%20final%202016-2017.pdf (Accessed September 2, 2018).
66. See the Canadian Federation of Humane Societies' Report, "2016 Animal Shelter Statistics for Canadian Humane Societies and SPCAs," https://d3n8a8pro7vhmx.cloudfront.net/cfhs/pages/1782/attachments/original/1524169616/Cats_In_Canada_2017-FINAL-EN-LRs.pdf?1524169616 (Accessed September 2, 2018).
67. Of the 6 to 8 million cats and dogs who entered shelters in the United States between 2012 and 2013, only some were deemed sufficiently healthy to be adopted. Of those cats and dogs, approximately 2.7 million were killed. The RSPCA figures obtained from the United Kingdom show that between 2008 and 2013, 46% of the total number of animals who entered the charity's shelters were killed. While some of the animals killed were killed for their own sakes, many of them were killed because there was simply no room for them in shelters (Nick Craven and Lynne Wallis, "Revealed: RSPCA Destroys HALF of the Animals that it Rescues—Yet Thousands Completely Healthy," in *The Daily Mail* (online edition), 8 January 2013, www.dailymail.co.uk/news/article-2254729/RSPCA-destroys-HALF-animals-rescues--thousands-completely-healthy.html (Accessed September 2, 2018). During 2016–2017 financial year, Australia's RSPCA killed approximately 21,000 of the 100,000 dogs and cats who entered their shelters (see RSPCA Australia's "National Statistics 2016–2017" Report). And in Canada, it is estimated that 21,000 of the 120,000 dogs and cats who entered shelters in 2016 were killed (see the Canadian Federation of Humane Societies' Report, "2016 Animal Shelter Statistics for Canadian Humane Societies and SPCAs").
68. See AnimalSheltering.org's "Pets by the Numbers," www.animalsheltering.org/page/pets-by-the-numbers (Accessed September 2, 2018).
69. Supporting "puppy mills" (e.g., by purchasing animals from pet stores) is an especially egregious moral failing. This is because almost all puppy mills treat their dogs—especially the breeding females—atrociously. The dogs tend to be confined to tiny, crowded, and squalid cages, and deprived of both human affection and the opportunity to play and exercise. As a result, many of the dogs who come from puppy mills suffer from a variety of (very serious) physical and emotional conditions.
70. I would like to thank David Benatar and Bob Fischer for their very helpful comments on an earlier draft of this chapter.

24
THE ETHICS OF KEEPING PETS

Jessica Pierce

Pet-keeping practices are extraordinarily diverse and constantly evolving, and they run the gamut from ethically benign to profoundly disturbing. Individual pet owners face numerous quandaries: Is it ethical to "own" an animal and keep him in my home? How long can I leave my dog home by herself? Should my cat be allowed outside? What is an ethical and sustainable diet for my dog or cat? For every conscientious pet owner, there is another who is cruel, neglectful, and physically or emotionally abusive, raising the possibility that we ought to object to the entire pet-keeping endeavor if it opens the door to so much animal suffering. Cultural pet-keeping practices are also worth careful moral scrutiny, such as the abandonment of animals to shelters, where millions are killed each year to keep supply and demand in balance, and the increasing popularity of exotic pets, such as sloths and fennec foxes.

Pet keeping is an increasingly pressing moral issue, as the practice continues to grow in popularity around the globe and the number of animals bought, sold, and kept as pets mushrooms. Millions—perhaps even hundreds of millions—of animals' lives are at stake. In the US, there are approximately 470 million animals being kept in homes as pets. This means that the number of pet animals well exceeds the number of people (Pierce 2016: 3). In the UK, just over half of all households own a pet, adding up to some 9.3 million dogs, 1.1 million rabbits (People's Dispensary for Sick Animals 2017: 4), 400,000 lizards, and 300,000 snakes sharing people's homes (Williams and Jackson 2016: 59).

Most pet animals experience some degree of compromised welfare, and many live under conditions more challenging than those faced by animals in industrial farming venues, zoos, and research laboratories. There are no legal or regulatory standards of care for pet animals and no oversight for pet keepers. The fact that pet animals are kept in the private sphere of the home means that their lives are mostly hidden from public scrutiny. Anyone of any age can buy an animal and keep this animal in his home, garage, or backyard, and she can do so without any training or knowledge of the animal's environmental, nutritional, and behavioral needs. All this adds up to bad news for animals.

What Is a "Pet"?

"Pet" is a socially constructed category. Any animal can be labeled a pet, even an animal who may have initially have been labeled "food." A pet is generally taken to be an animal kept in the domestic or home setting whose sole function is companionship, amusement, or personal curiosity. Following this definition, James Serpell (1989) suggests that pet is a category of animal with no economic or utilitarian function, setting the pet into a wholly separate moral category from animals raised for food, research, hard labor, or public display. Our relationships with pets, he says, are affiliative not utilitarian, and rather than being mere things, pet animals enjoy a moral status verging on quasi-personhood. Jessica Pierce (2016), in contrast, has argued that pet animals are not exempt from exploitation but are objectified and commodified in ways very much like animals raised for food. Pets are emotional commodities but commodities nonetheless.

There has been some attempt to shift the language of pet keeping to better reflect the non-exploitive nature of the human–animal relationship. Rather than talk of "pet" and "owner," we might use the language of *companion animal*, or even *animal companion*, with the emphasis no longer on the function the animal serves for us. And we might describe ourselves as *guardian* or pet *parent*, rather than "owner," to shift emphasis toward the relational aspects of human–animal friendships. Some philosophers argue that until our behavior changes, we need to continue using language that reflects our actual (and not our ideal) relationship to animals and that by using friendly terms such as *companion* and *guardian* we simply obscure the issues we are hoping to address (Bok 2011).

Is It Ethical to Keep Animals as Pets?

Perhaps the most basic ethical question in the realm of pet keeping is whether it is wrong to hold an animal captive for the sake of our own entertainment or pleasure. Even taking a conventional utilitarian view of animal ethics—that it is morally acceptable for us to use and even to kill animals, if we are fulfilling some urgent need that cannot be otherwise satisfied—pet keeping seems to come up short. Those defending the use of animals in biomedical research, for example, point to the important potential benefits of new medicines and treatments. Does pet keeping fulfill such an "urgent need," or is it merely a frivolous enterprise? How strong are the interests of animals in not being held captive when weighed against the desire of a human to have a furry friend in the house?

This type of benefit–harm moral reasoning is very common in animal ethics. Unfortunately, animals always get the losing card in this game because we begin from the assumption that animals are here for us to use, however beneficently; their lives are ours for the taking. We then move on to decide which uses are acceptable and which offend our moral sensibilities. Yet we have overlooked the most basic harm: denying them their basic freedom to live their own lives under their own terms.

Some of the strongest arguments against pet keeping eschew this benefit–harm model and take a principled approach: "using" animals, no matter what for, is always a form of exploitation and is always wrong. The "relationship" between human and pet is fundamentally one-sided and exploitive, making pets akin to slaves. They are someone's property, have no say in whether they become a pet, whose house they will live in and for how long, who they may interact with socially and when, what they will eat, whether and when they will reproduce, and even whether they will get to keep their sex organs. Whether or not individual pet owners are responsible and conscientious makes no difference: the entire enterprise of owning pets is ethically corrupt, however well it is done. As Gary Francione and Anna Charlton (2017: 3) write, "We oppose domestication and pet ownership because these violate the fundamental rights of animals. . . . Non-human animals have a moral right not to be used exclusively as resources, irrespective of whether the treatment is 'humane.'"

Unfortunately, these deeper ethical challenges to pet keeping have remained at the margins, even within philosophy, with few questioning the appropriateness of keeping animals as pets in human homes. This lack of engagement may reflect the common assumption that pet keeping is not a serious subject for scholarly analysis or the (mistaken) belief pet animals are not in need of moral concern because they enjoy pampered lives as part of human families, sharing our homes and beds. It may also be that discussions of pet keeping tend to focus on the human side of the equation—how pets benefit us and whether the benefits outweigh the risks and costs—rather than seeing things from the animals' perspective.

Most scholarly work in the realm of pet keeping has instead focused on understanding and enhancing the human–animal bond and on improving the welfare of the animals we keep as pets. So the question is shifted from "Is pet keeping ethical?" to "What do I need to do to fulfill my moral responsibilities to my pet?" A growing literature on the human–animal bond suggests that humans and animals can form lasting friendships that are reciprocal, mutually respectful, and enriching. This literature also tells us that the bond is tenuous and can easily be frayed or broken. For example, a mismatch between what a dog owner expects from her animal and what the animal is willing or able to provide often leads to feelings of dissatisfaction on the part of pet keepers. (The pets themselves likely feel dissatisfied, too.)

The Harm of Captivity

Pets are not generally seen as captive animals in the same way that, for example, animals in zoos or research laboratories are. But any animal who must live most of his or her life behind bars is captive. Even dogs and cats—who may have the freedom to roam around the house and perhaps even the backyard—are captive in important respects. They are not the masters of their own lives and don't have substantive control over their environment, social interactions, or daily choices.

The fact that there is an entire literature dedicated to so-called captivity effects should leave us in no doubt that captivity itself is a moral problem. This literature has enormous reach, from behavioral problems observed in various species held in captive environments to evidence of neurobiological and physiological changes induced by captive conditions to the human psychology literature, including a robust data set on the harms of captivity to human prisoners of war and to those under conditions of incarceration. These captivity effects include long-term activation of the hypothalamic–pituitary–adrenal axis, changes to immune function, brain morphology, reproductive behaviors, circadian rhythms, and so on and so on. This is important because we need to be alert to the compromises that captivity imposes, and if we choose to have pets, we have a responsibility to offset these compromises as best we can.

Which Animals Make Appropriate Pets?

Do some animals make better pets, ethically speaking, than others? Probably so, and the answer to "Which animals?" depends on two things. First, what burdens do captivity and confinement impose on a given species? And, second, what are the possibilities for reciprocal and mutually enhancing relationships with humans? Using these criteria, we might rank species in a rough order, from most ethically acceptable to least: dogs, cats, small domesticated mammals (rabbits, rats, hamsters), fish, birds, reptiles and amphibians, and wild animals.

Because dogs and cats can live in human environments with a large measure of freedom, and because both species have evolved in close relationship with humans and have "chosen" (evolutionarily speaking) to form companionable bonds with us, they may be the most likely candidates for "ethical pets" (see Pierce 2016). A dog, for example, may be confined to the house for part of each day but may also have ample opportunities to get outside, run free, follow her nose, and share in mutually enjoyable activities with her human, such as lounging on the couch, eating cheese and crackers, or going for a hike. In contrast, most reptiles must always be caged. Reptiles tend to dislike human handling and probably don't form companionable bonds with us. Indeed, ethologist Gordon Burghardt (2013) believes that no captive environment can be adequately stimulating for reptiles and amphibians, and the best we can hope for is a life of "controlled deprivation."

The zoo-animal welfare literature may also help shed some light on which species of animal might do well or poorly pets. Species with more behavioral plasticity tend to do better in captive environments (dogs and cats have more flexible behaviors than reptiles). Some species-specific characteristics, such as timidity, should serve as red flags. For timid species, the presence of stimuli perceived as threatening will over-activate the endocrine stress responses; these stress responses will be "turned on" far more than is normal or healthy. This may explain why prey species such as hamsters appear to sleep all the time: they have "shut down" emotionally because the barrage of fearful stimuli is too great.

Cats raise a unique ethical problem revolving around what measure of freedom they should enjoy. Some cat behaviorists, as well as a good number of cat owners, argue that cats need to be able to roam outside to be happy and have optimal welfare. They consider it cruel to keep a cat indoors. On the other side, some behaviorists and veterinarians worry that outdoor cats are exposed to unacceptable levels of danger, from cars, coyotes, and eagles. And many worry about the impact of outdoor cats on populations of songbirds and small mammals. Cats are highly motivated to hunt and are effective killers, no matter how much food they are provided by their owners. Depriving them of the opportunity to engage in this highly motivated behavior may lead to stress and frustration; allowing them to engage leads to significant collateral damage.

Welfare Problems for Pet Animals in the Home

The basic welfare needs of pet animals are the same as the needs of animals in other venues of human use: access to nutritionally appropriate food and clean water; comfortable and species-appropriate housing, with plenty of room to move around; ample opportunities to engage in normal species-specific behavior (e.g., preening for a bird, digging and gnawing for a rodent, scratching and hunting for a cat); opportunities for social interaction with others of their own kind; and timely access to veterinary care (Bekoff and Pierce 2017). Meeting these five basic needs requires a strong commitment on the part of pet owners, and unfortunately, many pets are lacking in one or several of these critical areas.

From US data, we know that 25% of dogs and more than a third of cats will never see a veterinarian in their lifetime. This means that potentially painful diseases are left untreated, and basic preventive care may be lacking. As an example of the former, over 10 million dogs in the US are thought to suffer from degenerative joint disease, and only a fraction of these are ever treated for their pain. Seventy percent of cats and 80% of dogs over the age of three are likely to have some form of oral disease, including tooth decay, gum inflammation, and periodontal disease. As in humans, gum disease in dogs has been linked to endocarditis (Glickman et al. 2009). Many of the commercial dog and cat foods on the market are nutritionally inadequate and leave animals both fat and undernourished. At least half of all dogs and cats in the US and the UK are obese, which can cause long-term chronic health problems likely to shorten their life span. In contrast, only 15% of the same owners think their dog is overweight, suggesting a gap in awareness and education about ideal weight and the health and welfare problems caused by obesity. Very few dogs and cats are given the opportunity for adequate physical exercise.

Perhaps the most widespread welfare problem for pet animals, and one that has thus far received relatively little attention, is boredom. Over the past decade, scientists have recognized that boredom is a particularly acute welfare risk. Although the research is still young, we now know that animals can and do experience boredom under similar conditions as humans—conditions of confinement, monotony, and predictability—and that boredom causes both physical and especially psychological harm. Boredom is related to but distinct from frustration, stress, depression, and apathy. As Charlotte Burn (2017: 141) wrote in a paper on animal boredom, "[c]hronic inescapable boredom can be extremely aversive, and understimulation can harm neural, cognitive, and behavioural flexibility. Wild and domestic animals are at particular risk in captivity, which is often spatially and temporally monotonous." We can begin to address boredom by providing animals with interesting opportunities and environments and by limiting our pet keeping to species of animal who do not need to be confined in cages or tanks.

A second unacknowledged welfare risk is loneliness. Many pet animals are kept as singletons, even those who are known to be highly social and for whom solitary life is stressful. Rabbits, guinea pigs, and rats, for example, are all social animals who live in complex social groupings in the wild. They suffer when kept alone. Loneliness is also a significant problem for dogs, evidenced by the high numbers of dog owners seeking behavioral advice for separation anxiety. Dogs have been genetically selected for their friendliness toward and desire to have contact with humans, and contact with humans has become a requirement for their well-being. Deprivation of adequate social contact with us can be, for many dogs, a serious welfare concern. Although there is loose consensus among dog behaviorists that four hours of alone time a day is the maximum we should ask of our dogs, a recent UK report found that almost 20% of pet dogs were left home alone every day for at least five hours (People's Dispensary for Sick Animals 2017: 11). Although cats have long been characterized as aloof and solitary, new research on cat behavior is challenging this stereotype, and there is growing concern that cats, like dogs, suffer from too much time alone.

Lack of appropriate training is also a risk factor, particularly for dogs. Many people who purchase a dog know little to nothing about how to appropriately care for and teach their dog to be a well-functioning member of the household. In one large owner study, researchers found that 1.2 million dog owners did no research whatsoever before bringing home their dog, and 1.2 million dogs receive absolutely no training (PDSA 2017: 11). At the same time, at least 60% of surveyed dog owners reported unwanted behavior in their pet. The roots of "misbehavior" typically point toward human inattention or incompetence, not the dog: unclear communication, boredom, and frustration on the part of the dog, lack of training or confusing training, and mismatch between who dogs are and who we expect them to be. Many popular training methods and books rely on punitive techniques or are based on misunderstandings of dog behavior, with unsurprisingly poor results for dog and human alike.

Special Issues of Ethical Concern Within Pet Keeping

Critics of pet keeping note that animals are victims of profound mistreatment by humans, who seem to view animals as disposable objects. You can buy a dog for about the same price as a pair of shoes, these critics say, and the shoes are likely to get better care. Humans are too often neglectful, cruel, and abusive toward pets, and readily abandon an animal on a street corner or slip them into the animal shelter's night box. The most serious areas of concern are abandonment, killing in shelters, and physical, emotional, and sexual abuse.

Abandonment

Many people seem to view getting a pet as akin to checking out a library book. If you pick one that you don't like, you can simply return it and get a different one. There are many faults in this logic, but the most obvious is ignoring the fact that dogs and cats have feelings and form attachments. Indeed, the social attachments dogs form with humans have been described, using theoretical models developed in the human psychology literature, as parallel to the attachment a child forms with a parent. The owner/parent is an attachment figure, and to have this person taken away is scary and distressing. Each time a dog or cat is adopted and then returned, the animal's chances of emotional breakdown increase, and so do their chances of winding up dead. Although no precise data exist, we have good evidence that large numbers of cats and dogs—probably the majority—don't live in one home their entire life. They are relinquished to shelters when they become inconvenient for their human owner, whether because they bark or meow too much, poop too much, and scratch the furniture or the family is moving and doesn't want to take the animal.

Shelters and Killing

Pets often end up part of a macabre dance, moving into and out of homes and into and out of shelters and rescues. Discarded and abandoned animals may or may not get a new home, depending on what kind of shelter they wind up in and where, and how many other dogs and cats are competing for new homes, what degree of psychological trauma they have experienced, and how resilient they are. In the US, only about 20% of people acquiring a new pet will adopt from a shelter, so the number of placements is quite limited. More than 7 million pets enter the shelter system each year, and only about 2.7 million leave. Of these remaining animals, 3 million will be killed each year, and the rest will simply bide their time. Although 3 million is the lowest number of shelter killings the US has seen in years, this is offensively high and should be of profound moral concern to all who care about animals.

Physical and Emotional Abuse

One of the strongest arguments against pet keeping—second only to the fact that millions of surplus pets are killed each year to keep supply and demand in relative balance—is that pet animals are uniquely vulnerable to human abuse and exploitation. The Humane Society of the United States estimates that there are hundreds of thousands of animal cruelty cases a year, which is likely a conservative estimate since many cases go undetected. The primary groups of abusers are domestic violence batterers, people with personality disorders, children and teens (primarily male and preschool age), hoarders (75% of whom are women), adult men and women with unrealistic expectations about how an animal should behave, and violent offenders (e.g., rapists). Dogs are the most common target of abuse, with cats a close second. Although some cruelty rises to the level of felony criminal animal abuse, much abusive behavior toward pets takes place on a less dramatic but more insidious level, such as the daily use of harsh "training techniques" such as spanking, shock collars, and "stringing up."

Sexual Abuse

Talking about sexual exploitation of animals tends to make people very uncomfortable, but greater awareness of the vulnerabilities of animals and the willingness of humans to take advantage of their position of power is essential. Zoophilia, or bestiality, is more common than most people realize. Although only a small number of violent sexual assaults on animals get reported, there are perhaps millions of "zoos" out there. According to one statistic, each "zoo" likely knows, on average, about 90 other people who engage in zoophilic activity. There are internet forums dedicated to sharing stories and exchanging advice, sex "farms" where a group of animals is kept ready for the sexual pleasure of paying guests, YouTube videos aplenty of homemade "zoo-porn," and an entire industry of professional animal-based pornography (Pierce 2016: 128–135).

According to Gieri Bollinger and Antoine Goetschel's (2005) typography, there are five basic forms of sexual activity with animals: genital acts (anal and vaginal penetration), oral–genital acts (fellatio, cunnilingus), masturbation (of animal by human), frottage (rubbing up against), and voyeurism. Many animals wind up with serious injuries or are killed by the sexual encounter or by their abuser, after the fact. Dogs are the most frequent victim of sexual assault, with horses coming in second, but no animal is safe from the human sexual imagination.

A great many pet owners refrain from sexually abusing their animals, but a great many don't. And this is one of the significant problems with pet keeping as it is practiced: animals are vulnerable to our whims and fantasies, whatever form these take. Pets are kept in the private sphere and our interactions with them occur behind the curtains. The high levels of sexual exploitation of pet animals are in themselves a strong argument against pet keeping.

Breeding

One cannot talk about the ethics of pet keeping without giving some attention to the way animals are "produced" for our consumption. We make dogs and cats our "babies" but only after their own mother has conceived, birthed, and raised them—and then had them abruptly taken away, often well before natural weaning has occurred. Animal advocates have two central concerns about breeding. The focus is generally on dogs, but other species face similar issues. The first concern is what are often called "puppy mills," the label given to commercial breeding operations with the worst animal cruelty violations. Bitches are often kept in small wire cages (which are extremely painful for a dog's paws), forcefully inseminated (artificially or not), and tasked with birthing one litter of pups after another until they are reproductively "spent," at which time they are killed. Even the commercial and family breeders with very high standards of care still raise ethical concerns because they are commodifying animals and are adding to the supply of desirable dogs while healthy animals in shelters are being killed.

The other concern about breeding is selective breeding of dogs for appearance rather than health, which has led to a large population of dogs with disabilities and lifelong physical discomfort. Perhaps the most striking examples of this are the brachycephalic dogs like pugs, bulldogs, and boxers. The shape of their skull has been altered to such a degree that their brains have shrunk; their foreshortened noses almost guarantee that they will have difficulty breathing; their bulging eyes are frequently infected; and their joints are malformed and often painful.

Keeping Exotic and Wild Animals

Arguably, one of the most damaging and least ethically justifiable pet-related practices is the keeping of wild and exotic animals as pets. There is no countervailing benefit to keeping wild animals as pets, either to animals or people, other than the staggering amount of money that some people are making by collecting, breeding, transporting, and selling wild animals.

There is no precise definition of "exotic pet." By some definitions, an exotic is any pet other than a dog, cat, or horse and can include familiar pets such as leopard geckos, hamsters, parrots, and ball pythons. "Exotic" can also refer to a nondomesticated, nontraditional, or non-native animal. Exotics can be captive-bred or wild-caught. Whatever the definition, and wherever they are sourced, these animals share one common feature: they are unsuitable as pets.

Unlike dogs and cats, exotic pets spend nearly all their time in a cage or tank. They are not domesticated and thus are fearful in the company of humans, and they have environmental and nutritional needs that are complex and beyond the knowledge and skills of most pet owners. Because these animals are often unfamiliar, it is more challenging for pet owners to determine signs of positive and negative welfare. Pet stores and Internet brokers offer little, if any, education for prospective owners, and information available online runs the gamut from quite good to awful and misleading. Very few veterinarians will treat exotic pets, and even fewer exotic pet owners avail themselves of veterinary specialists, and so pain and illness tend to go untreated. Owner satisfaction with these species tends to be relatively low, perhaps because of a misalignment between owner expectations and the behavior of the animal. For example, many parents buy geckos as "starter pets" for their children, but geckos dislike handling and may strike children as boring (because they have nothing to do in a barren cage and simply lay in one place all day).

The welfare of exotic pets is very often compromised, and mortality rates tend to be quite high (e.g., see Bush et al. 2014). One estimate suggests, for example, that at least 70% of captive reptiles will die before they reach the pet store shelf, and about 75% will not survive past their first year of being a pet (Warwick 2014: 79). Most exotic pets are kept as singletons, but for some species of animal, this kind of social isolation can cause severe psychological trauma. Wild parrots, for example, are a prey species and protect themselves in the wild by flocking together. The behavioral motivation to flock is present whether a parrot is in a forest or in a cage in someone's house; when deprived of the possibility for this behavior, the bird suffers distress. Stereotypic behaviors such as feather plucking and head bobbing—both pathological signs of stress—are common among captive parrots (Grant et al. 2017). Exotics often suffer from some combination of thermal stress, inappropriate handling, poor diet, restricted movement, and trauma from attempting to escape.

All manner of creatures are plucked from the wild to be marketed and sold in US and European markets. American pet dealers import and sell 225 million live animals annually, and 11 million fish are sucked out of the oceans to fill the aquariums of collectors and hobbyists. The pet trade is driving many species toward extinction, including the radiated and ploughshare tortoises from Madagascar, the slow loris from Southeast Asia, the red line torpedo barb from Asia, and the myna from Bali (Conniff 2017). Scientists are increasingly worried about the wild pet trade and have noticed correlations between loss of species and spikes in popularity of keeping that particular kind of animal as a pet. A 2015 study published in *Conservation Biology*, for example, showed that the price of a popular Indonesian songbird called the white-rumped shama jumped 1,500% between 2013 and 2015; during this same period, the species nearly vanished from the wild (Harris et al. 2015). Some conservation biologists now consider the pet trade an even more serious threat to ecosystem stability than habitat loss.

Captive breeding of exotic animals has its own set of problems: high mortality, difficulty in learning how to successfully breed particular species, and the need to continually replenish "stock" from the wild (see Tensen 2016; Lyons and Natusch 2011).

The exotic pet trade and pet-keeping practices have impacts on ecosystems that stretch well beyond the disappearance of the animals themselves. Often, methods of collecting animals are crude and damage the surrounding ecosystem and its living creatures. For example, cyanide bombs are set off in coral reefs to paralyze and then capture fish. Another problem is the release of pets when they have grown too large or have become a nuisance. Pet pythons have been released in high numbers in the Everglades and thrive in this environment; they also eat the local fauna and disrupt the ecological balance. Collateral damage occurs when novel bacteria or fungi are released into an ecosystem. One of the saddest examples of this is the decimation of salamander populations in Europe and the UK by a chytrid fungus, likely spread through the pet trade in amphibians from Southeast Asia.

Connecting the Dots

The burgeoning research into animal cognition and emotions is encouraging a moral reassessment of how humans use and interact with animals in many different venues. The more intelligent and emotional we recognize other animals to be, and the more empirical data we have on how profoundly animal welfare is compromised by some of our common practices, the stronger the moral arguments for abolition become. For example, in the realm of animals raised and killed for food, awareness of what pigs, cows, and chickens think and feel has prompted increased attention to animal welfare and ethical scrutiny of intensive farming practices. It has also inspired many people to choose a plant-based diet. Similarly, a greater understanding of animal feelings has led to reassessment of whether zoos and aquariums are ethically acceptable, particularly whether captivity imposes unacceptable levels of harm to highly intelligent and sensitive animals such as elephants, orcas, dolphins, and wolves.

This science is encouraging us to reassess our relationships to companion animals as well, by suggesting that pets suffer many of the same welfare problems as animals used in agriculture, research, or entertainment. Yet perhaps even more important than helping us identify welfare challenges, the expanding knowledge of who animals are can help us improve the lives of those animal companions who already share our homes. For example, the science of dog cognition is reshaping the way we understand the needs of our canine friends. We have a better understanding of how to effectively communicate with dogs and what kinds of experiences are important to them. Knowing, for instance, that dogs explore the world largely through their nose, we can make small changes, such as giving an on-leash dog ample time to stop and investigate smells when we go for a walk together.

Links

Ethical responsibilities and attitudes toward animals are often considered in isolation, apart from our responsibilities toward each other. But animal and human ethics are linked, and nowhere does this association become more evident than in the realm of pet keeping.

There is a robust literature on the links between cruelty toward animals and toward people. Criminal offenders are statistically more likely to have a history of animal abuse, and adolescents who abuse animals are at increased risk of delinquency. Domestic violence and animal abuse often co-occur. In families where animals are treated cruelly, interpersonal violence is often also present—which is why there has been some movement toward requiring veterinarians to report suspected animal abuse to social services. Animals are drawn into the vortex of violence both as victims and as weapons.

Violence toward animals stems from the same impulse as violence toward our fellow humans. On the flip side, compassion for animals and humans is also linked, and so our efforts to address the roots of violence toward animals will have broader reverberations.

Can Pet Keeping Be Ethical?

There is no doubt that an individual relationship between a person and a companion animal can be mutually beneficial, reciprocal, and loving. And pet keeping often rises to this level. But is there still an element, however faint, of exploitation in the relationship? This is a question that each individual person needs to contemplate as they consider acquiring a pet.

Even if we acknowledge the bonds of friendship and love that can form between human and animal, a broader question remains: Is the cultural practice of pet keeping so damaging to animals that we ought to move away from it altogether? Can you engage in pet keeping without supporting an industry that commodifies animals? Are there ways to protect animals adequately, such that pet keeping could be practiced without unacceptable levels of collateral damage?

Pet keeping will never be completely free of ethical compromise, but animal advocates generally agree that a few basic changes would vastly improve the situation for animals:

- *Appropriate species.* We could narrow the range of species we keep as pets to domesticated animals who have evolved in close connection with humans, are well adapted to living in human environments, and can live in mutually enhancing and somewhat voluntary relations with us. The most likely candidates here are dogs and cats. Wild animals and exotics should simply not be kept as pets.
- *Ethical sourcing.* Until there are no animals languishing in shelters or being euthanized as surplus, all pets should be sourced from this pool of "surplus" animals. It is our collective and individual responsibility to assure that these animals have a good home before turning to other sources such as pet stores or breeders. A public education and myth-busting campaign could increase shelter adoptions by helping the public understand that dogs and cats adopted from shelters have fewer behavioral issues than animals bought from stores or breeders. Although rescued animals may be "damaged goods" in the sense that they have suffered neglect or abuse, dogs and cats are remarkably resilient and usually adapt very well to a new home. The public also needs to understand that mixed-breed dogs and cats are healthier and have fewer psychological issues than purebred animals.
- *Better attention to welfare needs of pet animals.* Pet owners need to become educated about who their animal is and what their animal needs, including understanding and creating opportunities

for species-specific behaviors and providing comfortable housing, nutritionally appropriate food, timely veterinary care, mental enrichment, emotional fulfillment, and plentiful opportunities for social interaction and engagement. Pet owners can find many ways to increase an animal's sense of freedom and self-determination, for instance, letting a dog have time off-leash every day or respecting a dog's decisions about when and if she wants to interact with people or other dogs.

When we bring an animal into our home as a pet, the quality of this creature's life is entirely in our hands—and this is a momentous responsibility. Unlike our children, who will grow up and become independent beings, our pet animals will rely on us from birth to death and everything between. Once we have made the decision to bring an animal into our home, we have an obligation to do the very best we can to provide this animal with a good life. Empathy is taking the time to see what it feels like to be somebody else. We need to take time to see what it might feel like to be a pet.

Bibliography

Bekoff, M., and Pierce, J. (2017) *The Animals' Agenda: Freedom, Compassion, and Coexistence in the Age of Humanity*, Boston, MA: Beacon Press.

Bok, H. (2011) "Keeping Pets," in T. L. Beauchamp and R. G. Frey (eds) *The Oxford Handbook of Animal Ethics*, New York, NY: Oxford University Press.

Bollinger, G., and Goetschel, A. (2005) "Sexual Relations with Animals (Zoophilia): An Unrecognized Problem in Animal Welfare Legislation," in A. M. Beetz and A. M. Podberscek (eds) *Bestiality and Zoophilia: Sexual Relations with Animals* (pp. 23–45), West Lafayette, IN: Purdue University Press.

Burghardt, G. M. (2013) "Environmental enrichment and cognitive complexity in reptiles and amphibians: Concepts, reviews, and implications for captive populations," *Applied Animal Behaviour Science* 147: 286–298.

Burn, C. C. (2017) "Bestial boredom: A biological perspective on animal boredom and suggestions for its scientific investigation," *Animal Behaviour* 130: 141–151.

Bush, E. R., Baker, S. E., and MacDonald, D. W. (2014) "Global trade in exotic pets, 2006–2012," *Conservation Biology* 28: 663–676.

Conniff, R. (2017) "Loved to death," *Scientific American* 317(4): 40–45, October.

Francione, G. L., and Charlton, A. E. (2017) "The case against pets," *Aeon* [digital magazine], (http://aeon.co/essays/why-keeping-a-pet-is-fundamentally-unethical).

Glickman, L. T., Glickman, N. W., Moore, G. E., Goldstein, G. S., and Lewis, H. B. (2009) "Evaluation of the risk of endocarditis and other cardiovascular events on the basis of the severity of periodontal disease in dogs," *Journal of the American Veterinary Medical Association* 14(4): 486–494.

Grant, R., Montrose, V. T., and Wills, A. P. (2017) "ExNOTic: Should we be keeping exotic pets?" *Animals* 7: 47.

Harris, J. B. C., Green, J. M. H., Prawiradilaga, D. M., Giam, X., Giyanto, H. D., Putra, C. A., and Wilcove, D. S. (2015) "Using market data and expert opinion to identify overexploited species in the wild bird trade," *Conservation Biology* 187: 51–60.

Lyons, J. A., and Natusch, D. J. D. (2011) "Wildlife laundering through breeding farms: Illegal harvest, population declines and a means of regulating the trade of green pythons (*Morelia viridis*) from Indonesia," *Biological Conservation* 144(12): 3073–3081.

People's Dispensary for Sick Animals (2017) "PAW report," www.pdsa.org.uk/media/3290/pdsa-paw-report-2017_online-3.pdf.

Pierce, J. (2016) *Run, Spot, Run: The Ethics of Keeping Pets*, Chicago, IL: University of Chicago Press.

Serpell, J. (1989) "Pet keeping and animal domestication: A reappraisal," in J. Clutton-Brock (ed.) *The Walking Larder: Patterns of Domestication, Pastoralism, and Predation*, London: Unwin Hyman.

Tensen, L. (2016) "Under what circumstances can wildlife farming benefit species conservation?" *Global Ecology and Conservation* 6: 286–298.

Warwick, C. (2014) "The morality of the reptile 'pet' trade," *Journal of Animal Ethics* 4: 74–94.

Williams, D. L., and Jackson, R. (2016) "Availability of information on reptile health and welfare from stores selling reptiles," *Open Journal of Veterinary Medicine* 6: 59–67.

25
THE ETHICS OF COMPANION ANIMAL EUTHANASIA

Christine Overall

When people talk about euthanizing companion animals, they often resort to euphemisms. They "put their dog down"; they "put her[1] out of her misery"; they have her "put to sleep." These terms function to reduce people's distress about killing animals who are important parts of their lives. Companion animal euthanasia is often difficult and emotionally fraught.

The first step in discussing the ethics of companion animal euthanasia is to get clear about what the term *euthanasia* means. Based on the word's Greek roots, *euthanasia* means "good death," but Jessica Pierce (2013: 476) points out that the term is frequently used much more widely:

> The AVMA [American Veterinary Medical Association 2007] Guidelines use the term 'euthanasia' to cover the whole range of deliberate killing of animals, from gassing unwanted dogs in shelters, to pithing frogs in a classroom, to decapitating mice in a laboratory, to crushing the brains of cows in a slaughterhouse with a captive bolt gun. The word 'euthanasia' is asked to do too much, to obscure far too much moral nuance, since only a small portion of animal deaths could really be considered good and merciful.

This broad usage is inappropriate because it includes deaths that are clearly *not* good. James Yeates (2010b: 70) argues that "animal euthanasia" should signify "killing an animal in its [*sic*] interests." Typically, the word is used even more precisely to denote deliberate action taken, usually by a medical professional (a veterinarian) and usually near the end of life, to cause the death of an animal—in as pain-free a manner as possible—for beneficent reasons. This is how the term will be used in this chapter.

In discussions of euthanasia of and for human beings, it is customary to make a distinction between voluntary euthanasia, which is chosen by the individual, and nonvoluntary euthanasia, which is undertaken for the sake of the individual's interests but neither chosen nor rejected by the individual. Although most human companions want to enable their companion animals to "exercise meaningful control over what matters to them, to live 'on their own terms,' and to be agents in shaping" their futures (Donaldson and Kymlicka 2015: 65), the euthanasia of companion animals is inevitably nonvoluntary, "the beneficent killing of a being neither in concert with, nor contrary to, its [*sic*] consent" (Cholbi 2017: 266). For animals obviously cannot choose euthanasia. They lack the kind of cognition that would enable them to understand the concept of a death deliberately undertaken for the sake of their interests. That fact makes it all the more important to be clear about the moral justification of animal euthanasia. Indeed, it should make us cautious about choosing to euthanize companion animals, just as we should be hesitant to engage in the nonvoluntary euthanasia of human beings.

Yet as Michael Cholbi (2017: 264) points out, the euthanasia of companion animals "is far more the norm" than the euthanasia of human beings. We euthanize animals in cases where we would never euthanize people. Pierce (2012: 180–181) gives examples of a dog with bone cancer; a cat with inappropriate urination; a boxer with anxiety; even two animals raised together who are euthanized at the same time just because one of them must be "put down." Sometimes animals are euthanized simply because their behavior has become difficult for human beings to handle (Pierce 2012: 182; Haug 2011).

Nonetheless, it's commonplace to claim that, with respect to euthanasia, "we treat animals *better* than we treat humans." This is often said when people are discouraged about the lack of socially supported euthanasia for human beings (Bachelard 2002: 131–133; Pierce 2012: 191). The thought is that society should make euthanasia more readily available to human beings, as it is to animals. But is it indeed true that with respect to the end of life, we treat our companion animals better than we treat human beings? If some people hesitate to provide euthanasia for their dads, then why not for their dogs?

This chapter surveys various conditions and criteria that have been proposed for justifying the euthanasia of companion animals. There is no formulaic answer to ethical questions about animal euthanasia; the morality of the practice is more complex and difficult than the common platitude recognizes. And the evidence does not support the belief that, when it comes to the termination of life, animals are treated better than human beings.

Can Death Be Bad for Animals?

Some people believe that death is neither good nor bad for animals. James Yeates (2010a) calls this the "death is not a welfare issue" premise. This premise means, he writes,

> that, while death in the sense of the process ('dying') is perceived as a welfare issue in that it may involve suffering, death in the sense of the termination of life or post-mortem non-existence ('deadness') is not a matter of harm or benefit in animal welfare terms.
>
> *(Yeates 2010a: 229–230)*

As Yeates points out, adopting this premise would imply that killing an animal is not and cannot be a harm to her, provided that the method of killing causes no pain or suffering.

This is precisely the point of view held by David Velleman, who defends it by saying that animals (he uses the example of cows but believes the argument applies to other animals) cannot conceive of themselves as enduring beings. What they cannot conceive of, they cannot care about. Thus, animals cannot care about "the sequences of momentary goods" they may experience (Velleman 1993: 354). Velleman concludes that because animals cannot care about extended periods in their lives, the totality of their lives has no value for them. Hence, death takes nothing from the animal, and there is no time at which an animal is badly off because of her death (Velleman 1993: 357).

Elsewhere I have critiqued this argument in detail (Overall 2017: 251–253).[2] I do not repeat my arguments here but simply make a few general remarks. First, as I discuss in the following, there is some evidence that animals *do* have a sense of at least a limited future and hence of themselves as enduring beings. But even if they do not, it is hard to see why depriving an animal of a future of *unconnected* pleasures is not bad for her. It is impossible to believe that cutting short, through death, the many moments of pleasure that an animal could otherwise have experienced does not constitute a real loss for her.

Moreover, given that, according to Velleman (1993: 356), "the various benefits accruing to a cow at different moments must not add up to anything at all, not even to zero," then presumably a life of various pains accruing to the cow at different moments also do not add up to anything, not even to zero. But this view implies that a life of years of pleasure is no better for an animal than a life of misery and that there is no difference, for the animal, between a future full of suffering and a future full of enjoyment. It implies that there is literally no difference whatsoever—*from the animal's point of view*—between being treated kindly throughout her life and being treated cruelly. And if, as Velleman believes, there is no time when death is bad for an animal, then ending the animal's life after six months of extreme suffering is no worse for her than ending her life after a week of it.

These implications are both counterintuitive and untenable. If death were always only neutral for animals, then it simply would not matter whether or when we terminate a companion animal's life. On the contrary, it is only because we assume, correctly, that it is important to minimize the duration of a companion animal's suffering and promote the duration of her enjoyment that euthanasia becomes a moral issue. Death can therefore be good or bad for an animal, depending on its timing.

The Right Time

Cholbi argues that the justification of companion animal euthanasia rests on ascertaining the *right time* for the animal to die. If the animal is killed too soon or too late, then the euthanasia is not justified (Cholbi 2017: 268). "[E]uthanizing Ridge [a dog] is warranted to the extent that at the time of the act of euthanasia, Ridge does not stand to gain by living longer, but also loses nothing by dying at that point" (Cholbi 2017: 269). Likewise, Pierce (2012: 168) writes that although we cannot know *precisely* the right time for an animal's death, we must "find a golden mean between too soon and too late, between premature and overdue." Yeates (2010a: 237) (referring to farmed animals rather than companion animals) says,

> If the presence of a life would have positive value overall then death is a harm; if it would have negative value overall then death is a benefit (and if compared to a life neither worth living nor worth avoiding, then death might be neutral).

Presumably, if an animal's life has "positive value overall," then the animal has something to lose by dying now, and if the animal's life has "negative value overall," then the animal has nothing to lose by dying now.

This criterion for the justification and timing of animal euthanasia raises the question of how to evaluate when an animal has nothing to gain by continuing to live and nothing to lose by dying. Some might suggest that we should appeal to a hypothetical: What would the animal herself want? But it is not always easy to know what animals want, and asking what they want may not be the right question in this case.

The question is not, Would I, a human being, prefer to live or die under these circumstances and facing this future? Anthropomorphic approaches are of limited value, because animals have considerably less knowledge and understanding of the future than adult human beings do. While a human may, for example, understand his own illness and its prognosis, an animal does not. Animals can't understand the reasons for pain they may experience or for undergoing surgery or other difficult treatments; the animal has no theoretical understanding that some pains may be ongoing, and others may be temporary and lead to relief later on. It is implausible to suppose that an animal could conceptualize the idea "I am suffering and therefore I want someone to end my life" or even "I am suffering and I no longer want to live."

Nonetheless, we cannot justify euthanizing an animal merely because her grasp of the future is limited compared to that of adult human beings. For it is not thought morally justified to euthanize humans who, whether because of extreme youth or cognitive impairment, do not understand their futures and their deaths. In fact, we are justifiably *more* hesitant to end the lives of human individuals who are very young or cognitively impaired precisely because we see them as more vulnerable. From this perspective, it would seem that we should be similarly hesitant with animals, who are extremely vulnerable with respect to us and lack the capacity to resist or elude human dominance.

Animal Suffering

Rather than using the hard-to-determine criterion of what an animal might want, it makes more sense to appeal to the animal's well-being, her states of enjoyment, and, above all, her suffering. Most people would agree that if an animal is suffering and there is no feasible way to relieve that suffering, then euthanasia is justified (Pierce 2012: 167). Given that animals have no way of understanding their suffering, no way of putting it into context or hoping for better days, it would seem that when they feel great pain, that is all they are aware of (Pierce 2013: 473), and that fact makes relieving their pain even more urgent.

However, evaluating an animal's interior experiences raises epistemological problems: "How do we judge the quality of life for someone who cannot communicate in language?" (Pierce 2013: 477). Cats, for example, are notoriously stoical about enduring pain. Most human companions would probably say that close observation of the animal and the background provided by the ongoing relationship between the human and the animal provide good insights into the animal's quality of life. Tony Milligan (2009) argues that understanding an animal's pain involves more than merely in-the-moment observation. He says that what matters in evaluating an animal's suffering are not only quantitative considerations of the intensity and duration of the animal's pain (Milligan 2009: 406) but also "how pains are situated within a life" (Milligan 2009: 407). We should therefore ask ourselves what Milligan calls "[t]he life-role question"; that is, "What role will just this kind of pain or suffering play within the life of our dependent non-human companion when that life is considered as a whole?" (Milligan 2009: 407). In other words, we should consider the animal's situation within her own life narrative.

Milligan gives the example of two dogs, Rover and Rex, both suffering from painful arthritis. While Rover has had a life of "care and attention," Rex "was rescued by his guardian from years of cruelty and mistreatment" (Milligan 2009: 408). This latter factor, says Milligan, should play a role in any decision about the amount of pain Rex should be expected to endure:

> [T]he re-entry of pain into the life of Rex is a more serious matter than it is in the (already serious) case of Rover; and . . . a special sort of care and attention is in order to ensure that Rex should never again suffer either intentionally and maliciously or unintentionally.
>
> *(2009: 408)*

Perhaps, says Milligan, we might "intervene" (i.e., provide euthanasia) at the same point in both animals' lives, but our deliberations about each one will be different, and in other cases, such deliberations could result in different decisions for each animal.

It might be objected that animals lack the ability to have a narrative appreciation of their own lives; as Milligan (2009: 409) puts it, "[t]hey do not indulge in autobiography." But Milligan notes that even when a human being loses the capacity for a narrative understanding of his own life, we do not therefore think that "the obligations, care and love *which emerge out of our shared past* thereby cease" (Milligan 2009: 409, emphasis in original). Thus, we should "bring *our* narratives about *them* [companion animals] into deliberation about the ending of *their* lives" (Milligan 2009: 410, emphasis in the original). Milligan (2009: 411) continues:

> Animal guardians can (and often will) be epistemically privileged participants in end-of-life deliberations because they can (and often will) be the people who are best placed to bring

> the relevant narrative of a pet's life into view. In the absence of some appreciation that there is a narrative to be told, a creature cannot be treated as a unique particular companion. And in the absence of the narrative that the animal's guardian is best placed to supply there will be no way to make sense of the overall significance or life-role of a particular animal's pain and suffering.

It might be objected that human beings are no better at comprehending and evaluating an animal's narrative than they are at directly understanding and evaluating her subjective experiences. And it is true that no human being can have a perfect understanding of an animal (any more than he can have a perfect understanding of another human being). However, what is significant about an animal's narrative is that it is an account, not only of her immediate subjective experiences as best we can intuit them, but of her life events—including her medical history, her preferences and aversions, her needs and vulnerabilities, her interactions with other animals, and her relationships with her human companion(s). Her narrative permits the human companion to understand what the animal is undergoing not merely in the moment, but as a part of her ongoing history. The human can understand the animal's suffering and thereby make a good decision with respect to her death because appreciating her narrative requires a past and ongoing relationship with the animal that appreciates the animal as a complete being.

A Worthwhile Future?

Despite the limits on their knowledge and understanding of what will happen, there are reasons to think—*contra* Velleman—that animals are connected in important ways to their own futures. Bernard Rollin (2009: 1081) remarks,

> [I]t seems obvious that they [nonhuman animals] definitely anticipate the short-term future, as when a cat waits outside a mouse hole or a lion intercepts a gazelle. Anticipating the long-term future seems more problematic . . ., but it is far from clear where one draws the line between short and long term, and thus we may well be aborting future projects when we kill an animal.

Thus, the potential euthanasia of an animal is morally significant because it may well "abort" the animal's "future projects." These might include the rewards of ongoing relationships with humans and other animals, the delights of eating, or the pleasures of walking, playing, and sleeping.

So Cholbi is correct to suggest that we need to think about what an animal might lose through her death. Evaluating when euthanasia is justified requires thinking carefully about the likely future the companion animal can have. As Milligan (2009: 407) says, both guardians and vets should try to ensure that the companion animal has "something that approximates to *whatever is a good life for an animal of the relevant type*" (emphasis in original). That good life will vary both by the animal's species and by her individual personality (Cholbi 2017: 270). And when an animal is ill, disabled, or aging, it is important to consider what a good life might mean for that animal *at that animal's stage of life*. Thus, Yeates (2010b: 71) suggests that human companions should "generate a list of activities the animal appears to enjoy while remembering that older animals may have new, compensatory pleasures."

The notion of "compensatory pleasures" suggests that, even if the animal's life is less good than it once was, it may not always be wrong to keep her alive. It can't necessarily be assumed that a life different from the one lived in her prime is not acceptable to the animal (Pierce 2012: 145). The question is not whether the in-her-prime companion animal would want to live as an old-and-dying companion animal (something that presumably an animal cannot conceptualize). The question is whether, once she reaches that stage, the animal finds it worthwhile.

Frances Kamm (2017: 738) suggests that, whereas a life lived with good experiences just above the zero line might not be valuable for a human being, perhaps it would be valuable for a dog or a cat. While a noticeable decline in capacities and abilities might be bad for a human being, perhaps it would not be bad for a dog or a cat. Interestingly, Pierce (2012) thinks humans and animals may be more *alike* in this respect than Kamm believes. Pierce (2012: 193) remarks,

> Research has found that it is people who are thriving who most strongly advocate for the legal option of assisted death; once a person becomes seriously ill, their perspective often changes, and they put up with far more than they would have predicted. Yet with our animals, we assume that their level of tolerance for pain and suffering is much smaller than ours and that death will be a welcome relief. We think about their dying in the same way a healthy able-bodied person thinks about her own assisted death, not like a dying person, whose perspective on the process is certainly going to be different and, of course, in many ways more authentic.

In other words, when a healthy human being contemplates a future in which he is ill or seriously disabled, he thinks it not worth living—and he is likely to impose that same viewpoint on his animal companion. But both human and animal may well encounter life stages when they still find their existence worthwhile even with significant limitations on their capacities and perhaps some pain.

However, Jeff McMahan (2002) suggests that there are important ways in which animal experience is different from human experience. On the one hand, he says, animals are not capable of some of the forms of "higher dimensions of wellbeing" that are accessible to humans. These include "deep personal relations, aesthetic experience, achievement through the exercise of complex skills, and so on" (McMahan 2002: 203). But on the other hand, "animals can get considerably closer to the lowest depths of human misery than they can to the heights of human well-being" (McMahan 2002: 203). So according to McMahan animal experience is so constituted as to give the depths of suffering a greater weight in the assessment of well-being than they might have for human beings.

This difference, he suggests, explains "the common view that death is more often preferable to a prospect of significant suffering in the case of animals than in the case of persons." An animal such as a stray dog or cat who "faces a prospect of suffering" has the potential to undergo severe pain and misery, while her prospects for "compensating goods" may be lower. "This is because the animal may simply be incapable, within its [*sic*] expected life span, of experiencing sufficient good to outweigh the suffering of which it is capable and which it stands a serious chance of experiencing" (McMahan 2002: 203; cf. 487).

Is McMahan correct about this difference? Certainly no cat will ever play the flute, and no dog has the capacity to appreciate works by Van Gogh. The question is whether the absence of these abilities makes suffering more easily outweigh enjoyment. Although an animal may not be capable of what McMahan regards as the "higher dimensions of wellbeing," she may still be capable of enjoying significant relationships, both with human beings and with other animals. She may still experience the joys of eating her favorite food, taking a long rest, or simply being petted and cuddled. Indeed, I have argued that because animals can enjoy the same sort of pleasure over and over, and they are unlikely to engage in worry about the future or regret about the past, they may in important respects be capable of *greater* enjoyment than adult human beings are (Overall 2017: 253). Provided an animal retains the ability to experience and enjoy at least some of these aspects of her physical existence, it's not clear why her suffering should so readily be assumed to outweigh her pleasures.

McMahan (2002) goes on to say that the reasons not to cause or allow the animal to suffer in the future derive both from her "present time-relative interest in avoiding future suffering" and from her future time-relative interests. In other words, the animal has interests in avoiding future suffering both in the present, before the suffering occurs, and in the future, when it could occur and could compromise her future enjoyments; these two different kinds of interests give human beings reasons not to cause or allow the animal's suffering. If the animal's future suffering would be both "intense and protracted," then the reasons not to cause that suffering may be stronger than the reasons not to kill the animal (McMahan 2002: 487–488).

This seems correct. But he makes a further argument that is troubling. On the one hand, according to McMahan, the reason to kill the animal is made stronger by the animal's future time-relative interest in avoiding suffering, but on the other hand, the reason *not* to kill her "cannot be strengthened" by the animal's future time-relative interest in having goods.

> For to kill an animal is to ensure that it [*sic*] will have no future time-relative interests. Thus one can prevent possible later time-relative interests from constraining one's present action by preventing them from arising—in this case by killing their potential bearer now.
>
> *(McMahan 2002: 487–488)*

In other words, the animal's future interests can count in favor of killing her, but the animal's future interests cannot count *against* killing her—because as a result of being killed, she will have no interests at all.

Most human companions will probably find this argument implausible. If the animal's future interests give us reasons for acting in the present, why shouldn't her interests in having future benefits be just as significant as her interests in avoiding future suffering? Why shouldn't her future enjoyments give us a reason to keep an animal alive, just as much as her future suffering gives us a reason not to keep her alive (McMahan 2002: 489)?

McMahan's reasons for rejecting any symmetry between future interests in avoiding suffering and future interests in enjoying benefits are related to other theoretical concerns he has, independent of his discussion of animal welfare. He believes that treating these interests the same would undermine his case in support of late abortion, and it would support preventing spontaneous miscarriages for the sake of the fetuses rather than the future parents (McMahan 2002: 489–490). McMahan concludes that "[t]he positive justification for animal euthanasia therefore remains elusive. An animal's possible future suffering clearly matters but exactly how or why it matters is hard to say" (McMahan 2002: 493).

I don't think those who have animal companions need to worry, either about the implications that trouble McMahan or about the difficulty of saying how or why their companion's actual or potential suffering matters. Unlike a fetus, an animal companion is an existing being in the world, one with whom one or more human beings are in a relationship. As their human companions, they inevitably are concerned about their animals' suffering, whether present or potential, and want to prevent it. That is *part of what being a caring companion means*. And in addition, they inevitably want their animal companions to have what is good for them, whether it is food, exercise, sleep, play, or cuddling. Someone who is worried about the right time to euthanize his dog will not be reassured by being told that the dog's potential for future enjoyment does not count against euthanizing her. The animal's potential for future good matters to her human companion; concern for her future good cannot be extinguished simply by terminating the animal's life. Whatever lies in the future for one's cat or dog inevitably and justifiably weighs on the human's responsibilities for her.

Relationships and Intertwined Interests

The human/animal relationship both creates and helps to define the human's moral responsibilities to the animal. But how much of a role should that relationship play in the decision whether to euthanize the animal? Milligan (2009) suggests that, given that the human wants his animal companion to have as good a life as possible, the human's own interests will not likely diverge from those of the animal. In the small chance that they do, "the animal's interests ought to come first" (Milligan 2009: 409). Yeates (2012: 611) goes further and says that for an animal, a life worth living is *not* a function of the interests of human beings. But there are reasons to doubt that the interests of animal companions can be defined so separately from those of their humans.

The reason is that the animal's interests are constituted, in part, relationally; that is, what is good or bad for the animal has a deep connection to what is happening in her human companion's life. In particular, the human's socioeconomic, medical, and psychological condition may affect his ability to care appropriately for the animal. Thus, as Cholbi (2017: 276) writes, "[m]edical care for diseased animals can be astonishingly costly, and guardians at least have the right to take that in account when determining when animals should be euthanized." Even Yeates (2010b: 71), who says that a worthwhile life for an animal is not a function of human interests, regards as morally significant whether the human companion can "cope physically and emotionally with nursing" an animal who is seriously ill. Given the dependence of the animal on the human being, it may not be realistic or morally justified to consider the companion animal's interests entirely independently of those of the human.

However, caution is necessary because emotions play a central role in the relationship of a human being to an animal. Thus Cholbi (2017: 270) notes the danger that a guardian might keep an animal alive too long out of "strong emotional attachment." And Milligan (2009: 409) considers the possibility that the interests of a child in the family might favor delaying euthanasia—for example, so that the child can say good-bye to the animal.[3] Indeed, I have talked with people who worry that, out of attachment to their animal and an unwillingness to let her go, they may have delayed euthanasia too long and hence caused the animal inappropriate additional pain. I have also talked with others who fear that they may have euthanized their cat or dog too early out of fear both of what the animal might otherwise undergo and of trepidation that they might not be able to endure the animal's pain. These very real and deep emotions may complicate and even, in some cases, undermine the making of the best decisions about animal euthanasia. Thus, although a companion animal's interests are inevitably intertwined with those of her human owner, care is necessary to ensure that human emotions do not get in the way of meeting the animal's true needs at the end of life.

Convenience

Unfortunately, human interests tend to dominate in many cases of animal death. The deliberate killing of companion animals is not always undertaken solely to relieve the animal's present suffering or obviate inevitable future suffering. Often, it is as much (or more) for the convenience of human companions as for any "compassion" for the animal. For example, the human is moving or is traveling or is very busy. As a result, in such cases, some veterinarians agree to "euthanize" an animal not because they think doing so is morally unproblematic but because they fear that the human will try to end the animal's life himself, relinquish the animal to a shelter, or simply move on to another veterinarian (Pierce 2012: 180).

Ending an animal's life for the convenience of the human companion is, of course, possible because animals are considered to be property, not autonomous beings deserving of respect in their own right. Annamaria Passantino et al. (2006) point out the contradiction, exemplified in their nation of Italy but common elsewhere, between laws that treat animals as property—so that "owners" of animal property can request "euthanasia" for the animal even when the animal is healthy—and laws that recognize that animals, unlike all other property, are capable of experiencing pain and pleasure (Passantino et al. 2006: 492, 493). Killing an animal to suit the human's convenience does not deserve to be called euthanasia.

In addition, so-called euthanasia at animal shelters may be undertaken not so much in the interests of the animals themselves but, rather, because the shelter houses so many animals that homes cannot be found for all of them. Unfortunately, Western society appears to have a fair amount of complacency about the killing of shelter animals. As McMahan (2002: 199) points out,

> even among those who are willing to devote their time and energy to caring for animals, most approve of the painless killing of stray animals for whom homes cannot be found. The

> apparent assumption is that death is preferable, for the animal's own sake, to even the quite limited amount of suffering it [*sic*] might experience as a stray.

Some ethicists think that these shelter killings are justified and qualify as euthanasia. According to Cholbi (2017: 277),

> there will be a point at which the continued life of a shelter animal is not a benefit to it [*sic*], at which time euthanasia would be morally required on the grounds that the animal's optimal life span has been reached.

Yeates (2010b: 71) agrees:

> [I]t might be unrealistically optimistic to keep an unwanted animal alive if it [*sic*] is unsuitable as a pet, if rehoming centres are full, if it is likely to end up being kenneled for a long time or if it needs major surgery or medical treatment, even if it could have a good life in an ideal world.

Hence, Yeates distinguishes between "absolutely justified euthanasia where an animal cannot avoid having a life of overall negative welfare except by dying," and "contextually-justified euthanasia where an animal could have a life worth living in an ideal world" (Yeates 2010b: 71), but such a life is not possible within our current reality.

But other commentators are less certain of the justification of euthanizing shelter animals. Clare Palmer (2003: 571) asks whether continued life really would be worse for these animals than death. Perhaps it depends on which possible alternative life the animal's existence in the shelter is being compared to. Is it life as a stray (with the attendant possibilities of hunger, ill health, and danger from cars and people) or life with bad adoptive parents or simply indefinite life in the shelter? Palmer (2003: 575) argues that, on the one hand, "it is questionable whether painless killing is an appropriate way of discharging responsibilities to unwanted but dependent animals humans have themselves created," and, on the other hand, where animals such as dogs and cats are living independently of human beings, scavenging or hunting for themselves, "being taken to an animal shelter for painless killing seems to be a denial of their *lack* of relationship with particular human beings" (emphasis in original). Thus, "the ethical responsibilities of the creation of dependence where it exists should be taken more seriously; [and] relative independence where it exists should be respected" (Palmer 2003: 576–577).

Lee Anne Fennell (2003: 7) goes further, pointing out the general societal context in which the lives of shelter animals are ended:

> [E]uthanasia would seem to be morally problematic insofar as it cuts off the life that a pet could have enjoyed had he not been born into a society whose collective acts and omissions have resulted (we will assume for present purposes) in its inability to provide him with a life that is worth living.

The animal pays—with her life—for the moral failures of a society that brings animals into existence (or allows them to come into existence) but fails to provide an environment in which the animals can flourish. Fennell adds, "[T]he fact that painless killing emerges as the preferable alternative does not—and cannot—turn it into an act without moral significance" (Fennell 2003: 8). Killing animals in animal shelters, then, may not be so readily justified as many people assume.

Alternatives

The discussion of the shelter context shows that making morally realistic comparisons of different times for euthanasia depends upon knowing the kinds of conditions in which the animal could and would have lived, had she survived. This also applies in cases where animals are in a relationship with a human being. Pierce (2013: 477) writes, "Too many companion animals are killed prematurely because adequate palliative care is unavailable and because euthanasia is so deeply entrenched in the cultural narrative of pet ownership." This is an important area in which animal euthanasia is clearly different from euthanasia for humans, and animals do not fare well in this respect. Pierce (2012: 134) suggests that we should reject the limited dichotomy of either pursuing active treatments or providing euthanasia. Instead, we should seek to create viable alternatives to the current "treat or kill" status quo (Pierce 2012: 154), alternatives that would, in at least some cases, make the animal's life worth continuing, if only for a while.

The World Health Organization (n.d.) defines *palliative care* in relation to human beings, but most of the features of the definition would also be relevant to palliative care for companion animals. That is, palliative care for companion animals would provide relief from pain and other distressing symptoms, enhance the animal's quality of life, enable the animal to live as actively as possible within the limits of her health condition, and provide a support system for the animal's family.

Palliative care may be offered even when the animal is not nearing death. Hospice, on the other hand, provides care for terminally ill or dying animals. It is a way to "prolong not length of life itself but length of quality life" (Pierce 2012: 135). Hospice takes place at home, with visits to the veterinarian when needed.

> The key practical aspects of animal hospice care include relieving pain and discomfort (which can involve administration of drugs, massage, physical therapy) and maximizing pleasure for the animal (e.g., through social interaction, companionship, mental stimulation, play, walks, human touch, delicious foods).
>
> *(Pierce 2012: 132)*

If palliative care and hospice were available to companion animals, they would help to provide, says Pierce (2012: 155), a "Goldilocks approach to animal death: not too soon, not too late, but just right." In this respect, we need to move the care of animals who are seriously ill or very old closer to the kind of care we seek for human beings.

Conclusion

Death may be good or bad for an animal; its value depends on its timing. Hence, the justification of euthanasia in particular cases depends on ascertaining how much enjoyment the animal is likely to experience in the future and how much suffering the animal is undergoing and will undergo. Evaluating the animal's experience in these respects requires understanding the animal within the context of her own life narrative. A companion animal's interests cannot be completely separated from the interests of her human companion, but the animal should not be subjected to ongoing suffering merely to assuage the human's emotions. Unfortunately, some, perhaps many, cases of so-called animal euthanasia, including those at shelters, are not undertaken in the interests of the animals but arise from reasons of human convenience or even just the inability of human beings to provide good environments for companion animals. Finally, the option of palliative care and hospice for companion animals would provide a much-needed addition, for aging and ill animals, to the existing situation in which only continuing active treatment or euthanasia are available.

As the discussion in this chapter has shown, there is no easy formula for determining the justification and timing of companion animal euthanasia. And although it is commonplace to remark that, with respect to the end of life, we treat animals better than human beings, this is often not the case. There are good reasons to rethink when and why we euthanize companion animals.

Notes

1. Throughout this chapter, for clarity, I refer to individual animals using feminine pronouns and to individual human beings using masculine pronouns.
2. I argue that life is good for companion animals, that a longer life is better for them, and that—other things being equal—it is worthwhile to prolong their lives (Overall 2017).
3. However, Milligan (2009: 407–408) says that considering what role the animal's suffering plays within her life in no way commits the human companion to ignoring the intensity of the animal's pain.

Recommended Readings

Cholbi, M. (2017) "The euthanasia of companion animals," in C. Overall (ed.) *Pets and People: The Ethics of Our Relationships with Companion Animals*, New York, NY: Oxford University Press, pp. 264–278.

(Suggesting that the ethical foundations for euthanizing companion animals are different from the foundations for euthanizing human beings, this chapter argues that animal euthanasia is justified when the animal has reached her optimum life span.)

Meijer, E. (2018) "The good life, the good death: Companion animals and euthanasia," *Animal Studies Journal* 7(1): 205–225.

(The author argues for new policies governing animal euthanasia based on an acknowledgment of animal subjectivity and uniqueness.)

Milligan, T. (2009) "Dependent companions," *Journal of Applied Philosophy* 26(4): 402–413.

(The article argues that an animal's human companion has an "epistemically privileged" insight into end-of-life decisions for his animal, derived from his understanding of the animal's life narrative.)

Overall, C. (2017) "Throw out the dog? Death, longevity, and companion animals," in C. Overall (ed.) *Pets and People: The Ethics of Our Relationships with Companion Animals*, New York, NY: Oxford University Press, pp. 249–263.

(The chapter argues—contrary to the claims of many philosophers—that death is, in general, bad for companion animals; that their death is not necessarily less bad than it is for human beings; and that dying prematurely is bad for them.)

Pierce, J. (2012) *The Last Walk: Reflections on Our Pets at the End of their Lives*, Chicago: University of Chicago Press.

(This book is both a memoir and a work of applied philosophy. The author discusses her dog Ody's old age, dying, and death and reflects on the choices she makes for his care at each stage.)

Bibliography

Bachelard, S. (2002) "On euthanasia: Blindspots in the argument from mercy," *Journal of Applied Philosophy* 19(2): 131–140.

Cholbi, M. (2017) "The euthanasia of companion animals," in C. Overall (ed.) *Pets and People: The Ethics of Our Relationships with Companion Animals* (pp. 264–278), New York: Oxford, University Press.

Donaldson, S., and Kymlicka, W. (2015) "Farmed animal sanctuaries: The heart of the movement?" *Politics and Animals* 1: 50–74.

Fennell, L. A. (2003) "Killing with kindness: An inquiry into the routinized destruction of companion animals," *Between the Species,* August: 1–10, http://digitalcommons.calpoly.edu/cgi/viewcontent.cgi?article=1053&context=bts.

Haug, L. I. (2011) "Treat or euthanize: Helping owners make critical decisions regarding pets with behavior problems," *Veterinary Medicine* 106(11): 564–569.

Kamm, F. M. (2017) "The purpose of my death: Death, dying and meaning," *Ethics* 127(3), April: 733–761.

McMahan, J. (2002) *The Ethics of Killing: Problems at the Margins of Life*, New York, NY: Oxford University Press.

Milligan, T. (2009) "Dependent companions," *Journal of Applied Philosophy* 26(4): 402–413.

Overall, C. (2017) "Throw out the dog? Death, longevity, and companion animals," in C. Overall (ed.) *Pets and People: The Ethics of Our Relationships with Companion Animals*, New York, NY: Oxford University Press.

Palmer, C. (2003) "Killing animals in animal shelters," in S. J. Armstrong and R. G. Botzler (eds.) *The Animal Ethics Reader*, 2nd ed, New York, NY: Routledge.

Passantino, A., Fenga, C., Morciano, C., Morelli, C., Russo, M., Di Pietro, C., and Passantino, M. (2006) "Euthanasia of companion animals: A legal and ethical analysis," *Annali dell'Istituto Superiore di Sanità* 42(4): 491–495.
Pierce, J. (2012) *The Last Walk: Reflections on Our Pets at the End of Their Lives*, Chicago, IL: University of Chicago Press.
Pierce, J. (2013) "The dying animal," *Journal of Bioethical Inquiry* 10(4): 469–478.
Rollin, B. E. (2009) "Ethics and euthanasia," *Canadian Veterinary Journal* 50(10): 1081–1086.
Velleman, J. D. (1993) "Well-being and time," in J. M. Fischer (ed.) *The Metaphysics of Death*, Stanford, CA: Stanford University Press.
World Health Organization (n.d.) "WHO definition of palliative care," www.who.int/cancer/palliative/definition/en/.
Yeates, J. W. (2010a) "Death is a welfare issue," *Journal of Agricultural and Environmental Ethics* 23(3): 229–241.
Yeates, J. W. (2010b) "Ethical aspects of euthanasia of owned animals," *In Practice* 32: 70–73.
Yeates, J. W. (2012) "Quality time: Temporal and other aspects of ethical principles based on a 'life worth living,'" *Journal of Agricultural and Environmental Ethics* 25: 607–624.

26

LINKS BETWEEN VIOLENCE AGAINST HUMANS AND NONHUMAN ANIMALS

Examining the Role of Adverse Family Environments

Shelby Elaine McDonald

Introduction

Associations between animal maltreatment[1] and other antisocial behaviors, such as aggression and interpersonal violence, have been well documented in research on children, adolescents, and adults (Ascione et al. 2018; Felthouse and Bernard 1979; Tapia 1971). The clinical significance of animal maltreatment behaviors as an indicator of maladjustment was formally recognized in the revised third edition of the *Diagnostic and Statistical Manual of Mental Disorders* (*DSM*) in 1987, in the *DSM-III-R* (American Psychiatric Association 1987), which included animal cruelty as a symptom of conduct disorder. Later editions of the *DSM* (IV, IV-TR, 5; American Psychiatric Association 1994, 2000, 2013) retained children's physical cruelty to animals in a set of indicators of conduct disorder labeled, "aggression to people and animals"; other behaviors in this category included physical fighting, use of a weapon to inflict harm on others, and bullying (Ascione et al. 2018; Signal et al. 2013). Since the inclusion of animal cruelty as a symptom of conduct disorder in the *DSM*, relations between human- and nonhuman animal-related violence have received increasing attention in the research literature. In particular, associations between psychological disorders, crime, and animal maltreatment have been the focus of significant research (Ascione et al. 2018). This chapter provides an overview of current empirical knowledge on the intersection of violence toward people and nonhuman animals to highlight sociological factors and affective processes that may play a role in associations between animal maltreatment and interpersonal violence. In particular, we highlight literature on adverse family environments (violent households) to illustrate various factors that may be involved in the onset, maintenance, and intergenerational transmission of these antisocial behaviors. The chapter concludes with an overview and discussion of the practical implications of research in this area, current gaps in knowledge, and opportunities for future research.

Links Between Animal Maltreatment and Violence Against Humans

There is increasing recognition of links between violence against human and nonhuman animals. However, Patterson-Kane's (2016) recent meta-analysis of studies examining animal maltreatment among violent offenders documents a relatively weak link between animal maltreatment and human-directed violence. Specifically, the authors highlight that the majority of those who mistreat animals do not go on to perpetrate violent acts against humans. Still, collectively, studies suggest that animal maltreatment may be a risk factor for later aggressive and criminal behavior. In particular, animal maltreatment in childhood is a complex form of antisocial behavior, and recent studies highlight the utility of this behavior as a clinical marker for risk of long-term maladjustment, including interpersonal violence.

Childhood maltreatment of pets has been linked to childhood externalizing behaviors, such as conduct problems, delinquency, and callous/unemotional behaviors (Hartman, Hageman, Williams, and Ascione 2019; Plant et al. 2016; McDonald et al. 2018); adulthood psychopathy such as antisocial personality disorder, obsessive-compulsive and histrionic personality disorders, and pathological gambling (Vaughn et al. 2009); and criminal offending (Haden et al. 2018; Hensley et al. 2009; Knight et al. 2014). For example, Luk et al. (1999) compared 141 clinic-referred children with a history of animal maltreatment with a comparison group of children with no history of animal maltreatment and found increased severity of conduct disorder symptoms and poorer outcomes among children who were cruel to animals. Similarly, retrospective investigations of criminal offenders have found significantly higher levels of childhood animal maltreatment among individuals who have committed aggressive and violent crimes than by those who committed nonaggressive crimes (e.g., property and drug-related crimes; Kellert and Felthous 1985; Merz-Perez et al. 2001).

Although longitudinal research in this area is limited, a 10-year prospective study of 363 families found that children's animal maltreatment behaviors were significantly related to conduct disorder diagnoses, referral for violent antisocial offending, and marital violence in the home (Becker et al. 2004). Across studies, it is suggested that perpetrators' aggressive tactics against animals in childhood often mirror the tactics used against human victims in adulthood (Wright and Hensley 2003). Such findings lend support to the association between childhood maltreatment of animals, conduct problems, and antisocial behavior involving violence toward humans.

Intimate Partner Violence and Animal Maltreatment

> Right before Christmas I had called my brother in Mexico to wish them happy holidays when he walked in the door and heard me. He got so upset he started pushing me and punching the wall. He said since one of the most important things to me was my dog he would burn it by tying it up to the grill and turning it on in the back yard so that I learned my lesson to not ever call my brother again.
>
> —*Survivor of IPV*

Perhaps the most extensively documented link between violence toward human and nonhuman animals is the association between animal maltreatment and intimate partner violence (IPV; also known as "domestic violence"). IPV refers to behaviors within an intimate relationship that are intended to exert power and control over another individual; IPV often involves psychological, physical, and/or sexual harm by a current or former intimate partner or spouse (Brownridge et al. 2011). Although there is a growing body of literature indicating that men and people of other gender identities and expressions, as well as individuals in nonheterosexual partnerships, are victims of IPV at rates of notable significance, the majority of research in this area has focused on the statistically more prevalent scenario of a female victim and male partner. Recent, nationally representative research reports that 24% of women in the United States experience severe physical aggression by an intimate partner during their lifetime (Breiding et al. 2014). Nearly 16% of US children 17 years or younger report having witnessed physical intimate partner violence during their lifetime (Finkelhor et al. 2015); co-occurring physical aggression toward children by parents is estimated at 40% (Appel and Holden

1998; Jouriles et al. 2008). IPV perpetrators use numerous tactics of coercion and domination to control and harm their partners and their children (Langhinrichsen-Rohling 2010; Lindhorst and Tajima 2008). Psychological forms of coercion are used more frequently than direct physical and/or sexual abuse against female partners (Coker et al. 2000). As such, studies document that perpetrators of IPV who live in households with pets often use threats and violence against animals as a coercive tactic to intimidate, coerce, and/or retaliate against their partner(s) and children (Ascione et al. 2007; Barrett et al. 2017; Campbell et al. 2018; Collins et al. 2018; McDonald 2018; Newberry 2017).

Rates of IPV and concomitant animal maltreatment have varied extensively across studies of pet-owning intimate partner violence survivors. It is estimated that between 25% (Simmons and Lehmann 2007) and 71% (Ascione 1998) of IPV survivors with pets have experienced violence toward an animal by their abusive partner. Discrepancies in rates of co-occurring animal maltreatment and IPV across studies may be due to several factors including differences in methodology, sample selection, and cross-cultural variations in attitudes toward nonhuman animals. For example, in a study of IPV survivors in the US, Hartman et al. (2018) reported that Hispanic perpetrators of IPV, particularly those who relocated to the US from Mexico, were less likely to harm pets than non-Hispanic perpetrators of IPV. There is some evidence that intimate partner violence perpetrators may be more likely to utilize animal maltreatment as a coercive tactic when their partner has a strong emotional bond or attachment with the pet (Collins et al. 2018). For example, Faver and Cavazos (2007) reported that 88% of participants who experienced animal maltreatment by their partner identified the pet as a "very important" source of support. Among women who did not report maltreatment of pets by their partner, only 51% indicated that their pet was an important source of support.

There is evidence that IPV perpetrators who abuse pets are also more likely to engage in other forms of severe IPV behaviors, such as stalking, sexual assault, emotional abuse, and frequent physical injury of their partner (Simmons and Lehmann 2007; Walton-Moss et al. 2005). IPV survivors who experience maltreatment of pets are not only at risk of more severe and frequent forms of IPV; recent studies suggest that concomitant exposure to animal maltreatment may also exacerbate the negative effects of IPV on their physical and psychological health and safety (Collins et al. 2018; McDonald et al. 2015; McDonald, Cody et al. 2017). Barrett et al. (2017) reported that survivors who had experienced animal cruelty by their abusive partner were at greater risk for more severe and frequent IPV. Similarly, a recent study examining reports of police officers present at IPV scenes indicated that women reporting IPV by a suspect with a history of animal maltreatment were significantly more likely to report being strangled (76%) and forced to have sex (26%) when compared to survivors whose abuser did not engage in animal maltreatment (Campbell et al. 2018). This study also found higher rates of other lethality risks in households where animal maltreatment was present, such as threats to kill children and easy access to a firearm(s).

Children Exposed to Animal Maltreatment and Family Adversity

> The dog had multiple cracked teeth from him [partner] punching her in the face. He would stand on the dog's neck and stomp her ribs. He would pick her up with her excess skin and swing her over his head and slam her back on the ground. [. . .] He would lock cats in the kennel and shake it. The dog had a litter and he broke one of the puppies' paws by throwing it. He would rub the dog's snout in her pee until her nose was raw. He would make the dog eat her own fecal matter if she went [to the bathroom] in the house. He also had two gerbils that turned cannibalistic because of neglect and, after leaving a carcass in the cage for three months, he drowned the other and its babies. She [daughter] was in the room- a foot from him- during a lot of the [animal] abuse. She was a toddler when most of it was happening.
>
> —*Mother/Survivor of IPV, describing her child's exposure to the maltreatment of family pets*

Given the overlap between violence toward human and nonhuman animals, studies exploring risk factors and developmental pathways relating to IPV perpetration and concomitant maltreatment of animals are needed to advance prevention and intervention efforts with survivors and perpetrators of IPV and pet maltreatment. Adverse childhood experiences, such as witnessing IPV and animal maltreatment, are associated with increased risk for a wide range of short- and long-term social and emotional problems, including childhood maltreatment of animals and interpersonal violence. Through a series of studies funded by the National Institute of Child Health and Human Development (5R01HD066503–04: Principal Investigators: J.H. Williams and F. R. Ascione), my colleagues and I have found that children in adverse family environments live within a duality of experiencing both potential benefits (e.g., bonds, comfort, companionship, stress reduction) and risks associated with pets in the household (Collins et al. 2018; McDonald, Collins et al. 2015; McDonald, Collins et al. 2017) and that the interplay between these factors may account for the development of antisocial behavior involving violence to people and nonhuman animals. In the following, risk factors and processes that may underlie links to violence against human and nonhuman animals are discussed in the context of adverse family environments.

Childhood Adversity

Across numerous epidemiological studies, the prevalence of exposure to adverse childhood events is estimated at approximately 50% in the US population (Green et al. 2010; Kessler et al. 1997; McLaughlin et al. 2010; McLaughlin et al. 2012). Such experiences include a diverse set of exposures that have occurred prior to age 18 that negatively impact health and psychosocial well-being. These include victimization (e.g., exposure to IPV, child maltreatment, exposure to community violence) and nonvictimization adversities (e.g., neglect, household drug use, poverty). It is widely documented that childhood adversities frequently co-occur, and for the majority of children and adolescents, these events occur as multiple rather than single experiences (Shin et al., 2018). Children exposed to adverse events experience greater odds of developing psychopathology, including anxiety, mood, and disruptive behavior disorders, compared with individuals without such exposure. Furthermore, the strength of association between childhood adversity and psychopathology increases as exposure to adversity increases (Green et al. 2010; McLaughlin et al. 2010; McLaughlin et al. 2012; Shin et al. 2018).

Exposure to Animal Maltreatment in Childhood

Although exposure to animal maltreatment has largely been ignored in research on childhood adversity, it is well documented that exposure to animal maltreatment is prevalent and frequently co-occurs with other forms of adversity. Retrospective studies have identified prevalence rates of child and adolescent exposure to animal maltreatment ranging from 22% to 57% (Degue and DiLillo 2009; Flynn 1999, 2000; Henry 2004a, 2004b; Thompson and Gullone 2006). The likelihood of animal maltreatment exposure is higher among families in which other forms of family violence, such as child maltreatment and IPV, have occurred. For example, Ascione et al. (2007) found that 61.5% of pet-owning IPV victims residing at a domestic violence shelter(s) reported that their children had been exposed to animal cruelty in the home, whereas only 2.9% of women in a nonvictimized comparison group reported this exposure among their children. Deviney et al. (1983) examined animal abuse in families with substantiated reports of child abuse and found that in 88% of households where child physical abuse occurred, animal maltreatment was also present. A recent study of Child Protection Workers in Canada found that 94% of participants reported that they had observed evidence of animal neglect during investigations in the past year; 44% reported that they had observed an animal be physically abused during an investigation in the past year (Girardi and Pozzulo 2012).

There is emerging evidence that animal maltreatment exposure is associated with short- and long-term child health and developmental outcomes and accounts for variance in internalizing and externalizing psychopathology over and above other co-occurring childhood adversities (McDonald et al. 2016; Girardi and Pozullo 2015). As previously discussed, several studies have documented associations between childhood exposure to animal maltreatment and externalizing behaviors during child and adulthood such as delinquent behavior, bullying, and animal cruelty (Becker et al. 2004; Currie 2006; Girardi and Pozzulo 2012). Fewer studies have examined relations between animal maltreatment exposure and other socioemotional outcomes. In a recent study of children recruited from IPV services, we found that children exposed to maltreatment of pets were more than five times more likely than children not exposed to maltreatment of pets to have severely maladjusted profiles of adjustment, including clinical levels of multiple behavioral health problems (i.e., behavior problems, attention problems, social problems, and callous/unemotional traits; McDonald et al. 2016). In later analyses, we found that exposure to pet maltreatment was the strongest predictor of children's internalizing symptoms in a model that adjusted for the severity of children's IPV exposure, maternal abuse, maternal education, and other household risk factors (McDonald, Dmitrieva et al. 2017). Relatedly, a retrospective study by Girardi and Pozullo (2015) found a significant interaction effect between participants' level of bonds with pets and animal cruelty exposure, controlling for the effect of experiencing co-occurring emotional abuse in childhood. Among participants with medium-level bonds, those who were exposed to animal cruelty had significantly higher depression and anxiety scores than those who were not exposed to animal cruelty. Among participants who were not exposed to animal cruelty, those with medium-level bonds had lower depression and anxiety scores than those with low-level bonds.

Through mixed-methods research on children exposed to IPV, my colleagues and I found evidence of differences in animal maltreatment exposure when comparing children with asymptomatic profiles of adjustment to those with emotional and behavioral difficulties (McDonald, Cody, et al. 2017). Specifically, children with emotional and behavioral difficulties described exposure to severe forms of animal maltreatment (often resulting in injury and/or death of an animal) perpetrated with the intent of coercing the child or his or her maternal caregiver. They were also more likely to report experiencing direct victimization by a parent following an effort to protect a pet. We hypothesize that these children may be more likely to go on to engage in concomitant violence against humans and nonhuman animals. In contrast, children with low levels of emotional and behavioral problem symptoms experienced animal maltreatment that was less severe, most often reporting exposure to physical violence intended to punish or discipline the pet for misbehavior. They also rarely reported intervening in incidents of animal maltreatment and/or trying to protect pets. These findings suggest that specific types of animal maltreatment exposure may differ in their effects on children's outcomes and that children's bonds with pets and/or emotion regulation may influence these relations.

Given the co-occurrence of animal maltreatment and other forms of adversity, it is not surprising that research points to the utility of childhood maltreatment of animals as a clinical marker for child maltreatment and maladjustment. However, further work is needed to establish the extent to which specific types of animal maltreatment exposure differ in their relations with socioemotional adjustment in the context of co-occurring adverse events. Do specific types of animal maltreatment exposure predict subsequent animal cruelty and interpersonal violence? There is some evidence that the severity and type of witnessed animal cruelty, as well as the child's relationship to the perpetrator, have implications for understanding subsequent behavior problems. Thompson and Gullone (2006) found higher levels of animal cruelty perpetration among participants who had witnessed a family member or friend abuse an animal, compared to witnessing animal cruelty perpetrated by a stranger. Similarly, Hensley and Tallichet (2005) found that inmates who observed a friend or family member harm animals, compared to those who witnessed animals harmed by another individual, reported hurting animals more frequently.

Processes Underlying Aggression Toward Human and Nonhuman Animals

Cultural Norms and Violence Exposure

When considering processes that contribute to dysfunctional human–animal relationships, such as animal cruelty, it is important to consider the role of cultural norms in this context. A recent article by Henry (2018) provides an informative overview of broad models that describe the origins of animal cruelty within the context of cultural norms that serve as determinants of human–animal interactions. Henry (2018) states:

> The broadest models of human-animal interactions describe the origins of animal cruelty within the context of cultural norms and standards. Several authors have highlighted the role that culture plays in defining acceptable types of human-animal interactions (Flynn, 2001; Arluke, 2002; Pagani, Robustelli, and Ascione, 2010). For example, Flynn (2001) pointed out that, within Western societies, the relationship between humans and animals is contextualized by the Judeo-Christian ethical system within which humans are 'given dominion' over animals. Consequently, people may view animals as property, without a moral standing of their own. The construal of animals as property frees humans to engage in behaviors that cause suffering and death. As Flynn (2001) notes, the patterns of behavior that produce the most harm to animals (e.g. factory farming) are entirely legal in the United States. Pagani, Robustelli, and Ascione (2010) argue that value systems often include 'mixed messages' about our relationships with animals—we protect and embrace some (e.g. dogs and cats) while we kill and eat others (e.g. cows and pigs.) Some behaviors, such as slaughtering and eating a cow, are acceptable, while the exact same behavior is unacceptable if the target is different (e.g. a dog.) Thus, cultural and subcultural contexts define the human-animal interactions that are either acceptable or are 'animal cruelty.'
>
> *(pg. 459)*

As previously noted by Henry (2018) and McDonald et al. (2018), such broad models do not meaningfully contribute to our understanding of individual differences in relationships with animals, especially the development and persistence of animal cruelty. Social learning theory is another framework often used to explain maltreatment of animals in childhood, given that many children who engage in maltreatment of animals appear to be reproducing learned violent behaviors (McEwen et al. 2014). Thus, exposure to models of antisocial behavior, such as witnessing family (Ascione et al. 2003; Baldry 2005; Currie 2006; McEwen et al. 2014) and community violence (Boat et al. 2011; Gullone and Robertson 2008), may underlie links between violence toward human and nonhuman animals. As previously mentioned, children who experience bullying and are exposed to IPV and child abuse are at increased risk for perpetrating animal maltreatment. Baldry (2005) found that children between the ages of 8 and 12 years who had witnessed IPV were three times more likely to engage in maltreatment of animals. Similarly, Currie (2006) found that children between the ages of 5 and 17 years were 2.95 times more likely to report having engaged in maltreatment of animals if they had been exposed to IPV compared to children who had not been exposed to IPV. Among the limited longitudinal research in this area, McEwen et al. (2014) reported that U.K. children who were physically maltreated by their caregivers were 3.32 times more likely to engage in harm toward animals compared to children who had not been mistreated by caregivers.

Motivations for Maltreatment of Animals

There is a paucity of research examining motivations for animal cruelty across the life span. Still, a small body of research has examined motivations for maltreatment of animals in childhood. For example, Ascione (1997) found that children described harming animals for a wide variety of reasons, including curiosity, peer pressure, excitement, exploration, as a means of self-injury (when a child bothers an animal until it attacks them), or when motivated to imitate observed cruel behavior. Hensley and Tallichet (2005) explored retrospective accounts of animal maltreatment among incarcerated adults and found that child and adolescent cruelty toward animals perpetrated out of anger was most common, occurring in 48% of participants. Interestingly, participants who engaged in multiple acts of animal maltreatment were three times more likely to report that they were motivated to perpetrate harm in order to control the animal when compared with participants who engaged in a singular act of animal cruelty.

Emotion Regulation and Cognitive Appraisals

I, along with colleagues at the University of Denver (Ascione) and Arizona State University (Williams), have studied children's explanations of their animal maltreatment behaviors and found that differences in impulsivity and anger regulation may interact with beliefs supporting violence toward animals to influence children's animal-related conduct. Similar to Hensley and Tallichet's (2005) findings, many children in our study who engaged in maltreatment of pets said that they did so because they felt angry and wanted to punish the animal (45%). For example, one child in our study stated:

> I would throw the cat at the wall and sit on her. But I stopped now, there was just too much anger in my life and I didn't want to take it out on anyone else. I did grab her really hard once too. Back in Florida, I did choke my dog to death on accident.

A large body of work on externalizing behaviors has theorized that anger regulation (or dysregulation) mediates the relation between beliefs supporting aggression and aggressive behavior (Roberton et al. 2012). As such, anger-motivated maltreatment of pets may reflect anger dysregulation, including both under- and overregulation of emotions, and such processes may underlie links between violence against human and nonhuman animals. Further research in this area is needed and studies would do well to test whether under- and/or overregulation of emotions may influence children's ability to control intense emotional responses to animals and/or reduce inhibitions against animal maltreatment (McDonald et al. 2018), particularly in contexts when aggression and violence toward animals is normalized and justified. For example, a child in our study explained:

> [E]veryone in my family has hurt the cat at least once. Because she would scratch and bite us for no reason. Sometimes then, dad and me and my sisters, basically my entire family [would hurt her] because she would bite us so we would scruff her—grab her by the back of her neck where there's extra neck skin.

To the author's knowledge, no studies to date have explored how child emotion regulation influences decision making in the context of interactions with animals and how beliefs supporting violence to animals may interact with emotion regulation to influence aggression toward pets. A large body of research indicates that children with beliefs that aggressive behavior is normative and/or socially acceptable are more likely to act aggressively (Crick and Dodge 1994). Children in adverse family environments experience disproportionate exposure to multifarious forms of violence. When maltreatment of pets is validated and/or reinforced by family members, children may be more likely to engage in such behaviors. Although social information processing models are frequently employed as a framework for understanding children's antisocial behavior, they are rarely applied to understanding animal maltreatment. Social information processing models assume that social cognitions are a major driving force in aggressive behavior. Thus, beliefs supporting maltreatment of animals may account for individual differences in the link between exposure to animal maltreatment and interpersonal violence and subsequent aggression toward human and nonhuman animals (McDonald et al. 2018).

As previously mentioned, there is also some evidence that callous/unemotional traits (Dadds et al. 2008; Hartman et al. 2019b) and empathy deficits may predispose children to engage in maltreatment of animals (Hartman et al. 2019b; Plant et al. 2016). Callous/unemotional personality traits, which are characterized by a general lack of empathic and emotional arousal (Dadds et al. 2016; Dadds et al. 2006), are used to discern conduct disorder that is characterized by shallow affect and a lack of concern for others. In addition, the presence of these traits characterize a subgroup of antisocial youth that are more likely to have deficits in the processing of negative emotional stimuli (Blair et al. 2001; Loney et al. 2003), use premeditated violence (Frick et al. 2003), and exhibit more severe and aggressive antisocial behavior (Frick and Dickens 2006). There is some evidence that early socialization plays an important role in the emergence of callous/unemotional (CU) traits, and it is hypothesized that abnormal affective experiences in early development may play a role in this process. Therefore, children who experience exposure to adversity, such as IPV and animal maltreatment, may be at risk for developing CU traits (McDonald, Dmitrieva et al. 2017; McDonald 2018), which may underlie the overlap between violence toward human and nonhuman animals.

Indeed, there is evidence that witnessing and perpetrating animal maltreatment is linked to callous/unemotional traits (Hartman et al. 2018; McDonald, Dmitrieva et al. 2017; McDonald et al. 2016). In this manner, maltreatment of animals may reflect psychological risk and individual traits that are associated with increased risk of antisocial behavior, as well as their interplay with environmental context (exposure to models of animal abuse and interpersonal violence) and genetic risk (Ascione et al. 2018; McDonald et al. 2018). In our study of child survivors of IPV, a child with high levels of CU traits, as reported by their maternal caregiver, explained his motivations for animal maltreatment as follows:

> I tortured them. Sometimes I threw stuff at it. Maybe like socks and stuff. I did that when I was torturing it. Just like when she was being really bad. I enjoy it and I think the animal deserves to be hurt. I do still torture it when it's mean to me. I do like dad does. He throws it away, and I do this sometimes when it's mean to us. We only do it when it's mean to us or chewing on the cord. I felt kind of funny and I liked it.

Interestingly, in our study of children who engaged in animal maltreatment (McDonald et al. 2018), a smaller subset of the sample engaged in cruelty toward animals that was motivated by curiosity. These behaviors ranged in severity from mild (e.g., poking frog with stick) to severe (e.g., pulling legs off frogs multiple times). Children who engaged in animal maltreatment rooted in curiosity about animal behavior and/or anatomy did not express that negative emotions were a motivating factor driving their animal maltreatment behaviors. This finding highlights that it's important to examine the motivations for violence toward animals and humans as factors underlying the onset and maintenance of these antisocial behaviors, particularly when they co-occur.

Beliefs and Attitudes About Animals

As described by Hawkins and Williams (2016: 504),

> [h]uman–animal interactions are affected by beliefs about animal mind (Davis and Cheeke 1998). Beliefs about animal mind, that is, believing that nonhuman animals have the ability to think, feel, and experience emotions, is arguably the most important cognitive domain

> (affective empathy being the most important emotional domain) influencing the moral status of animals (Sorabji 1995), attitudes toward animals, and subsequent animal welfare (Ellingsen et al. 2010). [. . .] However, we still have little understanding regarding the role of beliefs about animal minds in children—animal relationships.

Hawkins and Williams (2016) also note that beliefs that animals are sentient are commonplace across cultures; however, beliefs may differ depending on species type across cultures and settings. Beliefs about animal minds have been linked to caring behaviors (Ellingsen et al. 2010), empathy toward and attitudes about animals (Hawkins and Williams 2016; Hills 1995; Knight et al. 2004), and concern for animals (Hawkins and Williams 2016; Herzog and Galvin 1997). Among adults, beliefs about animal minds are positively associated with concern for animal welfare. Less is known about children's beliefs, however. In a recent study reporting on a sample of 6- to 13-year-old U.K. children, Hawkins and Williams (2016) found that beliefs about animal minds (children's perceptions of animal sentience) correlated with humane behaviors and compassion toward animals. As discussed by Hawkins and Williams, children who believe animals are *insentient* may be more likely to engage in negative behaviors toward animals (Knight et al. 2004) and be more likely to compromise animal welfare through curiosity and play behaviors (McDonald et al. 2018). Alternatively, some hypothesize that children and youth with callous/unemotional traits may be *more likely* to engage in cruelty if they believe animals are sentient.

Given that individuals hold different beliefs about animal minds according to the type or species of animal (and in relation to their cultural context), it is also important to consider the role of animal type in the development of animal cruelty. Signal et al. (2018) found support for prior assertions linking human-directed empathy and attitudes toward animals; however, their findings suggest that the type of animal (i.e., category of animal: pet pest or profit animal) has an impact on the strength of this relationship. For example, in their large, community-based study of 1,606 Australian adults, human-directed empathy was strongly associated with pro-pet attitudes. This is not surprising given that pet animals are often considered members of the family within multispecies households (Signal et al. 2018). Interestingly, their study also found that elevated levels of anxiety (personal distress) in interpersonal situations (thought to be a dimension that moderates empathic responding) might be related to more positive attitudes toward pests (feral animals) and profit animals (animals used for food) but not pet animals. These findings are interesting to consider and reinforce the importance of evaluating the intersection of beliefs about animal minds and attitudes toward different types of animals among children and adults when considering the development and persistence of animal cruelty across the life span.

Social-Cognitive Models of Animal Cruelty

In my earlier work, I have drawn on social-cognitive perspectives on interpersonal aggression to discuss our team's findings related to animal cruelty and make recommendations for future research and intervention work (e.g., distinguishing between reactive and instrumental animal cruelty, considering hostile attribution biases that influence children's perceptions of animal behavior, implications of children's anger dysregulation and beliefs about aggression to pets). A recent paper by Henry (2018) portrays similar and complementary arguments and offers comprehensive and helpful recommendations for the adoption of a "social-cognitive model of animal cruelty." Although further elaboration on the topic is outside the scope of this chapter, I encourage readers interested in this topic to review Henry's detailed and informative review of social cognitive perspectives on animal cruelty. Henry provides a promising framework for animal cruelty research that (a) mirrors themes in children's narratives about their animal maltreatment behaviors (see McDonald et al. 2018) and (b) centers the study of animal cruelty within the broader literature on aggression to inform testable hypotheses that may lead to a greater understanding of individual differences in animal cruelty. Specifically, his proposed model highlights how social information processing, beliefs about aggression, and schemas may be underlying, interactive mechanisms through which aggression toward animals can be understood and addressed.

Gaps in the Literature and Directions for Future Research

There is a substantial need for longitudinal research aimed at understanding the onset, maintenance, and intergenerational transmission of violence against people and nonhuman animals. Identification of malleable risk and protective factors that can be targeted in intervention programs that aim to treat and prevent these overlapping forms of antisocial behavior will help to address these public health issues. Although exposure to animal maltreatment may pose risk to child development and well-being, positive aspects of human–animal interaction may confer protective effects in the context of environmental adversity. As such, there is great need for research to determine whether and the extent to which the associations between animal maltreatment exposure and child adjustment are moderated by positive human–animal interaction. Recent theoretical work suggests that pets may benefit child socioemotional health by improving social regulation of emotion and enhancing executive functioning (i.e., inhibitory control). It is hypothesized that these effects may be particularly pronounced among individuals who lack supportive social networks and nurturing environments (Brown and Coan 2016; Freund et al. 2016; Ling et al. 2016). To the author's knowledge, no published studies to date have directly examined how children's interactions with their pets influence relations between adverse childhood experiences and socioemotional development. However, I, along with colleagues at Virginia Commonwealth University and Greater Richmond SCAN (Stop Child Abuse Now), were recently awarded a grant from the National Institute of Child Health and Human Development to explore the relations among childhood adversity, human–animal interaction, and child outcomes (1R21HD097769–01; Principal Investigator: McDonald). Current evidence suggests that pet ownership and related interactions may promote emotional and social health in both normative and vulnerable populations of children and adolescents (Purewal et al. 2017). For example, cross-sectional studies have reported that pet ownership is associated with a decreased likelihood of general anxiety among children in a primary care sample and lower rates of depression and loneliness in homeless youth (e.g., Gadomski et al. 2015; Rhoades et al. 2015).

Previous studies also suggest that pet bonding and attachment may be stronger determinants of positive child development than pet ownership and that strong bonds with pets are associated with higher scores in domains of social competence, self-esteem, empathy, and perspective-taking abilities (e.g., Gadomski et al. 2015; Maruyama 2011; Poresky and Hendrix 1989; Rhoades et al. 2015; Triebenbacher 1998). In a recent study, my colleagues and I found that 7- to 12-year-old children exposed to IPV had high levels of positive engagement with pets (McDonald, Vidacovich et al. 2015). Most reported frequently engaging in play and caretaking behaviors, as well as seeking out pets during negative emotional states (e.g., crying with pets when sad). We also found that children's reports of positive engagement with pets were negatively correlated with parental reports of callous/unemotional traits, which are predictive of subsequent antisocial behavior involving violence against human and nonhuman animals. Still, there is inconsistent support in the broader literature for the emotional and social benefits of interactions with household pets, which may be partly attributable to the inconsistent methodological and measurement approaches employed across studies (Arambašić et al. 1999; McDonald, Vidacovich et al. 2015; Mathers et al. 2010).

Conclusion and Practical Applications

Literature examining the intersection of violence against humans and animals suggests that maltreatment of animals is a risk factor for interpersonal violence and related maladjustment across the life span. Exposure to childhood adversity, including maltreatment of animals, increases children's risk of antisocial behavior including violence against human and nonhuman animals. Callous/unemotional traits, compromised emotion regulation, and psychopathology, such as conduct disorder and antisocial personality disorder, are associated with early exposure to adversity, and these individual risk factors may play a role in the development of antisocial behavior involving violence against people and nonhuman animals. As a whole, literature on the risk and protective effects of pets among vulnerable children underscore the need for research to examine how animal maltreatment exposure and positive aspects of human–animal interaction (e.g., children's bonds and attachments with pets) interact and influence children's development in the context of co-occurring adversities. This is particularly important in the context of understanding the onset and maintenance of antisocial behavior involving violence to people and animals. The majority of research to date has been descriptive in design. Important questions remain as to the developmental processes and risk and protective factors involved in the etiology of interpersonal violence and concomitant animal maltreatment across the life course.

Although this area of research requires much more development, the information presented in this chapter has many implications for various professions, including mental health professionals (psychology, social work, counseling, therapists); educators, child development researchers, and specialists; professionals in law enforcement and forensic settings; and animal welfare workers. Here, I'd like to highlight recent arguments by Buck and Rauscher (2019) regarding the ethics of competency in human–animal relationships for helping professionals interacting with individuals across settings. The authors argue that:

> [t]herapists, social workers, psychologists, counselors, and helping professionals of all kinds have an ethical obligation to be competent to practice in complex and changing sociocultural settings. In the United States, there have been significant population shifts in race, ethnicity, marital status, and gender identity. Other changes include life expectancy, income disparity, and an increase in female participation in STEM fields. One sociocultural change that has received little attention is the shifting relational dynamic of pets in the lives of their owners. These companion animals play increasingly significant roles in American households, resulting in both an increase in the number of pet-owning households and an important shift in the human-animal dynamic from occupational to relational.
>
> *(pg. 535)*

The current chapter has reported on key issues related to animal maltreatment and demonstrated the importance of competency and knowledge related to animal cruelty and its correlates both in practice and research settings and across professional contexts. In my own field, which is the profession of social work, it is important to highlight that no code of ethics from *any* Anglo-American social work profession include reference to nonhuman animals (Taylor et al., 2014, as cited in Buck and Rauscher 2019). Indeed, in a national survey, Risley-Curtiss (2010) found that nearly 63% of social workers reported that their training did not include content on human–animal interaction. Buck and Rauscher argue that

> [l]acking formal training in the role of companion animals in the lives of individuals, families, and communities, compounded by its absence in codes of ethics and a paucity of training opportunities, therapists are, at best, ill-equipped to appropriately serve their clients when pet issues arise and, at worst, at risk of harming clients when the lack of knowledge presents in the form of bias.
>
> *(pg. 537)*

One could argue that this is the case for any helping profession intersecting with children, families, and animals and that the ability to identify and evaluate animal abuse as a potential indicator and predictor of human violence, neglect, and polyvictimization is tied to these professions' ethical responsibilities to engage in competent practice.

Finally, it is the author's hope that the research highlighted in this chapter demonstrates the importance, and need for, interdisciplinary partnerships between agencies and programs that provide services related to human and/or nonhuman animal welfare, as well as the need for policies and programs related to animal cruelty (particularly in child welfare and IPV settings). Currently, only 13 states in the US have laws that mandate cross-reporting of IPV, child abuse, and animal cruelty (Buck and Rauscher 2019). There is a substantial need for policy advocacy in this area. In many states, animal control officers are required to report suspected child abuse; however, typically there are no mandatory reporting procedures for animal abuse encountered by child welfare professionals. Moreover, there are no formal policies in place for police officers who encounter children who have witnessed animal abuse. Cross-reporting mandates are needed and should include animal, child, and adult protection workers; IPV workers; veterinarians; and law enforcement. Such procedures might help us prevent and intervene in incidents of animal maltreatment and more effectively promote the well-being of human and nonhuman animals.

Note

1. In this chapter, the term *animal maltreatment* is used to refer to "non-accidental, socially unacceptable behavior that causes pain, suffering, or distress to and/or the death of an animal" (Ascione and Shapiro 2009: 570). In the broader literature, animal maltreatment is more commonly known as "animal cruelty" and "animal abuse"; the terms are used interchangeably in this text. The author has elected to use the term *animal maltreatment* as it is a more comprehensive term that encompasses a broad range of behaviors that harm nonhuman animals.

Bibliography

American Psychiatric Association (1987) *Diagnostic and Statistical Manual of Mental Disorders: DSM-III-R*. Washington, DC: Author.

American Psychiatric Association (1994) *Diagnostic and Statistical Manual of Mental Disorders: DSM-IV*, Washington, DC: American Psychiatric Association.

American Psychiatric Association (2000) *Diagnostic and Statistical Manual of Mental Disorders: DSM-IV-TR*, Washington, DC: American Psychiatric Association.

American Psychiatric Association (2013) *Diagnostic and Statistical Manual of Mental Disorders: DSM-5*. Washington, DC: American Psychiatric Association.

American Veterinary Medical Association (2007) *Pet Ownership and Demographics Sourcebook*, Schaumburg, IL: American Veterinary Medical Association.

Appel, A. E., and Holden, G. W. (1998) "The co-occurrence of spouse and physical child abuse: A review and appraisal," *Journal of Family Psychology* 12(4): 578.

Arambašić, G. K. L., Kuterovac-Jagodić, G., and Vidović, V. V. (1999) "Pet ownership and children's self-esteem in the context of war," *Anthrozoös* 12(4): 218–223.

Ascione, F. R. (1997) "Battered women's reports of their partners' and their children's cruelty to animals," *Journal of Emotional Abuse* 1(1): 119–133.

Ascione, F. R. (1998) "Battered women's reports of their partners' and their children's cruelty to animals," *Journal of Emotional Abuse* 1(1): 119–133. https://doi.org/10.1300/J135v01n01_06.

Ascione, F. R., Friedrich, W. N., Heath, J., and Hayashi, K. (2003) "Cruelty to animals in normative, sexually abused, and outpatient psychiatric samples of 6- to 12-year-old children: Relations to maltreatment and exposure to domestic violence," *Anthrozoös* 16(3): 194–212. https://doi.org/10.2752/089279303786992116.

Ascione, F. R., McDonald, S. E., Tedeschi, P., and Williams, J. H. (2018) "The relations among animal abuse, psychological disorders, and crime: Implications for forensic assessment," *Behavioral Sciences and the Law*. Advance Online Publication. https://doi.org/10.1002/bsl.2370.

Ascione, F. R., and Shapiro, K. (2009) "People and animals, kindness and cruelty: Research directions and policy implications," *Journal of Social Issues* 65: 569–587.

Ascione, F. R., Weber, C. V., Thompson, T. M., Heath, J., Maruyama, M., and Hayashi, K. (2007) "Battered pets and domestic violence: Animal abuse reported by women experiencing intimate violence and by nonabused women," *Violence Against Women* 13(4): 354–373.

Baldry, A. C. (2004) "The development of the PET scale for the measurement of physical and emotional tormenting against animals in adolescents," *Society and Animals* 12(1): 1–17.

Barrett, B. J., Fitzgerald, A., Stevenson, R., and Cheung, C. H. (2017) "Animal maltreatment as a risk marker of more frequent and severe forms of intimate partner violence," *Journal of Interpersonal Violence*. doi: 10.1177/0886260517719542.

Becker, K. D., Stuewig, J., Herrera, V. M., and McCloskey, L. A. (2004) "A study of firesetting and animal cruelty in children: Family influences and adolescent outcomes," *Journal of the American Academy of Child & Adolescent Psychiatry* 43(7): 905–912.

Blair, R. J. R., Colledge, E., Murray, L., and Mitchell, D. G. V. (2001) "A selective impairment in the processing of sad and fearful expressions in children with psychopathic tendencies," *Journal of Abnormal Child Psychology* 29(6): 491–498.

Boat, B. W., Pearl, E., Barnes, J. E., Richey, L., Crouch, D., Barzman, D., and Putnam, F. W. (2011). Childhood cruelty to animals: Psychiatric and demographic correlates. *Journal of Aggression, Maltreatment & Trauma, 20*(7), 812–819.

Breiding, M. J., Chen, J., and Black, M. C. (2014) *Intimate Partner Violence in the United States – -2010*. Atlanta, GA: National Center for Injury Prevention and Control, Centers for Disease Control and Prevention.

Brown, C., and Coan, J. (2016) "Human-animal interaction and the development of executive functions," in L. S. Freund, S. McCune, L. Esposito, N. R. Gee, and P. McCardle (eds.) *The Social Neuroscience of Human-animal Interaction*, Washington, DC: American Psychological Association.

Brownridge, D. A., Taillieu, T. L., Tyler, K. A., Tiwari, A., Chan, K. L., and Santos, S. C. (2011) "Pregnancy and intimate partner violence: Risk factors, severity, and health effects," *Violence Against Women* 17(7): 858–881.

Buck, P. W., and Rauscher, K. A. (2019) "The ethics of competency in human-animal relationships," in *Clinician's Guide to Treating Companion Animal Issues*, London, UK: Academic Press. https://www.amazon.com/Clinicians-Treating-Companion-Animal-Issues/dp/012812962X/ref=asc_df_012812962X/?tag=hyprod-20&linkCode=df0&hvadid=312519927002&hvpos=1o1&hvnetw=g&hvrand=14896970390491768096&hvpone=&hvptwo=&hvqmt=&hvdev=c&hvdvcmdl=&hvlocint=&hvlocphy=9008458&hvtargid=pla-569044479266&psc=1

Campbell, A. M., Thompson, S. L., Harris, T. L., and Wiehe, S. E. (2018) "Intimate partner violence and pet abuse: Responding law enforcement officers' observations and victim reports from the scene," *Journal of Interpersonal Violence*. https://doi.org/10.1177/0886260518759653.

Coker, A. L., Smith, P. H., McKeown, R. E., and King, M. J. (2000) "Frequency and correlates of intimate partner violence by type: Physical, sexual, and psychological battering," *American Journal of Public Health* 90(4): 553.

Collins, E. A., Cody, A. M., McDonald, S. E., Nicotera, N., Ascione, F. R., and Williams, J. H. (2018) "A template analysis of intimate partner violence survivors' experiences of animal maltreatment: Implications for safety planning and intervention," *Violence Against Women* 24(4): 452–476.

Crick, N. R., and Dodge, K. A. (1994) "A review and reformulation of social information processing mechanisms in children's social adjustment," *Psychological Bulletin* 115(1): 74. https://doi.org/10.1037/0033-2909.115.1.74.

Currie, C. L. (2006) "Animal cruelty by children exposed to domestic violence," *Child Abuse & Neglect* 30(4): 425–435.

Dadds, M. R., Gale, N., Godbee, M., Moul, C., Pasalich, D. S., Fink, E., and Hawes, D. J. (2016), "Expression and regulation of attachment-related emotions in children with conduct problems and callous—Unemotional traits," *Child Psychiatry & Human Development* 47(4): 647–656.

Dadds, M. R., Hunter, K., Hawes, D. J., Frost, A. D., Vassallo, S., Bunn, P., . . . El Masry, Y. (2008) "A measure of cognitive and affective empathy in children using parent ratings," *Child Psychiatry and Human Development* 39(2): 111–122. https://doi.org/10.1007/s10578-007-0075-4.

Dadds, M. R., Whiting, C., and Hawes, D. J. (2006) "Associations among cruelty to animals, family conflict, and psychopathic traits in childhood," *Journal of Interpersonal Violence* 21(3): 411–429.

Davis, S. L., and Cheeke, P. R. (1998) "Do domestic animals have minds and the ability to think? A provisional sample of opinions on the question," *Journal of Animal Science* 76(8): 2072–2079.

DeGue, S., and DiLillo, D. (2009) "Is animal cruelty a "red flag" for family violence? Investigating co-occurring violence toward children, partners, and pets," *Journal of Interpersonal Violence* 24(6): 1036–1056.

Deviney, E., Dickert, J., and Lockwood, R. (1983) "The care of pets within child abusing families," *International Journal for the Study of Animal Problems* 4: 321–329.

Ellingsen, K., Zanella, A. J., Bjerkås, E., and Indrebø, A. (2010) "The relationship between empathy, perception of pain and attitudes toward pets among Norwegian dog owners," *Anthrozoös* 23(3): 231–243.
Faver, C. A., and Cavazos, A. M. (2007) "Animal abuse and domestic violence: A view from the border," *Journal of Emotional Abuse* 7(3): 59–81.
Felthouse, A. R., and Bernard, H. (1979) "Enuresis, firesetting and cruelty to animals: The significance of two-thirds of this triad," *Journal of Forensic Sciences* 24: 240–246.
Finkelhor, D., Turner, H. A., Shattuck, A., and Hamby, S. L. (2015) "Prevalence of childhood exposure to violence, crime, and abuse: Results from the national survey of children's exposure to violence," *JAMA Pediatrics* 169(8): 746–754.
Flynn, C. P. (1999) "Exploring the link between corporal punishment and children's cruelty to animals," *Journal of Marriage and the Family* 61(4): 971–981.
Flynn, C. P. (2000) "Woman's best friend: Pet abuse and the role of companion animals in the lives of battered women," *Violence Against Women* 6(2): 162–177.
Freund, L. S., McCune, S. E., Esposito, L. E., Gee, N. R., and McCardle, P. E. (eds.) (2016) *The Social Neuroscience of Human-animal Interaction,* Washington, DC: American Psychological Association.
Frick, P. J., Cornell, A. H., Barry, C. T., Bodin, S. D., and Dane, H. E. (2003) "Callous-unemotional traits and conduct problems in the prediction of conduct problem severity, aggression, and self-report of delinquency," *Journal of Abnormal Child Psychology* 31(4): 457–470.
Frick, P. J., and Dickens, C. (2006) "Current perspectives on conduct disorder," *Current Psychiatry Reports* 8(1): 59–72.
Gadomski, A. M., Scribani, M. B., Krupa, N., Jenkins, P., Nagykaldi, Z., and Olson, A. L. (2015) "Pet dogs and children's health: Opportunities for chronic disease prevention?" *Preventing Chronic Disease* 12: 1–10.
Girardi, A., and Pozzulo, J. D. (2012) "The significance of animal cruelty in child protection investigations," *Social Work Research* 36(1): 53–60.
Girardi, A., and Pozzulo, J. D. (2015) "Childhood experiences with family pets and internalizing symptoms in early adulthood," *Anthrozoös* 28(3), 421–436.
Green, J. G., McLaughlin, K. A., Berglund, P. A., Gruber, M. J., Sampson, N. A., Zaslavsky, A. M., and Kessler, R. C. (2010) "Childhood adversities and adult psychiatric disorders in the national comorbidity survey replication I: Associations with first onset of DSM-IV disorders," *Archives of General Psychiatry* 67(2): 113–123.
Gullone, E., and Robertson, N. (2008) "The relationship between bullying and animal abuse behaviors in adolescents: The importance of witnessing animal abuse," *Journal of Applied Developmental Psychology* 29(5): 371–379.
Haden, S. C., McDonald, S. E., Booth, L. J., Ascione, F. R., and Blakelock, H. (2018) "An exploratory study of domestic violence: Perpetrators' reports of violence against animals," *Anthrozoös* 31(3): 337–352.
Hartman, C. A., Hageman, T., Williams, J. H., and Ascione, F. R. (2019a) "Intimate partner violence and animal abuse in an immigrant-rich sample of mother—child dyads recruited from domestic violence programs," *Journal of Interpersonal Violence* 33(6): 1030–1047.
Hartman, C., Hageman, T., Williams, J. H., Mary, J. S., & Ascione, F. R. (2019b). Exploring empathy and callous–unemotional traits as predictors of animal abuse perpetrated by children exposed to intimate partner violence. *Journal of Interpersonal Violence*, 34(12), 2419–2437.
Hawkins, R. D., and Williams, J. M. (2016) "Children's beliefs about animal minds (Child-BAM): Associations with positive and negative child—animal interactions," *Anthrozoös* 29(3): 503–519.
Hawkins, R. D., McDonald, S. E., O'Connor, K., Matijczak, A., Ascione, F. R., and Williams, J. H. (2019) "Exposure to intimate partner violence and internalizing symptoms: The moderating effects of positive relationships with pets and animal cruelty exposure," *Child Abuse & Neglect*, 98, Advance Online Publication, https://doi.org/10.1016/j.chiabu.2019.104166
Henry, B. C. (2004a) "Exposure to animal abuse and group context: Two factors affecting participation in animal abuse," *Anthrozoös* 17(4): 290–305.
Henry, B. C. (2004b) "The relationship between animal cruelty, delinquency, and attitudes toward the treatment of animals," *Society & Animals* 12(3): 185–207.
Henry, B. C. (2018) "A social-cognitive model of animal cruelty," *Psychology, Crime & Law* 24(5): 458–478.
Hensley, C., and Tallichet, S. E. (2005) "Animal cruelty motivations: Assessing demographic and situational influences," *Journal of Interpersonal Violence* 20(11): 1429–1443.
Hensley, C., Tallichet, S. E., and Dutkiewicz, E. L. (2009) "Recurrent childhood animal cruelty: Is there a relationship to adult recurrent interpersonal violence?" *Criminal Justice Review* 34(2): 248–257.
Herzog, H. A., and Galvin, S. (1997) "Common sense and the mental lives of animals: An empirical approach," in R. W. Mitchell (ed.) *Anthropomorphism, Anecdotes and Animals*, Albany, NY: State University of New York Press.

Hills, A. M. (1995) "Empathy and belief in the mental experience of animals," *Anthrozoös* 8(3): 132–142.
Jouriles, E. N., McDonald, R., Slep, A. M. S., Heyman, R. E., and Garrido, E. (2008) "Child abuse in the context of domestic violence: Prevalence, explanations, and practice implications," *Violence and Victims* 23(2): 221.
Knight, K. E., Ellis, C., and Simmons, S. B. (2014) "Parental predictors of children's animal abuse: Findings from a national and intergenerational sample," *Journal of Interpersonal Violence* 29(16): 3014–3034.
Kellert, S., and Felthous, A. (1985) "Childhood cruelty toward animals among criminals and noncriminals," *Human Relations* 38: 1113–1129.
Kessler, R. C., Davis, C. G., and Kendler, K. S. (1997) "Childhood adversity and adult psychiatric disorder in the US National Comorbidity Survey," *Psychological Medicine* 27(5): 1101–1119.
Knight, S., Vrij, A., Cherryman, J., and Nunkoosing, K. (2004) "Attitudes toward animal use and belief in animal mind," *Anthrozoös* 17: 43–62.
Langhinrichsen-Rohling, J. (2010) "Controversies involving gender and intimate partner violence in the United States," *Sex Roles* 62(3–4): 179–193.
Lindhorst, T., and Tajima, E. (2008) "Reconceptualizing and operationalizing context in survey research on intimate partner violence," *Journal of Interpersonal Violence* 23(3): 362–388.
Ling, D. S., Kelly, M., and Diamond, A. (2016) "Human-animal interaction and the development of executive functions," in L. S. Freund, S. McCune, L. Esposito, N. R. Gee, and P. McCardle (eds.) *The Social Neuroscience of Human-animal Interaction*, Washington, DC: American Psychological Association.
Loney, B. R., Frick, P. J., Clements, C. B., Ellis, M. L., and Kerlin, K. (2003) "Callous-unemotional traits, impulsivity, and emotional processing in adolescents with antisocial behavior problems," *Journal of Clinical Child and Adolescent Psychology* 32(1): 66–80.
Luk, S. L., Staiger, P. K., Wong, L., and Mathai, J. (1999) "Children who are cruel to animals: A revisit," *Australian and New Zealand Journal of Psychiatry* 33: 29–36.
Maruyama, M. (2011) *The Effects of Animals on Children's Development of Perspective-Taking Abilities*, Ann Arbor, MI: ProQuest Information & Learning.
Mathers, M., Canterford, L., Olds, T., Waters, E., and Wake, M. (2010) "Pet ownership and adolescent health: Cross-sectional population study," *Journal of Paediatrics and Child Health* 46(12): 729–735.
McDonald, S. E. (2018) "Animal maltreatment in households experiencing family violence," in J. L. Ireland, C. A. Ireland, and P. Birch (eds.) *International Handbook on Aggression* (pp. 183–193), London, UK: Routledge.
McDonald, S. E., Cody, A. M., Booth, L. J., Peers, J. R., Luce, C. O. C., Williams, J. H., and Ascione, F. R. (2018) "Animal cruelty among children in violent households: Children's explanations of their behavior," *Journal of Family Violence*, 1–12.
McDonald, S. E., Cody, A. M., Collins, E. A., Stim, H. T., Nicotera, N., Ascione, F. R., and Williams, J. H. (2017) "Concomitant exposure to animal maltreatment and socioemotional adjustment among children exposed to intimate partner violence: A mixed methods study," *Journal of Child & Adolescent Trauma*, 1–13.
McDonald, S. E., Collins, E. A., Maternick, A., Nicotera, N., Graham-Bermann, S., Ascione, F. R., and Williams, J. H. (2019) "Intimate partner violence survivors' reports of their children's exposure to companion animal maltreatment: A qualitative study," *Journal of Interpersonal Violence*. Advance online publication. https://doi.org/10.1177/0886260516689775.
McDonald, S. E., Collins, E. A., Nicotera, N., Hageman, T. O., Ascione, F. R., Williams, J. H., and Graham-Bermann, S. A. (2015) "Children's experiences of companion animal maltreatment in households characterized by intimate partner violence," *Child Abuse & Neglect* 50: 116–127.
McDonald, S. E., Dmitrieva, J., Shin, S., Hitti, S. A., Graham-Bermann, S. A., Ascione, F. R., and Williams, J. H. (2017) "The role of callous/unemotional traits in mediating the association between animal abuse exposure and behavior problems among children exposed to intimate partner violence," *Child Abuse & Neglect* 72: 421–432.
McDonald, S. E., Graham-Bermann, S. A., Maternick, A., Ascione, F. R., and Williams, J. H. (2016) "Patterns of adjustment among children exposed to intimate partner violence: A person-centered approach," *Journal of Child & Adolescent Trauma* 9(2): 137–152.
McDonald, S. E., Vidacovich, C., Ascione, F. R., Williams, J. H., and Green, K. E. (2015) "The children's treatment of animals questionnaire: A Rasch analysis," *Anthrozoös* 28(1): 131–144.
McEwen, F. S., Moffitt, T. E., and Arseneault, L. (2014) "Is childhood cruelty to animals a marker for physical maltreatment in a prospective cohort study of children?" *Child Abuse and Neglect* 38(3): 533–543.
McLaughlin, K. A., Conron, K. J., Koenen, K. C., and Gilman, S. E. (2010) "Childhood adversity, adult stressful life events, and risk of past-year psychiatric disorder: A test of the stress sensitization hypothesis in a population-based sample of adults," *Psychological Medicine* 40(10): 1647–1658.
McLaughlin, K. A., Green, J. G., Gruber, M. J., Sampson, N. A., Zaslavsky, A. M., and Kessler, R. C. (2012) "Childhood adversities and first onset of psychiatric disorders in a national sample of US adolescents," *Archives of General Psychiatry* 69(11): 1151–1160.

Merz-Perez, L., Heide, K. M., and Silverman, I. (2001) "Childhood cruelty to animals and subsequent violence against humans," *International Journal of Offender Therapy and Comparative Criminology* 45(5): 556–573.

Newberry, M. (2017) "Pets in danger: Exploring the link between domestic violence and animal abuse," *Aggression and Violent Behavior* 34: 273–281.

Patterson-Kane, E. (2016) "The relation of animal maltreatment to aggression," in L. Levitt, G. Patronek, and T. Grisso (eds.) *Animal Maltreatment: Forensic Mental Health Issues and Evaluations* (pp. 140–158), New York, NY: Oxford University Press.

Plant, M., van Schaik, P., Gullone, E., and Flynn, C. (2016) "It's a dog's life": Culture, empathy, gender, and domestic violence predict animal abuse in adolescents—Implications for societal health," *Journal of Interpersonal Violence*. https://doi.org/10.1177/0886260516659655.

Poresky, R. H., and Hendrix, C. (1989, April) *Companion Animal Bonding, Children's Home Environments, and Young Children's Social Development*. Paper presented at the 21st National Biennial Meeting of the Society for Research in Child Development, Kansas City, MO.

Purewal, R., Christley, R., Kordas, K., Joinson, C., Meints, K., Gee, N., and Westgarth, C. (2017) "Companion animals and child/adolescent development: A systematic review of the evidence," *International Journal of Environmental Research and Public Health* 14(3): 234.

Rhoades, H., Winetrobe, H., and Rice, E. (2015) "Pet ownership among homeless youth: Associations with mental health, service utilization and housing status," *Child Psychiatry & Human Development* 46(2): 237–244.

Risley-Curtiss, C. (2010) "Social work practitioners and the human—companion animal bond: A national study," *Social Work* 55(1): 38–46.

Roberton, T., Daffern, M., and Bucks, R. S. (2012) "Emotion regulation and aggression," *Aggression and Violent Behavior* 17(1): 72–82.

Shin, S. H., McDonald, S. E., and Conley, D. (2018) "Patterns of adverse childhood experiences and substance use among young adults: A latent class analysis," *Addictive Behaviors* 78: 187–192.

Signal, T., Ghea, V., Taylor, N., and Acutt, D. (2013) "When do psychologists pay attention to children harming animals?" *Human-Animal Interaction Bulletin* 1(2): 82–97.

Signal, T., Taylor, N., and Maclean, A. S. (2018) "Pampered or pariah: Does animal type influence the interaction between animal attitude and empathy?" *Psychology, Crime & Law* 24(5): 527–537.

Simmons, C. A., and Lehmann, P. (2007) "Exploring the link between pet abuse and controlling behaviors in violent relationships," *Journal of Interpersonal Violence* 22(9): 1211–1222.

Sorabji, R. (1995) *Animal Minds and Human Morals: The Origins of the Western Debate*, Ithaca, NY: Cornell University Press.

Tapia, F. (1971) "Children who are cruel to animals," *Child Psychiatry and Human Development* 2(2): 70–77.

Thompson, K. L., and Gullone, E. (2006) "An investigation into the association between the witnessing of animal abuse and adolescents' behavior toward animals," *Society & Animals* 14(3): 221–243.

Triebenbacher, S. L. (1998) "Pets as transitional objects: Their role in children's emotional development," *Psychological Reports* 82(1): 191–200.

Vaughn, M. G., Fu, Q., DeLisi, M., Beaver, K. M., Perron, B. E., Terrell, K., and Howard, M. O. (2009) "Correlates of cruelty to animals in the United States: Results from the National Epidemiologic Survey on Alcohol and Related Conditions," *Journal of Psychiatric Research* 43(15): 1213–1218.

Walton-Moss, B. J., Manganello, J., Frye, V., and Campbell, J. C. (2005) "Risk factors for intimate partner violence and associated injury among urban women," *Journal of Community Health* 30(5): 377–389.

Wright, J., and Hensley, C. (2003) "From animal cruelty to serial murder: Applying the graduation hypothesis," *International Journal of Offender Therapy and Comparative Criminology* 47(1): 71–88.

PART V

Wild Animals

Conservation, Management, and Ethics

In philosophical circles, people tend to think of 1975 as the beginning of the contemporary conversation about animals. That's when Peter Singer published *Animal Liberation*, where he argued that our tendency to discount animal interests is usually due to speciesism, as well as for the view that we should stop using nonhuman animals for food, fashion, and research. Singer didn't have much to say about wild animals; it was really only the critics who did. In response to various pro-animal arguments, they would point out that if animals really do have morally relevant interests (or rights or what have you), then we don't simply need to worry about the way we treat pigs and cows. We also need to worry about the many ways in which wild animals suffer, whether at our hands or due to natural causes. But since that's absurd, proponents of animal liberation must be wrong about our obligations to the animals on farms and in labs.

For a long time, defenders of animals resisted criticisms of this kind. But eventually, some of them began to embrace the basic idea—namely, that we ought to do far more for wild animals than we previously thought. This has been one of the more interesting developments in the animal ethics literature over the last decade or so.

At the same time, the environmental movement has grown in influence. There is now more conversation than ever about how best to conserve and manage wild animal populations, particularly in light of climate change and habitat loss. But conservationists tend to focus on wild animals at the species level: they are concerned to preserve finches, not any particular finch, and lions, but not any particular lion. This creates a deep tension between environmentalists and animal ethicists, with the former concerned about animals as parts of ecosystems, important in much the same way that grasses and streams are important, and the latter concerned about animals as individuals. It's difficult to see how, if at all, these two perspectives on animals can be reconciled, as what's good for the species may not be good for the individual—and vice versa.

We find this tension in the debate about zoos, where some think that conservation is an excellent reason to keep some animals in captivity while others deny this. We find it in debates about managing wildlife, where one camp is, and another isn't, willing to kill some animals (ideally in painless ways, although rarely so in practice) to preserve a certain balance in an ecosystem. Consider cases where culling some white-tailed deer would help the herd, but it seems to be bad for the deer we kill. We find it in debates about when animals should be classified as invasive, or as overpopulated, given that classifying animals in these ways is usually understood as a justification for harming them. We find it in discussions about duties of assistance to wild animals, where some think that we ought to be trying to intervene in nature to minimize suffering—say, by providing wild animals with vaccines against certain diseases—while others think that we should let nature take its course.

Here, large questions are unavoidable. What kind of world are we trying to create? How do we understand animals within that vision? To what degree are we entitled to re-create the world in our own image, imposing our values on nature? Should we think of ourselves as members of an interspecies political community, of which both we and nonhuman animals are citizens? Should we regard morality as a construct that's useful for regulating human interactions, but that generates unhelpful results when extended to animals? Or are we in a position to know how things ultimately ought to be, such that surprising result about animals are just that—surprising, but not implausible for that reason? And, of course, regardless of how we answer these questions, we face complex empirical questions. What can we actually accomplish with zoos? To what degree can we improve things for wild animals? What are the best strategies for managing wild animal populations, if we decide that we need to manage them? Plainly, we shouldn't expect any easy answers.

27

ZOOS AND AQUARIUMS COMMITTING TO INTEGRATED SPECIES CONSERVATION

Markus Gusset

Introduction

Ensuring the well-being of other species is essential if humans are to ensure their own. The quality of the land, air, and water not only affects wild populations of animals and plants but will eventually determine humanity's fate as well. Quick and effective action must be taken to deal with the profound anthropogenic issues that confront natural ecosystems, such as growing human populations, continued pollution and over-exploitation of natural resources, and climate change (Tilman et al. 2017). Human actions and lifestyle choices are threatening the planet and the life-forms that inhabit it. To preserve the diversity of the world's wildlife, humans must change how they live.

However, it has proved extremely difficult to mobilize and sustain the social and political will necessary to change behavior for the benefit of wildlife and wild places. While many believe that species and habitat conservation are innately valuable, others need to be convinced of the material importance of conserving living fauna and flora. The key strategy for achieving the required revolution in attitudes and behaviors will be reconnecting the public with nature. People must be inspired to understand that life on earth is fragile, that the species that make up life on the planet depend on each another to survive, and that human survival is reliant on the species populations in natural ecosystems. It must also be made clear that species conservation has economic value: the richer the diversity of life, the greater the opportunity for medical discoveries, economic development, and adaptive responses to the ominous impacts of global climate change.

Zoos and aquariums (accredited or otherwise designated members of the professionally recognized zoological community) are uniquely poised to contribute to the successful conservation of species and ecosystems. Many zoos and aquariums are working to make sure that the range of species they care for is supported by meaningful conservation actions linked to the survival of species in the wild (Zimmermann et al. 2007; Dick and Gusset 2010; Conde 2013). While resources may not extend to providing support for every species, conservation actions taken for the most threatened populations will have a positive impact on all species within that habitat (Conde et al. 2015; Funk et al. 2017). A global appraisal of the contribution of the world zoo and aquarium community to wildlife conservation (Gusset and Dick 2010) showed that the evaluated projects are helping to improve the conservation status of high-profile threatened species and habitats in biodiversity-rich regions of the world.

The collective social, political, and financial power of zoos and aquariums as a community, as well as the potential impact of such vast audiences, can be potent. Zoos and aquariums enjoy wide-ranging levels of public credibility and trust and provide fun and intellectually stimulating destinations for visitors of all ages (Gusset and Chin 2016). Every year, an estimated 700 million visits are made to zoos and aquariums that are members of national or regional associations around the world (Gusset and Dick 2011a). While money and donations do not always translate into quality conservation efforts, funds are still an essential requirement for the implementation of conservation action. It is estimated that US$350 million are raised annually for direct support of wildlife conservation by zoos and aquariums in organized associations around the world (Gusset and Dick 2011a).

Instilling in all visitors a strong sense of excitement about and a desire to care for life on earth will create a solid platform for fulfilling the promise to care for and conserve wildlife (Fa et al. 2011; Rees 2011; Dick and Gusset 2013; Hosey et al. 2013). Zoological facilities are uniquely positioned to use a social-science, evidence-based approach to influence pro-environmental behavior (Gusset and Lowry 2014). A global evaluation of the educational impacts of visits to zoos and aquariums found that a significant number of people end their visit with higher biodiversity understanding and a greater knowledge of actions to help protect biodiversity (Moss et al. 2014, 2015, 2017a), with a potentially long-lasting effect (Jensen et al. 2017). The challenge, not only for zoological facilities but also for all environmental education providers, is to translate such biodiversity knowledge gains into the social and behavior change required to achieve greater biodiversity conservation (Moss et al. 2017b).

Integrated Species Conservation

To accomplish this, and to increase the effectiveness of global conservation efforts, zoos and aquariums are increasingly adopting a One Plan Approach (Figure 27.1). This conservation planning framework brings together experts from the global zoo and aquarium fraternity, local community representatives, governmental agencies, wildlife managers, conservation organizations, scientists, and others in developing conservation strategies to achieve the common goal of viable populations of the species thriving in healthy ecosystems. Through the One Plan Approach, all available resources are engaged in producing one comprehensive conservation plan for each target species. This integrated approach will result in more comprehensive actions, promote innovation in species conservation, cultivate greater collaboration between zoological facilities and with other conservation organizations, and allow for greater adaptability in the face of climate change.

Zoos and aquariums can and must become models of integrated conservation (Gusset and Dick 2013). As animal-care specialists, conservationists, educators, communicators, wildlife advocates and scientists, zoo and aquarium professionals must also become powerful agents of change and encourage the widespread application of the One Plan Approach, as exemplified in *Second Nature* (CBSG

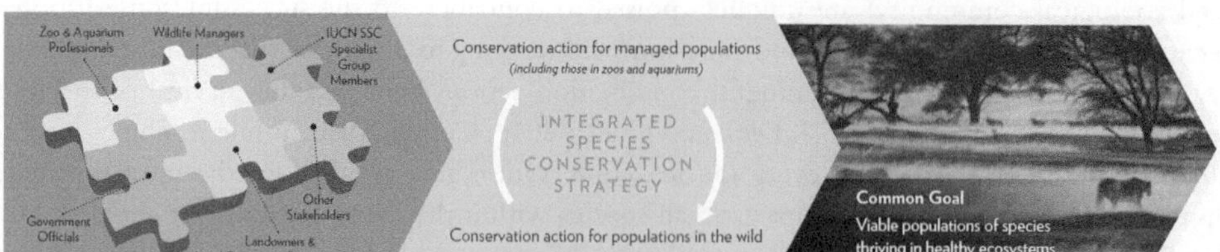

Figure 27.1 The One Plan Approach—a term coined by the IUCN SSC Conservation Breeding Specialist Group—refers to integrated species conservation planning that considers all populations of the species (inside and outside the natural range), under all conditions of management, and engages all responsible parties and resources from the start of the conservation planning initiative

Source: Illustration used with permission from Barongi et al. (2015).

2017). Their institutions must embrace the role of professional conservation organizations that operate sustainably. Fulfilling this responsibility has never been more essential.

The One Plan Approach also mandates that animals maintained in zoological facilities play a conservation role that benefits wild counterparts. The One Plan Approach links researchers in zoos and aquariums with scientists and conservationists working directly with wild populations. Likewise, education and capacity-building efforts must start with zoos and aquariums and expand to influence behavior change for conservation in society. Zoological facilities must work together, and be effective at partnering and collaborating with other conservation organizations to evaluate impacts and advocate for conserving biodiversity.

The World Association of Zoos and Aquariums (WAZA), the unifying organization for the world zoo and aquarium community (Penn et al. 2012), defines conservation in its *Committing to Conservation: The World Zoo and Aquarium Conservation Strategy* (Barongi et al. 2015: 12) as "securing populations of species in natural habitats for the long term." Therefore, this chapter focuses on direct contributions of zoos' and aquariums' core expertise in population management to species conservation within the One Plan Approach framework. "Population management" in this context refers to the human act of altering behavioral, reproductive, genetic, health-related, and welfare-related aspects of animal populations.

Sustainable Population Management

Global biodiversity targets (e.g., those by the United Nations Convention on Biological Diversity [CBD]) aim primarily to preserve biodiversity in natural habitats. However, because human impacts now affect all ecosystems, a rising number of species will benefit from, and increasingly require, intensive population management. This trend emphasizes the need for zoos and aquariums to be more directly involved in the intensive management of an increasing number of species both in zoological facilities and in the wild (Conway 2011). As zoos and aquariums engage in increased conservation breeding for the purpose of preserving biodiversity, careful species selection should be used to focus limited resources on those for which a long-term and broadly protective difference can be made.

The International Union for Conservation of Nature (IUCN), the world's largest and most diverse environmental network, has recognized that conservation breeding by zoos and aquariums has played a role in the recovery of one-quarter of the 64 vertebrate species whose threat status was reduced according to *The IUCN Red List of Threatened Species* (www.iucnredlist.org; Gusset and Dick 2012; also see Gilbert et al. 2017). Breeding animals in human care followed by reintroducing them into the wild as part of a coordinated recovery plan was one of the most frequently cited conservation actions that led to improvements in the IUCN Red List status. For birds, conservation breeding and reintroduction helped prevent the extinction of 6 out of 16 species that would probably have been lost in the absence of conservation measures. For mammals, conservation breeding and reintroduction have been more successful in improving conservation status than other conservation actions and contributed to the genuine improvement in the IUCN Red List status of at least nine species.

With more than 20,000 species in their care, zoos and aquariums have assumed increased leadership and responsibility for conservation-breeding programs over the years. No other group of institutions has the scientific knowledge and practical experience to keep and breed thousands of animal species, thereby offering enormous potential for contributing to wildlife conservation (Kleiman et al. 2010; Irwin et al. 2013). These zoo- and aquarium-based skills and resources are most effective for achieving conservation outcomes when applied through extensive and cross-disciplinary partnerships.

To fulfill the full suite of conservation roles required, wild-animal populations in human care must be demographically robust, the animals must be behaviorally competent and genetically representative of wild counterparts, and the breeding program must be able to sustain these characteristics for the future (Gusset and Dick 2011b). Individuals making up viable populations should be healthy in every respect, including a positive animal-welfare state (Maple and Perdue 2013; Manteca Vilanova 2015), and be sourced legally, sustainably and ethically.

Small populations are rarely sufficient for securing long-term persistence of a species. Conservation-breeding programs at the regional or global level can help form larger populations, if needed (Conde et al. 2013). Most programs are managed at the regional level for logistical and regulatory reasons. A new way of fostering collaboration inter-regionally is being tested through Global Species Management Plans (GSMPs). A GSMP involves the management of a particular species with a globally agreed set of goals while building upon and respecting existing regional processes.

International and regional studbooks provide the data that can help facilitate the coordination of such conservation-breeding efforts across zoological institutions. Studbooks are repositories of pedigree and demographic data on animals managed internationally or regionally. ZIMS is an application of Species360 (www.species360.org) that keeps track of individual animals throughout their lives. New features have been added to ZIMS to help studbook keepers, and well-run and up-to-date studbooks will improve the animal and population data ZIMS offers within the application. As members of Species360 enter data into ZIMS, they contribute to efficient population management across the zoological community. Furthermore, applying this type of living records system to small populations in reserves could advance the One Plan Approach and make a direct contribution to sustaining wildlife in nature (Schwartz et al. 2017).

It is vital to recognize that space for holding and breeding larger populations of many species is the greatest impediment to building long-term sustainability (Lacy 2013). This issue over available space was recognized in the 1980s, yet it remains a critical need in building sustainable populations today, with a demand for caring for more species in zoos and aquariums (Gusset et al. 2014). Another crucial matter is the difficulty that zoological professionals encounter in moving animals (or gametes) for breeding purposes. Regulatory hurdles continue to make transregional movement of animals difficult (Gusset and Dick 2011b). This threatens the successful implementation of GSMPs and other collaborative interregional programs. In addition, it thwarts cooperative management of species maintained in different regions whose collective population would be sustainable, if individuals in the isolated, regional populations could be moved predictably for breeding purposes.

Continuum of Management Intensity

A clear link should be established and communicated between field conservation, and the conservation work carried out in zoos and aquariums. In line with this objective, WAZA's *Committing to Conservation: The World Zoo and Aquarium Conservation Strategy* (Barongi et al. 2015) postulates the dawning of the era of increasing focus on a more holistic approach to integrated species conservation, which works along a continuum of management intensity (Figure 27.2) (Redford et al. 2012). This new paradigm—the One Plan Approach—has been recognized in the *IUCN SSC Guidelines for Species Conservation Planning* (IUCN SSC 2017). For illustration (see Figure 27.2), Kihansi spray toads (*Nectophrynoides asperginis*) are found almost exclusively in human care as an intensively managed population and many populations of African elephants (*Loxodonta africana*) require conservation interventions in order to persist, whereas the peregrine falcon (*Falco peregrinus*) is an example of a recovered, currently self-sustaining species. Furthermore, in order to build sustainable populations, zoological facilities must commit to supporting and training the staff who implement cooperative population management.

Increasingly, as a result of habitat loss, and habitat and population fragmentation, many wild populations have similarities to populations in human care—small in size, fragmented, and with limited gene flow between them. For example, animals reintroduced into relatively small, fenced reserves

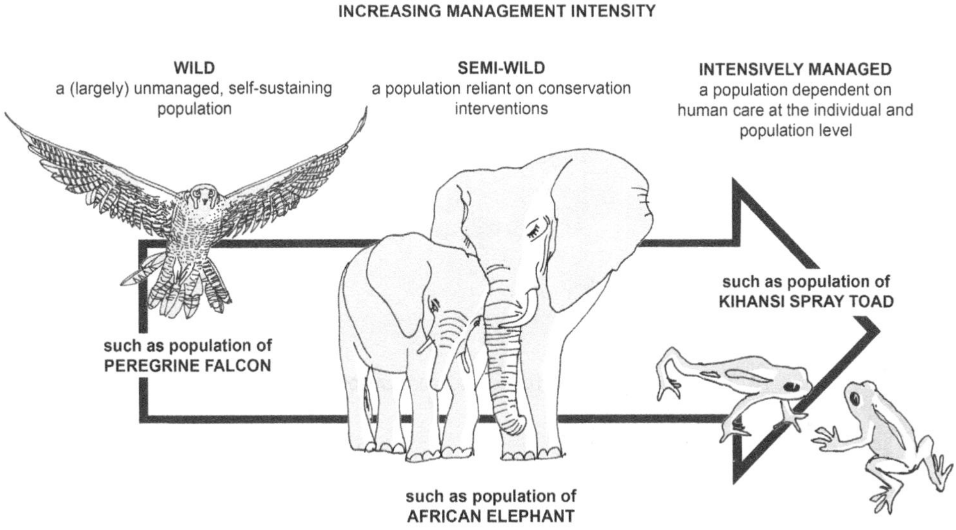

Figure 27.2 Integrated species conservation works along a continuum of management intensity, including little, if any, human intervention in wild populations all the way to intensively managed populations in some reserves and in zoos and aquariums

Source: Illustration by Júlia Hanuliaková/Zoo Design Inc.

have necessitated periodic translocation of individuals to mimic natural dispersal and maintain gene flow. This model is referred to as a managed metapopulation (Figure 27.3), as natural metapopulation processes such as dispersal are subject to human intervention.

Genome resource banks, as an additional component of metapopulation management (see Figure 27.3), serve to cryopreserve genetic material of animals (e.g., blood, gametes, embryos, tissues). Long-term population viability often requires transfers of animals (or gametes) for breeding. Traditionally, this included the exchange of animals between holders of the population in human care, import of animals from the wild to either bolster existing or establish new populations in human care, and export of animals from populations in human care to the wild. These transfers can be combined under one umbrella of interactive exchanges of animals (or gametes) between populations in the wild and in human care for achieving coincident conservation outcomes (Gusset and Dick 2013), as illustrated in Figure 27.3 by the various types of transfers. This greatly enhances the capacity to sustain viable populations both in zoological facilities and in the wild. For the sake of effective population management, legislation at national and international levels (including regulations by the Convention on International Trade in Endangered Species of Wild Fauna and Flora [CITES]) should be adapted and enforced to provide opportunities for such interactive exchanges.

The science of managing small populations in human care is of direct relevance to field-conservation programs that require intensive wildlife-management techniques. For example, fencing can be highly effective for preventing human—wildlife conflicts in wild-animal populations adjacent to settled areas. However, fenced populations will require human intervention to be viable in the long term. Similarly, fragmented and small populations may require translocation of animals among the few remaining sites to restore gene flow. As land-use change and, increasingly, climate change progress habitat fragmentation, deterioration, and destruction, translocation is likely to become an increasingly important conservation tool, as illustrated in Figure 27.3 by the transfer of animals between semi-wild populations. Strategic guidance is provided in the revised *IUCN SSC Guidelines for Reintroductions and Other Conservation Translocations* (IUCN SSC 2013). An example of an

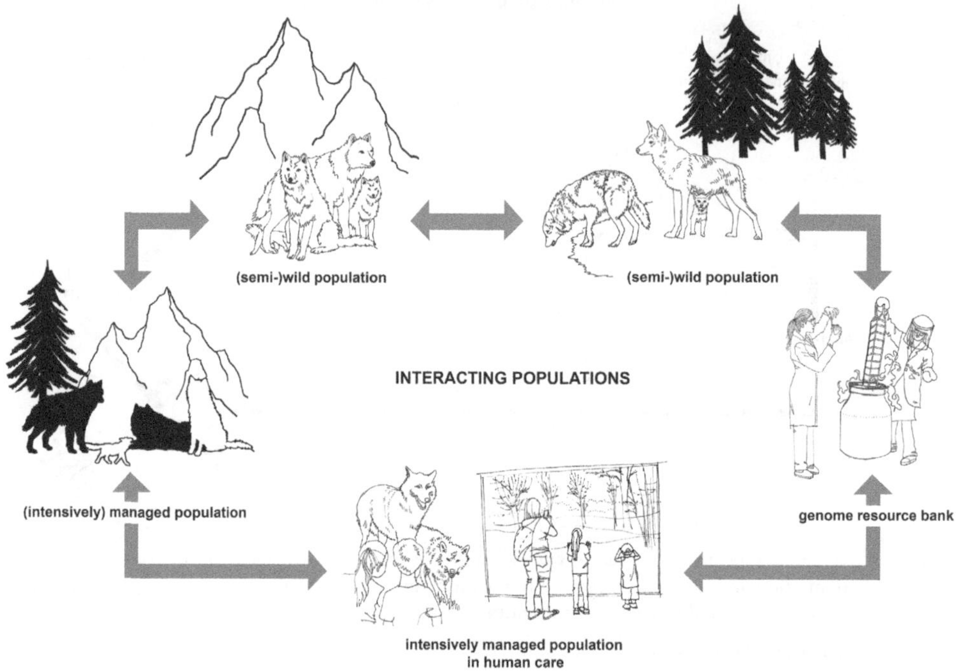

Figure 27.3 Metapopulation management involves managing a set of interacting populations under a common conservation goal. Its components may include multiple regional populations managed in human care (including range-country breeding programs), multiple (semi-)wild populations (including reintroduced populations) and genome resource banks. Arrows indicate transfers of animals (or gametes)

Source: Illustration by Júlia Hanuliaková/Zoo Design Inc.

integrated species conservation strategy, which involves conservation interventions along the continuum of management intensity and applies a metapopulation management approach, is provided in Box 27.1.

Box 27.1 Integrated Conservation of the Red Wolf in the United States

The red wolf (*Canis rufus*) is one of the world's most threatened canids (Figure 27.4). Once common throughout the eastern and south-central United States, the species was decimated by the early part of the 20th century as a result of predator control and habitat destruction. The red wolf was designated an endangered species in 1967. The establishment of a managed zoo population in the early 1970s allowed persistence of the species and a dedicated conservation-breeding program (split among multiple zoological facilities) facilitated a reintroduction program (initiated in the mid-1980s) that supported two wild populations, one of which persists today in North Carolina. This restoration effort was the first time in US history that a federally listed species was reintroduced to its historical range from which it had been extirpated. The US Fish and Wildlife Service set the explicit objective of managing red wolves via the interactive exchange of animals between populations in human care and in the wild. As of 2017, about 200 red wolves live in zoological facilities, and about 50 roam freely. The red wolf thus provides

Figure 27.4 Red wolf at Point Defiance Zoo & Aquarium

Source: Photograph by John Froschauer.

an example where a species once extinct in the wild benefits from an integrated conservation strategy (Figure 27.1) that contains conservation actions for both the managed zoo population and the (small, reintroduced) wild population. This strategy involves conservation interventions along the continuum of management intensity (Figure 27.2) and applies a metapopulation management approach (Figure 27.3). Nevertheless, anthropogenic mortality continues to threaten the persistence of the red wolf in its historical range, highlighting the need for intensive population management (human intervention) both in zoological facilities and in the wild (for more information, see www.fws.gov/redwolf/index.html).

As the biodiversity crisis intensifies, an increasing number of species will likely require some form of intensive population management in order to avoid extinction. Guidance on if and when activities in zoos and aquariums can be a beneficial component of an overall species-conservation strategy is provided in the revised *IUCN SSC Guidelines on the Use of* Ex Situ *Management for Species Conservation* (IUCN SSC 2014; McGowan et al. 2017). These guidelines outline a five-step decision-making process that defines potential conservation roles that populations in human care may play, the type of activities needed to fulfill those roles, and the feasibility, risks, and likelihood of success. Population management can be used more effectively as a conservation tool if the specific ways in which it can improve population viability or prevent extinction are identified and critically evaluated as part of an integrated approach to species-conservation planning.

In addition to advancing tools for the behavioral, reproductive, genetic, health-related, and welfare-related management of intensively managed populations, innovative approaches are needed to enhance the capacity to sustain viable populations both in human care and in the wild (Gusset et al. 2014). There are existing challenges that also need attention, such as the management of group-living species, low reproductive success, metapopulation management and adaptation to being kept in human care. Research and new technological advances (e.g., genomics) are emerging that have the potential to significantly change and improve how populations are managed. There will be the need to develop new ways and software tools to incorporate these findings and technologies into population management. Developing sustainable, genetically diverse populations is an obligation that serves field conservation and conservation work carried out in zoological facilities and animal-welfare goals.

Conclusion

Animals in zoos and aquariums act as ambassadors that, if leveraged effectively, can provide impact and reach to the support accredited zoo and aquarium communities give to wildlife conservation. It is essential to provide visitors with clear explanations about the conservation impact their everyday behavior is having on wild populations, both locally and globally, and to focus behavioral-change campaigns on the behavior changes that will be most positive for biodiversity conservation (Gusset and Lowry 2014).

It is imperative that all zoos and aquariums increase their contribution to and impact on saving species in the wild, including the provision of skills and technical and financial resources (Fa et al. 2014). Creating a clear connection between a live animal in a zoological facility and a conservation project in the field should be integrated into every master-planning process to make certain that adequate support is generated for saving species in the wild.

Zoological institutions are already playing a major role in the global conservation of species (Conde et al. 2011), and this will grow as their conservation missions are integrated into every aspect of operations. The One Plan Approach builds on the strengths and motivations to link all the skill sets and experience of zoo and aquarium staff to individuals and organizations working in the field (Figure 27.1). Advances in animal care and research with intensively managed small populations in zoological facilities are being applied to larger global issues.

Sustainable population management is one of the most critical issues for modern zoos and aquariums (Gusset and Dick 2011b; Gusset et al. 2014), and visitors may find it difficult to differentiate between the needs of an individual animal (animal welfare) and the conservation needs of a species (population management). Population management within zoological facilities regularly requires animal transfers, mate selection, social-group composition, euthanasia or contraception (Hildebrandt et al. 2017), and these requirements should be clearly explained to all stakeholders in conservation and welfare terms.

Excellent animal welfare is fundamental to achieving a shared wildlife-conservation goal. While conservation of wildlife is the core purpose of zoological facilities, positive animal welfare is their core activity (Gusset and Dick 2015). To that end, a commitment statement has been adopted as a part of WAZA's *Caring for Wildlife: The World Zoo and Aquarium Animal Welfare Strategy* (Mellor et al. 2015: 15): "Zoos and aquariums have a responsibility to achieve high standards of animal welfare in support of their goals as modern conservation organizations."

Further Reading

Barongi, R., Fisken, F. A., Parker, M., and Gusset, M. (eds.) (2015) *Committing to Conservation: The World Zoo and Aquarium Conservation Strategy*, Gland: WAZA Executive Office.
(The way forward for zoos and aquariums in wildlife conservation, population management and environmental education.)

Fa, J. E., Funk, S. M., and O'Connell, D. (2011) *Zoo Conservation Biology*, Cambridge: Cambridge University Press.
(Textbook providing an overview of how zoos and aquariums contribute to conservation.)

Hosey, G., Melfi, V., and Pankhurst, S. (2013) *Zoo Animals: Behaviour, Management, and Welfare*, 2nd ed., Oxford: Oxford University Press.
(Textbook providing an overview of how and why to keep animals in zoos and aquariums.)

Maple, T. L., and Perdue, B. M. (2013) *Zoo Animal Welfare*, Berlin: Springer-Verlag.
(Textbook providing an overview of how the welfare of animals in zoos and aquariums can be ensured.)

Mellor, D. J., Hunt, S., and Gusset, M. (eds.) (2015) *Caring for Wildlife: The World Zoo and Aquarium Animal Welfare Strategy*, Gland: WAZA Executive Office.
(The way forward for zoos and aquariums in animal welfare.)

Bibliography

Barongi, R., Fisken, F. A., Parker, M., and Gusset, M. (eds.) (2015) *Committing to Conservation: The World Zoo and Aquarium Conservation Strategy*, Gland: WAZA Executive Office.

CBSG (2017) *Second Nature: Changing the Future for Endangered Species*, St. Paul, MN: IUCN SSC Conservation Breeding Specialist Group.

Conde, D. A. (2013) "Role of zoological gardens," in N. MacLeod, J. D. Archibald, and P. Levin (eds.) *Grzimek's Animal Life Encyclopedia: Extinction,* Vol. 1, Detroit, MI: Gale/Cengage Learning.

Conde, D. A., Colchero, F., Güneralp, B., Gusset, M., Skolnik, B., Parr, M., Byers, O., Johnson, K., Young, G., Flesness, N., Possingham, H., and Fa, J. E. (2015) "Opportunities and costs for preventing vertebrate extinctions," *Current Biology* 25: R219–R221.

Conde, D. A., Colchero, F., Gusset, M., Pearce-Kelly, P., Byers, O., Flesness, N., Browne, R. K., and Jones, O. R. (2013) "Zoos through the lens of the IUCN Red List: A global metapopulation approach to support conservation breeding programs," *PLoS One* 8: e80311.

Conde, D. A., Flesness, N., Colchero, F., Jones, O. R., and Scheuerlein, A. (2011) "An emerging role of zoos to conserve biodiversity," *Science* 331: 1390–1391.

Conway, W. G. (2011) "Buying time for wild animals with zoos," *Zoo Biology* 30: 1–8.

Dick, G., and Gusset, M. (eds.) (2010) *Building a Future for Wildlife: Zoos and Aquariums Committed to Biodiversity Conservation*, Gland: WAZA Executive Office.

Dick, G., and Gusset, M. (2013) "Conservation biology," in M. D. Irwin, J. B. Stoner, and A. M. Cobaugh (eds.) *Zookeeping: An Introduction to the Science and Technology*, Chicago, IL: University of Chicago Press.

Fa, J. E., Funk, S. M., and O'Connell, D. (2011) *Zoo Conservation Biology*, Cambridge: Cambridge University Press.

Fa, J. E., Gusset, M., Flesness, N., and Conde, D. A. (2014) "Zoos have yet to unveil their full conservation potential," *Animal Conservation* 17: 97–100.

Funk, S. M., Conde, D., Lamoreux, J., and Fa, J. E. (2017) "Meeting the Aichi targets: Pushing for zero extinction conservation," *Ambio* 46: 443–455.

Gilbert, T., Gardner, R., Kraaijeveld, A. R., and Riordan, P. (2017) "Contributions of zoos and aquariums to reintroductions: Historical reintroduction efforts in the context of changing conservation perspectives," *International Zoo Yearbook* 51: 15–31.

Gusset, M., and Chin, S. A. (eds.) (2016) *WAZA Magazine 17: Future of Zoo and Aquarium Design*, Gland: WAZA Executive Office.

Gusset, M., and Dick, G. (2010) "'Building a future for wildlife'? Evaluating the contribution of the world zoo and aquarium community to *in situ* conservation," *International Zoo Yearbook* 44: 183–191.

Gusset, M., and Dick, G. (2011a) "The global reach of zoos and aquariums in visitor numbers and conservation expenditures," *Zoo Biology* 30: 566–569.

Gusset, M., and Dick, G. (eds.) (2011b) *WAZA Magazine 12: Towards Sustainable Population Management*, Gland: WAZA Executive Office.

Gusset, M., and Dick, G. (eds.) (2012) *WAZA Magazine 13: Fighting Extinction*, Gland: WAZA Executive Office.

Gusset, M., and Dick, G. (eds.) (2013) *WAZA Magazine 14: Towards Integrated Species Conservation*, Gland: WAZA Executive Office.

Gusset, M., and Dick, G. (eds.) (2015) *WAZA Magazine 16: Towards Positive Animal Welfare*, Gland: WAZA Executive Office.

Gusset, M., Fa, J. E., Sutherland, W. J., and the Horizon Scanners for Zoos and Aquariums (2014) "A horizon scan for species conservation by zoos and aquariums," *Zoo Biology* 33: 375–380.

Gusset, M., and Lowry, R. (eds.) (2014) *WAZA Magazine 15: Towards Effective Environmental Education*, Gland: WAZA Executive Office.

Hildebrandt, G., Perret, K., Eulenberger, K., Junhold, J., and Luy, J. (2017) *Understanding Zoos—A Multidisciplinary Consideration about Ex-Situ-Conservation, Surplus Animals and Euthanasia*, Münster: Schüling Verlag.

Hosey, G., Melfi, V., and Pankhurst, S. (2013) *Zoo Animals: Behaviour, Management, and Welfare*, 2nd ed., Oxford: Oxford University Press.

Irwin, M. D., Stoner, J. B., and Cobaugh, A. M. (eds.) (2013) *Zookeeping: An Introduction to the Science and Technology*, Chicago, IL: University of Chicago Press.

IUCN SSC (2013) *Guidelines for Reintroductions and Other Conservation Translocations*, Version 1.0, Gland: IUCN Species Survival Commission.

IUCN SSC (2014) *Guidelines on the Use of* Ex Situ *Management for Species Conservation*, Version 2.0, Gland: IUCN Species Survival Commission.

IUCN SSC (2017) *Guidelines for Species Conservation Planning*, Version 1.0, Gland: IUCN Species Survival Commission.

Jensen, E. A., Moss, A., and Gusset, M. (2017) "Quantifying long-term impact of zoo and aquarium visits on biodiversity-related learning outcomes," *Zoo Biology* 36: 294–297.

Kleiman, D. G., Thompson, K. V., and Kirk Baer, C. (eds.) (2010) *Wild Mammals in Captivity: Principles and Techniques for Zoo Management*, 2nd ed., Chicago, IL: University of Chicago Press.

Lacy, R. C. (2013) "Achieving true sustainability of zoo populations," *Zoo Biology* 32: 19–26.

Manteca Vilanova, X. (2015) *Zoo Animal Welfare: Concepts and Indicators*, Barcelona: Multimédica Ediciones Veterinarias.

Maple, T. L., and Perdue, B. M. (2013) *Zoo Animal Welfare*, Berlin: Springer-Verlag.

McGowan, P. J. K., Traylor-Holzer, K., and Leus, K. (2017) "IUCN guidelines for determining when and how ex situ management should be used in species conservation," *Conservation Letters* 10: 361–366.

Mellor, D. J., Hunt, S., and Gusset, M. (eds.) (2015) *Caring for Wildlife: The World Zoo and Aquarium Animal Welfare Strategy*, Gland: WAZA Executive Office.

Moss, A., Jensen, E., and Gusset, M. (2014) "Zoo visits boost biodiversity literacy," *Nature* 508: 186.

Moss, A., Jensen, E., and Gusset, M. (2015) "Evaluating the contribution of zoos and aquariums to Aichi Biodiversity Target 1," *Conservation Biology* 29: 537–544.

Moss, A., Jensen, E., and Gusset, M. (2017a) "Impact of a global biodiversity education campaign on zoo and aquarium visitors," *Frontiers in Ecology and the Environment* 15: 243–247.

Moss, A., Jensen, E., and Gusset, M. (2017b) "Probing the link between biodiversity-related knowledge and self-reported proconservation behavior in a global survey of zoo visitors," *Conservation Letters* 10: 33–40.

Penn, L., Gusset, M., and Dick, G. (2012) *77 Years: The History and Evolution of the World Association of Zoos and Aquariums 1935–2012*, Gland: WAZA Executive Office.

Redford, K. H., Jensen, D. B., and Breheny, J. J. (2012) "Integrating the captive and the wild," *Science* 338: 1157–1158.

Rees, P. A. (2011) *An Introduction to Zoo Biology and Management*, Oxford: Wiley-Blackwell.

Schwartz, K. R., Parsons, E. C. M., Rockwood, L., and Wood, T. C. (2017) "Integrating *in-situ* and *ex-situ* data management processes for biodiversity conservation," *Frontiers in Ecology and Evolution* 5: 120.

Tilman, D., Clark, M., Williams, D. R., Kimmel, K., Polasky, S., and Packer, C. (2017) "Future threats to biodiversity and pathways to their prevention," *Nature* 546: 73–81.

Zimmermann, A., Hatchwell, M., Dickie, L., and West, C. (eds.) (2007) *Zoos in the 21st Century: Catalysts for Conservation?* Cambridge: Cambridge University Press.

28
THE EDUCATIONAL VALUE OF ZOOS
An Empirical Perspective

Nancy Staus

Introduction

Although they have been in existence for over 3,000 years, zoos as we know them now are fairly recent phenomena (Alexander 1979). Originally established as private menageries to indulge the nobility, the first modern zoos that allowed the public to view wild exotic animals originated in Europe in the 18th century, followed by the first American zoos in the 1870s (Jamieson 1985). By the 1990s, there were reportedly more than 10,000 zoos worldwide found in almost every country in the world (Mason 2000). Zoos attract hundreds of millions of visitors each year, which makes watching wildlife in captive settings one of the most popular leisure activities in the world (Carr and Cohen 2011; Tribe and Booth 2003).

As the leisure and recreational needs and expectations of society have changed over time, as well as concern over captive animals' welfare, zoos have evolved to meet them. Once considered primarily places of entertainment, amusement, and recreation, zoos have repositioned themselves as centers of research and conservation, particularly with regards to the preservation of threatened or endangered species (Miller et al. 2004; Shani and Pizam 2010). More recently, zoos have begun to emphasize an educational agenda, positioning themselves as places where the public can learn about animals and their habitats, enhancing visitors' understanding of wildlife conservation and potentially increasing pro-environmental behaviors (Carr and Cohen 2011; Moss and Esson 2013; Ogden and Heimlich 2009).

Yet despite their obvious popularity and potential importance in promoting conservation education for large numbers of people, zoos and other animal-based attractions have been persistently criticized by animal rights and welfare advocates for keeping wild animals in captivity where they may suffer physically and/or mentally from their confinement (Jamieson 1985; Mason 2000; Esson and Moss 2013). Conway (1995: 2) referred to this as the "zoo paradox," in which zoos aim to "inspire public interest in wild creatures and nature, to provide ecological education, and to help save wild species from extinction, but in doing so they confine wild animals away from nature and manage their lives." That being the case, critics argue that providing evidence of the educational impact that zoos have claimed for themselves is essential to justify keeping wildlife in captivity for public display (Shani and Pizam 2010; Wagoner and Jensen 2010).

In this chapter, I present an empirical perspective of the educational value of zoos. I begin with a discussion about the educational goals of zoos, the efforts they make to reach those goals, and the current understanding about the learning that takes place in informal, free-choice environments such as zoos. Finally, I present the current evidence both for and against zoos' claims regarding their educational outcomes and conclude with a discussion about alternatives to zoos as a means to learn about wildlife and environmental conservation.

The Education Agenda of Zoos

Currently, all major zoo accreditation organizations contain strong statements of educational goals as being central to their missions (Moss and Esson 2013). For example, the World Association of Zoos and Aquariums (WAZA), which includes more than 1,200 member zoos and aquariums, states that the educational role of zoos is to "attract, inspire and enable people from all walks of life to act positively for conservation" (WAZA 2005: 35). Similarly, the Association of Zoos and Aquariums (AZA) in the United States claims to educate "over 180 million visitors . . . about wild animals, their habitats, their related conservation issues, and the ways in which they can contribute to their preservation" (AZA 2018). Patrick et al. (2007) found that education was mentioned in 131 out of 136 US zoo mission statements that they analyzed, appearing more frequently even than the theme of conservation. Clearly, education appears to be seen by zoos as a central component of their missions by providing a means to heighten visitor appreciation of the value of biodiversity and a feeling of connectedness with the natural world (Moss and Esson 2013).

Until recently, educational programs at zoos focused primarily on cognitive goals, such as teaching animal facts to school-aged children or increasing visitors' overall natural history knowledge (Balmford et al. 2007; Ogden et al. 2004). However, as zoos have begun to position themselves as conservation organizations, there has been an accompanying shift in their education goals to include a focus on affective and behavioral outcomes as well. For example, zoos seek to instill caring, empathy, and connectedness with animals. Such affective outcomes are important because emotional connections with nature are highly predictive of pro-environmental behaviors—much more so than knowledge alone (Kals et al. 1999; Hinds and Sparks 2008). In addition, visitors' affective responses to observing wildlife are related to short- and long-term cognitive outcomes such as environmental knowledge and attitudes (Ballantyne et al. 2011; Falk and Gillespie 2009; Staus and Falk 2017). Thus, promoting more positive emotional outcomes along with increased knowledge of conservation issues may help zoos meet their ultimate goal of influencing visitors' long-term conservation-related behaviors.

Zoos as Places for Free-Choice Learning

In order to meet the cognitive, affective, and behavioral goals described earlier, zoos deliver their educational messages in a number of ways. Many zoos offer formal, educator-led, curriculum-linked teaching to school groups, particularly children (Andersen 2003; Moss and Esson 2013; Wagoner and Jensen 2010). However, the vast majority of the 180 million annual visitors to zoos in the United States learn in a more informal manner by way of signage and other interpretation, interactions with staff or docents, and animal demonstrations (Fraser et al. 2009; Mony and Heimlich 2008; Povey and Rios 2002). Each visitor navigates the zoo in an unstructured way, guided by their own interest and visit agenda. Collectively, these informal, visitor-directed experiences are referred to as "free-choice learning" (Falk and Dierking 2002).

Learning outcomes of zoo visits by free-choice learners, like those to other educational leisure settings such as museums, science centers, or aquariums, are complex and unique to a particular visitor (Dierking et al. 2013). Learning researchers have come to appreciate that information is not simply transferred wholesale to learners. In fact, learning is a process of constructing meaning based on one's past experiences, existing knowledge, interests, and motivations; because everyone's mental foundations are unique to their life experiences, no two people will learn the same information in the same way (Falk and Staus 2013; Falk and Storksdieck 2005; Fosnot and Perry 2005). Therefore, it cannot be assumed that all zoo visitors will focus on or learn the information that is presented or taught to them during their visit. In other words, although many visitors "take away" the conservation information provided by the zoo through signage or presentations (and many others do not), their learning outcomes are a unique and highly individual combination of what they saw, read, heard, and felt throughout the experience (Ballantyne and Packer 2011; Ballantyne et al. 2010).

Despite their highly individual and unpredictable nature, learning outcomes are strongly influenced by the visitor's motivation and agenda which directs the visit in ways that can influence the quality and extent of learning outcomes (Falk 2006, 2009). For example, people visit zoos for a variety of reasons, not all of which are related to education or learning. Falk (2006) identified five such identity-related motivations which strongly influenced visitors' science learning goals: Explorers, Facilitators, Professional/Hobbyists, Experience Seekers, and Rechargers. His research showed that Explorers, who are driven by a general curiosity about science, and Professional/Hobbyists, with specific content-related interests, tend to enter zoos with distinct learning goals. In contrast, Facilitators are socially motivated and primarily enable the learning of others, particularly children. Experience Seekers, who are looking for fun or entertainment, and Rechargers who seek a restorative experience, are the least likely to actively pursue learning goals. In a three-year nationwide study of the educational impacts of zoos and aquariums in the United States, Falk et al. (2007) found that most visitors identified as Facilitators and Explorers, with Experience Seekers composing less than 10% of their sample.

In contrast, other researchers have found that zoos are perceived as leisure sites by most visitors, who do not come prepared to be educated but instead have an expectation of being entertained. In their assessment of the benefits and services sought by zoo visitors, Tomas et al. (2003) found that visitors to Fort Worth Zoo were more socially and entertainment-oriented than learning-oriented, rating family togetherness and wildlife enjoyment as most important, while the education domain was rated lowest in importance. Similarly, Rajack and Waren (1996) reported that visitors to Edinburgh Zoo in Scotland wanted to visit with friends, have fun, and be entertained; only 4% of visitors sampled were there to be educated.

These findings, if widespread, would seem to create a challenge for zoos in which education is a central element in their mission statements. However, Falk et al. (2007) reported that only half of visitors entered the zoo with a single, dominant identity-related motivation (e.g., entertainment). The rest held multiple reasons for visiting, including to learn about animals and their conservation. Furthermore, even for visitors who don't come with deliberate intentions to learn, most are drawn into an experience that incorporates learning (Packer 2006). Other researchers have reported similar results from a variety of zoological parks in the United States (Andereck and Caldwell 1994; Holzer et al. 1998; Milan and Wourms 1992), suggesting that a mix of education and entertainment at zoos is not necessarily problematic because enjoyment motives can be situated in experiences that support the advancement of a conservation agenda (Clayton et al. 2009). In fact, visitors may be seeking experiences of discovery and exploration rather than more traditional learning outcomes (e.g., increased factual knowledge) that are associated with the word *education*. This engagement in a learning experience purely for its enjoyment value is a phenomenon that Packer (2006) refers to as "learning for fun" and is a major aspect of free-choice learning.

Promoting Free-Choice Learning in Zoos

Since many visitors come to the zoo for entertainment and recreation purposes, the challenge for zoos is to provide learning opportunities that are also fun so that the experience of learning is seen as an enjoyable form of entertainment (Ballantyne et al. 2007; Packer 2006; Sickler and Fraser 2009). Researchers in a variety of educational leisure settings, including zoos, have identified several conditions under which learning for fun is likely to happen, which could be used to inform the development of successful learning experiences in zoos. These included creating a sense of discovery or fascination, presentations that appeal to multiple senses, and the ability to choose from a variety of experiences (Eisenberger 1999; Falk and Dierking 2000; Freedman 2000; Packer 2006). Similarly, Sickler and Fraser (2009) found that adult zoo visitors enjoyed experiences of discovery or tranquility and the ability to learn about and connect with the animals.

These findings suggest a number of ways for zoos to provide fun learning opportunities for their visitors. For example, interactive and interpretive exhibits are entertaining for visitors while also providing a stimulus for learning (Carpenter 1992). Touch tables that contain interpretive materials such as feathers, fur, or bones allow visitors to engage in exploratory discovery experiences using a variety of senses (Lindermann-Matthies and Kamer 2005). Animal demonstrations or keeper talks are often exciting and engaging, which can have an impact on visitors' knowledge and attitudes about an animal or a conservation issue (Povey and Rios 2002; Staus 2012; Staus and Falk 2017; Swanagan 2000). Furthermore, these types of experiences may increase empathy or feelings of connection with animals that could motivate greater concern and willingness to devote resources to protecting the animals and their environment (Allen 2002; Clayton 2006).

In addition to the learning experiences described above, the animals themselves can be considered a primary educational component of zoo exhibits (Churchman 1985). Key factors that seem to positively influence visitors' cognitive and affective learning outcomes include greater visibility of the animals, observable active animal behaviors, up-close viewing of animals, and more natural-looking habitats (Bitgood et al. 1988; Luebke et al. 2016). Visitors pay greater attention to more animated animal behaviors and spend more time viewing animals that are active, both of which increase opportunities for visitors to learn about the animal and its natural habitat (Altman 1998; Moss and Esson 2010). In addition, the act of viewing live animals can elicit positive emotions such as caring, connectedness, and empathy for animals and nature (Clayton et al. 2009; Myers et al. 2004). Zoos have responded to these findings by enhancing the naturalism of wildlife exhibits to replicate the ecosystems of the exhibited species and incorporating behavioral enrichments to encourage more natural behavior (Shani and Pizam 2010). Zoo professionals believe that naturalistic exhibits increase the affective impact on visitors, leading to increased interest in conservation after visiting such exhibits (Coe 1985; Derwin and Piper 1988).

However, there is a potential conflict between the goals of entertainment and education and the welfare of the animals on display. For example, while greater visibility and interaction is desirable for engaging visitors, the presence of large numbers of people can cause greater stress for animals, which can lead to inactivity, abnormal behavior (e.g., pacing), or aggression, especially when the animals have no way to escape from viewers (Anderson et al. 2003; Carlstead and Brown 2005; Mallapur and Chellam 2002). These effects are especially acute for primates, particularly when zoo visitors try to interact with the animals (Hosey 2000). Therefore, zoos must balance the prospective benefits of providing entertaining educational experiences for visitors with possible negative effects on the behavior and well-being of resident animals (Fernandez et al. 2009).

Zoo Visitors' Educational Outcomes

Unlike other educational leisure institutions such as museums and science centers, zoos command substantially larger audiences, with more than 600 million visitors annually throughout the world, nearly a third of which occur in the United States (AZA 2018; Kellert 1997; WAZA 2005). In addition to sheer numbers, zoo visitors also are more demographically diverse than at other types of institutions, including people in a variety of racial, ethnic, and socioeconomic categories (Bruni et al. 2008). Therefore, zoos have significant potential to educate the public about wildlife and its conservation (Balmford et al. 2007).

As discussed earlier, educational efforts in these organizations seek to focus on both the cognitive domain by increasing awareness and knowledge of conservation issues and the affective domain by increasing feelings of caring and empathy and promoting emotional connections to animals, which they claim will lead to positive behavioral outcomes. To meet these goals, many zoos have invested in a variety of learning opportunities for visitors above and beyond traditional signage, including interactive interpretation, public talks at exhibits, educator interventions, and animal demonstrations (Moss and Esson 2013). However, the quantity of such education-related outputs is not a necessarily reliable indicator of successful educational outcomes. In order to satisfy their critics, it is necessary for zoos to provide evidence to substantiate their claims as effective conservation education providers.

Perhaps because of the relative youth of conservation education efforts in zoos, until recently very few studies have attempted to quantify whether or how zoo visits impact people's conservation-related knowledge, attitudes, or behavior (Balmford et al. 2007). In fact, although a large proportion of zoos do conduct visitor research, most focus on measures of operational performance rather than those concerning mission-related educational learning outcomes (Luebke and Grajal 2011).

Despite many zoos' claims regarding their success as conservation education providers, the research to date reveals mixed and sometimes contradictory results. In this section, I present a review of the literature regarding learning outcomes at zoos, organized by the three major education goals: cognitive, affective, and behavioral outcomes.

Cognitive Outcomes

Historically, zoos, like other educational leisure providers, have approached visitor education by framing the problem as one of overcoming visitors' knowledge deficits in the hopes that greater knowledge about conservation issues would lead to pro-environmental actions (Falk and Staus 2013). Although this notion has been replaced by more sophisticated models of visitor learning (e.g., Ballantyne et al. 2007), cognitive outcomes—such as learning about animals and their natural habitats or increasing knowledge about conservation issues—remain a primary educational goal for many zoos (AZA 2018).

A number of researchers have evaluated the extent to which visitors demonstrated changes in knowledge and attitudes in response to particular experiences at zoos, and several found that zoo visitors reported increases in knowledge about a specific animal (e.g., tigers or apes) as a result of their visits (Broad and Smith 2004; Broad and Weiler 1998; Lukas and Ross 2005). Similarly, a recent multi-institutional study of 26 zoos and aquariums in 19 countries indicated that visitors' understanding of biodiversity and associated actions increased significantly over the course of their visit (Moss et al. 2015). Wagoner and Jensen (2010) found similar increases in biodiversity-related learning for schoolchildren visiting London Zoo. In contrast, others found very little evidence of changes in conservation knowledge or concern following an informal zoo visit (Anderson et al. 2003; Balmford et al. 2007; Gutierrez de White and Jacobson 1994). In an investigation of the impact of informal education at zoos on a variety of educational outcomes including knowledge and attitudes, Kellert and Dunlap (1989) reported no observable increase in either factual or conceptual knowledge of animals. Likewise, in a large-scale study of five zoos in the United Kingdom, Balmford et al. (2007) found little evidence for the effects of a zoo visit on adults' conservation knowledge or ability to engage in associated behaviors.

In an earlier study, Kellert (1979) reported that zoo visitors were actually less knowledgeable about animals than other outdoor recreationists such as backpackers and hunters. In contrast, Falk et al. (2007) found that zoo visitors had a greater than expected level of knowledge about basic ecological concepts, which may have explained the lack of significant gains in conservation knowledge and understanding that they reported after their visit. However, although visitors' knowledge didn't increase, zoos did appear to support and reinforce visitors' existing knowledge and attitudes.

These contradictory results may be explained by the nature of free-choice learning itself, in which visitors engage in self-directed exploration guided by their own interests and agenda. Although people recognize the zoo as a place for education, the primary motivation for most zoo visitors is actually a desire to enjoy themselves and spend time with their families. While zoo visitors may be receptive to the opportunities for learning provided by the zoo, they engage in them only insofar as it fits within their own personal agenda (Clayton et al. 2009). For example, Packer (2006: 336) reported that many visitors to educational leisure settings such as zoos "are drawn into an experience of learning, even if that is not consciously part of their stated agenda for their visit, or indeed they do not recognize or refer to it as 'learning.'" Alternatively, it is sometimes argued that attempting to educate visitors in an entertainment setting leads to mixed messages in which the conservation meaning is overwhelmed by the reality of "animal routines that resemble a circus show" (Swanagan 2000: 30).

Whichever view is correct, it is quite likely that the effects of free-choice learning experiences are not immediately apparent but that they take time for reflection or repeated visits to manifest themselves (Balmford et al. 2007; Falk and Dierking 2002; Falk et al. 2007). In addition, since free-choice learning outcomes are unique to each visitor, it is unlikely that they will be discerned by any one survey or interview question. Therefore, it is almost certain that learning is occurring but in ways that are not necessarily captured by traditional research instruments.

However, critics argue that even if visitors do learn from zoos, they are learning the wrong things. Because the animals are encountered outside of their natural context, visitors experience "a twisted view of wildlife and animal behavior" (Shani and Pizam 2010: 290). For instance, Farinato (2001: 146) asserts that "viewing an orangutan sitting in a grassy, moated outdoor yard . . . teaches nothing about the nature of the animal or its role in the non-zoo environment." Hancocks (2001) points out that zoos overrepresent mammals in comparison to other taxonomic groups such as reptiles and amphibians, which conveys a distorted view of their real proportions in nature. Finally, Jamieson (1985: 117) claims that what zoos ultimately teach us is "a false sense of our place in the natural order" in which animals "are there at our pleasure, to be used for our purposes." Indeed, Kellert (1997: 99) found that "many visitors leave the zoo more convinced than ever of human superiority over the natural world."

Affective Outcomes

In addition to promoting cognitive learning outcomes, zoos have recently positioned themselves as centers of caring with a goal of connecting people with wildlife through learning experiences that elicit feelings of connectedness and empathy for animals and nature (Bruni et al. 2008; Luebke and Grajal 2011; Rabb and Saunders 2005). Promotion of such feelings is an important visitor outcome because emotional affinity toward nature is part of a sequence of responses that are triggered by observing animals in zoos or wildlife tourism that eventually result in pro-environmental behavioral responses (Ballantyne et al. 2010).

In recent years, a number of researchers have found that the viewing of live animals promoted feelings of connectedness and caring in zoo visitors (e.g., Bruni et al. 2008; Clayton et al. 2009; Falk et al. 2007; Hayward and Rothenberg 2004; Myers et al. 2004; Schultz and Tabanico 2007). For example, studies in a variety of naturalistic exhibits indicated a significant increase in implicit connectedness with nature (a measure of unconscious cognitive associations) in visitors that increased with the duration of their stay, suggesting that zoo experiences can have a measurable impact on visitors' connections with nature (Bruni et al. 2008; Schultz and Tabanico 2007). However, no effects were observed for self-ratings of explicit connectedness with nature, indicating that the affective outcomes of the visit were not apparent to visitors at the time. In contrast, Falk (2007) and colleagues found that visitors explicitly reported that they experienced a stronger connection to nature as a result of a zoo visit. Although critics have argued that forming such connections to nature is not truly possible in a zoo environment, which is a constructed view of how humans wish nature to be (Berger 1980), or that zoo visitors are already more predisposed to care about animals than nonvisitors (Kellert 1996), the fact remains that many people perceive zoo experiences as connecting them with the natural world, which may lead to concern and care for nature (Bruni et al. 2008).

Other studies have shown that zoo visits can enable an experience of shared identity between humans and nonhuman animals, particularly primates (e.g., "That gorilla looks just like Daddy!") (Clayton et al. 2009: 393), and that these behavioral or conceptual connections increase their concern for the well-being of the animals (Myers et al. 2004). Conversely, Milstein (2009) argued that such encounters are inherently unequal because the confinement of the animal constitutes a critical difference between the visitor and the animal that encourages feelings of mastery and dominion rather than harmony and connection.

Feelings of connection and caring aside, it is becoming increasingly clear that a variety of affective elements are significant in the learning process and in influencing environmental attitudes and behaviors (Staus and Falk 2013). For example, many experiences at zoos and other animal-based attractions trigger feelings of emotional arousal or excitement that have been found to be strongly associated with positive short- and long-term learning outcomes (Falk and Gillespie 2009; Staus 2012). However, Smith (2008) reported a negative relationship between very high levels of emotional arousal and learning outcomes when visitors were too excited by an exhibit (a wild bird presentation) to pay attention to the associated conservation messages, suggesting that zoos need to manage experiences in such a way that they are exciting enough to elicit attention and learning but not too exciting for visitors to attend to key educational messages.

Others have suggested that conservation-learning outcomes are a consequence of having highly pleasurable experiences as these positive feelings significantly influence the meaning visitors take away from their visit (Clayton et al. 2009; Luebke et al. 2016). In contrast, Staus and Falk (2017) showed that feelings of displeasure were significant predictors of short-term learning in the context of an informal science learning experience (viewing a video of a near-predation event), even more so than arousal. Similarly, Esson and Moss (2013) reported significant levels of visitor engagement and reflection at a disturbing exhibition about the effects of irresponsible human behavior on the environment, suggesting that zoos could potentially include more challenging environmental themes to promote learning and behavior change.

Behavioral Outcomes

While increasing visitors' knowledge and affective outcomes are important educational objectives for zoos to pursue, ultimately, the overarching goal of education programs is to create new patterns of pro-environmental behavior in individuals and society by inspiring and enabling visitors to take meaningful conservation actions (Luebke and Matiasek 2013; Povey and Spaulding 2005; WAZA 2005).

Although a number of researchers have attempted to examine the role of zoos in influencing visitor behavior, there is little evidence for long-term impact from a single zoo experience (Smith and Broad 2008). Several studies indicated that commitment to conservation-related behavioral changes increased immediately after the zoo visit but reverted to pre-experience levels several months later (Adelman et al. 2000; Dierking et al. 2004; Manubay et al. 2002). While Esson and Moss (2014) reported some success in influencing more persistent behavior changes, these were in the context of a specially devised intervention consisting of workshops and focus groups rather than a single informal zoo visit. Therefore, it may require intense management of the zoo experience or, at minimum, repeated visits to realize the goal of long-term behavior change.

It is possible that rather than initiating *changes* in behavior, zoo visits may have a role in *reinforcing* pro-environmental behaviors with which visitors are already familiar or engaging in. For example, Smith et al. (2008) reported that the majority of visitors they sampled from a birds of prey demonstration at a zoo in Australia simply increased their commitment to an existing pro-environmental behavior (recycling) rather than undertaking a new behavior (removing roadkill from the road to prevent carrion-feeding birds from being hit by cars), despite making a commitment to engage in the new behavior immediately after their visit.

There are several plausible explanations for the apparent inability of zoos to realize their goal of influencing visitor behavior. First, it is likely that a single visit is insufficient for convincing people to change entrenched behavior, and therefore follow-up interventions will be required to reinforce conservation messages. As Milstein (2009: 42) notes,

> it may not be possible for the zoo visitor, who, according to zoo researchers, spends an average of five seconds at a zoo animal exhibit, to deeply encounter other animals and themselves in ways that influence perceptions or behaviors, or help visitors grapple with or transform the structural eco-social forces in which they play a part.

A second explanation is that exhibits are not designed or delivered in ways that adequately communicate the desired behaviors and outcomes or how they relate to the conservation issue, often promoting simplistic actions to address complex, systemic environmental problems (Milstein 2009: 39). Results of her study showed that visitors understood the conservation messages they encountered but were confused about how they could make a difference because the zoo did not provide a way to make a personal connection with the issue. For example, "visitors wonder whether recycling will lead to helping save tigers" (Milstein 2009: 40).

Finally, the conservation messages of zoos often focus on environmental problems in exotic places that are far removed from the everyday lives of the visitors. For example, zoos ask visitors to care about forest destruction in distant rainforests while local forestry practices remain unexamined. Likewise, Conway (2003: 12) states that

> it is time for zoos to stop arguing that exciting children in New York or Tokyo about the plight of Gorillas in Cameroon or the Democratic Republic of the Congo is responsive conservation. This process is too indirect, too slow, too far away and too unlikely to affect the real issues.

In other words, while the suffering of animals in remote places may elicit sympathy, attention to wildlife issues in the local ecosystems where visitors' roles in environmental issues are more directly relatable may be more likely to prompt conservation action (Milstein 2009).

Alternatives to Zoos for Conservation Education

In light of the mixed results regarding the success of zoos' educational aims, some have argued that zoos have become educationally redundant and are no longer necessary given the prevalence of alternative means to learn about wildlife and conservation education (Margodt 2000). In particular, critics of zoos suggest that the same learning outcomes could be achieved through educational media such as nature programs on television, wildlife documentaries, books, magazines, and the internet (Hancocks 2001; Margodt 2000; Sommer 1972), and there is some empirical evidence to support this assertion, at least in terms of cognitive outcomes. A recent study examining how adults learn about science indicated that free-choice learning experiences such as reading books and magazines about science, using the internet, and watching science-related documentaries were significant predictors of science knowledge (Falk and Needham 2013). Others have reported that documentaries did not elicit feelings of connectedness in viewers, although they did have a somewhat positive effect on donation behavior (Arendt and Matthes 2016). However, few studies have directly compared the roles of zoos and the media in providing information about wildlife.

To address this issue, Smith and Broad (2008) evaluated visitor learning in an Australian zoo to determine if people learned conservation information from their visit or if they had learned the information previously from other sources. Results indicated that although 60% of visitors did not learn any new information, the zoo played an educational role in reinforcing existing knowledge gained from other sources, particularly television and nature documentaries. The authors concluded that rather than being educationally redundant, zoos provide an important source of complementary information to that available in the media, and the two should work together to reinforce each other's messages (Smith and Broad 2008).

As another alternative, it is increasingly suggested that noncaptive wildlife tourism experiences could fulfill similar functions as those provided by zoos (Ballantyne et al. 2007). Numerous researchers have demonstrated the positive impacts of wildlife tourism on participants' environmental knowledge, attitudes, awareness of environmental issues, connectedness to nature, and desire to engage in pro-environmental behaviors (Ballantyne and Packer 2002, 2009; Lee and Moscardo 2005; Tisdell and Wilson 2005; Zeppel and Muloin 2008). For example, Ballantyne and Packer (2002) found that field trips to natural areas to observe and experience wildlife aroused feelings of empathy in schoolchildren, which, in turn, had an impact on their behavioral intentions toward conserving the environment and wildlife. Likewise, Tisdell and Wilson (2005) reported that visitors to a turtle-watching site showed increases in environmental knowledge and indicated a desire to engage in pro-environmental behaviors after seeing turtles.

Despite their potential as zoo alternatives, wildlife tourism experiences have their own set of challenges. Because they are often located in remote areas, they can be prohibitively expensive for many people, may require air travel which contributes to global climate change, and are much less accessible to visit than zoos. In addition, ecotourism is not necessarily a benign activity in terms of animal welfare as it can disrupt natural wildlife behaviors and compromise the ecological values of the area (Ballantyne et al. 2009; Ryan and Saward 2004). These issues have prompted some researchers to suggest that zoos themselves be designed and managed to serve as ecotourism destinations, although it is unclear how this would be done or in what way they would differ from zoos as they currently exist (Mason 2000; Ryan and Saward 2004).

Finally, some zoo critics have suggested ways to modify zoos to mitigate their negative effects on animals while promoting the learning outcomes desired by zoos and still appealing to the visiting public. Milstein (2009) describes a potential "non-zoo" experience based on contemporary wildlife rescue models which would house injured or orphaned animals while they recover and rehabilitate. Visitors could witness volunteers caring for the animals or even volunteer themselves, thus promoting real feelings of caring and connectedness. Such encounters would teach visitors about the effects of human activities on wildlife and de-emphasize the entertainment aspects that are so problematic in modern zoos. Similarly, Acampora (2005: 81) suggests keeping only native animals to "strip the zoo of its exoticism." In addition, instead of keeping animals in enclosures, he advocates establishing areas where native wildlife can be viewed in their natural habitat (e.g., penguins on an island in Australia) while engaging in their normal activities.

Conclusion

Despite the claims made by many zoos that they play a significant role in public conservation education, the research undertaken thus far reveals mixed and sometimes contradictory results which have not been universally accepted as an effective demonstration of zoos' positive impact on visitor learning outcomes (Moss and Esson 2013). For example, Acampora (2005: 80) states that the conservation and education outcomes of zoos "are neither necessary nor sufficient conditions for the existence of zoos." Likewise, Milstein (2009: 42) concludes, "in the end, it is indeed questionable whether the possible payoffs of a three-hour zoo visit outweigh the coercion that keeps animals in captivity for empty, monotonous lifetimes."

There is some evidence that zoos are viewed as unacceptable visitor attractions by a segment of the public as well. In a study of latent zoo visitors in the United Kingdom, Turley (1999) found that 40% of respondents associated traditional zoos with bars and unnatural conditions, and more than one third indicated that they did not visit zoos because they disliked seeing animals in captivity. Milstein (2009) also described the feeling of discomfort reported by many visitors who were conflicted about visiting zoos. Furthermore, Malamud (1998) notes that in popular literature, nearly all stories involving zoos portray them as places of cruelty and suffering, indicating that such interpretations are part of the cultural zeitgeist.

However, the fact remains that large segments of the public do enjoy visiting zoos and the evidence suggests that zoos have some positive educational impacts on visitors. In particular, zoos seem to be successful in promoting feelings of caring and connection with wildlife, although efforts to increase visitor knowledge and pro-environmental behaviors appear to have been less effective. This may be the case because despite zoos' efforts to position themselves as sites of conservation, research, and education, the reality is that by and large the public still views them as places of entertainment. Therefore, it is argued that "rather than providing a true learning experience about the value of and need for conservation, zoo educational experiences, particularly informal ones, may actually do little more than provide a socially acceptable veneer to the entertainment on offer" (Carr and Cohen 2011: 11).

Nevertheless, the enormous popularity of zoos throughout the world and across a variety of cultures strongly suggests that they are fulfilling a powerful human need, prompting some researchers to theorize that the behaviors referred to pejoratively as entertainment or amusement may actually be an indication of an innate biological need for interspecies animal interactions and encounters (Acampora 2005). Wilson (1984: 1) describes a wide range of human behaviors in relation to nature under the term *biophilia*, which is "the innate tendency to affiliate with life or life-like processes" that may underlie our strong desire for spectatorship of other animals. As an instinctive and evolutionarily adaptive behavior, biophilia is considered a sign of mental and physical health that should be supported and encouraged, particularly in a world in which humans are increasingly separated from wildlife and the natural ecosystems that support them and us (Orr 1993).

If this is the case, then it is incumbent on critics of zoos to develop other experiences that allow people, particularly those in urban environments with little access to nature, to cultivate their innate biophilic drives to encounter animals in ways that are more educationally effective and less ethically challenging than zoos. Until then, it is likely that controversies over the educational value of zoos, as well as zoos themselves, are here to stay.

Further Reading

Jamieson, D. (1985) "Against zoos," in P. Singer (ed.) *In Defense of Animals*, Oxford: Blackwell.
(Classic critique providing the foundation of the debate questioning the educational outcomes of zoos.)

Chiszer, D., Murphy, J., and Iliff, W. (1990) "For zoos," *The Psychological Record* 4: 3–13.
(Rebuttal of Jamieson's argument against zoos.)

Bibliography

Acampora, R. (2005) "Zoos and eyes: Contesting captivity and seeking successor practices," *Society & Animals* 13(1): 68–88.

Adelman, L., Falk, J., and James, S. (2000) "Impact of national aquarium in Baltimore on visitors' conservation attitudes, behavior, and knowledge," *Curator* 43(1): 33–60.
Alexander, E. P. (1979) *Museums in Motion: An Introduction to the History and Functions of Museums*, Nashville, TN: American Association for State and Local History.
Allen, S. (2002) "Looking for learning in visitor talk: A methodological exploration," in G. Leinhardt, K. Crowley, and K. Knutson (eds.) *Learning Conversations in Museums*, Mahwah, NJ: Lawrence Erlbaum.
Altman, J. D. (1998) "Animal activity and visitor learning at the zoo," *Anthrozoös* 11(1): 12–21.
Andereck, K., and Caldwell, L. (1994) "Motive-based segmentation of a public zoological park market," *Journal of Park and Recreation Administration* 12(2): 19–31.
Andersen, L. L. (2003) "Zoo education: From formal school programmes to exhibit design and interpretation," *International Zoo Yearbook* 38: 75–81.
Anderson, U. S., Kelling, A. S., Pressley-Keough, R., Bloomsmith, M. A., and Maple, T. L. (2003) "Enhancing the zoo visitor's experience by public animal training and oral interpretation at an otter exhibit," *Environment and Behavior* 35: 826–841.
Arendt, F., and Matthes, J. (2016) "Nature documentaries, connectedness to nature, and pro- environmental behavior," *Environmental Communication* 10(4): 453–472.
[AZA], Association of Zoos and Aquariums (2018) Conservation education home page, www.aza.org/conservation-education.
Ballantyne, R., and Packer, J. (2002) "Nature-based excursions: School students' perceptions of learning in natural environments," *International Research in Geographical and Environmental Education* 12(1): 1–19.
Ballantyne, R., and Packer, J. (2009) "Future directions for research in free-choice environmental learning," in J. H. Falk, J. E. Heimlich, and S. Foutz (eds.) *Free-choice Learning and the Environment*, Lanham, MD: Alta Mira Press.
Ballantyne, R., and Packer, J. (2011) "Using tourism free-choice learning experiences to promote environmentally sustainable behaviour: The role of post-visit 'action resources,'" *Environmental Education Research* 17(2): 201–215.
Ballantyne, R., Packer, J., and Falk, J. (2011) "Visitors' learning for environmental sustainability: Testing short- and long-term impacts of wildlife tourism experiences using structural equation modeling," *Tourism Management* 32(6): 1243–1252.
Ballantyne, R., Packer, J., and Hughes, K. (2009) "Tourists' support for conservation messages and sustainable management practices in wildlife tourism experiences," *Tourism Management* 30(5): 658–664.
Ballantyne, R., Packer, J., Hughes, K., and Dierking, L. (2007) "Conservation learning in wildlife tourism settings: Lessons from research in zoos and aquariums," *Environmental Education Research* 13(3): 367–383.
Ballantyne, R., Packer, J., and Sutherland, L. A. (2010) "Visitors' memories of wildlife tourism: Implications for the design of powerful interpretive experiences," *Tourism Management* 32: 770–779.
Balmford, A., Leader-Williams, N., Mace, G., Manica, A., Walter, O., West, C., and Zimmerman, A. (2007) "Message received? Quantifying the impact of informal conservation education on adults visiting UK zoos," in A. Zimmermann, M. Hatchwell, L. Dickie, and C. West (eds.) *Zoos in the 21st Century: Catalysts for Conservation?* Cambridge, UK: Cambridge University Press.
Berger, J. (1980) "Why look at animals?" in J. Berger (ed.) *About Looking*, New York, NY: Pantheon.
Bitgood, S., Patterson, D., and Benefield, A. (1988) "Exhibit design and visitor behavior: Empirical relationships," *Environment and Behavior* 20(4): 474–491.
Broad, S., and Smith, L. (2004) "Who educates the public about conservation issues? Examining the role of zoos and the media," *Proceedings from the International Tourism and Media Conference*, Tourism Research Unit, Melbourne, pp. 15–23.
Broad, S., and Weiler, B. (1998) "Captive animals and interpretation: A tale of two tiger exhibits," *Journal of Tourism Studies* 9(1): 14–27.
Bruni, C. M., Fraser, J., and Schultz, P. W. (2008) "The value of zoo experiences for connecting people with nature," *Visitor Studies* 11(2): 139–150.
Carlstead, K., and Brown, J. L. (2005) "Relationships between patterns of fecal corticoid excretion and behavior, reproduction, and environmental factors in captive Black (*Diceros bicornis*) and White (*Ceratotherium simum*) rhinoceros," *Zoo Biology* 24: 215–232.
Carpenter, E. K. (1992) "Museums, zoos, and the Epcot factor," *Print* 46: 88–97.
Carr, N., and Cohen, S. (2011) "The public face of zoos: Images of entertainment, education, and conservation," *Anthrozoös* 24(2): 175–189.
Churchman, D. (1985) "The educational impact of zoos and museums: A Review of the Literature," *Resources in Education* 20(12): 133–161.
Clayton, S. (2006) *Extending the Boundaries of the Self: Environmental Identity and Justice for Animals*. Paper presented at the meeting of the International Society for Justice Research, Berlin.

Clayton, S., Fraser, J., and Saunders, C. D. (2009) "Zoo experiences: Conversations, connections, and concern for animals," *Zoo Biology* 28(5): 377–397.
Coe, J. C. (1985) "Design and perception: Making the zoo experience real," *Zoo Biology* 4(2): 197–208.
Conway, W. (1995) "Zoo conservation and ethical paradoxes," in B. G. Norton, M. Hutchins, E. F. Stevens, and T. L. Maple (eds.) *Ethics on the Ark: Zoos, Animal Welfare, and Wildlife Conservation*, Washington, DC: Smithsonian Institution Press.
Conway, W. (2003) "The role of zoos in the 21st century," *International Zoo Yearbook* 38: 7–13.
Derwin, C. W., and Piper, J. B. (1988) "The African rock kopje exhibit: Evaluation and interpretive elements," *Environment and Behavior* 20(4): 435–451.
Dierking, L. D., Adelman, L. M., Ogden, J., Lehnhardt, K., Miller, L., and Mellen, J. D. (2004) "Using a behavior change model to document the impact of visits to Disney's Animal Kingdom: A study investigating intended conservation action," *Curator* 47(3): 322–343.
Dierking, L. D., Falk, J. H., and Storksdieck, M. (2013) "Learning from neighboring fields: Conceptualizing outcomes of environmental education within the framework of free-choice learning experiences," in R. Stevenson, M. Brody, J. Dillon, and A. Wals (eds.) *International Handbook of Research on Environmental Education*, Washington, DC: AERA/Routledge.
Eisenberger, R. (1999) "The museumgoer's motives: The social and the sublime," *Visitor Studies Today* 2(3): 1–4.
Esson, M., and Moss, A. (2013) "The risk of delivering disturbing messages to zoo family audiences," *Zoo Biology* 32: 13–18.
Esson, M., and Moss, A. (2014) "Zoos as a context for reinforcing environmentally responsible behaviour: The dual challenges that zoo educators have set themselves," *Journal of Zoo and Aquarium Research* 2(1): 8–13.
Falk, J. H. (2006) "An identity-centered approach to understanding museum learning," *Curator* 49: 151–166.
Falk, J. H. (2009) *Identity and the Museum Visitor Experience*, Walnut Creek, CA: Left Coast Press.
Falk, J. H., and Dierking, L. D. (2000) *Learning from Museums: Visitor Experiences and the Making of Meaning*, Walnut Creek, CA: AltaMira Press.
Falk, J. H., and Dierking, L. D. (2002) *Lessons Without Limits: How Free-Choice Learning is Transforming Education*, Lanham, MD: AltaMira Press.
Falk, J. H., and Gillespie, K. L. (2009) "Investigating the role of emotion in science center visitor learning," *Visitor Studies* 12(2): 112–132.
Falk, J. H., and Needham, M. D. (2013) "Factors contributing to adult knowledge of science and technology," *Journal of Research in Science Teaching* 50(4): 431–452.
Falk, J. H., Reinhard, E. M., Vernon, C. L., Bronnenkant, K., Deans, N. L., and Heimlich, J. E. (2007) *Why Zoos and Aquariums Matter: Assessing the Impact of a Visit to a Zoo or Aquarium*, Silver Spring, MD: American Association of Zoos & Aquariums.
Falk, J. H., and Staus, N. L. (2013) "Free-Choice Learning and Ecotourism," in R. Ballantyne and J. Packer (eds.) *International Handbook on Ecotourism*, Cheltenham, UK: Edward Elgar Publishing Limited.
Falk, J. H., and Storksdieck, M. (2005) "Using the contextual model of learning to understand visitor learning from a science center exhibition," *Science Education* 89: 744–778.
Farinato, R. (2001) "Another view of zoos," in D. J. Salem and A. N. Rowan (eds.) *The State of the Animals 2001*, Washington, DC: Humane Society Press.
Fernandez, E. J., Tamborski, M. A., Pickens, S. R., and Timberlake, W. (2009) "Animal-visitor interactions in the modern zoo: Conflicts and interventions," *Applied Animal Behaviour Science* 120: 1–8.
Fosnot, C. T., and Perry, R. S. (2005) "Constructivism: A psychological theory of learning," in C. T. Fosnot (ed.) *Constructivism: Theory, Perspective and Practice* (pp. 9–38), 2nd ed., New York: Teachers College Press.
Fraser, J., Bicknell, J., Sickler, J., and Taylor, A. (2009) "What information do zoo and aquarium visitors want on animal identification labels?" *Journal of Interpretive Research* 14: 7–19.
Freedman, G. (2000) "The changing nature of museums," *Curator* 43(4): 295–306.
Gutierrez de White, R., and Jacobson, S. (1994) "Evaluating conservation education programs at a South American zoo," *Journal of Environmental Education* 25(4): 18–22.
Hancocks, D. (2001) "Is there a place in the world for zoos?" in D. J. Salem and A. N. Rowan (eds.) *The State of the Animals 2001*, Washington, DC: Humane Society Press.
Hayward, J., and Rothenberg, M. (2004) "Measuring success in the Congo gorilla forest conservation exhibition," *Curator* 47: 261–282.
Hinds, J., and Sparks, P. (2008) "Engaging with the natural environment: The role of affective connection and identity," *Journal of Environmental Psychology* 28: 109–120.
Holzer, D., Scott, D., and Bixler, R. (1998) "Socialization influences on adult zoo visitation," *Journal of Applied Recreation Research* 23(1): 43–62.
Hosey, G. R. (2000) "Zoo animals and their human audiences: What is the visitor effect?" *Animal Welfare* 9: 343–357.

Jamieson, D. (1985) "Against zoos," in P. Singer (ed.) *In Defense of Animals*, Oxford: Blackwell.
Kals, E., Schumacher, D., and Montada, L. (1999) "Emotional affinity toward nature as a motivational basis to protect nature," *Environment and Behavior* 31: 178–202.
Kellert, S. R. (1979) Zoological parks in American society. Presentation at the American Association of Zoological Parks and Aquaria, St. Louis, MO.
Kellert, S. R. (1996) *The Value of Life: Biological Diversity and Human Society*, Washington, DC: Island Press.
Kellert, S. R. (1997) *Kinship to Mastery: Biophilia in Human Evolution and Development*, Washington, DC: Island Press.
Kellert, S. R., and Dunlap, J. (1989) *Informal Learning at the Zoo: A Study of Attitude and Knowledge Impact*, Philadelphia, PA: Zoological Society of Philadelphia.
Lee, W. H., and Moscardo, G. (2005) "Understanding the impact of ecotourism resort experiences on tourists' environmental attitudes and behavioural intentions," *Journal of Sustainable Tourism* 13(6): 546–565.
Lindermann-Matthies, P., and Kamer, T. (2005) "The influence of an interactive educational approach on visitors' learning in a swiss zoo," *Science Education* 90(2): 296–315.
Luebke, J. F., and Grajal, A. (2011) "Assessing mission-related learning outcomes at zoos and aquaria: Prevalence, barriers, and needs," *Visitor Studies* 14(2): 195–208.
Luebke, J. F., and Matiasek, J. (2013) "An exploratory study of zoo visitors' exhibit experiences and reactions," *Zoo Biology* 32: 407–416.
Luebke, J. F., Watters, J. V., Packer, J., Miller, L. J., and Powell, D. M. (2016) "Zoo visitors' affective responses to observing animal behaviors," *Visitor Studies* 19(1): 60–76.
Lukas, K. E., and Ross, S. R. (2005) "Zoo visitor knowledge and attitudes toward gorillas and chimpanzees," *Journal of Environmental Education* 36: 33–48.
Malamud, R. (1998) *Reading Zoos: Representations of Animals and Captivity*, New York: New York University Press.
Mallapur, A., and Chellam, R. (2002) "environmental influences on stereotypy and the activity budget of Indian Leopards (*Panthera pardus*) in four zoos in Southern India," *Zoo Biology* 21: 585–595.
Manubay, G., Smith, J. C., Houston, C., Schulz, K., Dotzour, A., and De Young, R. (2002) "Evaluating exhibits that promote conservation behavior: Developing a theoretical framework," proceedings of the 31st Annual North American Association for Environmental Education Conference, Boston, MA, August 6–11.
Margodt, K. (2000) *The Welfare Ark: Suggestion for a Renewed Policy in Zoos*, Brussels, Belgium: VUB University Press.
Mason, P. (2000) "Zoo tourism: The need for more research," *Journal of Sustainable Tourism* 8(4): 333–339.
Milan, L. M., and Wourms, M. K. (1992) "A zoological park is not just another museum," *Curator* 35(2): 120–135.
Miller, B., Conway, W., Reading, R. P., Wemmer, C., Wildt, D., Kleiman, D., and Hutchins, M. (2004) "Evaluating the conservation mission of zoos, aquariums, botanical gardens, and natural history museums," *Conservation Biology* 18(1): 86–93.
Milstein, T. (2009) "Somethin' tells me it's all happening at the zoo: Discourse, power, and conservationism," *Environmental Communication* 3(1): 25–48.
Mony, P. R. S., and Heimlich, J. E. (2008) "Talking to visitors about conservation: Exploring message communication through docent—visitor interactions at zoos," *Visitor Studies* 11: 151–162.
Moss, A., and Esson, M. (2010) "Visitor interest in zoo animals and the implications for collection planning and zoo education programmes," *Zoo Biology* 29: 715–731.
Moss, A., and Esson, M. (2013) "The educational claims of zoos: Where do we go from here?" *The Journal of Environmental Education* 44(2): 79–96.
Moss, A., Jensen, E., and Gusset, M. (2015) "Evaluating the contribution of zoos and aquariums to Aichi biodiversity target 1," *Conservation Biology* 29(2): 537–544.
Myers, O. E., Saunders, C. D., and Birjulin, A. A. (2004) "Emotional dimensions of watching zoo animals: An experience sampling study building on insights from psychology," *Curator* 47: 299–321.
Ogden, J., Gentile, C., and Revard, B. (2004) "Trends in conservation education: A primer," *AZA Communiqué*, August: 18–20.
Ogden, J., and Heimlich, J. E. (2009) "Why focus on zoo and aquarium education?" *Zoo Biology* 28(5): 357–360.
Orr, D. W. (1993) "Love it or lose it: The coming biophilia revolution," in S. R. Kellert and E. O. Wilson (eds.) *The Biophilia Hypothesis* (pp. 415–440), Washington, DC: Island Press.
Packer, J. (2006) "Learning for fun: The unique contribution of educational leisure experiences," *Curator* 49(3): 329–344.
Patrick, P. G., Matthews, C. E., Ayers, D. F., and Tunnicliffe, S. D. (2007) "Conservation and education: Prominent themes in zoo mission statements," *The Journal of Environmental Education* 38(3): 53–60.
Povey, D. D., and Rios, J. (2002) "Using interpretive animals to deliver affective messages in zoos," *Journal of Interpretation Research* 7: 19–28.

Povey, K., and Spaulding, W. (2005) *Message Design for Animal Presentations: A New Approach*. Paper presented at the American Zoo and Aquarium Association Annual Conference, Chicago, IL, September 13–18.
Rabb, G. B., and Saunders, C. D. (2005) "The future of zoos and aquariums: Conservation and caring," *International Zoo Yearbook* 39: 1–26.
Rajack, L., and Waren, N. (1996) "The modern zoo: How do people perceive animals," *Applied Animal Behavior Science* 47: 109–118.
Ryan, C., and Saward, J. (2004) "The zoo as ecotourism attraction—visitor reactions, perceptions and management implications: The case of Hamilton zoo, New Zealand," *Journal of Sustainable Tourism* 12(3): 245–266.
Schultz, P. W., and Tabanico, J. J. (2007) "Self, identity, and the natural environment: Exploring implicit connections with nature," *Journal of Applied Social Psychology* 37: 1219–1247.
Shani, A., and Pizam, A. (2010) "The role of animal-based attractions in ecological sustainability," *Worldwide Hospitality and Tourism Themes* 2(3): 281–298.
Sickler, J., and Fraser, J. (2009) "Enjoyment in zoos," *Leisure Studies* 28: 313–331.
Smith, L. (2008) *The Zoo Proposition: An Examination of the Role of Emotional Arousal in Influencing Pro-Environmental Behaviour*, unpublished doctoral dissertation, Monash University, Australia.
Smith, L., and Broad, S. (2008) "Comparing zoos and the media as conservation educators," *Visitor Studies* 11(1): 16–25.
Smith, L., Broad, S., and Weiler, B. (2008) "A closer examination of the impact of zoo visits on visitor behaviour," *Journal of Sustainable Tourism* 16(5): 544–562.
Sommer, R. (1972) "What do we learn at the zoo," *Natural History* 81: 26–27.
Staus, N. L. (2012) *Crossing the Cartesian Divide: An Investigation into the Role of Emotion in Science Learning*, unpublished doctoral dissertation, Oregon State University, Corvallis, OR.
Staus, N. L., and Falk, J. H. (2013) "The role of emotion in ecotourism experiences," in R. Ballantyne and J. Packer (eds.) *International Handbook on Ecotourism*, Cheltenham, UK: Edward Elgar Publishing Limited.
Staus, N. L., and Falk, J. H. (2017) "The role of emotion in informal science learning: Testing an exploratory model," *Mind, Brain, and Education* 11(2): 45–53.
Swanagan, J. (2000) "Factors influencing zoo visitors' conservation attitudes and behavior," *The Journal of Environmental Education* 31(4): 26–31.
Tisdell, C., and Wilson, C. (2005) "Perceived impacts of ecotourism on environmental learning and conservation: Turtle watching as a case study," *Environment, Development and Sustainability* 7: 291–302.
Tomas, S. R., Crompton, J. L., and Scott, D. (2003) "Assessing service quality and benefits sought among zoological park visitors," *Journal of Park and Recreation Administration* 21(2): 105–124.
Tribe, A., and Booth, R. (2003) "Assessing the role of zoos in wildlife conservation," *Human Dimensions of Wildlife* 8(1): 65–74.
Turley, S. K. (1999) "Exploring the future of the traditional UK zoo," *Journal of Vacation Marketing* 5(4): 340–355.
Wagoner, B., and Jensen, E. (2010) "Science learning at the zoo: Evaluating children's developing understanding of animals and their habitats," *Psychology & Society* 3(1): 65–76.
[WAZA], World Association of Zoos and Aquariums (2005) *The World Zoo and Aquarium Conservation Strategy: Building a Future for Wildlife*, Bern, Switzerland: WAZA Executive Office.
Wilson, E. O. (1984) *Biophilia*, Cambridge, MA: Harvard University Press.
Zeppel, H., and Muloin, S. (2008) "Conservation benefits of interpretation on marine wildlife tours," *Human Dimensions of Wildlife* 13(4): 280–294.

29

MORAL ARGUMENTS AGAINST ZOOS

Karen S. Emmerman

Introduction

In North America, zoos are a staple of elementary school field trips, a destination for families with young children, and a place where urban dwellers immerse themselves in a landscape far removed from the hard, concrete surfaces of their daily lives. What began as caged menageries kept by despots with no regard for animal welfare have developed into substantial zoological gardens staffed by zoo professionals with expertise in animal care. Over decades, in response to growing concerns that holding animals captive for life as entertainment for human beings is frivolous and morally questionable, zoos have shifted away from positioning themselves as sites of recreation. Repackaged as vital parts of the conservation puzzle during these ecologically troubled times, zoos have hoped to ward off moral concerns by underscoring their crucial role in educating the public and contributing to worldwide conservation efforts.

Despite these efforts, the existence of zoos prompts deep moral questions about captivity, power, rights, welfare, and care. Thinking about the moral arguments against zoos forces us all to reflect on who we are as a species, our role in creating the permanent loss of other animal species, and how we can appropriately respond to the conservation crisis. Zoo proponents see zoos as a biologically necessary and morally defensible part of this response. Zoo critics remain skeptical about both zoos' efficacy in achieving their education and conservation missions and zoos' ability to defend their existence on moral grounds. In this chapter, we consider moral arguments against zoos with attention to critiquing zoos' claims regarding education and conservation. In the end, we will see that zoos are not morally defensible.

Before we begin, here are some clarifications: First, "zoos" refers only to those zoological facilities that have been accredited either by the Association of Zoos and Aquariums (AZA) or the European Association of Zoos and Aquariums (EAZA). I restrict the discussion to this relatively small class of facilities because they represent what some people call "the best zoos," or "the good zoos." Accredited zoos must meet certain standards for animal welfare to achieve accreditation. There is a consensus among those who study zoos that unaccredited facilities likely are not meeting the necessary welfare requirements to even be considered as potentially ethical operations. Only 10% of zoos in the United States are accredited, leaving thousands of animal exhibitors that may or may not comply with the AZA's standards (Braverman 2013: 126, 129). Second, I use *zoos* as shorthand for "zoos and aquariums." Comparatively little has been written about the morality of aquariums, particularly regarding their professed contributions to public education and conservation efforts, but the arguments laid out here apply as much to them as to zoos. Third, I will use *animal* as shorthand for "other-than-human animals" for efficiency. Finally, I refer to animals living in zoos as "zooed" animals rather than "zoo" animals. I do this in much the same way many of us working in animal ethics use "farmed" animals rather than "farm" animals. No animal exists naturally in a zoo and referring to animals living in zoos as "zoo animals" reduces them to their instrumental value for humans.[1]

I end this introduction with a brief discussion of the long-standing debate in environmental ethics regarding whether moral value sits with individual animals or with species. In the second section, I provide an overview of moral arguments against zoos including rights-based, welfare-based, and ecofeminist considerations. In the third and fourth sections, I evaluate the two most common justifications for zoos: education and conservation. The chapter ends with a plea for epistemic humility and a reminder that humans cannot avoid our shame and grief at having wreaked havoc on the natural world by permanently incarcerating individual animals.

Before beginning the overview of moral arguments against zoos, we need to understand an important ethical debate that exists in the background of all discussions about zoos. The debate centers on whether moral value is properly understood as belonging to individual animals or to species. For many conservationists and environmental ethicists, species have more moral significance than individual animals because species are integral to the sustainability and integrity of ecosystem health whereas specific individuals are not. This viewpoint gives rise to culling practices, captive breeding programs, and other practices that sacrifice the interests and lives of individual animals with an eye toward species health and ecosystem preservation (Lacy 1995). Tom Regan, one of the most well-known animal rights theorists, famously referred to this viewpoint as "environmental fascism" (Regan 2004: 362). Value, he insists, rests with individuals, and individuals cannot be used as mere means to an end. Individual animals' interests can be sacrificed only if doing so accrues some benefit to them.

In a famous essay detailing the nuances of this debate, Mark Sagoff averred that the two positions are irreconcilable (Sagoff 1984). One can be an animal liberationist or a conservationist, but one cannot be both. Proponents of both positions are staunch. Those on the conservationist side believe animal liberationists are antienvironment, stubborn, and morally misguided. Those on the animal liberationist side believe conservationists are anti-animal, stubborn, and morally misguided. Dale Jamieson once asked, "Is it really better to confine a few hapless Mountain Gorillas in a zoo than to permit a species to become extinct?" Is it permissible to sacrifice a gorilla in service of the Gorilla (Jamieson 2002a: 173)? It is important to recognize how this debate functions in the background of the arguments for and against zoos. We cannot hope to settle it here, but it is worth noting that both positions stem from a desire to avoid moral tragedy. Animal liberationists wish to prevent the tragedy of holding zooed animals captive with all the moral implications that go along with captivity. Conservationists wish to prevent the tragedy of extinction. Unfortunately for both positions, tragedy is already on us. The question, therefore, is not how to prevent a tragedy from occurring but, rather, how we can face this tragedy of our own making without exploiting other animals.

Moral Arguments Against Zoos

Rights-based arguments against zoos are rooted in the theoretical position that animals, like humans, are rights-bearers. As such, there are ways of treating animals that are morally indefensible regardless of whatever positive consequences could be brought about by violating animals' rights. Just as it would be wrong to treat a human rights-bearer as having merely instrumental value, it is wrong to treat animals as tools for human use. In this way (as in others), the rights view distinguishes itself from standard utilitarian moral views in its strong abolitionist position. On the rights view, no welfare improvements could justify the violation of animals' rights. Animal rights theorists want empty cages, not more spacious cages.

According to some versions of the rights view, restricting animals' liberty is morally defensible in temporary situations where an animal's well-being is improved by time spent in captivity (Regan 1995: 45–46). Some may argue that life in captivity is beneficial to animals insofar as they are free of predation, starvation, thirst, and violent death—all conditions of life in the wild. Indeed, wild animals face innumerable potential sources of suffering. Still, rights-based zoo opponents argue that zoos cannot be said to forward the interests of the individuals spending their entire lives as zoo captives (Regan 1995). Through captivity, zooed animals are deprived of, at a minimum, their liberty rights. Through the various forms of treatment zooed animals experience, their rights to bodily integrity, reproductive liberty, and free association are also violated since zooed animals are rarely given an opportunity to consent either to physical manipulation or inter-zoo transfers that break up their family and social groups. As a result, zoos are morally indefensible from the rights perspective.

Dale Jamieson has argued that "there is a moral presumption against keeping wild animals in captivity" (Jamieson 2002a: 167). Like Regan, Jamieson notes that this presumptive liberty right for wild animals can be overridden in cases where captivity would impart important benefits on the animals in question. In a later essay on zoos, Jamieson added that animals born in captivity share this presumptive liberty right (Jamieson 2002b: 183). Moreover, it could be that animals born in captivity have a greater presumptive right to liberty than their wild-captured counterparts as the captive-born have never been afforded the opportunity to experience life in the wild (ibid.).

There is, however, disagreement among animal rights theorists regarding whether animals have a right to liberty. While Regan and Jamieson defend animals' liberty rights, Alisdair Cochrane (2009) rejects the notion that animals have a right to liberty because, he argues, animals have only an instrumental interest in freedom. If animals' well-being can be provided for in captive environments, Cochrane avers, then the denial of their freedom need not necessarily set back their interests. Cochrane notes that the instrumental interest in liberty, while robust, can be satisfied through mechanisms aimed at reducing captive animals' suffering. Liberation may not be obligatory. Whether animals can be said to have a liberty right is important when thinking about zoos, but even discounting a presumptive right to liberty, zoos remain in violation of other rights related to reproductive liberty, freedom of association, and rights not to be harmed.

On the rights view, improvements in exhibit space and attentiveness to animals' well-being cannot mitigate the fundamental wrong being done to zooed animals: the persistent violation of their rights. Of course, most animals currently living in zoos would not benefit from release into the wild. Due to the conditions of their captive lives, they are ill equipped for life outside the confines of human care. While the rights view advocates for the abolition of zoos as we know them, care would need to be taken to attend to the needs of zooed animals who cannot survive in the wild. They would live out their days under some form of captivity, be it in sanctuaries or wildlife refuges.

In addition to rights-based objections to zoos, there are welfare-based objections. These objections were perhaps more striking when the average city zoo consisted of small, barred cages with concrete floors and no opportunities for enrichment. Over time, through the concerted efforts of animal advocates, ethologists, and forward-thinking zoo professionals, it became clear that these barren, caged environments were not meeting the welfare needs of the animals held within their bars. These animals were frequently bored and stressed to the point of turning to self-mutilation. They had no opportunities for play, utterly inappropriate living environments, and inadequate nutrition, and they were often denied the companionship of their conspecifics (Carlstead et al. 2013; Clubb and Mason 2007; Halberstadt 2014; Tullis 2015).[2]

Admittedly, modern accredited zoos—with their naturalistic exhibits and attention to species-specific nutritional and enrichment needs—are better at meeting the needs of individual animals than either their predecessors or roadside zoos and attractions. Yet, despite these advancements in care and exhibit design, concerns about animal welfare persist. To begin with, zooed animals remain prisoners in their environments, perpetually subject to the gaze and control of their captors

(Acampora 2008; Gruen 2014c; Pepper 2017). Simply being captive for life results in a serious reduction in welfare even under the best conditions. Much has been written regarding the problems associated with captivity, and the issues are far too rich and complex to do them justice here. Hence, I set aside a general discussion of the ethics of captivity. Readers interested in learning more can start with Acampora 2008, Gruen 2014a; Marino et al. 2009.

There are additional welfare worries when one factors in how little time zooed animals spend in the larger, naturalistic exhibits so appealing to zoo-goers. Zooed animals used in zoo education programs are often housed in off-exhibit facilities (Maple et al. 1995: 227). Even the animals not used in education programs spend very little time in the naturalistic exhibit space. According to one group of zoo professionals, "many animals spend the majority of the twenty-four-hour day in a deprived holding area, and this may actually encompass the most active hours of the day for many species" (Maple et al. 1995: 228). Behind-the-scenes footage posted by zoos in North America reveals the nature of these holding areas: steel bars, small spaces, and concrete floors. Indeed, they look exactly like the zoo exhibit spaces that have been relegated to the past precisely because they cannot meet animals' welfare needs. These holding areas are typically inaccessible to the public due to concerns around safety and disease prevention (Braverman 2013: 85), but zoos also do not want zoo-goers to see these spaces as doing so would likely raise the very welfare concerns the zoos seek to mitigate with naturalistic exhibit spaces.

Another line of argument against zoos comes from ecofeminist theory. Ecofeminism, an ethical framework that extends feminist ethics to include consideration of nature (and, for the ecofeminism I am describing here, animals) offers a unique lens through which we can view moral objections to zoos. Ecofeminists do not speak with one voice. Hence, what follows represents one ecofeminist perspective on zoos. Many ecofeminists would share the concerns about zoos discussed earlier regarding rights violations and welfare but would augment them with others. These additional concerns include highlighting the immorality of separating bonded pairs, social groups, and families through inter-zoo transfers and captive breeding programs.[3] As theorists who consider relationships of love and care central to moral life, ecofeminists strongly object to the ways zooed animals' relationships with their conspecifics are regularly broken up to serve zoos' interests.

Ecofeminists also point to intersections in social justice concerns that manifest at zoos. In addition to the moral implications regarding animals, zoos are places of concern insofar as they have a history of demonstrating colonial control through the power white Europeans had to take animals out of their home countries as well as through underscoring the idea that urban spaces are "places of high culture" as compared to "the natural realm of the colonies and the hinterlands and their inhabitants" (Deckha 2012: 540). Zoos are also implicated in exoticizing non-Western people both through their history of displaying colonized humans (Braverman 2013: 73) and through their current culturally themed exhibits. Insofar as zoos are sites where the domination and exploitation of animals, as well as humans, is on display, they are of deep concern to ecofeminists. As Kelly Struthers Montford puts it, "[g]iven this history, the zoo should be considered an institution embedded in and productive of racist, sexist, ableist, and speciesist orderings of life" (Montford 2016: 79).

Furthermore, zoos are problematic manifestations of capitalism and rampant consumerism.[4] Consider the dissonance between zoos' mission to inspire care about the devastating loss of natural habitat and species worldwide and the zoo gift shop where you can purchase a stuffed animal or mug manufactured thousands of miles away. Or consider the hot dog from factory-farmed cows that zoo-goers munch on as they shake their heads in dismay while reading the zoos' placards about the destruction of the Amazon. Since industrialized agriculture is a leading contributor to habitat destruction worldwide, this underscores a disturbing disconnect between zoos' purported mission and the consumption-driven behaviors they encourage among zoo-goers.

Finally, ecofeminists object to zoos because they reify deeply problematic hierarchies. Through controlling, breeding, and exhibiting animals, as well as treating them as property, zoos perpetuate the idea of human exceptionalism and superiority. Through determining which animals are more charismatic and interesting and thus deserving of preservation and exhibition, zoos perpetuate the idea that some animal lives are more valuable than others. Ecofeminists are deeply suspect of hierarchies, and zoos run afoul of the ecofeminist commitment to respecting the lives and individuality of all animals. For these reasons, and others that remain beyond the scope of this chapter, ecofeminists are opposed to zoos.

While the arguments presented in this section are not an exhaustive representation of the moral arguments against zoos, they are sufficient for understanding why zoos are the object of critique from animal advocates. In response to critique, zoo advocates typically invoke zoos' education and conservation missions as reasons why the concerns raised above must be overlooked in favor of greater, more pressing concerns. Vast areas of the planet are devastated by human activity, and the animals who live in those areas are dying, resulting in catastrophic loss of species diversity. Given the dire nature of the predicament in which we find ourselves, zoo advocates argue for the relevance and urgency of zoos' education and conservation missions. In the following section, we consider how well these justifications stand up to moral scrutiny.

Evaluating Education as a Justification for Zoos

The education and conservation justifications offered on behalf of zoos are really two parts of the same justification. Zoos see themselves as providing education about conservation issues by emphasizing the importance of species preservation and informing the public about habitat destruction. Nevertheless, there are, importantly, different issues at stake in evaluating the education and conservation missions, so I take each in turn, beginning with the argument that zooed animals' captivity is justified due to the vital importance of zoos' educational mission.

Arguments that zoos are important sites of public edification abound. Hutchins et al. (2008: 516) argue that zoos and aquariums have a "direct connection to the public" through which they can raise awareness, and Maple et al. (1995: 230) deem conservation education "the undisputed cornerstone of zoo programs." Braverman (2013: 28) notes that as zoos have shifted away from recreation and toward conservation and education as their primary mission, they "have taken upon themselves (and have been publicly entrusted with) the responsibility of caring for animals and of educating the public to care." Zoos' education mission is twofold: to provide zoo-goers with scientific information regarding animals and their natural environments and to educate zoo-goers into a caring posture regarding animals and their habitats. For this reason, Maple et al. (1995: 230) talk about the "motivational zoo" as one that "realizes that its mission is to conserve wildlife and natural habitats through changing the attitudes of its visitors."

Do zoos succeed in educating the public? Answering this question proves quite difficult. A 2007 study undertaken by the AZA purported to demonstrate zoos' positive impact on zoo-goer attitudes towards other animals. The researchers claim their study shows that visits to accredited zoos and aquariums result in visitors' better understanding of humans' role in environmental degradation, that visitors feel a stronger connection to nature postvisit, and that zoos and aquariums "are enhancing public understanding of wildlife and conservation of the places animals live" (Falk et al. 2007: 4). The methodological validity of that study came under intense scrutiny by Marino et al. (2010), which prompted a defensive reply from the study's authors (Falk et al. 2010). Given the nature of the dispute about methodology, as well as the paucity of any other research in this area, it is difficult to know what to conclude about the AZA's 2007 study. Where there has been other research, the results are less than promising. Ian Parker notes that a World Association of Zoos and Aquariums global survey found that "zoo visits made people seventeen per cent less committed to take action on habitat protection and creation, and nine per cent less likely to act against pollution and climate change" (Parker 2017: 46). Indeed, the literature on zoos reveals that both zoo critics and zoo professionals question zoos' ability to meet their educational mission. At a minimum, it is clear more research is necessary before we can know if zoos succeed in educating the public.

Even if zoos were able to provide demonstrable proof of pedagogical success, important questions would remain. To begin with, concerns that zoos violate animals' liberty rights will not be assuaged by claims that zoos succeed in educating the public. Beyond concerns about rights violations, however, we might worry that zoos are teaching the wrong lessons. Concerns regarding the potential ill effects of zoo education come from both zoo critics and zoo professionals. Among the former, there are worries that, by breeding and exhibiting animals, zoos reinforce pernicious attitudes of human superiority and exceptionality. As Dale Jamieson (2002b: 179) put it, "the profound message of zoos is that it is permissible for humans to dominate animals." When we breed and display animals for humans' delight and education, we take for granted the permissibility of human domination over animals' lives. This undermines the very mission the zoo purports to support, namely, that of teaching zoo-goers about the ill effects of human domination over the natural world.

Zoo critics are also concerned about zoos' failure to undertake any kind of meaningful structural critique of the systems that make habitat destruction so widespread, such as capitalism's insatiable appetite for profit at the expense of all other values. Tema Milstein (2009) observes that one zoo she visited provided a display of actions zoo-goers can take to aid conservation efforts, but

> the display stops short of clearly explaining the embedded complexity of structural issues that visitors could, if informed, work to oppose, or of clearly linking causes, what's working, and what visitors can do to the destructive agency of large-scale, transnational political economic processes and specific actors that visitors, if informed, could lobby, boycott, or attempt to transform . . .
>
> *(Milstein 2009: 39–40)*

Zoos' failure to engage in structural critique undermines their education mission by leaving a crucial element of conservation education off the table. Zoo-goers are left with the impression that making more ethical consumer choices, such as purchasing shade-grown coffee, will help preserve habitat. While personal consumer choices certainly matter in the struggle to preserve Earth's disappearing landscapes, we do a disservice when we suggest that people's responsibility and agency begin and end with personal consumer choices (Maniates 2001). Critique of the structures upholding behaviors of domination and exploitation is vital to the work of educating and empowering the public to demand more of their governments and corporations with respect to conservation efforts. The problem may go beyond teaching the public the wrong things. Insofar as zoos rely on corporate donors, they are complicit in the very structures that make exploitation of nature possible.

Zoo professionals also have concerns about zoos potentially teaching people the wrong lessons. One worry is that zoos negatively impact the public's understanding of conservation by separating zoological gardens from botanical gardens and museums dedicated to other areas of the natural sciences. David Hancocks (1995a: 32), former director of the Woodland Park Zoo, claims that these disparate institutions are too piecemeal to help the public understand the story of ecosystem collapse. William Conway, former president of the Wildlife Conservation Society and former director of the Bronx Zoo, argues that zoos are far too slow-working to effectively educate children to make a difference in a conservation crisis that is as time-sensitive and pressing as the one we face.

Other zoo professionals point out that zoos' focus on exotic species undermines zoos' educational mission. Some have argued that, by undertaking more regional specialization, zoo exhibits and educational programming would more directly bring home the conservation message to local audiences (Hancocks 1995a: 35; Keulartz 2015: 345). This is a hard sell for zoos because exotic, charismatic megafauna, such as pandas, elephants, and tigers, are perceived as critically important for public outreach and keeping zoos economically viable. Whether this is the case is a matter of some dispute, but a glance at most zoos' websites will demonstrate the extent to which zoos rely on exotic animals as public relations tools.

Some of the concerns discussed earlier, particularly those of zoo professionals, could, in theory, be mitigated through a radical redesign of zoos and their focus. Although it seems unlikely, one can imagine zoos moving away from exhibiting exotic animals and transitioning to more regionally specialized exhibits, as well as bringing together scientific information regarding ecosystems. Nevertheless, ethical concerns regarding zoos' role in violating animals' rights and reinforcing deeply problematic beliefs regarding human superiority and exceptionalism cannot be overcome by overhauling zoos. Morally problematic practices are sewn into the fabric of zoos. Breeding animals, holding them captive for life, and using them as tools to inspire care for their wild counterparts is at the heart of what zoos do. Even if they could succeed in teaching zoo-goers something important about conservation, these moral concerns would remain.

Evaluating Conservation as a Justification for Zoos

The second and most oft-cited justification for holding animals captive in zoos is that doing so is necessary for species preservation. Faced with the tragedy of species extinction, zoos are posited as the new Noah's Ark—keeping individual species members safe while their wild counterparts succumb to the consequences of pollution, habitat destruction, poaching, wars, and so on, until the day when all members of the species can live free. Zoo defenders point to zoos' role in breeding and managing *in situ* (zoo-based) populations as vital to the project of maintaining *ex situ* (field-based) populations. Moreover, they note that for species that can no longer survive in the wild, zoos are the only hope of continued existence.

I begin the discussion of the conservation justification by sketching several logistical considerations that cast doubt on zoos' conservation mission and then turn to a discussion of the moral considerations regarding zoos and conservation. One concern about zoos' claim to function as the modern Ark is that the metaphor itself is fundamentally misguided. Ships pull into ports where passengers disembark. Given the speed of habitat destruction and limited success at preserving existing habitat, there are good reasons to worry that if zoos are arks, they will forever navigate the world, carrying their cargo on an endless journey. Without a significant shift in global priorities, there will be no final, wild destination for zooed animals. Dale Jamieson (2002b: 187) was right when he said the animals on the ark are "like passengers on a voyage of the damned, never to find a port that will let them dock or a land in which they can live their lives in peace and freedom." Zoo proponents would argue that, even with low probability of success, it is better to try to preserve species given the seriousness of what is at stake with mass extinction. Perhaps some hope is better than no hope. Given the complications faced by zoos' conservation mission articulated in what follows, however, it is not immediately clear that zoos can succeed as arks. Moreover, the cost to individual animals for even limited success will be tremendous.

Some worry that the metaphorical Ark isn't merely perpetually floating without a destination but is, in fact, sinking (Lees and Wilcken 2009: 8; cf., Keulartz 2015: 340). Maintaining captive populations is complex. Breeding programs require significant financial and technical resources, and housing captive-bred animals requires space, which poses a significant problem for zoo conservation programs. Zoos do not have the space to care for sustainable captive populations for the long term (Lindburg and Lindburg 1995: 197–198; cf., Conway 2011: 3–4; Hancocks 1995a: 35; Maple et al. 1995: 232; Wagner 1995: 213). Maintaining sufficiently genetically diverse captive populations requires large numbers of animals, all of whom require housing and care. Indeed, the AZA has shifted captive breeding priorities away from the most "vulnerable and threatened species" in favor of prioritizing Species Survival Plans (SSPs) for sustainable species (Braverman 2013: 184), thereby underscoring a conflict between the value of conservation and the value of sustainability. The species most in need of conservation are sometimes the least sustainable to maintain through captive breeding. This undermines zoos' claims of being critically important players in conserving the planet's most besieged animal species. David Hancocks (1995a: 35) says that "the task of saving the world's wildlife is too great for even the combined efforts of all the world's zoos."

There is considerable disagreement even among zoo professionals regarding zoos' role in conservation efforts (Braverman 2013: 177–180). The debate between zoo professionals regarding whether zoos should focus on local, regional collections rather than charismatic megafauna is one manifestation of this disagreement. The debate over whether zoos should focus on smaller species rather than charismatic megafauna is another (Hancocks 1995a; Keulartz 2015). Some zoo professionals question captive breeding programs and current conservation priorities and practices. This all points to reasons to worry that zoos cannot reasonably fulfill the conservation mission that forms the foundation of their justification for keeping animals captive for life.

One justification often cited by zoo proponents is that zoos are morally justified in keeping captive animals because reintroducing captive-bred animals to the wild is their ultimate end. The ark will indeed find a port, and the animals aboard will transition to life in the wild as soon as it is viable for them to do so. This claim flies in the face of reality. Reintroduction efforts are very rare and have a high failure rate. Most species for whom there are SSPs are never returned to the wild (Braverman 2013: 64). They generally spend their lives in captivity, used as tools for captive breeding efforts or designated as surplus and killed. Hancocks (1995b: 181) argues that zoos are not properly equipped to ready animals for reintroduction to the wild and that display areas are "rarely tolerable for sustaining natural behaviors." Moreover, the way zoos care for animals "in no way prepares their skills for survival" (Hancocks 1995b: 182; cf., Keulartz 2015: 241).

While some people writing on zoos remain optimistic about the prospects for reintroduction (Geist 1995; Wagner 1995; Hutchins 2007), Keulartz's skepticism is echoed by other zoo and conservation professionals (Shepherdson 2003; Lindburg and Lindburg 1995). Through a careful discussion of various reintroduction programs that achieved varying levels of success, Robert Loftin (1995) argues that we cannot generalize about the viability of reintroduction efforts but instead need to determine on an individual basis whether a proposed captive breeding program genuinely holds out the promise of success. Among the reintroductions Loftin considers, most successes were the result of considerable progress in habitat restoration. Unfortunately, in most cases, there is insufficient natural habitat to make reintroduction possible and the very problem that caused the need for reintroduction has not been remedied. Given the difficulties of successfully reintroducing captive-bred animals into the wild, it is disingenuous for zoos to pin the moral justification for holding animals captive on their (or their descendants') future reintroduction to the wild.

The cornerstone of zoos' conservation efforts is their captive breeding programs. It is through breeding and managing *ex situ* populations that zoos hope to contribute to conservation efforts. Zoos tout their captive-breeding successes publicly and rely on captive-bred charismatic species to generate public interest in their mission. Unfortunately, zoos' captive breeding efforts are morally problematic for several reasons.

One area of concern is the questionable veracity of zoos' claims to be aiding *ex situ* conservation. With SSPs based on sustainability rather than prioritizing threatened animals, there are reasons to wonder whether zoos are making conservation-minded decisions when they allocate resources to captive breeding programs for specific species. There are also relatively few species that have SSPs, and most SSPs are for mammals and charismatic megafauna rather than smaller and nonmammalian species who may be more integral to restoring ecosystemic order (Hancocks 1995a: 34–35). Conway (2011: 4) notes that managing vanishing species has been a low priority for zoos despite zoo rhetoric to the contrary. Moreover, zoos tend to devote resources to captive breeding programs for animals who help zoos inspire public interest and raise funds. The suggestion that zoos take space previously designated for charismatic megafauna and devote it to smaller, more sustainable species is often met with resistance from zoo professionals because zoos consider the charismatic flagship species necessary for conservation education and conservation support (Baker 2007: 147; cf., Braverman 2013: 178).

Additionally, while zoos portray themselves as large contributors to *ex situ* conservation efforts, the numbers do not support their claims. Although some zoos make a concerted effort to raise funds for *ex situ* conservation (Conway 2011: 4), Keulartz (2015: 343) cites a 2011 report showing that less than 5% of zoo income is spent on conservation, and Loftin (1995: 177) argues that with habitat loss playing the central role in species endangerment, zoo resources would be best spent on preserving natural habitat (cf., Kheel 2008: 230). While zoos cultivate the appearance of serving a public function, most are commercial entities in the business of staying in business. Despite advances in SSP priorities, captive breeding programs often reflect this reality.

Setting aside worries about resource allocation and breeding priorities, we find that captive breeding programs themselves are deeply morally questionable. To begin with, except in rare cases where reintroduction is possible, the purpose of captive breeding programs is to produce more captive animals. As was discussed in the introduction, there are multiple moral objections to permanent captivity for zooed animals, including claims that a captive life violates their rights, that captivity significantly thwarts their species-typical behaviors and needs, and that humans exert a morally unjustifiable degree of control over animals' lives in captive environments.[5] Captive breeding programs generate animals fated to become passengers on the voyage of the damned.

Creating genetically sustainable *ex situ* populations requires a careful focus on genetic diversity (Lees and Wilcken 2009; Conway 2011). Without diverse genetic pools, captive populations become inbred and unviable. Consequently, zoos exert complete control over zooed animals' reproductive lives, choosing who can mate with whom and who is genetically surplus and therefore removed from breeding programs. There is disagreement in the zoo community regarding the best methods for dealing with animals designated as genetically surplus.

Animals are also deemed surplus when zoos do not have the space to house them and when they do not have the money to support their care (Lindburg and Lindburg 1995: 197). Surplus animals cannot be released into the wild due to the difficulties of reintroduction. Often, they cannot be transferred to other zoos since most zoos face space and economic constraints and transfer will not mitigate the issue for genetically surplus zooed animals. This leaves zoos with a moral dilemma regarding the lives of surplus animals. In response to this dilemma, zoos in the United States favor contracepting such animals, though culling does occur (Parker 2017: 51–52), while zoos in Europe prefer breed-and-cull policies where animals are permitted to mate but their offspring are killed (Braverman 2013: 176; Parker 2017: 45–46). Both approaches are morally problematic.

For now, it is enough to recognize that captive breeding programs reduce animals' value to their genetic contributions and reproductive capacities. Animals are seen as tools—manipulable entries in a database who have no special value beyond that of the other tools of reproduction, such as test tubes and ultrasound machines. For example, Chai, an elephant formerly at the Woodland Park Zoo, was subjected to a shocking 112 attempts at artificial insemination (Doyle 2014: 45). The consequences of seeing zooed animals as mere means to an end are that their lives are manipulated and their individual interests are overridden. Because zooed animals are a small population, captive-breeding programs in the United States are managed through the AZA and require zoo-to-zoo transfers that break apart bonded pairs, social groups, and families. Zoos wrangle over ownership of the animals' offspring (Braverman 2013: 11), underscoring the lack of "ownership" the animals have over their own progeny. The zoos determine whose offspring will be taken away, when, and where they will go. Animals' reproductive interests are subordinate to the priorities of captive-breeding programs. There are, therefore, good reasons to question the morality of captive-breeding programs on rights and welfare grounds.

In addition to those considerations, some ecofeminists would add concerns about the ways captive breeding programs fail to recognize the moral value of relationships of love and care, both between animals and between animals and their human caregivers. Relationships of all kinds are disvalued in the zoo paradigm. Additionally, invasive control and exploitation of animals' sexuality and reproductive lives are profoundly problematic insofar as it reifies long-standing themes of domination through reproductive control. Finally, because animals often resist human efforts to control them, novel technologies are developed to overcome this resistance, which, in turn, results in more management (Braverman 2013: 171). Rather than viewing animals' recalcitrance as a manifestation of their will and questioning the primacy of captive breeding programs, zoo professionals merely invent innovative technologies to overwhelm animals' resistance.

Facing Regret and Embracing Humility

All this suffering and exploitation is carried out for a conservation mission that is uncertain in its efficacy and future viability. Some zoo proponents argue that where there exists a chance to save species (however slim), we must do so; to condemn a species to certain death is itself an unethical choice (Geist 1995). This is compelling in its way. None of us wants to face a world without the charismatic megafauna we cherish, and many of us mourn the idea of losing species integral to the proper functioning of ecosystems.

At the same time, we have arrived at the point where we must embrace a degree of humility about how much we can mitigate the harm our species has done, about the toll our managerial and technological solutions take on animals, and about what we know about what animals want, what helps them thrive, and how our genetic tinkering has an impact on their long-term prospects.[6] Jamieson (1995: 72) emphasized the importance of humility when he wrote,

> It may be that many species will not survive without human intervention. But this does not mean that they will survive with human intervention. Indeed, our track record as planetary managers is deplorable: we generally make things worse when we try to make them better.
>
> *(Jamieson 1995: 72)*

Moreover, Jamieson rightly reminds us that

> killing or confining an animal to preserve a species is not a conflict between the interest of the individual animal and the interest of the species; it is a conflict between the interest of the animal and the human desire to preserve the species.
>
> *(Jamieson 1995: 72)*

How we set up the conflict we face has profound moral implications. Most theorists talking about the conservation dilemma between the treatment of individual animals and the survival of the species lose sight of the fact that prioritizing the survival of a species is a decision humans have made, just as prioritizing economics and political power at the expense of the planet's various lifeforms is also a human decision. It is neither necessary nor inevitable that we imprison thousands of animals in the name of preserving species. It is a choice we make and a profoundly questionable one at that.

Still, the point about the necessity of humility runs both ways. Conway (2003: 11) asks, "[W]hat great prophet among us has sufficient foresight to predict which species will never again have an opportunity to love at liberty, or something approaching it? Who will assume the authority for instructing us in the rules for allowing a species to die?" These are fair questions: while we cannot ensure our management and intervention in animals' lives will stave off mass extinction, we also cannot know with certainty that they will not.

Yet, as compelling as success stories are, they are also stories of captivity, control, exploitation, and manipulation. Where there is success, there is also tragedy. Success stories are also a rarity in the world of conservation insomuch as there is very little reason to hope that most of the species focused on in zoos will ever find freedom. Zoo critics are attentive to the tragedy piece of the tale while holding space for the successes. Some would prefer to see species go extinct than have individual animals captured from the wild and manipulated by humans to produce offspring (Kheel 2008: 230). Others would suggest that zoos be reshaped into megaparks and refuges with very different policies and programs regarding breeding and conservation (Jamieson 2002b: 188; Conway 2003: 10). None of us are equipped to be Conway's prophet. While we do well to remember that, we also do well to remember that there comes a time when we must face our grief and shame rather than sacrifice animals' interests and lives in a last-ditch effort to avoid experiencing the tragic consequences of our species' choices.[7] The central question of zoos is not whether individual animals should be sacrificed to ensure future members of their own species. The central question of zoos is whether individual animals should be sacrificed to ensure humans can delay facing profound regret caused by the knowledge that we have wreaked irrevocable devastation on the natural world. For many of us, the answer is a resounding "no."

Notes

1. Lisa Kemmerer (2006) has written a helpful discussion of the importance of language as a tool for calling out oppressive and hierarchical structures.
2. Thousands of animals remain in these conditions in unaccredited zoos, roadside attractions, and personal collections all over the world. Hawthorne (2013: chapter 6) offers an accessible and detailed discussion of conditions for zooed animals as well as the connection between stereotypies and captivity.
3. Readers interested in animal sociability and their interests in developing and maintaining social groups and pairings can look to both empirical research on the topic and reports from animal sanctuaries. There is a tendency to think interests in social connection are related to higher levels of cognition. I resist this tendency in recognition of the limited human capacity to measure and fully comprehend animal cognition (de Waal 2016).
4. For a detailed discussion of consumption at the zoo, see Braverman (2013).
5. Although animal sanctuaries differ importantly from zoos in both mission and practices, these concerns about control and the implications of captivity are pressing for sanctuaries as well (see Emmerman 2014; Jones 2014).
6. See Fouts (1995), Jamieson (1995), Kheel (2008), and Loftin (1995) for further discussion of arrogance and humility in zoological and conservation science.
7. Lori Gruen (2014b) writes eloquently on the importance of facing the grief in our efforts to mitigate harm humans have done to animals.

Bibliography

Acampora, R. (2008) "Zoos and eyes: Contesting captivity and seeking successor practices," in S. J. Armstrong and R. G. Botzler (eds.) *The Animal Ethics Reader*, 2nd ed. (pp. 501–506), New York: Routledge.

Baker, A. (2007) "Animal ambassadors: An analysis of the effectiveness and conservation impact of *ex situ* breeding efforts," in A. Zimmerman, M. Hatchwell, L. Dickie, and C. West (eds.) *Zoos in the 21st Century: Catalysts for Conservation?* (pp. 139–154), Cambridge: Cambridge University Press.

Braverman, I. (2013) *Zooland: The Institution of Captivity*, Stanford, CA: Stanford University Press.

Carlstead, K., Mench, J. A., Meehan, C., and Brown, J. L. (2013) "An epidemiological approach to animal welfare in zoos: The elephant welfare project," *Journal of Applied Animal Welfare Science* 16(4): 319–337.

Clubb, R., and Mason, G. J. (2007) "Natural behavioural biology as a risk factor in carnivore welfare: How analysing species differences could help zoos improve enclosures," *Applied Animal Behaviour Science* 102: 303–328.

Cochrane, A. (2009) "Do animals have an interest in liberty?," *Political Studies* 57: 660–679.

Conway, W. (2003) "The role of zoos in the 21st century," *International Zoo Yearbook* 38: 7–13.

Conway, W. (2011) "Buying time for wild animals with zoos," *Zoo Biology* 30: 1–8.

Deckha, M. (2012) "Toward a postcolonial, posthumanist, feminist theory: Centralizing race and culture in feminist work on nonhuman animals," *Hypatia* 27(3): 527–545.

de Waal, F. (2016) *Are We Smart Enough to Know How Smart Animals Are?* New York, NY: W.W. Norton and Company.

Doyle, C. (2014) "Captive elephants," in L. Gruen (ed.) *The Ethics of Captivity*, Oxford: Oxford University Press.

Falk, J., Heimlich, J. E., Vernon, C. L., and Bronnekant, K. (2010) "Critique of a critique: Do zoos and aquariums promote attitude change in visitors," *Society and Animals* 18: 415–419.

Falk, J., Reinhard, E., Vernon, C. L., Bronnenkant, K., Heimlich, J. E., and Deans, N. L. (2007) *Why Zoos and Aquariums Matter: Assessing the Impact of a Visit to a Zoo or Aquarium*, Silver Spring, MD: Association of Zoos and Aquariums.

Fouts, R. (1995) "Science in zoos: Arrogance of knowledge versus humility of ignorance," in B. G. Norton, M. Hutchins, E. F. Stevens, and T. L. Maple (eds.) *Ethics on the Ark: Zoos, Animal Welfare, and Wildlife Conservation* (pp. 277–288), Washington, DC: Smithsonian Institution Press.

Geist, V. (1995) "Noah's Ark II: Rescuing species and ecosystems," in B. G. Norton, M. Hutchins, E. F. Stevens, and T. L. Maple (eds.) *Ethics on the Ark: Zoos, Animal Welfare, and Wildlife Conservation*, Washington, DC: Smithsonian Institution Press.

Gruen, L. (2014a) *The Ethics of Captivity*, Oxford: Oxford University Press.

Gruen, L. (2014b) "Facing death and practicing grief," in C. J. Adams and L. Gruen (eds.) *Ecofeminism: Feminist Intersections with Other Animals & the Earth*, New York, NY: Bloomsbury Academic.

Gruen, L. (2014c) "Dignity, captivity, and an ethics of sight," in L. Gruen (ed.) *The Ethics of Captivity* (pp. 231–247), Oxford: Oxford University Press.

Halberstadt, A. (2014) "Zoo animals and their discontents," *The New York Times Magazine*, July 3, www.nytimes.com/2014/07/06/magazine/zoo-animals-and-their-discontents.html.

Hancocks, D. (1995a) "Lions and tigers and bears, Oh No!" in B. G. Norton, M. Hutchins, E. F. Stevens, and T. L. Maple (eds.) *Ethics on the Ark: Zoos, Animal Welfare, and Wildlife Conservation* (pp. 31–37), Washington, DC: Smithsonian Institution Press.

Hancocks, D. (1995b) "An introduction to reintroduction," in B. G. Norton, M. Hutchins, E. F. Stevens, and T. L. Maple (eds.) *Ethics on the Ark: Zoos, Animal Welfare, and Wildlife Conservation* (pp. 181–186), Washington, DC: Smithsonian Institution Press.

Hawthorne, M. (2013) *Bleating Hearts: The Hidden World of Animal Suffering*, Alresford, England: Changemakers Books.

Hutchins, M. (2007) "The animal-rights-conservation debate: Can zoos and aquariums play a role?" in A. Zimmerman, M. Hatchwell, L. Dickie, and C. West (eds.) *Zoos in the 21st Century: Catalysts for Conservation?* (pp. 92–109), Cambridge: Cambridge University Press.

Hutchins, M., Smith, B., and Allard, R. (2008) "In defense of zoos and aquariums: The ethical basis for keeping wild animals in captivity," in S. J. Armstrong and R. G. Botzler (eds.) *The Animal Ethics Reader*, 2nd ed. (pp. 513–521), New York: Routledge.

Jamieson, D. (1995) "Wildlife conservation and individual animal welfare," in B. G. Norton, M. Hutchins, E. F. Stevens, and T. L. Maple (eds.) *Ethics on the Ark: Zoos, Animal Welfare, and Wildlife Conservation* (pp. 69–73), Washington, DC: Smithsonian Institution Press.

Jamieson, D. (2002a) "Against zoos," in *Morality's Progress*, Oxford: Oxford University Press.

Jamieson, D. (2002b) "Zoos revisited," in *Morality's Progress*, Oxford: Oxford University Press.

Jones, M. (2014) "Captivity in the context of a sanctuary for formerly farmed animals," in L. Gruen (ed.) *The Ethics of Captivity* (pp. 90–101), Oxford: Oxford University Press.

Keulartz, J. (2015) "Captivity for conservation? Zoos at a crossroads," *Journal of Agricultural and Environmental Ethics* 28(2): 335–351.

Kheel, M. (2008) *Nature Ethics*, Lanham, MD: Rowman & Littlefield.

Lacy, R. (1995) "Culling surplus animals for population management," in B. G. Norton, M. Hutchins, E. F. Stevens, and T. L. Maple (eds.) *Ethics on the Ark: Zoos, Animal Welfare, and Wildlife Conservation* (pp. 187–194), Washington, DC: Smithsonian Institution Press.

Lees, C. M., and Wilcken, J. (2009) "Sustaining the ark: The challenges faced by zoos in maintaining viable populations," *International Zoo Yearbook* 43: 6–18.

Lindburg, D., and Lindburg, L. (1995) "Success breeds a quandary: To cull or not to cull," in B. G. Norton, M. Hutchins, E. F. Stevens, and T. L. Maple (eds.) *Ethics on the Ark: Zoos, Animal Welfare, and Wildlife Conservation* (pp. 195–208), Washington, DC: Smithsonian Institution Press.

Loftin, R. (1995) "Captive breeding of endangered species," in B. G. Norton, M. Hutchins, E. F. Stevens, and T. L. Maple (eds.) *Ethics on the Ark: Zoos, Animal Welfare, and Wildlife Conservation* (pp. 164–180), Washington, DC: Smithsonian Institution Press.

Maniates, M. F. (2001) "Individualization: Plant a tree, buy a bike, save the world?" *Global Environmental Politics* 1(3): 31–52.

Maple, T., McManamon, R., and Stevens, E. (1995) "Defining the good zoo: Animal care, maintenance, and welfare," in B. G. Norton, M. Hutchins, E. F. Stevens, and T. L. Maple (eds.) *Ethics on the Ark: Zoos, Animal Welfare, and Wildlife Conservation* (pp. 219–234), Washington, DC: Smithsonian Institution Press.

Marino, L., Bradshaw, G., and Malamud, R. (2009) "The captivity industry: The reality of zoos and aquariums," *Best Friends Animal Society Magazine*, March–April: 25–27.

Marino, L., Lilienfeld, S. O., Malamud, R., Nobis, N., and Broglio, R. (2010) "Do zoos and aquariums promote attitude change in visitors? A critical evaluation of the American zoo and aquarium study," *Society and Animals* 18: 126–138.

Milstein, T. (2009) "Somethin' tells me it's all happening at the zoo: Discourse, power, and conservationism," *Environmental Communication* 3(1): 25–48.

Montford, K. S. (2016) "Dehumanized denizens, displayed animals: Prison tourism and the discourse of the zoo," *philoSOPHIA* 6(1): 73–91.

Parker, I. (2017) "The culling," *The New Yorker*, January 16: 42–53.

Pepper, A. (2017) *Glass Panels and Peepholes: Nonhuman Animals and the Right to Privacy*. Paper presented at the SSEA Pre-RoME Workshop, Boulder, August 9.

Regan, T. (1995) "Are zoos morally defensible?" in B. G. Norton, M. Hutchins, E. F. Stevens, and T. L. Maple (eds.) *Ethics on the Ark: Zoos, Animal Welfare, and Wildlife Conservation* (pp. 38–51), Washington, DC: Smithsonian Institution Press.

Regan, T. (2004) *The Case for Animal Rights*, 2nd ed., Berkeley: University of California Press.

Sagoff, M. (1984) "Animal liberation and environmental ethics: Bad marriage, quick divorce," *Osgoode Hall Law Journal* 22(2): 297–307.

Shepherdson, D. J. (2003) "Environmental enrichment: Past, present, and future," *International Zoo Yearbook* 38: 118–124.

Tullis, T. (2015) "The Bronx zoo's loneliest elephant," *The New York Times*, June 26, www.nytimes.com/2015/06/28/nyregion/the-bronx-zoos-loneliest-elephant.html

Wagner, F. (1995) "The should and should not of captive breeding: Whose ethic?" in B. G. Norton, M. Hutchins, E. F. Stevens, and T. L. Maple (eds.) *Ethics on the Ark: Zoos, Animal Welfare, and Wildlife Conservation* (pp. 209–218), Washington, DC: Smithsonian Institution Press.

30

DEFENSIBLE ZOOS AND AQUARIUMS

Clare Palmer, Hamish Morrin, and Peter Sandøe

Introduction

Zoos and aquariums are institutions that exhibit collections of animals from different species, usually to the public. Holding collections of animals in captivity for show is not new; menageries were widely found in the ancient world, while in medieval and early modern Europe, royal families collected wild beasts for their own entertainment and to display their social dominance (Hancocks 2001). The nineteenth century, however, saw the founding of institutions, such as London Zoo, that, like modern zoos, took scientific research and principles seriously and aimed to educate the public about the animals in their collections.

In their early days, zoos and aquariums were relatively uncontroversial (although the Bronx Zoo stirred a furious debate in 1906 by caging Ota Benga, a young African man from the Congo, in the primate house, as supposed "evidence" of earlier stages of human evolution). However, since the 1970s, zoos and aquariums have generated significant ethical controversies, both concerning whether they should exist at all and about the ways in which they supply and manage their animal collections in practice.

We begin our discussion of the question raised in the title by considering why it might be claimed that zoos and aquariums are ethically unacceptable *in principle*, and we argue against this view. However, that zoos and aquariums are ethically acceptable in principle does not mean that they are in all respects acceptable *in practice*, and they face significant ethical problems in terms of their management. To illustrate this, we introduce two recent ethical controversies that led to fierce and sustained ethical debates. Using these cases to inform our discussion, we spend the rest of the chapter, more pragmatically, considering what might count as an "ethically defensible" zoo or aquarium and exploring the ethical challenges facing such institutions.

Are Zoos and Aquariums Wrong in Principle?

Those who object to the existence of zoos and aquariums in principle often maintain that sentient animals kept in zoos have *rights* that are violated by being kept in captivity and/or on display. In order to understand what underpins this objection, we need to unpack it a little.

Sentient animals are beings with the capacity for conscious experience, including negative experiences such as suffering and frustration, and positive experiences such as pleasure and excitement. It's widely accepted that vertebrates, including mammals, birds, reptiles and amphibians, are sentient (although there has been some debate about fish). There's also evidence that some invertebrates, such as the cephalopods (which include the octopus family), may be sentient. However, it's not widely accepted that insects, mollusks, and jellyfish are sentient, so the arguments here would not apply to some insectariums, butterfly houses and aquarium exhibits.

For some philosophers, then, sentience entitles an animal to basic rights, where rights are understood to be very strong protections against certain sorts of human-inflicted harms. Among those harms is keeping sentient animals in captivity and on display.

Several closely related arguments can be made here, some explicitly based on animals' rights and others, on the kind of human–animal relationship captivity implies. The first is that sentient animals have a right never be used purely instrumentally, merely as means to some other end. Zoos and aquariums, it's maintained, treat animals merely as means to human ends, whether that end is human entertainment or a more obviously serious purpose, such as the preservation of the species to which they belong. Tom Regan (1995), for instance, argues that confining animals normally fails to recognize their basic right to be treated with respect; captivity is only morally permissible when it is best for the animal concerned (for instance, to provide medical care).

A related argument focuses on what is required to protect animals' basic interests. In a seminal paper, James Rachels (1976: 214) argues that "liberty is necessary for many nonhuman animals if they are to live the sorts of lives, and thrive, in ways that are natural to them," noting that this applies more to a wild animal such as a lion and less to a domesticated animal such as a laying hen. Similarly, representatives of a group called the Nonhuman Rights Project have argued that keeping elephants in zoos interferes with elephants' fundamental autonomy, because these animals "want to be able to live as wild elephants do" (NPR 2017).

A third argument relates not just to captivity, but also to the fact that the animals are on display. Captivity and display, it's argued, is both demeaning and undignified to the animals *and* it teaches people that dominating animals is an acceptable way to relate to them. Lori Gruen (2016) for instance, argues that "certain features of current zoo practices are fundamentally dignity-denying" because they deny animals' freedom to make basic decisions, such as with which fellow species members they want to associate.

A fourth argument is made by Dale Jamieson (1985: 117) who maintains that displaying animals in zoos is wrong, because by presenting animals as separate from and inferior to humans, and suggesting that they can be used for our pleasure and our purposes, zoos teach us "a false sense of our place in the natural order."

We think, however, that there are good reasons to be skeptical about these arguments. First, it's at least possible in principle that animals *could* be treated as beings that are valuable in their own right while being kept in zoos and aquariums; they need not be treated *merely* as means to some other end, as beings to be used *only* "for our pleasure and our purposes," as Jamieson says. Animals could be cared for respectfully, with a quality of life far better than they could expect in the wild while at the same time being useful, for instance by contributing to preserving their species, just as people can be useful while still being respected for what they are in themselves. An obvious objection here, however, is that to respect wild animals requires them to be free. This could be formulated in terms of the argument that wild animals have a right to liberty. But here we need to look a little closer.

Rights are often extended to animals on the grounds of basic interests that need protecting. So, if animals have a right to life, it's because they have a basic interest in continuing to live; if they have a right not to be tortured, it's because they have a basic interest in not having serious suffering inflicted on them. For them to have a right to liberty, they would need to have a basic interest in liberty. Rachels (1976 p. 210) argues that they do have this kind of basic interest—that "the interests of many other species [besides humans] are also harmed by a loss of freedom." However, if we look more closely at Rachels's argument, his concern is that without freedom, wild animals do not thrive; captivity negatively affects animals' welfare, causing reproductive failure and premature death.

Rachels is surely right that captivity in zoos and aquariums can cause serious suffering, although many zoos have made major changes to respond to the behavioral needs of zoo animals in the 40 years since Rachels published his paper. But the question is whether animals have a right to liberty that's *independent* of a right not to be made to undergo serious suffering. If animals in captivity are not caused suffering—say they have lives much better than they would have in the wild, where their freedom will typically also be limited in many ways—are their rights really being violated? Suppose a zoo has a reptile enclosure holding lizards in captivity, but that the enclosure is the size of the lizards' normal range; they have access to their normal foodstuffs, they can choose with whom they want to associate and mate, and they are protected from predators. Do these lizards have a basic interest in freedom, grounding a right to liberty? We don't think so; all their basic interests are being met.

In our view, what matters for animals is not liberty in itself, but the important things liberty normally allows animals to obtain, such as meeting their needs and, in particular, not being made to suffer (a position carefully defended by Alasdair Cochrane [2012]). If animals can obtain these things without liberty, then liberty is not a basic interest. For humans, unjustified captivity is a rights violation because most people care a lot about their liberty and therefore have a basic interest in obtaining it. Animals, on the other hand, while they can suffer from the *effects* of being in captivity, don't suffer directly from *being in* captivity, so being free in itself is not a fundamental interest.

Similar responses, we suggest, apply to the argument that captivity is undignified and demeaning to animals in zoos and aquariums. After all, animals themselves don't have a sense of their own dignity and of being demeaned. This isn't to say that animals are not sometimes distressed by human attention in zoos; Gruen (2016) comments on animals' discomfort from "being exposed to gawking, often irresponsible, humans." If the behavior of visitors in zoos causes animals discomfort, then that should motivate a change in the animals' housing (or what human behavior is permitted). But this isn't a reason to object to zoos *in principle*.

Here, we assume that since sentience grounds these animals' needs for special protection, for something to be of moral significance, it must be of significance to the animals themselves in virtue of their sentience. So, any problem for the animals must actually be perceived or experienced by them—either directly by the animals perceiving frustration, pain or other kinds of suffering, or indirectly by the animals missing out on something they would otherwise have enjoyed. It's sometimes argued, however, that animals can be harmed by things that don't appear to matter to them. So, for instance, Webster (2011: 29) claims that the "fundamental biological and psychological essence of an animal" can be "insulted," even if the animal itself doesn't undergo negative experiences (or lose out on positive experiences) as a result. However, as two of us have argued elsewhere (Sandøe et al. 2014), we can't see plausible reasons for believing that something we do to an animal is wrong if it does not matter to the animal (assuming it also has no significant implications for other animals or humans).

Objections to the very existence of zoos and aquariums maintain that displaying captive sentient animals is morally wrong and would be wrong *even if* the displayed animals had very good lives and *even if* significant conservation benefits resulted. Even though we (the authors of this chapter) have somewhat different ethical views about animals, all three of us agree that zoos and aquariums should not be (as it were) taken off the ethical table just because they limit the freedom of the animals on display. If animals in zoos and aquariums have good lives and if significant other benefits, especially for conservation, flow from their existence, then zoos and aquariums may be ethically acceptable, even though the animals in them are not free.

Before moving on, though, there's one further thing to note: Even those who object to zoos and aquariums in principle have to accept that they are not going away any time soon. Comtemporary studies of public perceptions of zoos and aquariums are limited and national differences may occur, but overall, based on published polls there seems to be public support of zoos and aquariums in the Western world. For example, a recent poll from YouGov in the US produced the following results (and very similar results have been found in a UK poll (YouGov UK 2017)):

Table 30.1 Attitudes of US public to zoos

YouGov NY *Zoos* *US nationally representative sample: April 27–28, 2017*						
YouGov *What the work! thinks*	*Total*	*Gender*		*Age*		
		Male	*Female*	*18 to 34*	*35 to 54*	*55+*
In general, do you support or oppose the existence of zoos and aquariums?						
Unweighted base	1,110	489	621	282	346	482
Base: All US adults	1,096	541	555	334	327	435
Strongly support	34%	36%	32%	25%	37%	39%
Somewhat support	39%	37%	41%	34%	38%	43%
Somewhat oppose	12%	12%	12%	16%	13%	9%
Strongly oppose	4%	4%	4%	4%	5%	4%
Don't know	10%	10%	10%	20%	8%	5%
Net: Support	73%	73%	73%	60%	75%	82%
Net: Oppose	17%	17%	17%	21%	17%	13%
How, if at all, have your feelings on zoos and aquariums changed in the past decade?						
Unweighted base	1,110	489	621	282	346	482
Base: All US adults	1,096	541	555	334	327	435
I'm more in favor of zoos and aquariums today than I was a decade ago	17%	18%	16%	15%	24%	13%
I'm more opposed to zoos and aquariums today than I was a decade ago	25%	24%	26%	34%	23%	20%
My feelings are no different today as they were a decade ago	48%	48%	47%	34%	41%	62%
Don't know	10%	9%	12%	16%	12%	5%

Source: Reproduced with permission from YouGov Plc US (2017).

Note: All figures, unless otherwise stated, are from YouGov Plc. Total sample size was 1096 adults. Fieldwork was undertaken between 27–28 April 2017. The survey was carried out online. The figures have been weighted and are representative of all US adults (aged 18+).

In this representative survey, 73% of US residents supported the existence of zoos and aquariums, 17% opposed their existence, and 10% were undecided; the clear majority therefore thought that zoos and aquariums were at least acceptable in principle. Having said this, these statistics do also show that significantly more people have moved away from support for zoos and aquariums than have moved in the opposite direction, and more young people than older people oppose them. But it's clear that zoos and aquariums are still popular institutions. Given that they will continue to exist for the time being, even people who are opposed to them in principle have reason to be concerned that they are managed in an ethically acceptable way.

This brings us to consider *in-practice* ethical issues related to zoos and aquariums. If zoos are not unacceptable in principle, what would they need to look like to stand up to ethical scrutiny? What goals would they have to endorse, and how far would they have to meet these goals? What kinds of animals should they hold in their collections? How should zoos and aquariums manage the acquisition of animals, their access to space, their reproduction, their lives and deaths? In order to illustrate what is at stake and to give our discussion a more concrete focus, we now introduce two recent highly controversial cases about zoo management.

Two Cases

Importing Wild Elephants Into the US

In 2015, a consortium of three US zoos applied to import eighteen wild elephants into the United States from Swaziland. The importing zoos—Dallas Zoo in Texas; Sedgewick County Zoo in Wichita, Kansas; and Henry Doorly Zoo and Aquarium in Omaha, Nebraska—claimed that the elephants would otherwise be culled in Swaziland, due to an extended drought and because the elephant population had grown too large for the space available, and was "changing forests into barren landscapes" (Sedgewick County Zoo 2018). The elephants could not be moved elsewhere in Africa on account of heavy poaching, and they were competing for habitat with endangered white and black rhinos.

However, importing wild elephants into the US was met with controversy; a letter of objection was signed by "scientists, conservationists, elephant care, animal welfare and policy experts" and widely circulated in the media (PAWS 2015). The signatories argued that the reasons given by the zoos for importing wild elephants at best lacked substantiation and at worst were false. The zoos, they argued, were not genuinely concerned about the welfare of elephants as individuals nor about species conservation, but rather wanted to restock "a dwindling zoo elephant inventory" and fill their emptying elephant exhibits. Zoo elephants in the US, they maintained, "do not thrive," and it is "archaic and unethical" to bring more wild elephants to the US given the challenges to their "health and viability."

The importing zoos and the American Zoological Association (AZA) which accredits all three zoos and supported importing the elephants, responded that while the elephants would be used in a breeding program, the aim of the program was to maintain a genetically viable African elephant population in the US so that welfare and other problems caused by inbreeding were avoided. Second, elephant welfare in the right zoo conditions could be perfectly good, and unlike in their wild habitat, these elephants would receive veterinary care and be safe from poaching. And third, keeping these elephants in zoos in the US would both benefit wild rhinos (by freeing up space for them) and wild elephants (by providing an educational and fundraising vehicle to support remaining wild elephants in Africa).

The zoos were granted permission by the US Fish and Wildlife Service to import the elephants. Sixteen are at the time of writing (August 2018) in residence in the three US zoos (one died in captivity before the elephants were imported, and another, Warren, died in Henry Doorly Zoo in 2017, while being anesthetized for a procedure to protect his cracking tusks).

Killing Marius, the Giraffe

In the spring of 2014, a healthy juvenile male reticulated giraffe that the keepers called Marius was killed by staff at Copenhagen Zoo, publicly dissected, and his remains fed to zoo lions. The giraffes at Copenhagen Zoo are used for a breeding program run by the European Association of Zoos and Aquaria; successful breeding programs require genetically diverse participants to avoid inbreeding. But Marius's genes were very similar to those of other giraffes in the breeding program, and there was limited space in Copenhagen Zoo and a risk of overpopulation, so he was regarded as "surplus," using a place that could be taken by a more genetically valuable giraffe. It might have been possible to move Marius elsewhere, but Copenhagen Zoo was not satisfied that any of them could guarantee Marius's long-term welfare. So, the zoo decided to euthanize him, and the zoo director, Bengt Holst, made the following statement:

> We see this as a positive sign and as insurance that we in the future will have a healthy giraffe population in European zoos. The same type of management is used in deer parks

> where red deer and fallow deer are culled to keep the populations healthy. The most important factor must be that the animals are healthy physically and behaviourally and that they have a good life whilst they are living whether this life is long or short.
>
> *(Holst 2014)*

That Marius's killing should be taken as a "positive sign" wasn't a view widely shared among popular and social media sources (Smith 2014), where there were accusations of "cruelty" and creating a "horror show." There were also more considered moral objections to killing a healthy young animal just because his genes were "surplus" to requirements (Bekoff 2018). Some argued that he should not have been born in the first place; various options, including sterilization, the separation of sexes, or contraception, were mooted. Others argued that even if his future welfare could not be absolutely guaranteed, it would still be better for him to be alive in another facility than to be killed on the cusp of adulthood. And yet others argued that while it might have been acceptable had Marius been quietly euthanized, the public display of dissection and feeding him to the lions sent entirely the wrong message about the disposability of animal lives (Cohen and Fennell 2016).

If Zoos and Aquariums Are Not Wrong in Principle, What Would Count as an Ethical Zoo in Practice?

We've argued above that zoos and aquariums are not wrong in principle. However, we've also provided two cases in which managing zoos and aquariums gives rise to ethical controversy. How should issues like these be dealt with by zoos? What features would a zoo or aquarium need to possess to be ethical in practice? Here, we outline two principles that we think could serve as a basis for an ethically defensible zoo or aquarium at the present time. These principles follow naturally from the earlier discussions, in that the first principle focuses on zoos and aquariums serving a good purpose in ethical terms, and the second focuses on the welfare of the animals that they keep.

In fact, we are not attempting to work this out alone. Precisely because of the controversies involving zoos and aquariums in the past couple of decades, national and international zoo and aquarium accreditation associations such as the World Association of Zoos and Aquariums (WAZA), the AZA and similar organizations in other parts of the world, as well as individual zoos, have developed ethical codes as a form of self-regulation. The best of these provide excellent guidance as to what an ethical zoo or aquarium might look like. The difficulty lies in the ability of individual institutions to live up to these principles. As our two case studies have illustrated, ethical controversies persist even about accredited zoos. In part, as we suggest, that's because the two key principles that shape ethical zoos can be in tension or even in conflict with one another.

Principle 1: Zoos and Aquariums Ought to Make a Significant Practical Contribution to Promoting Conservation

Many zoos and aquariums, and major zoo associations, have in the past three decades committed themselves to the conservation of wild species and their habitats. Attempts to cash out this commitment, however, can take different forms and may be direct or indirect. *Direct* conservation involves breeding and reintroducing endangered species and fundraising and research to support wild populations and habitats. *Indirect* conservation involves public engagement and environmental education aimed at positively changing knowledge, attitudes, and behavior with respect to conservation (WAZA 2015).

The conservation of wild species and habitats can be justified from a number of different ethical perspectives. The habitats in question may contribute to ecosystems that provide people with services such as foods and fibers, that are important in storing carbon, purifying water, and regulating other natural cycles, that are beautiful, or that are of historic and cultural value. On some ethical views, it's also argued that we can have direct moral responsibilities to maintain ecosystems in particular states (such as integrity or flourishing) independently of any uses they may have to people (e.g., Johnson 1993). Animal populations and species may be highly valued for their important contributions to ecosystems or because they are charismatic, beautiful or fascinating, and some ethicists also argue that species have intrinsic value, which gives us direct moral responsibilities not to endanger them (e.g., Smith 2016). Where species are valued for qualities independent of their contribution to ecosystems, their continued existence is important even if there's no obvious way of reintroducing them to wild ecosystems.

If a zoo or aquarium succeeds in making significant practical contributions to the protection of such important values, doing so would provide an ethical justification for their existence—unless these values are outweighed by other ethical concerns. So, the key questions are whether and how far they succeed in contributing to conservation, and whether they do so without significant compromise on the second key principle: animal welfare.

Principle 2: Zoos and Aquariums Ought to Protect and Promote Good Animal Welfare

Animal welfare concerns how the lives of sentient animals are going, for the sake of, and from the perspective of, the animals themselves (Palmer and Sandøe 2018). In recent years ensuring good animal welfare has become an increasingly prominent commitment of zoo associations; for instance, WAZA published a major report on animal welfare, *Caring for Wildlife*, in 2015.

Traditionally, good zoo animal welfare was understood as avoiding *negative* experiences (so, for instance, ensuring that animals were free from hunger, injury, and disease), but more recently, the importance of what's called *positive* welfare has been emphasized—ensuring animals have the opportunity to feel pleasure, stimulation, excitement, enjoyment. It is widely assumed that these things can be achieved by ensuring that animals in captivity are able to exercise important natural behaviors, for example related to feeding and care for offspring.

That animals in zoos and aquariums have good welfare in all these senses is important from a number of different ethical positions. For instance, good welfare is important from utilitarian ethical views, where the goal is to maximize happiness. It's also important on the kind of interest-based animal rights position, such as that mentioned earlier, as well as on ethical views that emphasize special obligations to care for animals that we have made vulnerable to and dependent on us.

If zoos and aquariums can make a significant contribution to the conservation of species and habitats while achieving, protecting, and promoting good animal welfare, we maintain that there's ethical justification for their existence. However, there are significant difficulties in reaching these goals in practice. One problem is that zoos and aquariums often fall short of making significant contributions to conservation, or any contribution they do make is extremely difficult to measure, and in some cases or with some species, they also have significant problems in achieving good animal welfare. A second concern is that the goals of conservation and of good animal welfare can be in tension with one another. The elephant importation case discussed earlier illustrates a conflict between conservation values and animal welfare. The case of Marius concerns conservation versus not killing a young healthy animal. Even though a quick, painless death may not pose a welfare issue in terms of causing suffering, it may still be argued that killing Marius involved a loss in potential life-years of good welfare.

How Far Do and Could Zoos and Aquariums Contribute to Conservation in Practice?

In thinking about zoos' and aquariums' potential contribution to conservation, we distinguished between *direct* and *indirect* conservation. Direct conservation involves practices with measurable effects on species or habitats, for instance, breeding populations of threatened species for the purpose of reintroduction; indirect conservation primarily relates to the effects zoos and aquarium have on visitors' attitudes and behaviors. In both cases, how much do zoos achieve in practice?

Direct Conservation

One obvious way in which zoos and aquariums could contribute to conservation is by keeping threatened species in their collections. But while zoos have recently increased the number of threatened species that they hold, some studies have argued that they should be doing better. Conde et al. (2013: 3), for instance, compared the species actually held in zoos with those listed as threatened by the International Union for the Conservation of Nature, and concluded that, with a few exceptions, "most [zoo] collections are not distinguishable from what would be expected if the species were selected at random." Other studies point out that while around a quarter of threatened *mammal* species can be found in zoos, only a tiny percentage of globally threatened *amphibian* species are held in zoo collections. This suggests that, if zoos are to make stronger contributions to conservation, as Keulartz (2015) argues, they should increase the number of threatened reptiles, amphibians and invertebrates they hold, in preference to keeping large, charismatic, but less threatened mammals. A further issue concerns how far keeping populations of endangered species really contributes to conservation, if those species are not or cannot be successfully reintroduced to the wild. To date, there have not been many successful reintroductions from captive-bred zoo populations, and in addition, appropriate habitats are becoming increasingly scarce due to human incursion and changing climate.

A further difficulty for captive breeding programs is that many populations of threatened species in captivity are small and distributed between zoos and aquariums (Conde et al. 2013). To ensure sufficient genetic diversity in breeding programs, these institutions therefore need to exchange either animals or gametes, or to bring in new individuals from the wild. But as we've already seen in both the African elephant and the Marius case, breeding programs to ensure genetic diversity can conflict with individual animals' welfare. We'll return to this shortly.

Zoos can also directly contribute to conservation by tethering zoo exhibits to fundraising for specific projects in the wild, though for this to be really effective, it needs to generate new money (rather than merely diverting donations that would have been received anyway). WAZA, for instance, includes fundraising in its "One Plan Approach," which aims to integrate species conservation efforts "in or out of the natural habitat" (Gusset and Dick 2013: 1). Zoo and aquarium fundraising for wild populations has had some success: "US$350 million is raised annually for direct support for wildlife conservation by zoos and aquariums in organized associations around the world" (WAZA 2015: 33)—although we don't know how much of this is "new" money. However, most of these funds currently come from a small number of institutions—there is certainly room for significant increase. Barongi (2018) estimates that if all WAZA members donated 5% of their budget, $1 billion (compared to currently less than a third of this) a year could be raised by zoos and aquariums to support wild conservation projects.

Zoos and aquariums are definitely making some contributions to direct conservation, but even based on the brief discussion here, it's clear that they could do more—especially where, as we discuss shortly, compromise in animal welfare is permitted in order to pursue a conservation goal.

Indirect Conservation

Globally, visits to zoos are estimated at 700 million a year (Gusset and Dick 2011). However, understanding the actual or potential effects of these visits on the visitors is extremely difficult. In particular, to measure conservation effects, studies need to be able to identify lasting changes not only to visitors' *attitudes* but also to their *behavior* over time. But as soon as longer time spans are introduced, there's potential for other factors than the zoo visit to influence attitudinal and behavioral outcomes. In addition, there's not only significant variation between studies in terms of what is being measured with respect to "pro-environmental" or "conservation" attitudes and behaviors but also disagreement about what these terms should actually be taken to mean (Nygren and Ojalammi 2018).

Perhaps the best-known research here is a study produced by Falk et al. (2007) on behalf of the AZA. This study aimed "to find out if zoos and aquariums successfully promote conservation" (p. 3); it took place at 12 US zoos over three years and used questionnaires, interviews, and tracking studies to follow visitors before and after their visits. The study found, positively, that a zoo/aquarium visit "has a measurable impact on the conservation attitudes and understanding of adult visitors" (p. 9), strengthening attitudes and values regarding conservation and connection to nature.

However, the study has been methodologically controversial (Marino et al. 2011), in part because it depended on subjects self-reporting (many studies show, for instance, that people exaggerate their own socially responsible behaviors when self-reporting). Other studies, in contrast, have come to much less positive results with respect to visitor learning and behaviour change. A recent qualitative meta-analysis of 31 papers studying empirical visitor research concludes that "[t]he overall evidence that the visitors learn about conservation and biodiversity, and even more importantly, that this learning results in behavioral changes, remains quite weak" (Nygren and Ojalammi 2018 [early view, not paginated])

It should be stressed that the absence of evidence for an effect of zoos on attitudes and behavioral change is not evidence of absence. We do not have evidence for saying that zoos and aquariums do *not* have a positive effect on visitor attitudes and behaviors; nor can we say, to return to Jamieson (1985), that their impact is obviously negative and contributes to "a false sense of the human place in nature."

However, we do need more well-designed studies, not only to indicate what effects zoos and aquariums are currently having on visitors but whether redesigned communications and collections could have more positive effects. Recent research by Clayton and Le Nguyen (2018), for instance, found that encouraging visitors to develop feelings of connection to and shared identity with animals, to think about how animals are similar to them, and to understand how what humans do threatens animals in the wild, provided a mixture of both information and motivation "to create concern and a tendency towards supportive behavior" (p. 211). Interestingly, this focus—on developing connections to and empathy with individual animals—to some degree bridges concern for the conservation of whole species and the welfare of individual animals.

How Far Do—and Could—Zoos Protect and Promote Animal Welfare?

Earlier, we noted that the goal of good animal welfare in zoos and aquariums could be supported from a number of different ethical perspectives. Obviously, however, some existing zoos and aquariums do not achieve good welfare for some or all animals (and may make little or no contribution to conservation either). These institutions can't be ethically justified. Accredited zoos and aquariums—those that are members of organizations such as WAZA and the AZA—all make commitments both to conservation and to animal welfare. But there are still many cases where it's reasonable to doubt that particular animals have good welfare.

We should clarify here, however, by saying what we mean by "good welfare." Earlier, we suggested that *good* welfare could both be understood as freedom from bad experiences/behavioral constraints, and more positively, as good experiences achieved through opportunities to manifest natural behaviors such as play and social interaction. This is picked up by WAZA (2015: 20) where good welfare for animals is described in terms of the "Five Domains" model. The model divides welfare into four physical or functional domains and one mental domain. Among the physical domains are nutrition, environment, physical health, and behavior. Animals should be free, for instance, from malnutrition, from disease or injury, from significant environmental challenges, and from behavioral restriction; they should have access to appropriate food, an environment that affords them opportunity and choice, they should be fit and able to express natural behavior. These four domains all serve to underpin the mental domain. Thus, the ultimate goal of the model is to ensure that zoo animals are minimally affected by unpleasant emotional states, such as fear, boredom, and distress, and able to engage in positive activities and experiences such as contentment and affectionate companionability.

Taking this as the kind of "good welfare" zoos and aquariums should be aiming at has several general implications. Some species of animals should not be kept in zoos and aquariums at all, if there's evidence that they can't have good welfare in captivity—this probably includes some species, such as cheetahs and forest duikers (Mason 2010) and orcas.

This restriction also raises questions about whether elephants should be kept in zoos, currently a debate that divides even those who are generally supporters of zoos. In 2005, Detroit Zoo retired its two African elephants to a sanctuary, noting that the zoo was unable to provide the elephants with sufficient space, that the coldness of the winter meant the elephants were confined for long periods, and that elephants needed to live in complex social groups, not just as a twosome (Detroit Zoo 2018).

Other zoos, as we've seen, have recently acquired African elephants from the wild to expand the size of their elephant groups. However, there are serious questions about whether zoo elephants can have sufficient space and the right kinds of social group for good welfare. Certainly, the persistence of high levels of stereotypies (apparently purposeless repetitive behaviors), foot problems, a poor reproductive record, and limited longevity suggests that captive elephants are not currently achieving good welfare, and that they should not be kept in zoos (although some form of captivity may nonetheless be the only available options for the *existing* African elephants held in zoos; Mason and Veasey 2010; Dow et al. 2011; Meehan et al. 2016). Relatedly, there are some species that can have good welfare in zoos and aquariums but that have complex or highly specialized needs, and many institutions are unable to meet them. In these cases, only some specialist institutions should keep such species.

A special problem about the elephant case was the need to bring animals from the wild into a zoo, generating concern for the welfare of the animals (compared to animals bred in captivity). A similar problem seems to exist on a much larger scale for fish in aquariums. Although it is difficult to find detailed information on the scale of marine species that are captive-bred, a large majority of species on exhibit in public aquariums seems to be wild-caught, likely due to a lack of economic incentives and of knowledge about optimal breeding facilities (Palmer 2014). It is estimated that more than 1,800 marine species are wild-caught in the marine aquarium trade (Rhyne et al. 2014) and that almost 55% of the marine species that are imported into the US are the same species that are on exhibit at public aquariums (Rhyne et al. 2012). However, the fishing techniques that are employed in the capture of wild fish can have a negative impact on the welfare of the animals, as well as on their habitats. For example, the illegal practice of cyanide fishing is widely used to capture live reef fish by poisoning and temporarily stunning the targeted fish, but it also kills invertebrates and non-targeted fish (Calado et al. 2014). The transportation of tropical wild caught fish also seems to have significant consequences. The mortality rate for some species in the aquarium trade reaches 80%, due to factors such as physical injuries and extreme changes in water quality and temperatures (Livengood and Chapman 2007). In light of these problems, many high-end aquariums make demands regarding form of capture, packing, and transport before buying from an exporter. Also, aquariums may require that fish caught for their collections be as small/juvenile as possible, creating the lowest pressure on wild fish populations.

However, there's also a converse point worth making. It might be that some species have excellent welfare and positively flourish in zoos on all possible indicators mentioned earlier. If such a species is likely to be very popular with visitors and might help with promoting conservation (in terms of education or fundraising), there could be a justification for keeping it in zoos, even if the species is not *itself* of high conservation value.

Put in this way, ethical difficulties about managing zoos don't sound so troubling. After all, there are many threatened species, so zoos could focus on keeping animals in their collections that are both threatened and that can have good welfare, and they could also keep animals from some popular species with lower conservation value, if these animals have good welfare and can support species of higher conservation significance. But nonetheless, as the Marius case illustrated, tensions between conservation and animal welfare (including avoiding culling "surplus" animals) persist in zoo and aquarium management.

Persistent Tensions Between Conservation and Animal Welfare

The two cases we considered earlier involving the introduction of animals from the wild with fresh, diverse genes, and the killing of genetically "surplus" animals, highlights some of the most challenging ethical problems in zoo management. There are many occasions where the value of some aspect of conservation (e.g. preservation of an endangered species through breeding) comes into conflict with some aspect of animal welfare. For instance, to encourage reproduction between genetically diverse individuals, transportation between different zoos or aquariums may be required. But this may be distressing and frightening for the animals concerned, and may require confinement and disruption of social groups—undermining many of the elements of WAZA's (2015) account of animal welfare.

In these kinds of value conflicts, even those who support zoos and work in them will disagree—just as there are currently disagreements in the zoo community about keeping elephants. The case of the giraffe Marius is obviously divisive, and it is not an isolated incident; killing "surplus" animals is a policy in many zoos across the world, including the US (Penfold et al. 2014). In fact, the authors of this chapter, who agree on the permissibility of zoos in principle, disagree about "surplus" animals. One of us (Palmer) takes the view that killing healthy young animals is a serious harm that should be avoided, since it frustrates the most basic interests and desires of the animal concerned, deprives the animal of positive future experiences, and may rupture social relationships—even though this may mean other kinds of welfare compromise in terms of contraception, sex separation, moving animals around, or shifting the composition of zoo species to avoid this situation arising (reticulated giraffes, after all, are of low conservation value). Sandøe, a (self-described!) hard-nosed consequentialist, argues that the best overall outcomes in terms of welfare involve allowing giraffes to breed and have "natural" family lives, to try to ensure healthy populations, and to painlessly kill adolescents where they are "surplus"—especially where a better welfare overall across the affected animals is brought about by their deaths. Morrin also would support the use of management euthanasia in some cases, arguing that breeding and offspring rearing behaviors are important not only to welfare but also to the maintenance of viable captive populations and to the possibility of future reintroduction attempts.

Both these population management strategies have ethical justifications and can be arrived at by a process of reasoning and reflection. We are not going to attempt to recommend a "best policy" here. However, what we do think is important is that debates about these issues are undertaken publicly and openly, by zoos and other stakeholders in the debate. In the Marius case, Copenhagen Zoo was open about its policy and the reasons for adopting such a policy. Even though this led to vilification in some quarters, it also allowed for debate about its reasons and discussions of the implications of adopting alternative policies.

The important thing for us is that ethical debates with a focus on both conservation and animal welfare should be publicly undertaken, and that zoos are open about how they deal with the issue (a way in which Copenhagen Zoo obviously excels).

Conclusion

In this chapter, we've argued that there is no good objection in principle to the existence of zoos and aquariums. This conclusion is based on ethical premises with which some people, for example those who subscribe to strong animal rights positions, will disagree. However, for the time being, attitudes that favor zoos and aquariums prevail, and therefore, anyone who cares about zoo animals should agree with our effort to formulate and promote principles of best ethical practice for zoos and aquariums. We formulate two principles that an ethically defensible zoo or aquarium must live up to: it should make a significant practical contribution to promoting conservation, and it should in practice protect and promote good animal welfare. This means that zoos and aquariums that don't make meaningful contributions to conservation and/or that have poor animal welfare are unacceptable, ethically speaking. In addition, zoos and aquariums that, broadly, meet these criteria may need to reconsider some of their practices: how they acquire animals for their collections, which species they keep, what they do about reproduction, how they house their animals, and so on. Even so, there will be practices over which those who think zoos are permissible will disagree—such as in cases such as Marius's. Much more informed ethical reflection and debate are needed with respect to all these issues in terms of how zoos and aquariums are managed over the next few decades.

Acknowledgment

Thanks are due to Sara Kondrup for help in searching literature and in editing the references.

Bibliography

Barongi, R. (2018) "Committing to conservation: Can zoos and aquariums deliver on their promise?" in B. A. Minteer, J. Maienschein, and J. P. Collins (eds.) *The Ark and Beyond: The Evolution of Zoo and Aquarium Conservation* (pp. 107–121), Chicago, IL: University of Chicago Press.

Bekoff, M. (2018) "Zoothanasia: The cruel practice of killing healthy zoo animals," [position statement], www.salon.com/2018/02/03/zoothanasia-the-cruel-practice-of-killing-healthy-zoo-animals_partner/

Calado, R., Leal, M., Vaz, M., Brown, C., Rosa, R., Stevenson, T., . . . Thornhill, D. (2014) "Caught in the act: How the U.S. lacey act can hamper the fight against cyanide fishing in tropical coral reefs," *Conservation Letters* 7(6): 561–564.

Clayton, S., and Le Nguyen, K. (2018) "People in the zoo," in B. A. Minteer, J. Maienschein, and J. P. Collins (eds.) *The Ark and Beyond: The Evolution of Zoo and Aquarium Conservation* (pp. 204–211), Chicago, IL: University of Chicago Press.

Cochrane, A. (2012) *Animal Rights Without Liberation*, New York, NY: Columbia University Press.

Cohen, E., and Fennell, D. (2016) "The elimination of Marius, the giraffe: Humanitarian act or callous management decision?" *Tourism Recreation Research* 41(2): 168–176.

Conde, D., Flesness, N., Colchero, F., Jones, O., and Scheuerlein, A. (2013) "Zoos can lead the way with ex situ conservation," *WAZA Magazine* 12: 26–29.

Detroit Zoo (2018) "Winky and Wanda: A tale of two elephants," https://detroitzoo.org/about/your-detroit-zoo/elephants/

Dow, T., Holaskova, I., and Brown, J. (2011) "Results of the third reproductive assessment survey of North American Asian (*Elephas maximus*) and African (*Loxodonta africana*) female elephants," *Zoo Biology* 30(6): 699–711.

Falk, J., Reinhard, E., and Vernon, C. (2007) *Why Zoos & Aquariums Matter: Assessing the Impact of a Visit to a Zoo or Aquarium*, Silver Spring, MD: Association of Zoos and Aquariums.

Gruen, L. (2016) *Shifting Towards an Ethics of Sanctuary*, [contributor Response], Center for Humans & Nature, www.humansandnature.org/shifting-toward-an-ethics-of-sanctuary

Gusset, M., and Dick, G. (2011) "The global reach of zoos and aquariums in visitor numbers and conservation expenditures," *Zoo Biology* 30(5): 566–569.

Gusset, M., and Dick, G. (2013) [editorial], *WAZA Magazine* 14: 1.

Hancocks, D. (2001) *A Different Nature: The Paradoxical World of Zoos and Their Uncertain Future*, Berkeley, CA: University of California Press.

Holst, B. (n.d.) *Copenhagen Zoo, Q & A to Scientific Director Bengt Holst* [Interview], www.zoo.dk/en/about/enrichment-and-management/enrichment-and-management-of-the-animal-population/.

Holst, B. (2014) *Q&A to Scientific Director Bengt Holst* [Interview], https://www.nbcnews.com/news/world/zoo-kills-unwanted-giraffe-marius-feeds-carcass-lions-n25726

Jamieson, D. (1985) "Against zoos," in P. Singer (ed.) *In Defense of Animals* (pp. 132–143), New York, NY: Basil Blackwell.

Johnson, L. (1993) *A Morally Deep World*, London: Cambridge University Press.

Keulartz, J. (2015) "Captivity for conservation? Zoos at a crossroads," *Journal of Agricultural and Environmental Ethics* 28(2): 335–351.

Livengood, E. J., and Chapman, F. A. (2007) *The ornamental fish trade: An introduction with perspectives for responsible aquarium fish ownership*, Department of Fisheries and Aquatic Sciences, UF/IFAS Extension, http://edis.ifas.ufl.edu/pdffiles/FA/FA12400.pdf

Marino, L., Lilienfeld, S., Malamud, R., Nobis, N., and Broglio, R. (2011) "Strong claims, feeble evidence: A rejoinder to Falk et al," *Society & Animals* 19(3) (2010): 291–293.

Mason, G. J. (2010) "Species differences in responses to captivity: Stress, welfare and the comparative method," *Trends in Ecology & Evolution* 25(12): 713–721.

Mason, G., and Veasey, J. (2010) "What do population-level welfare indices suggest about the well-being of zoo elephants?" *Zoo Biology* 29: 256–273.

Meehan, C. L., Mench, J. A., Carlstead, K., and Hogan, J. N. (2016) "Determining connections between the daily lives of zoo elephants and their welfare: An epidemiological approach," *PLoS One* 11(7): e0158124. https://doi.org/10.1371/journal.pone.0158124

[NPR] National Public Radio, Inc (2017) "The nonhuman rights project advocates for zoo animals," [interview with Steven Wise of the Nonhuman Rights Project], www.npr.org/2017/11/18/565025405/

Nygren, N. V., and Ojalammi, S. (2018) "Conservation education in zoos—A literature review," *TRACE, Finnish Journal for Human-Animal Studies* 4, [early view, not paginated].

Palmer, C. (2014) "An investigation into the UK's largest public aquarium chain," *Captive Animals' Protection Society*, http://sea-lies.org.uk/wp-content/uploads/2014/04/An-Investigation-into-the-UKs-Largest-Public-Aquarium-Chain-C.Palmer-CAPS-20142.pdf

Palmer, C., and Sandøe, P. (2018) "Welfare," in L. Gruen (ed.) *Critical Terms for Animal Studies* (pp. 424–438), Chicago, IL: University of Chicago Press, [in press].

[PAWS] Performing Animal Welfare Society (2015) *Letter Opposing Import of Elephants from Swaziland*, October 26, www.pawsweb.org/documents/SWAZILAND_EXPERTS_UPDATED.pdf

Penfold, L. M., Powell, D., Traylor-Holzer, K., and Asa, C. S. (2014) "'Use it or lose it:' Characterization, implications, and mitigation of female infertility in captive wildlife," *Zoo Biology* 33: 20–28.

Rachels, J. (1976) "Do animals have a right to liberty?" in T. Regan and S. Singer (eds.) *Animal Rights and Human Obligations* (pp. 205–223), Englewood Cliffs, NJ: Prentice-Hall.

Regan, T. (1995) "Are zoos morally defensible?" in B. G. Norton, M. Hutchins, E. F. Stevens, and T. L. Maple (eds.) *Ethics on the Ark: Zoos, Animal Welfare, and Wildlife Conservation*, Washington, DC: Smithsonian Institution Press.

Rhyne, A. L., Tlusty, M. F., and Kaufman, L. (2014) "Is sustainable exploitation of coral reefs possible? A view from the standpoint of the marine aquarium trade," *Current Opinion in Environmental Sustainability* 7: 101–107.

Rhyne, A. L., Tlusty, M. F., Schofield, P. J., Kaufman, L., Morris, J. S., and Bruckner, A. (2012) "Revealing the appetite of the marine aquarium fish trade: The volume and biodiversity of fish imported into the United States," *PLoS One* 7(5): e35808.

Sandøe, P., Hocking, P. M., Collins, S., Forkman, B., Haldane, K., Kristensen, H. H., and Palmer, C. (2014) "The blind hens' challenge: Does It undermine the view that only welfare matters in our dealings with animals?" *Environmental Values* 23(6): 727–742.

Sedgewick County Zoo (2018) "Get to know the elephants," www.scz.org/animal_exhibits-four.php

Smith, R. (2014) "Giraffe killing at Copenhagen zoo sparks global outrage," *National Geographic*, https://news.nationalgeographic.com/news/2014/02/140210-giraffe-copenhagen-science/

Smith, I. (2016) *The Intrinsic Value of Endangered Species*, New York, NY: Routledge.

[WAZA] World Association of Zoos and Aquariums (2015) *Committing to Conservation: The World Zoo and Aquarium Conservation Strategy*, www.waza.org/files/webcontent/1.public_site/5.conservation/conservation_strategies/committing_to_conservation/WAZA%20Conservation%20Strategy%202015_Landscape.pdf

Webster, J. (2011) "Husbandry and animal welfare," in J. Webster (ed.) *Management and Welfare of Farm Animals: The UFAW Farm Handbook*, 5th ed., Chichester: Wiley Blackwell.

YouGov UK (2017) "If the Scottish independence referendum were held tomorrow, who would win? Plus, headscarves and zoos results," [poll], https://yougov.co.uk/opi/surveys/results#/survey/199d5070-089c-11e7-80fa-d2249ae0b02d

YouGov PLC US (2017) "Zoos," *YouGov NY*, [poll], https://d25d2506sfb94s.cloudfront.net/cumulus_uploads/document/pb3xz5vhmy/Results%20for%20YouGov%20NY%20(Zoos)%20086%2005.08.2017.pdf

31

KILLING FOR CONSERVATION

Ethical Considerations for Controlling Wild Animals

Sara Dubois

In a pristine and natural world, self-regulating populations of wildlife would roam in large undisturbed habitats. They would be subject only to natural processes such as disease, predation, weather events, and geological changes that over time cause evolutionary shifts in species diversity and abundance. In the Anthropocene, however, humans drive the decline or loss of species and habitat, through both direct (e.g., deforestation) and indirect actions (e.g., climate change) that may cause intentional or unintentional harms to wild animals (Fraser and MacRae 2011). While conservation measures are increasingly needed to curtail the effects of rapid environmental changes on wildlife, many suggest the costs of conservation should also include animal welfare (Paquet and Darimont 2010) and an evolved ethic of compassion for individual animals that goes beyond their value as a member of their species (Ramp and Bekoff 2015).

Humans are responsible for many actions that harm and help wildlife. Conservation, which aims to protect and increase biodiversity of species and populations through applied science and value-based decisions, can do both. A significant tool in these efforts is wildlife control, used internationally to manage populations of both wanted and unwanted wildlife. Although many types of control exist, direct lethal efforts are the most controversial despite being extensively used. Often the language used to describe killing for conservation is softened into terms such as *culling*, *dispatching*, *eradication*, *harvesting*, and *mitigation*, but it is important to be clear and transparent in describing these actions to ensure that the analysis used to make control decisions is not blurred by such policy language.

As an animal welfare scientist with a wildlife biology background, one might expect that I would focus this discussion on the "how" of wildlife control, as suffering and quality of life can be measured to determine the humaneness of specific intervention methods. Yet throughout this chapter, the "what" and "why" of wildlife control is the focus of ethical scrutiny. When framing the problem itself, special attention should be paid to the language used to describe both the actions and the animals and to what is "humane," given these concepts are often products of certain institutional and political situations. Furthermore, intervention justification requires a comprehensive and transparent evaluation beforehand, one that can be applied across global contexts (i.e., International Consensus Principles for Ethical Wildlife Control; Dubois et al. 2017).

"Pests" or invasive species are a frequent target for wildlife control, but, as this chapter describes, iconic and native species, such as elephants, kangaroos, badgers, monkeys, deer, wolves, and owls, plus many other bird species, reptiles, and amphibians, are also killed by the masses as a conservation "solution." Frequently, humane considerations are lacking in wildlife control as the definition of "humane" varies greatly between individuals and communities based on their values and experience. Although the methods of a control action can be evaluated through animal welfare science measuring relative humaneness (Littin et al. 2014; Sharp and Saunders 2011), this may not be included in the criteria of a control program where removal "by any means necessary" is mandated. Furthermore, the ethics of interventions, why it is necessary to undertake action, may not be considered at all in some cases.

The absence of formal ethical consideration is not surprising as only a handful of biology departments in North America and Europe teach animal ethics courses. This means future biology professionals, wildlife managers, and conservation practitioners are not being taught to consider the lives and experiences of animals in their work. Respect for animals may be inherent through culture and religion in other parts of the world, but models of wildlife management and conservation are frequently exported from Eurocentric views despite alienating local communities (Jones and Braun 1996). Ethical conversations in these professions tend to center on scientific rigor, confidentiality, and data use, rather than whether or not wild animals have self-interests or if proxy considerations for consent should be made to protect the inherent value of wild animals' lives. These same students are taught through secondary and tertiary education that they need to cut up animals to study them, despite the technology that has since enabled equal or better learning outcomes using nonanimal alternatives (Ormandy 2015). Where those animals came from, what their life and death were like, what their cultural significance is, and the environmental implications of their disposable biohazardous bodies, are not often part of the curriculum or course learning objectives.

Conservation problems are wicked. They are complex, dynamic, and massive in scale and occur in diverse political, economic, and cultural contexts. Add sentience and animal well-being as layers of interest, and effective solutions with social license seem impossible to achieve. There is a role for practical ethics here, as "theory-based approaches sometimes fail to address the ethical concerns of conscientious people facing complex, real-life problems of animal ethics" (Fraser 2012: 722). Integrating ethical and consistent decision-making in tough conservation or human–wildlife conflict dilemmas, such as wildlife control, can be achieved with stepwise and widely agreed-on principles (Dubois et al. 2017). When applying these in practice, mistakes will be made as human nature is fallible and even science can be unpredictable, but sharing and learning from these mistakes can only improve future study designs, reduce duplication of studies and use of animals, increase focus on effective solutions, and make the best use of often very limited funding.

Harm in the Name of Conservation

As human populations have grown and expanded across the globe over the past centuries, animals have been hunted for food, clothing, medicine, and other commerce or killed when considered a threat to human safety or property. Killing and capture have also been turned into leisure activities, as wild animals provided opportunities for recreation or entertainment. Then, often to further scientific study of animals, humans have also observed, chased, caught, captured, displayed, experimented on, killed, dissected, and taxidermied wildlife. Both the fields of wildlife management and conservation biology have evolved in response to declining populations of wildlife, and although the disciplines have historical overlap, it is important to distinguish their philosophical and practical differences (Peterson and Nelson 2017). Managing game species for sustainable use under the doctrine of public trust (wildlife management) and applying ecological principles to conserve nature and maximize biodiversity (conservation biology) can even work against each other; for example, when a species is introduced because it is desirable for hunting, this has significant impact on conservation of other species and habitats (e.g., introductions of feral pigs, fallow deer, carp fish).

Conservation as an applied science situates policy recommendations in value-based decision-making and often requires tough choices to be made. Conservation ethics draw from various values for nature such as aesthetic, intrinsic, and instrumental value but remains primarily focused on utilitarian goals, maximizing good for the greatest number of beings. Decisions about what species and populations ought to be conserved, and what species and populations ought to be controlled, and how any of this should be done, are still debated. Agreement among conservationists can be challenging to achieve, particularly when trying to balance the benefits and harms to the environment, species, populations, individual animals—and, often, humans.

Many traditional conservation practices directly harm certain animals for the greater perceived good. For example, studying animals themselves: from the extreme of removing body parts for identification and sampling to collaring, tagging, marking, and branding, plus all the chasing, darting, and capture methods that are deemed necessary to get access to animals for identification and sampling (e.g., Arnemo et al. 2006; Bloch and Irschick 2005; Dennis and Shah 2012; May 2004; Sikes et al. 2016; Trefry et al. 2013; Wilson and McMahon 2006). Unintentional harms that can occur during captivity projects include capture stress, injury or self-injury, forced breeding in unnatural conditions, deprivation, and mortality (e.g., Jule et al. 2008; Spotswood et al. 2012; Wilson and McMahon 2006). If eventually released back to the wild, postcapture harms or death can result (e.g., Blomberg et al. 2018; Cattet et al. 2008). Moving animals for long distances by translocation, reintroduction or even short distance relocation, also involves harms related to capture, group separation, temporary captivity stress, disorientation in new areas without known food and water sources, and increased predation (e.g., Germano et al. 2015; Massei et al. 2010; Tarszisz et al. 2014). Often these projects are viewed as successful if a portion of the animals survive to release, but few projects are funded for long-term monitoring of reproductive success to know if they will genetically contribute to the next generation.

When it comes to killing animals in the name of conservation, even if methods can cause an instantaneous death without pain or suffering, the consequences of removing that individual as a member role of its social structure, its genetic contribution, or its trophic role in the ecosystem can lead to harms to other animals and ecosystems. Killing for conservation, especially for sampling purposes alone, needs greater scrutiny and humane, nonlethal alternatives must be developed to obtain conservation data (Hammerschlag and Sulikowski 2011). Despite conservation's ethical underpinnings in environmental ethics, the ethical costs to individual animals are not adequately considered in conservation, even in the presence of animal care ethics committees or peer review (Vucetich and Nelson 2007). Science can help predict, model, and measure outcomes related to conservation measures, but it cannot dictate policy (Darimont 2017), and most important, science does not give permission to harm animals.

Defining Wildlife Control and Using Labels

Wildlife control is a challenging topic to study and communicate, as the numbers of animals involved are not well documented and estimates could be in the billions annually in Canada or the US alone. Limiting the definition to vertebrate wildlife control, it can be summarized as the poisoning, trapping, hazing, deterring, exclusion, relocation, translocation, and/or killing of wild animals implemented in an effort to restrict animal activity to address perceived or actual human–wildlife conflict (i.e., public health and safety, property or crop protection, nuisance) or as a conservation strategy. Control actions can be directed at native wild animals, introduced animals and feral populations of domesticated animals. "Pest control" is likely the most commonly known form of wildlife control, but it is only one type, often associated with commensal rodents in residential and commercial contexts. The idea of counting the number of mice killed intentionally every year by people globally—let alone any other species controlled—is so abstract that wildlife control becomes inconceivable to quantify; perhaps this is the reason why it is not a subject raised frequently as an ethical dilemma to reconcile, as where does one begin?

Many terms have been used to describe wild animals that are the targets of control, and most have negative connotations: *pest, nuisance, unwanted, invasive, alien, exotic, introduced, hyperabundant,* or *feral.*

Such labels are fluid, vary by location and culture, and reflect human attitudes, not inherent properties of the animals themselves. Those who advocate for compassionate conservation emphasize that the de-categorization of animals is necessary to improve how we treat them (Draper et al. 2015), as "bad" animals tend to receive less moral consideration (Bekoff 2009). Furthermore, these labels can perpetuate myths and undermine empathy development; in extreme cases, such as when bounties are encouraged, the culture of killing is celebrated and may desensitize children and adults to acts of violence (The Guardian 2017; Williams 2018).

Humans intentionally or unintentionally enabled the movement of wildlife to novel environments and altered natural predation and feeding. As a result, these labels reflect our own influences on creating unnatural or out of balance ecosystems and not the worth of the animal. Perhaps these labels create the false moral distance to allow for intentional and painful practices, such as poisoning, glue boards, leghold traps, and snares, to exist as common and legal control methods despite being indiscriminate, unmonitored, and inhumane. As vertebrate animals of similar cognitive and emotional complexity can be expected to have similar capacities for suffering (Mellor et al. 2009), greater consideration for humaneness, and its enforcement, in wildlife control policy is necessary, regardless of any given label.

What Is Humane When It Comes to Controlling Wildlife?

If asked, what is "humane" for animals to a classroom of university students, one can expect to hear different answers on a continuum of "doesn't hurt" to "nonlethal" and "don't touch at all." A poll of 800 residents in British Columbia (BC) asked just this, "What does humane mean to you in terms of pest control?" The answers varied from using nonlethal actions only (38%), not using poison (40%), trapping and releasing animals (55%), and quick killing (55%) to not injuring trapped animals (70%; BC SPCA 2015). Given that the word *humane* differs among people, their values, and their experiences, it is hard to standardize what humane wildlife control is. The definition also varies between organizations and legal systems. Yet the term is widely used to describe common practices, guidelines, standards, and terms in international agreements. For example, the Agreement on International Humane Trapping Standards (AIHTS), among Canada, Russia, and the US, is a trade agreement describing permissible traps for certain wild animals with fur that is marketable (AIHTS 1997). However, one does not have to be a scientist to know that the AIHTS's allowable suffering from struggling in a killing snare or body-gripping device for 300 seconds for at least 80% of caught animals is not humane and would not be considered an acceptable form of death in farm, companion, or research animal standards (AVMA 2013; CCAC 2010; Iossa et al. 2007).

There are at least three distinct definitions of *humane*: a legal definition, a scientific definition, and a societal definition. Legally, many activities that cause pain and suffering to wildlife are still allowed under many federal and provincial laws that permit certain "generally acceptable practices of industry" to continue under animal cruelty legislation and wildlife regulations (which determine species, demographics, seasonality, and numbers allowed). Yet, scientifically, the humane test is whether sentient animals are caused distress (physical and psychological), for how long, and at what intensity. Thus, an animal can be legally trapped using a snare for example, but this may not meet the scientific definition of *humane* if the animal is left to suffer until the trap is checked days later. And then, many will ask, "Does the animal need to be trapped in the first place?" This is where the societal definition of humane asks that on top of the methods, what are the ethics, justifications, and alternatives of the situation. These varying definitions present a challenge to animal protection agencies who have to enforce what is legally humane, while animal welfare science set outs to measure what is scientifically humane, and the greater community decides what is societally acceptable as humane.

There are currently few assurances that wildlife control practices are taking place ethically and no clear standards for "humane wildlife control." Freezing, drowning, and suffocation across all species are totally unacceptable methods yet are still used by the public and government agencies. Where such clear animal welfare science exists, the bar for legal methods should be raised to provide both the public and enforcement agencies with evidence-based guidance. Developing scientific standards for wildlife control is an ongoing need, and novel programs, such as Animal-Kind accreditation (BC SPCA 2018) for pest control, are forging that path. However, the methods themselves are only one aspect of an overall assessment in society's evaluation of the humaneness of such actions. With many complex considerations for when and why wildlife control is used, the how should be much easier to create enforceable criteria for, yet it lacks legal and regulatory attention as government agencies themselves are often leading inhumane control programs (Green and Rohan 2012; Parr 2018).

Controlling National Icons

Certain charismatic wild animals enjoy national reverence through tourism, symbolism, and cultural importance, and there is no bigger animal that garners such admiration and at the same time angst than elephants. As national icons and cultural deities in some African and Asian countries, elephants roam savannahs and forests in diverse protected and human-occupied areas. Relationships with people vary as elephants are working animals, tourist attractions, or mysterious wild creatures to some, while others fear losing their livelihood and/or fear for their personal safety living among elephants. When highly managed and living in restrictive protected areas, elephant populations can become locally high and they may cause environmental and property damage when intentionally limited in range. Previous culls and continued removal of young elephants for off-shore sales to decrease local populations have been extremely scientifically and ethically controversial and have not achieved long-term management goals to reduce conflict (Fernando et al. 2008; Scholes and Mennell 2008; Van Aarde and Jackson 2007).

A national icon cared for in wildlife rehabilitation hospitals and visited at local game parks, kangaroos are widely distributed across the Australian landscape; they are also killed at a rate of more than a million and a half per year for various control reasons (Boom et al. 2012). Concerns range from hyperabundance causing ecosystem and agricultural impacts to vehicle collisions and nuisance activities (BBC News 2017). Under commercial permits, kangaroos are hunted on private lands, contributing to a multimillion-dollar meat and pet food industry; plus, sport killing and illegal poaching is a well-known activity for the iconic perceived pest. Serious welfare issues center on the lack of training and monitoring of how to kill adults, plus orphaning of pouched young and dependent joeys and the effects of hunting with dogs (Croft 2004; Descovich et al. 2015). Further concerns remain for the overall culture of entrenched interests that have created an economy through the exploitation of the nationally important species (Ramp 2013).

Culling kangaroos is only one control program that raises national controversy down under, as seen with the recent Australian government's announcement to kill 2 million cats by 2020 (Australian Government 2015; Tharoor 2015). The "war" on cats stems from the desire to save vulnerable native wildlife from this introduced and efficient predator. Aside from animal welfare concerns about poisoning methods, the sheer distribution of feral cats on the landscape and diverging views about management make achieving a measurable conservation goal unrealistic unless targeted collaboratively to high conservation-priority areas (Peterson et al. 2012). Without addressing other community issues, such as homeless cats, sterilization, managed colonies, and confinement, the ongoing need to kill cats could be never-ending and create extreme suffering for little conservation gain (Peterson et al. 2012). Proposals to introduce dingoes to control cats show the complexity of conservation decisions and the need for an ethical framework for wildlife control, as our ongoing desire to manage nature has previously led to killing predators, such as dingoes and foxes, which resulted in unintended and perverse outcomes of lethal predator control (Hunter et al. 2018).

Killing One Species to Save Another

One of the most well-known wildlife control controversies pits two native species of raptors against each other, the habitat-sensitive Spotted Owl, which faces extinction in the Pacific Northwest of Canada and US, versus the generalist Barred Owl. The Barred Owl has expanded its range and competes directly with Spotted Owls, who can only reside in old-growth forests (Peterson and Robins 2003). As logging has diminished old-growth habitat, Spotted Owls struggle to survive has been imperiled by its cousin and they may be forever lost. Governments on both sides of the border have taken up culling Barred Owls but have not successfully stopped the endangered Spotted Owl decline or miraculously restored old-growth habitat. Investment in the captive breeding of Spotted Owls has been significant with little success to date (Pynn 2016), and some suggest too little regard for habitat has left it too late to save the species (Moore 2013).

The same wildlife control strategy, which distracts from habitat protection, is currently being used to justify the killing of wolves to save caribou in BC and Alberta. Critical habitat loss has resulted in several caribou herds dwindling to "functionally extinct" numbers (Robbins 2018), despite endangered species status and federally mandated recovery plans. Cull targets have been placed on wolves who are guilty only for performing natural predation behavior, yet this taxpayer-funded wildlife control is quicker and cheaper than protecting habitat from natural resource development or recreation (Proulx et al. 2017). The social acceptability of killing one species to save another is lacking when habitat loss and other human-driven ecosystem imbalances are the root cause of species declines (Dubois and Harshaw 2013), as causing more harm does not fix the problem.

Conflict on the Farm

Agricultural conflicts with wildlife bring about their own set of control issues on many levels, some conservation-related and others strictly to protect livestock and crops. This is the origin of predator control, whereby all carnivores on the landscape are killed to protect cattle, sheep, and other range species. Predator control examples internationally range from wild canids and felids to mesopredators, such as raccoons and mustelids; decades of culling these species without changing livestock management practices are finally being challenged as economically, environmentally, and ethically unsound. Many other species are also affected by crop-protection measures. In Mauritius, for example, flying foxes (a species of fruit-eating bat) are a threatened species that feed on fruit and nectar. Despite opposition from conservationists, control measures have been ramped up to cull flying foxes to protect local crop industries (Florens 2015). Culling wildlife on a massive scale is not uncommon in agriculture, as birds such as European Starlings are known to cause more than a billion dollars of economic damage to crops in the US annually (Homan et al. 2017) and vast numbers are killed every year. However, in the case of flying foxes, killing a threatened species that is an important seed disseminator is contrary to both conservation and agricultural goals.

Perhaps one of the most well-known wildlife control programs internationally is the badger cull in the UK. Badgers are native species in the UK that have a key ecosystem role as an omnivorous predator of insects, small mammals, reptiles, and amphibians. They are also carriers of bovine tuberculous (TB) but not generally affected by the disease. Farmers in the UK have targeted badgers as being responsible for TB in their cattle, and for four decades have lobbied governments to kill thousands of badgers despite no slowdown of the disease spread (Cassidy 2017). Pilot culls were undertaken in a number of counties and monitored but revealed that culling badgers actually increased incidence of bovine TB in surrounding areas (Donnelly et al. 2006). Furthermore, illegal small-scale culls would also elevate TB risks to cattle (Bielby et al. 2014). Yet other significant TB vectors go unresolved, such as cattle-to-cattle transmission and farm staff, supplies, and vehicles traveling between farms without proper biosecurity (Bourne et al. 2007). Recent proposals to pay a bounty for every badger shot could lead to 100,000 badgers being killed by 2020 and will likely fail to reduce TB occurrence due to factors described earlier (Leake 2018).

Keeping the Skies Safe

Most people do not think of bird control when they are about to travel by plane; rather, arriving on time and with their baggage is top of mind. However, occasionally a local story will catch media attention and a midair incident with a bird will remind travelers that not all flight technical challenges are a result of mechanical issues and that nature can also be unpredictable. Hundreds of millions of dollars are spent annually keeping passengers safe from bird strikes at airports and military bases internationally and to offset the potential cost in billions for plane damage, plus flight delays and cancellations (Allan 2000). From working dogs, falconry, habitat modification, deterrents and barriers, and visual and sound effects to capture and killing programs, many wildlife control tactics are needed to ensure success when the stakes are so high (Belant and Martin 2011; Bradbeer et al. 2017). Wildlife control in aviation aims to modify the behavior of animals to reduce their numbers within the active areas of the flight paths. Bird control at airports is regulated in countries such as the US, Canada, Australia, and most European countries; however, in some developing nations, such requirements do not exist (Allan 2000).

Birds are not the only concern for airports, but mammal and reptile management plans are also necessary. Small terrestrial mammals, although not a strike concern, can attract raptors and cause damage to fields and infrastructure. Alligators, bats, bears, caimans, coyotes, deer, foxes, moose, sloths, snakes, and wild cats are additional species of concern (e.g., Novaes et al. 2016; Scheideman et al. 2017). Effective and ethical approaches to wildlife control at airports require detailed knowledge of local wildlife ecology and an understanding of animal physiology and behavior (DeVault et al. 2017). Lethal measures should be a last resort as ongoing control and mitigation will be continually needed, while detailed documentation of management practices is critical to adapting and learning from operational plans (DeVault et al. 2017).

Don't Forget the Reptiles and Amphibians

Wildlife control is not only applicable to large, charismatic furred and feathered species, as examples of introduced reptiles and amphibians in many parts of the world demonstrate the need to address losses to ecosystems and compromising the welfare of other animals. For example, in Florida in the US, exotic crocodilians, frogs, lizards, snakes, and turtles are all breeding and thriving in this subtropical climate with few predators (Engeman and Avery 2016). Strategies to address the unprecedented concerns for local wildlife and habitats have been proposed based on risk, scale and stage of invasion and practical outcomes for control (Engeman and Avery 2016). With so many harms occurring to other wildlife and ecosystems as a result of the new species, it is hard to reconcile which to animals to control first, what will have the greatest reduction in harm, and what will be sustainable.

An international example of amphibian wildlife control is the American bullfrog; native to southeastern US and Canada, it is now widespread in these countries and has been introduced in western Europe, Central and South America, and parts of Asia (Adams and Pearl 2007). The American bullfrog is a significant predator and competitor to native species and may spread disease to native frogs. These behaviors, not just their nonnative status, make them an invasive species worthy of control attention. Eradication is an option for small isolated ponds and should be emphasized when local endangered species with vulnerable natural history characteristics are present (Adams and Pearl 2007). Attention should also be paid to culling seasonally at key developmental stages to decrease populations as alternative timing and removal of tadpoles and adults may be counterproductive (Govindarajulu et al. 2005). Managing habitat and community characteristics not only are important for reducing populations but can also be altered to promote coexistence where eradication is unlikely (Adams and Pearl 2007). Caution should be used when considering the introduction of other native predators (Louette 2012), as this could just create new problems to control in the future.

International Consensus Principles for Ethical Wildlife Control

In an attempt to find common solutions to complex wildlife issues, an international meeting of academics, animal welfare organizations and conservation practitioners involved in wildlife control was held in July 2015 at the University of British Columbia (BC SPCA 2016). Over two days, shared experiences of flawed, failed and inhumane control projects were discussed to acknowledge institutional and agency approaches that have rarely resulted in achieving project goals, often at significant cost and controversy to local communities. Sparked by the desire to define humane wildlife control criteria for animal-cruelty enforcement purposes, the meeting reviewed global approaches, unacceptable methods, and previous decision-making frameworks (BC SPCA 2016) to propose the first set of international principles reached by consensus for determining ethical wildlife control (Dubois et al. 2017). The principles help to fill a gap where a lack of standards and regulations exist for wildlife control and more broadly establish a common framework for decision-making.

The ethical wildlife control principles are designed to be evidence-based and serve as a practical check for proponents designing projects and provide a transparent mechanism for external stakeholders to gauge the validity and appropriateness of wildlife control (Dubois et al. 2017). Although some conservation projects will be required to complete institutional animal care reviews, the principles go beyond recommending project refinements and can be used by exempt government agencies and private companies who should also be held to the same ethical standards. The ethical principles can be summarized into seven questions to ask before undertaking control efforts, as shown in Box 31.1.

Box 31.1 Ethical Wildlife Control (Dubois et al. 2017)

1. Can the problem be mitigated by changing human behavior?
2. Are the harms serious enough to warrant wildlife control?
3. Is the desired outcome clear and achievable, and will it be monitored?
4. Does the proposed method carry the least animal welfare cost to the fewest animals?
5. Have community values been considered alongside scientific, technical, and practical information?
6. Is the control action part of a systematic, long-term management program?
7. Are the decisions warranted by the specifics of the situation rather than a negative categorization of the animals?

A useful example to apply the seven international consensus principles for ethical wildlife control is to contrast projects that meet, or fail to meet, the criteria. As an example, this can be seen in the different approaches to managing unwanted deer in the province of BC. In the federal park Gwaii Haanas National Park Reserve, Sitka deer were introduced for hunting purposes more than a hundred years ago and have thrived on many of the islands, causing significant loss of biodiversity in otherwise undisturbed ecosystems (Parks Canada 2018). With no natural predators, populations increased measurably despite local hunting efforts, and eradication was proposed on several islands where full removal would be possible to restore native plants, birds, and other dependent species. The community supported the removal by professional shooters (and subsequent distribution of meat locally) on certain islands but wanted to maintain a hunted population on large human-occupied islands. The well-funded conservation project involved the local First Nation (Haida Nation), scientists, park staff, and professional international control expertise in the multiyear project planning, implementation, and monitoring (Hudson 2017). All the ethical principles were met in this case given the scale of the project, which realistically could eradicate the deer from the desired islands.

In contrast, in the urban district of Oak Bay adjacent to Victoria, BC, the population and movement of the native black-tailed deer species are unknown as they move throughout the greater region. As natural predators, such as cougars, are removed anytime they are observed in the urban area, vehicle impacts are one of the few mortality sources for the nonhunted population. Some deer had also become a nuisance to some residents as a result of unintentional and intentional feeding in neighborhood gardens. There was significant community disagreement over how to manage the deer, little municipal funding dedicated, and there was no plan for long-term monitoring following a onetime cull of 11 animals in 2015 (Cleverley and Harnett 2015). The animals were killed by captive bolt gun after being cage trapped in secret locations on private properties, and bodies were given to a local First Nation for meat. Almost all the ethical principles were not adhered to in this case: there was no enforcement of deer feeding or speed limit reductions in high collision areas, no scientific deer population count, no specific end goal for management, no plan for postcull monitoring, and social license was lacking.

These examples contrast a project with a conservation goal to that with a human-wildlife conflict, but as seen in previous examples, it does not mean that all conservation projects will meet the ethical principles, nor will all human–wildlife conflicts fail to do so. Yet the principles allow for consistent decision-making for both types of projects through the systematic evaluation of their rationale and methodology. Because so many wildlife control projects have been initiated internationally over the past decades, often by government agencies not subject to peer review or animal ethics committees, transparent external reporting of the process and outcomes is key to learning from past mistakes.

The Academic Journal That Will Never Be Published, But Should: Mistakes With Wildlife

Mistakes happen: in work, in school, in training, and in between. Yet learning about mistakes with wildlife through publications is rare as errors are glossed over as study design issue, ineffective methods or improper modeling, failed equipment, a lack of resources or time, and other objective variables. Rarely will an acknowledgment of human error be cited. Bad decisions by an individual or agency can happen—maybe it was the choice of a release location, time of day, poor trap placement, lack of consideration for other animals or people, or even a missing decimal or program step. Used too frequently, "more research is needed" is a catch-all phrase for many wildlife control projects, which needs to be teased apart to ensure that additional animal suffering is not occurring unchecked.

Furthermore, there is a lack of published studies on mistakes made in the fields of conservation and wildlife management, which itself is an ethical dilemma—few researchers want to share when animals have accidentally died in a trap, been accidentally killed by a predator or conspecific, injured so significantly they had to be killed, or, in an attempt to kill, were not fully rendered unconscious or were only critically injured. Unnecessary suffering or death of animals is alarming and regrettable, but it is also tragic to allow a biased project that has no measurable objective and realistic chance to reduce overall harms, which cannot be sustained and lacks social license, to proceed and intentionally harm animals through wildlife control. There is an urgent need to learn from mistakes so that others do not follow the same path and cause pointless and unnecessary pain, suffering, or death.

Learning from mistakes is also difficult when the overall effectiveness of interventions is not evaluated on a timely basis. In the case of killing wolves to save caribou in BC, progress from government control programs is to be reviewed every five years despite "adaptive management" (course-correcting from lessons learned) being a fundament tenet of conservation programs (BCMFLNRO 2014). This culling program, which lacked external ethical review, has not published or demonstrated that, after four years and more than 500 wolves killed, endangered caribou populations are better off (MacLeod 2017); in fact, they are worse (Cox 2018).

If the international consensus principles (Dubois et al. 2017) had been applied, the first principle of changing human behavior would have prioritized habitat protection and reduced industrial impacts on caribou, and control measures might have looked much different. Although the risk of losing caribou populations merited serious intervention, restoring populations (and not habitat) as the sole outcome meant that the wolves as natural and effective predators were the main target, and only narrow values informed management decisions. Ariel shooting from helicopters and the use of Judas wolves (Alberts 2016) were blatantly inhumane but necessary according to government biologists (Parr and Paquet 2017). Almost all the principles for ethical wildlife control were not met, and a lack of systematic, long-term management plans other than killing as many wolves as possible fails to solve cumulative impacts like habitat loss and degradation, forest fires, climate change, caribou health, disease, and reproduction.

Understandably, no one wants to reveal that a project failed to reduce human–wildlife conflict or solve the conservation problem, as they might not get funded, be published, or, for government agency-driven projects, be elected again. Yet, human error, human ignorance, ego, and fear of failure or judgment from peers inhibit our ability to learn from each other. Mistakes in planning wildlife control can be reduced using these objective ethical principles, but there is also a need to acknowledge when and why wildlife control measures do not achieve their goals despite proper planning and implementation. The approval of ineffective and unethical wildlife control projects has to stop, as the conservation crisis, with declining populations of wild animals and dwindling wild spaces, does not have time to waste limited efforts and resources.

In addition to mistakes with wildlife and the total number of animals killed for control purposes, we need to better document successful wildlife control programs and continue to refine the ethical consensus principles. Evidence on the effectiveness of interventions is key to share through accessible, open-source, and online platforms to reduce the unnecessary suffering and killing of animals and spur innovation to solve wicked problems faced internationally in managing our coexistence with our wild neighbors.

Bibliography

Adams, M. J., and Pearl, C. A. (2007) "*Problems and opportunities managing invasive bullfrogs: Is there any hope?*" in F Gherardi (ed.) *Biological Invaders in Inland Waters: Profiles, Distribution and Threats*, Dordrecht, The Netherlands: Springer.

AIHTS Agreement on International Humane Trapping Standards (1997) "Agreement on international humane trapping standards between the European Community, Canada and the Russian Federation," http://eur-lex.europa.eu/legal-content/EN/TXT/PDF/?uri=CELEX:21998A0214(02)&from=EN.

Alberts, E. C. (2016) "Wolf kept alive year after year so government can kill his friends," *The Dodo*, March 1, www.thedodo.com/judas-wolf-canada-1634797811.html.

Allan, J. R. (2000) "The costs of bird strikes and bird strike prevention," *Human Conflicts with Wildlife Economic Considerations* 18: 147–153.

Arnemo, J. M., Ahlqvist, P., Andersen, R., Berntsen, F., Ericsson, G., Odden, J., Brunberg, S., Segerström, P., and Swenson, J. E. (2006) "Risk of capture-related mortality in large free-ranging mammals: Experiences from Scandinavia," *Wildlife Biology* 12(1): 109–113.

Australian Government (2015) "FAQs: Talking feral cats and their impacts," *Department of Environment and Energy*, www.environment.gov.au/system/files/resources/bb591b82-1699-4660-8e75-6f5612b21d5f/files/factsheet-tackling-feral-cats-and-their-impacts-faqs.pdf.

AVMA American Veterinary Medical Association (2013) *AVMA Guidelines for the Euthanasia of Animals: 2013 Edition*, Schaumburg, IL, www.avma.org/KB/Policies/Documents/euthanasia.pdf.

BBC News (2017) "Why kangaroo culling divides Australia," *BBC News*, February 15, www.bbc.co.uk/news/world-australia-38964535.

BCMFLNRO BC Ministry of Forests, Lands and Natural Resource Operations (2014) *Management Plan for the Grey Wolf (Canis lupus) in British Columbia*, Victoria, BC: B.C. Ministry of Forests, Lands and Natural Resource Operations, p. 48, www.env.gov.bc.ca/fw/wildlife/management-issues/docs/grey_wolf_management_plan.pdf.

BC SPCA British Columbia Society for the Prevention of Cruelty to Animals (2015) "What is 'humane' pest control?" *BC Resident Poll Results*, September, https://spca.bc.ca/faqs/what-is-humane-pest-control/.

BC SPCA British Columbia Society for the Prevention of Cruelty to Animals (2016) "A common approach to wildlife control for animal welfare and protection organisations: Expert forum," July 27–28, 2015 Report. University of British Columbia, Vancouver, Canada, https://spca.bc.ca/wp-content/uploads/UBC_AWP__BC_SPCA_Expert_Forum_on_Wildlife_Control_-_Final_Report_Endorsed_-_27_June_2016.pdf.

BC SPCA British Columbia Society for the Prevention of Cruelty to Animals (2018) *AnimalKind Wildlife & Rodent Control Accreditation Standards*, www.animalkind.ca.

Bielby, J., Donnelly, C. A., Pope, L. C., Burke, T., and Woodroffe, R. (2014) "Badger responses to small-scale culling may compromise targeted control of bovine tuberculosis," *Proceedings of the National Academy of Sciences* 111(25): 9193–9198.

Bekoff, M. (ed.) (2009) *Encyclopedia of Animal Rights and Animal Welfare*, 2nd ed., Santa Barbara, CA: Greenwood Publishing.

Belant, J. L., and Martin, J. A. (2011) *Bird Harassment, Repellent, and Deterrent Techniques for Use on and Near Airports*, Vol. 23, Washington, DC: Transportation Research Board.

Bloch, N., and Irschick, D. J. (2005) "Toe-clipping dramatically reduces clinging performance in a pad-bearing lizard (Anolis carolinensis)," *Journal of Herpetology* 39(2): 288–293.

Blomberg, E., Davis, S., Mangelinckx, J., and Sullivan, K. (2018) "Detecting capture-related mortality in radio-marked birds following release," *Avian Conservation and Ecology* 13(1), article 5.

Boom, K., Ben-Ami, D., Croft, D. B., Cushing, N., Ramp, D., and Boronyak, L. (2012) "Pest and resource: A legal history of Australia's kangaroos," *Animal Studies Journal* 1(1): 17–40.

Bourne, F. J., Donnelly, C. A., Cox, D. R., Gettinby, G., McInerney, J., Morrison, I., and Woodroffe, R. (2007) *"Bovine TB: The Scientific Evidence," Final Report of the Independent Scientific Group on Cattle TB*, London: Independent Scientific Group on Cattle TB, http://webarchive.nationalarchives.gov.uk/20081108133322/www.defra.gov.uk/animalh/tb/isg/pdf/final_report.pdf.

Bradbeer, D. R., Rosenquist, C., Christensen, T. K., and Fox, A. D. (2017) "Crowded skies: Conflicts between expanding goose populations and aviation safety," *Ambio* 46(2): 290–300.

Cassidy, A. (2017) "Badger-human conflict: An overlooked historical context for bovine TB debates in the UK," Chapter 4 in C. M. Hill, A. D. Webber, and N. E. C. Priston (eds.) *Understanding Conflicts About Wildlife: A Biosocial Approach*, Oxford: Berghahn Books.

Cattet, M., Boulanger, J., Stenhouse, G., Powell, R. A., and Reynolds-Hogland, M. J. (2008) "An evaluation of long-term capture effects in ursids: Implications for wildlife welfare and research," *Journal of Mammalogy* 89(4): 973–990.

CCAC Canadian Council on Animal Care (2010) *CCAC Guidelines On: Euthanasia of Animals Used in Science*, Ottawa, ON, Canada, www.ccac.ca/Documents/Standards/Guidelines/Euthanasia.pdf.

Cleverley, B., and Harnett, C. (2015) "Nothing learned from Oak Bay deer cull: B.C. SPCA," *Times Colonist*, March 10, www.timescolonist.com/news/local/nothing-learned-from-oak-bay-deer-cull-b-c-spca-1.1788416.

Cox, S. (2018) "B.C. approved 83 logging cut blocks in endangered caribou habitat in last six months," *The Narwhal*, October 23, https://thenarwhal.ca/b-c-approved-83-logging-cut-blocks-in-endangered-caribou-habitat-in-last-six-months/.

Croft, D. B. (2004) "Kangaroo management: Individuals and communities," *Australian Mammalogy* 26(1): 101–108.

Darimont, C. (2017) "Trophy hunting: Science on its own can't dictate policy," *Nature* 551(7682): 565.

Dennis, T. E., and Shah, S. F. (2012) "Assessing acute effects of trapping, handling, and tagging on the behavior of wildlife using GPS telemetry: A case study of the common brushtail possum," *Journal of Applied Animal Welfare Science* 15(3): 189–207.

Descovich, K. A., McDonald, I. J., Tribe, A., and Phillips, C. J. C. (2015) "A welfare assessment of methods used for harvesting, hunting and population control of kangaroos and wallabies," *Animal Welfare* 24(3): 255–265.

DeVault, T. L., Blackwell, B. F., Belant, J. L., and Begier, M. J. (2017) Wildlife at Airports, Wildlife Damage Management Technical Series. Fort Collins, CO: USDA, APHIS, WS National Wildlife Research Center, p. 19.

Donnelly, C. A., Woodroffe, R., Cox, D. R., Bourne, F. J., Cheeseman, C. L., Clifton-Hadley, R. S., Wei, G., Gettinby, G., Gilks, P., Jenkins, H., and Johnston, W. T. (2006) "Positive and negative effects of widespread badger culling on tuberculosis in cattle," *Nature* 439(7078): 843–846.

Draper, C., Baker, L., and Ramp, D. (2015) "Why compassionate conservation can improve the welfare of wild animals," Conference Poster. UFAW International Animal Welfare Science Symposium, Zagreb, Croatia, www.researchgate.net/publication/282664567_Why_Compassionate_Conservation_Can_Improve_the_Welfare_of_Wild_Animals.

Dubois, S., and Harshaw, H. W. (2013) "Exploring 'humane' dimensions of wildlife," *Human Dimensions of Wildlife* 18(1): 1–19.

Dubois, S., Fenwick, N., Ryan, E. A., Baker, L., Baker, S. E., Beausoleil, N. J., Carter, S., Cartwright, B., Costa, F., Draper, C., Griffin, J., Grogan, A., Howald, G., Jones, B., Littin, K. E., Lombard, A. T., Mellor, D. J., Ramp, D., Schuppli, C. A, and Fraser, D. (2017) "International consensus principles for ethical wildlife control," *Conservation Biology* 31(4): 753–760.

Engeman, R., and Avery, M. L. (2016) "Prioritizing management and research actions against invasive reptiles in Florida: A collaboration by an expert panel," *USDA National Wildlife Research Center—Staff Publications* (1862).

Fernando, P., Kumar, M. A., Williams, A. C., Wikramanayake, E., Aziz, T., and Singh, S. M. (2008) *Review of Human-Elephant Conflict Mitigation Measures Practiced in South Asia*, Gland, Switzerland: WWF.

Florens, F. V. (2015) "Flying foxes face cull despite evidence," *Science* 350(6266): 1325–1325.

Fraser, D. (2012) "A 'practical' ethic for animals," *Journal of Agricultural and Environmental Ethics* 25(5): 721–746.

Fraser, D., and MacRae, A. M. (2011) "Four types of activities that affect animals: Implications for animal welfare science and animal ethics philosophy," *Animal Welfare* 20(4): 581–590.

Germano, J. M., Field, K. J., Griffiths, R. A., Clulow, S., Foster, J., Harding, G., and Swaisgood, R. R. (2015) "Mitigation-driven translocations: Are we moving wildlife in the right direction?" *Frontiers in Ecology and the Environment* 13(2): 100–105.

Govindarajulu, P., Altwegg, R., and Anholt, B. R. (2005) "Matrix model investigation of invasive species control: Bullfrogs on Vancouver Island," *Ecological Applications* 15: 2161–2170.

Green, W., and Rohan, M. (2012) "Opposition to aerial 1080 poisoning for control of invasive mammals in New Zealand: Risk perceptions and agency responses," *Journal of the Royal Society of New Zealand* 42(3): 185–213.

The Guardian (2017) "New Zealand's possum war: 'barbaric' drowning of babies at school fair sparks outcry," July 5, www.theguardian.com/world/2017/jul/05/new-zealands-possum-war-barbaric-drowning-of-babies-at-school-fair-sparks-outcry.

Hammerschlag, N., and Sulikowski, J. (2011) "Killing for conservation: The need for alternatives to lethal sampling of apex predatory sharks," *Endangered Species Research* 14(2): 135–140.

Homan, H. J., Johnson, R. J., Thiele, J. R., and Linz, G. M. (2017) *European Starlings*, Wildlife Damage Management Technical Series. Ft. Collins, CO: USDA, APHIS, WS National Wildlife Research Center, 26p.

Hudson, A. (2017) "Restoring balance project closing in on 'deer zero,'" *Haida Gwaii Observer,* September 7, www.haidagwaiiobserver.com/home/restoring-balance-project-closing-in-on-deer-zero/.

Hunter, D. O., Lagisz, M., Leo, V., Nakagawa, S., and Letnic, M. (2018) "Not all predators are equal: A continent-scale analysis of the effects of predator control on Australian mammals," *Mammal Review* 48(2): 108–122.

Iossa, G., Soulsbury, C. D., and Harris, S. (2007) "Mammal trapping: A review of animal welfare standards of killing and restraining traps," *Animal Welfare* 16(3): 335–352.

Jones, B. T., and Braun, E. (1996) "Reversing marginalization private community-based conservation in Namibia," *Wilderness Worldwide* 2(2): 39–42.

Jule, K. R., Leaver, L. A., and Lea, S. E. (2008) "The effects of captive experience on reintroduction survival in carnivores: A review and analysis," *Biological Conservation* 141(2): 355–363.

Leake, J. (2018) "Badger cull goes national to fury of animal campaigners," *The Sunday Times*, May 27, www.thetimes.co.uk/article/badger-cull-goes-national-to-fury-of-animal-campaigners-pshccnrmx.

Littin, K., Fisher, P., Beausoleil, N. J., and Sharp, T. (2014) "Welfare aspects of vertebrate pest control and culling: Ranking control techniques for humaneness," *Revue scientifique et technique* (International Office of Epizootics) 33(1): 281–289.

Louette, G. (2012) "Use of a native predator for the control of an invasive amphibian," *Wildlife Research* 39: 271–278.

MacLeod, A. (2017) "Government to restart winter wolf cull," *The Tyee*, November 17, https://thetyee.ca/News/2017/11/17/Government-Restart-Winter-Wolf-Kill/.

Massei, G., Quy, R. J., Gurney, J., and Cowan, D. P. (2010) "Can translocations be used to mitigate human-wildlife conflicts?" *Wildlife Research* 37(5): 428–439.

May, R. M. (2004) "Ecology: Ethics and amphibians," *Nature* 431(7007): 403.

Mellor, D., Patterson-Kane, E., and Stafford, K. J. (2009) *The Sciences of Animal Welfare*, Oxford: John Wiley & Sons.

Moore, D. (2013) "In B.C., to keep one owl alive, another species must die," *The Globe and Mail*, January 27, www.theglobeandmail.com/news/british-columbia/in-bc-to-keep-one-owl-alive-another-species-must-die/article7902769/.

Novaes, W. G., Grossman, N. V., Pimentel, D. S., and Prada, M. (2016) "Terrestrial mammal and reptile hazards in an airport in the Brazilian amazon," *Human-Wildlife Interactions* 10(1): 122–127.

Ormandy, E. (2015) "Alternatives to animals in teaching: Literature review," *Animals in Science Policy Institute*, www.animalsinscience.org/projects/completed/ubcalternatives/literature-review/.

Paquet, P. C., and Darimont, C. T. (2010) "Wildlife conservation and animal welfare: Two sides of the same coin," *Animal Welfare* 19(2): 177–190.

Parks Canada (2018) "Llgaay gwii sdiihlda—restoring balance project," Gwaii Haanas National Park Reserve, National Marine Conservation Area Reserve, and Haida Heritage Site, www.pc.gc.ca/en/pn-np/bc/gwaiihaanas/nature/conservation/restauration-restoration/retablir-restore.

Parr, S. (2018) "There's nothing legitimate about killing Canada's iconic wolves," *Huffington Post*, January 8, www.huffingtonpost.ca/sadie-parr/theres-nothing-legitimate-about-killing-canadas-iconic-wolves_a_23322128/.

Parr, S., and Paquet, P. (2017) "B.C. government scientists admit wolf cull is inhumane, then propose to expand it," *National Observer*, February 1, www.nationalobserver.com/2017/02/01/news/bc-government-scientists-admit-wolf-cull-inhumane-then-propose-expand-it.

Peterson, A. T., and Robins, C. R. (2003) "Using ecological-niche modeling to predict Barred Owl invasions with implications for Spotted Owl conservation," *Conservation Biology* 17(4): 1161–1165.

Peterson, M. N., Hartis, B., Rodriguez, S., Green, M., and Lepczyk, C. A. (2012) "Opinions from the front lines of cat colony management conflict," *PLoS One* 7(9): e44616.

Peterson, M. N., and Nelson, M. P. (2017) "Why the North American model of wildlife conservation is problematic for modern wildlife management," *Human Dimensions of Wildlife* 22(1): 43–54.

Proulx, G., Alexander, S., Barron, H., Bekoff, M., Brook, R., Bryan, H., Darimont, C., Dubois, S., Lukasik, V., McCrory, W. P., and Paquet, P. (2017) "Killing wolves and farming caribou benefit industry, not caribou: A response to Stan Boutin," *Nature Alberta* 47(1): 4–11.

Pynn, L. (2016) "Wing and a prayer: Northern Spotted Owls on the brink despite captive breeding Program," *Vancouver Sun*, September 11, http://vancouversun.com/news/local-news/wing-and-a-prayer-northern-spotted-owls-on-the-brink-despite-captive-breeding-program.

Ramp, D. (2013) "Bringing compassion to the ethical dilemma in killing kangaroos for Conservation," *Journal of Bioethical Inquiry* 10(2): 267–272.

Ramp, D., and Bekoff, M. (2015) "Compassion as a practical and evolved ethic for Conservation," *BioScience* 65(3): 323–327.

Robbins, J. (2018) "Gray ghosts, the last caribou in the lower 48 states are 'functionally extinct,'" *New York Times*, April 14, www.nytimes.com/2018/04/14/science/gray-ghost-caribou-extinct.html.

Scheideman, M., Rea, R., Hesse, G., Soong, L., Green, C., Sample, C., and Booth, A. (2017) "Use of wildlife camera traps to aid in wildlife management planning at airports," *Journal of Airport Management* 11(4): 408–419.

Scholes, R. J., and Mennell, K. (eds.) (2008) *Elephant Management: A Scientific Assessment of South Africa*, Johannesburg, South Africa: Wits University Press.

Sharp, T., and Saunders, G. (2011) *A Model for Assessing the Relative Humaneness of Pest Animal Control Methods*, Santa Barbara, CA: Department of Agriculture, Fisheries and Forestry, Greenwood Publishing.

Sikes, R. S., Gannon, W. L., and Animal Care and Use Committee 1 of the American Society of Mammalogists (2016) "Guidelines of the American society of mammalogists for the use of wild mammals in research," *Journal of Mammalogy* 92(1): 235–253.

Spotswood, E. N., Goodman, K. R., Carlisle, J., Cormier, R. L., Humple, D. L., Rousseau, J., Guers, S. L., and Barton, G. G. (2012) "How safe is mist netting? Evaluating the risk of injury and mortality to birds," *Methods in Ecology and Evolution* 3(1): 29–38.

Tarszisz, E., Dickman, C. R., and Munn, A. J. (2014) "Physiology in conservation translocations," *Conservation Physiology* 2(1): cou504.

Tharoor, I. (2015) "Australia actually declares 'war' on cats, plans to kill 2 million by 2020," *The Washington Post*, July 16, www.washingtonpost.com/news/worldviews/wp/2015/07/16/australia-actually-declares-war-on-cats-plans-to-kill-2-million-by-2020.

Trefry, S. A., Diamond, A. W., and Jesson, L. K. (2013) "Wing marker woes: A case study and meta- analysis of the impacts of wing and patagial tags," *Journal of Ornithology* 154(1): 1–11.

Van Aarde, R. J., and Jackson, T. P. (2007) "Megaparks for metapopulations: Addressing the causes of locally high elephant numbers in southern Africa," *Biological Conservation* 134(3): 289–297.

Vucetich, J. A., and Nelson, M. P. (2007) "What are 60 warblers worth? Killing in the name of Conservation," *Oikos* 116(8): 1267–1278.

Williams, T. (2018) "Coyote carnage: The gruesome truth about wildlife killing contests," *Yale Environment 360*, May 22, https://e360.yale.edu/features/coyote-carnage-the-gruesome-truth-about-wildlife-killing-contests.

Wilson, R. P., and McMahon, C. R. (2006) "Measuring devices on wild animals: What constitutes acceptable practice?" *Frontiers in Ecology and the Environment* 4(3): 147–154.

32
ETHICAL DIMENSIONS OF INVASIVE ANIMAL MANAGEMENT

Tim S. Doherty and James C. Russell

Animals occurring outside of their native range are disproportionately likely to cause environmental, economic and social impacts. Those causing excessive negative impacts ('invasive animals,' defined later) are typically managed to avoid, remedy, or mitigate such harm. This often involves lethal control, which necessarily raises concerns about animal welfare and the ethics of killing animals. While animal welfare and rights are prominent ethical dimensions of invasive species management, others include human responsibility to resolve anthropogenic impacts, the pain and suffering inflicted by the invasive animals on other animals, and the broader consequences for collectives (e.g. populations, species, and ecosystems). In this chapter, we first outline the philosophical dimensions of invasive animal management, followed by a discussion of the ethical perspectives and issues around managing invasive animals.

What Are Invasive Species?

There are multiple overlapping terms used to describe animals existing outside of their native range, including invasive, introduced, exotic, alien, pest, and migrant, amongst others (Colautti and MacIsaac 2004). These terms are often conflated or used interchangeably, which creates confusion. The two main criteria used to classify non-native species are biogeographical or geopolitical origin and relative impact (positive, negative, or benign). Species that occur outside of their native range—irrespective of impacts—are considered *non-native*, which is synonymous with alien and exotic. Most operational definitions use the term *invasive* to refer to non-native species that generate negative impacts (i.e. *invasive species* or *invasive alien species*). This is the definition we adopt here, although we acknowledge that debate exists about what constitutes an invasive species and their value in non-native ecosystems (Simberloff 2003; Cassey et al. 2005; Brown and Sax 2005; Davis et al. 2011; Simberloff et al. 2011; Courchamp et al. 2017). Although it can sometimes be difficult to distinguish native from non-native species in historical records (e.g. if ancient humans were a vector of transportation outside the native range), the operational definition of an invasive species is nonetheless practical in application to biodiversity conservation and other sectors today, in an era of massively increased human mobility (Seebens et al. 2017). It need not be necessary to have a single definition of what constitutes an invasive species; however, the particular definition being invoked should be clearly stated.

The negative impacts of invasive species on ecosystems, economies, and humans can occur through myriad mechanisms. The Global Invasive Species Database classifies the environmental impacts of invasive species into 13 mechanisms: competition, predation, hybridisation, disease transmission, parasitism, poisoning/toxicity, biofouling, grazing/herbivory/browsing, rooting/digging, trampling, flammability, interaction with other invasive species, and other mechanisms not captured elsewhere (Blackburn et al. 2014). Many of these mechanisms also apply to economic activities and humans.

Invasive animals have altered biodiversity worldwide and are a leading cause of species decline and extinction (Bellard et al. 2016). For instance, invasive mammalian predators, such as rats *Rattus* spp. and cats *Felis catus* (Figure 32.1), have contributed to more than 50% of bird, mammal and reptile extinctions worldwide (Doherty et al. 2016). Invasive freshwater fish, such as carp *Cyprinus carpio* (Figure 32.1), negatively impact ecosystems by preying on and competing with native species, spreading disease, and depleting native vegetation (Cambray 2003; Simon and Townsend 2003). Invasive birds, such as the common myna *Acridotheres tristis* (Figure 32.1), can aggressively compete with native birds for nesting sites and food resources (Charter et al. 2016). The impacts of invasive animals on humans span health and safety, property, and livelihoods. The economic costs of invasive animals are primarily incurred in agriculture, forestry, and public health and include not only the costs associated with reduced yields but also significant amounts spent on invasive species control. Invasive insects are estimated to cost the global economy at least US$70 billion per year, plus an additional US$6.9 billion of associated health costs (Bradshaw et al. 2016). Invasive animals can spread diseases to humans, and domestic and wild animals, such as dogs *Canis familiaris* spreading rabies in India and elsewhere (Menezes 2008), Australian brushtail possums *Trichosurus vulpecula* spreading bovine tuberculosis in New Zealand (Ryan et al. 2006), and cats spreading toxoplasmosis worldwide

Figure 32.1 Examples of invasive animal species: feral cat *Felis catus* (top left), black rat *Rattus rattus* (top right), carp *Cyprinus carpio* (bottom left) and common myna *Acridotheres tristis* (bottom right)

Source: Top layer © Tim Doherty (left) and James Russell (right); bottom layer courtesy of succo/pixabay (left) and smarko/pixabay (right).

(Tenter et al. 2000). These wide-ranging impacts necessitate that invasive animal management be a major concern globally.

Animal Ethics and Human Responsibilities

Despite there being a large body of evidence demonstrating that invasive species cause harm to humans, ecosystems and their components and economies, debates persist on the need to manage invasive animals (Simberloff 2003; Russell and Blackburn 2017; Ricciardi and Ryan 2018). Part of the reason this occurs is due to contrasting perspectives that differentially prioritise individuals or collectives (e.g. populations or species; Fraser 2010; Parke and Russell 2018). The main philosophical perspectives related to invasive animal ethics are anthropocentrism, animal rights, welfarism, and biocentrism (Singer 1997; Kawall 2017). In brief, the first focuses primarily on how animals can be useful to humans, the second and third challenge this view and focus on the primacy of the rights or welfare of individual animals, while the fourth shifts the focus from individuals to ecological collectives. Welfarism primarily focuses on "sentience" as the trait which confers moral standing, which is a term variously described as the ability to perceive or feel things (Oxford English Dictionary 2015), experience positive and negative affective states (Duncan 2006), and experience pain and suffering (Jones 2013). Most operational definitions of sentience include all vertebrates, cephalopods, and some crustaceans; see Jones (2013) for further discussion. Animal welfare principles promote the humane and responsible use of animals and can also be embedded in an anthropo- or biocentric approach, but contrast with the animal rights view which opposes any use of animals by humans. Biocentric and eco-centric approaches are similar, although the latter includes consideration of non-living components of nature. Kawall (2017) and Bekoff (2010) provide more detailed assessments of the history and characteristics of these different viewpoints.

Singer (1997) argued that where invasive animals harm ecosystems, ethical debates are triangular because they include conflicting views related to human interests, individual animal rights and collective effects (e.g. ecosystems). Eggleston et al. (2003) also identified three similar sets of values for consideration in the management of vertebrate pests: anthropocentric concerns, animal welfare and rights, and ecological protection. The vexed issue of feral pig *Sus scrofa* management in the Hawaiian Islands provides a good example of this. Pigs cause massive damage to Hawaiian ecosystems (Nogueira-Filho et al. 2009), and land managers attempt to reduce these impacts through hunting, trapping, baiting, and fencing (Stone and Loope 1987; Campbell and Long 2009). Some animal rights groups have opposed lethal control of pigs, especially inhumane methods such as snaring (Knickerbocker 1994; Maguire 2004). Pig control has also been opposed on cultural grounds because pigs have been culturally adopted as part of indigenous Hawaiian culture; they are hunted customarily, their meat is considered traditional food, and they feature in social festivities (Adler 1995; Pfeiffer and Voeks 2008). Complex situations such as this demonstrate how consideration of human, animal and ecosystem values can be critical to an effective ethical framework for managing introduced animals.

More recently, some have advocated for incorporating compassion into wildlife management (Bekoff 2013; Ramp and Bekoff 2015; Wallach et al. 2015). Proponents argue that compassionate conservation is not solely an animal rights position (Ramp and Bekoff 2015) but, rather, "a 3-tiered conservation ethic that encompasses the welfare of individuals, populations, and ecosystems" (Wallach et al. 2015: p. 1481). However, in practice advocates have gone on to suggest that we should "embrace [introduced] cats as part of Australia's environment" and rename them "Australian wildcats" (Wallach and Ramp 2015). This is despite cats being a primary driver of Australia's extinction crisis (Ziembicki et al. 2015; Woinarski et al. 2015), where they kill an estimated 1 million birds and 1.7 million reptiles per day (Woinarski et al. 2017; Woinarski et al. 2018). Such an application of compassionate conservation therefore seems at odds with its intent to be a broad environmental ethic. The decision to not conduct lethal control of introduced predators is questionable where it disregards the deaths of animals caused by introduced predators that facilitate species declines and extinctions (Russell et al. 2016). If the motivation for not conducting lethal control is the view that animal deaths are unacceptable, this is firmly an animal rights position, whose application conflicts with the large number of animal deaths caused by the predators themselves. This is particularly relevant where introduced predators engage in surplus killing of excess prey without eating it (Short et al. 2002; Major and Jones 2005).

Hutchins (2007, 2008) argued that animal rights and conservation ethics are incompatible because the former places equal value on common and threatened species, while a conservation approach prioritises species at risk of extinction. This conflict is evident in the case of grey squirrel *Sciurus carolinensis* invasion in Italy, where it threatens native red squirrels *Sciurus vulgaris* (Genovesi and Bertolino 2001). A trial eradication that began in 1997 was halted prematurely due to legal action by animal rights groups opposing the use of euthanasia. The legal battle continued for three years while the grey squirrel continued to expand such that eradication was no longer considered feasible (Genovesi and Bertolino 2001). Between 1970 and 2010, red squirrels declined and were replaced by grey squirrels in 62% of their range within the invaded area (Bertolino et al. 2014). Whether the animal rights proponents saw the loss of red squirrels as an acceptable cost of not harming the grey squirrels, or whether they were simply concerned with not having humans directly killing animals, is unclear.

Further ethical challenges arise where introduced pest animals are threatened species in their native range (Marchetti and Engstrom 2016). Examples include European rabbits *Oryctolagus cuniculus* introduced to New Zealand and elsewhere (Lees and Bell 2008); bantengs *Bos javanicus* which are introduced in Australia and threatened in Indonesia (Bradshaw et al. 2006); and bovids (*Ammotragus lervia* and *Ovis orientalis*) that are introduced in the Canary Islands and threatened in their native ranges (Garzón-Machado et al. 2012). Control actions to protect an invaded ecosystem from an invasive species may conflict with conservation efforts elsewhere if that species is classed as threatened. Having extant non-native populations of a threatened species may be better than allowing it to become extinct, but this decision is difficult where the invasive species has negative impacts on an ecosystem, including threatening other species with extinction (Marchetti and Engstrom 2016). Conservation practitioners and land managers typically operate at local scales and may not consider they have a duty towards, or simply may be unaware of, the status of an invasive species in its native range. Discussion of these issues has only recently gained momentum (Lees and Bell 2008; Clavero 2014; Díez-León et al. 2015; Marchetti and Engstrom 2016; Gibson and Yong 2017), and there are no clear guidelines on how to manage them.

Welfare Principles for Managing Invasive Animals

Integrating animal welfare into wildlife management is a complex endeavour and many sets of principles and guidelines have been developed to facilitate this process (e.g. Eggleston et al. 2003; Littin et al. 2004; Littin and Mellor 2005; Yeates 2010; Invasive Animals CRC 2014; Hadidian 2015; Dubois et al. 2017). For example, Littin et al. (2004) outlined six principles for ethically sound vertebrate pest control that primarily drew on the issues of animal welfare and evidence-based management. Other guidelines have focused purely on operational or practical aspects of pest animal control, with no reference to animal welfare (Invasive Animals CRC 2014). The most recent and comprehensive set of principles was developed by Dubois et al. (2017) which include both the operational and animal welfare components, as well as the addition of human considerations. Their seven step-wise principles posed as questions are the following:

1. Can the problem be mitigated by changing human behaviour?
2. Are the harms serious enough to warrant wildlife control?
3. Is the desired outcome clear and achievable, and will it be monitored?

4. Does the proposed method carry the least animal welfare cost and to the fewest animals?
5. Have community values been considered alongside scientific, technical, and practical information?
6. Is the control action part of a systematic, long-term management program?
7. Are the decisions warranted by the specifics of the situation rather than negative labels applied to the animals?

Below we consider some of these principles in more detail, including numbers 4 and 5 under *Animal welfare and control techniques* and numbers 2, 3, 5 and 6 under *Management ethics, design and monitoring*. Much of this discussion focuses on animal welfare as this is the most well-developed viewpoint in terms of practical application to invasive animal management. We acknowledge that other positions, particularly rights, may advocate for different management approaches.

When invasive animal management will take place, the most immediate ethical issue relates to animal welfare during control operations. *Animal welfare* has many interpretations (Fisher 2009), but perhaps the most widely adopted definition and the one that we use here is "the physical and mental state of an animal in relation to the conditions in which it lives and dies" (World Organisation for Animal Health 2019: 333). This concept represents a continuum from poor welfare (i.e. suffering) to good welfare (i.e. well-being), and the preceding definition goes on to say:

> An animal experiences good welfare if the animal is healthy, comfortable, well nourished, safe, is not suffering from unpleasant states such as pain, fear and distress, and is able to express behaviours that are important for its physical and mental state. Good animal welfare requires disease prevention and appropriate veterinary care, shelter, management and nutrition, a stimulating and safe environment, humane handling and humane slaughter or killing. While animal welfare refers to the state of the animal, the treatment that an animal receives is covered by other terms such as animal care, animal husbandry, and humane treatment.
>
> (*World Organisation for Animal Health 2019)*

It is important to note from the outset that an invasive animal may be in poor welfare regardless of any direct human intervention, beyond its original introduction.

Approaches for managing invasive animal populations include both lethal and non-lethal measures. Non-lethal approaches include fencing, repellents and fertility control, and lethal methods include biological control, poisoning, shooting, and trapping, amongst others. Both lethal and non-lethal approaches can cause pain and distress to both target and non-target animals. There is typically a strong expectation from the public that the methods used are humane, adverse impacts are reduced or avoided, and animal welfare standards are upheld. Indeed, an animal being an invasive species does not mean that welfare standards should be disregarded.

Risks to welfare during lethal control occur when target animals are killed, or are not killed but are otherwise harmed (e.g. sub-lethal poisoning), and when non-target animals are negatively impacted by a control technique, either directly (e.g. poisoning), or indirectly (e.g. changes to food availability). The welfare costs (WC_{Total}), but not benefits, of a control or eradication operation can be described using the following formula (Cowan and Warburton 2011; Warburton et al. 2012):

$$WC_{Total} = (WC_{TL} \times N) + (WC_{TSL} \times N) + (WC_{NTL} \times N) + (WC_{NTSL} \times N),$$

where
WC_{TL} = welfare cost to target species that are killed,
WC_{TSL} = welfare cost to target species that are sub-lethally poisoned or injured,
WC_{NTL} = welfare cost to non-target species that are killed,
WC_{NTSL} = welfare cost to non-targets that are sub-lethally poisoned or injured, and
N = the number of animals in each of these categories.

Using this formula, the total welfare costs are minimised where N is minimised and the techniques used have the lowest welfare costs. Such a model is appealing when striving to minimise welfare costs in a control program, but importantly, it does not also account for welfare benefits (i.e. increases in animal welfare to surviving target and non-target species) and avoidance of future welfare costs (i.e. unborn invasive animals), such as in complete eradication.

Eradication is a special case of minimising welfare costs. Although the aim in an eradication is to remove all individuals of a species from an area (i.e. kill the maximum individuals present), by doing so it minimises long-term welfare costs through removing the need to kill future unborn individuals if only control (not eradication) is undertaken. Thus, consideration of time horizon is important for eradications because welfare costs are minimised when current and future deaths are both accounted for. Minimising the number killed of a target species in eradications can be achieved by conducting the operation as quickly as possible to minimise recruitment (while still ensuring success) and targeting control in the low point of a population cycle.

Warburton et al. (2012) used modelling to show how alternative management strategies can minimise the number of introduced possums, ship rats *Rattus rattus* and wallabies *Macropus rufogriseus* killed in pest control operations in New Zealand. Here, we discuss both the efficacy and cost of alternative strategies as the latter is a major factor determining pest control operation implementation and design. For possums, an initial 90% knockdown followed by annual maintenance control at this level resulted in the least deaths, but this strategy was also the most expensive (Warburton et al. 2012). The strategy with the highest number killed but lowest cost was 90% knockdown followed by maintenance control once the population returned to 50% carrying capacity (Warburton et al. 2012). For ship rats, aerial baiting immediately prior to population irruptions killed fewer individuals and was cheaper than annual ground-based baiting during the bird breeding season (Warburton et al. 2012). The situation was more complex for wallabies, with the number killed increasing until target density was 60% of carrying capacity and then declining, while costs increased as density declined (Warburton et al. 2012). Warburton and Anderson (2018) provide a more detailed discussion of integrating ethics, ecology, and economics in wildlife management.

Other approaches for reducing welfare impacts during animal control operations include choosing control tools that have the least harm to animals (i.e. Question 4), improving the humaneness of existing control tools, and developing new control tools that are more humane. Various frameworks have been produced for assessing the humaneness of control tools, with the most comprehensive being that of Sharp and Saunders (2011). A primary advantage of this approach over others is that the assessment process produces a score of relative humaneness that can be compared between any lethal and non-lethal approaches. Also, the assessment tool can be used for any species, whereas earlier tools were designed for specific species, taxonomic groups, or control tools (e.g. Iossa et al. 2007; Baker et al. 2012). This assessment tool has been used widely in Australia and New Zealand, primarily for introduced mammals but also some fish and bird species (Littin et al. 2014).

The Sharp and Saunders (2011) model is a two-part process. Part A assesses the impact on overall welfare and the duration of this impact and can be applied to any technique, whereas Part B is used only for lethal techniques and assesses the level and duration of suffering until the animal becomes insensible (Figure 32.2). Impacts in Part A are classified as either no impact, mild, moderate, severe or extreme in each of five domains: nutrition, environment, health, behaviour, and mental state (Sharp and Saunders 2011; Littin et al. 2014; Figure 32.2, left). The degree of impact is then combined with the duration (seconds to weeks) to give a numerical score for Part A (Figure 32.2, left). For lethal techniques, Part B is similarly applied to assess the level of suffering until insensibility (Figure 32.2, right), and the scores are then combined to give an overall welfare impact score (e.g. 4D). Periodic evaluation of different techniques is important because best practice approaches can be superseded by new methods that emerge from research or technological advances. For instance, a fertility control

Overall impact on welfare	Duration of impact				
	Immediate/ seconds	Minutes	Hours	Days	Weeks
Extreme	5	6	7	8	8
Severe	4	5	6	7	8
Moderate	3	4	5	6	7
Mild	2	3	4	5	6
No impact	1	1	1	1	1

Level of suffering	Time until insensibility				
	Immediate/ seconds	Minutes	Hours	Days	Weeks
Extreme	E	F	G	H	H
Severe	D	E	F	G	H
Moderate	C	D	E	F	G
Mild	B	C	D	E	F
No impact	A	A	A	A	A

Figure 32.2 Scoring matrix for Part A: overall welfare impact (top); scoring matrix for Part B: assessment of mode of death (bottom)

Source: Reproduced from Sharp and Saunders (2011).

drug has been tested which shows marked success in reducing rat population size at small scales (Witmer et al. 2017).

A framework such as this helps remove subjectivity from management decisions. For instance, opposition to lethal control often carries an implicit assumption that non-lethal methods are more humane, but this is not always the case. Two examples of this are the management of feral horses *Equus caballus* in the Australian Alps and trap–neuter–release programs for free-ranging cats. Horses introduced to the Alps have significant negative impacts on alpine ecosystems, including many threatened plant and animal species (Cherubin et al. 2019; Foster and Scheele 2019). Attempts at population control via aerial shooting have been strongly opposed by special interest groups, including animal rights advocates, who instead advocate trapping and rehoming (Driscoll et al. 2019). This is despite an analysis using the Sharp and Saunders (2011) model which showed that trapping and rehoming would have inferior welfare outcomes compared to aerial shooting due to stress and suffering during capture, holding and transportation (Humaneness Assessment Panel 2015). Similarly, trap–neuter–release, whereby cats are captured, sterilized, and returned to the environment, is commonly advocated for the management of free-ranging cats instead of euthanasia, especially in the US (Longcore et al. 2009). However, this approach neglects the fact that free-ranging cats often have poorer welfare outcomes than owned, indoor cats, including higher rates of disease and traffic accidents

(Rochlitz 2004; Goldkamp et al. 2008; Marston and Bennett 2009). Furthermore, returning cats to the environment, even when fed, does not end their predation of native wildlife (Hawkins et al. 2004; Longcore et al. 2009). These examples demonstrate the importance of considering the entire pest management chain, including the consequences of a management action after it has finished.

Management Ethics, Design, and Monitoring

Technological advances have undoubtedly improved welfare outcomes for individual animals, but there has been less focus on the ethical and welfare implications of control operations that fail to meet their biodiversity objectives (Warburton and Norton 2009). Littin and Mellor (2005) argued that the ethics of pest animal control can be assessed by asking (1) whether an action is necessary and (2) whether it is justified. Implicit in these questions is another question, 'Do the benefits outweigh the harms?' For lethal control to be justified, it must reduce the impacts of invasive animals on the assets or values of interest (i.e. a benefit) beyond that achieved by non-lethal control (Warburton and Norton 2009). But invasive animal management is often laden with uncertainty about the magnitude of impacts and the outcomes of control actions. Adaptive management is advocated as the gold standard for dealing with uncertainty in pest animal management, while less desirable approaches include trial and error, dogmatism, risk aversion, and deferred action (Parkes et al. 2006).

To ensure that their actions are ethically defensible, pest animal managers have a responsibility to measure the response of target assets (e.g. a threatened species) to control actions, but many do not do this. In Australia, Reddiex and Forsyth (2006) found that 88% of 1,915 control operations of introduced foxes *Vulpes vulpes*, dogs, rabbits, pigs, cats, or goats *Capra hircus* did not include monitoring of the resource they were trying to protect. Similarly, in New Zealand, only 16% of 83 pest plant and animal control programs involved 'outcome monitoring' of the system values aiming to be protected (Clayton and Cowan 2010). In contrast, for projects controlling introduced rats on islands, 62% monitored rats, and 58% monitored biodiversity (Duron et al. 2017).

Control operations that fail to meet their objectives and do not include outcome monitoring may unnecessarily compromise the welfare of animals. At this point, it is important to acknowledge that poor monitoring and an ineffectual control program are not codependent. Managing invasive animals is logistically challenging and external factors can render programs ineffective. But if monitoring is involved, this provides a learning opportunity to improve future management actions, even if the primary objective is not achieved (Parkes et al. 2006). If no monitoring is involved and biodiversity objectives are not met, animal deaths may be in vain.

Conclusion

Managing invasive animals is complex and contentious, but necessary to reduce or avoid their negative impacts on humans, other individual animals, and larger ecological collectives. Animal ethics, particularly animal welfare, is a major area of concern. We have highlighted key ethical issues related to managing invasive animals, namely the ethics of killing, human responsibility to resolve anthropogenic impacts, the pain and suffering inflicted by invasive species on other animals, and the paradox of invasive species that are threatened in their native range. Common ethical and moral perspectives used to navigate these issues are animal rights, welfarism, anthropocentrism, and biocentrism. How invasive animals are managed often depends on which of these ethical viewpoints is considered paramount, whether it be an explicit value-based decision, or something more implicit. In many cases, simultaneous consideration of animal, human and environmental interests is necessary, and this will likely become increasingly important as the issue of invasive animal management persists in this era of global change.

Bibliography

Adler, P. S. (1995) "Pig wars: Mediating forest management conflicts in Hawaii," *Negotiation Journal* 11: 209–215.

Baker, S. E., Ellwood, S. A., Tagarielli, V. L., and Macdonald, D. W. (2012) "Mechanical performance of rat, mouse and mole spring traps, and possible implications for welfare performance," *PLoS One* 7: e39334.

Bekoff, M. (2010) *Encyclopedia of Animal Rights and Animal Welfare*, Santa Barbara, CA: Greenwood Press.

Bekoff, M. (2013) *Ignoring Nature No More: The Case for Compassionate Conservation*, London: The University of Chicago Press.

Bellard, C., Genovesi, P., and Jeschke, J. M. (2016) "Global patterns in threats to vertebrates by biological invasions," *Proceedings of the Royal Society B: Biological Sciences* 283: 20152454–20152459.

Bertolino, S., Cordero, N., and Martinoli, A. (2014) "A grey future for Europe: *Sciurus carolinensis* is replacing native red squirrels in Italy," *Biological Invasions* 16: 53–62.

Blackburn, T. M., Essl, F., Evans, T., Hulme, P. E., Jeschke, J. M., Kühn, I., Kumschick, S., Marková, Z., Mrugała, A., Nentwig, W., Pergl, J., Pyšek, P., Rabitsch, W., Ricciardi, A., Richardson, D. M., Sendek, A., Vilà, M., Wilson, J. R. U., Winter, M., Genovesi, P., and Bacher, S. (2014) "A unified classification of alien species based on the magnitude of their environmental impacts," *PLoS Biology* 12: e1001850.

Bradshaw, C. J. A., Isagi, Y., Kaneko, S., Bowman, D. M. J. S., and Brook, B. W. (2006) "Conservation value of non-native banteng in Northern Australia," *Conservation Biology* 20: 1306–1311.

Bradshaw, C. J. A., Leroy, B., Bellard, C., Roiz, D., Albert, C., Fournier, A., Barbet-Massin, M., Salles, J. M., Simard, F., and Courchamp, F. (2016) "Massive yet grossly underestimated global costs of invasive insects," *Nature Communications* 7: 12986.

Brown, J. H., and Sax, D. F. (2005) "Biological invasions and scientic objectivity: Reply to Cassey *et al.*," *Austral Ecology* 30: 481–483.

Cambray, J. A. (2003) "Impact on indigenous species biodiversity caused by the globalisation of alien recreational freshwater fisheries," *Hydrobiologia* 500: 217–230.

Campbell, T. A., and Long, D. B. (2009) "Feral swine damage and damage management in forested ecosystems," *Forest Ecology and Management* 257: 2319–2326.

Cassey, P., Blackburn, T. M., Duncan, R. P., and Chown, S. L. (2005) "Concerning invasive species: Reply to Brown and Sax," *Austral Ecology* 30: 475–480.

Charter, M., Izhaki, I., Ben Mocha, Y., and Kark, S. (2016) "Nest-site competition between invasive and native cavity nesting birds and its implication for conservation," *Journal of Environmental Management* 181: 129–134.

Cherubin, R.C., Venn, S. E., Driscoll, D. A., Doherty, T. S., and Ritchie, E. G. (2019) "Feral horse impacts on threatened plants and animals in sub-alpine and montane environments in Victoria, Australia," *Ecological Management and Restoration* 20: 47–56.

Clavero, M. (2014) "Shifting baselines and the conservation of non-native species," *Conservation Biology* 28: 1434–1436.

Clayton, R., and Cowan, P. (2010) "Management of animal and plant pests in New Zealand—patterns of control and monitoring by regional agencies," *Wildlife Research* 37: 360–371.

Colautti, R. I., and MacIsaac, H. J. (2004) "A neutral terminology to define 'invasive' species," *Diversity and Distributions* 10: 135–141.

Courchamp, F., Fournier, A., Bellard, C., Bertelsmeier, C., Bonnaud, E., Jeschke, J. M., and Russell, J. C. (2017) "Invasion biology: Specific problems and possible solutions," *Trends in Ecology & Evolution* 32: 13–22.

Cowan, P., and Warburton, B. (2011) "Animal welfare and ethical issues in island pest eradication," in C. Veitch, M. Clout, and D. Towns (eds.) *Island Invasives: Eradication and Management*, Gland, Switzerland: IUCN.

Davis, M. A., Chew, M. K., Hobbs, R. J., Lugo, A. E., Ewel, J. J., Vermeij, G. J., Brown, J. H., Rosenzweig, M. L., Gardener, M. R., Carroll, S. P., Thompson, K., Pickett, S. T. A., Stromberg, J. C., Tredici, P. Del, Suding, K. N., Ehrenfeld, J. G., Philip Grime, J., Mascaro, J., and Briggs, J. C. (2011) "Don't judge species on their origins," *Nature* 474: 153–154.

Díez-León, M., Miranda, R., Ariño, A. H., and Galicia, D. (2015) "Setting priorities for existing conservation needs of crayfish and mink," *Conservation Biology* 29: 599–601.

Doherty, T. S., Glen, A. S., Nimmo, D. G., Ritchie, E. G., and Dickman, C. R. (2016) "Invasive predators and global biodiversity loss," *Proceedings of the National Academy of Sciences* 113: 11261–11265.

Driscoll, D. A., Worboys, G. L., Allan, H., Banks, S. C., Beeton, N. J., Cherubin, R. C., Doherty, T. S., Finlayson, C. M., Green, K., Hartley, R., Hope, G., Johnson, C. N., Lintermans, M., Mackey, B., Paull, D. J., Pittock, J., Porfirio, L. L., Ritchie, E. G., Sato, C. F., Scheele, B. C., Slattery, D. A., Venn, S., Watson, D., Watson, M., and Williams, R. M. (2019) "Impacts of feral horses in the Australian Alps and evidence-based solutions," *Ecological Management & Restoration* 20: 63–72.

Dubois, S., Fenwick, N., Ryan, E. A., Baker, L., Baker, S. E., Beausoleil, N. J., Carter, S., Cartwright, B., Costa, F., Draper, C., Griffin, J., Grogan, A., Howald, G., Jones, B., Littin, K. E., Lombard, A. T., Mellor, D. J., Ramp, D., Schuppli, C. A., and Fraser, D. (2017) "International consensus principles for ethical wildlife control," *Conservation Biology* 31: 753–760.

Duncan, I. J. H. (2006) "The changing concept of animal sentience," *Applied Animal Behaviour Science* 100: 11–19.

Duron, Q., Shiels, A. B., and Vidal, E. (2017) "Control of invasive rats on islands and priorities for future action," *Conservation Biology* 31: 761–771.

Eggleston, J. E., Rixecker, S. S., and Hickling, G. J. (2003) "The role of ethics in the management of New Zealand's wild mammals," *New Zealand Journal of Zoology* 30: 361–376.

Fisher, M. W. (2009) "Defining animal welfare—does consistency matter?" *New Zealand Veterinary Journal* 57: 71–73.

Foster, C. N., and Scheele, B. C. (2019) "Feral-horse impacts on corroboree frog habitat in the Australian Alps," *Wildlife Research* 46: 184–190.

Fraser, D. (2010) "Toward a synthesis of conservation and animal welfare science," *Animal Welfare* 19: 121–124.

Garzón-Machado, V., Del-Arco-Aguilar, M. J., and Pérez-de-Paz, P. L. (2012) "Threat or threatened species? A paradox in conservation biology," *Journal for Nature Conservation* 20: 228–230.

Genovesi, P., and Bertolino, S. (2001) "Human dimension aspects in invasive alien species issues: The case of the failure of the grey squirrel eradication project in Italy," in J. McNeely (ed.) *The Great Reshuffling: Human Dimensions of Invasive Alien Species*, Gland, Switzerland and Cambridge, UK: IUCN.

Gibson, L., and Yong, D. L. (2017) "Saving two birds with one stone: Solving the quandary of introduced, threatened species," *Frontiers in Ecology and the Environment* 15: 35–41.

Goldkamp, C. E., Levy, J. K., Edinboro, C. H., and Lachtara, J. L. (2008) "Seroprevalences of feline leukemia virus and feline immunodeficiency virus in cats with abscesses or bite wounds and rate of veterinarian compliance with current guidelines for retrovirus testing," *Journal of the American Veterinary Medical Association* 232: 1152–1158.

Hadidian, J. (2015) "Wildlife in U.S. cities: Managing unwanted animals," *Animals* 5: 1092–1113.

Hawkins, C. C., Grant, W. E., and Longnecker, M. T. (2004) "Effect of house cats, being fed in parks, on California birds and rodents," in W. Shaw, L. Harris, and L. VanDruff (eds.) *Proceedings of the 4th International Symposium on Urban Wildlife Conservation*, Tuscon, AZ: University of Arizona.

Humaneness Assessment Panel (2015) *Assessing the humaneness of wild horse management methods. Kosciuszko National Park Wild Horse Management Plan. A report on the outcomes of a humaneness assessment panel assembled on behalf of the Independent Technical Reference Group*, Sydney, Australia: State of NSW and the Office of Environment and Heritage.

Hutchins, M. (2007) "The limits of compassion," *The Wildlife Professional,* Summer: 42–44.

Hutchins, M. (2008) "Animal rights and conservation," *Conservation Biology* 22: 815–816.

Invasive Animals CRC (2014) "Principles of pest animal management," www.pestsmart.org.au/principles-of-pest-animal-management/.

Iossa, G., Soulsbury, C. D., and Harris, S. (2007) "Mammal trapping: A review of animal welfare standards of killing and restraining traps," *Animal Welfare* 16: 335–352.

Jones, R. C. (2013) "Science, sentience, and animal welfare," *Biology and Philosophy* 28: 1–30.

Kawall, J. (2017) *A History of Environmental Ethics*, eds. S. M. Gardiner and A. Thompson, Oxford: Oxford University Press.

Knickerbocker, B. (1994) "Snaring of Hawaii's feral pigs angers animal activists," *The Christian Science Monitor,* www.csmonitor.com/1994/0105/05072.html

Lees, A. C., and Bell, D. J. (2008) "A conservation paradox for the 21st century: The European wild rabbit *Oryctolagus cuniculus*, an invasive alien and an endangered native species," *Mammal Review* 38: 304–320.

Littin, K. E., Fisher, P., Beausoleil, N. J., and Sharp, T. (2014) "Welfare aspects of vertebrate pest control and culling: Ranking control techniques for humaneness," *Revue scientifique et technique (International Office of Epizootics)* 33: 281–289.

Littin, K. E., and Mellor, D. J. (2005) "Strategic animal welfare issues: Ethical and animal welfare issues arising from the killing of wildlife for disease control and environmental reasons," *Revue scientifique et technique (International Office of Epizootics)* 24: 767–782.

Littin, K. E., Mellor, D. J., Warburton, B., and Eason, C. T. (2004) "Animal welfare and ethical issues relevant to the humane control of vertebrate pests," *New Zealand Veterinary Journal* 52: 1–10.

Longcore, T., Rich, C., and Sullivan, L. M. (2009) "Critical assessment of claims regarding management of feral cats by trap-neuter-return," *Conservation Biology* 23: 887–894.

Maguire, L. A. (2004) "What can decision analysis do for invasive species management?" *Risk Analysis: An Official Publication of the Society for Risk Analysis* 24: 859–868.

Major, H. L., and Jones, I. L. (2005) "Distribution, biology and prey selection of the introduced rat *Rattus norvegicus* at Kiska Island, Aleutian Islands, Alaska," *Pacific Conservation Biology* 11: 105–113.

Marchetti, M. P., and Engstrom, T. (2016) "The conservation paradox of endangered and invasive species," *Conservation Biology* 30: 434–437.

Marston, L. C., and Bennett, P. C. (2009) "Admissions of cats to animal welfare shelters in Melbourne, Australia," *Journal of Applied Animal Welfare Science* 12: 189–213.

Menezes, R. (2008) "Rabies in India," *Canadian Medical Association Journal* 178: 564–566.

Nogueira-Filho, S. L. G., Nogueira, S. S. C., and Fragoso, J. M. V (2009) "Ecological impacts of feral pigs in the Hawaiian Islands," *Biodiversity and Conservation* 18: 3677–3683.

Oxford English Dictionary (2015) *Oxford English Dictionary Online*, http://dictionary.oed.com/.

Parke, E. C., and Russell, J. C. (2018) "Ethical responsibilities in invasion biology," *The Ecological Citizen* 2: 17–19.

Parkes, J. P., Robley, A., Forsyth, D. M., and Choquenot, D. (2006) "Adaptive management experiments in vertebrate pest control in New Zealand and Australia," *Wildlife Society Bulletin* 34: 229–236.

Pfeiffer, J. M., and Voeks, R. A. (2008) "Biological invasions and biocultural diversity: Linking ecological and cultural systems," *Environmental Conservation* 35: 281–293.

Ramp, D., and Bekoff, M. (2015) "Compassion as a practical and evolved ethic for conservation," *BioScience* 65: 323–327.

Reddiex, B., and Forsyth, D. M. (2006) "Control of pest mammals for biodiversity protection in Australia. II. Reliability of knowledge," *Wildlife Research* 33: 711–717.

Ricciardi, A., and Ryan, R. (2018) "The exponential growth of invasive species denialism," *Biological Invasions* 20: 549–553.

Rochlitz, I. (2004) "Clinical study of cats injured and killed in road traffic accidents in Cambridgeshire," *The Journal of Small Animal Practice* 45: 390–394.

Russell, J. C., and Blackburn, T. M. (2017) "The rise of invasive species denialism," *Trends in Ecology & Evolution* 32: 3–6.

Russell, J. C., Jones, H. P., Armstrong, D. P., Courchamp, F., Kappes, P. J., Seddon, P. J., Oppel, S., Rauzon, M. J., Cowan, P. E., Rocamora, G., Genovesi, P., Bonnaud, E., Keitt, B. S., Holmes, N. D., and Tershy, B. R. (2016) "Importance of lethal control of invasive predators for island conservation," *Conservation Biology* 30: 670–672.

Ryan, T. J., Livingstone, P. G., Ramsey, D. S. L., De Lisle, G. W., Nugent, G., Collins, D. M., and Buddle, B. M. (2006) "Advances in understanding disease epidemiology and implications for control and eradication of tuberculosis in livestock: The experience from New Zealand," *Veterinary Microbiology* 112: 211–219.

Seebens, H., Blackburn, T. M., Dyer, E. E., Genovesi, P., Hulme, P. E., Jeschke, J. M., Pagad, S., Pyšek, P., Winter, M., Arianoutsou, M., Bacher, S., Blasius, B., Brundu, G., Capinha, C., Celesti-Grapow, L., Dawson, W., Dullinger, S., Fuentes, N., Jager, H., Kartesz, J., Kenis, M., Kreft, H., Kuhn, I., Lenzner, B., Liebhold, A., Mosena, A., Moser, D., Nishino, M., Pearman, D., Pergl, J., Rabitsch, W., Rojas-Sandoval, J., Roques, A., Rorke, S., Rossinelli, S., Roy, H. E., Scalera, R., Schindler, S., Štajerova, K., Tokarska-Guzik, B., Van Kleunen, M., Walker, K., Weigelt, P., Yamanaka, T., and Essl, F. (2017) "No saturation in the accumulation of alien species worldwide," *Nature Communications* 8: 1–9.

Sharp, T., and Saunders, G. (2011) *Assessing the Humaneness of Commonly Used Invasive Animal Control Methods*, 2nd ed., Canberra: Australian Government Department of Agriculture, Fisheries and Forestry.

Short, J., Kinnear, J. E., and Robley, A. (2002) "Surplus killing by introduced predators in Australia—evidence for ineffective anti-predator adaptations in native prey species?" *Biological Conservation* 103: 283–301.

Simberloff, D. (2003) "Confronting introduced species: A form of xenophobia?" *Biological Invasions* 5: 179–192.

Simberloff, D., Alexander, J., Allendorf, F., Aronson, J., Antunes, P. M., Bacher, S., Bardgett, R., Bertolino, S., Bishop, M., Blackburn, T. M., Blakeslee, A., Blumenthal, D., Bortolus, A., Buckley, R., Buckley, Y., Byers, J., M., R., Campbell, F., Campbell, K., Campbell, M., Carlton, J. T., Cassey, P., Catford, J., Celesti-Grapow, L., Chapman, J., Clark, P., Clewell, A., Canning Clode, J., Chang, A., Chytrý, M., Clout, M., Cohen, A., Cowan, P., Cowie, R. H., Crall, A. W., Crooks, J., Deveney, M., Dixon, K., Dobbs, F. C., Cameron Duffy, D., Duncan, R., Ehrlich, P. R., Eldredge, L., Evenhuis, N., Fausch, K. D., Feldhaar, H., Firn, J., Fowler, A., Galil, B., Garcia-Berthou, E., Geller, J., Genovesi, P., Gerber, E., Gherardi, F., Gollasch, S., Gordon, D., Graham, J., Gribben, P., Griffen, B., Grosholz, E. D., Hewitt, C., Hierro, J. L., Hulme, P., Hutchings, P., Jarošík, V., Johnson, C., Johnson, L., Johnston, E. L., Jones, C. G., Keller, R., King, C. M., G. J. Knols, B., Kollmann, J., Kompas, T., Kotanen, P. M., Kowarik, I., Kühn, I., Kumschick, S., Leung, B., Liebhold, A., MacIsaac, H., Mack, R., G. McCullough, D., McDonald, R., Merritt, D. M., Meyerson, L., Minchin, D., Mooney, H. A., Morisette, J. T., Moyle, P., Müller-Schärer, H., Murray, B. R., Nehring, S., Nelson, W., Nentwig, W., Novak, S. J., Occhipinti, A., Ojaveer, H., Osborne, B., Ostfeld, R. S., Parker, J., Pederson, J., Pergl, J., Phillips, M. L., Pyšek, P., Rejmánek, M., Ricciardi, A., Ricotta, C., Richardson, D., Rilov, G., Ritchie, E. G., Robertson, P. A., Roman, J., Ruiz, G., Schaefer, H., Schaffelke, B., Schierenbeck, K. A., Schmitz, D. C., Schwindt, E., Seeb,

J., David Smith, L., Smith, G. F., Stohlgren, T., Strayer, D. L., Strong, D., Sutherland, W. J., Therriault, T., Thuiller, W., Torchin, M., van der Putten, W., Vilà, M., Von Holle, B., Wallentinus, I., Wardle, D., Williamson, M., Wilson, J., Winter, M., Wolfe, L. M., Wright, J., Wonham, M., and Zabin, C. (2011) "Non-natives: 141 scientists object," *Nature* 475: 36.

Simon, K. S., and Townsend, C. R. (2003) "Impacts of freshwater invaders at different levels of ecologocial organisation, wtih emphaiss on salmonids and ecosystme consequences," *Freshwater Biology* 48: 982–994.

Singer, P. (1997) "Neither human nor natural: Ethics and feral animals," *Reproduction, Fertility and Development* 9: 157–162.

Stone, C. P., and Loope, L. L. (1987) "Reducing negative effects of introduced animals on native biotas in Hawaii: What is being done, what needs doing, and the role of national parks," *Environmental Conservation* 14: 245–258.

Tenter, A. M., Heckeroth, A. R., and Weiss, L. M. (2000) "*Toxoplasma gondii*: From animals to humans," *International Journal for Parasitology* 30: 1217–1258.

Wallach, A. D., Bekoff, M., Nelson, M. P., and Ramp, D. (2015) "Promoting predators and compassionate conservation," *Conservation Biology* 29: 1481–1484.

Wallach, A. D., and Ramp, D. (2015) "Let's give feral cats their citizenship," *The Conversation*, https://theconversation.com/lets-give-feral-cats-their-citizenship-45165.

Warburton, B., and Anderson, D. (2018) "Ecology, economics and ethics: The three Es required for the sustainable management of wild sentient species," in S. Sarkar and B. Minteer (eds.) *A Sustainable Philosophy—The Work of Bryan Norton* (pp. 237–252), Cham, Switzerland: Springer International Publishing.

Warburton, B., and Norton, B. G. (2009) "Towards a knowledge-based ethic for lethal control of nuisance wildlife," *Journal of Wildlife Management* 73: 158–164.

Warburton, B., Tompkins, D. M., Choquenot, D., and Cowan, P. (2012) "Minimising number killed in long-term vertebrate pest management programmes, and associated economic incentives," *Animal Welfare* 21: 141–149.

Witmer, G. W., Raymond-Whish, S., Moulton, R. S., Pyzyna, B. R., Calloway, E. M., Dyer, C. A., Mayer, L. P., and Hoyer, P. B. (2017) "Compromised fertility in free feeding of wild- caught Norway rats (*Rattus Norvegicus*) with a liquid bait containing 4-vinylcyclohexene diepoxide and triptolide," *Journal of Zoo and Wildlife Medicine* 48: 80–90.

Woinarski, J. C. Z., Burbidge, A. A., and Harrison, P. L. (2015) "Ongoing unraveling of a continental fauna: Decline and extinction of Australian mammals since European settlement," *Proceedings of the National Academy of Sciences* 112: 4531–4540.

Woinarski, J. C. Z., Murphy, B. P., Legge, S. M. S. M., Garnett, S. T. T., Lawes, M. J. J., Comer, S., Dickman, C. R., Doherty, T. S., Edwards, G., Nankivell, A., Paton, D., Palmer, R., and Woolley, L. A. A. (2017) "How many birds are killed by cats in Australia?" *Biological Conservation* 214: 76–87.

Woinarski, J. C. Z., Murphy, B. P., Palmer, R., Legge, S. M., Dickman, C. R., Doherty, T. S., Edwards, G., Nankivell, A., Read, J. L., and Stokeld, D. (2018) "How many reptiles are killed by cats in Australia?" *Wildlife Research* 45: 247–266.

World Organisation for Animal Health (2019) *Terrestrial Animal Health Code*, Paris: World Organisation for Animal Health.

Yeates, J. (2010) "What can pest management learn from laboratory animal ethics?" *Pest Management Science* 66: 231–237.

Ziembicki, M. R., Woinarski, J. C. Z., Webb, J. K., Vanderduys, E., Tuft, K., Smith, J., Ritchie, E. G., Reardon, T. B., Radford, I. J., Preece, N., Perry, J., Murphy, B. P., Legge, S. M., Leahy, L., Lawes, M. J., Kanowski, J., Johnson, C. N., James, A., Griffiths, A. D., Gillespie, G. R., Frank, a. S. K., Fisher, A., and Burbidge, A. A. (20150) "Stemming the tide: Progress towards resolving the causes of decline and implementing management responses for the disappearing mammal fauna of northern Australia," *Therya* 6: 169–225.

33

PROPERTY, REGULATION, AND ENDANGERED SPECIES CONSERVATION

Steven McMullen

One of the most important environmental challenges facing policy-makers is the preservation of animal populations and biodiversity. As a result of the growing human population, increased land use, and hunting/fishing, there has been a steady and alarming decrease in the populations of free-living animals and an increase in species extinctions. While documented rates of extinction vary greatly, careful estimates indicate that we are losing species at between 10 and 1,000 times the rate that extinctions would naturally occur. Moreover, while a few species have increased their population, many different kinds of animals are far less numerous as a result of human hunting and displacement.

While there is wide agreement about this problem, conservation strategies and policies can be far more contentious. This chapter examines the logic and effectiveness of two broad approaches to conserving endangered species. One approach is based on government regulation and limits and is favored by many environmental advocates. The second approach is based on assigning private ownership of animal populations and key habitats and is favored by a smaller but significant number of social scientists and policy-makers. Underlying these policy debates is a deep disagreement about why animal populations should be preserved and what progress would look like. Those that argue in anthropocentric terms—considering only the benefits that animal populations provide to humans—often favor the property approach. In contrast, those that argue in terms of the health of ecosystems and the well-being of animals tend to favor the more traditional regulation approach.

This divide is partially due to the fact that these two approaches conserve different things, and solve different problems. Commercial property-based approaches tend to work well with "tragedy of the commons" problems. The paradigmatic examples are fish stocking and the growing market for farmed North American bison. In contrast, the regulation-based approach is better suited for species endangered by habitat loss, pollution, or endangered species. These approaches have often been necessary to preserve local populations of endangered animals. The paradigmatic examples here include the revival of the North American populations of the peregrine falcon and the bald eagle, both of which were severely depleted by the use of insecticides containing DDT. What is often lost in policy debates is that animal welfare, autonomy, and ecosystem preservation may all be sacrificed when using the property-based approaches. For animal ethics scholars who are concerned about the lives of individual animals and human violence toward animals, this will often make property-based approaches unattractive.

Endangered Animals

The growth of human populations has consistently displaced animals, often leading to the extinction of large birds and mammals. While this process can be traced as far back as the Pleistocene epoch, the rate at which animals have become extinct has accelerated, with a majority of the extinctions of the last 2,000 years happening in since 1900 (Regenstein 1985). While there is significant debate about the numbers, there is broad agreement that the current rate of human-caused animal extinctions is at least 10 times the natural rate of extinction and possibly as high as 1000 times the natural rate (Chivian and Bernstein 2008; Pimm et al. 2014). Many experts have thus argued that species preservation and biodiversity should be a primary global environmental goal. Fortunately, we have sufficient evidence to be confident that species preservation is possible through active management of hunting/fishing and avoiding habitat destruction.

Estimating the rate of extinctions, and the degree of biodiversity loss, is complicated by the fact that scientists are still in the process of documenting existing species, particularly for insects. There are a variety of different estimates of current animal populations and undocumented species and therefore extinction rates. In the absence of regular and accurate documentation of species populations, scientists often model the rate at which species will become extinct as their habitat is diminished (Pearce 2015). To illustrate the difficulty, note that about 800 species extinctions have been documented over the last 400 years (about 2 species losses each year) but that models of species extinctions indicate losses substantially higher than this. The best estimates indicate that many of the species that are likely to go extinct in the near future are undocumented or understudied and thus difficult to count (Pimm et al. 2014).

Apart from extinction, the livelihood of nonhuman animals is substantially threatened. Recent models of animal populations illustrate the scope of the problem. Prior to human expansion, free-living mammals amounted accounted for 4% of total planetary biomass. Today, free-living nonhuman mammals account for less than 0.5% of total biomass, whereas humans and farmed animals account for 16% (Bar-On et al. 2018). This massive replacement of free-living mammals with farmed animals and human habitats has pushed many species of animals to the margins. Scientists estimate that at least one third of amphibian species, one eighth of bird species, one fifth of fish species, and one fifth of reptiles worldwide are threatened with extinction (Wake and Vredenburg 2008; McCallum 2007; Chivian and Bernstein 2008).

The nature of the conservation problem varies, but for the purposes of this chapter, it is necessary to outline two categories of human-caused species endangerment. The first is the classic tragedy of the commons (Hardin 1968), in which humans overhunt or overfish an animal to the point where the population cannot sustain itself. Free-living animals whose bodies are in direct demand by humans, either as food or for their bones, hoofs, horns, hair, or feathers, are particularly susceptible. The mechanics of a commons problem work as follows: (1) individual people each have access to the animal through hunting or fishing without limit or with inadequate limits, and (2) total demand for the animal is beyond what the population can replace. In this case, the benefits of overhunting or overfishing are individual (revenue from the animal body). The cost of overexploitation, by making the animals scarce, is shared by all people who may want to hunt or fish or is borne by the animal/ecosystem. Individuals can be "free riders" by overhunting or overfishing and allowing others to bear most of the costs. Often, depletion becomes inevitable, and in the race to get access to the animals, everyone has an incentive to hunt/fish beyond the sustainable rate.

While there are a number of different ways societies have avoided these commons problems, there is a long history of failure. Humans have hunted/fished many species to extinction. Two examples of species extinctions caused by this kind of problem will illustrate the scenario. First, Stellar's sea cow, a relative of the manatee, was discovered by Europeans in the 1700s in the North Pacific. The large slow animal was prized for food and for its blubber and skin and was hunted to extinction within 30 years of its discovery by Europeans. While there was no organized effort to hunt down the sea cows, the race by fisherman to find them eliminated the species very quickly. A second example is the North American passenger pigeon, once the most numerous bird species in North America and possibly in the world. Accounts of the flocks of pigeons that darkened the sky were common in the 18th and 19th centuries. The birds were easily hunted for food, however, and were often viewed as pests. Their survival, moreover, depended on large flocks, and after a dramatic decline in the late 1800s, their population was reduced to a few captive flocks, finally dying out in 1914 (Yeoman 2014). Both these species were hunted to extinction before the modern environmental movement started shaping conservation policy. Without a collective effort to monitor, manage, and limit hunting, there was no check on individual behavior.

The tragedy of the commons continues to endanger animals subject to hunting and fishing. Elephants and rhinoceroses killed for ivory/horns are common victims. Most species of fish that have become scarce in recent years have been overfished as a result of these kinds of incentives, in part because fish can be subject to a double-commons problem: there is free riding not only by individual fishing companies but also by countries. Limits on the number of fish that can be caught often need to be the subject of international agreements to be effective, and even these agreements might not be successful if countries place the short-term profitability of fishing companies above the long-term health of a species.

While commons problems are well-understood, the majority of endangered species are not and never were in demand by humans and were never subject to hunting or fishing. Instead, many animals are threatened as a negative side effect of some other human activity. This is the second broad type of human-caused species endangerment that is considered here. There are many different ways that human activity can threaten a population of animals. For example, the habitat of a species might become scarce as a result of agriculture or logging, or alternatively, a nonnative invasive species might be introduced to an ecosystem, killing off or crowding out a native species. These cases must be separated from the cases of overhunting or overfishing because the cause and solutions differ.

Two famously endangered North American bird species illustrate these non-commons problems. First, the California condor, famed for having the longest wingspan of any current living North American bird, has been brought to the brink of extinction in the 20th century. Three factors contributed to the birds' decline. First, the birds are scavengers and were often poisoned by lead bullets that remained in the animals they ate (Starin 2015; Church et al. 2006). Second, due to misconceptions, they were often hunted by farmers and ranchers, who believed that the birds killed the livestock that they would be found eating. Finally, the animals suffered from loss of habitat, the loss of eggs from DDT poisoning, and numerous other challenges. The story of the peregrine falcon is similar. The use of the insecticide DDT thinned the eggs of falcons, resulting in a dramatic decline in North America and Europe. Over much of the birds' North American range it remains locally extinct. While each species has its own complex history, these non-commons problems likely contribute to more species declines than do hunting and fishing. One attempt at aggregating different causes of extinction noted that, of those species whose extinction was well understood, only 27% were caused by overexploitation, whereas 34% of extinctions were due to habitat loss, and 39% due to invasive species (Van Kooten and Bulte 2000: 280). Protection policies must be different in each case.

Conservation Through Regulation

In order to preserve a species that is threatened or endangered, there are two kinds of intervention that are often required. First, the human action causing the species decline needs to diminish or stop entirely. This could mean protecting existing habitat, eliminating a practice that has a side effect of hurting the animal, or limiting hunting/fishing. Second, if the species has fallen below the sustainable population level, active steps must be taken to aid in the breeding and reestablishment of animals. Both actions can be taken by public or private entities, and both can be very costly but in different ways. The two main policy models for preserving species—regulation and property approaches—differ substantially in how the costs and benefits of conservation are apportioned.

First, in a traditional regulatory approach, as is exemplified by the Endangered Species Act in the United States, a government enacts legal protections for the endangered animal. This protection can take a number of forms. If the animal is endangered through overexploitation, then killing the animal may become illegal or severely curtailed. For example, wolf hunting was made illegal in the United States outside of Alaska and has only resumed in a handful of states as the population of wolves has recovered. If a person in the United States intentionally kills an animal on the US endangered species list, the punishment can include fines up to $50,000 and/or imprisonment for up to a year. Similarly, if the habitat of an animal is threatened, the law may limit the ability of landowners to develop or alter their property. This has caused some conflict in the United States, particularly around wetlands and forests that lose market value when an endangered species is discovered nearby. Finally, if a species is endangered through the use of a chemical, an invasive species, or some other practice, the law can impose restrictions on activities that cause the problem.

Some of the most famous global animal protections take the form of this kind of regulation. There are strict limitations on the hunting of rhinoceroses, some kinds of elephants, all kinds of tigers, and the Migratory Bird Treaty Act and related treaties protect birds throughout North America in cooperation with other countries. In many cases, laws also include bans on the trade of endangered animals, although illegal trade in bear and tiger body parts, shark fins, rhino horns, and ivory remains an ongoing concern. In many cases, the black-market prices for these illicit goods are high enough that poachers can afford to use high-tech methods to poach and ship goods.

While these kinds of interventions are often sufficient to slow the decline of a population, for critically endangered species recovery often requires additional costly interventions. The whooping crane, brought to the brink of extinction, has recovered somewhat but only with the extensive and expensive efforts of academics, conservationists, and governments over a period of 50 years. The whooping crane is a particularly charismatic species, and so it has been easy to garner public support for intensive investment. Other bird species, such as the bald eagle and peregrine falcon, have benefited from similar public support. Less charismatic and similarly endangered species, however, are less likely to inspire recovery funding. While the US Endangered Species Act requires the creation of a recovery plan, there is often insufficient government funds to even approve recovery plans for many species and limited overall investment in reestablishment (Brown and Shogren 1998; Heinzerling 2011). Recent evidence indicates that 80% of the public funds allocated to species protection in the United States are spent on just 5% of species, particularly those that have great popular appeal or are game animals (Evans et al. 2016; Gold 2017). The endangered species act, overall, is an ambitiously conceived legislative program, but the strict monetary constraints limit its implementation.

While regulatory approaches to species conservation are the norm, there are many critics of this approach. First, private landowners often bear a disproportionate amount of the financial burden of protecting endangered species. The costs can be unexpected and steep if an endangered species is discovered on private land in advance of development or logging. Many scholars and activists have argued that the incentives created by the laws are perverse. If a nearby species is likely to be listed as endangered, landowners have preemptively destroyed habitat or harvested timber. Moreover, if landowners find an endangered species on their land, there is a high incentive to illegally "shoot, shovel, and shut up" rather than face the economic cost of regulatory land-use limits (Adler 2008a; Evans et al. 2016; Brown and Shogren 1998; Adler 2008b).

Estimates of the cost of species protections to private parties vary. Researchers investigating the cost of saving the spotted owl, which inhabits large areas of valuable forests in the northwestern United States, found that increasing the survival odds of the owl to 91% (which would involve preserving 1,900 pairs of owls) would cost $33 billion (Montgomery et al. 1994). In other cases, researchers estimating the economic impact of conservation efforts on other regions with critical habitats found extremely small costs (Brown and Shogren 1998; Shogren 1997). Because these costs are sometimes high and the government does not compensate private parties for losses from the regulation, there has been a regular stream of ultimately unsuccessful lawsuits challenging the law on constitutional grounds (Mank 2007).

There is substantial disagreement about how successful this regulatory approach has been at protecting species. The broad enactment of these regulations in the United States has resulted in very few species recoveries. Those that have recovered, such as the grizzly bear, wolf, bald eagle, and peregrine falcon, have often been species with high levels of investment. While few have gone extinct after being listed, the vast majority have remained on the list and have seen little growth or have continued decline (Evans et al. 2016). At the same time, even these results might be a significant success. One evaluation of the endangered species act estimated that 227 species in North America would have gone extinct if these regulations had not been passed and enforced (Scott et al. 2005).

Conservation Through Ownership and Markets

In contrast to the traditional regulatory approach to species preservation, a number of scholars have argued, instead, for market and property-based mechanisms to encourage private conservation efforts. A famous example of this approach looks back to the 19th-century preservation of the North American Bison. Once numbering in the tens of millions on the great plains, bison were hunted to the brink of extinction in the 1880s. While the US government attempted to save the animal, their efforts were ultimately unsuccessful. Instead, a handful of ranchers saw an opportunity to farm the species and preserved some small herds (Anderson and Hill 2004; Hill 2012). These profit-seeking individuals succeeded in breeding the animals. Bison populations have since recovered from a few hundred animals in the late 1800s to more than 500,000 today, most of which are farmed on private land.

The broad goal of property-based conservation approaches is to give private parties an incentive to preserve the long-term viability of a threatened resource. The most common way to do this is either to give them private ownership over a population of animals or to give them a tradeable "right" to a portion of the free-living population. Doing so will create a functioning market for the right to breed or harvest the animals. If the ownership is secure, the owners will profit from careful preservation and limited use of the resource. In short, the free-rider problem that plagues common resources is reversed or eliminated. Where previously individuals had an incentive to overexploit a population or ignore broader environmental care, property rights reverse the incentives, giving owners a stake in the future viability of the ecosystem and the population of animals.

The most successful examples of this approach have been in the regulation of fisheries. Overfishing has been a chronic problem worldwide. Scholars have estimated that current fish populations are, at least, an order of magnitude smaller than they were before the introduction of industrial fishing methods (Myers and Worm 2003; Jackson et al. 2001). In the face of these declines, many states have imposed strict limits on how many fish can be removed, but the tragedy of the common remains. Fishing in some waters is poorly regulated, and fishing limits often require the agreement of multiple governments. Even the United States and Canada, countries with generally good relations, have found it difficult to agree on the shares of Pacific salmon that each country will take each year (*The Economist* 1997). More important, simple regulations will often do a poor job protecting fish populations (Grafton et al. 2006; Costello et al. 2008). New approaches to sustainably managing fisheries often take the form of individual transferrable quotas or catch shares. These approaches give fishing companies a tradable right to remove a certain share of the allowed catch from a given area. Rather than racing against other companies to catch a larger share of a quota, companies in these new arrangements have an incentive to manage the fishery sustainably in order to preserve the future value of their share. The results of these property- or rights-based approaches have been quite positive. Fisheries have increased in value, populations of fish have recovered, and companies often have incentives to more take more care to preserve the broader ecosystem (Arnason 2012; Grafton et al. 2006; Adler 2012).

The reasons for the success of these property-based approaches go beyond the elimination of the tragedy of the commons. Often, the political conflict between regulators and owners can be eliminated. Rather than fighting for less regulation, fishing companies that own a catch share will instead push for tighter limits to preserve the long-term value of their stake (Adler 2012). Additionally, individual owners with the incentive to preserve a population of animals, often bring private capital to the conservation project. Because the success of the population is central to their commercial success, owners will often invest their time and money in the protection of the habitat and the health of the species. Given that public conservation projects tend to be chronically underfunded, the infusion of private money into local conservation efforts is often necessary.

Most important, perhaps, ownership creates the opportunity for positive innovation and competition. Owners will work to find more efficient and effective ways to prevent poaching, improve animal populations, and make more targeted interventions in the ecosystem. This often requires local knowledge of a particular place and species, knowledge that is hard to collect and use at the state level but can be key to the success of a local ecosystem. The creation of a market in the property rights, moreover, means that if another company develops a better way to preserve the species or manage the ecosystem, they can purchase the right to do so. Producers who do not care to invest in a fishery or ecosystem or use outdated harvesting methods will not be able to afford to continue in business.

These property-rights based approaches are likely to be preferable to public regulatory approaches when the public investment in conservation is low or when the public institutions are not able to consistently enforce their policies. This is often the case around the world and particularly where poaching is a substantial problem. It is for this reason that some have argued that legal hunting of endangered rhinoceroses, elephants, and lions might be an essential part of the conserving these species. Allowing trophy-hunting gives landowners a reason to open their land to the animals, dramatically increasing the available habitat (Goldman 2014; Leader-Williams et al. 2005; Leader-Williams et al. 2001). Partly as a result of controlled hunting, populations of white rhinoceroses have recovered substantially since the 1970s (Paterniti 2017). The banning of hunting of elephants in Kenya, moreover, preceded a steep decline in the county's elephant population (Sugg and Kreuter 1994). Where hunting is allowed, landowners have an incentive to help control poaching in order to preserve the population for paying customers. Local people that may view large animals as pests to be eliminated are much more likely to live with the animals if they provide a source of income, either from hunters or tourists. The customers, moreover, may help fund, and provide pressure for conservation efforts (Lindsey et al. 2006).

There is no consensus, however, that hunting is the best way to conserve species. The money from hunters does not always support conservation efforts, and without the right institutional setup, hunting can be more of a problem than a solution. Furthermore, the market for trophies can create perverse incentives, encouraging the "farming" of lions, raised in poor conditions, only to be released for the hunter to kill (Paterniti 2017). This outcome is similar to the way animals are treated in agricultural markets and points to one of the most significant weaknesses of property-based approaches to animal conservation.

Markets for Animals

While creating markets for animal ownership can sometimes align the incentives of owners with conservation goals, this approach has only worked well for those species that have a market value. It is notable that every case in which property-based conservation approaches have worked, there is a tragedy of the commons problem of overexploitation. This is true of excessive fishing, the hunting of large mammals and birds, and hunting/poaching of rhinos and elephants for horns/tusks. Without a strong market demand for the animals, the incentives created by property rights can actually do more harm than good. Because the majority of endangered animals are not endangered as a result of overexploitation but are instead threatened by habitat loss or invasive species, there is a hard limit on the number of cases in which these policy approaches can be used. Even if, hypothetically, the US government had issued hunting rights for peregrine falcons during their DDT-caused decline, there would not have been enough demand for the birds to warrant a commercial movement to eliminate the use of DDT as a pesticide. It is very unlikely, moreover, that hunting demand would have been sufficient to fund the recovery programs that reintroduced the birds. In this case, the regulatory approach was the only option.

Even more important, creating a market for animals and habitats can backfire if the market value of the animal population declines. By one historical account, even if the hunting of the American bison had not been supported by the US government at the time and even if the land on the plains had been parceled out and assigned strict ownership, the bison would likely have been wiped out. Given the technology of the time, farming bison was difficult and expensive, whereas farming cattle was cheap. Ranchers would have had a strong incentive to populate their land with cattle instead of bison on the basis of market incentives (Anderson and Hill 2004; Hill 2012). Similarly, if the lives of the remaining spotted owls had been auctioned off, rather than having their habitat protected through regulation, lumber companies may have quickly purchased all the rights and killed the owls, preferring to profit from logging the forests where the owls lived.

Markets are not, in general, good at protecting biodiversity. Wherever markets for animals dominate, the degree of biodiversity plummets. The most profitable and in-demand species will quickly replace native species, because of the incentives created by the market. Thus, when lakes or game preserves are stocked with animals for hunting and fishing, the few most popular game species dominate, often replacing native species that are in less demand (McMullen and Molling 2015; Wilson 2010). In agriculture, where the lives of billions of animals are determined by market forces, the amount of genetic diversity is minimal. In short, allowing markets to put a price on an animal can incentivize conservation behavior, but it can also incentivize the elimination of animal populations if there is a better avenue for profit.

Similarly, the record of property-regimes protecting ecosystems is mixed. There is some evidence that fisheries regulated through catch-shares are better preserved, and some kinds of property-based institutions can incentivize reduced bycatch (Miller and Deacon 2017). At the same time, fish farming practices, which are characterized by strong property rights, often depend heavily on wild-caught fish. Aquaculture also imposes a heavy environmental burden (Naylor et al. 2000). Similarly, animal farming has a steep environmental cost (The Pew Environment Group 2013; Steinfeld et al. 2006). It is only in a few cases that ownership of animals and the environment predictably results in better conservation outcomes.

Pushing the logic of markets one step further, it is important to note that markets do not reliably protect animals as much as they protect the marketable characteristics of animals. Those animals that have been the longest victims and beneficiaries of human ownership and market incentives are farmed animals. Predictably, chickens, pigs, and cows have been carefully genetically selected to produce meat, eggs, and milk. The incentives created by competitive markets will push farmers to adopt the production equipment, genetic selection, and slaughter technique that maximizes profit, not the ones that are good for the animal (McMullen 2015; Lusk and Norwood 2011). Sometimes animals benefit from these incentives, but often the result is harm to animals. The chickens that have been carefully bred to produce meat grow unnaturally fast and suffer substantial pain in movement by the time they are killed (Norwood and Lusk 2011). If the commercial care of white rhinos, along with the harvesting of their horn, was protected solely by the market incentives, it may well be that the well-being of the species would quickly be pushed aside in favor of breeding to find animals that would regrow their horns more quickly, eat less, and move more slowly.

Why Save Endangered Species?

The policy debate about whether to protect species by assigning property rights or through regulation reveals a deep disagreement about the goals of conservation policy. The values ungirding the creation of markets for animals are deeply anthropocentric. While individual policy-makers may introduce property rights into fishery regulation out of a desire to preserve animals, the market mechanisms efficiently communicate human preferences only. It is humans who participate in markets as buyers and producers, and so their preferences and welfare drive the prices and incentives. More important, the preferences of individual humans drive markets for animals on the basis of their willingness to pay for the animal bodies or the goods they produce. Human preferences for animal well-being are rarely incorporated into the market, even when they exist (McMullen 2015). Pushing even further, there is good evidence that the act of participating in markets shapes and filters people's preferences so thatthey are less likely to reflect ethical concern for animals (McMullen 2016; Sagoff 2007b; Falk and Szech 2013).

For most of the proponents of market-based environmental policy[1] this anthropocentric frame extends into their theory. The problem that conservation efforts are trying to solve, in economic terms, is a failure of the market to efficiently maximize human preference satisfaction, as when a commons problem causes a market to collapse from overexploitation. In these terms, an animal that has no market value, especially if people are not willing to pay money to save the species, is not worth saving (De Allessi 1998; Pennington 2005). While this may sound extreme to some academic ears, it is not unusual for public policy arguments about animals and the environment to be framed almost entirely in terms of costs and benefits to humans. Similarly, in legal disputes, liability for harm to animals is often measured in terms of the lost market value of the animal and paid to the animals' owner (Francione 2001; Kelch 1998).

Environmental ethics scholars are likely to reject this anthropocentric framing, instead desiring to preserve species on the basis of the intrinsic value of ecosystems as a community of interdependent living beings (Nash 1989; McCauley 2006). In this framework, endangered species are a symbolic piece of a larger environmental crisis and may present the prospect of an irreversible loss of a key part of a morally valuable community. For these scholars, species conservation through property and markets is often viewed with suspicion for a host of reasons. First, there is concern that market incentives tend to atomize and separate people and animals from their larger environmental connections (Meyer 2009; Daly and Cobb 1994). Second, and most fundamentally, environmental scholars are likely to see economic logic as the underlying problem causing environmental crises and thus be skeptical that moral concerns can be adequately protected via economic incentives (Sagoff 2007a, 1992; Barry 1999).

While animal ethics scholars have not given as much attention to endangered species as have environmental ethicists,[2] many of the concerns will be shared. Where environmental scholars will often focus on the value of an ecosystem, animal ethics literature has instead focused on the value of individual animals. The accounts of the intrinsic value of animals differ, with some placing value of the experience of pain or pleasure (Singer 2009) and others valuing each animal as moral subjects and holder of rights (Regan 2005; Francione 1995). The broad thrust of work in animal ethics, however, has focused on pushing the boundaries of moral consideration beyond humans and rejecting anthropocentric thinking.

It is common throughout the animal ethics literature, moreover, to identify human ownership of animals as part of a larger system of legal oppression that enables exploitation. This property-based approach, for most ethicists, will appear contradictory. Such an approach places the animals more fully into an economic system where they are valued precisely because of human's desire to do them violence. Even those animal ethics scholars that do not call for the abolition of ownership would be quick to point out the degree to which human ownership can disregard the legitimate interests of other animals (Sunstein 2001; McMullen 2016; Cochrane 2012). There is a significant ethical problem with market-based conservation approaches that explicitly encourage harm to animals, even if these approaches aim to conserve the species.

Conclusion

There is broad agreement that humans have caused a rapid increase in the rate of species extinctions, and the dramatic reduction in the population and range of many species, even if they are not categorized as "endangered." There is also agreement that biodiversity is valuable and should be protected through public policy. The specifics of conservation policy, however, are subject to a significant amount of debate. Traditional regulatory policies, which place strict limits on human activities that might directly or indirectly harm endangered species, fit much more neatly with the ethical frameworks common among environmental and animal ethicists. New market-based approaches, which assign property rights to animals and environmental goods, in contrast, fit more closely with anthropocentric environmental goals and with economic thinking.

The two approaches are often contrasted, but for many endangered species, the regulatory approach is the only option, and in other cases, the approaches are creatively combined. Property rights can only be effective, however, if there are markets for animals that can be mobilized to fund and motivate conservation efforts. Those animals that are threatened by overexploitation, then, are candidates for this kind of approach. Most endangered species, however, are not threatened by overhunting or overfishing, and most will not garner wide investment because of their popularity. For these species, a broad public commitment and investment must come from outside the market.[3]

Additionally, the kind of conservation that occurs through these property-based approaches does not align well with the goals of most environmental or animal advocates. These markets create an incentive to more efficiently exploit animals for use by humans. It is only in rare cases that animal ownership results in the pursuit of animals' interests. In most cases, market competition and ownership only ensure that, on the margin, animal interests are ignored (McMullen 2015).

This does not mean, however, that market-based approaches to conservation should not be pursued. In some cases, like the regulation of hunting and fishing activities, these approaches can have clear administrative benefits and better results for animals and the environment. When facing a severely underfunded public conservation program, and the prospect of the collapse of communities of animals or whole species, it may be that auctioning off limited fishing and hunting rights is the best policy approach.

Notes

1. This movement of scholars and policy-makers often works under the label of "free market environmentalism." See, for example, Anderson and Leal's classic text by that name (Anderson and Leal 2001).
2. There are some notable exceptions, such as Regenstein (1985), and Bekoff (2010).
3. It might be possible even in these cases to create a market in which participants entrepreneurially bid for contracts to save species. This could create market-like incentives to innovate toward conservation goals. This is not, however, the kind of market solution usually discussed in this literature.

Bibliography

Adler, J. (2008a) "Money or nothing: The adverse environmental consequences of uncompensated land use control," *Boston College Law Review* 49(2): 301.

Adler, J. (2008b) "Perverse incentives and the endangered species act," *Resources for the Future Policy Commentary* (blog), August 4, www.rff.org/blog/2008/perverse-incentives-and-endangered-species-act.

Adler, J. (2012) "Property rights and fishery conservation," *The Atlantic*, May 24, www.theatlantic.com/business/archive/2012/05/property-rights-and-fishery-conservation/257604/.

Anderson, T. L., and Hill, P. J. (2004) *The Not So Wild, Wild West: Property Rights on the Frontier*, 1st ed., Stanford, CA: Stanford Economics and Finance.

Anderson, T. L., and Leal, D. R. (2001) *Free Market Environmentalism*, New York, NY: Palgrave Macmillan.

Arnason, R. (2012) "Property rights in fisheries: How much can individual transferable quotas accomplish?" *Review of Environmental Economics and Policy* 6(2): 217–236. https://doi.org/10.1093/reep/res011

Bar-On, Y. M., Phillips, R., and Milo, R. (2018) "The biomass distribution on earth," *Proceedings of the National Academy of Sciences*, May: 201711842. https://doi.org/10.1073/pnas.1711842115

Barry, J. (1999) *Rethinking Green Politics: Nature, Virtue and Progress*, London: Sage Publications. https://doi.org/10.4135/9781446279311.

Bekoff, M. (ed.) (2010) "Wild animals and ethical perspectives," in *Encyclopedia of Animal Rights and Animal Welfare*, Santa Barbara, CA: ABC-CLIO.

Brown, G. M., and Shogren, J. F. (1998) "Economics of the endangered species act," *Journal of Economic Perspectives* 12(3): 3–20. https://doi.org/10.1257/jep.12.3.3

Chivian, E., and Bernstein, A. (eds.) (2008) *Sustaining Life: How Human Health Depends on Biodiversity*, illustrated ed., Oxford and New York: Oxford University Press.

Church, M. E., Gwiazda, R., Risebrough, R. W., Sorenson, K., Page Chamberlain, C., Farry, S., Heinrich, W., Rideout, B. A., and Smith, D. R. (2006) "Ammunition is the principal source of lead accumulated by California condors re-introduced to the wild," *Environmental Science & Technology* 40(19): 6143–6150. https://doi.org/10.1021/es060765s

Cochrane, A. (2012) *Animal Rights Without Liberation: Applied Ethics and Human Obligations*, New York, NY: Columbia University Press.

Costello, C., Gaines, S. D., and Lynham, J. (2008) "Can catch shares prevent fisheries collapse?" *Science* 321(5896): 1678–1681. https://doi.org/10.1126/science.1159478

Daly, H. E., and Cobb, J. B. (1994) *For the Common Good: Redirecting the Economy Toward Community, the Environment, and a Sustainable Future*, 2nd updated ed., Boston, MA: Beacon Press.

De Allessi, L. (1998) "Property rights as the basis for free-market environmentalism," in P. J. Hill and R. E. Meiners (eds.) *Who Owns the Environment?* (pp. 1–36), Lanham, MD: Rowman & Littlefield Publishers.

The Economist (1997) "Salmon war on two fronts," June 26, www.economist.com/node/91873.

Evans, D. M., Che-Castaldo, J. P., Crouse, D., Davis, F., and Epanchin-Niell, R. (2016) *Species Recovery in the United States: Increasing the Effectiveness of the Endangered Species Act*, 20. Issues in Ecology, Washington, DC: Ecological Society of America, www.esa.org/esa/wp-content/uploads/2016/01/Issue20.pdf.

Falk, A., and Szech, N. (2013) "Morals and markets," *Science* 340(6133): 707–711. https://doi.org/10.1126/science.1231566

Francione, G. L. (1995) *Animals, Property, and the Law*, Philadelphia, PA: Temple University Press.

Francione, G. L. (2001) *Introduction to Animal Rights: Your Child or the Dog?* 1st ed., Philadelphia, PA: Temple University Press.

Gold, S. (2017) "The endangered species act won't save animals: It's not designed to," *Slate*, May 30, 2017, www.slate.com/articles/health_and_science/science/2017/05/the_endangered_species_act_wasn_t_meant_to_save_the_animals.html.

Goldman, J. (2014) "Can trophy hunting actually help conservation?" *Conservation Magazine*, January 15.

Grafton, R. Q., Arnason, R., Bjørndal, T., Campbell, D., Campbell, H. F., Clark, C. W., Connor, R. et al. (2006) "Incentive-based approaches to sustainable fisheries," *Canadian Journal of Fisheries and Aquatic Sciences* 63(3): 699–710. https://doi.org/10.1139/f05-247

Hardin, G. (1968) "The tragedy of the commons," *Science* 162(3859): 1243–1248. https://doi.org/10.1126/science.162.3859.1243

Heinzerling, L. (2011) "The law of big problems," *The New York Times*, April 24, online ed., www.nytimes.com/roomfordebate/2011/04/20/saving-endangered-species-as-the-climate-changes/the-law-of-big-problems.

Hill, P. J. (2012) "The non-tragedy of the bison commons," *The PERCOLATOR* (blog). January 2, www.perc.org/blog/non-tragedy-bison-commons.

Jackson, J. B. C., Kirby, M. X., Berger, W. H., Bjorndal, K. A., Botsford, L. W., Bourque, B. J., Bradbury, R. H. et al. (2001) "Historical overfishing and the recent collapse of coastal ecosystems," *Science* 293(5530): 629–637. https://doi.org/10.1126/science.1059199

Kelch, T. (1998) "Toward a non-property status for animals," *NYU Environmental Law Journal* 6: 531.

Leader-Williams, N., Milledge, S., Adcock, K., Brooks, M., Conway, A., Knight, M., Mainka, S., Martin, E. B., and Teferi, T. (2005) "Trophy hunting of black rhino diceros bicornis: Proposals to ensure its future sustainability," *Journal of International Wildlife Law & Policy* 8(1): 1–11. https://doi.org/10.1080/13880290590913705

Leader-Williams, N., Smith, R. J., and Walpole, M. J. (2001) "Elephant hunting and conservation," *Science* 293(5538): 2203–2204. https://doi.org/10.1126/science.293.5538.2203b

Lindsey, P. A., Alexander, R., Frank, L. G., Mathieson, A., and Romañach, S. S. (2006) "Potential of trophy hunting to create incentives for wildlife conservation in Africa where alternative wildlife-based land uses may not be viable," *Animal Conservation* 9(3): 283–291. https://doi.org/10.1111/j.1469-1795.2006.00034.x

Lusk, J. L., and Norwood, F. B. (2011) "Animal welfare economics," *Applied Economic Perspectives and Policy*, November. https://doi.org/10.1093/aepp/ppr036

Mank, B. C. (2007) "After Gonzales v. Raich: Is the endangered species act constitutional under the commerce clause?" *University of Colorado Law Review* 78: 375–465.

McCallum, M. L. (2007) "Amphibian decline or extinction? Current declines dwarf background extinction rate," *Journal of Herpetology* 41(3): 483–491. https://doi.org/10.1670/0022-1511(2007)41[483:ADOECD]2.0.CO;2

McCauley, D. J. (2006) "Selling out on nature," *Nature* 443(7107): 27–28. https://doi.org/10.1038/443027a

McMullen, S. (2015) "Is capitalism to blame? Animal lives in the marketplace," *Journal of Animal Ethics* 5(2): 126–134.

McMullen, S. (2016) *Animals and the Economy* (pp. 21–40), London: Palgrave Macmillan, Palgrave Macmillan Animal Ethics Series.

McMullen, S., and Molling, D. (2015) "Environmental ethics, economics, and property law," in M. D. White (ed.) *Law and Social Economics: Essays in Ethical Values for Theory, Practice, and Policy*, New York, NY: Palgrave Macmillan.

Meyer, J. M. (2009) "The concept of private property and the limits of the environmental imagination," *Political Theory* 37(1): 99–127. https://doi.org/10.1177/0090591708326644

Miller, S. J., and Deacon, R. (2017) *Mobilizing Markets to Reduce Bycatch in Marine Fisheries*, 57. PERC Policy Series, Bozeman, MT: Property and Environment Research Center, www.perc.org/2017/04/20/mobilizing-markets-to-reduce-bycatch-in-marine-fisheries/.

Montgomery, C. A., Brown, Jr. G. M., and Darius, M. A. (1994) "The marginal cost of species preservation: The Northern spotted owl," *Journal of Environmental Economics and Management* 26(2): 111–128. https://doi.org/10.1006/jeem.1994.1007

Myers, R. A., and Worm, B. (2003) "Rapid worldwide depletion of predatory fish communities," *Nature* 423(6937): 280–283. https://doi.org/10.1038/nature01610

Nash, R. F. (1989) *The Rights of Nature: A History of Environmental Ethics*, Madison, WI: University of Wisconsin Press.

Naylor, R. L., Goldburg, R. J., Primavera, J. H., Kautsky, N., Beveridge, M. C. M., Clay, J., Folke, C., Lubchenco, J., Mooney, H., and Troell, M. (2000) "Effect of aquaculture on world fish supplies," *Nature* 405(6790): 1017–1024. https://doi.org/10.1038/35016500

Norwood, F. B., and Lusk, J. L. (2011) *Compassion, by the Pound: The Economics of Farm Animal Welfare*, 1st ed., New York, NY: Oxford University Press.

Paterniti, M. (2017) "Should we kill animals to save them?" *National Geographic*, September 20, www.nationalgeographic.com/magazine/2017/10/trophy-hunting-killing-saving-animals/.

Pearce, F. (2015) "Global extinction rates: Why do estimates vary so wildly?" *Yale E360*, August, https://e360.yale.edu/features/global_extinction_rates_why_do_estimates_vary_so_wildly.

Pennington, M. (2005) "Liberty, markets, and environmental values: A Hayekian defense of free-market environmentalism," *The Independent Review* 10(1): 39–57.

The Pew Environment Group (2013) *Big Chicken: Pollution and Industrial Poultry Production in America*, Washington, DC: The Pew Charitable Trusts, www.pewtrusts.org/~/media/legacy/uploadedfiles/peg/publications/report/PEGBigChickenJuly2011pdf.pdf.

Pimm, S. L., Jenkins, C. N., Abell, R., Brooks, T. M., Gittleman, J. L., Joppa, L. N., Raven, P. H., Roberts, C. M., and Sexton, J. O. (2014) "The biodiversity of species and their rates of extinction, distribution, and protection," *Science* 344(6187): 1246752. https://doi.org/10.1126/science.1246752

Regan, T. (2005) *Empty Cages: Facing the Challenge of Animal Rights*, Lanham, MD: Rowman & Littlefield Publishers.

Regenstein, L. (1985) "Animal rights, endangered species and human survival," in P. Singer (ed.) *In Defense of Animals*, 1st ed. New York, NY: Basil Blackwell.

Sagoff, M. (1992) "Free-market versus libertarian environmentalism," *Critical Review* 6(2–3): 211–230.

Sagoff, M. (2007a) *Economic Theory and Environmental Law*, eds. David Campbell and Robert Lee, International Library of Environmental Law and Policy Series, Aldershot, UK and Burlington, VT: Ashgate.

Sagoff, M. (2007b) *The Economy of the Earth: Philosophy, Law, and the Environment*, 2nd ed., Cambridge and New York: Cambridge University Press.

Scott, J. M., Goble, D. D., Svancara, L. K., and Pidgorna, A. (2005) "By the numbers," in D. D. Goble, J. Michael Scott, and F. W. Davis (eds.) *The Endangered Species Act at Thirty: Vol. 1: Renewing the Conservation Promise* (pp. 16–35), None ed., Vol. 1. Washington, DC: Island Press.

Shogren, J. F. (1997) "Economics and the endangered species act," *Endangered Species Update* 14(1–2), www.umich.edu/~esupdate/library/97.01-02/contents.html.

Singer, P. (2009) *Animal Liberation: The Definitive Classic of the Animal Movement*, New York: Harper Perennial.

Starin, D. (2015) "Condors or lead ammunition? We can't have both," *The Ecologist*, January 21, /2015/jan/21/condors-or-lead-ammunition-we-cant-have-both.

Steinfeld, H., Gerber, P., Wassenaar, T., Castel, V., Rosales, M., and de Haan, C. (2006) *Livestock's Long Shadow*, Rome, Italy: Food and Agriculture Organization of the United Nations, www.fao.org/docrep/010/a0701e/a0701e00.HTM.

Sugg, I. C., and Kreuter, U. P. (1994) *Elephants and Ivory: Lessons from the Trade Ban*, London: IEA Environment Unit.

Sunstein, C. R. (2001) "Slaughterhouse jive," *The New Republic*, January 29, 2001, www.newrepublic.com/article/books-and-arts/slaughterhouse-jive.

Van Kooten, G. C., and Bulte, E. H. (2000) *The Economics of Nature: Managing Biological Assets*, 1st ed., Malden, MA: Wiley-Blackwell.

Wake, D. B., and Vredenburg, V. T. (2008) "Are we in the midst of the sixth mass extinction? A view from the world of amphibians," *Proceedings of the National Academy of Sciences*, August. https://doi.org/10.1073/pnas.0801921105

Wilson, E. O. (2010) *The Diversity of Life*, Cambridge, MA: Belknap Press of Harvard University Press.

Yeoman, B. (2014) "Why the passenger pigeon went extinct," *Audobon Magazine*, April 17, www.audubon.org/magazine/may-june-2014/why-passenger-pigeon-went-extinct.

34

THE *LAISSEZ-FAIRE* VIEW

Why We're Not Normally Required to Assist Wild Animals

Clare Palmer

Introduction

Life for wild animals in Yellowstone National Park can be tough. Take bighorn sheep, for example. Bighorn sheep live in small populations all over the park, and they are highly vulnerable to outbreaks of serious disease. A pinkeye epidemic in 1981–1982 had significant impacts on bighorn sheep populations in Yellowstone, reducing them by 60% (National Park Service, n.d., b). Pinkeye is not itself fatal to bighorn sheep, but it significantly impairs their vision. Since they spend their lives clambering around precipices in the high, steep mountains, not being able to see means that they end up falling, and injuries sustained during falls are likely, sooner or later, to kill them.

After the pinkeye epidemic, populations of bighorn sheep took a long time to recover, and, in 2012, another disease began to manifest itself in bighorn populations: pneumonia. By 2015, pneumonia could be found in bighorn sheep throughout the park, as researchers at the Bighorn Sheep Disease Research Consortium (n.d.) found out:

> When a bighorn sheep population is initially infected, often as many as a third, and sometimes up to 90%, of the herd may die of pneumonia. Most survivors are apparently immune, but their lambs are not and usually die before weaning.

While bighorn sheep can live for more than a decade, diseases such as pinkeye and pneumonia cut their lives short. Sometimes they don't even make it to adulthood. And dying can be a long, painful process—of starvation, for instance, if a broken limb from a fall due to pinkeye prevents an animal from reaching food. And, it's worth noting, there's nothing special about bighorn sheep in this regard. Most wild animals don't die from old age. They are torn apart by predators, burned up in wildfires, frozen, drowned, starved, run over, or infected with deadly diseases. And leaving aside how they die, during their lives they can suffer from intestinal parasites, skin lesions, tick and other insect infestations, frequent hunger and thirst, overheating and overcooling, and episodes of utter exhaustion and extreme terror. That's not to say there aren't some good things too: times when animals are not hungry (perhaps having just eaten another animal!) not sick and not threatened by cold or extreme weather. But still, there's clearly a lot of pain and suffering in wild animal lives (see Tomasik 2015 for a fuller account.) Indeed, some wild animal lives may be "not worth living," understood to mean that suffering and negative experiences significantly outweigh any positive experiences over the course of an animal's life.

In the human case, we normally think such suffering is bad and that, if we can, we should do something to prevent or to relieve it. If a wildfire were threatening a town and we were in a position to rescue some of those threatened by it, we'd think it our duty to do so. If we saw a stranger being chased by an angry grizzly bear and we were carrying bear spray and could use it, we'd think we should rip the bear spray open as fast as we could. What's more, most people think we owe something like this to our companion animals, too. If a pet cat had what looked like a painful disease, most owners would think they had a responsibility to take it to a vet. During Hurricane Katrina, many owners refused to be rescued because the rescuers would not take their pets too; thousands of animals, whose owners reluctantly left them, starved or drowned. Public outcry about this led, in 2006, to the US Pets Evacuation and Transportation Standards Act (PETS Act), a bipartisan initiative that among other things "authorizes FEMA [the Federal Emergency Management Agency] to provide rescue, care, shelter, and essential needs for individuals with household pets and service animals, and to the household pets and animals themselves following a major disaster or emergency" (American Veterinary Medical Association, n.d.). While the chief motivation for this legislation may have been preventing *people* dying because they resist separation from their pets, it does seem to be widely agreed that we should "provide rescue, care, shelter and essential needs" for companion animals, if we can.

However, this view doesn't, in practice at least, extend to cover our relations with wild animals. We are undoubtedly surrounded by wild animal pain all the time: in our yards, in local parks, on the streets; where the hawk catches the mouse or the sudden spring freeze kills the nestling chicks. We watch wildlife documentaries in which animals are disemboweled and dismembered, starved by droughts and drowned in floods. But even these episodes of suffering don't generally evoke the kinds of moral response or practical action that drowning dogs or starving cats frequently do. There's no general acceptance of the idea that we should "provide rescue, care, shelter, and essential needs" for *wild* animals. Instead, many people appear to adopt a *laissez-faire* approach to wild animal pain and suffering; we allow such suffering to take its course, and we don't intervene to relieve it.

Here, I consider whether there's anything morally problematic about taking this kind of *laissez-faire* approach to wild animal suffering. Is it morally inconsistent, for instance, to maintain that we have a moral responsibility to assist suffering companion animals while taking a hands-off approach to wild animal suffering? I suggest here that a broadly *laissez-faire* ethical approach to wild animal suffering can be defended—not so much in the sense that we *must never* assist wild animals but, rather, that we're not generally *required* to do so, although (as I also argue) there are nonetheless some circumstances where we at least have moral reason to assist them. This approach to wild animal suffering is controversial, and there are several alternative ways of thinking about the issue. I begin by considering some of these alternative approaches to wild animal pain and suffering as a way to locate the position I propose here.

Animal Suffering Doesn't Matter!

First things first: some people don't accept that (nonhuman) animals matter at all in a moral sense; we don't owe them anything directly. On one version of this view, being able to feel pain is a condition for mattering morally, and it's believed that no nonhuman animals can feel pain, so none of them matter. However, vanishingly few people currently maintain this view, and with good reason, since there's little empirical support for the claim that no nonhuman animals feel pain. An alternative version of this view holds that even though animal pain and suffering do exist, pain and suffering aren't sufficient conditions for mattering morally. Some versions of what's called *ethical contractarianism*, for instance, don't regard animals as being of direct moral relevance. The ethical sphere is understood to be a kind of mutually beneficial, reciprocal deal constructed between humans. Since animals can't understand or reciprocate such a deal, they fall outside its protection. Companion animals may still matter on this view; however, this isn't because the companion animals in themselves have high moral significance but, rather, because people are emotionally attached to them, and *people* matter. Since individual wild animals rarely attract such emotional attachment, their suffering falls completely outside the moral sphere.

But again, this view of animal pain and suffering is not very widely held (even among ethical contractarians, let alone among those who take other theoretical approaches to ethics). Most people agree that animal pain and suffering *does* matter ethically. Accepting this—as Zamir (2007) has effectively argued—does not require a commitment to anything like an egalitarian view on which the suffering of humans and that of other animals matter *equally*. It's entirely possible to think that humans are much more important than other animals (for instance, based on additional valuable capacities humans are believed to possess) and still to think that animal suffering matters in itself. So, given how widely the moral significance of animal suffering is accepted, I don't discuss the "Animal suffering doesn't matter!" view further here.

There is one further point to make here, however: there is still uncertainty about *which* species of animals are capable of feeling pain and suffering. While there are interesting and important debates about this, these aren't the focus of this chapter, so I'll take a shortcut. I concentrate here on animals about whom there is little controversy in terms of pain and suffering: mammals, birds, reptiles, and amphibians.

Always Assist: Wild Animal Suffering Is Still Suffering!

Now let's return, specifically, to *wild* animal suffering. An intuitively plausible view here would surely be that suffering is suffering, whatever species of animal or context is involved. Wild animal suffering is just as bad as the suffering of a companion animal, if it's of the same severity, and so should be responded to in the same way, at least in principle. On this view, a *laissez-faire* approach to wild animal suffering, of the kind outlined earlier, therefore misses something morally important—that like cases should be treated alike.

I added "at least in principle" here due to one obvious objection to this view: that there's a dissimilarity between the *outcomes* of assisting wild animals and caring for one's own companion. Unlike taking a companion animal to the vet, for instance, relieving one wild animal's suffering could just shift the suffering from one animal to another. For example, in the simplest of terms, preventing a hungry lion from attacking a zebra because the zebra would suffer is likely to protect the zebra at the expense of causing the lion to suffer from hunger. So, in intervening to relieve wild animal suffering, it's clearly important to avoid a situation where more suffering results from the intervention than is relieved by it. The goal must be to reduce *aggregate* or *overall* suffering in the wild.

Although different theoretical approaches to ethics could underpin the view that we should intervene in the wild when doing so would reduce overall suffering, the most obviously relevant theory here is *hedonistic utilitarianism*. Utilitarians, in general, aim to bring about the best consequences; hedonistic utilitarians understand the best consequences in terms of maximizing happiness and minimizing suffering. So, if it were likely that a particular intervention would make the lives of wild sentient animals go better and reduce their suffering, hedonistic utilitarians would generally be in favor of it (unless there were some alternative but more efficient way of relieving equal or greater amounts of suffering elsewhere).

It's helpful, in thinking about overall wild suffering, to distinguish between the *scales* of different kinds of wild interventions. Some interventions might be on a small, local scale—perhaps rescuing a wild animal that's fallen through ice or putting a fallen chick back in its nest. These interventions are unlikely to have a significant knock-on effect in the sense of harming other animals, but they are also unlikely to decrease wild animal suffering much overall. More significant decreases in suffering would require more systematic interventions. Reducing suffering from wild animal diseases, for example, pinkeye or pneumonia, could require vaccination campaigns, veterinary treatment in the wild, and/or the segregation of populations to prevent the spread of disease. Such interventions could have a significant impact in terms of preventing suffering among animals that otherwise would have caught these diseases and reducing the suffering of those who still succumb. But interventions such as this could also have broader ecological impacts with the potential to increase suffering. For instance, we can imagine vaccination campaigns that could lead to population growth, resulting in a lack of available resources to support the expanded population and thereby to individual animals undergoing hunger or even starvation. In cases such as these, utilitarians would have to judge whether intervening or not intervening would be most likely to bring about the best consequences over time.

On a broader scale still, some who are concerned about wild animal suffering have suggested the possibility of harnessing new technologies to help. Here, the goal is not so much to relieve existing suffering as to prevent it in the first place. So, for instance, it might be possible to gene edit species that produce very high numbers of offspring, few of whom make it to adulthood and many of whom die painfully, to produce fewer offspring, or to genetically modify predators to reduce or eliminate their hunting instincts while either providing them with alternative foodstuffs or additionally modifying them so that they can eat plant matter instead. (For discussion of both these proposals, see Delon and Purves 2018). These kinds of proposals obviously involve transforming the "natural" world in pursuit of the goal of reducing wild animal suffering. Clearly, many environmentalists and those concerned about preserving Earth's remaining wildness would object strongly to such proposals. But those who defend these methods argue that there's nothing morally significant about what just happens to have evolved and is "natural"; to use Nussbaum's (2007: 399) terms, they maintain that we should try to achieve a gradual "supplanting of the natural by the just."

Of course, as with wild animal vaccination cases, such transformative proposals also raise concerns about generating more suffering overall over time. Delon and Purves (2018), in their discussion of these proposals, accept that we do have a *pro tanto* obligation to relieve wild animal suffering—that is, we have such an obligation provided it isn't outweighed by other considerations. But, they maintain, large-scale interventions such as the gene editing mentioned earlier can't currently be morally justified, because at present we lack the knowledge and ability to judge whether such gene editing would or would not *actually* reduce wild animal suffering.

So, to summarize this approach, suffering is intrinsically bad, in the animal case as well as the human case, and in the wild animal case as well as the companion animal case. We should try to prevent or relieve wild animal suffering where we can, if we can do so without creating more suffering. There are likely to be disagreements in practice about the cases to which this actually applies and how much we need to know about the likely outcomes of any intervention before we make it; there's more likely to be agreement about small scale, local cases than large-scale cases. But there's no *in-principle* difference between what's owed to wild and companion animals; whether to intervene or not is contingent on the likely outcomes of any particular intervention in terms of impacts on suffering. I consider some problems with this view later.

Another cluster of views takes a very different perspective. Here, it's argued that a difference in context makes a difference in what's owed to animals. We *don't* have the same obligations to animals living independently in the wild as we do to our pets, for instance. However, each of these views has a somewhat different flavor. Some of them argue that we *should not* assist wild animals because there's something important about wildness. I call these the "Never assist!" views. Others maintain that we're not required to assist wild animals because we don't have the kind of relationship to them that requires us to help out. These views take a particular kind of contextual approach, which I call a *laissez-faire* view.

Never Assist: Protect Wildness!

Protect Natural Processes

One view here is based on the protection of *wildness* or *naturalness* understood in terms of human-independent ecological processes. It's argued that wildness is valuable and that humans should constrain their actions to allow human-independent ecological and evolutionary processes to continue without interference. Wild animal pain and suffering are baked into such natural processes, so, as Rolston (1989: 134) argues, "we have no obligations to help wild animals; we are obliged to leave them alone." On this view, then, we don't have any kind of obligation to prevent or relieve wild animal suffering, not even a *pro tanto* one; quite the contrary, we have an obligation *not* to intervene in the natural world, even where acute suffering is occurring, in order to allow for the continuance of wild, natural processes.

Protect Wild Animals' Wild Natures

An alternative view here is not so much about protecting natural ecological processes, but about respecting wild animals' natures. Take wild deer, for instance. Everett (2001: 54) argues that a deer is a wild animal and "as such, the sort of creature whose flourishing is generally thought incompatible with widespread human intervention." What's critically important for wild animals such as deer is that they "flourish according to their natures." Showing respect for wild animals means that we should respect the wildness of their natures, and we should not intervene to assist them, for instance, against predators, since such intervention would fail to respect their wildness.

Some Problems With These Wildness Arguments

These two wildness arguments raise difficulties. The first argument takes the position that there's something valuable about the human independence of particular ecological processes, that something like "autonomy" in nature matters. If all that's meant here is that some people *actually do* value autonomy in nature, then this appears indisputable. But while some people valuing independent nature may give us *a* reason for not intervening, it doesn't by itself seem a sufficiently strong reason to support the position that therefore we shouldn't intervene when intervention would prevent massive amounts of suffering. The claim may instead be, however, that we *should* respect human-independent natural processes (whether or not we actually do). But then we need an argument to justify why we *should* respect these processes, and that's more difficult to develop. Of course, it's widely agreed that autonomy is important for individual people, but that's because people can reason about what to do, make choices, and have preferences of their own. There are also ways in which sentient animals may be described as autonomous: they clearly have preferences for one thing or state over another; when permitted, they can make decisions and choices about what they do with their lives. But processes such as evolution or speciation (to name two of the ecological processes Rolston discusses) don't have the capacity to be autonomous in anything like this sense; they can't make choices or have preferences. They just change in one direction or another. So it's not clear why we would respect or protect their independence at all.

The second argument maintains that animals should be free to "flourish according to their natures," and this certainly sounds plausible. But this raises the question how of "wildness" could be part of an animal's nature. Take the bighorn sheep, for instance. It's easy to make sense of someone saying that it's part of the nature of bighorn sheep to eat grasses and sedges, for instance, or that it's part of their nature to jump along steep rock faces. But "wildness" isn't an ability or a behavior of this kind; it's a concept about a (lack of) relationship with or control by human beings. One possible interpretation is that "wildness" here should be taken to mean something like "being free to manifest natural capabilities without being inhibited by human activity" (Palmer 2010: 81). But this interpretation could equally provide a justification for assisting wild animals as refusing to assist them. After all, treating an animal suffering from pinkeye or pneumonia could restore the animal's ability to manifest natural capabilities and behaviors such as jumping from rock to rock that disease would inhibit, and vaccination could protect the freedom to manifest such capabilities over time. So it's at least as plausible to argue that human interventions can allow wildness in this sense to flourish as the reverse.

In sum, "Never assist" arguments seem difficult to justify, whether the wildness that should be protected is attributed to the ecological context or to the nature of wild animals themselves. But together with the "Always assist (if it would really minimize suffering)!" view, these form the two poles between which the alternative position, the one described in this chapter's title, falls. This is the *laissez-faire* view: we are not generally *required* to assist wild animals (unlike the "Always assist!" view), but that doesn't mean doing so is *impermissible* (unlike the "Never assist!" view). In fact, as I go on to suggest, there may be some occasions when we should assist wild animals, even on the *laissez-faire* view, but these result from special relations rather than being general obligations to all wild animals.

The *Laissez-Faire* View

First, some clarifications. I take the *laissez-faire* view to be one of a family of broadly *contextual* approaches to animal ethics. These contextual approaches share some important things in common. First, they accept a distinction between *harming* and *assisting*, and second, they maintain that factors *other* than a being's capacities are important to moral decision-making. I'll briefly explain these in turn.

Harming and Assisting

A *laissez-faire* position means that we aren't normally obliged to prevent or relieve wild animal suffering. However, that doesn't mean that we're free to *cause* wild animal suffering; this isn't a view on which animal suffering just doesn't matter! That we're not morally required to vaccinate wild bighorn sheep against pink eye does not make it acceptable to inject blindness serum into their eyes. Not having obligations to *assist* wild animals doesn't mean that we can *harm* them with impunity.

This distinction between harming and assisting is controversial in some schools of ethical thought. For instance, most utilitarians reject it, arguing that there's no substantial moral difference between causing suffering and failing to prevent or relieve it, if you could have done; in either case, you are responsible for its persistence, and this view is reflected in the "Always assist!" approach. However, those who take a *laissez-faire* view understand instances of harm differently from instances of non-assistance. There are different ways of thinking about this, but one well-known explanation has been developed by Kamm (2007). She argues that we can think of harms as impositions by some people on others, impositions that make people (or in this case, animals) worse off, changing the status quo by depriving those who are harmed of some good they would otherwise have had. Assistance, on the other hand, as Kamm (2007) argues, is a burden on the one who assists, rather than on the one who benefits from the assistance. While what we do shouldn't harm others, on the *laissez-faire* view, we don't have general duties to improve the situation of others at our own expense (Palmer 2010). Other contextual views, as I outline later, defend more significant duties of assistance. Still, these duties to assist lack the stringency of our duties not to harm.

Special Relationships

A second feature shared by contextual views, including the *laissez-faire* approach to wild animal suffering, is that it's not only *capacities* (such as being able to feel pain) that matter morally. Other factors also matter in terms of thinking about our obligations to others, factors that relate to context, community membership, relationship, or historical entanglement. These kinds of morally relevant, special relationships are commonly accepted in human ethics. For instance, it's widely agreed that parents have special obligations toward their own children that they don't have to children in general, even though all children are morally considerable and have similar morally relevant capacities. That particular kinds of contexts and relationships can generate *special*, rather than *general*, responsibilities, ones that apply to particular individuals or are group-differentiated, is currently being widely explored in animal ethics. And it can provide a consistent foundation for accepting a *laissez-faire* view with respect to assisting wild animals, while simultaneously insisting that we have special obligations to help domesticated animals, such as our animal companions.

Explaining the Laissez-Faire *View*

Given this background, what would *a laissez-faire* argument with respect to wild animal suffering look like? I begin by sketching a simple outline: wild animals, like the bighorn sheep, live independently of human beings. They have their own relationships, communities, and social structures; they find their own food and water; they are born, mate, reproduce, and die independently of us. We are not responsible for their genetic constitution, their health, or where they live. If they are hosting painful parasites, contracting painful diseases, or torn down by predators, that's not due to humans. Compare this situation to that of a pet poodle, for instance. Poodles live entirely dependent on human beings. Their social structures are largely located within human families; they frequently don't live with other dogs at all. We provide them with food and water; poodles are not equipped to care for themselves or to go out hunting in savage poodle packs. They mate and reproduce largely only with human permission, if they do either at all (after all, many are spayed or neutered), and their time of death is often chosen by us. Their bodies have been significantly genetically shaped by human choices; their state of health is, at least to some degree, dictated by human decisions (diet, exercise, veterinary care). They live in environments and buildings that we created and built for our own purposes and then found a way of fitting poodles in.

In all these ways, the situation of poodles is very different from the situation of bighorn sheep. Seen in this context, it seems strange to think that what we owe to bighorn sheep would be like what we owe to poodles. Poodles, for instance, have been made dependent and vulnerable because of the ways we have created and located them; they have been bred in shape and temperament to be unable to provide for themselves and confined in environments that would make it impossible for them to hunt and forage, even if, in principle, they were able to. And this state of affairs underpins the argument that we have special responsibilities to provide for poodles, since *we made them* unable to provide for themselves, in ways that are not true of bighorn sheep in Yellowstone, who are self-sufficient without us.

The basic claim, then, of the *laissez-faire* view is that while duties not to harm are general, applying to all sentient animals, duties to assist arise out of contexts and relationships, especially those where humans have created dependent vulnerability in animals. We breed and keep domesticated animals in ways that make their self-sufficiency largely impossible. This means we acquire special obligations to care for them. We haven't generally had these kinds of impacts on wild animals. So we don't generally have special obligations to assist them. This doesn't mean that we are not *permitted* to assist wild animals, a view that might be insisted on were the goal to protect wildness or, more generally, not to interfere with human-independent processes, as in the "Never assist!" view discussed earlier. The argument here is just that we're not normally *required* to do so; it's not normally wrong if we don't assist suffering wild animals.

Implications and Human Environmental Impacts

The *laissez-faire* view I've outlined earlier has some potentially useful implications, especially in relation to its current main alternative, "Always assist!" although it also faces problems, as I'll suggest below. First, it is significantly less demanding than a view on which we're required to try to eliminate as much animal suffering as we can, including the suffering of wild animals. Given that wild animal suffering is ubiquitous and can be acute, a general requirement to assist wild animals would impose heavy burdens on us—burdens that, it seems reasonable to say, it's psychologically implausible that many people would be willing to bear and act on (for a discussion of psychological plausibility in animal ethics, see Kasperbauer 2017). In this sense, the *laissez-faire* view at least has some practical advantages. Second, there's no ethical reason, in this view, for engaging in a radical reshaping of natural landscapes or wild animals' natures in order to find ways to reduce predation, disease, and the harshness of wild conditions. This also means that the *laissez-faire* approach can fit more harmoniously with environmentalist views and those that value naturalness/wildness, since it's not an implication of the view that we should "supplant the natural by the just"—taking what's "just" here as acting to produce a world with less wild animal suffering.

While (from some perspectives, at least!) these implications are advantages, the *laissez-faire* view also brings some difficulties. I consider just two of these here. First, I discuss the difficulty raised by human impacts on the environments of even wild animals, meaning that they may no longer be independent in morally relevant ways. And, second, I consider the possibility that we do have some kinds of positive duties to help suffering wild animals in need, though these duties are much less strong than are supposed in the "Always assist!" view.

Human Impacts on Wild Animals' Environments

Let's go back to the bighorn sheep. So far, I've focused on populations of bighorn sheep in Yellowstone. But a number of western US states, including Utah, Nevada, and Arizona, also have populations of desert bighorn sheep. Studies by the National Park Service (n.d., a) suggest that these desert bighorn sheep populations are likely to be negatively impacted by anthropogenic (human-caused) climate change. Climate change appears to be warming up and drying out the desert bighorn sheep habitats, making drinking water less easily located. Since lactating bighorn sheep, in particular, need to drink every day, lower water availability is likely to restrict their movements and to cause them discomfort or suffering, and it may kill some of them. Populations already seem to be declining. Does the *laissez-faire* approach imply we have no obligation to do anything here since these are wild animals?

Cases like this require us to look back at one of the underlying justifications for the *laissez-faire* approach: that we don't normally have moral responsibilities to assist animals where there's no kind of causal responsibility for their vulnerability or suffering. Generally speaking, in the case of wild animals, we're not responsible for their suffering: we don't cause predation or mudslides or parasites. Unlike companion or laboratory animals, for instance, we haven't made them dependent or vulnerable by (for instance) breeding them in ways that make them unable to fend for themselves.

But in the case of the climate threat to desert bighorn sheep, human activities *are* having a negative impact on wild animals; they are being made vulnerable to thirst, their movements are being restricted, and they are being caused to suffer by climate change. This means that the justification for adopting a *laissez-faire* approach no longer seems to apply; if humans are significantly causally responsible for increased vulnerability or harms to wild animals, then they are also morally responsible for giving assistance, if doing so successfully is possible.

I say "if possible," for this is a case in which help will be difficult. Climate change is not a straightforward problem with a relatively simple origin and a clear set of possible solutions. Compare it, for instance, with a proposal to build a road through bighorn sheep habitat, a project that would have the possible effect of increasing sheep–vehicle collisions, as well as fragmenting sheep habitat. In a road-building case, we're clear who is involved with constructing the road, and there are many possible options for reducing or removing potential harms to sheep: not building the road at all, rerouting the road through less sensitive areas, and/or adding extensive mitigation, such as wildlife crossing places to the road, both to reduce collisions and to maintain connectivity across bighorn sheep habitat. In the *laissez-faire* view, all these should be given serious consideration because humans are introducing the hazard into bighorn sheep territory, making them vulnerable to increased suffering.

Climate change, however, is a much more complicated issue in terms of negative impacts on human beings, let alone wild animals. Climate change lacks any easily identifiable, directly causally responsible individuals, local governments, or businesses; it poses a hazard that's gradual and time-lagged and where there's considerable uncertainty about exact effects on landscape and water supplies over time; and it's difficult to determine who should do what as an ethical response. It doesn't seem as though ordinary individuals have direct responsibilities to bighorn sheep in this kind of case, anyway; it's not as though we should all head out into the desert with buckets of water. It's much more plausible to maintain that those who are most responsible for managing the relevant land areas—such as in this case the National Parks Service—should develop strategies and policies for responding to wild animal suffering caused by climate change. In any case, there are obviously many more difficulties involved in trying to help desert bighorn sheep in the context of climate change than would be the case in mitigating the negative impacts of a particular road; it may not be possible to help much at all. But still, the basic point here is, I think, clear: while in the *laissez-faire* view I've described, there are no *general* duties to assist wild animal suffering as a consequence of factors such as predation, native diseases, and parasites, special obligations arise due to human actions that harm or threaten wild animals. So the *laissez-faire* view becomes less *laissez-faire* the more that the vulnerability and suffering faced by wild sentient animals becomes a human causal responsibility.

This contrasts both with an "Always assist!" and a "Never assist!" approach. In the former view, we should try to reduce *all* animal suffering, independently of the cause of the suffering. In the latter view, we should normally avoid interventions that compromise the wildness of ecological processes and animals, even if there's a human-produced threat of considerable wild animal suffering. So, for instance, suppose a possible policy response to the loss of water sources from climate change for desert bighorn sheep was to create artificial troughs and pools in the desert, filled by artesian water. This could be ethically justified on both the "Always assist!" and the *laissez-faire* view, although different reasons would be given by each for doing so. However, constructing artificial water sources is likely to be seen from the "Never assist!" position as undermining the natural processes of the desert, affecting its place-wildness, and (perhaps) as turning the desert into something more like a zoo or safari park, where humans decide how animals' needs are to be met.

Now that I've explained the *laissez-faire* view in more detail, further obvious objections might arise: Isn't it too restrictive with respect to our obligations to assist? Why insist that we don't need to help wild animals that are suffering unless humans caused it? One doesn't have to go all the way to an "Always assist!" view to think that we should help wild animals that suffer, for instance, if we could do so easily, on a small scale, without any obvious ecological ramifications. If I see a fawn who has fallen through the ice on a pond, and I could easily save him or her, is there no obligation, not even a weak one, that I should help out, if I'm not myself at risk? If the baby bird is fluttering on the ground at my feet while the parent bird is agitated above, is it really just fine for me to walk away, provided that I don't stamp on the chick in the process?

Don't We Have Stronger Duties to Assist Than the Laissez-Faire *View Suggests?*

This objection suggests that, without going so far as to accept that we should always intervene to prevent wild suffering if we can do so without making matters worse, we have at least some duties to assist wild animals, even if these are relatively weak and can be easily outweighed. Hedberg (2016), for instance, defends what he calls a "gradient view." In this view, as in the *laissez-faire* approach, we have duties to assist wild animals that have been adversely affected by our actions, but *in addition*, we have "a weak presumptive duty to assist ... which typically arises when their needs are great and costs to us are minimal" (Hedberg 2016: 436). If wild animals matter morally, Hedberg argues, we should relieve their suffering, especially if it is extreme and we can do so easily, even if we are not responsible for causing it. However, the closer our relationships to particular animals are, the stronger our duties to assist them become. So we have much stronger obligations to assist our animal companions than we do wild animals that we have never previously encountered. But still, we should assist wild animals in some circumstances.

Hedberg's gradient view falls into the broad category of contextual approaches to animal ethics. Like those who defend the *laissez-faire* view, he accepts that there's a distinction between harming and assisting. But, he maintains, we may have both very strong duties not to harm wild animals *and* weak duties (rather than no duties) to assist them. The gradient view, then, also maintains the moral significance of special relationships and contexts while making these somewhat less central than they are to the *laissez-faire* view. Special relationships strengthen our duties to assist, but these duties exist in weakened form even where no special relationship holds. Presumably, such a position would not only suggest that we should relieve severe wild animal suffering if we easily can, but also that we should prevent it—for instance, by vaccination campaigns or even by the use of technology—provided we can do so fairly easily. In this sense, the gradient view takes some steps toward "Always assist!" and away from the *laissez-faire* view. Indeed, one difficulty with this approach is that an explanation is needed as to why, if the duties to assist are just based on mattering morally and not on the closeness or distance of a relationship, they are weaker than duties not to harm in the first place. Hedberg himself (2016: 440) says that "consequentialist reasoning"—something like the "Always assist!" view—provides the simplest and most compelling account of our moral duties. So, although in some ways this is an appealing amendment to the *laissez-faire* view, a much more comprehensive argument is still needed to support something like the gradient view.

Conclusion

The *laissez-faire* view presented here defends the position *that we are not generally required to assist wild animals*. Obligations to assist, unlike duties not to harm, depend on the existence of some kind of morally relevant relation or context. Thus, the *laissez-faire* view rejects the position that we should always assist wild animals, that we have the same responsibilities to prevent and relieve animal suffering wherever it's found and however it's caused. The *laissez-faire* view is also not based on the idea that we should not interfere with natural processes or reduce wildness. In the *laissez-faire* view, assisting wild animals is generally *permissible*, even though it is not generally required. In addition, we may have duties to assist wild animals if we are responsible in some way for harming them, assuming any assistance would actually be successful in helping the animals concerned.

This view—like all positions in animal ethics!—does present significant difficulties, only some of which are discussed here (see Palmer 2010 for further discussion of some of these). However, in arguing that we don't have moral responsibilities to redesign the natural world to prevent animal suffering if we could, while at the same time accepting that animals we've harmed deserve our help, it puts forward at least a plausible approach for addressing wild animal suffering.

Bibliography

American Veterinary Medical Association (n.d.) "PETS act (FAQ)," www.avma.org/KB/Resources/Reference/disaster/Pages/PETS-Act-FAQ.aspx.

Bighorn Sheep Disease Research Consortium (n.d.) "About pneumonia," http://bighornhealth.org/about-pneumonia.

Delon, N., and Purves, D. (2018) "Wild animal suffering is intractable," *Journal of Agricultural and Environmental Ethics* 31(2): 239–260.

Everett, J. (2001) "Environmental ethics, animal welfarism and the problem of predation: A Bambi lover's respect for nature," *Ethics and the Environment* 6(1): 42–67.

Hedberg, T. (2016) "Animals, relations, and the laissez-faire intuition," *Environmental Values* 25(4): 427–442.

Kamm, F. M. (2007) *Intricate Ethics: Rights, Responsibilities, and Permissible Harm*, Oxford: Oxford University Press.

Kasperbauer, T. J. (2017) *Subhuman: The Moral Psychology of Human Attitudes to Animals*, Oxford: Oxford University Press.

National Park Service (n.d. a) "The adverse effects of climate change on desert bighorn sheep," www.nps.gov/articles/desertbighornsheepresearch.htm.

National Park Service (n.d. b) "Bighorn sheep," *Yellowstone National Park*, www.nps.gov/yell/learn/nature/bighorn-sheep.htm.

Nussbaum, M. C. (2007) *Frontiers of Justice: Disability, Nationality, Species Membership*, Cambridge, MA: Harvard University Press.

Palmer, C. (2010) *Animal Ethics in Context*, New York, NY: Columbia University Press.

Rolston, H. (1989) *Environmental Ethics: Duties to and Values in the Natural World*, Philadelphia, PA: Temple University Press.

Tomasik, B. (2015) "The importance of wild-animal suffering," *Relations: Beyond Anthropocentrism* 3(2): 133–152. doi: 10.7358/rela-2015-002-toma.

Zamir, T. (2007) *Ethics and the Beast: A Speciesist Argument for Animal Rights*, Princeton, NJ: Princeton University Press.

35
WELFARE BIOLOGY

Catia Faria and Oscar Horta

Introduction

Welfare biology is a proposed cross-disciplinary field of research that can be defined as the study of sentient living beings with respect to their positive and negative well-being. It was initially suggested in 1995 by Yew-Kwang Ng (1995, see also Carpendale 2015), and since then, others have argued in favor of its creation.[1] The main purpose of welfare biology is to determine the circumstances that affect the well-being of animals living in the wild such that their lives go well or badly. Applied welfare biology would be the study of the ways to improve in practice the situation of animals and prevent the harms they suffer.

When most people think of the harms suffered by animals in the wild, those that typically come to their mind are the ones that humans inflict on them. However, as we will see in the following, they are many others that they suffer due to natural causes such as hostile weather conditions, a lack of food and water, natural disasters, accidents, attacks, parasites and infections, and so on. What makes welfare biology a novel field of research is that it would be concerned with all the harms affecting nonhuman animals, including those not caused by humans. In fact, as the latter have not been given any significant consideration until now, their study would be the most archetypal and important tasks for this new field.

Welfare biology would differ very significantly from conservation biology, whose point is to gain the knowledge necessary to inform those measures aiming at the conservation of biological entities such as species, landscapes, biocenoses, or ecosystems. Conservation biology sees animals as parts of these wider and more abstract entities, caring about their interests only as long as this is instrumentally valuable for their continued existence. This explains why it can defend methods that can be harmful to them, ranging from the confinement and breeding of certain animals in zoos to the killing of others whose populations are intended to be reduced or eradicated in certain areas (Sagoff 1984; Soulé 1985; Hargrove 1992). In contrast, welfare biology would be concerned with sentient animals as individuals and would assess changes to ecosystems or species only in light of how it may affect animals positively or negatively.[2]

In the following sections, we will see the reasons to promote welfare biology and what seem to be its most promising starting points, together with the main objections that can be presented against it. The second section presents the reasons why the development of welfare biology is significantly important, by explaining the evidence leading to conclude that animals in the wild typically endure lives with huge amounts of suffering. The third section considers the reasons why welfare biology has not been developed yet. The fourth section explains the starting points for welfare biology to develop. Then, the fifth section presents ways in which animals in the wild are currently being helped at a short scale, and the sixth section introduces ways to help them at a larger scale. The seventh section then considers some normative objections that can be presented against welfare biology, such as that we should not be concerned with nonhuman animals at all, that doing so would be too demanding, that such concerns should be overridden by the pursue of environmental aims, or that we should let animals to their own devices. The eight section assesses pessimistic objections claiming that any efforts carried out to make a change for animals in the wild will either fail or bring about counterproductive consequences. Finally, the chapter ends with some concluding remarks.

The Importance of Welfare Biology

The importance of welfare biology can be appraised by examining the available evidence and plausible inferences concerning the well-being of animals living outside human control. Many people among the public (and among animal advocates as well) have an idyllic view of nature, according to which it is assumed that animals living in the wild generally have good lives, that is, lives with net positive value—if not with high levels of well-being (Balcombe 2006). However, data from animal population dynamics and life history theory give us sufficient grounds to reject this claim. For the majority of animals that come into existence, death intervenes before they reach sexual maturity. This is mainly due to the prevalent reproductive strategy, which consists in producing large numbers of offspring. Given the finite resources available in their environments, animals that follow such strategy have very low survival rates, with most individuals dying shortly after coming into existence (Lack 1954; MacArthur and Wilson 1967; Roff 1992; Stearns 1992; Cappuccino and Price 1995; Reznick et al. 2002). Examples range from amphibians and fish to invertebrates and mammals, including small rodents. Some of them, such as some amphibians and arthropods, may lay tens of thousands of eggs. Others, such as some fishes, may lay millions of them (Fraser-Bruner 1951; McAuliffe 1978; Mendi 1988). If they were to survive to adulthood, the population of these animals would soon be multiplied by millions. However, this is not the case. To be sure, populations of different animals can fluctuate dramatically. But on average, each adult animal is substituted in the next generation only by one individual. This is relevant from the point of view of the animals' well-being since it means that all animals that are born—minus one per parent—die prematurely due to different natural events. This typically causes them to experience significant pain. Some may be killed by starvation, disease, or parasites. Some may be eaten by predators. Some may freeze to death or die by dehydration and so on. As a result, their lives are short and include few opportunities for positive well-being. Due to this, their short lives are likely to contain more suffering than well-being. In some cases, they may even consist almost entirely of suffering. Thus, a massive number of animals in the wild experience more suffering than well-being. Moreover, since most animals reproduce in this way, this suggests that the majority of animals actually experience this fate. On this basis, it has been argued that suffering and premature death are actually the norm in wild ecosystems and that these disvalues are likely to prevail over happiness in nature (Ng 1995; Horta 2010a, 2017; Faria and Paez 2015; Tomasik 2015a [2009]; Faria 2016).

In addition, animals that do survive to adulthood face many threats to their health and physical integrity, which entail a great amount of suffering, including among others hunger and thirst; physical injuries; cold, heat, and other forms of discomfort; disease; and, in the case of many animals, psychological stress and fear (Cooper 1999 [1982]; White 2008; McCue 2010; Wobeser 2013). These circumstances not only can also cause huge amounts of suffering to the animals but can also bring about their deaths by direct or indirect means (by increasing their susceptibility to other harmful events). All this reinforces the idea that suffering and untimely death are widespread in nature. Welfare biology could provide us with a more detailed account of the extent to which this is so by interpreting the information we already have from different fields in ecology about how the lives of animals are and by making more precise estimates of which of them are sentient at the time they die.

By failing to study these issues there is an important part of what happens in ecosystems that we fail to be aware of, which leaves our knowledge about the world around us fundamentally incomplete. Still, in addition to this, there are also powerful practical reasons at stake here. If what happens to sentient beings is morally significant, then we have reasons to take action for their own sake.

This is particularly important considering the number of sentient animals living in the wild. According to estimations made by Tomasik (2015b [2009]) it can be up to 10^{18} to10^{22}. Considering this, as well as the current neglectedness of this research topic, we can conclude that the development of studies in welfare biology is of the utmost importance.

Any view concerned with obtaining a better situation for nonhuman animals will have reasons to support this conclusion. This is so in the case of deontological approaches defending positive rights or duties of assistance to those in need of aid. It also seems that character-based views should support having a caring or a benevolent attitude toward those who need our help. As for consequentialist approaches, it also seems clear that for them a scenario with less suffering and other harms in the wild is preferable to one with more. If we consider certain views in particular, we can also find that they can have further reasons to agree with this. Egalitarian, prioritarian, and sufficientarian approaches will support intervening to change a certain state of affairs if there is a group of individuals in it suffering a very bad situation. They would thus support taking action to improve the situation of animals in the wild even if, instead of most animals, only a small part of them were suffering very bad lives. Negative approaches, which hold that reducing disvalue (such as suffering) takes priority over promoting positive things (such as pleasure), will also agree on this. Utilitarianism will also be in favor changing the bad situation in which animals are in the wild, as it would mean an increase in the level of aggregate positive minus negative well-being, especially given the extent to which suffering may prevail over happiness in the wild. Perfectionist views will also have to acknowledge that the prevalent conditions in the wild make it impossible for the majority of animals to fulfill their potential and flourish. Other views can reach similar conclusions.

Why Is Welfare Biology Not Yet Underway?

One may wonder why welfare biology is still underdeveloped. One main reason for this is that we live in speciesist societies. Factors affecting human well-being have already been studied in much detail. Arguably, if this has not happened regarding the well-being of other animals is simply because their interests are not taken seriously enough. However, this is not the full explanation. Concern for nonhuman animals has increased significantly in the past decades around the world, and so has the study of the well-being of animals exploited by humans. But, unlike what has happened in the case of these animals, no social pressure has taken place to promote measures to protect animals from the harms they suffer in the wild (unless they are anthropogenic). This is due to the idyllic view of nature mentioned earlier, which is widespread among the public and animal advocates.

Life scientists may not share this rosy view, but here the lack of consideration for nonhuman animals appears again. Ecologists, zoologists, and veterinary scientists typically work on topics that are either directly or indirectly relevant to the promotion of human interests or, in some cases to the promotion of conservationist ends, for which what happens to individual sentient animals is important only instrumentally, not in itself (Schmidt 1990; The Wildlife Society 2011). This lack of concern for nonhuman animals explains why there has been little interest in developing this new field thus far. Otherwise, it seems that the issue would have received some attention, however little. The fact that we have not been interested at all in this question, while we have been so curious about other aspects within ecosystems, shows not only that the issue has not been considered pressing but also how little attention has been given to the interests of nonhuman animals.

Despite all this, there have been some researchers who have researched on the well-being of animals within ecosystems (Kirkwood et al. 1994; Ng 1995; Kirkwood and Sainsbury 1996; Morris and Thornhill 2006; Horta 2010a, 2017; Pearce 2015; Tomasik 2015a [2009]; Faria 2016; Ryf 2016; Alonso and Schuck-Paim 2017; Brennan 2017). In addition, some work on the reasons to aid animals in the wild has been also carried out by animal ethicists (Sapontzis 1987; Clement 2003; Cowen 2003; Fink 2005; Hadley 2006; Morris and Thornhill 2006; Nussbaum 2006; Horta 2010a, 2017; Donaldson and Kymlicka 2011; Faria and Paez 2015; McMahan 2015; Torres 2015; Faria 2016; Johannsen 2017; Soryl 2019; a review of the literature is available in Dorado 2015). Much of this work has addressed metatheoretical issues and is not actual research in biology. Still, it can help to encourage the development of welfare biology, by challenging the idyllic view of nature and, particularly among scientists, the view that individual animals do not matter.

Prospects for the Development of Welfare Biology

Welfare biology may develop from the combination of two disciplines that are already well established. One is the science of animal welfare.[3] Work under the label of "wild animal welfare" has been carried already (Sainsbury et al. 1995; Kirkwood 2013), although it has been focused mostly on the situation of animals affected by human action. Work on the well-being of other animals would just constitute a logical expansion of animal welfare science, in order to cover the study of nonanthropogenic harms to animals. The other discipline from which welfare biology would develop is ecology, the study of the relations of organisms with each other and their environments. While ecology has many different subfields (such as population ecology, community ecology, systems ecology, evolutionary ecology, landscape ecology, behavior ecology, etc.), we do not have yet a discipline focused on the study of the animals' well-being in different ecosystems and on how environmental relations can affect it. However, much of the knowledge that has already been gathered in the field can give us important insights about these issues.

To develop welfare biology, a quite logical starting point is the work on interventions in the wild that have already been carried out and that can benefit nonhuman animals. The following sections discuss some examples of these forms of intervention at a small and at a large scale.

Current Ways of Helping Animals on a Relatively Small Scale

Cases of animals being rescued from mud ponds, frozen lakes, or natural disasters appear in the media every now and then (see, for instance, Associated Press 2009; Hartman 2011; Shrinivasa 2018). All these interventions are instrumental in saving a significant number of animals every day. But other measures have been carried out that can benefit more animals (Animal Ethics 2016). The following are two examples of this.

Rescue Centers

There are sheltering and rescue centers for orphan, sick, or injured wild animals in different countries (Bovenkerk et al. 2003). Often, they are carried out for conservationist aims exclusively, meaning they only target certain animals and leave others unattended. Yet, contingently, they may have positive consequences for the animals affected.

Food and Water Provision

Feeding or providing water to animals can save many hungry or dehydrated animals from dying. This is not an uncommon practice, even though it should be carried out carefully in order not to increase animal populations. Some national parks have carried out feeding programs when certain animal populations were at risk of being significantly reduced or disappear, such as in droughts or heavy snows (Reuters 2002).

Helping Animals on a Large Scale

There are other forms of intervention that have aided or may aid a much more significant number of animals. The following are some promising areas of investigation.

Vaccination

Research on vaccination programs against different diseases has been taking place for the last decades. The implementation of these programs has saved many animals from certain death, and often from significant suffering. Rabies has been eradicated from large areas, by distributing them in baits with the vaccine from helicopters or in other ways. Other programs have been also implemented to fight diseases such as tuberculosis, swine flu, or Ebola (Koenig et al. 2007; Fausther-Bovendo 2012; Rupprecht et al. 2003).

In most cases, these programs have been implemented with the aim of preventing the targeted diseases from spreading to humans or to domesticated animals. Yet these measures show that this is a perfectly feasible way to save a significant number of animals from natural harms in the wild.

Urban Welfare Ecology

When people think of an ecosystem, they tend to visualize wild or semi-wild ones, but many ecosystems appear very different. There are urban and suburban ecosystems, as well as those that have been created in industrial areas, the study of which is the task of urban ecology. Research on the populations and the lives of the animals living in these places has been carried out for a long time.[4] This has often been done with the aim of eliminating the negative impact they may have in the promotion of human interests, by keeping animal populations at low levels. In other cases, it has been carried out to preserve certain species of animals (VanDruff et al. 1994; McCleery et al. 2014; Adams 2016). This knowledge could also be applied to reduce the harms those animals typically suffer. In fact, urban or industrial areas seem to be excellent places to carry out interventions for the sake of nonhuman animals, given how controlled these ecosystems are. The results of such programs could be useful to gather further knowledge about how to best assess the well-being of animals in other ecosystems.

Protecting Large Herbivores

Environmental management efforts are usually carried out to further human interests or, in some cases, conservationist aims. However, they could take place with the aim of reaching the best possible situation from the point of view of nonhuman interests. Sometimes that may coincide with conservationist efforts. An example of this is the protection of large herbivores such as elephants. These animals tend to have very high survival rates. They only have an offspring at a time, investing a lot of parental care, and often reaching an old age. Most important, they consume large amounts of biomass, thus significantly reducing the amount available to other animals (Guldemond and Van Aarde 2008; Guldemond et al. 2017). This prevents the presence of small animals, who would otherwise have existed in great numbers and with large progenies. It also stops large trophic chains from emerging. Thus, ecosystems where elephants are present tend to exhibit lower amounts of suffering and premature death. Due to this, welfare biology research with a potential to inform policies of this kind can have a significant positive impact not just for the animals these policies target directly (in this case, elephants) but also for the rest of the animals indirectly affected.

To conclude, both at a small and at a large scale, there are interventions in ecosystems already carried out that show that it is perfectly feasible to act in ways that can benefit nonhuman animals and, what matters most, that can benefit them overall, that is, that can bring about a more positive scenario considering the interests of all the animals affected.

This is compatible with the claim that sometimes we cannot be completely certain about the net benefit of a certain intervention to all animals affected (particularly, when considering predator species), since there are many other interventions (particularly, those that tackle herbivores) that strongly suggest that we can not only benefit animals but that we can also benefit them on balance. At any rate, the concern with the net benefit of interventions to help animals does not cast doubt on the development of welfare biology. Instead, it points out precisely to the need to further work on the field in order to develop increasingly feasible and net beneficial ways of intervening in the future.

Normative Objections to Welfare Biology

We can now consider the main objections that can be presented against the development of welfare biology. This section addresses normative objections, based on moral positions, while the next one examines practical objections regarding welfare biology's chances to succeed.

One way to claim that welfare biology is a pointless enterprise is by arguing that we should not be concerned, or only very little, with nonhuman animals. However, as a number of theorists have argued extensively in the literature, there are strong reasons to claim that no defense of this view is sound (see, for instance, Gompertz 1997 [1824]; Pluhar 1995; Cavalieri 2001; Bernstein 2015). It has been argued that this view is an instance of speciesism, an unjustified unequal consideration or disadvantageous treatment of those who do not belong to a certain species (Horta 2010b). Rejecting speciesism implies treating equal interests equally, independently of the species whose interests they are. As sentient individuals, nonhuman animals share with human beings fundamental interests in being free from suffering and in enjoying their lives, which should be of similar moral concern. Taking these interests into account would imply not only refraining from harming nonhuman animals but also, and crucially, finding ways to improve their well-being, such as happens in the human case. Accordingly, those rejecting speciesism have decisive reasons to support the development of welfare biology.

Moreover, as we have seen, the number of sentient animals living lives with significant amounts of suffering, and often with net-negative levels of well-being, is huge. The weight of their aggregate interests is thus vast. Due to this, we can conclude that even those holding speciesist views should also be committed to support welfare biology to the extent that they consider that nonhuman interests matter, even if only a little.

Another objection points out that accepting that we have reasons to prevent animals from suffering nonanthropogenic harms implies an overly demanding moral viewpoint. However, this objection would be problematic if humans were in the situation of nonhuman animals. Consider some scenario in which most human beings were dying in terrible pain for natural causes. It would not only make total sense to claim that we should make significant efforts to help them; it would actually also seem completely intolerable not to support those efforts. If we hold a different view in the case of nonhuman animals, it is because we are undervaluing their interests. But there are reasons to regard such a position as speciesist.

Environmentalist views may give rise to a different type of objection. Those holding such views can claim that promoting wild animal well-being will threaten other more important values such as the preservation of ecological wholes, the wilderness or the natural (Sagoff 1984; Callicott 1989;

Rolston 1992; Katz 1992). This would be because the harms animals suffer within ecosystems are a direct result of ecosystemic relations and trying to improve the situation of animals could potentially disrupt them.

It is worth noting about this that these concerns would not apply in the case of the proposal made earlier regarding animals living in areas that have been already significantly transformed by human beings. Those areas include at least urban and industrial areas, and may well include also others such as suburban and agricultural ones. At any rate, after closer examination, it could also be argued that these environmentalist objections are based on questionable normative theses (Johnson 1981; Vinding 2014; Faria 2016). If the preservation of environmental values is a compelling reason to oppose the promotion of well-being, it should also be a reason to oppose the promotion of human well-being. Most people would find this rather unacceptable. Again, in order to block this implication environmentalists may only retreat to some form of anthropocentric speciesism, according to which only human interests should be given priority over the promotion of other alleged values. Alternatively, they may accept that environmentalist values should not be given priority over well-being (human or nonhuman) and thus acknowledge the importance of welfare biology. For their part, those rejecting the idea that we should further environmental values will also reject this objection. Accordingly, whether certain natural entities such as ecosystems are to be preserved will be so insofar as they are instrumental to the promotion of well-being, something that welfare biology will help us to assess. This view makes sense if we consider that what makes it possible for us to be harmed or benefited is just the capacity to have mental states.

Another way to lessen the importance of welfare biology is by claiming that we should not worry about wild animal suffering since our reasons to assist others in need are only generated when there is a causal link between an individual's particular situation of exposure to a harm and previous human action. Since there is no causal relation between present wild animal suffering and past human action, we have no obligations to prevent or alleviate it (Palmer 2010, 2015). It can be argued that from the point of view of animals in need of aid, this view is not based on something relevant for how good or bad their situation is. Note, too, that it would also exclude distant human beings in need suffering from natural causes—at the very least, those with whom we fail to engage in any significant relationship (Faria 2015).

There is a final objection that can be presented, which is that by doing so, we would be interfering with their way of life and failing to respect their capacity to live as they prefer (Donaldson and Kymlicka 2011). This objection would only apply in the case of significant interventions making radical changes in present ecosystems. But it can also be pointed out that the objection is based on the assumption that animals succeed in meeting the challenges they face in nature. While this happens in some cases, it is not so in the case of the overwhelming majority of them. As we saw earlier, in the wild, most animals die in their early youth, thus being unable to live as they would like to due to natural reasons. If these animals were able to make an informed decision about it, they would not support the present situation in which they endure such terrible lives. They would plausibly support forms of intervention that can achieve the best possible situation for them. That is exactly the kind of interventions that welfare biology would inform.

Practical Objections Against Welfare Biology

A different kind of challenge to the development of welfare biology arises from the idea that the magnitude of wild animal suffering and death would make this discipline unmanageable as a field of inquiry. According to this objection, any knowledge that might be gathered about the factors affecting wild animal well-being will always be incomplete, given the complexity of natural ecosystems. Therefore, welfare biology would never provide us with a reliable guide to inform actual policies aiming at improving the situation of these animals. Moreover, there is the risk that welfare biology would not succeed in making things substantially better for animals living in the wild or even proves to be perverse. This would render the effort of developing applied welfare biology either futile or counterproductive.

However, this objection seems misguided. Most scientific fields deal with complex phenomena, yet the difficulty in reaching a total understanding of them is not something that renders their work valueless or inapplicable. In addition, there are already feasible ways of improving wild animal well-being, which are not merely cosmetic from the point of view of the animals (as explained in earlier regarding helping animals on a small and a large scale). In most cases, these forms of intervention only aid a limited number of animals, and research is needed to assess whether they may have negative effects on others. But it is clear that some animals are being helped by them. Moreover, some measures (such as the protection of elephants) appear to have a wider impact, resulting in a net reduction of the harms suffered in the ecosystem where they are implemented.

Moreover, those claiming that intervention in the wild can have bad effects are being too pessimistic regarding the results of interventions. In addition, they are being too optimistic about the current situation of animals in the wild. The argument seems to assume that the current situation is slightly bad and that intervention may improve it at a high risk of turning it very bad. But the present situation is not just moderately bad. It is an extremely serious situation, where suffering and premature death are the norm, not radically different from the one that critics would deem as catastrophic as a result of intervention. This does not mean that the situation might not become worse. But it means that a certain failed intervention may have very bad effects, yet those effects not significantly differ from the present state of affairs. This setting aside, of course, that informed interventions might avoid those bad effects.

Another version of this objection claims that, even if feasible, the task of ending the harms suffered by those animals is helpless, as there will always be significant suffering and premature death in natural ecosystems. This claim is right, but it misses the point against welfare biology. The fact that wild animal suffering is colossal does not give us reasons against trying to understand it more accurately and try to reduce it or to make it increasingly more feasible to do so in the future. The case here is not substantially different from other applied fields of knowledge that can reduce, even if not end up altogether, serious problems affecting the well-being of both humans and nonhuman animals. Being unable to stop all the harms is not a compelling reason not to prevent as many as possible from occurring.

Conclusion

Welfare biology is the study of sentient animals with respect to their well-being. As a discipline, welfare biology could partly develop from animal welfare science. It would not be identical to it, however, since it must include the study of all animals not living under human control. These range from free-living animals in urban, suburban, and industrial environments to animals within natural ecosystems. Welfare biology would also develop from the science of ecology, although it would grow beyond its current purview by focusing on how the relations of organisms with their environments affect animal well-being.

The development of welfare biology is of utmost importance, especially due to the very low levels of wild animal well-being. Due to wasteful reproductive strategies, most animals that live in the wild have premature deaths and lead lives of net suffering. Animals who make it to adulthood also suffer for a number of reasons. The idyllic view of nature, according to which animals in the wild live great lives, is thus wrong.

The magnitude of wild animal suffering, combined with the current neglectedness of the topic, may constitute a challenge to the development of welfare biology as a workable field of inquiry. Yet, there are already some promising starting points of research. These include vaccination or feeding programs targeting certain wild animal populations, studies carried out in urban ecology, and specific conservation policies that potentially contribute to the aggregate well-being of animals within ecosystems. This shows that the development of these programs is far from being something unrealistic and that policy makers may well be interested in promoting them.

Widespread speciesist attitudes and environmentalist beliefs among the public and life scientists may challenge the development of this discipline by minimizing the relative importance of animal well-being. Nevertheless, independently of the weight assigned to human well-being or other alleged environmental values, it follows from the most important ethical positions that animal well-being matters. Given the data already available about the well-being of animals beyond human control, this provides us with compelling reasons to try to understand and improve their situation.

Finally, our lack of knowledge regarding the complexity of ecosystems could also cast doubt on the development of the discipline, insofar as we may view this effort as futile or counterproductive. However, this downplays the impact of feasible measures already being carried out. The fact that we do not yet have accurate data to assess the net value of such measures is a strong reason for the development of welfare biology instead of a reason against it. In addition, as it happens in other applied fields of research, the assessment of welfare biology's expectable value should be made by considering the prospects of suffering reduction instead of suffering eradication. On the basis of such a criterion, the expected value of developing welfare biology is extremely high.

Notes

1. The term *welfare biology* has sometimes been used with a meaning different to the one employed here, namely, to refer to an applied field of ecology aimed at gaining the knowledge necessary to improve human well-being. Rigorously speaking, however, this field may be more aptly named "human welfare biology" (Wells 1993; Ghosh 1999).
2. Efforts are currently being carried out by a growing number of researchers and supporters of conservationism who reject its lack of concern for nonhuman animals as individuals. A label that has become standard in recent years to name this approach is that of "compassionate conservation" (Harrop 2011; Bekoff 2013). The aims of compassionate conservationist efforts differ from those of welfare biology, as its point is to carry out conservationist efforts in ways that minimize the harms suffered by nonhumans in the process, not to try to minimize the harms suffered by nonhuman animals due to all kind of reasons, including natural ones. Still, compassionate conservationists may be among the most receptive life scientists to the idea of carrying out work in welfare biology.
3. This field of research should not be understood as endorsing the claim that nonhuman animals can be used as long as the harms they suffer are reduced. This position is also known as "animal welfarism" and has been criticized by Francione (1995) and Haynes (2008).
4. Animals living in these areas have been referred to by Donaldson and Kymlicka (2011) as "liminal." It must be noted however that the distinction between these animals and those they consider truly "wild" is not morally relevant, as the interests of the animals are basically the same. They are also affected similarly by their relations to their environments. Indeed, all that was described earlier about why animals suffer and die prematurely in nature apply in the case animals in urban, suburban, and industrial areas. More important, if anything, the distinction here should be made between the different area's animals live and not between the animals themselves.

Bibliography

Adams, C. E. (2016) *Urban Wildlife Management*, Boca Raton: CRC press.

Alonso, W. J., and Schuck-Paim, C. (2017) "Life-fates: Meaningful categories to estimate animal suffering in the wild," *Animal Ethics*, http://animal-ethics.org/life-fates-essay-prize-2017.

Animal Ethics (2016) "Helping animals in the wild," *Wild animal suffering, Animal Ethics*, http://animal-ethics.org/helping-animals-in-the-wild.

Associated Press (2009) "Millions of animals feared dead in Australia," *NBC News*, November 2, http://nbcnews.com/id/29143295/#.W2GLmdIzY2x.

Balcombe, J. P. (2006) *Pleasurable Kingdom: Animals and the Nature of Feeling Good*, London: Palgrave Macmillan.

Bekoff, M. (ed.) (2013) *Ignoring Nature No More: The Case for Compassionate Conservation*, Chicago, IL: University of Chicago Press.

Bernstein, M. H. (2015) *The Moral Equality of Humans and Animals*, Basingstone: Palgrave Macmillan.

Bovenkerk, B., Stafleu, F., Tramper, R., Vorstenbosch, J., and Brom, F. W. A. (2003) "To act or not to act? Sheltering animals from the wild: A pluralistic account of a conflict between animal and environmental ethics," *Ethics, Place and Environment* 6: 13–26.

Brennan, O. (2017) "'Fit and happy': How do we measure wild-animal suffering?" *Wild Animal Suffering Research*, http://was-research.org/paper/fit-happy-measure-wild-animal-suffering.

Callicott, J. B. (1989) *In Defense of the Land Ethic: Essays in Environmental Philosophy*, Albany, NY: State University of New York.

Cappuccino, N., and Price, P. W. (eds.) (1995) *Population Dynamics: New Approaches and Synthesis*, San Diego: Academic Press.

Carpendale, M. (2015) "Welfare biology as an extension of biology: Interview with Yew-Kwang Ng," *Relations: Beyond Anthropocentrism* 3: 197–202.

Cavalieri, P. (2001) *The Animal Question: Why Nonhuman Animals Deserve Human Rights*, Oxford: Oxford University Press.

Clement, G. (2003) "The ethic of care and the problem of wild animals," *Between the Species* 10(3), http://digitalcommons.calpoly.edu/bts/vol13/iss3/2.

Cooper J. E. (1999 [1982]) "Physical Injury", in A. Fairbrother, L. N. Locke, and G. L. Hoff (eds.) *Noninfectious Diseases of Wildlife*, Ames, IA: Iowa State University Press.

Cowen, T. (2003) "Policing nature," *Environmental Ethics* 25: 169–182.

Donaldson, S., and Kymlicka, W. (2011) *Zoopolis: A Political Theory of Animal Rights*, Oxford: Oxford University Press.

Dorado, D. (2015) "Ethical interventions in the wild: An annotated bibliography," *Relations: Beyond Anthropocentrism* 3: 219–238.

Faria, C. (2015) "Disentangling obligations of assistance: A reply to Clare Palmer," *Relations: Beyond Anthropocentrism* 3: 211–218.

Faria, C. (2016) *Animal Ethics Goes Wild: The Problem of Wild Animal Suffering and Intervention in Nature*, PhD thesis, Pompeu Fabra University, Barcelona.

Faria, C., and Paez, E. (2015) "Animals in need: The problem of wild animal suffering and intervention in nature," *Relations: Beyond Anthropocentrism* 3: 7–13.

Fausther-Bovendo, H., Mulangu, S., and Sullivan, N. J. (2012) "Ebolavirus vaccines for humans and apes," *Current Opinion in Virology* 2: 324–329.

Fink, C. K. (2005) "The predation argument," *Between the Species* 13(5): 1–16, http://digitalcommons.calpoly.edu/bts/vol13/iss5/3/.

Francione, G. L. (1995) *Animals, Property and the Law*, Philadelphia, PA: Temple University Press.

Fraser-Bruner, A. (1951) "The ocean sunfishes (family Molidae)," *Bulletin of the British Museum (Natural History)*, 1: 89–121.

Ghosh, D. (1999) *Selected Essays on Welfare Ecology*, Calcutta: Centre for Sustainable Living.

Gompertz, L. (1997 [1824]) *Moral Inquiries on the Situation of Man and of Brutes*, Lewiston: Edwin Mellen.

Guldemond, R. A. R., Purdon, A., and van Aarde, R. J. (2017) "A systematic review of elephant impact across Africa," *PLoS One* 12(6): e0178935.

Guldemond, R. A. R., and Van Aarde, R. J. (2008) "A meta-analysis of the impact of African elephants on savanna vegetation," *Journal of Wildlife Management* 72: 892–899.

Hadley, J. (2006) "The duty to aid nonhuman animals in dire need," *Journal of Applied Philosophy* 23: 445–451.

Hargrove, E. C. (1992) "Foundations of wildlife protection attitudes," in E. C. Hargrove (ed.) *The Animal Rights/Environmental Ethics Debate, The Animal Rights/Environmental Ethics Debate: The Environmental Perspective* (pp. 151–184), Albany: State University of New York.

Harrop, S. (2011) "Climate change, conservation and the place for wild animal welfare in international law," *Journal of Environmental Law* 23: 441–462.

Hartman, K. I. (2011) "Deer rescued from middle of frozen lake in Canada," *Digital Journal*, December 15, http://digitaljournal.com/article/316149.

Haynes, R. P. (2008) *Animal Welfare: Competing Conceptions and Their Ethical Implications*, Dordrecht: Springer.

Horta, O. (2010a) "The ethics of the ecology of fear against the nonspeciesist paradigm: A shift in the aims of intervention in nature," *Between the Species* 13(10): 163–187, http://digitalcommons.calpoly.edu/bts/vol13/iss10/10/.

Horta, O. (2010b) "What is speciesism?" *Journal of Agricultural and Environmental Ethics* 23: 243–266.

Horta, O. (2017) "Animal suffering in nature: The case for intervention," *Environmental Ethics* 39: 261–279.

Johannsen, K. (2017) "Animal rights and the problem of r-strategists," *Ethical Theory and Moral Practice* 20: 333–345.

Johnson, E. (1981) "Animal liberation against the land ethic," *Environmental Ethics* 4: 265–273.

Katz, E. (1992) "The big lie: Human restoration of nature," *Research in Philosophy and Technology* 12: 231–241.

Kirkwood, J. K. (2013) "Wild animal welfare," *Animal Welfare* 22: 147–148.

Kirkwood, J. K., and Sainsbury, A. W. (1996) "Ethics of interventions for the welfare of free-living wild animals," *Animal Welfare* 5: 235–243.

Kirkwood, J. K., Sainsbury, A. W., and Bennett, P. M. (1994) "The welfare of free-living wild animals: Methods of assessment," *Animal Welfare* 3: 257–273.

Koenig, P., Lange, E., Reimann, I., and Beer, M. (2007) "CP7_E2alf: A safe and efficient marker vaccine strain for oral immunisation of wild boar against classical swine fever virus (CSFV)," *Vaccine* 25: 3391–3399.

Lack, D. (1954) *The Natural Regulation of Animal Numbers*, Oxford: Clarendon.

MacArthur, R. H., and Wilson, E. O. (1967) *The Theory of Island Biogeography*, Princeton, NJ: Princeton University Press.

McAuliffe, J. R. (1978) *Biological Survey and Management of Sport-Hunted Bullfrog Populations in Nebraska*, Lincoln: Nebraska Game and Parks Commission.

McCleery, R. A., Moorman, C. E., and Peterson, M. N. (eds.) (2014) *Urban Wildlife Conservation: Theory and Practice*, Dordrecht: Springer.

McCue, M.D. (2010) "Starvation physiology: Reviewing the different strategies animals use to survive a common challenge," *Comparative Biochemistry and Physiology A: Molecular and Integrative Physiology* 156: 1–18.

McMahan, J. (2015) "The moral problem of predation," in A. Chignell, T. Cuneo, and M. Halteman (eds.) *Philosophy Comes to Dinner: Arguments on the Ethics of Eating*, London: Routledge.

Mendi, M. (1988) "The effects of litter size variation on mother-offspring relationships and behavioural and physical development in several mammalian species (principally rodents)," *Journal of Zoology* 215: 15–34.

Morris, M. C., and Thornhill, R. H. (2006) "Animal liberationist responses to non-anthropogenic animal suffering," *Worldviews* 10: 355–379.

Ng, Y.-K. (1995) "Towards welfare biology: Evolutionary economics of animal consciousness and suffering," *Biology and Philosophy* 10: 255–285.

Nussbaum, M. C. (2006) *Frontiers of Justice: Disability, Nationality, Species Membership*, Cambridge, MA: Harvard University Press.

Palmer, C. A. (2010) *Animal Ethics in Context*, New York: Columbia University Press.

Palmer, C. A. (2015) "Against the view that we are usually required to assist wild animals," *Relations: Beyond Anthropocentrism* 3: 203–210.

Pearce, D. (2015) "A welfare state for elephants? A case study of compassionate stewardship," *Relations: Beyond Anthropocentrism* 3: 133–152.

Pluhar, E. B. (1995) *Beyond Prejudice: The Moral Significance of Human and Nonhuman Animals*, Durham: Duke University Press.

Reuters (2002) "Zambia distributes food to starving wildlife," *SOS Rhino*, October 18, http://sosrhino.com/news/rhinonews101802.php.

Reznick, D., Bryant, M. J., and Bashey, F. (2002) "*r*-and *K*-selection revisited: The role of population regulation in life-history evolution," *Ecology* 83: 1509–1520.

Roff, D. A. (1992) *Evolution of Life Histories: Theory and Analysis*, Dordrecht: Springer.

Rolston III, H. (1992) "Disvalues in nature," *The Monist* 75: 250–278.

Rupprecht, C. E., Hanlon, C. A., and Slate, D. (2003) "Oral vaccination of wildlife against rabies: Opportunities and challenges in prevention and control," *Developments in Biologicals* 119: 173–184.

Ryf, P. (2016) *Environmental Ethics: The Case of Wild Animals*, master thesis, University of Basel, Basel.

Sagoff, M. (1984) "Animal liberation and environmental ethics: Bad marriage, quick divorce," *Osgoode Hall Law Journal* 22: 297–307.

Sainsbury, A. W., Bennett, P. M., and Kirkwood, J. K. (1995) "Welfare of free-living wild animals in Europe: Harm caused by human activities," *Animal Welfare* 4: 183–206.

Sapontzis, S. F. (1987) *Morals, Reason, and Animals*, Philadelphia: Temple University Press.

Schmidt, R. H. (1990) "Why do we debate animal rights?" *The Wildlife Society Bulletin* 18: 459–461.

Shrinivasa, M. (2018) "Tusker rescued from muddy pond near Alur," *The Times of India*, March 23, http://timesofindia.indiatimes.com/city/bengaluru/tusker-rescued-from-muddy-pond-near-alur/articleshow/63279182.cms.

Soryl, A. A. (2019) *Establishing the Moral Significance of Wild Animal Welfare and Considering Practical Methods of Intervention*, master thesis, University of Amsterdam, Amsterdam.

Soulé, M. E. (1985) "What is conservation biology?" *BioScience* 35: 727–734.

Stearns, S. C. (1992) *The Evolution of Life Histories*, Oxford: Oxford University Press.
Tomasik, B. (2015a [2009]) "How many wild animals are there?" *Essays on Reducing Suffering*, http://reducing-suffering.org/how-many-wild-animals-are-there.
Tomasik, B. (2015b [2009]) "The importance of wild-animal suffering," *Relations: Beyond Anthropocentrism* 3: 133–152.
Torres, M. (2015) "The case for intervention in nature on behalf of animals: A critical review of the main arguments against intervention," *Relations: Beyond Anthropocentrism* 3: 33–49.
VanDruff, L. W., Bolen, E. G., and San Julian, G. J. (1994) "Management of urban wildlife," in T. A. Bookhout (ed.) *Research and Management Techniques for Wildlife and Habitats*, Bethesda: The Wildlife Society.
Vinding, M. (2014) *A Copernican Revolution in Ethics*, Los Gatos: Smashwords.
Wells, D. (1993) "Green politics and environmental ethics: A defence of human welfare ecology," *Australian Journal of Political Science* 28: 515–527.
White, T. C. R. (2008) "The role of food, weather and climate in limiting the abundance of animals," *Biological Reviews* 83: 227–248.
The Wildlife Society (2011) *Animal Rights Philosophy and Wildlife Conservation*, Bethesda: The Wildlife Society, http://wildlife.org/wp-content/uploads/2014/05/animal_rights_8.30.2011.pdf.
Wobeser, G. A. (2013) *Investigation and Management of Disease in Wild Animals*, Dordrecht: Springer.

36

WILD ANIMALS AS POLITICAL SUBJECTS

John Hadley

In what follows, the expression 'political subject' means something akin to *citizen* or, at least, legitimate stakeholder. To be a political subject is to have a certain status within a nation-state. To have status ordinarily means that the state must provide you with certain 'goods.' Goods can take the form of rights or interest protections. Ideally, rights and interest protections are legally enshrined guarantees. In effect, to be a political subject is to have a stake in the laws and institutions that serve to regulate dealings between citizens. Basically, being a political subject means that you are on the radar, so to speak. Should wild nonhuman animals (hereafter, animals) be on the political radar?

I begin by examining a possible negative answer to that question. I then use the question to frame an examination of how a number of influential political theories may be applied to animals. The conclusion is that, like so many philosophical questions concerning animals, answers to the political subject question turn importantly on facts about the psychological capacities of the animals themselves.

Political Obligation

Some might give a negative answer to the political subject question because animals cannot be held under political obligation. An individual held under political obligation is expected to obey the law (Dagger and Leftkowitz 2014). In line with this objection, because animals are not smart enough to be held morally responsible for their actions, there is no sense in which the law can apply to them.

A response to the objector could take the form of an argument from marginal cases (Dombrowski 1997). An argument from marginal cases aims to expose a double-standard by suggesting that animals and so-called marginal human beings should be treated alike. Paradigmatic examples of marginal human beings are people with advanced dementia and very young children. The rationale is that if one's opponent extends rights to marginal human beings, then consistency demands that they also extend such rights to animals. In line with the response, children and cognitively impaired people are also not smart enough to obey the law, yet they *are* political subjects, so a lack of knowledge and skills must be no barrier to animals being political subjects.

Marginal cases–style arguments are ineffective, however, when the opponent is prepared to bite the bullet and deny to marginal human beings the rights being withheld from animals. Accordingly, a viable response from the opponent might be to argue that the future agency of children and the prior agency of cognitively impaired human beings entitles them to honorary political subject status (Rawls 1999: 446). A proponent of this view denies honorary subject status to only a small class of human beings, namely, human beings that are permanently cognitively impaired and have been so since birth.

But perhaps the claim that animals do not qualify for honorary political subject status on the grounds that they are not smart enough to be held under obligation rests upon an unsound empirical claim? After all, some theorists are prepared to argue that animals have their own moral codes or sets of norms (Bekoff and Pierce 2009). Others argue that animals possess a range of moral emotions and show genuine concern for others (Rowlands 2012). While it may be a stretch to suggest animals understand that they must abide by codes or norms, it is common for people to ascribe a feeling of shame to certain companion animals. Perhaps implicit in the concept of animal shame is a mental state akin to the recognition of having done something wrong? From this point, it is not a big leap to argue that, say, dogs, deserve apprentice political subject status. Notice that this kind of theory would restrict political subjecthood only to animals that feel a sense of shame. This kind of view equates animals with young children whose subjecthood capacities are forever stuck in an early developmental stage.

Some might object, however, that an expectation to obey the law is not necessary for political subject status in the first place. In line with this view, an obligation to obey the law would be an unacceptable constraint on the autonomy of an individual (Wolff 1998 [1970]). One drawback of applying this argument to animals is that it changes how the core value of liberalism, freedom (Christman 2015), is understood. The significance of autonomy is tied to freedom or 'liberty.' The idea is that one's choices in life are *bona fide* choices. While it is easy to claim that animals may freely choose means, it is not so easy to say they freely choose ends. Arguably, the ends of animals are evolutionarily endowed (Agar 2001), which means their preferences are, in effect, preprogrammed into them. But some might argue that there is no difference between the autonomous agency of humans and the preference autonomy of animals because autonomy is a capacity that admits of degree (De Waal 2006). This is the kind of objection that could be made by a proponent of the theory of mind known as continuism. A continuist about animal minds might place the capacities and knowledge that underpin the autonomous agency of persons on a continuum.[1] Only the concept of a continuum, the continuist argues, can do justice to Darwin's signature claim that the differences between human and animal psychological capacities are differences of degree not kind (Darwin 2004 [1879]: 151). If this kind of strategy is successful, then the answer to the question of whether animals should be on the political radar must be yes. But, arguably, this answer is not credible because it would extend political subject status even to invertebrates. After all, if autonomy is a capacity that admits of degree, then the continuum will stretch from human persons and higher apes at one end to single-cell organisms at the other (Agar 2001).

Social Contract Theory

A further objection, however, is that debate over animal intelligence is irrelevant because there are no grounds for animals having allegiance to the state. In line with this objection, a political subject cannot be held under obligation unless the state can ensure their safety. The objection continues, as the state does not afford wild animals with safety; in fact, it facilitates the destruction of their habitat and permits the hunting of individual animals; there is no meaningful sense in which animals could owe the state their allegiance.

An objection of this kind could be made by someone that wants to apply social contract theory to animals (Rowlands 2002; Smith 2016; Talbert 2006). According to social contract theory, the state exists for the mutual benefit of citizens who consent to abide by the laws of the legislature in return for the maintenance of peace and security. Citizens and government alike have a bond of trust that breaks when the state fails to fulfill its contractual obligations (Locke 1990[1690]: 397–398, 411).

To be able to give consent a citizen must have the relevant conceptual knowledge and an ability to project oneself into the future and compare different ways of living. In hospital situations, a guardian makes decisions on the patient's behalf when they are not able to give consent. A similar system of third-party consent for animals would require a credible way of pairing animals with responsible guardians. While it is straightforward to assign guardianship to third parties when the animals concerned are domesticated, the wildness of wild animals makes it difficult to identify someone suitable. Guardians must be able to act in the best interests of animals and be suitably informed about their particular circumstances. Representatives from animal protection and environmental advocacy groups, lawyers, ecologists, and academics with expertise in the relevant species, could be suitable guardians (see Hadley 2015b: 17).

Guardianship helps to put animals on the political radar. Even though they are not smart enough to obey the law, animals can be represented by third parties. Third-party guardianship would enable animals to have a political subject status comparable to marginal human beings who are not morally responsible for their actions but nonetheless can choose how to satisfy their preferences.

Rawlsian Political Theory

Moral Pluralism

John Rawls (1921–2002) was an influential American political philosopher. Does Rawls's theory allow animals to be political subjects? Rawls (2005) defines a political subject (he uses the term 'citizen') as both reasonable and rational. Rawls defines reasonable as having a sense of justice and a willingness to abide by 'fair terms of cooperation' (p. 446). He defines rational as the capacity to revise one's way of life (p. 72). If we accept Rawls's definitions, animals, with the possible exception of higher apes, are neither reasonable nor rational because they cannot understand terms of cooperation and they lack an ability to reflect on their own lives. While it is logically possible to direct a marginal cases style argument in response to Rawls and in support of animals, the argument will be ineffective if the Rawlsian theorist excludes from among the class of the reasonable and rational human beings who have a permanent incapacity.

A number of theorists (Abbey 2007; Cochrane 2010; Garner 2013) identify the Rawlsian notion, moral pluralism, as the main obstacle to extending political subject status to animals. Roughly, moral pluralism is the idea that citizens ought to be allowed to choose their own ways of living and the state should only interfere if considerations of basic justice are at stake. In other words, people ought to be free to do what they want so long as they respect the rights of other citizens. The problem for animals, however, is that Rawls restricts the class of citizens to moral agents only, which means autonomous agency is the most important value. In line with Rawlsian theory, autonomous agency features in the development of principles and the design of institutions in the state. Accordingly, the implications for animals are not good because, as Garner (2013) observes, "moral pluralism would presumably include different attitudes toward the treatment of animals," which entails pluralism, "is likely to trump attempts to protect the interests of animals where such attempts conflict with the liberty of humans" (p. 50). Garner's point is that people use and abuse animals in the service of promoting their own well-being and in the process of finding meaning in life. Regrettably, it is just plainly obvious that many people strongly desire and take pleasure in doing things that cause animals to suffer. Some citizens identify first and foremost as farmers, rodeo riders, biomedical researchers, and so forth, and they find meaning in activities such as hunting, fishing, racing, and breeding animals. Under the protection of laws and norms which ensure the promotion of moral pluralism, the state must leave animal users and abusers to it (Garner 2013: 26).

Public Reason and the Interest Theory of Animal Rights

To suggest that animals are off the radar completely in Rawlsian theory, however, would be an overstatement (Abbey 2007; Garner 2013: 25). In *A Theory of Justice* Rawls condemns cruelty towards animals and acknowledges the ethical significance of sentience. He also enjoins 'duties of compassion and humanity' toward 'animals and the rest of nature' (1999: 448). While it is true that Rawls is pessimistic about attempts to extend the social contract to animals, a brief discussion in *Political Liberalism* opens the door for a means of enfranchising them, albeit through the advocacy of citizens. In a critical reevaluation of his theory of public reason, Rawls says that there are 'numerous political values' (1993: 245) to appeal to in attempts to have legislators pass animal protection and environmental laws.

Roughly, public reason is the idea that debate about political matters ought to proceed by appeal to widely shared beliefs and values (Larmore 2003; Quong 2018). The rationale of public reason is that citizens invariably disagree about how to live and find meaning in life. Citizens live by their own moral and religious codes that shape their sense of identity. Rawls's view is that it is too much to ask a fellow citizen to endorse a policy proposal that jeopardizes the viability of their own worldview.

The requirement to base policy proposals on terms acceptable to all suggests that animal protection measures consistent with mainstream values would be on firm Rawlsian ground. A case for animal property rights, for example, that appeals to basic needs and biodiversity (Hadley 2015b; Cook 2017; Bradshaw 2018) would be one such proposal. After all, property is a ubiquitous institution and few people would deny that animals have an interest in having their basic needs met. While it is true Rawls denied that animals were party to the social contract, and he certainly did not countenance that they could own property, the claim here is simply that public reason creates the space for policy proposals which may secure for animals a degree of protection. For example, animals might be afforded a right to negotiate at times when human land-use decisions put their lives at risk (Hadley 2015b: 19). A human guardian acting in the best interests of the animals would have a chance to persuade the landholder to amend or abandon their development plans. Nothing in a guardianship system like this poses a threat to any citizen's world view. Landholders are not being asked to give up the vocation that brings meaning to their lives.

Public Reason and the Inherent Value Theory of Animal Rights

Public reason would not be as amenable, however, to the inclusion of animals on orthodox animal rights grounds. The problem is that the signature concept of orthodox animal rights philosophy, inherent value (Francione 2000; Regan 1983), is metaphysically contentious (Hadley 2015a). In line with orthodox animal rights theory, inherent value is purportedly a categorical property that is based upon a raft of psychological capacities possessed by 'normal mammals of one year or more' (Regan 1983). A categorical property is a property that does not admit of degree—an individual either possesses the property or does not. To the scientifically minded, however, it is a mystery how a property that does not admit of degree can be based on psychological capacities that clearly do. To put it bluntly, inherent value looks logically similar to a religious idea like sanctity or ensoulment. In fact, recent social science research indicates that animal rights philosophy occupies the place of a functional religion in lives of animal rights advocates (Jamieson et al. 2003; Hadley 2017a). If the functional religion thesis is true, then proponents of animal rights will be subject to the same rules of public debate as any other religiously minded citizens. Public reason precludes appeal to elements of specific worldviews during public deliberation about the principles to govern society (Larmore 2003: 384). This will require activists to refrain from appeals to specialty concepts such as inherent value.

Lockean Political Theory

John Locke (1632–1704) was an influential seventeenth-century English political philosopher. Two important concepts in Lockean theory are self-ownership and the labor-mixing theory of property (Locke [1690] 1990). A number of theorists have argued for various forms of animal self-ownership (Favre 2000; Milburn 2017). Implicit in the concept self-ownership is the idea that the individual is the master of their own destiny and ought to be free from interference. For animals, self-ownership would translate into moral or legal rights to enable them to lead a species-typical life without being hunted or harassed by human beings.

An obvious objection to the extension of self-ownership to animals is to claim that the capacity for autonomous agency is necessary for self-ownership and animals are not autonomous agents. Autonomous agency is needed, the objection continues, to render meaningful the claim that a self-owning individual is genuinely a master of their own destiny. Animals, the objector concludes, are not masters of their own destinies because the shape of their lives is the product of evolutionary endowment and not freely chosen.

A response from the proponent of animal self-ownership is to concede that animals are not autonomous agents but argue they nonetheless have 'authority-over-self' or exhibit 'something resembling self-ownership' (Milburn 2017: 635). Josh Milburn makes a move like this in his recent attempts to politically enfranchise animals on distinctly Lockean grounds. Milburn (2016, 2017), first, claims that the scope of the Lockean proviso ought to extend to animals and, second, argues for animal property rights on the grounds of labor mixing.

Milburn's argument for extending the Lockean proviso to animals presupposes that they are rights bearers (2016: 283). It's safe to say that if, as Milburn presupposes, animals are rights-bearers, then the extension of the Lockean proviso to them is fairly straightforward. The rationale of the proviso is to accommodate the interests of nonowners that are made worse off by private appropriators of land and natural resources. If animals matter enough to be rights-bearers, and assuming that habitat destruction renders them worse off, then they will be among the relevant class of nonowners.

The Labor-Based Theory of Animal Property Rights

Milburn's argument for animal property rights on labor-based grounds, however, is more contentious. Labor mixing, Locke argued, was the means by which individuals could justly acquire exclusive ownership over discreet parcels of commonly held or unowned land or natural resources. The exercise of labor, in line with Lockean theory, metaphorically extends the boundaries of a person's body to include the property object. The idea that one's property is a part of one's body is consistent with the theory of self-ownership as Locke initially conceived it. The problem for Milburn's theory is that the 'root idea' underpinning the acquisition of property is desert (Becker 1977: 49–50; Hadley 2015b: 35–52), and it is a stretch to suggest that animals deserve property rights in return for the exercise of their labor.

The deservingness of labor is the product of either ingenuity or exertion and, arguably, animals are neither especially creative nor exert themselves in the service of a reflectively chosen end. To labor in the morally relevant sense is not simply to expend energy in the service of an evolutionarily endowed end but, rather, to also show ingenuity or engage in exertion that is experienced as some kind of sacrifice for a chosen end. Labor is undertaken as a means to the realization of one's own way of living. The moral significance of such an endeavor is a function of industriousness. Industriousness is a value consistent with the framing assumption of liberal property theory—the initial patrimony (Locke [1690] 1990: 205). The initial patrimony is the founding tenet of property theory that God gave the Earth to human beings in common for purposive use. Implicit in the initial patrimony is the idea that life is ultimately a collective endeavor and stakeholders are bound together by casual necessity. Locke's proviso that would-be appropriators should leave 'enough and as good for others' is an attempt to do justice to the other-regarding values implicit in the initial patrimony.

Milburn may respond by arguing that exertion alone can yield a right even in cases when the individual exerts themselves in the grip of an obsession. His point here is to suggest that freely choosing an end is not necessary for the kind of industriousness that entitles an individual to the fruits of their labor; instead, exertion itself is sufficient, and this opens the logical door to animals having property rights on labor-based grounds. But the deservingness of an obsessive whittler (Milburn 2017: 638), for example, is the creativity or ingenuity that the whittler demonstrates, and not the exertion. While we may say that an activity such as whittling is 'painstaking' and this implies that exertion is the basis of desert, in cases of obsession the laboring comes too easy and alternative grounds needs to be found. The alternative grounds, creativity or ingenuity, is not a good logical fit for animals because their skills are not the product of reflective development over time but instead 'blind' evolutionary endowment.

Apart from the problems relating to its ethical grounds, the labor-based argument for animal property rights also faces scope problems. The Lockean theory extends property rights only to especially active species like beavers or squirrels (Milburn 2017). In contrast, the basic needs theory of animal property rights would enfranchise any animals that use natural 'goods' to satisfy their basic needs. Moreover, the labor theory will extend protection to parcels of land that are smaller than the parcels protected under an interest-based animal property rights regime. As Milburn (2017) notes, laboring behavior only takes place in certain areas within an animal's territorial range. Territory-marking behavior, which Hadley and Cooke posit as the means of determining the boundaries of animal property, will mark larger parcels of land for protection than the parcels marked by laboring behavior. This difference alone is a good reason for preferring the interest-based regime over the labor-based regime as a means of promoting habitat protection and biodiversity conservation.

All habitat-focused property-based attempts to empower animals, however, face implementation problems. One problem is the multiple-owner problem (Hadley 2015b: 59–60). Many different individual animals from different species will have an interest in, or mix their labor with, the very same parcels of land. Whom among them is to be the rightful owner? The scope for conflict between multiple rights-bearers is considerable and a lawyer's banquet beckons.

Wild Animal Sovereignty

These kinds of practical difficulties also apply to the wild animal sovereignty theory. The wild animal sovereignty theory is a combination of orthodox animal rights theory and group rights theory. In line with the wild animal sovereignty theory, in addition to rights not to suffer and be killed, animals bear certain rights because they are members of 'distinct sovereign communities' (Donaldson and Kymlicka 2011; 2013a). Presumably, distinct communities will share overlapping sovereign territories, and some kind of mechanism needs to be found to distinguish one community from another. In this respect, the wild animal sovereignty theory is vulnerable to a multiple sovereign territory problem (Cochrane 2013; Ladwig 2015: 290).

The chief aim of the wild animal sovereignty theory is to facilitate the conditions for wild animal communities to flourish as self-governing and autonomous (Donaldson and Kymlicka 2011: 173). As proponents of sovereignty put it,

> [i]f wild animals are to be accorded sovereignty, we need to show that they are competent to take care of themselves, and to manage their communities independently, separate from humans.... What matters for sovereignty is the ability to respond to the challenges the community faces, and to provide a social context in which its individual members can flourish.
>
> *(Donaldson and Kymlicka (2011: 175)*

What is the best way to demonstrate that wild animals 'are competent to take care of themselves'? It's fair to say that both proponents of wild animal sovereignty and animal property rights are in furious agreement on the answer to this question. Whichever habitat protection theory is preferred, what is necessary is the strict curtailment of damaging human impact on the natural areas in which animals reside. In other words, what's needed is for humans to observe the general rule of traditional animal rights theory: 'leave nature be' (Regan 1983: 363). Perhaps, then, both wild animal sovereignty and animal property rights amount to little more than elaborate ways of restating the traditional hands-off (wild) animal rights position (Hadley 2015b: 93)?

A proponent of wild animal sovereignty may object that only sovereignty entails that humans provide material assistance to animals (Donaldson and Kymlicka 2011: 179). But such an objection makes the mistake of misinterpreting a rule of thumb for a lawlike prescription (Hadley 2015b: 87). The traditional animal rights injunction to 'leave nature be' should be understood *not* as enjoining the complete separation of humans from animals but, rather, the complete separation of humans from animals *to the extent that human impact is pernicious*. If, as is empirically possible, humans can interact with wild animals in ways that don't undermine their 'competence to look after themselves,' then such interaction is consistent with 'leaving them be.' The key question, then, is whether there are 'goods' that wild animal sovereignty theory affords which are not also provided by either traditional animal rights theory or animal property rights theory. This question is difficult to answer definitively because both sovereignty and property rights are logically malleable concepts.

Reparations for Past Injustices

A response from a proponent of sovereignty is that only a theory of sovereignty offers reparations to animals. After all, reparations are common in international law, and human-centered sovereignty theory is a rich source for many ideas in the wild animal sovereignty theory (Donaldson and Kymlicka 2011: 170, 172, 178; Donaldson and Kymlicka 2013b: 771–773). But, as Palmer (2010: 96–114) argues, there are a number of logical problems with extending reparations to groups of animals. The most serious difficulty is that the moral weight of reparations turns on the harm that only the spoken testimony of victims can render visible (Palmer 2010: 99). While it may be possible to gauge the impact of human actions on the well-being or basic rights of animals, the peculiar injustice of the legacy of past actions is a harm that is difficult to determine. There is no sense in which animals can feel aggrieved for past harms and understand the role played by deceased perpetrators and their living beneficiaries. Animals don't have a sense of justice that can be sated by reparations. If Palmer can be read as suggesting that the possession of a sense of justice is necessary in order to be the beneficiary of genuine reparations, then the claim that wild animal sovereignty affords positive goods that animals property rights theory cannot is false.

Interest Theory of Animal Rights

The interest theory does not make the possession of a sense of justice a necessary condition of political subject status; instead, sentience alone is sufficient (Cochrane 2012; Garner 2013). In line with the interest theory, interests are tied to well-being and, consistent with orthodox theories of well-being (Fletcher 2016), well-being is the measure of whether a life is going well or badly.[2] According to the interest theory, sentient animals have interests in certain goods if getting them, or not getting them, makes the animals better or worse off, respectively. An animal has an interest in food and water, for example, because without it they will suffer and eventually die. While an interest-based formula potentially allows for many different goods to be in an individual's interest, interest theorists will only give rights status to those interests which pass a threshold of moral importance. The interest in not suffering and the interest in continuing to live, for example, are purportedly 'important enough to impose a duty upon others' (Cochrane 2012: 43) and, therefore, are the grounds for a *prima facie* right not to suffer and be killed.

There is a fundamental conceptual tension, however, in the basic structure of the interest theory of rights. The tension is a product of bringing together two fundamentally distinct concepts: interests and rights. I will use a financial metaphor to explain this tension. Interests are like cash; rights are like fixed assets. Combining them in a single theory is like making a purchase by paying partly with cash and partly with, say, a block of land. There is nothing inherently wrong with making a purchase in this way but it makes the task more complicated than it needs to be. The difficulty is a product of meshing the liquidity of cash with the solidity of fixed assets; likewise, in the interest theory of rights, the difficulty is a result of meshing a concept (interests) that is designed to be traded away if need be with a concept (rights) that is meant to be held sacrosanct.

The so-called problem of rights inflation (Cochrane 2012: 44–46) illustrates the inherent difficulty of combining interests with rights. Recall that the interest theory ties interests to well-being. Well-being, in turn, is defined with reference to whatever makes life go better or worse. Given that having just about anything makes a person's life go better, then a person has an interest in just about everything. This means that, in theory, a person can have rights to all manner of things. For example, it is in my interest to have the bus run on time because my life goes better when it runs according to schedule. Does the interest theory entail a right to have the bus run on time? In line with the interest theory, there is no logical obstacle standing in the way of someone having an inalienable right to have the bus run on time; it all depends on the decision procedure used to determine whether and how interests translate into rights. But, as is well known in the moral epistemology literature, the credibility of ethical decision-making procedures is difficult to determine (Jamieson 1993). At times when basic moral ideas, such as rights and well-being, are at stake, decision procedures are notoriously open to the charge of being ad hoc or based on unreliable intuitions (Kagan 1998: 12–13). By bringing together two fundamentally different concepts, interests and rights, proponents of the interest theory of rights make an already fraught exercise even more difficult.

Non-ideal Theory

Proponents of the interest theory pride themselves on being sensitive to political reality and each employs a split-level theory to enhance the feasibility of the case for animal rights. Garner (2013) employs an ideal theory–non-ideal theory distinction; Cochrane (2012) invokes the distinction between *prima facie* and concrete rights. Implicit in the use of distinctions is the idea that the kind of ideal world thinking that characterizes orthodox animal rights theorizing delivers proposals that are utopian. The concern of theorists to produce non-ideal theory begs the questions, Why not produce theories that reflect reality more wholeheartedly? and Why not have the current property status of animals set the background of theoretical analysis (See O'Sullivan 2011; Smith 2012)? The task of the 'non-ideal' philosopher who is sympathetic to animals would then be to eke out a progressive theory by redescribing ordinary valuing practices in new ways (See Hadley 2013, 2017b: 999–1003).

Conclusion

In the preceding, I discussed the question of whether animals should be political subjects. I gave an indication of what follows for animals if they are accepted as political subjects. Some philosophers have gone so far as to argue that animals should have property rights over their habitat. Others have argued that important philosophical concepts such as self-ownership, sovereignty, and guardianship apply to animals just as much as to humans. Implicit in all these arguments is the idea that animals are sufficiently smart enough to be political subjects.

There are philosophical ideas that make it difficult to answer yes to the political subject question. If a *bona fide* political subject must be able to obey the law, then the answer to the question is no. Two further obstacles are Rawlsian concepts such as moral pluralism and public reason. The former rules out giving animals the kind of rights that would prevent people from harming them; the latter requires animal advocates to campaign in ways that they may find too restrictive (Humphrey and Stears 2006). Debate about whether animals should receive reparations is ongoing; so, too, is debate about the extent to which political theory should be sensitive to political realities.

Notes

1. David Peña-Guzmán (2017) makes a continuist argument for animal suicide.
2. The leading theories of well-being are hedonism, the desire theory, and the objective list theory.

Further Reading

Andrews, K (2015) *The Animal Mind: An Introduction to the Philosophy of Animal Cognition*. Oxon: Routledge. (A comprehensive and accessible introduction to the philosophical debate over animal minds.)

Hargrove, E (1992) *The Animal Rights/Environmental Ethics Debate: The Environmental Perspective*. Albany: State University of New York Press.
(The seminal introduction to the theoretical and practical issues involved in the debate between animal rights individualism and environmental holism.)

Svolba, D. (2016) Is there a Rawlsian Argument for Animal Rights? *Ethical Theory and Moral Practice* 19: 973–984. (An opinionated discussion of the extension of Rawlsian theory to animals with a detailed critique of Rowland's position.)

Bibliography

Abbey, R. (2007) "Rawlsian resources for animal ethics," *Ethics and the Environment* 12: 1–22.

Agar, N. (2001) *Life's Intrinsic Value*, New York, NY: Columbia University Press.

Becker, L. (1977) *Property Rights: Philosophic Foundations*, Boston, MA: Routledge and Keegan Paul Ltd.

Bekoff, M., and Pierce, J. (2009) *Wild Justice: The Moral Lives of Animals*, Chicago, IL: University of Chicago Press.

Bradshaw, K. (2018) "Animal property rights," *University of Colorado Law Review* 89: 809–861.

Christman, J. (2015) "Autonomy and liberalism: A troubled marriage," in S. Wall (ed.) *The Cambridge Companion to Liberalism* (pp. 141–162), Cambridge: Cambridge University Press.

Cochrane, A. (2010) *An Introduction to Animals and Political Theory*, London: Palgrave McMillan.

Cochrane, A (2012) *Animal Rights Without Liberation*, New York, NY: Columbia University Press.

Cochrane, A. (2013) "Cosmozoopolis: The case against group-differentiated animal rights," *Law, Ethics and Philosophy* 1: 127–141.

Cooke, S. (2017) "Animal kingdoms: On habitat rights for wild animals," *Environmental Values* 26: 53–72.

Dagger, R., and Lefkowitz, D. (2014) "Political obligation," in E. N. Zalta (ed.) *The Stanford Encyclopedia of Philosophy*, Fall 2014 ed., https://plato.stanford.edu/archives/fall2014/entries/political-obligation/.

Darwin, C., (2004 [1879]) *The Descent of Man*, London: Penguin Books.

De Waal, F. (2006) *Primates and Philosophers: How Morality Evolved*, Princeton, NJ: Princeton University Press.

Dombrowski, D. A. (1997) *Babies and Beast: The Argument from Marginal Cases*, Urbana, IL: University of Illinois Press.

Donaldson, S., and Kymlicka, W. (2011) *Zoopolis: A Political Theory of Animal Rights*, Oxford: Oxford University Press.

Donaldson, S., and Kymlicka, W. (2013a) "A defense of animal citizens and sovereigns," *Law, Ethics and Philosophy* 1: 143–160.

Donaldson, S., and Kymlicka, W. (2013b) "Reply: Animal citizenship, liberal theory and the historical moment," *Dialogue* 52: 769–782.

Favre, D. (2000) "Equitable self-ownership for animals," *Duke Law Journal* 50: 473–502.

Fletcher, G. (2016) *The Routledge Handbook of Philosophy of Well-Being*, London: Routledge.

Francione, G. (2000) *Introduction to Animal Rights: Your Child or the Dog*, Philadelphia, PA: Temple University Press.
Garner, R. (2013) *A Theory of Justice for Animals*, New York, NY: Oxford University Press.
Hadley, J. (2013) "Liberty and valuing sentient life," *Ethics and the Environment* 18(1): 87–104.
Hadley, J. (2015a) "A meta-level problem for animal rights theory," in Elisa Aaltola and John Hadley (eds.) *Animal Ethics and Philosophy: Questioning the Orthodoxy* (pp. 15–30), London: Rowman and Littlefield International.
Hadley, J. (2015b) *Animal Property Rights: A Theory of Habitat Rights for Wild Animals*, Lanham: Lexington Books.
Hadley, J. (2017a) "Religiosity and Public Reason: The Case of Animal Rights Advocacy," *Res Publica* 23(3): 299–312.
Hadley, J. (2017b) "From welfare to rights without changing the subject," *Ethical Theory and Moral Practice* 20(5): 993–1004.
Humphrey, M., and Stears, M. (2006) "Animal right protest and the challenge to deliberative democracy," *Economy and Society* 35(3): 400–422.
Jamieson, D. (1993) "Method and moral theory," in P. Singer (ed.) *A Companion to Ethics*, Oxford: Blackwell.
Jamieson, W. V., Wenk, C., and Parker, J. V. (2003) "Every sparrow falls: Understanding animal rights activism as a functional religion," in S. J. Armstrong and R. G. Botzler (eds) *The Animal Ethics Reader* (pp. 556–562), London: Routledge.
Kagan, S. (1998) *Normative Ethics*, Boulder, CO: Westview Press.
Ladwig, B. (2015) "Against wild animal sovereignty: An interest-based critique of *zoopolis*," *The Journal of Political Philosophy* 23(3): 282–301.
Larmore, C. (2003) "Public reason," in S. Freeman (ed.) *The Cambridge Companion to Rawls*, Cambridge: Cambridge University Press.
Locke, J. (1990 [1690]) *Two Treatise of Government*, ed. Peter Laslett, Cambridge: Cambridge University Press.
Milburn, J. (2016) "The demandingness of Nozick's 'Lockean' proviso," *European Journal of Political Theory* 15(3): 276–292.
Milburn, J. (2017) "Nonhuman animals as property holders: An exploration of the Lockean account," *Environmental Values* 26(5): 629–648.
O'Sullivan, S. (2011) *Animals, Equality and Democracy*, Basingstoke: Palgrave McMillan.
Palmer, C. (2010) *Animal Ethics in Context*, New York: Columbia University Press.
Peña-Guzmán, D. M. (2017) "Can nonhuman animals commit suicide? *Animal Sentience* 20(1).
Quong, J, (2018) "Public reason," in E. N. Zalta (ed.) *The Stanford Encyclopedia of Philosophy*, spring 2018 ed., https://plato.stanford.edu/archives/spr2018/entries/public-reason/.
Rawls, J. (1999) *A Theory of Justice, Revised Edition*, Oxford: Oxford University Press.
Rawls, J. (2005) *Political Liberalism, Expanded Edition*, New York, NY: Columbia University Press.
Rega, T. (1983) *The Case for Animal Rights*, Berkeley, CA: University of California Press.
Rowlands, M. (2002) *Animals Like Us*, London: Verso.
Rowlands, M. (2012) *Can Animals Be Moral?* Oxford: Oxford University Press.
Smith, K. (2012) *Governing Animals*, New York: Oxford University Press.
Smith, K. (2016) "A public philosophy for the liberal animal welfare state," in Robert Garner and Siobhan O'Sullivan (eds.) *The Political Turn in Animal Ethics* (pp. 69–84), London: Rowman and Littlefield International.
Talbert, M (2006) "Contractualism and our duties to nonhuman animals," *Environmental Ethics* 28: 201–215.
Wolff, R. P. (1998) *In Defense of Anarchism*, 3rd ed., Berkeley, CA: University of California Press.

PART VI

Animal Activism

Editor's Introduction

All the contributors to this section believe that there is something very seriously wrong about the way that human beings relate to nonhuman animals, and they're all keen to change human/animal relations in one way or another. Suppose, then, that we object to the way that humans use animals. How might we respond? Many people have thought that some form of activism is required, although there are many forms on which people focus. There are the techno-optimists, who think that we'll end animal farming by creating artificial meat, dairy, and eggs. There are the ecofeminists who think that we won't be able to address animal exploitation without addressing the underlying structures that produce all other forms of exploitation—racial, gender-based, economic, and so on. There are those who see individual dietary change—typically, going vegan—as morally obligatory for most people. There are some who are trying to build spaces that can serve as a model of how humans might relate to animals, namely, animal sanctuaries. Others are open to all sorts of strategies but insist that whatever we do, it ought to be whatever does the most to reduce animal suffering.

What are the pros and cons of these responses to the human use of animals? It's difficult to answer this question without understanding how social movements evolve, function, and try to change existing structures. How did animal activism come to take the various forms that it's taken? How do activist communities balance the goals of gaining supporters (which is sometimes easier if you water down the message) and preserving their identities (which dictates that you should stay true to your founding principles)? How can activists make changes from within the legal system we have? What are the lessons that we can learn from past activist movements, both in terms of what to imitate and what to avoid? How does our particular moment in history dictate what is and isn't politically possible?

This section addresses these and related questions. However, before we delve into that material, let's stand back for a moment. As I mentioned, there's significant agreement among the contributors to this section. Given as much, we might wonder how to think about animal activism if you *aren't* on board with the assumptions that motivate it. On such views, animal activism can seem extreme, especially in its more confrontational forms: think, for instance, of activists who try to release chickens from chicken sheds or who hold protests inside grocery stores and restaurants. How should we think about the ethics of activism if we aren't activists ourselves?

There are at least two ways that we might think about the ethics of activism. The first says that we can't separate the ethics of activism from the ethics of the cause. So if the cause is just, then many activist strategies are going to be morally permissible, and for some people in certain circumstances, perhaps even morally required. And if the cause isn't just, then many of the same tactics will be morally wrong. On this sort of view, until we know whether it's morally wrong for us to be raising chickens for food, we can't tell whether it's morally wrong to tell people to stop eating them—much less break into a chicken farm and try to release them. If it's morally wrong to be raising chickens, then acting on their behalf may well be morally obligatory. It isn't a violation of your privacy to try to convince you that your diet is morally objectionable. And if there's nothing wrong with raising chickens for food, then perhaps it's wrong to tell others to stop eating them. After all, it's none of their business what you purchase and consume. The upshot: if this view is correct, then there is no conversation about the ethics of activism—or, at least, not much of one—until we answer the kinds of questions that contributors explored in the rest of the book.

However, another view is available. On this approach, our thinking about the ethics of activism should be sensitive to "the burdens of judgment"—namely, all the factors that make it difficult for people to figure out what's true, even when they're doing their best. The idea, very roughly, is that being sensitive to these burdens should affect the standards to which we hold one another. What are some of the barriers to knowing what's true? There are many. For instance, we often don't have enough information: the evidence is insufficient to tell us what to believe. Or the information that we have is very difficult to interpret. Moreover, we have an array of biases that make it difficult for us to process even good information. Worse, the people around us do as well, making it even harder for us to reach the truth. For example, not only do we each have a confirmation bias—where we are strongly inclined to trust evidence that confirms what we already believe rather than to look for disconfirming evidence—but we are surrounded by people who operate in the same way and encourage us to trust only evidence that supports our preconceptions.

As a result, there's a lot of reasonable disagreement about moral matters, where people really are trying to do their best to think clearly about morality, but it's simply hard to do, and thoughtful people will reach different conclusions. This means that when we think about the ethical standards to which we hold others, we've got reasons to make those standards sensitive to the burdens of judgment. If we concede that there's room for a lot of reasonable disagreement, even if everyone is doing their best to figure out the moral truth, then we had better be careful about condemning forms of activism simply because we reject the moral assumptions on which they are based. First, that might set back our own objectives as advocates for particular causes. Second, if we were to say that it's wrong to agitate for change based on assumptions we reject, then that might prevent us from learning important moral truths. After all, if we are doing something very seriously wrong by eating animals, then it's important to know that. Unless we think it's essentially impossible that animal activists are correct in their views, it's good to have them out there making the case. Third, even when activists aren't presenting arguments but are simply acting on their beliefs—say, by engaging in civil disobedience—we can still think of that as a way of trying to convince us of what's true. These sorts of actions communicate that activists aren't simply saying what they say because it's expedient to do so; instead, they're willing to bear some costs for their views, which discourages us from writing them off as caught up in a fad. Even disruptive, confrontational activism can be part of a societal conversation about what's right and wrong.

If this second view is correct, then "activist" shouldn't be a label that we apply to dismiss someone, as it so often is in public discourse. And while we might disagree with their views, and so not think that we have any reason to act as they do, we won't condemn their actions on this basis. This isn't to say that there are no constraints at all on what activists can permissibly do. Perhaps it's always wrong to destroy property in pursuit of an activist agenda, or perhaps it's always wrong to physically harm other people in pursuit of such an agenda. After all, activists should also be sensitive to the ways that the burdens of judgment affect their audience! The point is just that there's plenty of room for discussion, and skepticism about what activists believe doesn't necessarily justify any criticisms of what activists do. And as we come to better understand the ways that activists think, perhaps even skeptics can see activists in a more favorable light.

37

THE HISTORY OF ANIMAL ACTIVISM

Intersectional Advocacy and the American Humane Movement

Janet M. Davis

In September 2017, People for the Ethical Treatment of Animals (PETA) kicked off a vegan billboard campaign in Detroit, Michigan: "Peace in our lunchtime! Choose peace. Choose vegan" (Baldas 2017). In local interviews, PETA president Ingrid New revealed that she and her colleagues chose Detroit for the campaign kickoff because the Federal Bureau of Investigation and WalletHub had recently ranked the city as the nation's most violent and unhealthiest: "We thought Detroit would be a good place to encourage residents to choose peace, nonviolence and compassion" (Baldas 2017). The local reaction was immediate. Sam Riddle, a civil rights activist and political director of the Michigan National Action Network, condemned PETA's emphasis on personal choice as implicitly racist:

> Detroit may be America's fattest and most violent city but to attribute that to us not eating our veggies is racist BS because Detroit is also America's blackest, most segregated and poorest city. . . . If PETA wants Detroit to improve our diet, PETA must address the historic roots of why we eat like we eat including slavery on plantations run by PETA ancestors.
>
> *(Jackman 2017)*

Fundamentally, Riddle flatly rejected PETA's narrowly construed call for dietary self-improvement, arguing that PETA's tactics willfully ignored the brutal consequences of centuries of trauma, structural inequality, and white supremacy.

The reception in Detroit marked PETA's latest controversial foray into intersectional interspecies politics. What I mean here is a politics in which systems of discrimination based on race, class, gender, and species are interconnected and inseparable forms of oppression. In 2005, the organization was criticized for its traveling exhibition "Are Animals the New Slaves?" which featured twelve photographic panels juxtaposing explicit violence toward African Americans and animals. One of the most shocking panels, titled "Hanging," showed a photograph of two African American lynching victims surrounded by a white mob next to a picture of a dead cow hanging in a slaughterhouse. The National Association for the Advancement of Colored People (NAACP) and the Southern Poverty Law Center swiftly dismissed the exhibition as a racially insensitive publicity ploy. NAACP spokesperson John White observed incredulously: "PETA operates by getting any publicity they can. They're comparing chickens to black people?" (*PETA Rethinks Slavery Analogy* 2005). When the exhibition traveled to New Haven, Connecticut, on August 8, 2005, Scot

X. Esdaile, NAACP president of the Greater New Haven Chapters, remarked, "Once again, Black people are being pimped. You used us. You have used us enough" (Wright 2011). After touring in seventeen cities to stinging criticism, PETA's leadership suspended the exhibition. Nonetheless, such campaigns continued. In 2009, the group picketed an American Kennel Club (AKC) dog show at Madison Square Garden (among other places) wearing authentic-looking mock Ku Klux Klan robes as a way to equate the AKC and its purebred registry with the white supremacist KKK and its "master race" ideologies. In 2011, PETA filed a lawsuit against SeaWorld, which argued (unsuccessfully) that the Thirteenth Amendment constitutionally protected a group of orcas from enslavement at SeaWorld (Fontaine 2009; Macon 2009; *PETA Sues SeaWorld for Violating Orcas' Constitutional Rights* 2011).

PETA leaders have long maintained that it is incumbent to connect the shared histories of racial oppression and animal cruelty—however shocking or uncomfortable—to underscore the totality of interspecies violence. By contrast, civil rights activists argue that such comparisons—however well intentioned—are unavoidably racist. Specifically, they contend that these associations resurrect a pernicious body of racist pseudoscience from the eighteenth, nineteenth, and early twentieth centuries that equated black people with monkeys, apes, and other animals to justify myriad forms of white supremacist dehumanization: enslavement, lynching, racial segregation, disenfranchisement, and the exhibition of the Congolese Mbuti man, Ota Benga, in the Monkey House at the Bronx Zoo in 1906 (Kim 2015; Bradford and Blume 1992). Critics have also charged that the ubiquity of white people in contemporary animal protectionism has amplified the movement's disconnection from the lived experiences of people of color. Civil rights activist Tim Wise, accordingly, has called modern animal advocacy an animal "whites" movement (Wise 2005).

This chapter explores the fraught history of American animal protectionism and intersectionality. Specifically, I examine the movement's historical relationship with African American human rights struggles—from antebellum abolitionism to twentieth-century Jim Crowism. Metaphors of enslavement and direct participation in abolitionism helped catalyze the organized animal protection movement in Great Britain and in the United States. I pay special attention to a historical phase of American animal protectionism starting in the late nineteenth century and ending after World War II that looked quite different from the contemporary animal "whites" movement. During this era, African American animal welfare leaders forged a dynamic and synergetic interracial and interspecies social justice movement dedicated to human and animal welfare. As a case study, I examine the remarkable career of Reverend Frederick Rivers Barnwell (1883–1958), an African American animal welfare leader who made animal advocacy inseparable from his civil rights career as a minister, teacher, NAACP chapter founder, Urban League member, and public health worker in Jim Crow Texas. Nonetheless, Barnwell's extraordinary career has been all but forgotten. I argue that the historical erasure of Barnwell's work—and that of other black animal advocates across the Jim Crow South—has contributed to a pervasive ideological climate of antagonistic bifurcation in the contemporary "animal whites" movement. Furthermore, I contend that key transformations in the movement's ideological scope contributed to this fragmentation: as the movement became increasingly dedicated to animal subjectivity and animal rights, it became less connected to the human rights networks and reform movements that originally brought it to life.

Abolitionism and Antebellum Reform: The Catalyst for an Organized Animal Protection Movement in the United States

In the late eighteenth century, the seeds of the animal protection movement crystallized out of the acceleration of transatlantic abolitionism and other social reform movements flowering during the Second Great Awakening (1790–1840). In this era of expansive religious revivalism and reform, the English abolitionist, evangelical, and member of Parliament, William Wilberforce helped found the Abolition Society in 1787 and dedicated his political career to abolishing slavery. In 1800, he helped a group of supporters introduce a bill in Parliament to ban bull-baiting. Although the bill was unsuccessful, Wilberforce and his colleagues persisted: in 1821, Wilberforce co-sponsored a bill in the House of Commons "to prevent cruel and improper treatment of Cattle" (broadly defined to protect laboring livestock), which also failed. But the bill's primary sponsor, Richard Martin, a member of Parliament for Galway, successfully reintroduced the legislation in 1822. Thereafter, organized institutional animal advocacy spread quickly in Great Britain and beyond (Turner 1980: 15, 21, 39–40). The Society for the Prevention of Cruelty to Animals (SPCA) was incorporated in 1824 and became the "Royal" SPCA in 1840 when Queen Victoria provided royal sponsorship for the organization. In a society powered by animal muscle, early British animal welfare legislation targeted acts of physical violence and neglect toward laboring animals. Animals used in blood sports, such as bear and bull baiting, cockfighting, dogfighting, and coursing, were also subjects of new protective legislation (Ritvo 1987; Shevelow 2008).

The metaphor of interspecies bondage was a powerful literary tool for antislavery writers during the antebellum era. Historian Joshua Kercsmar observes that British and American abolitionists frequently coupled human enslavement and animal cruelty to condemn the fundamental brutality of speculative capitalism (Kercsmar 2018; Kercsmar 2014). American antislavery activists, such as Lydia Maria Child, and formerly enslaved writers, such as Henry Bibb, created a new literary genre, the cruelty narrative, which told horrific stories of enslavement accompanied by stark illustrations of stakes, cuffs, chains, and shackles used to confine and torture people. Historically, these brutal devices of punishment and confinement were also used as a form of brutal dominion in livestock husbandry, a shared material history of violence that Marjorie Spiegel explores in *The Dreaded Comparison* through the strikingly similar visual images of shackles, chains, and whips used to enslave and to confine people and domestic animals (Spiegel 1996). Slaveholders also applied the same language to livestock and enslaved human beings, such as breaking, breeding, buying, branding, selling, and castration (Davis 2016: 41). The nation's most famous antislavery novel, *Uncle Tom's Cabin* (1852), included several examples of animal cruelty. Harriet Beecher Stowe portrayed the abolitionist character Mrs. Bird as a moral exemplar whose virtue was doubly reinforced when she whipped her wayward sons and sent them to bed without dinner for stoning a kitten with the neighbor boys. Vicious slaveholders, such as Simon Legree, mistreated animals, a detail that Stowe used to underscore the totality of their moral degradation (Stowe 1852: 88, 373–374). Throughout Stowe's long literary career, she infused animal welfare into her plotlines, including her collection of short stories *A Dog's Mission; or, the Story of the Old Avery House and Other Stories* (Stowe 1880).

Stowe also was in direct contact with one of the nation's most prominent future animal welfare leaders during the antebellum period. Her attorney was the Boston abolitionist, Samuel Sewell, who practiced law with a newly minted young attorney named George Thorndike Angell, the future founder and president of the Massachusetts Society for the Prevention of Cruelty to Animals (MSPCA) in 1868. Before the Civil War, Sewell and Angell's practice in Boston was a major center of abolitionist activity. Sewell provided William Lloyd Garrison with substantial financial backing in 1831 to launch his influential abolitionist paper, *The Liberator*. Sewell contested the constitutionality of the Fugitive Slave Law (1851) in representing Thomas Sims and Anthony Burns, both of whom had traveled to Massachusetts after escaping slavery and were court-ordered back to the South and re-enslaved. Furthermore, Sewell was in regular contact with John Brown during the 1850s and provided legal counsel to his large family in the aftermath of the raid on Harper's Ferry (Davis 2016: 43). George Angell's moral consciousness was deeply influenced by his direct contact with abolitionism at his law practice. In this bustling, intimate environment, Angell regularly met abolitionists, such as Garrison, who also used his newspaper platform to promote animal welfare issues. Other future animal welfare leaders, such as Caroline Earle White, founder of the Pennsylvania SPCA (1867) and Women's Pennsylvania SPCA (1869), were directly tied to the abolitionist movement. During this same period, new animal welfare laws were passed in New York (1829), Massachusetts (1836), and Philadelphia (1855), among other places.

Animals were legally considered property under the tenets of common law and police officers were granted exclusive powers of enforcement. Nonetheless, this flurry of legislation represented an important cultural and legal shift. Springing from a growing cultural emphasis on free moral agency and human perfectibility during the Second Great Awakening, abolitionism and animal protectionism emerged out of the same historical moment when social reform and moral reform were one and the same.

A New Social Movement Emerges After the Civil War

The Civil War hastened the institutionalization of the animal protection movement. The war spawned a Constitutional rights revolution through the ratification of the Thirteenth, Fourteenth, and Fifteenth Amendments, which abolished slavery, made manhood suffrage a constitutional right, and enshrined the constitutional right to equal protection under the law, respectively, among other provisions. During the Civil War, the recent technology of photography became the "Clio of the war," documenting a sprawling theater of unmitigated suffering through haunting images of dead soldiers and horses (often next to each other), and the physical violence of enslavement, the Civil War's root cause. One of the most famous wartime photographs was of man named Gordon (also referred to as Peter), who escaped slavery in 1863 and joined a Union encampment in Baton Rouge. In the series of photographs, Gordon turns his face to the camera from behind, and his back, which faces the viewer, reveals a thick tangle of raised scars—offering irrefutable visual evidence of the grisly nature of enslavement. First published in *Harper's Weekly* in 1863 and sold as *carte de visite* photographs, the images swiftly dispelled insidious proslavery propaganda about the benevolence of the "peculiar institution" (Faust 2008; Trachtenberg 1990). Bringing human and animal suffering to a national audience through the medium of photography and print, the Civil War helped create the cultural conditions for an organized animal welfare movement.

A year after the war ended, on April 10, 1866, the New York Legislature incorporated the American Society for the Prevention of Cruelty to Animals (ASPCA). The first formal state animal protection organization in American history, the ASPCA's charter of incorporation gave this pioneering organization an extraordinary degree of authority: its officers were vested with the powers of arrest, thus empowering them to become unofficial police officers who dressed the part in police-like uniforms and badges to enforce the state's anticruelty law. Journalist Horace Greeley observed that Henry Bergh, a shipping heir and the ASPCA's founder and first president, became the most recognizable person in New York City because he routinely prowled the streets and personally arrested people caught in the act of overloading or beating animals used primarily in haulage and transportation. Within days of incorporation, Bergh and his powerful allies successfully drafted a groundbreaking new state anticruelty law. While animals were still considered property under the new law, historian Susan Pearson observes that this legislation recognized animals as sentient beings who possessed the right to protection from cruelty and neglect (Pearson 2011: 78–80). A year later, Bergh and his supporters expanded the law to prohibit additional forms of cruelty, including blood sports and abandonment.

Within fifteen years, a majority of states and cities had passed new anticruelty laws. Far-flung new animal protection societies spanned the United States, including Pennsylvania (1867), Massachusetts (1868), San Francisco (1868), Cincinnati (1873), and Texas (1879). In the early 1900s, some form of animal welfare legislation existed in every state in the Union. In a muscle-powered society, animal advocates dedicated their efforts primarily to laboring livestock and food animals. They paid particular attention to the plight of urban horses as ubiquitous agents of haulage and transportation. Animal advocates tried to improve the abusive logistics of livestock transport by railroad from western rangelands to urban stockyards and slaughterhouses. They successfully lobbied Congress to pass the nation's first federal animal welfare law in 1873, which required livestock to receive food, water, and rest stops at least every twenty-eight hours. They shut down dogfights, rat baiting, and cockfights. They petitioned against the use of live animals for scientific experiments in laboratories and classrooms, and they routinely shot aged and infirmed workhorses to end lives of suffering. By 1874, the animal protection movement had expanded to include children after ASPCA officials prosecuted a case of child abuse in New York City. Consolidated new "humane societies" included children and animals as subjects of protection (Davis 2016).

George Angell, founder and president of the MSPCA, expanded the movement's formative historical nexus with abolitionism. The MSPCA's monthly magazine, *Our Dumb Animals*, often featured stories about President Lincoln, "the Great Emancipator," and his fierce devotion to animal and human welfare (Davis 2016: 45). In 1889, Angell founded a new branch of the MSPCA, the American Humane Education Society (AHES), which was dedicated to local grassroots animal welfare educational activities in schools, churches, and the missionary field—both at home in the United States and abroad in India, the Philippines, Cuba, Puerto Rico, Hawaii, and other sites of American evangelical proselytization. In 1890, Angell underwrote the publication of Anna Sewell's haunting biography of a horse, *Black Beauty*, which had been published in England in 1877 and became a major catalyst for banning the fashionable, physically painful checkrein, also called the bearing rein, which was attached to the bit and extended to a second attachment point on the harness or saddle, thus forcing the horse's head into a fixed position. Angell made his historical evocation of enslavement, abolition, and animal advocacy plain in advertising the American publication of *Black Beauty* as "the Uncle Tom's Cabin of the Horse" (Davis 2016: 42).

Reverend Frederick Rivers Barnwell—A Case Study of Intersectional Animal Advocacy in Jim Crow Texas

At the turn of the twentieth century, the AHES had grown into a transnational organization, whose activities and influence had expanded far beyond its Boston headquarters. Although the earliest SPCA leaders during the Gilded Age were white elites, a new generation of African American activists expanded the humane movement into the Jim Crow South in the late nineteenth and early twentieth centuries. Specifically, these animal advocates made this history within black institutions—specifically, churches and schools.[1] Starting in the 1910s, a group of southern black ministers, educators, and reformers worked as state representatives for the AHES. The organization's Boston headquarters supplied state field secretaries with copies of *Our Dumb Animals* for distribution and sale, as well as stereopticons, lantern slides, and other media materials for lectures, classroom activities, parades, plays, and pageants. State representatives also received copies of *Beautiful Joe* (1893), and *Black Beauty* (1877) for use in far-flung humane education activities at schools, churches, and youth group meetings (*Black Beauty: The Uncle Tom's Cabin of the Horse* 1890: 26). Like their white brethren in the humane movement, black activists were typically college-educated and well-to-do.

An in-depth exploration of the work of Reverend Frederick Rivers Barnwell illuminates the important ways in which African American animal advocates forged an intersectional social movement. Barnwell was a civil rights leader and Texas field representative for the American Humane Education Society beginning in 1915 (*More Work Started in the South* 1916: 190). He was a minister, a charter member of the Fort Worth Branch of the NAACP (1918), a member of the Fort Worth Urban League, the Young Men's Christian Association, and composer and chorus director for the Baptist Missionary and Education Convention of Texas. He was also a leader of the National Negro Health Movement and served as the director of Negro Health Education with the Texas Tuberculosis Association during the New Deal, among many other activities. Born on January 11, 1883, in South Carolina, Barnwell graduated from Lincoln University in Chester County, Pennsylvania, in 1908 and earned a graduate degree in Sacred Theology (STB) there in 1911. Founded in 1854 as the Ashmun Institute and renamed in memory of President Lincoln in 1866, Lincoln was the nation's first degree-granting black university. Known as the "Black Princeton" because of its similarly demanding classical curriculum and foundational ties to the Presbyterian Church, Lincoln University embodied W.E.B. DuBois's call for an intellectual "Talented Tenth" to lead a burgeoning civil rights movement. Later graduates included poet Langston Hughes (1929) and US Supreme Court Justice Thurgood Marshall (1930) (*History: Lincoln University* 2019). In 1942, Lincoln University awarded Barnwell an honorary doctorate in recognition of his distinguished career (*Dr. F. R. Barnwell Is Awarded An Honorary Degree* 1942). In activities spanning nearly a half century, Barnwell's animal advocacy and social justice work reveal symbiotic connections across three features of the black freedom struggle: physical mobility, the black church, and education, with a special emphasis on moral and physical health.

Barnwell's work for the AHES was dependent on travel by automobile across Texas. Reports published in *Our Dumb Animals* provide a detailed log of miles traveled, locations visited, and numbers of people who heard him speak. In March 1917, for example, Barnwell drove nearly one thousand miles, delivered seventy addresses and four sermons, visited twenty-five schools and four Sunday schools, and handed out more than 2,000 pieces of humane education literature to children and adults (*Our Month's Work* 1917: 10). Twenty years later, Barnwell was still driving extensively. According to the AHES's annual report for 1937, "Mr. Barnwell traveled nearly 10,000 miles to reach the colored people of Texas . . . where 22,733 persons were reached. He secured a large amount of press publicity and distributed nearly 5,000 pieces of literature" (*Great Activity in the South* 1938: 44).

Barnwell's extraordinary mobility highlights a critical facet of post-emancipation black life in Texas. The right to travel freely to reunite with family members, engage in courtship, or find decent work was a hallmark of hard-fought citizenship that also became a central theme in black literature, blues, and gospel music (Foner 1988; Davis 1999). Yet after the Congressional Compromise of 1877 terminated Reconstruction, white Texans, like other white southerners, enacted broad local anti-vagrancy laws that enabled the arrest of any black person who was either jobless and searching for work or already contracted to an employer and seeking a new job. This post-Reconstruction system of arrest, steep fines, incarceration, and enslavement through contract labor (which was constitutionally enshrined in the Thirteenth Amendment) persisted into the twentieth century (Barr 1995: 53–55, 148). In a Jim Crow society that routinely circumscribed black movement and mobility through vagrancy and segregation laws, in addition to outright violence, Barnwell's extensive automobile travel made its own powerful statement about black freedom.

Barnwell journeyed frequently to black churches, a cornerstone of black life after the Civil War, when emancipated African Americans founded their own religious institutions. In Texas, new branches of the Baptist and Methodist Church constituted the vast majority of post–Civil War black churches (Barr 1995: 67). Barnwell preached humane sermons to myriad denominations, using abundant Scriptural references to kindly biblical stewardship for "the least among us" (Matthew 25: 31–46). In 1921, he successfully convinced black Methodist conferences to adopt resolutions supporting "humane responsibility and stewardship" as part of new church-based humane education programs (*Colored Methodists* 1921: 138). When Barnwell was at home in Fort Worth, he served as an associate pastor at St. Andrew's Methodist Episcopal Church. But he was often on the road. April was a particularly busy travel month owing to the annual celebration of Be Kind to Animals Week, an event designed largely for schoolchildren launched nationally by the American Humane Association in 1915. Local Humane Sunday sermons opened and closed the festivities. In April 1917, Barnwell spent the week in Austin, where he gave nine sermons and five lectures at local black churches to 2,000 people (*Humane Sermons in Austin* 1917: 26).

Other sermons addressed specific strands of the animal protection movement. As part of a campaign to support warhorses sponsored by the American Red Star Animal Relief during World War I,

Barnwell devoted portions of his sermons to compassionate horse care in the field. On June 30, 1918, Barnwell preached to more than 1,100 black soldiers stationed at Camp McArthur in Waco in his dual capacity as an animal protectionist, minister, and field marshal for the Colored Branch of the Red Cross. The following Sunday, July 7, 1918, Barnwell returned to Camp McArthur, where he "addressed a large audience both white and colored" and distributed copies of "The Horse's Prayer" (*Mr. Barnwell in Camp* 1918: 58; *Program Arranged for Negroes' Red Cross Benefit* 1918). Barnwell reported how millions of Allied horses served in the Great War, primarily in haulage, pulling heavy artillery, food, and supplies through pouring rain and deep mud to the miserable trenches on the Western Front. After hearing Barnwell preach at Marlin that summer, Mrs. Eliza E. Peterson, a Texarkana-based superintendent of the Woman's Christian Temperance Union's Department of Work Among Colored People, wrote an unsolicited letter to *Our Dumb Animals* in praise of Barnwell: "He is accomplishing vast good for righteousness. Sentiment is being changed right along from indifference to consideration of the sub-human creatures.... I am glad to find Rev. Barnwell sincere and through and through a humane man" (*Appreciative Words* 1918: 62). In a transitional era when motor power was supplanting muscle power, Barnwell's powerful language of patriotism, service, and citizenship made an urgent case for the warhorse as an essential working member of society (Davis 2014).

Barnwell's pleas for equine kindness during World War I and its aftermath occurred amid escalating racist violence in Houston (1917), East St. Louis (1917), Chicago (1919), and Washington, D. C. (1919), which targeted black soldiers—whose very appearance in uniform challenged white supremacy. Mobs also attacked recent African Americans migrants who flocked to urban wartime centers in search of work and to escape debt peonage. Barnwell always brought complimentary copies of *Our Dumb Animals* for community leaders wherever he traveled. Because of its constant presence in Barnwell's presentations, the magazine amplified the ideological content of his work. *Our Dumb Animals* explicitly condemned racist violence in wartime America:

> Europe is finding it hard to reconcile the wild and savage outbreaks in the United States where law is trampled underfoot, with our proud claim of making the world safe against the might of brutal force. Not only have the lynchings of our colored citizens nearly doubled during the six months of 1918 as compared with the same six months of 1917, but the mob spirit, rapidly spreading.... We are reaping what we have sown. Too long we have tolerated this spirit. War has naturally inflamed it, and widened its fatal power.
>
> (The Mob *1918: 67)*

The minister and the magazine offered a synthetic intersectional critique of American violence that escaped censure among white audience owing to a seemingly apolitical focus on animals.

Throughout Barnwell's career, children were essential to his work. When Eliza Peterson, Superintendent for the Woman's Christian Temperance Union (WCTU), met Barnwell, she was impressed to see "a crowded house and more young people than old, which all was the better" (*Appreciative Words* 1918: 62). Peterson's observations tacitly acknowledged the humane education work that she and other African American WCTU activists had pioneered with children in Texas and across the South during the 1890s. In a society where women were still disenfranchised, WCTU activists engaged in moral suasion using their gendered authority as moral guardians of the home to press for social change. Two WCTU departments were especially important in bringing black WCTU animal protectionists together in a formal institutional setting—the Department of Work among Colored People and the Department of Mercy, which wedded the WCTU's broader mission of sobriety to moral uplift and kindness to people and animals. Mrs. Lucinda (Lucy) Simpson Thurman, a national superintendent of the Department of Work among Colored People and a Department of Mercy leader, was so effective as a WCTU organizer in Texas that African American members simply called their branches "Thurman WCTUs" (Davis 2016: 72–73).

Building on African American WCTU members' dynamic vision of children as harbingers of a just future, F. Rivers Barnwell regularly spoke at black schools—a foundational institution of black culture and society after emancipation. Starting in 1915, he created local youthful animal welfare groups at schools, churches, and elsewhere, known as Bands of Mercy, which were dedicated to proper pet and livestock care, feeding wildlife, bird identification, writing animal stories, making animal art, and sharing personal stories of direct intervention when encountering cruelty. Founded in 1882 by Massachusetts SPCA president George Angell and Thomas Timmins, an English minister, the Bands of Mercy movement was transnational in scope and figured prominently in Barnwell's work. *Our Dumb Animals* highlighted the number of new bands he founded around the Lone Star State each year. In 1936, for example, he organized 772 Bands of Mercy with 81,696 members during 15,560 miles of travel. He estimated that he reached over 100,000 people "by the spoken word"(*Colored Workers in South Carolina and Texas* 1937: 44).

Barnwell's animal protection activities, like his sermons, occasionally crossed the color line. In 1919, the Tarrant County Humane Society sponsored an interracial youth birdhouse building competition, with half of the prize money slated for African American children and the other half for white children. Barnwell's energetic leadership generated 150 African American entries, while only fourteen white children submitted birdhouses. Under these circumstances, black children received awards that would have otherwise gone to white children. All birdhouses subsequently were placed in city parks across Fort Worth—a mark of integration in a segregated city (*Bird-House Contest in Texas* 1919: 185).

Barnwell's youthful activities often took place in public spaces. His groups staged plays and pageants at local parks, and they marched in colorful street parades featuring exhortations to kindness on banners and posters, all of which gave public form and voice to the movement. In April 17, 1937, Barnwell staged a huge event at Greenway Park in Fort Worth, replete with awards for best humane posters and scrapbooks, speeches by local leaders, including W.A. Meacham, assistant superintendent of the Fort Worth Public Schools, and an elaborate pageant jointly held by African American students from across the city. "The Nations in Humane Ideals" paid special attention to Native Americans as exemplars of kindness. In the words of the pageant,

> If America had followed the example of the early Indian in his love and care for animals we would be more humane today. The Indians had no humane societies for the care and protection of animals but they loved them most devotedly. They didn't kill animals for sport but only when they needed them for food for their families.
>
> (Pageant in Fort Worth *1937: 93)*

While such representations offered a romantic and unitary view of diverse Native cultures, the pageant unabashedly criticized contemporary American animal practices.

Children's posters for Be Kind to Animals Week offer another connection to Reverend Barnwell's work in other fields of social justice. In a photograph of Barnwell with the Band of Mercy of Gay Street School in Fort Worth in 1927, children posed with "Be Kind to Animals" banners, as well as posters of bird nests commanding viewers to "Touch Not," and "Please Don't Throw Stones at Me," as well as images of bare trees captioned with the specter of starvation: "What You Waste Would Keep Them Alive All Winter" (*Band of Mercy of Gay Street School, Fort Worth, Texas* 1927: 110).

Speaking directly to bodily suffering and deprivation, the posters for Be Kind to Animals Week complemented Barnwell's concurrent work in black health care and education. Barnwell provided ministerial service during National Negro Health Week, an annual event stressing food security, hygiene education, neighborhood sanitation drives, and calls for more medical professionals in black communities, which contained only one black doctor per 3,000 residents in 1915. Booker T. Washington and the National Negro Business League launched the event just months before Washington's death in 1915 to combat the grim statistical reality of black life in the early twentieth century, which included an infant mortality rate for black children of 180.6 per 1,000 live births, or twice the rate of white babies (Quinn and Thomas 2001: 44–45). During the New Deal, Barnwell was appointed director of Negro Health Service with the Texas Tuberculosis Association. While Barnwell's work stressed individual responsibility for one's good health, he also pressed for state action to address the structural causes of poor health in black communities, which were often forced to locate in areas without clean running water and basic municipal services (*Texas Plans to Celebrate Health Week* 1936). In his dual capacity as health official and humane movement leader, Barnwell gladly embraced popular media, such as radio, newspapers, and the theater to educate black communities and to reach white audiences (Barnwell 1937: 565–571). Throughout his career, he never lost sight of the pernicious ways that structural racism did violence to black lives.

Barnwell, however, pressed for social and economic equality through nonconfrontational methods because he lived in Jim Crow Texas. In this way, his approach mirrored that of other southern African American animal protectionists, such as Reverend Richard Carroll of South Carolina and William Key of Tennessee. Historian Paula Tarankow contends that Carroll and Key's accommodationist tactics in the spirit of their associate Booker T. Washington garnered critical support from affluent, politically influential white patrons (Tarankow 2018). During the NAACP's national boycott movement of the racist feature-length movie *Birth of a Nation* in 1915, Barnwell told W.E.B. DuBois that staging an integrated picket line in Dallas and Fort Worth would be virtually impossible for safety reasons. Instead, he urged DuBois and the NAACP to write separately to the black and white departments of the City Welfare Board to "try to influence" the board's film censors through moral suasion (Stokes 2007: 145).

Nonetheless, as a founder of the Fort Worth chapter of the NAACP, Barnwell aligned himself more closely to DuBois than Washington. *Our Dumb Animals* vigorously supported the boycott of *Birth of a Nation* in Boston and beyond:

> This is what the photo-play known as 'The Birth of a Nation' does. . . . it appears to have been most skillfully and deliberately planned to arouse and widen in the North that prejudice against the Negro that has characterized the worst elements of the South . . . to permit the continuance of this exhibition in Boston, the home of Garrison and Phillips and Sumner, is a reproach to our city.
>
> (Race Prejudice *1915: 8)*

Consequently, each time that Barnwell gave a free copy of *Our Dumb Animals* to local officials during his humane education presentations, they also received a direct rebuke of white supremacy on the magazine's pages.

Barnwell retired from active field service with the AHES in 1942, but he remained a dedicated animal advocate and civil rights activist for the rest of his life (*Omega's Century Club is Born* c.1950: 4). He continued to participate in multiple vectors of black institution building, including leadership duties of the regional chapter of the historically black Omega Psi Phi fraternity. As chair of the Social Welfare Committee of the Texas Commission on Interracial Cooperation in the postwar period, Barnwell pushed for full implementation of President Truman's Fair Deal in the Lone Star State (*Rev. Barnwell Receives Nat'l. Omega Award* 1947). His activism ended with his death. On February 5, 1958, Barnwell died at home after suffering a heart attack. He was seventy-five (*Frederick Rivers Barnwell Death Certificate* 1958). Black media mourned his passing. Fort Worth's oldest black high school, I.M. Terrell High, honored his memory with the creation of the annual F. Rivers Barnwell Service Award, until its closure in 1973 when the city's schools were integrated (*Fort Worth Assembly to Present Eight Debutantes* 1960; Gilmore 2017). But outside of his community, Barnwell was largely forgotten. Even the physical traces of his vibrant neighborhood institutions vanished. In the 1960s, the construction of Interstate 35 demolished Barnwell's home at 1328 Louisiana Street, where he and his wife raised their seven children. The mammoth highway project also obliterated the Gay Street School, where Barnwell's students proudly displayed their banners for Be Kind to Animals Week (Google Maps 2018). Greenway Park, home to Barnwell's public Bands of Mercy pageants and plays, still exists but is encircled by I-35, a tangle of frontage roads, and grim industrial parks. It is inaccessible to pedestrians.

The physical erasure of Barnwell's home and the sites of his activism in Fort Worth mirror the broader erasure of his pioneering work from the historical memory of modern animal protectionism. Barnwell's generation of animal advocates was deeply involved in myriad social justice movements, which they deemed to be inseparable from their work with animals. Many of these movements were tied to church-based activism. Indeed, many of Barnwell's African American contemporaries in the AHES, including Seymour Carroll, Richard Carroll, and John W. Lemon, were ministers. During the twentieth century, however, the humane movement gradually secularized. Formative church-based alliances with the WCTU and other temperance organizations dissolved after Prohibition was repealed in 1933. Similarly, the totalizing "gospel of kindness" that wedded animal advocacy and child protectionism splintered as the new profession of social work used social science methodologies to build a new human-centered field. At the same time, the expansion of the Darwinian revolution helped scientists understand the enormous depth and complexity of animal sentience, intelligence, culture, and social life. Consequently, as the humane movement became more animal-centered, it gradually lost its connections to the human rights movements that originally catalyzed animal protectionism.

By the 1970s, animal protectionism had branched into two distinct, but interconnected movements. The first represented a direct ideological and programmatic through-line from the post–Civil War SPCA animal welfare movement with a focus on preventing and prosecuting cruelty and neglect, as well as sheltering and adoption. The second strand constituted a growing dedication to animal rights, rejecting legal definitions of animals as property and making veganism a litmus test of one's commitment to the movement. Like earlier phases of the movement, animal rights activists and their scholarly allies deployed the history and language of slavery and racism to mobilize their constituents. In his classic text *Animal Liberation* (1975), utilitarian philosopher Peter Singer posited that "speciesism," a term coined by Richard Ryder, is a form of oppression akin to racism and sexism (Singer 1990: 6; Ryder 1971: 41–82). In the 1990s, legal scholars Gary L. Francione and Anna Charlton formulated a position as "new abolitionists," a call for veganism and a commitment to nonviolence in all realms of life Francione and Charlton 2015, 1996).

Yet, as the opening of this essay suggests, the deployment of abolitionist language sits uneasily with many contemporary social justice movements. In the absence of deeper historical analysis giving equal time to human rights and animal rights, antiracist claims in defense of animals can remain potentially suspect, in part because they ultimately deemphasize the enduring pervasiveness of racial oppression. Political scientist Claire Jean Kim observes that PETA's unitary interspecies battle cry, "We Are All Animals," rests on a postracial elision, a teleological sleight of hand in which racism—ostensibly vanquished—serves as a lesson for animal liberation:

> ["We Are All Animals"] succinctly repackages and falsely truncates the story of anti-Blackness to serve the present purposes of animal liberation. The Black struggle is not the subject (or even one of the subjects) . . . it is symbol, tool, and prop.
>
> *(Kim 2015: 285)*

Yet if scholars and activists move beyond "symbol, tool, and prop" to consider the expansive work of civil rights activists, such as Reverend Frederick Barnwell, in tandem with their focus on animals, a more complex and more hopeful precedent for multispecies justice might emerge.

Yet the question remains: How might a shared movement take shape? A court battle in Iowa offers one potential path forward. In 2012, the Iowa state legislature passed a draconian law that criminalizes the act of recording and/or reporting any irregular activity at animal agricultural facilities, such as dairies, feedlots, and slaughterhouses. Five other agricultural states subsequently passed similar anti-whistle-blower legislation, commonly known as "Ag Gag" laws (Prygoski 2015). In Iowa, journalists, advocacy groups, employees, or anyone else working undercover to document abuses at these facilities face felony charges and up to a year in jail if convicted.

Industrial animal agriculture has long been a flashpoint for human rights activists and animal protectionists alike. Labor activists, immigrant rights groups, civil liberties organizations, and other human rights groups denounce the dangerous working conditions endemic to virtually all animal agricultural facilities, including long working hours for low wages, hazardous offal-slicked floors, the constant "speed up" of assembly-line butchering and processing, recurrent psychological trauma, and the horrific frequency of industrial accidents. In 2004, Human Rights Watch published a damning report on the abusive laboring conditions at modern factory farms: *Blood, Sweat, and Fear: Workers' Rights in U.S. Meat and Poultry Plants* (Human Rights Watch 2004; Genoways 2015; Foer 2010). Given the large number of undocumented immigrants employed in the animal agriculture industry, many workers fear termination and deportation if they seek redress for abusive laboring conditions. Similarly, animal protection groups, such as Farm Sanctuary, PETA, and the ASPCA, condemn the ubiquity of suffering at such operations, including confinement practices such as farrowing crates and battery cages, forced molting for laying hens, coerced repeated pregnancies among dairy cows and sows, and stunning and slaughter practices that prolong pain (Foer 2010).

In Iowa, human rights groups and animal advocates successfully joined forces to fight against the Ag Gag law as a violation of the First Amendment. In 2017, the American Civil Liberties Union (ACLU) of Iowa, the Animal Legal Defense Fund, the Center for Food Safety, and Public Justice filed a joint lawsuit on behalf of an equally diverse group of clients, including Bailing Out Benji, an Iowa-based group working to abolish puppy mills, PETA, and Iowa Citizens for Community Improvement (ICCI), an organization dedicated to racial justice, immigrant rights, environmental regulation, and workers' rights at factory farms. On January 9, 2019, the US District Court for the Southern District of Iowa ruled that the state's Ag Gag law was unconstitutional and thus could no longer legally be enforced. This victory came on the heels of similar decisions in Idaho and Utah (*Victory in Lawsuit against Iowa's 'Ag Gag' Law* 2019). According to Rita Bettis Austen, legal director for ACLU Iowa,

> Ag [G]ag . . . effectively silenced advocates and ensured that animal cruelty, unsafe food safety practices, environmental hazards, and inhumane working conditions go unreported for years. We are so pleased with the Court's order today and that the law has finally been held to be unconstitutional.
>
> (Victory in Lawsuit against Iowa's 'Ag Gag' Law *2019)*

Ultimately, the pluralistic plaintiffs joined forces out of a shared commitment to the constitutional right to bear witness in order to combat cruelty to animals and people. Their success in states where powerful agricultural lobbyists normally sway the political process represents an instructive example for future coalitions dedicated to multispecies justice.

Note

1. Similarly, historian Ava Purkiss has pushed back on the history of physical culture as a "white" movement by finding black women's exercise regimes and healthful activities in early and mid-twentieth-century black institutions.

Bibliography

"Appreciative words" (1918) *Our Dumb Animals* 51(4): 62.

Baldas, T. (2017) "PETA to Detroit: Stop eating animals. You're the most violent, fattest city in America," *Detroit Free Press*, September 27, www.freep.com/story/news/local/michigan/detroit/2017/09/27/peta-detroit-violence/708606001/

"Band of Mercy of Gay Street School, Fort Worth, Texas" (1927) *Our Dumb Animals* 60(7): 110.

Barnwell, F. R. (1937) "Health education of negroes provided by press, radio, and theatre," *Journal of Negro Education* 6(3): 565–571.

Barr, A. (1995) *Black Texans: A History of African Americans in Texas, 1528–1995*, Norman: University of Oklahoma Press.

"Bird-House contest in Texas" (1919) *Our Dumb Animals* 51(12): 185.

"Black beauty: The uncle Tom's cabin of the horse" (1890) *Our Dumb Animals* 23(3): 26.

Bradford, P., and Blume, H. (1992) *Ota Benga: The Pygmy in the Zoo*, New York: St. Martin's Press.

"Colored methodists act" (1921) *Our Dumb Animals* 53(9): 138.

"Colored workers in South Carolina and Texas" (1937) *Our Dumb Animals* 70(3): 44.

Davis, A. (1999) *Blues Legacies and Black Feminism: Gertrude "Ma" Rainey, Bessie Smith, and Billie Holiday*, New York, NY: Vintage.

Davis, J. (2014) *'Where Gasoline Can't Go': Equine Patriotism and the American Red Star Animal Relief Campaign during World War I*. London: School of Oriental and African Studies Publications, www.soas.ac.uk/history/conferences/war-horses-conference-2014/file94551.pdf.

Davis, J. (2016) *The Gospel of Kindness: Animal Welfare and the Making of Modern America*, New York: Oxford University Press.

"Death certificate for Frederick rivers Barnwell" (1958) State of Texas, Certification of Vital Record, Department of State Health Services, Vital Statistics Unit.

"Dr. F. R. Barnwell is awarded an honorary degree" (1942) *Chicago Defender*, May 9, p. 15.

Faust, D. G. (2008) *This Republic of Suffering: Death and the American Civil War*, New York: Knopf.

Foer, J. S. (2010) *Eating Animals*, New York: Back Bay Books.

Foner, E. (1988) *Reconstruction: America's Unfinished Revolution, 1863–1877*, New York: Harper & Row.

Fontaine, S. (2009) "PETA dresses up as KKK, equates dog-breeding to racism," *NewsOne*, February 12, https://newsone.com/110111/peta-dresses-up-as-kkk-equates-dog-breeding-to-racism/

"Fort worth assembly to present eight debutantes" (1960) *New Pittsburgh Courier*, December 24, 1960, News & Newspapers—ProQuest.

Francione, G., and Charlton, A. (1996) *Rain without Thunder: The Ideology of the Animal Rights* Movement, Philadelphia: Temple University Press.

Francione, G., and Charlton, A. (2015) *Animal Rights: The Abolitionist Approach*, Logan: Exempla Press.

Genoways, T. (2015) *The Chain: Farm, Factory, and the Fate of Our Food* (New York: Harper Paperbacks.

Gilmore, C. (2017) "The color of Texas classrooms and the first black high school in fort worth, NBC Dallas-fort worth, February 28, www.nbcdfw.com/news/local/The-Color-of-Texas-Classrooms-And-the-1st-Black-High-School-in-Fort-Worth-414953653.html

"Great activity in the south" (1938) *Our Dumb Animals* 71(3): 44.

"History: Lincoln University," www.lincoln.edu/about/history.

Human Rights Watch (2004) *Blood, Sweat, and Fear: Workers' Rights in U.S. Meat and Poultry Plants*, New York: Human Rights Watch.

"Humane sermons in Austin" (1917) *Our Dumb Animals* 50(2): 26.

Jackman, M. (2017) "PETA's Detroit vegan campaign draws charges of racism from Sam riddle," *Detroit Metro Times*, September 28, www.metrotimes.com/the-scene/archives/2017/09/28/petas-detroit-vegan-campaign-draws-charges-of-racism-from-sam-riddle

Kercsmar, J. (2014) *The Politics of Husbandry: Managing Animal and Slave Labor in England and America, 1550–1815*, PhD thesis, University of Notre Dame, South Bend.

Kercsmar, J. (2018) "Animals and critiques of capitalism in early antislavery writings," *American Historical Association Annual Meeting*, January 4, Washington, D.C.

Kim, C. J. (2015) *Dangerous Crossings: Race, Species, and Nature in a Multicultural Age,* Cambridge, UK: Cambridge University Press.

Macon, D. (2009) "Trivialize white supremacy by comparing its victims to abused animals," *Stuff White People Do Blog*, February 9, http://stuffwhitepeopledo.blogspot.com/2009/02/trivialize-victims-of-white-supremacy.html.

"Mr. Barnwell in camp" (1918) *Our Dumb Animals* 51(4): 58.

"The mob" (1918) *Our Dumb Animals* 51(5): 67.
"More work started in the south" (1916) *Our Dumb Animals* 48(12): 190.
"Omega's century club is born" (c. 1950) [*Omega Psi Phi*] *Oracle*, n.d., p. 4.
"One month's work" (1917) *Our Dumb Animals* 50(1): 10.
"Pageant in fort worth" (1937) *Our Dumb Animals* 70(6): 93.
Pearson, S. (2011) *The Rights of the Defenseless: Protecting Animals and Children in Gilded Age America*, Chicago: University of Chicago Press.
"PETA rethinks slavery analogy" (2005) *Los Angeles Times*, August 14, http://articles.latimes.com/print/2005/aug/14/nation/na-peta
"PETA sues SeaWorld for violating orcas' constitutional rights" (2011) *PETA Blog*, October 25, www.peta.org/blog/peta-sues-seaworld-violating-orcas-constitutional-rights/
"Program arranged for Negroes' Red Cross Benefit" (1918) *Fort Worth Star-Telegram*, August 11, Readex by NewsBank.
Prygoski, A. (2015) *Brief Summary of Ag-gag Laws, Animal Legal & Historical Center*, Lansing: Michigan State University, www.animallaw.info/article/brief-summary-ag-gag-laws
Purkiss, A. (2016) *'Mind, Soul, Body, and Race': Black Women's Purposeful Exercise in the Age of Physical Culture, 1900–1939*, PhD thesis, University of Texas at Austin.
Quinn, S. C., and Thomas, S. (2001) "The national negro health week, 1915 to 1951: A descriptive account," *Minority Health Today* 2(3): 44–45.
"Race prejudice" (1915) *Our Dumb Animals* 48(1): 8.
"Rev. Barnwell receives Nat'l. Omega Award" (1947) *Fort Worth Mind*, January 18, p. 1, Readex, a Division of NewsBank.
Ritvo, H. (1987) *The Animal Estate: The English and Other Creatures in the Victorian Age*, Cambridge: Harvard University Press.
Ryder, R. (1971) "Experiments on animals," in S. Stanley Godlovich, R. Godlovich, and J. Harris (eds.) *Animals, Men and Morals: An Enquiry into the Mal-Treatment of Non-Humans*, London: Victor Gollanoz.
Shevelow, K. (2008) *For the Love of Animals: The Rise of the Animal Protection Movement*, New York: Henry Holt.
Singer, P. (1990) *Animal Liberation*, 2nd ed., New York: New York Review of Books.
Spiegel, M. (1996) *The Dreaded Comparison: Human and Animal Slavery*, New York: Mirror Books.
Stokes, M. (2007) *D. W. Griffith's The Birth of a Nation: A History of the Most Controversial Motion Picture of All Time*, New York: Oxford University Press.
Stowe, H. B. (1852) *Uncle Tom's Cabin, or, Life among the Lowly*, New York: Signet Classic.
Stowe, H. B. (1880) *A Dog's Mission; or, the Story of the Old Avery House and Other Stories*, New York: Fords, Howard, & Hulbert.
Tarankow, P. (2018) *Race and the Romance of Reunion: African American Animal Advocates in the New South*, American Historical Association Annual Meeting, January 4, Washington, D.C.
"Texas plans to celebrate health week" (1936) *Chicago Defender*, March 14, p. 5, ProQuest Historical Newspapers.
Trachtenberg, A. (1990) *Reading American Photographs: Images as History, Mathew Brady to Walker Evans*, New York: Hill & Wang.
Turner, J. (1980) *Reckoning with the Beast: Animals, Pain, and Humanity in the Victorian Mind*, Baltimore: Johns Hopkins University Press.
"Victory in lawsuit against Iowa's 'Ag Gag' law" (2019) ACLU.org, January 9, www.aclu.org/news/victory-lawsuit-against-iowas-ag-gag-law
Wise, T. (2005) "Animal Whites: PETA and the politics of putting things in perspective," *Blog*, August 13, www.timwise.org/2005/08/animal-whites-peta-and-the-politics-of-putting-things-in-perspective/
Wright, D. (2011) "Another PETA exhibit compares animal cruelty to slavery," BET, July 21, www.bet.com/news/national/2011/07/21/another-peta-exhibit-compares-animal-cruelty-to-slavery.html

38

THE POLITICAL AND CULTURAL SOCIOLOGY OF ANIMAL ADVOCACY

Erin M. Evans

> A couple of interactions I had with veterinarians were very transformative for me because they modeled the doctor-patient relationship, not the researcher-subject relationship. They modeled the ability to identify pain and distress, the presence of disease, and that kind of interaction was very important to me. It was like I started to interact with these . . . or started to recognize these animals as having a predicament, a predicament of having pain and distress.
>
> *(Interview 2014)*

The chapter-opening quote is from an interview I conducted with a bioethicist in 2014. This reflection is part of his story of leaving his career as a laboratory scientist for a career in bioethics and animal welfare, where he remains today. Other people I interviewed had similar experiences with veterinarians, students, and fellow researchers before they uprooted their careers and abandoned their lives as animal researchers. This dramatic conversion from animal researcher to animal advocate isn't common, but what is common—and what is taken for granted by so many animal advocates—is talk of concern for animal welfare. With the increasing number of animals killed for food, science, entertainment, or clothing each year, it's difficult to prove that this concern influences practice, but as a person who remembers what it was like trying to find soy milk or tofu in the grocery store in the early 1990s, let alone a menu with "vegan" as a descriptor, I can attest that there have been marked changes.

This rhetoric or discourse of animal welfare is built partly by structural factors that both constrain and facilitate a shift in the logic of animal usage. The process through which a social movement's demand is incorporated into law and policy is called "institutionalization," and even when a law such as the Animal Welfare Act is passed, its implementation can be weak and largely symbolic. An important question and debate for both scholars and activists who focus on animal advocacy are whether animal advocates should pursue welfare reform laws that provide incremental changes in conditions for animals or whether they should pursue only abolition (Francione and Garner 2011). A debate of this form underpins virtually every major social movement, from civil rights to environmentalism to feminist movements. It also underpins the focus of this chapter: the sociology of becoming a compassionate for animals and/or of becoming an animal advocate. The difference between these attitude changes is a matter of degree.

I start with the three major theoretical frameworks that scholars use to explain why social movements like animal advocacy emerge. Then I introduce some of the theorizing on the nature of human–animal relationships and the boundaries involved, with a focus on how boundaries that determine whether an animal is worthy of moral concern are maintained or dismantled. Throughout the chapter, I use quotes from my research on animal research to illustrate the nuances of the personal traits, institutional conditions, and social movement tactics that factor into the growing power of the animal advocacy movement.

Why Do People Participate in Social Movements?

> *I think a big thing that kind of provoked my departure and transformation was a failure to rationalize, a failure to believe in ideological explanations for these actions. Instead I didn't go into it with another ideology, like an animal rights ideology, I went into it with an open mind and an awareness to kind of judge for myself what was right and what was wrong.*
>
> *(Interview 2014)*

In the preceding quotation, a long-term animal rights activist describes her path toward animal advocacy. Similar to the bioethicist I describe in the beginning of the chapter, she worked in an animal laboratory but as a student research assistant working toward a degree in neuroscience. As an assistant, she worked with rats doing procedures she described as "horrifying . . . surreal and nightmarish," like one where she implanted pieces of candy into rats' skulls as a training technique for the actual devices. Eventually, she was a whistleblower against the lab and started working as a full-time animal activist. I asked her, "What do you think made you feel empathetic towards the animals in your lab?"

> *That's a great question that I'm still trying to understand myself. I think a big part of it was simply an openness to the experience of emotional connection with other beings and I, in spite of everything I thought I knew or believed, I ended up empathizing with these animals.*

But there are others who experience or are exposed to videos and images of animal abuse and either don't feel empathy or don't feel it to the extent required to act on it. What sorts of conditions drive people such as this whistle-blower to empathize enough with animals to act on their empathy? How do these individual stories speak to broader societal changes animal advocates want to see? In the following, I provide a foundation for social movements theory by reviewing how scholars explain patterns of participation across social movements. Then I explore some of the sociological bases for an individual's discovery of a personal compassion for animals.

Theoretical camps reflect broad approaches scholars use to explain some element of our world. Those who use "macro" approaches investigate contextual conditions—such as cultural norms or political, economic, or legal systems—in search of patterns that could explain why people participate in movements and why those movements emerge. Others use "micro" approaches, which focus on individual interactions that might explain why individual people choose to participate in a movement and how those choices diffuse across groups of people. Macro approaches emphasize the role of structures that constrain behavior. Micro approaches emphasize agency, the ability of people to make individual choices with control and freedom. The following brief overview of social movement theory examines how participation in activism, like all social phenomena, involves the interplay of individual agency and structure. Here, I focus on three major theoretical camps that fall within the spectrum of macro and micro approaches: one uses resources to explain social movement processes and outcomes, the second uses messaging, and the third explains movements through political processes.

Early research specific to social movements was sparked by civil rights uprisings in the 1950s and 1960s. At the time, many scholars characterized these massive waves of protests as mob behavior as opposed to rationally calculated collective action by people without political power (Kornhauser 1959; see review in McAdam et al. 1996; Park 1972).[1] Pushing against this characterization, scholars emphasized that groups with less power had to use protest tactics to disrupt the status quo and force power holders to respond to their needs. Research slowly began emphasizing the use of organizing processes and collective action as resources that are used rationally (Lipsky 1968; McCarthy and Zald 1977; Piven and Cloward 1977). This shift in theoretical approach toward "resource mobilization" reflected a broader desire to de-stigmatize movements of the 1960s and dispel myths of activist irrationality. However, this theory left a gap in the literature on the ideational and cultural qualities of building public support, recruiting new activists, and sustaining movement participation (Goodwin and Jasper 1999).

Grievances can be long-standing before they are recognized and presented in ways that galvanize support. Resource mobilization was less useful for describing the micro-level interpretive work activists do to mobilize people and garner sympathy from the public. Meaning is constructed through individual interactions that diffuse and become embedded in society, and meaning informs how activists construct the presentation of their grievances and demands. This process is called "framing" (Goffman 1974; Simmel 1955; Snow and Benford 1988).

For instance, I could frame the grievance of meat production in several different ways. I could frame it as an environmental problem because factory farming is a major contributor to climate change. This would be useful for an audience of people who understand the meaning of environmental degradation and climate change that has been constructed through environmentalism. I could frame it as a problem of health because meat contains high levels of cholesterol, antibiotics, and growth hormones, which contribute to maladies such as heart disease and antibiotic resistance and "superbugs." The health frame is likely to resonate with larger audiences because most people care about their health. Meat production could also be framed as a problem of immigrant labor abuse because factory farms are the largest employer of immigrant laborers who endure unsafe and abusive working conditions. Or I could frame it as a problem of ethics and animal suffering because billions of animals live and die in horrible conditions for meat production. These frames reflect the varying ways animal advocacy issues can psychologically appeal to different individuals. Framing theory in social movements gained traction as more scholars began to consider individual-level processes that factor into social movement processes, and with this shift came a resurgence in research concerned with emotions (Benford and Snow 2000; Goodwin et al. 2001; Goodwin et al. 2009).

During this time scholars also explored the ways that political structures explained the emergence of movements (Eisinger 1973), in particular how the degree to which democratic processes were open or closed to citizen participation explained varying degrees of mobilization (Kitschelt 1986; McAdam 1982; Meyer 1990). Scholars explored how political systems offer opportunities that increase protest activity (Meyer and Imig 1993; Meyer 2006), how systems mediate movement demands (Amenta 1998; Amenta 2006), and how they change in response to movement cycles of claims-making and political concessions (Andrews 2004; Banaszak 1996; for a detailed review of the literature, see Snow et al. 2004). This contextual approach called for consideration of both cultural norms that constrain activists and those that offer them tools for building support (for a review, see Earl 2004; McCammon et al. 2007).

Overall, we know a lot more about how social movements emerge and persist than we know about their effects, especially movements that target the state (Amenta et al. 2010; Meyer and Lasch-ever 2015). We know less about how social movements affect nonstate targets, such as businesses and scientific institutions, such as laboratory research (Epstein 1996; Luders 2006; Moore 2007; Soule 2009; Wolfson 2001). The political outcomes of animal advocacy receive very little attention from scholars (Evans 2016; Francione and Garner 2011). In fact, even sociological research on emotions and culture in animal advocacy movements is quite limited despite its 100-plus years of mobilization (for a review, see Munro 2012).

Explaining how and why people experience an emotional shift to embrace animals as sentient and important beings can't be divorced from the political context (Cherry 2016; Einwohner 1999), and the ways policy reforms facilitate attitudinal shifts can help us understand the cultural and emotional realms we share with animals. Jasper and his colleagues (1998, 1999), and more recently Wren (2013) and Cherry (2010, 2016), analyze factors such as emotions, moral shocks, and personal connections in an effort to explain participation in animal advocacy. Here, I focus on how the construction and maintenance of boundaries guide how people relate to animals and to specific species of animals and whether they feel compelled to act on feelings of compassion through either lifestyle changes or participation in animal advocacy (Ellis 2014).[2] As opposed to discussing types of emotions we feel towards animals, given the wide range of emotional responses people can experience, let's focus on the boundaries between "human" and "animal" and how disruptions to those boundaries can be encouraged by laws and policies that are often not considered related to our emotional lives.

"Conversion" to Animal Advocacy

> There are measures in place to kind of dull that reaction. So, you probably heard that they give numbers to all these animals; they don't give them names. You're supposed to handle them in exactly the same way, but I saw early on that these different rodents had different personalities that (sic) I couldn't help but come to understand them as individuals. Their individuality came out, in those brief moments where I handled them, putting them in one box from another. So I don't know what primed me to be that way.
>
> *(Interview 2014)*

This preceding quote is from the same woman who left animal research and became an animal advocate, and the laboratory norm she describes here reflects the very deliberate maintenance of boundaries that guide animal researchers' emotional responses to their research subjects. If you see an animal as being an automaton, a robotic creature devoid of self-awareness and emotion, you won't attribute differences in behavior to personality. Rather, you'll attribute individual differences to chance or instinct. This reflects boundaries between "human" and "animal" and between species of animals. If you feel emotionally attached to a companion animal who is a dog, but you eat pigs and do not respond to videos and images of them suffering during slaughter, this reflects a boundary between an animal who deserves compassion and an animal who does not. That boundary is maintained by a variety of mental and emotional processes that, for some, can be disrupted.

The process of becoming an animal advocate, especially for those who take on vegan lifestyles, is often compared to religious conversion because of its moral underpinnings and the substantial behavior changes it can inspire (Austin and Flynn 2015; Jamison et al. 2000; Jasper and Nelkin 1992). Due to limited space, I focus on the moral underpinnings instead of the substantial lifestyle changes other authors discuss in this volume.

Moral boundaries often guide our emotions toward animals—sometimes whether we feel emotions towards them at all and especially whether we act on emotions that arise. Emphasizing emotional drivers for animal advocacy comes with substantial pushback because it can easily delegitimize animal advocates' claims as irrational. Coupled with this is the movement's female-dominant demographic, so this framing can reflect sexist stereotypes as well (Adams 1990; Groves 2001). I'm emphasizing emotions through moral boundaries that are constructed through the human identity and guide our emotional reactions toward animals, especially animals with whom we don't typically share day-to-day space, such as animals used for food, research, and entertainment. Whether the decision is more intellectually or more emotionally driven, acting on newfound compassion for animals reflects a disruption of the boundaries that inform beliefs and decisions about which entities deserve moral consideration.

Work on the moral status of animals has a rich lineage (for a review, see Beauchamp and Frey 2014; Calarco 2015), much of which hinges on whether animals are sentient, that is, whether they are able to engage with the environment and themselves with awareness and emotion as opposed to only acting on instinct. Many think of being human as being uniquely sentient, and in the US, it is widely accepted that humans deserve freedom so long as that freedom does not impede other humans' freedom. For instance, you don't have the freedom to kill or jail someone, unless you are the US state, which shows how, even for humans, this value is compromised by institutionalized racism, sexism, xenophobia, gender and sexual norms, and so on, but in terms of overt jurisprudence, all human life is supposed to be treated equally. This is why so many thinkers are concerned with the distinction between "human" and "animal" and whether animals are sentient. If animals or some species of animals were deemed sentient, then an important boundary between "human" and "animal" would be disrupted, and to impede on animals' freedom would be a tangential violation of an agreed-on moral code.

We're at a strange juncture with animals—welfare regulations legally incorporated an acknowledgment that animals feel pain, but we haven't institutionalized animals' sentience. There is an entire discipline of animal behavior, ethology, that's substantially devoted to the explicit purpose of alleviating animal suffering in captivity, and yet we have legal, systematic practices that inflict extreme forms of suffering on animals. (This is especially true for rodents and birds who are explicitly excluded from the Animal Welfare Act.) The debate is whether using these questionable political tools, such as welfare regulation, as incremental means towards animal liberation helps or hinders the end goal of establishing animals as morally important beings. The interplay between emotional cognitions, moral boundaries, and political tools is central to that debate.

Moral Boundaries, Emotion Work, and Animal Advocacy

> So in our in-vitro lab in my PhD lab we used frog egg extracts, so I have done that, which I was fine with. I don't know if they were any less happy than out getting chased by things in the wild. I didn't like going down in the animal facilities, and that was one of the reasons I decided I didn't want to do anything else with animals. Would it stop me if the most amazing important discovery is in front of me, I'd get a collaborator to do it, but doing the sacrificing myself, I couldn't do it.
>
> *(Interview 2013)*

This interview was with a researcher who used yeast in her lab, instead of animal "models," as they're called. This quote illustrates some of the reasoning that goes into maintaining moral boundaries around animals who do and do not deserve to suffer and whether people should participate in that suffering if it's for a worthy purpose. First, she was fine with using frog egg extracts, which involve invasive procedures to acquire the eggs, but with a qualification about their not being "any less happy." Then she mentioned that although she would participate in using other animals (she was referring to mice here), it would have to be for a major discovery, she would "have a collaborator do it," and she wouldn't kill the animals herself. ("Sacrificing" refers to killing laboratory animals, often abbreviated as "sac-ing.")

Animal advocates spend deliberate effort managing their emotions (Herzog 1993) and external expressions of them (Jacobsson and Lindblom 2013). This is what Hochschild (1983) calls "emotion work" in her research on people who work in caring professions (e.g., flight attendants, domestic workers, and nurses). People who work with animals in abusive institutions also do emotion work, sometimes deliberately, but often through latent "boundary labor" (Ellis 2014). Latent emotion work is the maintenance of boundaries that separate animals who do and do not deserve emotional consideration, and conditions attached that legitimize suffering for some greater good (Arluke 1992; Ellis 2014; Pachirat 2011). These boundaries may be maintained unconsciously until a trigger brings them into question, in turn requiring us to do more deliberate emotion work if we are to maintain the boundaries.

One form of emotion work is "moral disengagement," a psychological and social-psychological concept that can help explain why people participate in activities they consider morally wrong (Bandura 2007; Detert et al. 2008; Mitchell 2011). The most relevant mechanisms for moral disengagement from participation in animal abuse include justification, physical and linguistic distance, and deindividuation (Detert et al. 2008), all of which can be viewed through the sociological lens of emotional boundary work. These mechanisms are active in those who inflict pain on animals within institutions and in those who purchase, consume, or otherwise benefit from that pain. So the question of how to disrupt latent and/or deliberate emotional boundary work is relevant to animal advocates trying to change public norms. So, again, what brings moral disengagement to light in a way that sparks action?

A common story of conversion involves an event, a person, a companion animal, or a graphic image triggering personal reflection and, sometimes, movement participation (Austin and Flynn 2015; Evans 2015; Halfmann and Young 2010; Herzog 1993; Jacobsson and Lindblom 2013; Jasper and Poulsen 1995). This can come in the form of what Jasper and Nelkin (1992) and Jasper and Poulsen (1995) call a "moral shock . . . when an event or situation raises such a sense of outrage in people that they become inclined toward political action, even in the absence of a network of contacts" (Jasper and Poulsen 1995: 498). But while moral shocks work like this for some, others are exposed to graphic imagery of institutionalized animal abuse and don't experience a meaningful change, especially those who work inside the institution. There are also people who are critical of practices that aren't overtly violent, such as being caged, and a range of triggers that are often more subtle, like growing up with companion animals or watching movies with sympathetic characters who are animals. This kind of shift in perspective requires dismantling boundaries and/or an intellectual shift that can't justify keeping animals, like laboratory mice, in cages for the duration of their lives. This is where networks and exposure to other ways of relating to animals come into play.

When a person is tied to organizations, ranging from local groups such as Meetups to national groups such as Greenpeace, they're more likely to start participating in a social movement activity. And when people are part of multiple social networks, meaning that they have ties to multiple people from multiple social fields, they're also more likely to participate across multiple social movements (for a review, see Diani 2009). After recruitment, being a part of a collective identity is a strong factor in sustaining participation over time, where feelings of solidarity, support, and affection resonate with a person's sense of self and keep them involved in a movement (for reviews, see Hunt and Benford 2009; Jasper and McGarry 2015). Cherry (2006) didn't find this to be true for the animal rights activists she studied, where social networks, more than collective identity, sustained participation and their commitment to a vegan lifestyle.

Interpersonal connectedness exposes individuals to new ways of thinking that can shift the boundaries central to their perception of their relationship to animals (Cherry 2010). These symbolic boundaries can be tied to other communities and belief systems. For instance, in their study of the growing tendency to interpret Christian doctrine to embrace animal rights, Austin and Flynn (2015) proposed that the combination of social networks and message framing is influential in this growing shift within Christian churches. Our relationship to animals is linked with multiple culturally contingent affiliations, and a disruption of boundaries isn't always as immediate as a moral shock. Boundary disruption can be a slow recoding of moral and emotional schemas, where an element of a person's worldview is slowly brought into question because of triggers that range from intellectual to emotional and the space in between. The workplace, or labor sector, is a field where this can occur as well.

Professionals who work within animal industries aren't born unfeeling animal abusers, and some who work within abusive industries express a sense of care for animals. Veterinarians and husbandry experts, for instance, experience tension because of the dual and conflicting roles of protector and user of animal bodies (Arluke 1992; Carbone 2004; Tannenbaum 1993; Wilkie 2005). The context within which human–animal interaction takes place is important to understanding the emotional relationship between people and objectified animals with whom they work, where the roles people play within market-driven animal industries influence how they perceive those animals as sentient beings (Wilkie 2005). Ideological perspectives, especially political ideologies, often change in times of crisis (as we've seen with the newly elected administration and the surge of new activists that came with it), but change in individual ideas also occurs during times of stability (Carstensen 2011), for instance within the workplace, where incremental shifts can facilitate new ways of relating to animals.

Welfare Reform as Facilitating Transformative Change

We love conversion stories—accounts of the emotionally laden experience of realizing animals experience suffering and the accompanying changes in behavior. Unlearning patterns of socialization that reinforce objectification of others may not be a dramatic conversion, but rather a slow unlearning process of which a person may not be fully aware. Understanding the various processes through which people change their relationship to animals empowers activists, especially in light of the ongoing, contentious, and personally painful debate between welfare reform and abolition.

We also love binaries. We love the feeling that choices are right or wrong, good or bad; but committing to these binaries might be detrimental to those working so hard on the ground to make positive change for animals. The shift from using and consuming animals to acknowledging that they deserve moral consideration may be facilitated by policy reforms that normalize the practice of moral concern. Scholars who study the effects of policy reform and incremental change have focused largely on whether welfare regulations actually improve conditions for animals, how those reforms affect the public's perception of animal research, and the ramifications on animals' legal status as objects (Francione and Garner 2011). The effects of welfare reform on those who work within the institution of animal research hasn't received as much attention (Evans 2016; Pachirat 2011). The shifts in practices that regulations influence may also shift the norms and attitudes of the regulated.

Incremental policy reforms meant to improve conditions for animals can inadvertently facilitate a shift in individual perspective that calls into question what it means to be an "animal." This individual shift reflects a repercussion in meaning-making, where institutionalization of animal welfare slowly changes the construction of what it means to be an "animal." Some welfare reform requires veterinarians, inspectors, oversight committee members, or welfare specialists to be involved in regulated animal industries. These actors model a relationship to animals that can lead others to question assumptions about the internal lives of animals and their own relationships to those lives. This shift within animal industries is relevant to what animal advocates seek in the wider public, where schemas used to guide human relationships with animals are disrupted, creating a space where assumptions about animals can be reconstructed.

Within animal industries relationships between humans and animals involve mechanized interactions in which machines and forced routinized processes turn both humans and animals into automata devoid of time and agency to reflect on personal action (Pachirat 2011). These processes call to mind routine consumption behavior. Go to the store, buy meat, cook meat, go to work, earn barely enough money to go to the store, buy meat, buy cheap shampoo, clothes, anything as long as you can afford it—and don't think about what actually went into producing these cheap goods we're supposed to buy. Consumption and participation in market-driven industries are highly mechanized, and we are mostly removed from the process of producing the animals we consume (Ellis 2014). In this context, small gestures that remind us of the lives being used and consumed can be powerful triggers to the emotional boundary work that helps us justify the pain our dollar supports, or the pain one inflicts on animals within the industry.

The Animal Welfare Act, with all of its flaws, mandates that veterinarians and oversight committees, called Institutional Animal Care and Use Committees (IACUCs), oversee federally funded animal laboratories.[3] Although veterinarians and members of IACUCs mostly rubberstamp experiments and are beholden to institutional imperatives to keep research moving (Hansen 2012), I found that sometimes individual interactions with these individuals can cause a shift in attitudes. In addition to the bioethicist in the introduction of this chapter, I interviewed multiple people who were personally affected by veterinarians and IACUCs, including the same whistle-blower who experienced small moments of reflection that culminated in a transformative shift in her career, from research assistant to animal rights activist.

> *The veterinarians were so weird! ::laughs:: Anyway, so, the veterinarian who was responsible for training us on this technique, um, would occasionally, and during other techniques, would occasionally whisper to these animals. He would actually apologize if he did anything invasive, when he euthanized animals, I constantly noticed this guy apologizing to these animals. So it seemed like this guy was not taking this well. He hadn't fully adapted to the culture of alienation from these animals. So, uh, the vets, I think, you know were party to a lot of stuff that I felt ethically unsure about . . .*

Although welfare regulations (along with vegan food alternatives, spay and neuter programs, sanctuaries, and any of the other tactics activists use) don't end animal use immediately, although they are not the end goal, they may be an integral part of an end goal that is in sight. Animal liberation may not happen through dramatic conversion stories but through small changes that introduce a dramatically new way of relating to animals.

Notes

1. This early characterization is also a vestige of research on the Holocaust that sought to explain why people participated in Nazi atrocities against the Jewry.
2. For the sake of brevity, I use *animal* to mean "nonhuman animal" and *animal advocacy movement* to refer to the diverse range of campaigns, goals, strategies, and tactics that comprise the movement.
3. Contract laboratories and other profit-driven research facilities are free of most regulatory constraints, including record keeping and any other devices of oversight.

Bibliography

Adams, C. J. (1990) *The Sexual Politics of Meat: A Feminist-Vegetarian Critical Theory*, New York: Bloomsbury Academic.

Amenta, E. (1998) *Bold Relief: Institutional Politics and the Origins of Modern American Social Policy*, Princeton: Princeton University Press.

Amenta, E. (2006) *When Movements Matter: The Townsend Plan and the Rise of Social Security*, Princeton: Princeton University Press.

Amenta, E., Caren, N., Chiarello, E., and Su, Y. (2010) "The political consequences of social movements," *American Review of Sociology* 36: 287–307.

Andrews, K. T. (2004) *Freedom Is a Constant Struggle*, Chicago: University of Chicago Press.

Arluke, A. (1992) "Trapped in a guilt cage," *New Scientist* 134(1815): 33–35.

Austin, R. L., and Flynn, C. P. (2015) "Traversing the gap between religion and animal rights: Framing and networks as a conceptual bridge," *Journal of Animal Ethics* 5(2):144–158.

Banaszak, L. A. (1996) *Why Movements Suceed or Fail: Opportunity, Culture, and the Struggle for Woman Suffrage*, Princeton: Princeton University Press.

Bandura, A. (2007) "Impeding ecological sustainability through selective moral disengagement," *International Journal of Innovation and Sustainable Development* 2(1): 8–35.

Beauchamp, T. L., and Frey, R. G. (eds.) (2014) *The Oxford Handbook of Animal Ethics*, New York: Oxford University Press.

Benford, R. D., and Snow, D. A. (2000) "Framing processes and social movements: An overview and assessment," *Annual Review of Sociology* 26: 611–639.

Calarco, M. (2015) *Thinking through Animals: Identity, Difference, and Indistinction*, Stanford: Stanford University Press.

Carbone, L. (2004) *What Animals Want: Expertise and Advocacy in Laboratory Animal Welfare Policy*, New York: Oxford University Press.

Carstensen, M. B. (2011) "Ideas are not as stable as political scientists want them to be: A theory of incremental ideational change," *Political Studies* 59: 596–615.

Cherry, E. (2006) "Veganism as a cultural movement: A relational approach," *Social Movement Studies* 5(2): 155–170. doi: 10.1080/14742830600807543.

Cherry, E. (2010) "Shifting symbolic boundaries: Cultural strategies of the animal rights movement," *Sociological Forum* 25(3): 450–475. doi: 10.1111/j.1573-7861.2010.01191.x.

Cherry, E. (2016) *Culture and Activism: Animal Rights in France and the United States*, New York and London: Routledge.

Detert, J. R., Trevino, L. K., and Sweitzer, V. L. (2008) "Moral disengagement in ethical decision making: A study of antecedents and outcomes," *Journal of Applied Psychology* 93(2): 374–391.

Diani, M. (2009) "Networks and participation," in D. A. Snow, S. A. Soule, and H. Kriesi (eds.) *The Blackwell Companion to Social Movements*, Malden, MA: Blackwell Publishing.

Earl, J. (2004) "The cultural consequences of social movements," in D. A. Snow, S. A. Soule, and H. Kriesi (eds.) *The Blackwell Companion to Social Movements*, Malden, MA: Blackwell Publishing.

Einwohner, R. L. (1999) "Practices, opportunities, and protest effectiveness: Illustrations from four animal rights campaigns," *Social Problems* 46(2): 169–186.

Eisinger, P. (1973) "The conditions of protest behavior in American cities," *American Political Science Review* 81: 11–28.

Ellis, C. (2014) "Boundary labor and the production of emotionless commodities: The case of beef production," *The Sociological Quarterly* 55(1): 92–118.

Epstein, S. (1996) *Impure Science: Aids, Activism, and the Politics of Knowledge*, Berkeley: University of California Press.

Evans, E. (2015) "Bearing witness: How controversial organizations get the media coverage they want," *Social Movement Studies*, TBD. doi: 10.1080/14742837.2015.1060158.

Evans, E. (2016) "Stumbling blocks or stepping stones? The problems and promises of policy reform goals for the animal advocacy movement," *Sociological Perspectives* TBD.

Francione, G., and Garner, R. (2011) *The Animal Rights Debate: Abolition or Regulation*, New York: Columbia University Press.

Goffman, E. (1974) *Frame Analysis*, Cambridge, MA: Harvard University Press.

Goodwin, J., and Jasper, J. M., (1999) "Caught in a winding, snarling vine: The structural bias of political process theory," *Sociological Forum* 14(1): 27–54.

Goodwin, J., Jasper, J. M., and Polletta, F. (eds.) (2001) *Passionate Politics: Emotions and Social Movements*, Chicago: The University of Chicago Press.

Goodwin, J., Jasper, J. M., and Polletta, F. (2009) "Emotional dimensions of social movements," in D. A. Snow, S. A. Soule, and H. Kriesi (eds.) *The Blackwell Companion to Social Movements*, Malden, MA: Blackwell Publishing.

Groves, J. M. (2001) "Animal rights and the politics of emotion: Folk constructs of emotions in the animal rights movement," in J. Goodwin, J. M. Jasper and F. Polletta (eds.) *Passionate Politics: Emotions and Social Movements*, Chicago: University of Chicago Press.

Halfmann, D., and Young, M. P. (2010) "War pictures: The grotesque as a mobilizing tactic," *Mobilization: An International Quarterly* 15(1): 1–24.

Hansen, L. A. (2012) "Institution animal care and use committees need greater ethical diversity," *Journal of Medical Ethics* 39(3): 188–190.

Herzog, H. A. (1993) "'The movement is my life:' The psychology of animal rights activism," *Journal of Social Issues* 49(1): 103–119.

Hochschild, A. R. (1983) *The Managed Heart: Commercialization of Human Feeling*, Berkeley: University of California Press.

Hunt, S. A., and Benford, R. D. (2009) "Collective identity, solidarity, and commitment," in D. A. Snow, S. A. Soule, and H. Kriesi (eds.) *The Blackwell Companion to Social Movements*, Malden, MA: Blackwell Publishing.

Jacobsson, K., and Lindblom, J. (2013) "Emotionwork in animal rights activsm: A moral-sociological perspective," *Acta Sociologica* 56(1): 55–68.

Jamison, W. V., Wenk, C., and Parker, J. V. (2000) "Every sparrow that falls: Understanding animal rights activism as functional religion," *Society & Animals* 8(3): 305–330.

Jasper, J. M., and McGarry, A. (2015) "Introduction: The identity dilemma, social movements, and contested identity," in A. McGarry and J. M. Jasper (eds.) *The Identity Dilemma: Social Movements and Collective Identity*, Philadelphia: Temple University Press.

Jasper, J. M., and Nelkin, D. (1992) *The Animal Rights Crusade: The Growth of a Moral Protest*, New York: Free Press.
Jasper, J. M., and Poulsen, J. (1995) "Recruiting strangers and friends: Moral shocks and social networks in animals rights and anti-nuclear protests," *Social Problems* 42(493).
Kitschelt, H. P. (1986) "Political opportunity structures and political protest: Anti nuclear movements in four democracies," *British Journal of Political Science* 16: 57–85.
Kornhauser, W. (1959) *The Politics of Mass Society*, New York, NY: Free Press.
Lipsky, M. (1968) "Protest as a political resource," *The American Political Science Review* 62(4): 735–754.
Luders, J. (2006) "The economics of movement success: Business responses to civil rights mobilization," *American Journal of Sociology* 111(4): 963–998.
McAdam, D. (1982) *Political Process and the Development of Black Insurgency, 1930–1970*, Chicago: University of Chicago Press.
McAdam, D., McCarthy, J. D., and Zald, M. N. (1996) *Comparative Perspectives on Social Movements: Political Opportunities, Mobilizing Structures, and Cultural Framings*, Cambridge: Cambridge University Press.
McCammon, H. J., Muse, C. S., Newman, H. D., and Terrell, T. M. (2007) "Movement framing and discursive opportunity structures: The political successes of the U.S. women's jury movements," *American Sociological Review* 72: 725–749.
McCarthy, J. D., and Zald, M. N. (1977) "Resource mobilization and social movements: A partial theory," *American Journal of Sociology* 82(6): 1212–1241.
Meyer, D. S. (1990) *A Winter of Discontent: The Nuclear Freeze and American Politics*, New York: Praeger.
Meyer, D. S. (2006) *Politics of Protest*, Oxford: Oxford University Press.
Meyer, D. S., and Imig, D. R. (1993) "Political opportunity and the rise and decline of interest group sectors," *The Social Science Journal* 30(3): 253–270.
Meyer, D. S., and Laschever, E. (2015) "Social movements and the institutionalization of dissent in America," in R. Valelley, S. Mettler, and R. Lieberman (eds.) *The Oxford Handbook of American Political Development* (pp. 563–592), New York: Oxford University Press.
Mitchell, L. (2011) "Moral disengagment and support for nonhuman animal farming," *Society & Animals* 19: 38–58.
Moore, K. (2007) *Disrupting Science*, Princeton, NJ: Princeton University Press.
Munro, L. (2012) "The animal rights movement in theory and practice: A review of the sociological literature," *Sociology Compass* 6(2): 166–181.
Pachirat, T. (2011) *Every Twelve Seconds: Industrialized Slaughter and the Politics of Sight*, New Haven: Yale University Press.
Park, R. E. (1972) *The Crowd and the Public, and Other Essays*, Chicago: University of Chicago Press.
Piven, F. F., and Cloward, R. A. (1977) *Poor People's Movements*, New York: Vintage.
Simmel, G. (1955) "The web of group affiliations," in R. Bendix (ed.) *Conflict and the Web of Group Affiliations*, New York: Free Press.
Snow, D. A., and Benford, R. D. (1988) "Ideology, frame resonance, and participant mobilization," in B. Klandermans, H. Kriesi and S. Tarrow (eds.) *From Structure to Action: Social Movement Participation across Cultures* (pp. 197–218), Greenwich, CT: JAI Press.
Snow, D. A., Soule, S. A., and Kriesi, H. (2004) "Mapping the terrain," in D. A. Snow, S. A. Soule, and H. Kriesi (eds.) *The Blackwell Companion to Social Movements*, Malden, MA: Blackwell Publishing.
Soule, S. A. (2009) *Contention and Corporate Social Responsibility*, New York: Cambridge University Press.
Tannenbaum, J. (1993) "Veterinary medical ethics: A focus on conflicting interests," *Journal of Social Issues* 49(1): 143–156.
Wilkie, R. (2005) "Sentient commodities and productive paradoxes: The ambiguous nature of human-livestock relations in Northeast Scotland," *Journal of Rural Studies* 21: 213–230.
Wolfson, M. (2001) *The Fight against Tobacco: The Movement, the State, and the Public's Health*, New York: Aldine de Gruyter.
Wrenn, C. L. (2013) "Resonance of moral shocks in abolitionist animal rights advocacy: Overcoming contextual constraints," *Society & Animals* 21: 379–394. doi: 10.1163/15685306-12341271.

39

BEYOND COMPARE

Intersectionality and Interspeciesism for Co-liberation With Other Animals

Nekeisha Alayna Alexis

> In our movements, we have been organizing and theorizing around the *literal, physical,* bodies of the oppressed, rather than going to the root of those oppressions *conceptually* . . . What makes the physical violation of these bodies possible is their citizenship of the space of the Other, or the "subhuman." They were all smuggled into the hierarchy to bolster the superiority of the white ruling class.
>
> *(Ko 2017 : 89)*

> When they enter, we all enter.
>
> *(Crenshaw 1989 : 167)*

A common tactic in the effort to free animals used for food, experimentation, entertainment, and other anthropocentric purposes is to compare the violence they undergo to extreme past and present violence experienced by oppressed human communities. The intent is to communicate in a shorthand way that the abuses people commit on farms, in circuses, at research facilities, and in other arenas of terror are unjustifiable. These comparisons reflect animal advocates' belief that we are on the right side of history, that we can create a more just and compassionate world, and that future generations will eventually renounce these actions. Yet these attempts to inspire concern for other animals have significant flaws that outweigh any limited benefits.

Cursory analogies to slavery, the Holocaust and other human atrocities treat these incidents as props for animal causes, disregarding the distinct traumas cows, elephants, gorillas, rats, minks, and other nonhuman persons undergo as well as the particular violations that human victims have faced and continue to endure. This rhetorical device also disparages already marginalized groups, alienating the very people whose liberation depends, in part, on confronting the ideological and social constructs that objectify other animals. Furthermore, by foregrounding similarities between human and nonhuman animal suffering as a reason to act for other animals' well-beings, advocates reinforce the very supremacies we need to dismantle to bring about substantive change. Said differently, the strategy that is meant to center the plight of other animals and compel more people to act on their behalf ends up diminishing other animals in their own struggle and excluding those who are poised to be their strongest comrades. In light of this conundrum, I join other animal liberationists in considering more robust, possibility-rich forms of persuasion to grow our movements.

Specifically, I echo other scholars and activists who embrace intersectionality as an essential diagnostic tool for understanding the ill-logics that normalize subjugating other animals. Moreover, I conclude that this thicker comprehension provides additional insights into the workings of other oppressions in the aftermath of colonialism, which, in turn, calls animal liberationists to resist racism, sexism, and other toxic systems of power as a matter of also freeing other animals. By identifying how struggles on behalf of other animals integrate with the resistance of systemically disadvantaged people, we open the door to more compelling narratives than comparisons allow. In addition, we increase our ability to develop more just short- and long-term organizing practices that move us all—humans and other animals alike—toward new futures for every body.

With these ideas in mind, I have constructed this chapter in three waves. First, I review and identify the shortcomings of several animal advocacy campaigns that rely on comparisons to historical and current instances of human violence. Second, I summarize and present intersectionality as an alternative analytical tool for better understanding the systemic abuse that humans commit against other animals and against targeted peoples. Third, I present ideas for advocacy that operate from an intersectional, interspecies perspective for co-liberation.

Campaign Tales: On Rape and All Lives Matter

The black-and-white video begins with a young woman looking directly at the camera. She opens with the words "One man held me down;" then another young woman says, "While another man touched me." For the next 40 seconds, five women take turns testifying to what appear to be personal accounts of sexual assault as a piano plays softly in the background. They say phrases like "I was so scared," "They got me pregnant," and "This is not okay." Then one of the women leans toward the camera with tears in her eyes and says, "I am you," followed by another woman who says, "Only different." Here, the messaging takes a turn as the lone Black woman in the ad raises a photo of a caged cow in front of her face.

With the farmed animal connection established, the video continues. Two women hold up photos of a beleaguered sow as the following statements display center screen: "EVERY YEAR, BILLIONS OF ANIMALS ARE BORN INTO THE MEAT, EGG, AND DAIRY INDUSTRIES" and "ALMOST ALL OF THEM ARE A RESULT OF FORCIBLE ARTIFICIAL INSEMINATION." As the video fades to black, the word *RAPE* replaces "forcible artificial insemination," making a new sentence: "ALMOST ALL OF THEM ARE A RESULT OF RAPE." As the video nears its close, the Black woman reappears with the photo of the caged cow as the message, "DON'T PARTICIPATE. GO VEGAN," appears. The closing frame features the name and website address for PETA (People for the Ethical Treatment of Animals). The YouTube version of the video ends with a link to another video titled *How Cows Are Raped on Farms* (2016).

A day after the release of *Women Explain What Rape Feels Like for Animals in the Food Industry*, PETA published on its website a brief statement from president and co-founder Ingrid Newkirk. It read in part:

> We are talking about rape: It is rape when someone sticks their hand into a vagina or rectum without permission. That's the dictionary definition of rape. We believe that everyone should see *the reality* of dairy, meat, and turkey production and then, unless they're ethically blind, they will be appalled. Every decent person abhors and denounces sexual abuse of women but we cannot blithely accept the sexual abuse of other females who happen not to be human but have the same vulnerability to pain. Any woman (or man!) with a heart should scream bloody murder on behalf of the animals who can't scream it themselves.
>
> *(Newkirk 2016)*

The statement made it clear that PETA was self-consciously transgressing a powerful social boundary that reserves the word *rape* for a kind of sexual violence between humans that is usually, although not exclusively, associated with cisgender, heterosexual male abuse of women. By dramatically alluding to women's survivor narratives to say "rape-is-rape," regardless of the victim's species, the video sought to simultaneously confront people with grim facts about how flesh foods are produced, expose the hypocrisy of being against sexual violence in one arena while enabling it elsewhere, and challenge people to abstain from flesh foods as a matter of consistency. Newkirk (2016) further articulated this sentiment, saying, "Kindness is a virtue, empathy is something we have to have, it can't be all about me, me, me, my species, my gender, my narrow group."

While it's unclear how many people heeded the call to stop aiding and abetting rapists by becoming vegan, the video did cause widespread shock and anger. In her article, "Appalling new PETA ad compares rape victims to animals," Marie Solis states, "For PETA to appropriate the language of survivors' experiences and use it for animal rights, is as irresponsible as it is tasteless" (2016). In a similar story, "Outrageous New PETA Ad Compares Cows With Rape Victims," Rebecca Shapiro rounds up representative responses on Twitter, including, "no survivor of sexual assault should have to go through their trauma being compared to an animal's when we've already been dehumanized;" "This is awful, and insensitive. You continue to spit in the faces of margainilized [*sic*] and victimized groups to push your message;" and "Apparently Peta's [*sic*] new ad campaign is 'If we trigger rape&sexual [*sic*] assault victims, maybe they'll become vegans!'" (2016). In an opinion piece, Boston University student Sarah Burnstein explained the upset, saying,

> We live in a world where sexual assault is stigmatized, where rape victims cannot get the justice they deserve from our judicial system and where women and their bodies are objectified regularly. Comparing one injustice to the other does nothing for either case, and PETA's insistence that they never did anything wrong . . . is simply demeaning to rape victims. Besides, as an organization that promotes better morals and ethics, shouldn't PETA be one of the first organizations to respect the thoughts and contributions of sexual assault victims?
>
> *(2016)*

Although organizers insisted that "[a]cknowledging that animals are sexually abused for meat & dairy doesn't take away from seriousness of sexual abuse of humans" (Shapiro 2016), both the campaign and PETA's posture in the face of criticism was off-putting. As a result, they not only alienated countless women; they also made it difficult for women to take seriously the idea of sexual violence against other female animals, much less relate that suffering to their own.

Despite PETA's extensive reputation for using problematic comparisons, including a Save the Whales vegetarian billboard that shamed heavyset women and a "Holocaust on Your Plate" exhibit juxtaposing Nazi concentration camps with factory farms, the organization is not alone in this practice. Since the rise of Black Lives Matter in 2013, one of the more unfortunate developments among animal advocates has been repurposing the antagonistic phrase "All Lives Matter" for animal rights and vegan activism. The slogan can now be found on apparel combining silhouettes of farmed animals, pets and/or humans, and hashtags on social media. There are also derivatives for specific causes such as "Cat Lives Matter." Although the most readily available All Lives Matter image, an illustration of a cow, cat, pig, and dog was credited to artist Victoria Novak, the message does not seem to belong to any one group or campaign. Instead, the phrase appears to have spread at a grassroots level among predominantly White individuals and groups.

The most generous reading of "All Lives Matter," the vegan edition, is that it is another attempt to name the ill-logics of discriminating between species of other animals. For example, some people use it to point out the discrepancy between extending compassion to the animals we categorize as pets or see as otherwise endearing while annihilating other sentient creatures. All Lives Matter is also used to argue that human and nonhuman animals have more commonalities than people recognize and are basically "the same," especially when it comes to avoiding pain and seeking joyful, meaningful lives. This sentiment is also expressed in sayings such as "We are all animals," "We are all earthlings," or as *Women Explain What Rape Feels Like for Animals* put it, "I am you. Only different." Yet even if the goal is to expose the irrationality within socially constructed species barriers and to increase empathy across those divides, the use of "All Lives Matter" discourse among animal advocates is fundamentally anti-Black. As Black vegan philosopher Aph Ko notes in her reflection on "all vegans rock," a companion phrase used to dismiss Black and other people of color who consider veganism and animal liberation alongside racial identity and anti-oppression concerns, it is "a way to employ post-racial rhetoric to violently silence activists of color who are trying to organize around their own experiences" (Ko 2017c: 18).

"All Lives Matter," and derivatives such as "Cops Lives Matter" and "White Lives Matter" arose as part of White backlash against Black communities for disruptively and unapologetically protesting outbursts of White vigilantism, such as George Zimmerman's assault and shooting of Trayvon Martin, and Theodore Wafer's shooting of Renisha McBride, and police violence, such as in the shooting deaths of Rekia Boyd, Eric Garner, Aiyana Jones, and Walter Scott. The terminology is an attempt to dismiss personal testimonies and public examples of the peculiar vulnerability and disposability of Black people, and to silence Black resistance to this state of affairs. As Cory Wrenn, sociologist and scholar of social movements and human–nonhuman relations, explains:

> "All Lives Matter" is not alliance-building, it's alliance-destroying. It suggests that mobilization to improve the life chances and well-being of the Black community is somehow unwarranted or distracting. It erases difference, and when difference goes invisible, this invisibility supports systems of inequality that feed on difference.
>
> *(2016)*

A potent example of the ugly side of flattening appropriate distinctions is apparent in Twitter user Vegan Revolution's response to the Baltimore uprising against the police killing of 25-year-old Freddie Gray: "Black lives matter . . . more than Chickens or Cows lives . . . apparently" (Galarza 2015). However, as Wrenn continues,

> [e]spousing that Black lives matter does not mean that Nonhuman Animals do not matter, or that white lives do not matter, or that anyone else's lives do not matter. It only means that the systemic oppression faced by Blacks is abominable and must stop. White vegans have an obligation to support this effort, not to derail it.
>
> *(2016)*

Regardless of their intentions, vegans who use "All Lives Matter" demonstrate deep ignorance and insensitivity about, and sometimes open hostility toward the historical and present-day threats to Black thriving. To take up that refrain for any cause is to participate in racist contempt for this manifestation of Black liberation, no matter how useful the language may seem.

Slavery: A Dreadful Comparison

When liberationists fail to think critically about the complex ways domination of other animals interlocks with racism, sexism, heterosexism and other "isms," they often end up trivializing human experiences of violence. One of the ways this frequently occurs is in the comparison between captive, tortured food animals and human slavery, especially European/White enslavement of Africans and their descendants during the 15th to 19th centuries. For example, in his provocatively titled editorial "Meat: the slavery of our time," environmental blogger Jim Motavalli summarizes the disastrous outcomes of farming and consuming other animals, then argues that the rise of flesh-food-related illnesses and realities such as extreme water shortages and rapid deforestation will force an "inevitable vegetarian revolution" (2009). People might not want to forego eating carcasses, but abstention will be the only reasonable, available choice.

Throughout the article, Motavalli does not articulate how meat or meat production is like slavery, or even the type of slavery to which he is referring. Indeed, he does not mention slavery again until the final paragraph, where he insists:

> Meat will disappear—except as a luxury available to few—and the ethical issues will evolve, too. In the way that slavery, once a broad social norm, later became an unthinkable crime, we can expect to see a similar shift once meat-eating disappears from our planet. Perhaps, some day, the very idea of eating animal flesh will seem as remote as the idea of owning humans does now.
>
> *(2009)*

The comparison has the shock value of associating meat-eaters with slave owners and those who supported or tolerated the practice. The reference also rests on the assumption that the matter of owning other people is settled on the side of being "morally absurd and logistically impossible" (Motavalli 2009). An essential piece of the argument is the belief that, as an intolerable system becomes less workable, participants' desirability for and attitude about said practice will naturally change. However, if Motavalli was alluding to transatlantic slavery, it is important to note that the institution ended because of a number of complex factors, not least of which included open and covert acts of defiance by slaves and abolitionists. Legal, social, and political advocacy in the face of the pro-slavery establishment as well as a massive Civil War all contributed to slaves' emancipation. These confrontations bear little resemblance to a passive "revolution" based on unavoidable, external circumstances.

Motavalli's argument that attitudes and ethics change because of what is most practical is also unconvincing. In the case of slavery and emancipation, the White ruling class merely developed new socially acceptable forms of exploiting Black bodies and labor, such as sharecropping and arbitrary imprisonment. Today, a similar trajectory is apparent in the situation facing food animals. Although doomsday climate scenarios have led to a formidable rise in plant-based living, it has also generated rapid, substantial investments in animal-based alternatives, from genetically modified, environmentally friendlier cows to "post-vegan" options (Ito 2018) like meats made from cultured animal cells harvested from cows' fetuses. As of this writing, it takes "roughly 50 liters of serum and costs about $6,000 to produce a single beef burger" (Ito 2018). The continued use of pregnant cows, the extremeness of the procedure, and the price of the product as compared to the ease and cost-effectiveness of plant-based eating have not deterred interest in these foods.

Last, despite the horrors of the past, it is still not the case that slavery is "unthinkable." As late as 2017, reports surfaced of Libyan auctioneers selling migrants and refugees from neighboring countries to bidders seeking manual laborers. The purchase prices for these vulnerable humans ranged from US$400 to US$800. Some of the individuals had been beaten, mutilated, deprived of food, and otherwise violated by their handlers. As one reporter put it:

> [t]he auctions take place in a seemingly normal town in Libya filled with people leading regular lives. . . . But inside the slave auctions it's like we've stepped back in time. The only thing missing is the shackles around the migrants' wrists and ankles.
>
> *(Elbagir et al. 2017)*

Although this may seem like an extraordinary example, it is still the case that forms of enslavement persist in sex trafficking, textile production, and other industries (Deckha 2010: 40–42). Slavery might be socially taboo across the globe, but it remains part of humanity's social and economic fabric.

Motvalli's slavery comparison is not only factually incorrect: it also breaks down conceptually. First, it bears noting that there is a distinction between classifying African-descended/Black people as beings outside humanity to justify violence against them, on one hand, and justifying violence against farmed and other animals because they are a different animal species, on the other. A suitable parallel may be that, while there are social prohibitions against murder in general, a spouse who kills his or her partner without provocation has a different cast from a spouse who kills the stranger next door without provocation, even though the partner and the stranger are both human beings from the same neighborhood. Each killing is illegal and morally inexcusable. Yet, we recognize a difference in the tragedy when there are closer ties between the victim and victimizer. At the very least, the relational proximity between spouses or siblings or close friends gives a murder between those parties a higher shock factor than does killing someone who is less close.

Animal advocates' refusal to acknowledge this nuance—that although we are all biological animals, our relationships are prioritized in ways that generate varying degrees of impact when they are violated through violence—is a key reason why "this-is-that" comparisons between slavery and animal agriculture can be so loaded. White supremacy still rejects Black people's biological and social belonging to the category of human and "either, erases, or rejects, or diminishes the value of contributions offered by black people" (Ko 2017a: 3). Although enslavement of humans and of other animals are interrelated systems, and we are all "the same" in crucial and unappreciated ways, there remains a felt difference between exploiting other animals in ways that block them from *fully and freely existing in their own species worlds* while Black and other non-White people are still struggling for full recognition and power within the species category to which *we already technically belong*.

Motavalli's optimism also underestimates how thoroughly "anthropocentric humanism" (Weitzenfeld and Joy 2014: 5) governs Western and Western-influenced societies. Cultivated during the Enlightenment, this "ideological commitment to conceptualizing human being over and against animal being, and privileging human consciousness and freedom as the center, agent and pinnacle of history and existence" (Weitzenfeld and Joy 2014: 5) still governs normative relationships with other animals. The rise of new animal-based industries despite current and impending environmental calamities is one example of how the impulse to harness other animals for human purposes remains so long as this ideology has a hold. Furthermore, references to slavery and other atrocities unintentionally perpetuate this anthropocentric humanism by suggesting that other animals' suffering matters most when it mirrors human experiences. Instead of calling attention to the intrinsic value of animals, these comparisons cast doubt as to whether we believe their stories of trauma are compelling in themselves. This might partly explain why "lower animals"—such as fish, invertebrates, and silkworms, all of whom are also killed *en masse* but whose suffering cannot easily be construed in terms of our own—are usually absent from mainstream advocacy campaigns.

Finally, Motavalli's hope for worldwide vegetarianism reflects a misunderstanding about what the end of Transatlantic Slavery did and did not accomplish. While it cannot be overstated how tremendous it was to the physical, emotional, economic, political, spiritual, social, and psychological well-being of Black people to attain freedom, the Emancipation Proclamation did not stamp out slavery or White supremacy's attempts to devour and destroy Black life. On the contrary, in the years immediately following the declaration and at least until World War II, millions of Black people in the US found themselves subject to forced labor and brutality through various means of legal re-enslavement. Consequently, one of slavery's enduring lessons is that ending particular manifestations of violence does not eradicate the ill-logics that first conceived of the victims as acceptable targets. Subjugation shape-shifts wherever toxic hierarchies live on.

Enter Intersectionality

Motavalli, All Lives Matter, and PETA demonstrate a simple but important premise: how we understand an issue determines the approaches we use to address it, and how we address the issue affects how closely we can get to our ultimate goals. If we frame animal liberation primarily in terms of stopping cruelty, we will face a battery of new challenges where suffering is minimal but exploitation remains, from proposals to genetically modify farmed animals so they don't feel pain during captivity and slaughter (Shriver 2009), to the mass breeding of transgenic animals for nonlethal pharmaceutical uses (Potenza 2018). If we see convincing people that they are biologically or socially "the same" as other animals as the path toward animal liberation, we miss how the very same emphasis on similarity can also perpetuate domination. For it is precisely because of significant commonalities between humans and animals as varied as chimpanzees, rats, pigs, rabbits, and guinea pigs that military, scientific, and commercial industries use them in excruciating and lethal experiments for our benefit and against their interests. Without first establishing the belief that every animal is a unique, irreplaceable being who deserves dignity, compassion, freedom from violence, and freedom for self-actualization within their community, independent of human whims, "the similarity is a reason to exploit or violate" (Ko 2017: 109–110).

Finally, if we see animal liberation as an isolated issue that is merely analogous to other human oppressions rather than a fundamental, inextricable part of how human oppressions function, we're liable to continue making at least four costly mistakes. We will allow ideologies that harm people and other animals alike to persist and, as in the case of slavery, to resurface. We will repeat strategies that are not only ineffective for inspiring change, but will also create enemies of our best potential comrades. We will re-create racism, sexism, ableism, speciesism, and other problems in our movements. Lastly, we will miss opportunities to advance what Veganism of Color calls, "consistent antioppression" (www.veganismofcolor.com). As Syl Ko urges, "[w]e actually need to *believe* these things are related and convey this at the theoretical level several steps before we organize and take to the streets" (2017: 84). Intersectionality supports this pressing task.

Introducing Intersectionality

The term *intersectionality* was coined by Kimberlé Crenshaw to explain the peculiar challenges Black women faced as they sought redress for incidences of systemic discrimination. In "Demarginalizing the Intersection of Race and Sex: A Black Feminist Critique of Antidiscrimination Doctrine, Feminist Theory and Antiracist Politics," she examines how US antidiscrimination laws' approach to unfair treatment on the basis of race *or* sex undermined petitioners complaints based on their race *and* sex combined. In one instance, the court dismissed the plaintiffs' case because their employer had hired White women and Black men in positions that Black women sought, which voided their claims of exclusionary policies according to the law. In another case, the court ruled a Black female plaintiff could only present evidence of discrimination against qualified Black women in her suit, despite having ample evidence that her company did not promote women nor Black men into management positions. In a third case, the court prevented Black women from bringing a racial discrimination suit on behalf of themselves and their Black male coworkers because they were women. Although they were still able to prove discrimination occurred, Black men at the company were excluded from any settlements.

In each situation, the courts refused to recognize Black women as "a multiply-disadvantaged class" for fear that doing so would give them an unfair edge in proving their claims (Crenshaw 1989: 145). Crenshaw likened these decisions to an accident in the middle of a street intersection:

> Consider an analogy to traffic in an intersection, coming and going in all four directions. Discrimination, like traffic through an intersection, may flow in one direction, and it may

> flow in another. If an accident happens in an intersection, it can be caused by cars traveling from any number of directions and, sometimes, from all of them. Similarly, if a Black woman is harmed because she is in the intersection, her injury could result from sex discrimination or race discrimination. Judicial decisions which premise intersectional relief on a showing that Black women are specifically recognized as a class are analogous to a doctor's decision at the scene of an accident to treat an accident victim only if the injury is recognized by medical insurance. Similarly, providing legal relief only when Black women show that their claims are based on race or on sex is analogous to calling an ambulance for the victim only after the driver responsible for the injuries is identified. . . . In these cases the tendency seems to be that no driver is held responsible, no treatment is administered, and the involved parties simply get back in their cars and zoom away.
>
> *(Crenshaw 1989: 142)*

On top of facing institutional bias, Black women received little or no recourse from the very apparatus they expected to assist them. Worse still, the legal system was not the only space in which they encountered these kinds of obstacles: even initiatives dedicated to women's and to Black people's empowerment were inadequate for meeting their needs.

In her essay, Crenshaw also describes how mainstream feminists and struggles for Black liberation in the US often neglect the combined force of racism and sexism, and its particular impact on Black women. Single-issue feminists' neglect of race has resulted in a framework that examines patriarchy from White women's perspectives while presenting these experiences as universal to all women; overlooks the relative social advantages White women have due to their racial category; perpetuates systemic racism; and fails to critically address the distinct ways that sexism manifests across cultural and racial contexts, including in the lives of Black women. Meanwhile, single-issue "Black liberation politics" (Crenshaw 1989: 160) has also ignored the ways gender mitigates Black women's experiences of racism; encouraged Black women to forgo confronting sexism as a lesser or conflicting priority than confronting racism; and perpetuates patriarchy and sexism within the movement. As a consequence, feminism's "potential to broaden and deepen its analysis by addressing non-privileged women remains unrealized" (Crenshaw 1989: 154), while Black resistance to racism has participated in "the subordination of certain aspects of the Black female experience in order to ensure the security of the larger Black community" (Crenshaw 1989: 163).

In addition to illuminating the multidimensional nature of oppression, Crenshaw critiques top-down, single-axis responses to it. In this instance, she uses the analogy of a basement and escape hatch, saying,

> Imagine a basement which contains all people who are disadvantaged on the basis of race, sex, class, sexual preference, age and/or physical ability. These people are stacked-feet standing on shoulders-with those on the bottom being disadvantaged by the full array of factors, up to the very top, where the heads of all those disadvantaged by a singular factor brush up against the ceiling. Their ceiling is actually the floor above which only those who are not disadvantaged in any way reside. In efforts to correct some aspects of domination, those above the ceiling admit from the basement only those who can say that "but for" the ceiling, they too would be in the upper room. A hatch is developed through which those placed immediately below can crawl. Yet this hatch is generally available only to those who—due to the singularity of their burden and their otherwise privileged position relative to those below—are in the position to crawl through. Those who are multiply-burdened are generally left below unless they can somehow pull themselves into the groups that are permitted to squeeze through the hatch. As this analogy translates for Black women, the problem is that they can receive protection only to the extent that their

> experiences are recognizably similar to those whose experiences tend to be reflected in antidiscrimination doctrine.
>
> *(Crenshaw 1989: 151–152)*

This illustration provides another way of understanding the need for intersectionality and the problem of motivating for justice based on similarities. First, a focus on getting through the escape hatch obscures important nuances about how structural inequalities work against different groups. Second, it validates the assumption that those on the upper floor are the ideal, and reinforces the idea that fighting to be affiliated with, but never quite like, the propped-up class is the only hope for those under the ceiling. Finally, the escape hatch distracts from the need to dismantle the ceiling altogether. In contrast, Crenshaw advocates first supporting those who bear the most burden in the hierarchy and "restructuring and remaking the world where necessary" (Crenshaw 1989: 167) to relieve their conditions. It is this appeal to start from the bottom that tests the currently anthropocentric nature of intersectionality, and makes visible the need to recognize and incorporate species within the analysis.

Intersectionality and Species

While the concept of intersectionality was articulated out of the experiences of Black women and remains attendant to our specific needs, Crenshaw and others have also expanded the initial focus on race and gender to more broadly examine "how structures make certain identities the consequence of and the vehicle for vulnerability" (Crenshaw 2016a). In her TED Talk, "The Urgency of Intersectionality," she identified heterosexism, classism, xenophobia, and ableism as dynamics that, in addition to racism and sexism, create distinct constellations of harm for different communities in particular contexts (Crenshaw 2016b). However, one of the perceived "group memberships" (African American Policy Forum 2013) her model did not include—and that standard iterations of intersectionality miss—is that of species. This is a substantial oversight, as notions of species and animality are key for determining who is and isn't appropriate for violation and disposal: who is and isn't, as philosopher Judith Butler theorizes, "a grieveable life" (2015).

As previously noted, anthropocentric humanism posits human beings as the supreme and central figure among all earthly creatures. This idea involves separating "the human" from other animals by denying or minimizing animal characteristics in human beings, and diminishing or obscuring any qualities in other animals that are found in or have been attributed to humans. Black vegan philosopher Syl Ko summarizes this principle, saying:

> One of the clinchers introduced by colonial thinking is that we are not just different and special when measured against all of the other animals, but we are their *opposite*. . . . If the human is the definitive representation of value itself, then, following the golden rule of the human-animal opposition, the animal is the definitive representation of the *absence* of value itself.
>
> *(2017: 11)*

In this schema, humans are seen as autonomous subjects who are exclusively or the most cultural, rational, linguistic, sentient, intelligent, self-aware, and above instinct and nature. Meanwhile, as humanity's inverse, "the animal" is rendered as a dependent nonsubject/object that is lacking culture, irrational, non- or prelinguistic, disorderly, wild, dumb, and governed solely by instinct and nature. Envisioned as "merely in the world, bound by natural law," it follows that other animals "have no dignity to violate and, thus, are owed little to no direct moral consideration" (Weitzenfeld and Joy 2014: 5). Theirs are not grieveable lives. On the contrary, their not-humanness permits, if not invites, extermination and violence (Deckha 2010: 33–34).

One of the clarifications Ko makes about the human and animal division is that "the human" does not refer to all biological *Homo sapiens*. Instead, Enlightenment and colonial thinking imagined "the human" as a superior type of being: the European/White (Christian heterosexual, able-bodied, upper-class) male. Although the concept of "the human" became limited to this particular subset of *Homo sapiens*, Ko points out that the category of "the animal" pushed millions of vastly different non–*Homo sapiens* animals into a single, oversimplified, and absurd container. Because "the human" is reserved for a single group, it is also the case that "the animal" includes *Homo sapiens* who fall outside the ideal. For:

> [i]f "the human" is really an expression of whiteness as the ideal way of being homo sapiens, then "the animal" is supposed to express a *deviation* from this way of being. "The general "animal" then, . . . can also apply to those members of homo sapiens who deviate from the way whites look and/or behave, and what' values and commitments they hold, and so forth.
> *(2017: 23)*

In other words, on this side of colonization, the dominant concepts of "'humanity/human' and 'animality/animal' have been constructed along *racial* lines" (Ko 2017: 23). Consequently, "[t]he human-animal divide is the ideological bedrock underlying the framework of white supremacy,'" and, far from being peripheral to the matter, "[t]he negative notion of 'the animal' is the *anchor* of this system" (Ko 2017: 23).

In *Dangerous Crossings: Race, Species, and Nature in a Multicultural Age*, political scientist Claire Jean Kim helpfully sketches this historical, mutually reinforcing relationship between species and race. She especially focuses on the ways in which White philosophers, pseudo-scientists, and popular culture in the US situated Black, indigenous, and Chinese populations as being outside the realm of "the human," and ranked them according to their presumed proximity to the fictive notion of "the animal." She observes:

> The conventional wisdom is that powerful groups have used animalization to "dehumanize" less powerful groups, demoting them from category A (humans) into category B (animals) and thus stripping them of the entitlements, rights, and protections distinctively owed to humans . . . I argue . . . that racialized groups were never seen as fully human to begin with, so they did not have this status to lose. They were always already animal or animal-like.
> *(Kim 2015: 24)*

Rooted in European racial narratives, these White imaginations placed "the Negro," "the Indian," and "the Chinese" in conceptual borderlands alongside nonhuman animal species and other *homo sapiens* who were racialized and animalized outside "the human" sphere. Within this order, Black people were the aggressive, hypersexual, and hopelessly un-civilizable gap between "man" and "ape," and counted among the livestock (Kim 2015: 37–38). Meanwhile, just above "the Negro," indigenous tribes were seen as primitive, savage, "animal-like men" (Kim 2015: 44) and associated with wolves and other forest creatures. Finally, above "the Indian" and closest to Whites were Chinese immigrants, who, depending on the economic climate, were "depicted as almost every kind of animal—frogs, camels, horses, elephants, vipers, vermin, tigers, among others" (Kim 2015: 57). These centuries-old race and animal tropes still operate in the present with devastating results for those who live under them. As Kim demonstrates, they also complicate the work of liberationists, especially White activists speaking out against animal-based eating, entertainment, and ritual practices among non-White and non-Western cultural groups.

In addition to unpacking the intersection of race and species, theorists have also examined the role of species in constituting other oppressions. For example, feminist vegans like Carol J. Adams,

Marti Kheel, and Heather Fraser have shown how animality informs the male/female hierarchy and binary, and these forces' devastating effects for all identified as nonmale and nonhuman animals. Fellow scholars, such as Anthony Nocella, Hannah Monroe, and Sunaura Taylor, have discussed how animality functions within ableism to denigrate and marginalize disabled individuals and communities. Still other writers and thinkers like pattrice jones and James Stanescu have examined queerness and other animals, while self-identified transsexual Calvin Neufeld has called for a "trans-vegan alliance" (2015). Like Crenshaw, critical animal studies scholar and lawyer Maneesha Deckha has counseled against using a top-down, similarity approach to confronting these imbalanced relations and the "violence-producing category" of the subhuman that drives them (2010: 47). Instead of attempting to smuggle more nonhumans and less-than-humans into the protected "human" class, as in Crenshaw's ceiling analogy, Deckha suggests "a better strategy may be to minimize the human/nonhuman boundary altogether" (2010: 47).

Toward Interspeciesism and Co-liberation

Although there is a strong temptation to try and erase the human–animal division by comparing types of suffering, a non-anthropocentric, species-minded intersectionality makes visible the problem in ways that allow us to identify and attack the root. As Ko writes,

> in both the narrative of speciesism and the narrative of racism [and I would add sexism, ableism, heterosexism, etc.] the members of the losing side both fall short of the *real* human status and, as a result, their suffering and their deaths are mundane, normal and expected.... If we want to make a connection, *this* is the connection we should be making. We're really not "comparing" anything in this type of thinking. We're noting a common source.
>
> *(2017: 86–87)*

If we grasp this fully, we will also understand that (1) the seemingly passive work of reconceiving human being *is* activism on behalf of other animals and (2) resisting White supremacy, patriarchy, and other forms of oppression *from an interspecies perspective* is not a distraction from or a nuisance to our liberationist goals, but is instead inseparable from creating a more just world for every body. The difference between intersectional, interspecies models of centering the source versus the current approach of concentrating on comparisons is subtle but important. To make this point clearer, I want to suggest a few initial principles and examples of possible alternatives.

Expose the Oppressive Foundation

In order to illustrate what it might look like for animal advocates to expose the oppressive foundations of the suffering and violence we witness, let us revisit PETA's "Don't Participate. Go Vegan" campaign. Perhaps one of the biggest ironies of *Women Explain What Rape Feels Like for Animals* is that at no point does the video depict the sexual violence of forced artificial insemination. The video does not include footage of a farmer restraining a cow, inserting his forearm into her rectum and inserting the "gun" into her vagina against her will and best interests. There are no clips of breeders corralling thousands of hens per day into chutes, restraining each one—breasts down, legs down, and tail up—until their "vents" open and the terrified birds "beat their wings and struggled in panic" (Davis 2017). We do not see sexually aroused sows with men sitting on top of them or placing weights on their backs to ensure their uteruses are wide open for insertion of semen. If the goal was to show "the reality" of sexual, gendered, heteronormative violence in the flesh-food industry, the better strategy would have been to portray the experiences of the actual female victims and to make explicit that these systems are an extension of patriarchy and rape culture. With this point established, veganism could then be framed as an act of feminist resistance to reproductive tyranny and "interspecies sexual assault" (Davis 2017). Women do not need to stand in for or speak as female animals, nor is it necessary to argue that each group's experiences of rape are "the same" to get the message across. The sexual violation farmed female animals experience is its own testimony.

Seek (Genuine) Solidarity Across Intersections

Because the "negative notion of 'the animal'" is lethal for other animals and for animalized *Homo sapiens*, it follows that our liberationist visions and activism must respond to multiple group needs. In the case of PETA's "Don't Participate. Go Vegan." campaign, another appropriate way to tackle sexual violence in flesh-food production would be to call attention to sexual violence against female laborers in meat-packing plants. According to a yearlong *Frontline* investigation, women laborers are routinely harassed by their supervisors and, when they risk complaining to management, are consistently ignored, demoted, fired, or further harassed (Yeung and Rubenstein 2013). For these predominantly women of color, especially Latinx women, complicating factors such as language barriers, citizenship status, and level of dependence on their jobs create power imbalances and cycles of silence on which their abusers capitalize. An intersectional, interspecies campaign to truly confront sexual violence in the flesh-food industry would also address racism and White supremacy, sexism and patriarchy, classism, and xenophobic immigration policies, in addition to exposing and ending sexual violence against farmed animals. This work would also require acknowledging rampant sexual violence in plant-based industrialized agriculture, and challenging vegans to make plant-based choices that factor in anti-oppression concerns, such as purchasing organic, local, and fairly traded goods, wherever possible.

The same multidimensional approach would distinguish those who genuinely believe that all lives matter from those attempting to hijack or oppose the Black Lives Matter movement. If animal advocates truly believe that slavery relates to what we do to other animals, then they will also be present in tangible ways for Black liberation initiatives that combat the legacy of racism and White supremacy that results from that system. This can involve showing up to protests against police brutality, creating and providing platforms for Black vegan politics, working for food justice and security to support plant-based eating, and providing financial and legal support for low-income communities of color devastated by factory farming. It also means not using racist slogans such as "All Lives Matter" and ill-conceived slavery analogies, especially when Black vegans and other antiracists speak out against them. Intersectional, interspecies advocacy insists that we join other struggles for the sake of every body. It entails remembering that oppression and liberation of other animals are tangled up with oppression and liberation of animalized peoples and organizing ourselves accordingly.

Build and Support Interspecies Resistances

The above suggestions may seem daunting to the point of impossible, raising important practical questions about resource distribution, time constraints, and use of energy. With this in mind, I briefly offer Food Empowerment Project (FEP) as one example of this trajectory. Founded by animal rights movement veteran Lauren Orenelas, FEP's work includes promoting ethical veganism: including disseminating information via a vegan Mexican website; protesting at a local chicken slaughterhouse; collaborating with organizations in communities of color and low-income communities to support access to healthy food; organizing annual school supply drives for farmworkers; and leading a coalition to advance farmworker rights in California. Recognizing the need for vigilance among plant-based eaters, they also maintain a list of vegan chocolate producers who do not use child labor or slavery. All their materials are available in English and Spanish. When asked by Animal Charity Evaluators how she decides where to focus her energies, Ornelas responded,

> All of these issues are connected. In order to create positive change in all of the areas we work on, and to move towards building the type of world we envision, those of us fighting for a more just and compassionate world need to be bigger and stronger. The only way we can create this change is to be consistent in our ethics and to be united.
>
> (*Alonso 2017*)

FEP demonstrates that we need not see these varied tactics as detracting from each other. When viewed with intersectional, interspecies lenses, we can rest assured that they are multiple fronts of the same fight. Organizations such as Afro Vegan Society, Grow Where You Are, Hip Hop Is Green, Sanctuary Publishers, and Vine Sanctuary also exemplify this perspective.

Conclusion

Black lives, disabled lives, women's lives, indigenous lives, queer lives, and other Othered lives are expendable in part because we/they still occupy the borderlands with animals of other species. The struggle for co-liberation from this liminal space where we/they are acceptable targets of violence will mean deconstructing White and other supremacist notions and structures that have placed us/them there. We must rethink the concept of "the animal" and practice new relationships to actual other animals, as well as reenvision what it means to be "human" in order to move toward liberation for discounted *Homo sapiens* and other animals alike. Shifting from comparisons and similarities to intersectionality and interspeciesism within all our social justice movements is core to this demanding, ongoing, and lifegiving project.

Bibliography

African American Policy Forum (2013) *Intersectionality*, January 17, 2013, www.aapf.org/2013/2013/01/intersectionality.

Alonso, E. (2017) "ACE interviews: Lauren Ornelas," *Animal Charity Evaluators*, July 13, https://animalcharityevaluators.org/advocacy-interventions/advocacy-advice/learn-from-professionals/lauren-ornelas/.

Burnstein, S. (2016) "PETA cannot equate artificial insemination to rape," *The Daily Free Press*, November 9, https://dailyfreepress.com/blog/2016/11/09/burstein-peta-cannot-equate-artificial-insemination-to-rape/.

Butler, J. (2015) "Precariousness and grievability—when is life grievable?" *Verso*, November 16, www.versobooks.com/blogs/2339-judith-butler-precariousness-and-grievability-when-is-life-grievable.

Crenshaw, K. (1989) "Demarginalizing the intersection of race and sex: A black feminist critique of antidiscrimination doctrine, feminist theory and antiracist politics," *University of Chicago Legal Forum* 1989(1, Article 8): 139–166.

Crenshaw, K. (2016a) "On intersectionality," *Southbank Centre*, March 14, 2016, www.youtube.com/watch?v=-DW4HLgYPlA.

Crenshaw, K. (2016b) "The urgency of intersectionality," *TED*, December 7, 2016, www.youtube.com/watch?v=akOe5-UsQ2o.

Davis, K. (2017) "Interspecies sexual assault: A moral perspective," *Animal Liberation Currents*, June 12, 2017, www.animalliberationcurrents.com/interspecies-sexual-assault/.

Deckha, M. (2010) "The subhuman as a cultural agent of violence," *Journal for Critical Animal Studies* VIII(3): 28–51, www.criticalanimalstudies.org/wp-content/uploads/2012/09/JCAS-Special-Issue-Women-of-Color-November-FINAL-2010.pdf.

Elbagir, N., Razek, R., Platt, A., and Jones, B. (2017) "People for sale," *CNN*, November 11, https://edition.cnn.com/2017/11/14/africa/libya-migrant-auctions/index.html.

Galarza, D. (2015) "Vegan advocacy group compares 'black lives' to 'chickens' and 'cows,'" *Eater*, April 30, www.eater.com/2015/4/30/8521129/vegan-advocacy-group-compares-black-lives-chickens-cows-animals.

Ito, J. (2018) "Fake meat, served six ways," *Wired*, July 2, www.wired.com/story/ideas-joi-ito-on-fake-meat/.

Kim, C. J. (2015) *Dangerous Crossings: Race, Species, and Nature in a Multicultural Age*, New York: Cambridge University Press.

Ko, S. (2017a) "Black lives, black life," in *Aphro-ism: Essays on Pop Culture, Feminism, and Black Veganism from Two Sisters*, Brooklyn: Lantern Books.

Ko, S. (2017b) "We can avoid the debate about comparing human and animal oppressions, if we simply make the right connections," in *Aphro-ism: Essays on Pop Culture, Feminism, and Black Veganism from Two Sisters*, Brooklyn: Lantern Books.

Ko, S. (2017c) "#AllVegansRock," in *Aphro-ism: Essays on Pop Culture, Feminism, and Black Veganism from Two Sisters*, Brooklyn: Lantern Books.

Motavalli, J. (2009) "Meat: The slavery of our time," *Foreign Policy*, June 23, https://foreignpolicy.com/2009/06/03/meat-the-slavery-of-our-time/.

Newkirk, I. (2016) *PETA Statement: Re: Women Explain What Rape Feels Like For Animals In The Food Industry*, November 3, www.peta.org/media/news-releases/peta-statement-re-women-explain-rape-feels-like-animals-food-industry/.

Neufeld, C. (2015) "Trans veganism," *Geez Magazine*, August 14, Summer 2015, https://geezmagazine.org/magazine/article/trans-veganism/.

PETA (2016) "Women explain what rape feels like for animals in the food industry," November 2, www.peta.org/features/rape-milk-pork-turkey/.

Potenza, A. (2018) "Are these genetically engineered cows the future of medicine?" *The Verge*, December 3, 2016, www.theverge.com/2016/12/3/13819482/genetically-engineered-animals-drugs-sab-cows-pharming-future.

Shapiro, R. (2016) "Outrageous new PETA ad compares cows with rape victims," *Huffington Post*, November 3, 2017, www.huffingtonpost.com/entry/peta-ad-rape_us_581befcfe4b0aac624836a29.

Shriver, A. (2009) "Knocking out pain in livestock: Can technology succeed where morality has stalled?" *Springer Science + Business Media B.V.*, August 21, www.academia.edu/2560199/Knocking_Out_Pain_in_Livestock_Can_Technology_Succeed_Where_Morality_has_Stalled.

Solis, M. (2016) "Appalling new PETA ad compares rape victims to animals," *Mic*, November 3, https://mic.com/articles/158513/appalling-new-peta-ad-compares-rape-victims-to-animals.

Weitzenfeld, A., and Joy, M. (2014) "An overview of anthropocentrism, humanism, and speciesism in critical animal theory," in A. Nocella, II, J. Sorenson, K. Socha and A. Matsuoka (eds.) *Defining Critical Animal Studies: An Intersectional Social Justice Approach*, New York: Peter Lang.

Wrenn, C. (2016) "What black lives matter can teach white vegans," *Corey Lee Wrenn, Ph.D.*, July 12, www.coreyleewrenn.com/white_vegans/.

Yeung, B., and Rubenstein, G. (2013) "Rape in the fields: Female workers face rape, harassment in U.S. agriculture industry," *Frontline*, June 25, www.pbs.org/wgbh/frontline/article/female-workers-face-rape-harassment-in-u-s-agriculture-industry/.

40
POLITICAL LOBBYING FOR ANIMALS

Joanna Grossman

Introduction

A pro–animal testing article from the 1980s published in *The Scientist* leads with the derisory headline and subtitle: "More protection for lab animals? Unnecessary rules are already strangling medical research." In the piece, the author bitterly observes, "One U.S. congressmen told me that he receives more mail lobbying for animal rights bills than he gets on all other issues combined" (Guyton 1988). It would appear that, in this particular respect, things had not changed too much from the 1960s, when Congressman Leo O'Brien (D-NY) said about an animal welfare bill he was sponsoring, "I doubt that there is any subject on which Members [of Congress] receive more mail" (Shaffer 1966).

The same holds true today; legislators frequently hear from constituents on issues pertaining to animal welfare. None of this should come as too great of a surprise. After all, the "concerns of the animal advocacy movement have increasingly taken root in public consciousness" (Donaldson and Kymlicka 2011: 1). Simply put, public support for animal protection efforts is high—arguably at an all-time high and with no sign of abatement. A 2015 Gallup poll found—rather incredibly—that a third of Americans believe animals should have the same rights as people (Riffkin 2015). That might come across as a fairly radical position, and certainly not the more tempered view the majority of respondents espoused, but one key area of overwhelming consensus is that animals deserve protection. In the same Gallup poll, a mere 3% answered that "animals require little protection from harm and exploitation 'since they are just animals'" (Riffkin 2015). Predictably, then, the last few decades have seen "a remarkable number of federal statutes . . . enacted to protect species, animals, and animal welfare" (Sunstein 2004: 253).

Today in the United States, legislation is arguably the "primary vehicle" when it comes to the growing field of animal law—notwithstanding the reality that "court-based challenges have been extremely important in animal protection" (Waldau 2011: 90). Often change comes incrementally, with legislative proposals building upon existing protections (Waldau 2011: 90). Certainly, the United States is a long way from attaining constitution-level provisions safeguarding animal welfare, as in Germany, which amended its constitution in 2002 to include considerations for animal interests, or perhaps most notably, India's constitution, which states, "It shall be the duty of every citizen . . . to have compassion for living creatures" (Kolar 2005: 147; Kelch 2011: 286). Even so, the humane treatment of animals is "so entrenched in our moral culture" that the US legal system endeavors to "establish the principle . . . in animal welfare laws"—both through "general animal welfare laws, such as anticruelty laws" that prohibit the "infliction of suffering on animals without distinguishing between uses of animals" and through "specific animal welfare laws" that apply to a "particular animal use" (e.g., animals used in experiments) (Francione 2000: 7).

Overview of US Federal Law

Early Victories and Landmark Legislation

In terms of animal protection, "by far the most important measure is the Animal Welfare Act (AWA), which imposes on those who deal in or with animals a wide range of negative constraints and affirmative duties" (Sunstein 2004: 254–255). Among other areas, the statute governs "the humane handling, care, treatment, and transportation of animals by dealers, research facilities, and exhibitors" (Sunstein 2004: 254–255). With its broad scope, the AWA proposes an "ambitious set of safeguards against cruel . . . practices" and suggests "something not so . . . different from a bill of rights for animals" (Sunstein 2004: 254–255). Prior to the AWA's enactment in 1966, animal experimentation was largely unregulated (Pacelle 2011: 47). As the US Department of Agriculture—which is charged with enforcing the law—notes, the AWA provides an excellent case study of "how a law gets proposed and eventually passed" (USDA 2007). In this instance, the "tipping point" was a series of magazine exposés on the "procurement process for . . . delivering dogs for biomedical research" (USDA 2007). One article detailed the disappearance of a Dalmatian named Pepper; the dog's guardian recognized "his missing dog in a picture taken of an animal dealer's overcrowded truck" (USDA 2007). His family tried to retrieve the dog but were denied entrance to the premises where they suspected Pepper was being kept; Congressman Joseph Resnick (D-NY) intervened and was similarly rebuffed (USDA 2007). In the end, Pepper—who like countless other dogs had been stolen to be sold to laboratories—was killed; she died during a botched surgery to implant an experimental cardiac pacemaker (Engber 2009). Another article, titled "Concentration Camp for Dogs," featured "pictures of skeletal dogs" that investigative journalists discovered at a Maryland dog dealer's farm (USDA 2007). The public was outraged, and the outcry led to a congressional response. The Animal Welfare Institute's founder and president, Christine Stevens, helped push for legislative action in response to the scandals, and in the summer of 1966, President Lyndon Johnson signed the AWA into law (Guither 1998: 41).

Public demand (or, more bluntly, politicians getting an earful) is a recurring theme when it comes to legislative reform. Several years before the AWA's passage, President Eisenhower was asked at a press conference whether he would sign the Humane Methods of Livestock Slaughter Act (HMSA) into law. His response was telling: "If I went by mail, I'd think no one was interested in anything but humane slaughter" (Congressional Record 2001). Animal advocacy groups such as the Animal Welfare Institute and the Humane Society of the United States supported the legislation, but as will be detailed in a later section, loopholes often abound and prevent a law from reaching its full potential. Indeed, farm animals raised for food were (and continue to be) in desperate need of legal protections. Still, the HMSA—which requires that livestock be rendered unconscious prior to slaughter—underscored an important statement from the federal government in terms of the imperative to reduce suffering.

Other early and significant federal victories include several laws that were signed by President Nixon. The Marine Mammal Protection Act (1972) "forbids the harming of any whale or other marine mammal within American waters, and authorizes sanctions against nations that do harm them" (Scully 2002: 182). The Endangered Species Act (1973)—oft-referred to as one of America's bedrock environmental laws—continues to enjoy widespread public support as a mechanism to prevent extinction. Equines—both wild and domestic—received much-needed attention with the passage of the Wild Free-Roaming Horse and Burro Act (1971) and the Horse Protection Act (1970). Despite the latter's broad name, the Act deals primarily with penalizing the practice of "soring"—the abusive practice of using chemicals and painful objects on a horse's hooves to incite a higher gait during "walking horse" competitions. The former protects these animals from "capture, branding, harassment, or death" and allows for the continued existence of herds on public lands (P.L. 92–195). With each of these examples (and, of course, this is also true of earlier statutes that pertain to wildlife and conservation), legislators continue to consider provisions that would alter the impact and scope of these laws in myriad ways.

Recent Achievements and Ongoing Initiatives

By any conservative estimate, every session of Congress sees dozens of proposals pertaining to animals in some form or fashion—everything from standalone bills that garner media attention to tucked away "riders" aimed at undermining animal welfare in huge legislative vehicles that were perhaps never truly intended to see the light of day. The following are a few examples that are illustrative of the kinds of successes that pro-animal lobbying has achieved in recent years, as well as a consideration of the major factors that contribute to such success.

Emergencies and Disasters

The devastation and aftermath of Hurricane Katrina were like nothing else most Americans had ever witnessed. The human toll was almost unfathomable, and subsequent hurricanes like Sandy, Maria, and Harvey have only further underlined the importance of preparedness measures and evacuation plans. Not to make short shrift of the human lives that were lost, but during Katrina, there is no question many were struck by what was happening to the animals during the rescue efforts. The Associated Press's coverage of a little boy who was forcibly separated from his pet dog named Snowball, with the child crying until he vomited, understandably moved and upset many individuals who could not fathom being forced to abandon their cherished companion animals under such circumstances. As one journalist explained, "you saw pictures of dogs standing on roofs and cats swimming in these toxic waters, and there was a huge public outcry" (Brulliard 2017). Legislators responded with a bipartisan initiative, the Pets Evacuation and Transportation Standards (PETS) Act, which ensures that state and local emergency operational plans accommodate pets and animals during and following a major disaster.

Trapping

A perennial subject of debate in Congress is the use of body-gripping traps—something that "animal advocates have brought . . . to the forefront of public consciousness" through decades of activism (Fox 2002: 126). Charles Darwin decried steel-jaw leghold traps as violent devices that inflict enormous agony, but even so, this "gratuitous torture of wildlife" (and of pets unlucky enough to fall victim to the traps) continues throughout much of the United States (Darwin 1863; Scully 2002: 355). The National Trappers Association (NTA) "celebrates the 'valiant and heroic exploits' of its trade, which . . . consists in subjecting an estimated four million animals every year . . . to slow death by shock, bleeding, or strangling" (Scully 2002: 355). Although groups like the NTA represent a small and extreme fringe within the entire US population, progress on this front has been slow, in part because the broader community of sportsmen's groups has historically been opposed to any kind of trapping reforms. When the Senate Environment and Public Works Committee held a hearing in 2015 to consider the Refuge from Cruel Trapping Act—a measure to prohibit the use of body-gripping traps on national wildlife refuges—a wide range of "mainstream" hunting groups wrote to members of the committee in opposition to the bill. Still, the issue keeps coming up in large part because so many Americans oppose the practice:

> Public outcry prompted Congress to include language in the 1997 appropriations bill directing the U.S. Fish and Wildlife Service to convene a task force to "study the use of animal traps in the National Wildlife Refuge system . . . [and to] consider the humaneness' of such traps."
>
> *(Fox 2002: 126)*

More recently, the 2017 and 2018 omnibus spending packages included language directing the U.S. Fish and Wildlife Service to warn members of the public if a wildlife refuge permits visitors to engage in trapping. These are modest measures (that will perhaps serve as precursors to significant reform at the federal level) but they reveal the subject's staying power on Capitol Hill. Indeed, in 2016, the Library of Congress issued a report noting that well over 100 countries have already banned the use of steel-jaw leghold traps (Law Library of Congress 2016a, 2016b).

Wildlife Trafficking

When it comes to wildlife, legislators of course also look at international matters. With an increasingly connected world (and where animal protection issues rarely if ever occur in a vacuum), Congress often considers proposals that impact the welfare of animals abroad. Wildlife trafficking offers an example par excellence of the necessity to wed animal concerns to broader social, cultural, geopolitical, and, frankly, human concerns. In this case, the link between wildlife trafficking and national security (e.g., funding from the illicit trade going toward warlords and terrorist groups) helps to cement the interest of stakeholders who might not otherwise be inclined to support—let alone push—for animal protection measures.

In our post-9/11 world, questions of national security take on heightened significance; this reality coupled with decimated populations of already severely imperiled species like elephants (which could go extinct in our lifetimes if the rate of poaching continues) has led to legislation aimed at tackling the crisis. In 2016, President Obama signed into law the Eliminate, Neutralize, and Disrupt (END) Wildlife Trafficking Act—legislation that, among other provisions, directs the US State Department to identify countries that are major sources, transit points, or consumers of trafficked wildlife products and allows prosecutors to treat smuggling endangered species parts as a predicate offense under money laundering statues (which leads to increased penalties and greater facility in prosecuting crimes).

The END Wildlife Trafficking Act deals with a host of "species threatened by poaching and the illegal wildlife trade" (P.L. 114–231), but some further context about elephants is warranted since there are few animals that provoke such strong reactions from constituents and their elected officials. At least in recent years, perhaps nowhere was this more evident than in the fervor and backlash surrounding the Trump administration's decision to resume the importation of elephant trophies. President Trump initially seemed to back away from the unpopular decision, calling elephant hunting a "horror show" (a sentiment that has not tracked with the USFWS's stated intentions). With any decision pertaining to these iconic and remarkably intelligent animals comes intense scrutiny; there is a lot on the line when the fate of a species hangs in the balance. In advance of the 17th meeting of the Conference of the Parties to the Convention on International Trade in Endangered Species of Wild Fauna and Flora (a multilateral treaty signed by 183 countries), 54 members of Congress sent a letter supporting the highest level of protection for African elephants (Beyer 2016).

The Will of the People

Essential to each of these successes and numerous others at the federal level (e.g., criminalizing attending organized animal fights such as cockfighting rings or dogfighting "pits," shifting away from requiring animals for toxicity testing) is popular support. To be clear, some lawmakers are champions who are willing to go to bat for virtually any animal welfare cause regardless of its popularity (often the same members of Congress who join the bipartisan Animal Protection Caucus or the Congressional Endangered Species Act Caucus to fend off attacks against wildlife) but such zeal is scarcely the norm—if it were, many more explicitly pro-animal laws would be sent to the president's desk each year.

Public outcry often prompts action—that is to say, when enough of the public makes it clear that a particular abuse will not stand and/or some problem must be remedied (as with the passage of the AWA). A more run-of-the-mill manifestation of this dynamic in legislative advocacy is the use of polling data to convey to congressional offices that a position enjoys broad support (or widespread opposition) among constituents. For example, polls showing overwhelming opposition among Americans to horse slaughter (80%), help shore up congressional support for setting policies that reflect that reality (AWHC 2017; ASPCA 2012).

Similarly, by adhering to the view that a unified voice can be more effective than individual entities attempting to tackle an issue alone, advocates have been successful in forming large coalitions that drive change on Capitol Hill. During negotiations for the 2014 and 2018 farm bills (massive legislative vehicles dealing with food and agriculture policy), hundreds of organizations banded together to help defeat the "King Amendment," which could have nullified local- and state-level animal welfare laws (HSUS 2018).

Strategies for Success

Follow the Money

Animal welfare advocates have often met with success on the legislative front by addressing the funding angle. What this means in practice is (a) using spending bills (which Congress must pass in some form each fiscal year to fund the operations of the federal government) to reform, promote, or eliminate certain policies for the benefit of animals (and which can be tied to funding); (b) highlighting examples of government waste; or (c) advancing arguments that proposal *x* is ultimately a more efficient use of taxpayer dollars—and, of course, these routes are not mutually exclusive.

Going after the purse strings, for example, has proven successful in blocking the opening of horse slaughter facilities in the United States. The "defund" provision included in the final appropriations package specifies that the US Department of Agriculture (USDA) cannot spend money inspecting horse slaughter facilities; such language is, of course, not tantamount to a flat-out ban on horse slaughter, but without federal inspections (which are mandated by law), the practice cannot occur. Such provisions are not without controversy; in recent years, some members of Congress have come out in support of horse slaughter, and the House Appropriations Committee has, at times, voted against including the language. However, the Senate has consistently voted in favor of the defund amendment, and thus, the provision has successfully made it to the proverbial finish line in the final negotiated spending package.

Perhaps shockingly, the federal government remains the "largest funder of animal experiments" in the United States (Church 2002: 251). And while many are no doubt completely unaware that the US government devotes billions of dollars toward animal experimentation, as Peter Singer bluntly put it in *Animal Liberation*, we cannot "pretend that we have nothing to do with these practices" since animal experimentation is "paid for out of the taxes we pay" (Singer 1975: 22).

Although the National Institutes of Health is typically singled out given the massive sums the agency doles out "in grants to many institutions both inside and outside federal agencies" (Church 2002: 251), the fact is taxes underpin many of the practices that individuals might find objectionable—from lethal control methods used to "manage" wildlife at the behest of ranchers to experiments and necropsies on canines at the Department of Veterans Affairs (VA). Perhaps not surprisingly, such areas have found support among right-wing allies given the overlap with arguments about wasteful government spending (and a misuse of taxpayer dollars). The conservative-leaning Taxpayers for Common Sense has endorsed legislative proposals to curb USDA's Wildlife Services spending on lethal control measures. The agency within USDA kills an estimated 3 and 5 million animals each year using some of the most inhumane methods imaginable (e.g., aerial gunning operations that have gone awry and led to pilot deaths, chemical poisons, body-gripping traps). White Coat Waste—itself a right-leaning animal advocacy and watchdog group—has led successful efforts to defund VA experiments on dogs through provisions in spending bills. Of course, with any appropriations language, it is important to note that such hard-fought victories are temporary and apply only to the fiscal period the bill covers. Consequently, animal protection groups must constantly prepare themselves for the next cycle—the next legislative fight, so to speak—at least until a permanent fix is achieved through legislation that would outright prohibit the practice in question (e.g., the Safeguard American Food Exports [SAFE] Act, which would ban horse slaughter and prevent the export of horses into Canada or Mexico for slaughter).

A Good Defense

While proactive measures to protect animals are generally the desired and ideal outcome, a critical part of legislative advocacy is combating harmful proposals that would undermine animal welfare. Nowhere is this reality more apparent than in areas pertaining to wildlife, and particularly endangered and threatened species protections. Legislators put forward a slew of provisions each Congress that would chip away at environmental and conservation statutes that support biodiversity and species survival, particularly the ESA. In 2017, for example, Defenders of Wildlife tallied more than 160 legislative attacks against the ESA in a two-year period (and such a count does not include provisions unrelated to the ESA that would harm wildlife). The Center for Biological Diversity keeps a running tally as part of its "Politics of Extinction" campaign that reveals a marked uptick of legislative attacks in recent years. Animal protection, wildlife, and conservation groups have done an admirable job of holding the line in a charged and hyper-partisan political climate by ensuring that "poison pill" riders that would strip protections from imperiled species are not included in omnibus spending packages, for example. Similarly, animal protection groups regularly work to ensure that controversial bills such as the so-called sportsmen's packages that are perennially introduced and which include provisions targeting animals (e.g., expanding trapping on public lands, allowing the ivory trade to resume) are not folded into larger legislative vehicles. In 2016, for example, 20 senators formally objected to the inclusion of anti-wildlife riders (which originated in a House-passed sportsmen's bill) in the energy package that was being considered (Booker 2016).

Obstacles and Challenges

Dysfunction in the Political Process

The need to combat harmful measures hints at some of the difficulties that exist when engaging in the legislative process. Of course, anyone who follows the news even tangentially is likely familiar with some of the more unpopular aspects of Congress. In *Animal Liberation*, Peter Singer understandably takes a bleak view of US politics:

> Since moving to the United States . . . I have thought about why America lags so far behind Europe in its protection of animals . . . [One] reason for the difference . . . is that the U.S. political process is simply more corrupt. Elections are many times more costly . . . [allowing]

> the agribusiness industry far greater control over Congress than it can hope to have over the political processes in Europe.
>
> *(Singer 1975: xi, xii)*

Anyone seeking to advance animal welfare causes should be under no illusions about the entrenched special interests that have considerable firepower at their disposal—and which often stand in the way of even the most modest reforms. The meat, dairy, and egg industries spend millions of dollars each election cycle in direct or indirect lobbying expenditures (Open Secrets 2018a, 2018b, 2018c; Joy 2010: 89). Prominent agribusiness groups such as the American Farm Bureau Federation have used their political heft to oppose efforts to ban the slaughter of downer cows and prohibit the extreme confinement of farm animals on factory farms (Pacelle 2011: 299).

If you can imagine a type of animal abuse, chances are high some form of "lobbying apparatus" exists to advocate for its continued existence; big-name groups like the Safari Club International and the National Rifle Association lobby in support of trophy hunting, but even cockfighters, trappers, and fur "farmers" have national trade groups (Pacelle 2011: 283). When it comes to animal experimentation, "dozens of political action committees support animal research interests . . . [and] individual pharmaceutical companies have their own PACs [political action committees]" (Church 2002: 251). Unfortunately, since legislators and their staff simply do not have time to become experts in the wide range of fields that involve the use of animals, they "rely on what the 'experts' tell them" and industries that profit off of animals "are prepared to spend huge amounts of money to oppose legislation that will deprive them of their profitable markets" (Singer 1975: 93).

The influence that animal-use industries wield cannot be underestimated but I would be remiss not to address the more benign or run-of-the-mill obstacles to progress: gridlock and dysfunction. It is no secret that partisan bickering has become the norm in Washington. Every year staffers describe the environment on Capitol Hill as the most partisan they've seen. One telling indication of the level of dysfunction is that regular order for the budget process has not been followed for decades; the "last time all twelve appropriations bills were passed individually prior to the start of the fiscal year was in 1994" (Byrne and Mitchell 2018). Advocates engaging in lobbying—especially at the federal level—have to be in it for the long haul because progress through political channels can come at a snail's pace. Reflecting on his years trying to sway federal lawmakers, farmer-turned-animal rights activist Howard Lyman noted that "it's definitely not impossible, and it's always worth trying, but more and more, I've come to feel that the most fundamental changes in this country come . . . from the bottom up, and the game Washington plays best is catch-up" (Lyman 1998: 79).

Questions of Legal Status, Rights, and Interests (and the Interplay With Legislative Advocacy)

Although some countries include animal protection in their constitutions, instituting that kind of macro-level change would be a tall order in the United States (see the difficulties that broadly supported campaigns like the Equal Rights Amendment have encountered). Without question, legal traditions impact "what changes are possible in any society," and in the United States, a "concept of property rights that favors even minor human interests" colors all matters pertaining to animals (Waldau 2011: 105, 119). State and federal agencies charged with administering policies tend to operate under a framework of "resource utilization"—for example, exterminating certain animals that are considered pests for the sake of increasing "game" populations or so that cattle (themselves considered resources) have more room to graze (Waldau 2011: 105).

On the one hand, many people—and undoubtedly most, if not all self-professed animal lovers—view animals as individual beings with "desires, beliefs, and the ability to act in pursuit of their goals" (Regan 1983: 116). As Tom Regan explained, "animals have a kind of autonomy—preference autonomy" (Regan 1983: 116). It is no stretch to say that animals prefer to be free from harm and danger. But this is not how issues are typically framed legislatively since animals are considered property, and "owners"—for the most part—have the "right to do what they will with their private property" (Joy 2010: 103; Baur 2008: 189). Indeed, the fact that animals are legal property constitutes "one of the defining problems of pursuing farmed animal abuse charges" (Baur 2008: 189).

If "rights" are understood to be "legal protection against harm, then many animals already do have rights" (Sunstein 2004: 5). After all, our laws "go well beyond prohibiting beating, injuring, and the like, and impose affirmative duties on people with animals in their care" (Sunstein 2004: 5–6). But if we understand "rights" to mean that animals "should not be subject to human use and control," then certainly the United States falls short by any measure (Sunstein 2004: 10). The mere status as legal property precludes such an individualistic view of rights; ultimately, most laws fall under the construct of relief from suffering.

Limitations of the Law and Glaring Omissions

Even long-fought victories—which should rightly be celebrated—still rarely mean true success is achieved. The reason can be summed up in one word: loopholes. Many of the important animal welfare statutes that exist in this country are so riddled with loopholes that one can reasonably query how such fundamental protections can fall so short of their stated goals.

By its very name, the AWA suggests breadth and gravity. Incredibly, however, the law's provisions do not apply to the vast majority of animals used in research. As psychologist Hal Herzog bluntly put it,

> I find the congressional declaration that rats, mice, and birds are not animals to be bizarre, unnecessary, and even morally offensive. The rats/mice/birds exclusion intentionally obfuscates the true numbers of animal used in research. And it lends credence to the claims of animal rights groups that research institutions have something to hide.
>
> *(Herzog 2015)*

Any rational person would agree that mice, birds, and rats are animals; unfortunately, for the sake of convenience, profits, politics, concealment, or some combination of these factors, the AWA was severely limited in its scope. Senator Jesse Helms (R-NC) offered an amendment that specifically excludes "birds, rats of the genus Rattus, and mice of the genus Mus." Recent data reveal that an "astounding 98.8% of animals housed in biomedical research facilities in the United States are exempt from any coverage under the Animal Welfare Act" (Herzog 2015).

Of course, another major omission is that the AWA "does not apply to the treatment of animals raised for food or clothing" (Sunstein 2004: 254). Farm animals—who are specifically excluded from the AWA—fare poorly across the board given that no federal statute regulates their treatment, and "states typically exempt farm animals from anticruelty statutes" (Sunstein 2004: 254–255). Even the HMSA does not apply to the largest number of terrestrial land animals killed for food—poultry (Singer 1975: 152). This is, without question, a serious failing of the law. As with the AWA, a shockingly high percentage of the animals who might otherwise seem to fall under the law's purview are excluded. Chickens, turkeys, and fish—which "constitute 99 percent of the animals slaughtered for food" in the United States—do not receive even the minimal protection that they be rendered "insensible to pain" before being killed (Simon 2013: 47). Not surprisingly, in the absence of basic protections, the results are horrific, with birds often still conscious as they hang shackled upside down to have their throats slit (Foer 2009: 133).

Serious penalties for the abuse or mistreatment of farms animals are virtually nonexistent. In the case of the HMSA, the primary enforcement mechanism is merely to temporarily halt the operations of a slaughter plant (Pacelle 2011: 107). The fact is that

> conduct that would be considered cruelty to a dog or a cat—denial of food, water, shelter, and veterinary care, as well as being chained or confined . . . in condition[s] that cause both physical and psychological disorders—is often lawful and, alarmingly, standard practice in the treatment of billions of farm animals in the United States.
>
> *(Baur 2008: 188)*

And while "every U.S. state has a statute prohibiting cruelty to farm animals," the majority have "exceptions to their anticruelty statutes for farm animals"—what are commonly referred to as "customary farming exemptions" that "put no limits on how animals are treated . . . as long as what is done to them is [considered] 'customary'" (Simon 2013: 36; Baur 2008: 188). Some of the

> many common farming practices now deemed lawful under most states' anticruelty statues include: crushing or severing the testicles of unanaesthetized animals, slaughtering chickens while they are aware and alert, [and] killing unwanted male chicks or spent laying hens by suffocation, starvation, or disposal in a garbage can or wood chipper.
>
> *(Simon 2013: 37)*

Getting Involved

Direct Advocacy

Alterations that severely limit "the scope of anticruelty protection" can easily lead to a sense of hopelessness since what results are laws that often fail to help the animals they were intended to protect (Waldau 2011: 109). The reality is that legislative advocacy is an ongoing battle and even successes (e.g., blocking congressional attempts to thwart more humane standards for farm animal care under USDA's organic program) can be undone in the blink of an eye (e.g., the Trump administration announcing it would roll back the organic rules to enhance animal welfare). Politics is messy by nature and advocates have to take the long view. That said, there are countless opportunities for individuals to get involved at virtually every level of politics (and at the risk of sounding clichéd, to provide a voice for the voiceless).

One avenue closely tied to legislation is the opportunity to weigh in on regulations that shape myriad aspects of animal protection, wildlife management, habitat preservation, and other environmental policies that impact biodiversity or land use. Agencies often seek public input during the rulemaking process, giving individuals a chance to submit written feedback on proposals or in some cases attend public meetings to voice their opinions. In 2016, for example, the USDA held a series of public meetings to gather input on how to strengthen the Horse Protection Act regulations to protect horses from soring (a persistent problem despite the federal law designed to prevent it from happening). The rulemaking process (or more informal guidance that agencies also sometimes issue) represents a critical follow-up step to ensure that the details behind the intent of a bill are fleshed out through clear directives, and can be properly enforced. Administrative regulations are especially prevalent in matters relating to the "hunting, handling, or importation of wildlife" (Waldau 2011: 92).

If nothing else, remarks from elected officials (including a sitting president as when Eisenhower expressed surprise at the correspondence on slaughter) reveal the usefulness of direct constituent outreach. Calling your representative in advance of a vote or asking him or her to support a bill is a straightforward and expedient way to register support for animal welfare measures and signal to the legislator that constituents care and follow their positions. Staffers at the front desk will convey information about a high volume of calls asking a senator to vote no on a harmful measure, for example. Each congressional office has staff dedicated to responding to constituent inquiries and tracks which issues receive the most mail. Town halls tend to be relatively well publicized, but a legislator's staff will schedule meetings with constituents in a district office or in Washington, D.C., to listen to concerns; most offices have a point person who specifically handles animal welfare as part of his or her legislative portfolio. Scorecards such as the ones published by the Animal Welfare Institute and the Humane Society of the United States help to keep tabs on legislators and hold them accountable for positions on animal welfare issues (and provide a way for the public to see their record over time).

State and Local Initiatives

Justice Brandeis's famous notion of state and local governments serving as laboratories of democracy seems especially apt in matters pertaining to animal protection. Attempts at reform that have fared poorly in Congress (e.g., animal agribusiness groups opposing improvements for the housing conditions of egg-laying hens) have prevailed on Election Day ballots. In other words, when questions of animal welfare are put directly to the people, the public generally—and decisively—makes its preference for humane treatment clear.

Grassroots movements have been remarkably successful in getting propositions on the ballot at the state level. Advocates have worked to get the thousands of signatures required to place an initiative on the ballot (exact figures vary depending on state law in those jurisdictions that allow such initiatives, but the numbers are high—California, for example, requires hundreds of thousands of signatures to qualify). The last three decades in particular have seen a significant uptick in the number of ballot measures and popular initiatives pertaining to animal protection, perhaps corresponding with an increasing awareness of the plights myriad animals face in modern society (Waldau 2011: 91)

While federal lawmakers have yet to enact meaningful and substantive restrictions on the use of body-gripping traps in the United States, efforts to ban cruel traps have been successful in several states (Fox 2002: 126). Arizona, California, Colorado, Massachusetts, and Washington have all severely curtailed the use of body-gripping traps through ballot initiatives (Fox 2002: 128). And several other states have done so via legislation or regulations, serving as an important reminder that lobbying one's state legislators provides a means that can be every bit as useful for enacting change in the absence of federal action.

Of course, farm animal welfare is another area where federal law is sorely lacking, but here again, state-level initiatives with important ramifications for the rest of the country have been successful. In 2002, Florida passed a ballot measure banning gestation crates for pregnant sows. Arizona followed suit in 2006, banning gestation crates and veal crates. In 2008, California passed Proposition 2—a ballot measure phasing out battery cages, gestation crates, and veal crates. The victory of Proposition 2 sent a strong message in its own right, but it was followed with legislation banning the sale of eggs from hens living in battery cages, meaning if egg producers from other states wanted to sell their eggs in California, they would have to meet this higher welfare standard.[1] In 2016, Massachusetts voters passed Question 3 by a vote of 78% to 22%—a ballot measure that contained both an outright ban on extreme confinement (battery cages, gestation crates, and veal pens) and a prohibition on the in-state sale of products resulting from such confinement systems. In a certain sense, each of these initiatives has built upon the progress of earlier measures. And by focusing on some of the worst conditions and practices (e.g., pregnant pigs, veal calves, and egg-laying hens who live out their days in misery), advocates were able to call attention to the serious problems inherent to massive factory farms.

Ballot initiatives tend to be high-profile campaigns (and highlight the power of citizen engagement and direct democracy), but, of course, examples of positive animal-related reforms at the state and local levels occur through multiple means. In 2014, South Dakota became the 50th state to adopt felony-level penalties for animal cruelty (in 1990, fewer than 10 states had felony provisions; Berry 2014). Selecting a few state and local measures to highlight is difficult, but suffice it to say, examples of vibrant grassroots activism abound with advocates working tirelessly to institute changes through their state legislature, local ordinances, and by speaking with elected officials (or engaging with boards and commissions charged with overseeing the implementation of policies, as is often the case with wildlife management). The goal is to work on all fronts and leave no stone unturned.

In the past few years alone, San Francisco and Los Angeles became the largest US cities to ban fur sales; New York City, New Jersey, and Hawaii banned wild and exotic animals from circuses; Maryland passed a bill prohibiting the sale of dogs from "puppy mills" as well as legislation requiring research facilities to adopt out dogs used in scientific experiments; Rhode Island became the latest state to allow companion animals to be included in domestic violence protective orders; Illinois banned the sale of cosmetics tested on animals; Hawaii implemented perhaps the nation's most stringent wildlife trafficking ban; and Texas—like other states before—enacted a ban on the shark fin trade. The list could go on, but regardless of which particular examples one cites, these cases speak to a broader social awareness of animal welfare. The numerous local bans affecting the ability of circuses or traveling shows to utilize exotic animals (e.g., prohibitions on the use of "bullhooks" to train elephants by inflicting pain) sent shockwaves throughout this industry—most notably with Ringling Bros. and Barnum and Bailey Circus retiring its elephants and ceasing operations. One might argue the repercussions have also included SeaWorld ending its orca performances and pledging to stop breeding these cetaceans in captivity.

Conclusion: Pessimism Versus Optimism

The animal protection movement has become a far more prominent aspect of US society over the past few decades. In that sense, as Sue Donaldson and Will Kymlica write in *Zoopolis*, the "movement can be seen as a success." But Donaldson and Kymlica actually take a much bleaker view, and not without good reason:

> We would argue that the movement has largely failed . . . The relentless expansion in human population and development continues to take away habitat for wild animals. Our population has doubled since the 1960s while wild animal populations have dropped by a third. And the factory farm system keeps growing to meet . . . the demand for meat . . . to the point that humans today kill 56 billion animals per year for food (not including aquatic animals) . . . these global trends are truly catastrophic, dwarfing the . . . victories achieved through animal welfare reforms.
>
> *(Donaldson and Kymlicka 2011: 1–2)*

This is an eminently reasonable view to take. Legislation can be important and can signal a shift in consciousness, but legislative victories alone are unlikely to stem this tide. However, it is also worth reflecting on what it would mean to live in a society without laws aimed at protecting animals—in ways both large (outright bans on cruel practices with penalties against perpetrators) and small (modest reforms in the living conditions of farm and lab animals who might otherwise know only abject horror and deprivation). This is certainly not a world in which anyone opposed to animal abuse would want to live. That unanimity is perhaps one of the few things that truly unites individuals of all political stripes when push comes to shove. Charles Krauthammer advanced a rather remarkable comparison for a conservative icon when he wrote,

> We often wonder how people of the past, including the most revered and refined, could have universally engaged in conduct now considered unconscionable. Such as slavery . . . Surely some contemporary practices will be deemed equally abominable by succeeding generations. The only question is: Which ones? I've long thought it will be our treatment of animals.
>
> *(Krauthammer 2015)*

Particularly in a liberal democracy, lobbying for change is a vitally important means for creating a society that upholds those values that are widely held by its constituency. It can be easy to take a somber view of legislative advocacy since as Paul Waldau rightly notes, "there is always a certain asymmetry in legislative matters" and the "ratio of successes to failures . . . does not alone reveal whether a cause is gaining or losing ground" (Waldau 2011: 90). But even ostensible failures—or at a minimum, the inertia that is often the norm in Congress—can impact public discourse and lead to changes via other routes, for example, corporate policies, regulatory or executive actions, state, and local laws, or simply by affecting consumer behavior.

We do not know what laws pertaining to animals will look like in the years to come, but "dramatic change has taken place within the last century"—change that has opened minds to the very notion of animal rights (Waldau 2011: 101). Perhaps most importantly, legislative reforms have underscored that advocating for animal protection is not so radical after all. On the passage of a federal bill to clamp down on horrific "crush videos" where animals are tortured and mutilated, Matthew Scully observed,

> When we speak of the rights of animals, we are talking as a practical matter about legal protections within our power to grant or deny, and that is what such a law actually looks like. Was the moral world upended? Was radicalism loosed upon society? . . . No, a law was passed saying that you can't crush hamsters and mice . . . and puppies for fun or profit, and if you do then you are going to jail.
>
> *(Scully 2002: 293)*

And while protecting the comparatively small number of animals tortured in crush videos might seem like a drop in the bucket compared to the systemic and institutional abuses taking place (something that can be said of many animal welfare reforms), we should take every opportunity to remedy clear wrongs, protect innocent creatures, and work to ensure that momentum keeps building at every level of society and the institutions that govern it—from a local town board meeting to the halls of Congress.

Note

1. In 2018, California voters again passed a ballot measure pertaining to the confinement of farm animals. Proposition 12 passed with more than 60% of the vote. One key difference between the 2018 ballot initiative and Proposition 2, which passed in 2008, was that the new initiative set specific minimum space requirements for pregnant pigs, calves raised for veal, and egg-laying hens.

Bibliography

American Society for the Prevent of Cruelty to Animals (2012) "Research confirms Americans strongly oppose slaughter of horses for human consumption," *ASPCA Press Release*, February 1.

American Wild Horse Campaign (2017) *National Survey Results*, Raleigh, NC: Public Policy Polling, https://americanwildhorsecampaign.org/media/2017-national-poll

Baur, G. (2008) *Farm Sanctuary*, New York: Simon & Schuster.

Berry, C. (2014) "All 50 states now have felony animal cruelty provisions," *Animal Legal Defense Fund Press Release*, March 14.

Beyer, D. (2016) "Beyer leads house delegation to restore elephant protections," *Office of Congressman Beyer Press Release*, September 21.

Booker, C. (2016) "20 Senators urge committee on energy and natural resources leaders to reject extreme anti-wildlife measures as part of energy bill," *Office of Senator Cory Booker Press Release,* October 5.

Brulliard, K. (2017) "How the chaos of hurricane Katrina helped save pets from flooding in Texas," *The Washington Post,* August 31.

Byrne, B., and Mitchell, P. (2018) "Debt ceiling working group letter," *Offices of Congressman Byrne and Mitchell*, February 12.

Church, J. (2002) "The Politics of Animal Research," in K. Stallwood (ed.) *A Primer on Animal Rights*, New York: Lantern Books.

Congressional Record (2001) *Senate Concurrent Resolution 45—Expressing the Sense of Congress that the Humane Methods of Slaughter Act of 1956 Should Be Fully Enforced So as to Prevent Needless Suffering of Animals,* Washington, DC: Vol. 147, Pt. 7, p. 9879.

Darwin, C. (1863) "Trapping agony," *Gardeners Chronicle and Agricultural Gazette,* August.

Defenders of Wildlife (2017) "ESA under attack on Capitol Hill," *Press Release*, July 19.

Donaldson, S., and Kymlicka, W. (2011) *Zoopolis: A Political Theory of Animal Rights*, Oxford: Oxford University Press.

Engber, D. (2009) "Pepper: The stolen dog that changed American science," *Slate,* June 1.

Foer, J. S. (2009) *Eating Animals*, New York: Hachette Book Group.

Fox, C. (2002) "National wildlife refuges: Sanctuaries or killing fields," in K. Stallwood (ed.) *A Primer on Animal Rights*, New York: Lantern Books.

Francione, G. (2000) *Introduction to Animal Rights*, Philadelphia: Temple University Press.

Guither, H. (1998) *Animal Rights: History and Scope of a Radical Social Movement*, Carbondale: Southern Illinois University Press.

Guyton, A. (1988) "More protection for lab animals?" *The Scientist,* May 30. https://www.the-scientist.com/news/more-protection-for-lab-animals-62925

Herzog, H. (2015) "Congress should declare that mice are animals – now!" *Psychology Today*, April 20.

Humane Society of the United States (2018) *Diverse Opponents of Bills that Threaten State Laws,* December 6, https://drive.google.com/file/d/0B70TaAxEEA8BS1FVZS1QWUZvYlk/view

Joy, M. (2010) *Why We Love Dogs, Eat Pigs, and Wear Cows*, San Francisco: Conari Press.

Kelch, T. (2011) *Globalization and Animal Law: Comparative Law, International Law, and International Trade*, The Netherlands: Kluwer Law International.

Kolar, R. (2005) "Three Years of Animal Welfare in the Germany Constitution: The Balance from an Animal Welfare Perspective," *ALTEX* 22(Special Issue 2): 146–149.

Krauthammer, C. (2015) "Free willy," *The Washington Post,* May 7. https://www.washingtonpost.com/opinions/free-willy/2015/05/07/4d1a82f2-f4f2-11e4-b2f3-af5479e6bbdd_story.html

Law Library of Congress (2016a) *Animal Trap Laws in Selected Jurisdictions: Supplemental Report,* Washington, DC: Global Legal Research Center.

Law Library of Congress (2016b) *Laws on Leg-Hold Animal Traps Around the World,* Washington, DC: Global Legal Research Center.

Lyman, H. (1998) *Mad Cowboy*, New York, NY: Touchstone.

Open Secrets (2018a) *Dairy: Long-Term Contribution Trends.* Washington, DC: The Center for Responsive Politics, www.opensecrets.org/industries/totals.php?cycle=2018&ind=A04

Open Secrets (2018b) *Meat Processing & Products: Long-Term Contribution Trends,* Washington, DC: The Center for Responsive Politics, www.opensecrets.org/industries./totals.php?cycle=2018&ind=G2300

Open Secrets (2018c) *Poultry & Eggs: Long-Term Contribution Trends,* Washington, DC: The Center for Responsive Politics, www.opensecrets.org/industries/totals.php?cycle=2018&ind=A05

Pacelle, W. (2011) *The Bond: Our Kinship with Animals, Our Call to Defend Them*, New York: HarperCollins.

Regan, T. (1983) *The Case for Animal Rights*, Berkeley: University of California Press.

Riffkin, R. (2015) "In U.S., more say animals should have same rights as people," *Gallup News,* May 18, https://news.gallup.com/poll/183275/say-animals-rights-people.aspx

Scully, M. (2002) *Dominion: The Power of Man, the Suffering of Animals, and the Call to Mercy*, New York: St. Martin's Press.

Shaffer, H. B. (1966) *Treatment of Animals in Medical Research: Editorial Research Reports,* Vol. 1, Washington, DC: CQ Press.

Simon, D. (2013) *Meatonomics*, San Francisco, CA: Conari Press.

Singer, P. (1975) *Animal Liberation*, New York, NY: HarperCollins.

Sunstein, C., and Nussbaum, M. (eds.) (2004) *Animal Rights: Current Debates and New Directions*, Oxford: Oxford University Press.

U.S. Department of Agriculture National Agricultural Library (2007) *Legislative History of the Animal Welfare Act,* Beltsville, MD: Animal Welfare Information Center Resource Series, No. 41.

Waldau, P. (2011) *Animal Rights*, Oxford: Oxford University Press.

41
EFFECTIVE ANIMAL ADVOCACY

Jeff Sebo

Introduction

Imagine that you are a doctor, volunteering your time to save people in the aftermath of a natural disaster. There are many more people who need help than you have the ability to help. How should you approach your work? Many people find it natural to say that you should triage. That is, you should try to do the most good possible with your limited resources. If you have to choose between treating two people, one of whom has a major injury and the other of whom has a minor injury, then you should prioritize the person with the major injury, all else being equal. Similarly, if you have to choose between treating two people, one of whom requires relatively few scarce resources and the other of whom requires relatively many scarce resources, then you should prioritize the person who requires relatively few scarce resources, all else being equal. Granted, it might seem callous to prioritize lives this way. But in a state of emergency, we naturally understand that triage is an expression of compassion, not callousness. If we want to save the most lives possible or relieve the most suffering possible, then we need to think carefully about how best to use our limited resources so that we can achieve this aim.

However, as MacAskill (2015), Singer (2015), and many others have noted, many people seem to forget this point when it comes to addressing other, more chronic problems.[1] For example, when we make choices about advocacy and philanthropy, many of us do what feels personally meaningful rather than think carefully about how to save the most lives possible or relieve the most suffering possible. Indeed, many of us spend more time researching which appliances to buy than which organizations to support with our time, energy, and other resources.

This is where effective altruism (EA) comes in. EA is the project of using evidence and reason to do the most good possible. For an effective altruist (also EA), the world is in a chronic state of emergency, with many more individuals who need help than we have the ability to help. So, we should triage: We should try to do the most good possible with our limited resources. If we have to choose between supporting two organizations, one of which would save or spare more lives than the other as a result of our support, then we should prioritize the organization that would save or spare more lives, all else being equal.

Effective animal advocacy (EAA), then, is EA in the context of animal advocacy. An effective animal advocate (also EAA) attempts to answer questions like: Can we do more good overall if we work on farmed animal advocacy or companion animal advocacy? Can we do more good overall if we engage in moderate advocacy or radical advocacy? In each case, they attempt to use evidence and reason to answer these questions, and then they pursue the course of action that they estimate will do the most good overall.

This chapter discusses how EAAs attempt to use evidence and reason to do the most good possible in animal advocacy, as well as conceptual, principled, and practical issues that this approach to animal advocacy faces. While I mostly focus on summarizing these issues, I also share my own views about them. In particular, I claim that EAA sets the right goals for animal advocacy but that its pursuit of these goals is still a work in progress. Thus, I claim, if we want to do the most good possible in animal advocacy (as, I think, we should), then we should work to (a) improve our assessments of how we can do the most good possible and (b) supplement our use of these assessments with other, more indirect ways of doing the most good possible.

Conceptual Issues

EAs use the following framework to decide which issues to prioritize. First, they think about the *scale* of an issue. If one issue involves more suffering or death than another issue, then it has higher priority for an EA, all else equal. Second, they think about the *neglectedness* of an issue. If one issue is more neglected than another, then it has higher priority for an EA, all else equal. Third, they think about the *tractability* of an issue. If one issue can be addressed more effectively or efficiently than another, then it has higher priority for an EA, all else equal. Finally, they think about *personal fit*. If one issue is a better match for your talents and interests than another, then it has higher priority for you as an EA, all else equal (although this consideration matters less than the other three).

What do EAs see as the most important issues to be working on right now, considering all these questions? For many EAs, the top priority is the category of *existential risk*, that is, issues that risk greatly reducing the potential for flourishing in the world (Bostrom 2002). Examples include plagues, pandemics, biotechnology, nanotechnology, nuclear technology, artificial intelligence, and totalitarianism (Bostrom and Ćirkovic 2011). These issues are top priorities because, in each case, the harm at stake is massive (as well as potentially permanent), neglected, and tractable. Many EAs think that we should then prioritize global health and animal welfare, since, again, the harm at stake is massive (even if not potentially permanent), neglected, and tractable. Of course, many other issues are important too. However, many EAs see these as the most important issues to be working on in the present moment.[2]

What, then, do EAAs see as the most important issues to be working on right now in the context of animal advocacy, considering these questions? For many EAAs, the top priority is farmed animal welfare. To see why, compare farmed animal welfare with companion animal welfare and wild animal welfare, respectively.[3]

First, consider how farmed animal welfare compares with companion animal welfare. Farmed animal welfare ranks higher in scale, since many more farmed animals than companion animals suffer and die each year. Farmed animal welfare also ranks higher in neglectedness, since farmed animals receive less support than companion animals do. Finally, farmed animal welfare ranks higher in tractability, since it costs less time, energy, and money to prevent suffering or death for the average farmed animal than it does for the average companion animal. As a result, farmed animal advocacy ranks higher than companion animal advocacy in scale, neglectedness, and tractability. (I consider objections to this analysis later.)

Now consider how farmed animal welfare compares with wild animal welfare. Wild animal welfare ranks higher in scale, since wild animals experience even more suffering and death than farmed animals do, as a result of hunger, thirst, illness, injury, and predation. Wild animal welfare also ranks higher in neglectedness, since wild animals receive even less support than farmed animals do.[4] However, wild animal welfare currently ranks much *lower* in tractability, since we currently have no reliable way to help wild animals at all. For many EAAs, the upshot is that we should focus mostly on farmed animals at present while engaging in advocacy and research that lays the groundwork for helping wild animals in the future.[5] (I consider objections to this analysis later as well.)

What kinds of approaches do EAAs take to these issues? Consider farmed animal welfare. EAAs tend to take approaches such as consumer outreach, corporate outreach, and development of alternative food products such as plant-based meat. In contrast, they do not tend to take approaches such as rescue or sanctuary. Why the difference? Many EAAs think that when you take the former approaches, you are potentially helping more animals at once, and so you are potentially spending less time, energy, and money per animal. In contrast, they think that when you take the latter approaches, you are helping fewer animals at once, and so you are spending more time, energy, and money per animal. (I consider objections to this analysis later as well.)[6]

EAAs also aspire to take a pragmatic approach to this work. For example, when it comes to consumer outreach, many EAAs think that if we can help more animals by advocating for meat reduction than by advocating for meat elimination, then we should do so. When it comes to corporate outreach, many EAAs think that if we can help more animals by advocating for Meatless Mondays than by advocating for vegan menus, then we should do so. And, when it comes to the development of alternative food products, many EAAs think that if we can help more animals by partnering with conventional food producers than by maintaining independence, then we should do so.[7] (I once again consider objections to this analysis later.)

Over the past decade or so, a network of organizations has developed around the idea of EA in general and EAA in particular. With respect to EAA in particular, these organizations Animal Charity Evaluators (ACE), which is dedicated to finding and promoting the most effective ways to help animals. Among the organizations ACE recommends are Animal Equality, The Good Food Institute, The Humane League, the Albert Schweitzer Foundation, Compassion in World Farming, Faunalytics, L214, Otwarte Klatki, ProVeg International, The Nonhuman Rights Project, and Vegan Outreach. Some of these organizations are explicitly committed to EAA. Others are not, but they still do excellent work by EAA standards.

This is more or less the interpretation of EAA that many EAAs accept. However, this conception of EAA raises many conceptual questions. Consider two examples.

First, what is the relationship between EAA and different moral theories? Some people see a connection between EAA and utilitarianism, which holds that we morally ought to maximize well-being in the world. It makes sense that people would see this connection, since utilitarianism and EAA do share several core features. For example, they are both committed to *impartial benevolence*, that is, to promoting well-being for all sentient beings. They are both committed to *aggregation*, that is, to evaluating states of affairs in terms of how much well-being they contain. They are both committed to *maximization*, that is, producing the best possible state of affairs (although they might not agree on what counts as well-being and, thus, they might not agree on what counts as the best state of affairs). And more. In contrast, other moral theories reject some or all of these features. For example, Kantianism holds that we morally ought to treat others as ends independently of whether doing so produces the best possible outcome. Care theory holds that we morally ought to cultivate relationships of care independently of whether doing so produces the best possible outcome. Virtue theory holds that we morally ought to cultivate virtuous characters independently of whether doing so produces the best possible outcome. And so on.[8]

However, even if EAA has more in common with utilitarianism than with other moral theories, a nonutilitarian can take interest in EAA as well. There are at least two reasons for this. First, no matter what moral theory we accept, we might find that we have at least some reason to participate in EAA, within the moral limits set by this theory. After all, all moral theories hold that consequences are morally relevant, even if some moral theories hold that other factors are morally relevant too. For example, a Kantian, care theorist, or virtue theorist might think that we should, at least in part, promote a world in which everyone can treat each other as ends, be in relationships of care, or have virtuous characters. Moreover, even if these theories do not *require* us to pursue these goals, they might at least *permit* us to do so, within certain moral limits. Granted, each theory might require or permit the pursuit of different goals, might require more than the pursuit of these goals, and might prohibit some means of pursuing these goals. Still, insofar as they require or permit the pursuit of particular goals, people who accept these theories can use an EAA framework to engage in informed, rational, effective, efficient pursuit of the goals required or permitted by this theory, within the limits set by this theory.

The second reason that it is possible for a nonutilitarian to take interest in EAA is that it is possible (and increasingly common) to interpret EAA in relatively general, theory-neutral terms, so that proponents of different moral theories can accept one and the same interpretation of EAA. For example, on a relatively general interpretation, EAA endorses the goal of *doing the most good possible in animal advocacy within certain moral limits*, but it does not endorse any particular conception of the good (such as pleasure theory or desire theory), nor does it endorse the goal of doing the most good possible in animal advocacy *by any means necessary* (such as by violating the rights of human or nonhuman animals). This interpretation of EAA strikes a good balance between utilitarian and nonutilitarian values. It also coheres with how many people think about triage cases. For example, in a standard triage case, many people agree that one should attempt to do the most good possible without violating rights. But people do not agree about what it means to do the most good possible, nor do they agree about whether or not one should do the most good possible in ways that violate rights.[9]

So, which interpretation of EAA makes the most sense all things considered? This depends on how we answer a related conceptual question: What kind of role is EAA meant to play in our thinking about what to do? For example, is EAA meant to be a *moral theory* that tells us what makes right actions right in the context of animal advocacy? Is it meant to be a *project that we engage in as individuals*? Is it meant to be a *project that we engage in as a collective*? And so on. It can be tempting to think about EA, in general, and about EAA, in particular, in a way that combines all of these roles. Yet these roles are importantly different.

In particular, if EAA is a moral theory, then, as with any moral theory, we should evaluate it in terms of theoretical virtues such as simplicity, explanatory power, and prescriptive power. In this case, we would each select an interpretation of EAA that coheres with our moral theory of choice, such as a utilitarian EAA, Kantian EAA, care theoretic EAA, or virtue theoretic EAA. We would then debate which interpretation is correct by considering arguments from principles, arguments from thought experiments, and so on. This interpretation would probably be relatively specific, since a good moral theory will not leave important moral questions open.

In contrast, if EAA is an individual project, then, as with any individual project, we should evaluate it in terms of practical virtues such as: "If I engage in this project, will that help me to do the things that I want to do in my life?" In this case, we would each select an interpretation of EAA that can guide us in building and sustaining a life of effective animal advocacy, given everything that we know about ourselves as individual advocates. This interpretation would probably be more general than the previous one, owing to the need to accommodate our individual epistemic and practical limitations (as I discuss later).

Finally, if EAA is a collective project, then, as with any collective project, we should evaluate it in terms of practical virtues such as: "If we engage in this project together, will that help us to do the things that we want to do in this movement?" In this case, we would select an interpretation of EAA that can guide us in building and sustaining a movement of effective animal advocacy, given everything that we know about ourselves as a community. Once again, this interpretation would probably be more general than the previous one, owing to the need to accommodate not only our individual and shared limitations but also reasonable disagreement within our movement.

My own view, partly following Pummer and MacAskill (2019), is that EAA is best understood as a collective project, not as a moral theory or individual project. As a result, I think that we should accept a relatively general interpretation of EAA. In particular, I think that we should define EAA as the project of using evidence and reason to do the most good possible in animal advocacy, within certain moral limits. This interpretation strikes a good balance, since it creates a strong mission for the EAA movement, but it also allows the EAA movement to benefit from expanded membership, an expanded division of labor, and a healthy amount of epistemic humility. We can each then decide for ourselves, by considering our moral theory of choice (and our project of living a life that complies with that theory) to what degree and in what ways we should participate in this project.

Principled Issues

The principled critiques that I consider here are all based on the idea that EAA sets the wrong goals for animal advocacy. That is, they are all based on the idea that we are not morally permitted or required to try to do the most good possible in animal advocacy. Since many people interpret EAA in utilitarian terms, some of these critiques apply more to a utilitarian EAA than to other interpretations. Thus, as we will see, EAAs can reply to these critiques in at least three ways: They can accept that EAA has the relevant features and insist that these features are good. They can accept that these features are bad and deny that EAA has them. And, they can say that EAA is a project, not a moral theory, and so each of us can choose to participate in our own way.

Here, then, are four common principled issues that people have with EAA:

1. *We should not evaluate states of affairs in terms of aggregate well-being.* First, some critics take issue with the idea that we should evaluate states of affairs in terms of how much well-being they contain. In particular, if we evaluate states of affairs this way, then we will sometimes produce surprising results. For example, we might find that reducing farmed animal suffering or wild animal suffering by 1% is better than reducing companion animal suffering by 99%. After all, there are so many farmed animals and wild animals in the world that a minor reduction in suffering per farmed animal or wild animal outweighs a major reduction in suffering per companion animal in the aggregate. Similarly, we might find that reducing insect suffering by 1% is better than reducing human suffering by 99%. After all, there are an estimated 1 billion insects for every human in the world. Thus, if we suppose that the average insect experiences one millionth the amount of well-being that the average human does at any given time (a conservative assumption in my view), it follows that the ratio of insect well-being to human well-being in the world at any given time is still about 1,000 to one, in which case a 1% reduction of suffering per insect does, indeed, outweigh a 99% reduction of suffering per human in the aggregate. Yet one might find this kind of aggregation implausible. In particular, one might think that this kind of aggregation ignores the inherent moral value of individual human and nonhuman animals, as well as the inherent moral value of fairness, equality, and other such distributions of well-being in the world.[10]

2. *We should not harm the few for the sake of the many.* Second, some critics take issue with the idea that we should harm the few so that we can help the many. For example, suppose that we face a choice between promoting meat reduction and promoting meat elimination. If we stipulate that promoting meat reduction sometimes produces better outcomes overall, then we might think that the EAA framework implies that we should sometimes promote meat reduction, in spite of the fact that doing so will foreseeably harm some animals in the short term.[11] Similarly, suppose that we face a choice between using racism, sexism, and ableism in animal advocacy (e.g., by making reductive and appropriative comparisons between human and nonhuman oppressions) and refusing to do so. If we stipulate that the former approach sometimes produces better outcomes overall, then we might think that the EAA framework implies that we should sometimes take the former approach, in spite of the fact that doing so will foreseeably harm some humans in the short term. Yet one might find these results implausible. In particular, one might think that we should never knowingly and willingly harm some individuals for the sake of helping others, even if we know for a fact that we would be

doing the most good possible as a result.[12] (Note that I later question these assumptions about what does the most good possible.)

3. *We do not need to neglect the few for the sake of the many.* Third, some critics take issue with the idea that we should neglect a small number of individuals if we can help a large number of individuals instead. For example, suppose that we face a choice between supporting corporate outreach, on one hand, and supporting rescue and sanctuary, on the other hand. If we stipulate that supporting rescue and sanctuary sometimes produces worse outcomes overall, then we might think that the EAA framework implies that we should sometimes *not* support rescue or sanctuary, in spite of the fact that this activity would help many animals.[13] Similarly, suppose that we face a choice between multi-issue animal advocacy (i.e., animal advocacy that stands in solidarity with other movements) and single-issue animal advocacy (i.e., animal advocacy that does not). If we stipulate that a multi-issue approach sometimes produces worse outcomes overall, then we might think that an EAA framework implies that we should sometimes *not* take a multi-issue approach, in spite of the fact that this activity would help many humans. Yet one might find these results implausible too. In particular, one might think that we should feel free to support any good cause in our advocacy, even if we know for a fact that we would not be doing the most good possible.[14] (Note that, as earlier, I later question these assumptions about what does the most good possible.)

4. *We do not need to demand so much from ourselves.* Fourth, and relatedly, some critics take issue with EAA because they see it as too demanding, physically as well as psychologically. First, they see EAA as too *physically* demanding because, as we have seen, if our aim is to do the most good possible, then we need to apply this framework to every aspect of our lives. We need to apply it to what major we select in college, what job we accept after college, how we spend our nights and weekends, and so on. Moreover, while we might hope that living a happy life will allow us to produce better outcomes than living an unhappy life will, we also need to be prepared to sacrifice our own well-being for the sake of others if doing so is necessary, at least to a degree. Second, critics see EAA as too *psychologically* demanding because, as we have seen, if our aim is to do the most good possible, then what we should do depends on the product of a complicated calculation. We have to ask: Which action, of every action available to me, will do the most good possible from now until the end of time? This question is difficult if not impossible to answer with confidence. Thus, it is difficult if not impossible to know which of our actions, if any at all, are right and wrong according to an EAA framework.[15] Yet one might find these results implausible too. In particular, one might think that there is a limit to how much we should demand from ourselves, physically and psychologically. We should be able to be happy in life, and we should be able to know which of our actions, if any, are right and wrong.[16]

With respect to all four of these principled critiques, an EAA has at least three replies available, which are complementary. The first reply available to an EAA is to accept that EAA has (some of) these features and insist that these features are good. Granted, it might initially seem implausible that we should, say, sacrifice ourselves or others for the greater good. But, first, we might not have to make these sacrifices as often as we fear (for reasons that I explore in the next section). Second, insofar as we do have to make these sacrifices, this implication might seem more plausible once we remind ourselves that everyday life is an emergency situation that requires triage. Third, every framework for deciding what to do will have at least some implausible implications. This is especially true in the modern world, since this world is much different—with people much more connected across space and time—than the world that our ancestors lived in thousands of years ago, when our current moral intuitions evolved. The point of practical ethics, then, is to consider the pros and cons of every framework holistically so that we can select the framework that makes the most sense in our current context, given the problems that we now face.

The second reply available to an EAA is to accept that (some of) these features are bad and insist that EAA does not have these features. For example, we can say that EAA is committed to aggregation only above a certain threshold of well-being. If we do, this would allow EAA to avoid certain seemingly implausible results, for example concerning insect well-being. (I personally think that we should accept full aggregation, but this reply is available to people who disagree.) We can say that EAA is committed to doing the most good possible without violating rights. If we do, this would allow for moral limits within EAA. (I think this is a good idea.) And we can say that EAA is committed to any impartially benevolent conception of the good. If we do, this would allow for moral freedom within EAA. (I think this is a good idea too.) Insofar as we interpret EAA in these ways, EAA would not fully face the principled critiques that we have considered in this section. However, it might then face other principled critiques, such as the critique that it privileges some individuals over others and allows a great deal of preventable suffering and death in the world.

The third reply available to an EAA is to say, as discussed earlier, that EAA is a project, not a moral theory. Thus, as discussed previously, we should evaluate EAA not by considering whether this framework is, say, simple and powerful but, rather, by considering whether this framework will help us build and sustain an effective animal advocacy movement. We can each then decide for ourselves to what degree and in what ways we should participate in this project. So, for example, if we think that EAA is a good project, with the exception that it involves too much aggregation, too much moral sacrifice, or too little moral freedom for us, then we can always proceed accordingly. For example, we can decide to spend 10% of our discretionary time, energy, and money participating in EAA, and we can then decide to spend the rest of our discretionary time, energy, and money on other morally permissible activities. Moreover, insofar as we participate in EAA, we do not have to decide between EAA values and our own values, since we can focus on the work that fits with both. From this perspective, even if EAA is a demanding project, a life that involves participating in EAA does not need to be similarly demanding.

My own view is that a combination of these three replies is best. As I said in the previous section, I think that we should interpret EAA as the collective project of using evidence and reason to do the most good possible in animal advocacy, within certain moral limits. On this interpretation, we should accept that EAA involves aggregation, maximization, and impartial benevolence, and we should insist that these features are good. However, we should also accept that EAA is not identical to utilitarianism, since EAA does not require us to accept any particular conception of the good, nor does it permit us to do the most good possible by any means necessary. Finally, we should say that, since EAA is a project rather than a moral theory, we can all decide for ourselves to what degree and in what ways we should participate. I think that these replies, considered holistically, are compelling. They describe a project in animal advocacy with appropriately high standards while also describing a project that many different animal advocates can identify with and participate in.

Practical Issues

Whereas principled critiques of EAA attempt to show that EAA sets the wrong goals for animal advocacy, practical critiques attempt to show that EAA is not achieving its own goals. In particular, these critiques are all based on the idea that our current assessments of what will do the most good possible are not fully accurate. Thus, if we want to do the most good possible in animal advocacy (as, I think, we should), then we should work to (a) improve our assessments of how we can do the most good possible and (b) supplement our use of these assessments with other, more indirect ways of doing the most good possible.

In particular, consider four respects in which one might worry that EAA thinking (i.e., using evidence and reason to assess what will do the most good possible, with the intention of acting accordingly) is not always the best way to achieve EAA aims (i.e., doing the most good possible):

1. *Evaluating states of affairs in terms of aggregate well-being is not always optimal.* First, EAA thinking could lead to a focus on aggregate well-being rather than on rights or justice. Yet, one might worry, this focus on aggregate well-being can be self-defeating in some cases in practice, since it might be

that promoting rights or justice in the short term is necessary for promoting aggregate well-being in the long run. This could be true for many reasons. For example, it might be that if people see animals as mere "receptacles of utility" rather than as individuals who merit rights and justice, then people will be less likely to promote particular kinds of social, legal, and political changes for animals (e.g., they will be less likely to support moral personhood, legal personhood, and political membership for animals). And it might be that if we want to maximize aggregate well-being in the long run, then we should promote these kinds of social, legal, and political changes in the short term. This kind of reasoning, if correct, would support placing moral limits on aggregation in some contexts.[17]

2. *Harming the few for the sake of the many is not always optimal.* Second, EAA thinking could lead to a willingness to harm the few for the sake of helping the many all else equal. Yet, one might worry, this kind of moral math can be self-defeating in some cases in practice, since it might be that refusing to harm the few in the short term is a necessary means to helping the many in the long run. This might be true for many reasons. For example, it might be that if we often harm some animals to help other animals (e.g., by promoting meat reduction instead of meat elimination), then we will undermine efforts to promote rights or justice for animals, and, as a result, we will limit our ability to maximize aggregate well-being in the long run. Similarly, it might be that if we often harm humans to help nonhumans (e.g., by promoting animal rights in racist, sexist, or ableist ways), then we will undermine efforts to promote rights and justice for all (as well as to build coalitions across movements), and, as a result, we will limit our ability to maximize aggregate well-being in the long run. This kind of reasoning, if correct, would support placing moral limits on animal advocacy in some contexts.

3. *Neglecting the few for the sake of the many is not always optimal.* Third, EAA thinking could lead to a willingness to neglect the few for the sake of helping the many all else equal. Yet, one might worry, this kind of moral math can be self-defeating in some cases in practice, since it might be that helping the few in the short term is a necessary means to helping the many in the long run. This might be true for many reasons as well. For example, it might be that if we never provide support for individual animals (e.g., if we never support rescue or sanctuary), then we will undermine efforts to show humans a different, more equitable kind of relationship with nonhumans, and, as a result, we might limit our ability to maximize aggregate well-being in the long run. Similarly, it might be that if we never provide support for humans (e.g., if we never stand in solidarity with other social movements), then, as before, we will undermine efforts to promote rights and justice for all (as well as to build coalitions across movements), and as a result, we might limit our ability to maximize aggregate well-being in the long run. This kind of reasoning, if correct, would support embracing moral freedom within animal advocacy in some contexts.

4. *Demanding so much from ourselves is not always optimal.* Finally, EAA thinking can be extremely physically and psychologically demanding. Thus, one might worry, this kind of approach to animal advocacy might be self-defeating in some cases in practice, since it might be that spending at least some time, energy, and money on other activities is necessary for doing the most good possible in the long run. After all, we all have limited psychological resources, and so we will not always be able to assess what will do the most good possible and act accordingly—and if we try, we might make bad assessments. Moreover, we also have limited physical resources, and so we will not always be willing to assess what will do the most good possible and act accordingly—and if we try, we might burn out. This kind of reasoning, if correct, would support relaxing certain kinds of demands on animal advocacy (and life more generally). It would also support investing in self-care and community care as part of EAA, in spite of the fact that the benefits of such work can be difficult to measure.[18]

As discussed in the previous section, even if we accept these points in some cases, we might not accept them in all cases. That is, we might think that there are many cases where EAA thinking is optimal (and so we should attempt to do the most good possible directly) as well as many cases where EAA thinking is suboptimal (and so we should attempt to do the most good possible indirectly, by pursuing other goals directly instead). If so, then the challenge is to learn how to strike a virtuous balance in practice. I return to this point in a moment.

Why might one worry that EAA thinking will produce these kinds of suboptimal outcomes? There are at least two related reasons. First, there is arguably a risk of methodological bias in EAA thinking. In particular, if your aim is to use evidence and reason to do the most good possible, there is a risk that you will engage in cost–benefit analysis more often than you should (i.e., that you will use this decision procedure when other decision procedures would be better). And, insofar as you use cost–benefit analysis more often than you should, there is also a risk that you will develop a bias in favor of interventions whose benefits are relatively easy to measure or quantify and/or whose costs are relatively hard to measure or quantify and against interventions whose benefits are relatively hard to measure or quantify and/or whose costs are relatively easy to measure or quantify.

Second, there is arguably a risk of standpoint bias in the current EAA movement. In particular, the current EAA movement has a relatively high percentage of utilitarians and other consequentialists. It also exists at the intersection of the effective altruism and animal advocacy movements, both of which have a reputation (partly earned, partly not) of consisting of wealthy white people. Moreover, the EAA movement functions as a social community too. This has many benefits, since altruism is easier when done in community with others. However, it also has risks, since it can create a social barrier between the EAA movement/community and other movements/communities. This, in turn, can make it hard for the EAA movement/community to expand its membership, as well as for it to collaborate and communicate effectively with other movements/communities.

The worry, then, is that these methodological and demographic issues might conspire to create a risk of bias in favor of, for example, working within current systems to bring about moderate reforms. And, at least to a degree, the EAA movement does seem to have this priority. However, we need to be careful here. First, even if this priority makes sense in light of bias, that does not mean that it is a product of bias. Second, even if it is a product of bias, that does not mean that it is incorrect. Third, other movements and communities are susceptible to bias as well, so it is not as though the risk of bias is unique to EAA. Fourth, one virtue of EAA is a commitment to taking seriously the effects that biases and heuristics can have on our thinking about what to do. So insofar as EAAs detect a risk of bias, they will be motivated to try to mitigate that risk.

Supposing, then, that the risk of methodological bias and standpoint bias are real (as I think they are), what can EAAs do to preserve the benefits of this framework while mitigating the risks? Here are two complementary steps that they can take, both of which many are already taking.[19]

First, EAAs can work to *improve* their use of EAA thinking. They can strike a balance between quantitative and qualitative assessments, as well as between assessments of individual interventions and of sets of interventions. For example, they can study the history of social movements to see what sets of interventions tend to work well and what sets of interventions tend to work badly. Then, with respect to the sets of interventions that seem to be effective for other movements, they can then study the degree to which the animal protection movement is relatively similar and the degree to which the animal protection movement is taking a relatively similar approach. Of course, this will not eliminate the risk of bias. But insofar as EAAs strike a virtuous balance between quantitative and qualitative reasoning and between individualistic and holistic reasoning, it will at least reduce this risk of bias.

Second, and relatedly, EAAs can work to *supplement* their use of EAA thinking. They can strike a balance between cost–benefit analysis and other decision procedures in everyday life. In particular, insofar as we have reason to think that engaging in cost–benefit analysis will be useful, we should use this decision procedure. Insofar as we do not, we should use other decision procedures. For example, Kantian thinking can help us to establish valuable rules to follow. Care theoretic thinking can help us to cultivate valuable relationships within and outside the movement. Virtue theoretic thinking can help us to cultivate valuable character traits. EAAs may or may not think that these moral theories are correct. But even if they do not, they can still benefit epistemically and practically from using these theories as practical guides in certain cases.[20]

As part of these efforts, EAAs work to promote diversity, equity, and inclusion within the movement and links across movements. This can take the form of pursuing unity (working with people in other movements), solidarity (helping/not harming people in other movements), and mutual understanding (sharing information and arguments with people in other movements). This will allow EAAs to expand the demographics of the movement and build coalitions with other movements—all of which will benefit the EAA movement epistemically (by allowing it to incorporate a wider range of perspectives) and practically (by allowing it to incorporate a wider range of skills).[21] This, in turn, can help mitigate the risk of methodological bias and standpoint bias, since other methodologies and standpoints will start to become more prevalent within the movement.

Fortunately, many EAAs are working to improve and supplement their application of direct utilitarian reasoning in exactly these ways. For example, Animal Charity Evaluators, Faunalytics, Sentience Institute, and other such nonprofit organizations are conducting and promoting quantitative as well as qualitative research about particular interventions as well as sets of interventions. This includes research about farmed animal advocacy and wild animal advocacy, confrontation and conciliation, institutional change and individual change, and more. Moreover, organizations such as Encompass are working to promote diversity, equity, and inclusion within the EAA movement, such that it can retain its focus on impartially benevolent farmed animal advocacy while expanding and diversifying the range of people who hold positions of power within the movement.

Of course, there are risks involved with some of this work. For example, insofar as EAAs promote nonutilitarian decision procedures and promote viewpoint diversity within the movement, they will likely invite disagreement into the movement about what to do and how to do it. People will disagree more about facts (e.g., what kind of political system would do the most good) as well as about values (e.g., whether we should be trying to do the most good). And while disagreement can have benefits (e.g., it can help everyone learn), it can also have risks. For example, it can result in conflict. It can result in value drift, that is, in the EAA movement deviating from its current mission. It can also result in fragmentation, that is, in the EAA movement fragmenting into multiple movements with multiple missions. It can even result in dissolution, that is, in the EAA movement dissolving entirely.

How should we assess these risks and benefits? Especially given the possibility of bias, we should proceed carefully. On one hand, conflict, value drift, fragmentation, or dissolution would clearly have bad effects. On the other hand, they might not happen, and even if they did happen, they could also have good effects. Conflict could lead to necessary epistemic and practical change. Value drift could lead to a mission that preserves the current strengths of EAA while overcoming the current limitations. Fragmentation could lead to multiple movements that accomplish more collectively than a single movement could individually. Dissolution could lead current EAAs to do similar work in the context of animal advocacy in general, whether or not this work is still labeled as EAA.

I do not try to reach an overall assessment of these risks and benefits here. Instead, I simply express my own general hope for the future of EAA. In my view, the ideal EAA movement would resemble the current EAA movement in many ways, for example, it would aim to do the most good possible in animal advocacy, within certain moral limits, and it would use cost–benefit analysis to pursue these ideals whenever possible. However, it would also improve and supplement cost–benefit analysis in the ways discussed earlier so that the EAA movement can expand and diversify, can communicate and collaborate more effectively with other movements, and can allow for creativity, improvisation, and toleration of difference.

What kind of balance the EAA movement should strike at any given point is, in part, a function of how influential it is within the broader animal advocacy movement at that point. Ten years ago, EAAs had relatively little influence. Thus, it makes sense that they spent most of their time, energy, and money on approaches that seemed most likely to be effective, to complement the work that other people were doing. However, at present, EAAs influence an estimated 25% of donations in animal advocacy (Edgerton 2019). Thus, it makes sense that they would now strike more of a balance between (a) supporting approaches that seem most likely to be effective (insofar as EAAs are still complementing work that other people are doing) and (b) supporting the development of a broad, pluralistic animal advocacy movement that involves different people taking different approaches (insofar as EAAs are now influencing the shape of the movement as a whole). My hope is that as the EAA movement develops, we can continue to ask what kind of balance between direct and indirect approaches is best and proceed accordingly.

Conclusion

EAA is the project of using evidence and reason to do the most good possible in animal advocacy. As I have claimed in this chapter, I think that this is the correct goal for animal advocates to pursue. However, I also think that we should build certain moral limits and freedoms into our pursuit of this goal. Moreover, I think that, if we want to pursue this goal effectively, then we should work to (a) improve our assessments of how we can do the most good possible and (b) supplement our use of these assessments with other, more indirect ways of doing the most good possible. Among other things, this means supporting the development of a broad, pluralistic animal advocacy movement that includes many different people taking many different approaches.

Acknowledgments

Thanks to Aaron Call, Leah Edgerton, Bob Fischer, and Tyler John for helpful feedback and discussion.

Notes

1. MacAskill (2015) motivates the idea of effective altruism with a similar example (pp. 29–32). This introduction, as well as much of the rest of this piece, is indebted to his work.
2. For discussion of these cause areas from an effective altruist perspective, see the Open Philanthropy Project website: www.openphilanthropy.org/focus.
3. For discussion of these cause areas from an effective altruist perspective, see the Animal Charity Evaluators website: https://animalcharityevaluators.org/advocacy-interventions/prioritizing-causes/#detailed.
4. Note that I am not counting environmental conservation work as wild animal welfare work, since this work does not typically address the suffering of individual wild animals.
5. For discussion of wild animal suffering, see Horta 2010; Sebo forthcoming; and Tomasik 2015.
6. For discussion of these issues from an effective altruist perspective, see the Sentience Institute website: www.sentienceinstitute.org/foundational-questions-summaries
7. For arguments in favor of these approaches, see Leenaert 2017. For arguments against, see Taft 2016.
8. For discussion of these moral theories, see Schlottmann and Sebo 2018, Chapter 3.
9. Singer (1972) takes a similar approach by arguing that we have a moral obligation to prevent very bad things from happening if we can do so without sacrificing anything morally significant.
10. For discussion of the ethics of aggregation, see Voorhoeve 2014; and Temkin 1996. For discussion of insect suffering, see Horta 2010; Sebo forthcoming; and Tomasik 2015. For discussion of how to rigorously estimate nonhuman well-being, see Budolfson and Spears 2020.
11. For discussion of reducetarianism, see Kateman 2017.
12. For discussion of racism in animal advocacy, see Harper 2010. For discussion of sexism in animal advocacy, see Gaarder 2011. For discussion of ableism in animal advocacy, see Taylor 2017.
13. For discussion of the moral and political value of sanctuaries, see Donaldson and Kymlicka 2015.
14. For discussion of the ethics of multi-issue advocacy, see Sebo 2018.
15. For discussion, see Kagan 1991; Lenman 2005; Railton 1984; and Scheffler 1982.
16. For discussion of what life is like for people who try to live altruistically, see MacFarquhar 2016.
17. For discussion of a related conception of animal dignity, see Gruen 2014.

18. For discussion of self-care and community care in activism, see jones 2007.
19. The remainder of this section draws from, and overlaps with, the suggestions made in Sebo and Singer 2018.
20. For utilitarians and other philosophers who endorse this kind of indirect moral thinking, see Mill 2004; Sidgwick 1981; Hare 1982; and Sebo and Paul (2019).
21. For discussion of the epistemic benefits of diversity, see Kitcher 1990; and Strevens 2003.

Bibliography

Bostrom, N. (2002) "Existential risks," *Journal of Evolution and Technology* 9: 1.

Bostrom, N., and Ćirkovic, M. (eds.) (2011) *Global Catastrophic Risks,* Oxford: Oxford University Press.

Budolfson, M., and Spears, D. (2020) "Quantifying animal well-being and overcoming the challenges of interspecies comparisons," in B. Fischer (ed.) *The Routledge Handbook of Animal Ethics*, New York: Routledge.

Donaldson, S., and Kymlicka, W. (2015) "Farmed animal sanctuaries: The heart of the movement?" *Politics and Animals* 1: 50–74.

Edgerton, L. (2019) "Strategic considerations for the future of effective animal advocacy," *2019 EA Global*, San Francisco.

Gaarder, E. (2011) *Women and the Animal Rights Movement,* New Brunswick: Rutgers University Press.

Gruen, L. (2014) "Dignity, captivity, and an ethics of sight," in L. Gruen (ed.) *The Ethics of Captivity*, Oxford: Oxford University Press.

Hare. R. M. (1982) *Moral Thinking: Its Levels, Methods and Point,* Oxford: Oxford University Press.

Harper, B. (ed.) (2010) *Sistah Vegan: Black Female Vegans Speak on Food, Identity, Health, and Society,* New York: Lantern Books.

Horta, O. (2010) "Debunking the idyllic view of natural processes: Population dynamics and suffering in the wild," *Télos* 17: 73–88.

Jones, P. (2007) *Aftershock: Confronting Trauma in a Violent World,* New York: Lantern Books.

Kagan, S. (1991) *The Limits of Morality,* Oxford: Clarendon Press.

Kateman, B. (ed.) (2017) *The Reducetarian Solution,* New York: TarcherPerigee.

Kitcher, P. (1990) "The division of cognitive labor," *Journal of Philosophy* 87: 5–22.

Leenaert, T. (2017) *How to Create a Vegan World: A Pragmatic Approach,* New York: Lantern Books.

Lenman, J. (2005) "Consequentialism and cluelessness," *Philosophy & Public Affairs* 29(4): 342–370.

MacAskill, W. (2015) *Doing Good Better: How Effective Altruism Can Help You Make a Difference,* New York: Gotham Books.

MacFarquhar, L. (2016) *Strangers Drowning: Impossible Idealism, Drastic Choices, and the Urge to Help,* New York: Penguin.

Mill, J. S. (2004) *Utilitarianism and Other Essays,* London: Penguin.

Mill, J. S. (2008) *On Liberty and Other Writings,* Cambridge: Cambridge University Press.

Parfit, D. (1986) *Reasons and Persons,* Oxford: Oxford University Press.

Pummer, T., and MacAskill, W. (2019) "Effective altruism," in H. LaFollette (ed.) *International Encyclopedia of Ethics.*

Railton, P. (1984) "Alienation, consequentialism, and the demands of morality," *Philosophy and Public Affairs* 13(2): 134–171.

Scheffler, S. (1982) *The Rejection of Consequentialism,* Oxford: Oxford University Press.

Schlottmann, C., and Sebo, J. (2018) *Food, Animals, and the Environment: An Ethical Approach,* New York: Routledge.

Sebo, J. (2018) "Multi-issue food activism," in A. Barnhill, T. Doggett, and M. Budolfson (eds.) *The Oxford Handbook of Food Ethics,* Oxford: Oxford University Press.

Sebo, J. (forthcoming) "Animals and climate change," in M. Budolfson, T. McPherson, and D. Plunkett (eds.) *Philosophy and Climate Change,* Oxford: Oxford University Press.

Sebo, J., and Paul, L. A. (2019) "Effective altruism and transformative experience," in H. Greaves and T. Pummer (eds.) *Philosophy and Effective Altruism,* Oxford: Oxford University Press.

Sebo, J., and Singer, P. (2018) "Activism," in L. Gruen (ed.) *Critical Terms for Animal Studies,* Chicago: Chicago University Press.

Sidgwick, H. (1981) *The Methods of Ethics,* Indianapolis: Hackett.

Singer, P. (1972) "Famine, affluence, and morality," *Philosophy and Public Affairs* 1(1): 229–243.

Singer, P. (2015) *The Most Good You Can Do: How Effective Altruism is Changing Ideas About Living Ethically,* Yale: Yale University Press.

Strevens, M. (2003) "The role of the priority rule in science," *Journal of Philosophy* 100(2): 55–79.

Taft, C. (2016) *Motivational Methods for Vegan Advocacy: A Clinical Psychology Perspective,* Danvers: Vegan Publishers.
Taylor, S. (2017) *Beasts of Burden: Animal and Disability Liberation,* New York: New Press.
Temkin, L. (1996) "A continuum argument for intransitivity," *Philosophy & Public Affairs* 25(3): 175–210.
Tomasik, B. (2015) "The importance of wild animal suffering," *Relations* 3(2): 133–152.
Voorhoeve, A. (2014) "How should we aggregate competing claims?" *Ethics* 125(1): 64–87.

42
CULTURED MEAT
A New Story for the Future of Food

Max Elder

A feather slowly floats to the ground, falling painlessly from a chicken named Ian. Primary cells are isolated from that feather, placed in an environment that mimics Ian's body, and fed all the nutrients they need to grow. These cells do what they naturally do: double, over and over again. Eventually, they create tissue that is then formed into nuggets, battered, fried, and joyously consumed by a group of friends at a picnic bench situated on a perfectly manicured green lawn. Ian, the chicken from whom these cells came, clucks around the picnic bench, blissfully ignorant that his flesh is being consumed in the form of nuggets mere feet away. Ian remains unharmed, his fallen feather long forgotten.

Ian is a real chicken, and that picnic scene really happened in California in 2017. This meal was filmed as part of the food start-up JUST's (formerly known as Hampton Creek) video announcement that they were producing cultured meat, a new type of animal-based meat that is grown outside of animal bodies. Some call this food production 'carniculture,' a form of meat agriculture (Hopkins and Dacey 2008). Others call it 'cellular agriculture,' as it involves the domestication and growth of cells instead of animals. Cultured meat challenges a lot of fundamental questions about what meat is, about our relationship to animals as individuals and as sources of food, about the future of food production, and about the values we hold regarding our food system.

If this sounds like science fiction, that's because it was. There is a long history of forecasts related to the synthetic production of protein. Back in 1930, the British secretary for India, Frederick Edward Smith, Earl of Birkenhead, wrote:

> It will no longer be necessary to go to the extravagant length of rearing a bullock in order to eat its [sic] steak. From one 'parent' steak of choice tenderness, it will be possible to grow as large and as juicy a steak as can be desired. So long as the 'parent' is supplied with the correct chemical nourishment, it will continue to grow indefinitely, and, perhaps, internally.
>
> *(1930: 18–20)*

One year later, Winston Churchill spoke of synthetic foods that look "practically indistinguishable" from their animal-based counterparts. He famously stated that "we shall escape the absurdity of growing a whole chicken in order to eat the breast or wing, by growing these parts separately under a suitable medium" (1932: 397). The same year, Yale chemical engineer C. C. Furnas forecasted that the "independent farmer as we now know him probably is doomed" because "animals are laboratories where chemical compounds are made" and "[m]an is learning chemistry very rapidly, and he is also learning to put nature out of business" (1932: 47–8). These visions are from more than 80 years ago, yet they sound almost identical to what JUST and others are currently trying to build.

It's critically important to look back in order to look forward. Cultured meat may seem radical in its novelty, especially compared to current industrial animal agriculture. Looking back over our culinary and agricultural histories, however, provides a humbling and holistic perspective on how new animal agriculture really is. Humans have been around and consuming food for about the past 3 million years. Farming began when humans started domesticating plants and animals some 10,000 to 12,000 years ago. And the kind of industrialized animal agriculture that has come to dominate our modern foodscape is about 100 years old, or about 0.0005% of human history.

Considering agriculture's short past, historian James McWilliams asks us to question agriculture's long-term future. "Given its late arrival, and given its lack of precedent," he says, "agriculture by its nature raises an interesting question: Why should we think we got it right the first time around?" (2017: vii). It doesn't take much to be convinced that the first time around, we failed. A more challenging question is one that JUST chief executive officer Josh Tetrick poses often: "What would it look like if we started over?" (2015). This question is not about turning back our agricultural clock to zero but instead asking what a new clock would look like, totally redesigned.

Tetrick is one of many who believe we need to start over, with cultured meat as a key piece of the new design. Cultured meat offers us new choices for our food system. As Jonathan Safran Foer eloquently writes in his book *Eating Animals*,

> [i]f we are not given the option to live without violence, we are given the choice to center our meals around harvest or slaughter, husbandry or war. We have chosen slaughter. We have chosen war. That's the truest version of our story of eating animals.
>
> Can we tell a new story?
>
> *(2009: 244)*

We've always been telling stories about food systems. Even before we thought of harvest or slaughter at the systems level, there were religious prohibitions, rituals, taboos, and rules around what to eat and how to prepare it. This new story is not just a story about how our food system is designed but also a story about what we choose to put in our bodies and why.

A vast majority of eaters choose to put conventional meat in their bodies. Meat typically consists of a collection of muscle, fat, and connective tissues. Historically, animals have been our 'production platforms' for meat. Advances in tissue engineering and synthetic biology have opened up new possibilities to produce meat not inside animal bodies but instead inside bioreactors—fermentation tanks like you would find in a beer brewery. Cultured meat is an attempt to answer a fundamental question about food production: "Rather than raise an entire complex organism only to harvest these [muscle, fat, and connective] tissues, why not start at the basic unit of life, the cell, to produce meat?" (Datar and Bolton 2014: 153).

Growing meat from the cell level up—as opposed to the animal-level down—has a lot of potential benefits. In theory, it produces meat without any bacterial pathogens, without any antibiotic residue, without volatility of supply due to infectious diseases that have historically killed massive numbers of farmed animals, without a huge water footprint, without taking up large amounts of arable land, and without the rearing and slaughtering of more than a trillion sentient animals every year. Cultured meat could be produced with greater transparency, with complete traceability, with high-skilled laborers, with constantly improving technology, with more controllable quality, and perhaps even with cells from animals that no longer exist (dodo bird nuggets or woolly mammoth burgers, anyone?).

Many technological hurdles remain to truly produce humane and sustainable cultured meat at scale to meet such a hopeful vision of possibility. Some of these challenges involve removing animals from growth media in a cost-effective manner, engineering large-scale bioreactors in which to grow this meat, producing all the components of meat other than the muscle tissue (the fat, blood, etc.), isolating primary cell lines from which to culture meat, and determining how to speed up the duplication of cells so that costs are lowered and supply can meet demand, among many, many others. This chapter assumes that these challenges can and will be overcome—an assumption that many companies, investors, and activists in the space believe—in order to explore the ethical issues that emerge in a world where cultured meat is as readily available in certain parts of the world as traditional meat is.

In this new story, many ethical questions remain unanswered and even more remain unasked. How should we think about the opportunities and challenges of this new form of food production? How might eaters understand cultured meat as a product, and which moral (and nonmoral) considerations might govern their consumption choices? How might we create a compelling story to encourage people to eat cultured meat, and how might we accelerate the transition from factory farming as quickly as possible? This chapter aims to address these questions and provide the foundation for deeper investigation into a world in which synthetic biology fundamentally rewrites the story of our farms and our food.

What Is It?

Every new story starts with a name. In this case, we don't have one yet. Since this type of food production has only recently moved from science fiction novels to food research and development departments, there is disagreement over terminology. As sociologist Neil Stephens has said, the best description of in vitro meat is an "as-yet undefined ontological object" (2010: 400). While perhaps accurate, an 'undefined ontological object' isn't exactly a pill consumers will swallow, nor does it help us understand what cultured meat really is.

You may have heard people try to define this ontological object with descriptions such as 'synthetic,' 'fake,' 'Frankenmeat,' 'lab-grown,' 'clean,' 'in-vitro,' and even 'cell-based.' There are different arguments for each name and exploring this debate will help shine a light on how people are thinking about cultured meat, some of the implicit values eaters hold, and what this new food might mean for the future.

The open question about terminology goes beyond an academically-interesting etymological inquiry. The linguistic framing of these products will have implications on consumer acceptance. More important, it will set the stage for what is looming on the horizon: a high-stakes battle over standards of identity—the mandatory requirements for what different food products must contain in order to be called a certain name. If the current battles over eggless mayo and nut-based milks and cheeses have taught us anything, it's that big industries won't let innovators enter their product categories without a fight.

Terms such as *fake*, *synthetic*, or *Frankenmeat* are as appetizing as they are accurate—the cells used in cultured meat originate from a real animal and the tissue is far from an imitation. Technology is mimicking nature to create this food product. At one point in history, all ice was cut and shipped from naturally frozen ice in ponds. Today, we have a piece of consumer technology, the freezer, which produces ice. No one thinks the ice in their freezer is fake or synthetic. Cultured meat scientists are creating a different piece of technology to create meat outside of animal bodies, just like your freezer creates ice outside of the freezing outdoors.

Many journalists have called this meat 'lab-grown,' since start-ups currently developing the foundational technology for this meat wear white coats inside laboratories as opposed to aprons inside kitchens or straw hats on farms. While this meat is grown in labs today, it won't be for much longer. Instead, it will likely be grown in large bioreactors that are organized in a way that looks a lot more like a microbrewery than a research laboratory. Although not immediately apparent as you walk down your grocery store aisle, almost every food product on shelves today is first developed in a laboratory (see de Salcedo 2015; Lusk 2016). Our whole food system is 'lab-grown.' Advocates argue that we shouldn't call this meat 'lab-grown' since we don't call the cereal we have for breakfast or the frozen pizza we had for dinner 'lab-grown.'

Others call this type of meat 'in vitro' meat. The Latin *in vitro* literally means 'in glass' and refers to the process by which cells are studied in a glass test tube or petri dish, as opposed to inside a living organism (or 'in vivo'). Since cells have historically been cultured on a small scale for biomedical research, most of those experiments have been referred to in the academic literature as 'in vitro.' Sir Paul McCartney has famously said that "if slaughterhouses had glass walls, everyone would be a vegetarian" (PETA). 'In vitro' cell culture might prove McCartney wrong, offering a glass wall to meat production while enabling carnivores to still eat meat. However, once cell culture is commercialized and reaches the scale necessary for mass production, the actual culturing of meat will not take place 'in vitro' but instead in large, metallic bioreactors. As such, few find 'in vitro' a convincing name for the long-term future.

Proponents of the term *clean* argue that this meat is cleaner than its animal-based counterpart—especially in terms of its supply chain, carbon footprint, antibiotic use, and bacterial residue. Many actively involved with developing this technology believe meat grown via cell culture also produces a cleaner conscience for the consumer eating it. Some randomized trials have tested the consumer acceptance of terms such as *clean* versus *cultured*, with slightly higher preference for *clean* (Greig 2017). Proponents of *clean* also point to clean energy as analogously communicating the environmental benefits of this technology.[1] The religious distinction between clean and unclean animals, with some prohibitions around the consumption of the latter, complicates this picture. Proponents of 'clean' meat tend to think this distinction is more popularly characterized in terms such as kosher or halal as opposed to clean versus unclean and thus are not very concerned with this potential complication (see Leviticus 20: 24–25 for a discussion of 'clean beasts').

The term *de jour* at the time of writing this chapter is 'cell-based' meat, first proposed in a joint letter to the president of the United States by the cultured meat start-up Memphis Meats and the North American Meat Institute to urge a clear regulatory path for the technology (Watson 2018). Cell-based seems the least descriptive, as traditional meat from an animal body is technically just as cell-based as its cultured meat analog. Plant-based meat alternatives also are 'cell-based' insofar as both plants and animals contain cells. The term does little to meaningfully differentiate or describe anything other than what happens when competitors try to compromise.

Ironically, much of the meat consumed today, especially the cheap meat sold at many restaurants, would likely not hold up to most consumers' standards for meat; it's filled with artificial flavors and supplemented with commodity plants like soy in order to reduce costs. Some chains such as Taco Bell have come under scrutiny for selling meat products that contain only about one-third actual meat, so perhaps *fake* or *synthetic* are better terms for other meat products in our food system.

Some think the future will reveal branded cultured meat products that don't have any qualifying descriptor. The cultured meat on the shelves might simply be called 'Memphis meat' (cultured meat produced by the start-up Memphis Meats) or 'Finless Tuna' (cultured tuna produced by the start-up Finless Foods).

The most provocative name for cultured meat—and perhaps the most influential in the long-term—might be no name at all. This naming debate begs the question: why does cultured meat need to be called anything other than what it is? Meat.[2]

Since this debate is not yet settled—and it is unclear exactly what it will take for it to be settled—this chapter uses the term *cultured meat*. These cells are grown outside of an animal and inside a cell culture; the verb for producing these cells is *culturing*; and thus, the adjective *cultured* is a good default for all intents and purposes here.

Is It Natural?

The various terms used to describe cultured meat demonstrate a consumer tension at the heart of this new food technology: Is this food natural? Notions of naturalness have been at the center of many debates regarding new food technologies around the world, greatly influencing their acceptance or rejection (Siegrist 2008). How people answer this question about cultured meat will likely deter or drive adoption. One review of 72 studies conducted in 32 countries covering more than 85,000 people found that "for the majority of consumers, food naturalness is crucial" for adoption (Román et al. 2017: 44). Regardless of whether

> cultured meat may contribute to solving major ethical concerns with respect to livestock farming and animal slaughter for human consumption, and may contribute to the alleviation of hunger problems in the world, the technology for producing meat might as well be perceived as intervening and messing too much with nature.
>
> *(Verbeke, Sans, and Van Loo 2015: 287)*

The challenge is that there is no standard definition of natural, and no regulatory definition either.

One theory is that many consumers are only comfortable eating meat that is "formed on the body of the animal through a process of organic growth rather in a lab through the synthetic simulation of natural growth" (McWilliams 2017: 161). This could pose a serious barrier to acceptance, as consumer research has found initial reactions of cultured meat to be grounded in considerations of unnaturalness[3] (Verbeke, Sans, and Van Loo 2015).

Cultured meat proponents have a few reasons to be optimistic about this naturalness concern. Current practices of meat production look far from natural, involving tightly confined cage systems; use of hormones and antibiotics; feed filled with corn, soy, and waste products; and a level of industrialization that looks very little like nature. So, to reject cultured meat as unnatural would force us to reject most of the food in our fridges as facsimiles of natural products. Consumers have been good at rationalizing this away when it comes to conventional meat, so it is possible a similar rationalization could occur for cultured meat.

Our bias for the natural is full of contradictions and any link between naturalness and goodness is not immediately obvious. The same in vitro processes used for cultured meat are widely accepted in other cases that aren't food-related but nevertheless involve our bodies, such as medical applications and in vitro fertilization (Welin and van der Weele 2012). Synthetic antibiotics created in labs are unnatural but are considered good compared to the horrible, natural diseases they combat. Millions of people around the world happily use lifesaving insulin, all of which is synthesized. And those who go through in vitro fertilization to get pregnant likely don't think their children are any less natural than their friends' children, conceived 'naturally' during that drunken night on Valentine's Day.

In terms of the naturalness of animal-based food products, some research has found that people view the process of domestication as much less damaging to their notions of naturalness than, for example, the insertion of a gene from another species (Rozin 2005). Neither occurs 'naturally,' yet inserting genes tends to be viewed with more skepticism than domestication. Perhaps this partially explains why some frame the evolution of cultured meat as the next wave of domestication—but this time, we're domesticating cells instead of animals (Dance 2017).

Whether cultured meat is viewed as natural or unnatural will affect consumer acceptance, but it doesn't settle any ethical debate about whether cultured meat is good or bad for the world or its inhabitants. And the assumption that natural is good and unnatural is bad does not always hold in this context. Some argue that it is exactly the 'unnaturalness' of cultured meat that makes it such an attractive and necessary development. As Hopkins and Dacey (2008) posit,

> [i]f anything, the alleged "unnaturalness" of cultured meat may be precisely the thing we are looking for—some option that is not currently offered by the physical constraints of given nature. In this sense, cultured meat may be superior to what nature offers—humans can live out their natural propensity to eat meat while also sparing animals from the horrors of that propensity.

We should question whether our notions of naturalness are consistent and, when they aren't, ask what underlying values are really guiding our behavior. Do we really care about naturalness for its own sake, or are there deeper implicit values for which we use notions of 'naturalness' as a proxy? What are those values, and how might cultured meat fulfill or fall short of them?

Is It Better Than Conventional Meat?

Some stories require rewriting. The story of our agricultural systems is large enough that rewriting takes a lot of effort, time, and money. Are the problems of our current story large enough to justify such a revision?

Conventional meat production has a long list of environmental, economic, and social problems. How does cultured meat compare? Cultured meat offers the possibility of reducing the suffering of billions of sentient creatures every year, creating much more environmentally friendly meat products, dramatically reducing the amount of antibiotics used in our food system, and disrupting an industry fraught with poor labor practices. However, these are all still *possibilities*. Since we have yet to build cultured meat production facilities—so-called carneries—we don't yet fully understand the impact this new story will have on animals, humans, or the planet we co-inhabit.

Many thinkers, food activists, and venture capitalists are ready to tell us a new story about meat to drive adoption of an alternative. For example, the industrial farming of animals is, as historian Yuval Noah Harari says, "arguably one of the worst crimes in history" (Shapiro 2018: x). There are almost 136 billion farmed animals killed each year around the world.[4] To put that number into perspective, the estimated number of humans to have ever existed totals about 108 billion (Haub 2011), which means we kill more animals on farms around the world *every year* than all humans who have ever walked this earth in the history of *Homo Sapiens*. The life of these farmed animals is, to borrow a phrase from Thomas Hobbes, "poore, nasty, brutish, and short" (2004: 92). Others push back, arguing that conventional factory farming has produced cheap protein for a growing population that is hungry for it. Such efficiencies in productivity are the result of impressive investments in creating an efficient and global food system. Progress!

Cultured meat proponents are quick to ask: progress at what cost? One benefit of cultured meat is the possibility of truly humane meat that reduces suffering and relieves pressure from natural ecosystems by decoupling meat consumption from the farming of animals. Because of current meat consumption, almost 90% of all wild fish stocks are currently fully fished or overfished (FAO 2016). Traditional animal farming is a leading contributor to global deforestation, land degradation, water pollution, and desertification (Koneswaran and Nierenberg 2008). If we can write a new story about meat, the process might save both billions of animals and entire ecosystems.

One attempt to preserve the lives of farmed animals is to administer low-dosage pretherapeutic antibiotics. These stave off disease and keep animals healthy so they reach slaughter weight sooner. In the United States alone, about 80% of all antibiotics sold (by volume) are used by farmers (Sneeringer et al. 2015).[5] However, this practice has made animal agriculture a major contributor to antimicrobial resistance which, if left unchecked, could kill 10 million people a year by 2050—or one person every three seconds (O'Neill 2016). We are creating an environment in which relatively harmless bacteria such as *E. coli* can be deadly, all because we want cheap animal flesh for dinner. Many of the most severe *E. coli* outbreaks in 2018 have been traced back to water contamination from animal agriculture.

Current animal agriculture negatively impacts not only nonhuman animals but also human animals involved across almost the entire industry. Antibiotic use on farms is only one of many examples. The bad economics of animal farming, high rate of depression, and physically demanding farm work contribute to a suicide rate for farmers that is about twice the general population, leading one reporter to call farmer suicide an international crisis (Kutern 2014).

Slaughterhouses are also hidden far away from urban areas in geographically isolated rural towns. This is out of sight and out of mind for most meat consumers. The modern-day slaughterhouse is "physically hidden from sight by walls and socially veiled by the delegation of dirty, dangerous, and demeaning work to others tasked with carrying out the killing, skinning, and dismembering of living animals" (Pachirat 2011: 3–4). The workers at these slaughterhouses are often immigrants or resettled refugees—some undocumented—making labor abuses common and often invisible. Increasingly across the United States, judges are sentencing defendants to rehab working at slaughterhouses instead of prison, which is evidence enough for how poor the conditions are at these facilities (Harris and Walter 2017).

While large-scale cultured meat production has not yet been realized, it is certainly possible that this higher-skilled labor and cleaner production process will enable a new industry with higher labor standards and better wages. Some have imagined these facilities as located in urban centers, open to the public, and much more transparent than current animal production. Imagine the fun you've had at your local brewery, but replace the contents of the fermentation tanks with meat and you get one idea of what this might look like. It's by no means inevitable, but it is possible.

Farming animals "uses far more land resources than any other human activity" (FAO 2013: 2). One quarter of earth's land is used for ruminant grazing, and one third of arable land—land capable of being used to grow crops—is used to grow feed for farmed animals (FAO 2012). We use so much land to grow feed crops because animals are tremendously inefficient at converting the food that they eat into the flesh that we eat. In terms of protein, cows only convert a mere 4% of the protein in their feed into protein humans can eat. Other forms of meat aren't much better—pigs convert 15%, chicken 20%, and farmed fish 18% (Ranganathan et al. 2016: 43). It's no wonder that food policy researcher Frances Moore Lappé believes "we have turned cattle into protein disposal systems" (1991: 68). If you look at calories instead of protein, the conversion rates are even lower for all animals, so animals are energy disposal systems as well.

Given how resource-intensive animal farming is, it should be no surprise that the livestock sector is responsible for 18% of global greenhouse gas emissions, more than all emissions from transportation (FAO 2006). The rampant deforestation and monoculture crop production that both result from agriculture, in addition to the greenhouse gas (GHG) emissions, have led many to blame industrial animal farming for driving earth into a sixth mass extinction event. Our culinary preferences are not only killing the animals that end up on our plates but also entire species who are out of sight and out of mind.

Environmentally, there have been a few anticipatory life cycle analyses performed to explore the possible impact that cultured meat might have on our planet. Compared to conventionally produced meat, one initial life-cycle analysis found that cultured meat uses between 7% and 45% less energy, produces 78% to 96% fewer GHGs, utilizes 99% less land, and takes 82% to 96% less water (Tuomisto and Teixeira de Mattos 2011). However, many assumptions were made in this particular analysis about how the technology would work in the future and what the inputs would be, both of which remain open questions.

Some more recent scholarship has questioned whether these anticipatory benefits are justified. Alexander et al. (2017) analyzed alternatives to conventional animal products and determined that, although "cultured meat was found to have a lower land footprint than beef, it had a similar efficiency to poultry meat, but with substantially higher direct energy requirements" (28). As this technology continues to be developed, life-cycle analyses must be continually performed to determine cultured meat's environmental trajectory. Some evidence suggests we should be cautiously optimistic here, but to "assume that a transition away from livestock production in favor of *in vitro* meat is already fully understood is to oversimplify the interdependent nature of technology, society, and the environment" (Matticka et al. 2015: 253).

None of the potential benefits of cultured meat will be realized if no one eats it. One study of US consumers found that, while most respondents would be willing to try cultured meat, only one third of respondents said they "were definitely" or "probably" willing to eat cultured meat regularly or as a replacement for conventional meat (Wilks and Phillips 2017). If these numbers seem positive, recent examples from biotechnology and nanotechnology indicate that consumers might not embrace novel food technologies as much as expected during their initial development (Verbeke 2011). If consumers are only interested in experimenting with cultured meat but not actually adopting a diet in which cultured meat replaces conventional meat, the ability of this emerging food technology to solve the problems outlined earlier is greatly diminished.[6]

Are We Avoiding Our Moral Failures?

In his book on eating meat, Simon Fairlie (2010) argues that cultured meat opens up space for humans to turn our backs on nature. For some, cultured meat is "the dream product that lies at the end of this road" toward factory-produced and highly processed forms of protein (228). Fairlie's concern is that cultured meat may not be as innocuous as it appears as it could drive a wedge between ourselves and the natural world, a wedge that presumably is minimized by eating animals. He worries that cultured meat will detach us from our animal identities: "by eating animals we help to ensure that we ourselves remain animals" (Fairlie 2010: 231).

Fairlie's argument is flawed in a few critical ways. First, many animals walking this earth are naturally herbivorous, so it doesn't logically hold that remaining an animal means adhering to an omnivorous diet. Second, many humans are vegan or vegetarian and very few (one hopes) would consider them any less animal because of it. Furthermore, why should animality—at least conceptualized here—be preserved? We've designed social contracts and laws to ensure that we move beyond an animal nature that is red in tooth and claw. It is also not obvious where, if animal nature is to be preserved, humans get license to kill nonhuman animals to maintain human-animality. Finally, how does eating animals preserve animality? Such an idea is "comparable to a claim that eating human flesh will make us human" (Welin 2013: 31).

While the details of Fairlie's arguments may not be sound, he does unearth a broader concern about cultured meat and our relationship to technology and nature that is worth more seriously considering. Some might be concerned that cultured meat "fits in with an increasing dependence on technology, and the worry is that this comes with an ever greater estrangement from nature" (Welin 2013: 31). Others push back on such a claim, arguing that this technology won't estrange us from nature but instead allow us to move beyond our abuse and extractive relationship to nature, opening space for a more symbiotic interconnection.

Can technology actually solve the problem of meat in a meaningful way? Or "[a]re we, by turning to technology, evading the challenges in human—animal relations and giving up on animals?" (Welin 2013: 31). As philosopher Ben Bramble puts it,

> [t]he moral problem stems from the fact that we will likely switch over to lab-grown meat because it is cheap, or thanks to its benefits for human health or the environment. That is,

> we will do it for our own sake and not for the sake of animals. [. . .] If we switch over to lab-grown meat just for our own sake, and not for the sake of animals, then the morally dubious part of us that is responsible for our inaction on factory farming will remain intact. If this part of us has other bad consequences, then we might have lost a valuable opportunity to confront it and avoid those outcomes.
>
> *(2017)*

Cultured meat may solve for symptoms—environmental degradation, animal suffering, farm labor abuses, and more—while allowing humanity to avoid confronting our deeper problematic orientation toward sentient beings as things that can be used as a means to our own ends. If we don't account for these moral failures, we may continue to have similar, or worse, problems long into the future.

The mass production of cultured meat, or even simply the idea of it, "may not immediately change meat consumption, yet [. . .] it may help to change the moral landscape in which our thoughts and decisions about meat, or protein consumption, take place" (Van der Weele and Driessens 2014: 86). Cultured meat may remove animals from our food supply chain or simply redraw the landscape of animals in our moral imaginations, opening the opportunity for speciesism to continue to exist despite our food system largely no longer reinforcing the bias. The cultivated discipline of seeing meat products as animals, at one point not merely alive but also living *a life*, might not have a technological shortcut. Even if they appear cuttable, the corners of moral reasoning often need to be authentically navigated with time and persistence.

One counterargument to this concern can be found in Upton Sinclair's declaration that "[i]t is difficult to get a man to understand something, when his salary depends upon his not understanding it" (1935: 109). This insight may be as true for one's dinner as it is for one's salary. An understanding of the moral patienthood of farmed animals might be incredibly challenging as long as dinner—for omnivores, at least—depends on humanity not understanding it. Insofar as cultured meat removes the suffering of animals from our dinners, perhaps there's new ground for a deeper moral understanding of animal lives and their moral worth.

How Should We Think About Cultured Meat?

Cultured meat has captured the imaginations of many over the past few years. This should not be a surprise, as "the idea of cultured meat functions as a catalyst for wider imaginative thought on meat and its potential alternatives" (van der Weele and Driessen 2013: 651). Start-ups have raised millions to produce cultured meat, nonprofits are dedicated to advancing the field, and much of the animal welfare world believes it to be the key to unlocking a truly humane future of food. Given the current state of industrial animal agriculture around the world and the resources being invested into its commercialization, cultured meat might not be a question of if but when.

While the negative implications of traditional animal agriculture are numerous and well documented, the longer-term implications of cultured meat are less clear. As American futurist and researcher Roy Amara has said, "[w]e tend to overestimate the effect of a technology in the short run and underestimate the effect in the long run" (Searls 2012: 257). The long list of immediate benefits ought not distract from questioning the long-term implications. Although there's reason to be cautiously optimistic about the positive long-term effects of cultured meat, we should be hesitant of hype machines and skeptical of silver bullets. With that healthy skepticism in mind, there are real consequences to *not* revolutionizing our food system with the creation of a more healthy, humane, and sustainable future of food. Industrial animal agriculture is currently plagued with problems, and cultured meat might be one good answer, among many, to our question: What would this story look like if we started over?

Acknowledgments

This chapter benefited greatly from the thoughtfulness of Mary Allen, Quinault Childs, Bradley Kreit, Neil Stephens, Sarah Smith, Cor van der Weele, Yoni Wilkenfeld, and Ben Wurgaft.

Notes

1. Cultured meat requires energy to produce, so the extent to which this meat is 'clean' environmentally depends somewhat on the extent to which 'clean energy' is used in the production process.
2. It's worth noting a tension here in the ambiguity of what meat *is* exactly. Ethan Brown, chief executive officer of the plant-based company Beyond Meat, has said in a 2015 interview with PBS that "[m]eat is really made up of five constituent parts: the amino acids, lipids, carbohydrates, minerals and water. They're all actually present in plants. What we're doing is building a piece of meat directly from those plants, and so the compositions are basically the same. And in that case we are delivering meat." This raises deeper questions about how constituent parts (amino acids, lipids, etc.) must be combined to form what we think of as meat, and when that combination falls short. One potential solution can be found in a performance: the process "of things becoming food can be conceptualized as a 'performance' that is conducted through both producer and consumer (and other bodies in between), and by which certain imaginaries, perceptions, expectations, and realities are made to matter and not matter" (Sexton 2016: 68). The answer to the ambiguity of what meat is may be in the imagination, perception, expectations, and realities of this performance between food producers such as Brown and everyday eaters.
3. Some argue that the 'unnatural' aspects of cultured meat are the source of its redemptive qualities. Cultured meat is "superior to what nature offers—humans can live out their natural propensity to eat meat while also sparing animals from the horrors of that propensity" (Hopkins and Dacey 2008: 587).
4. This number comes from an analysis by Lewis Bollard, Farm Animal Welfare Program Officer at the Open Philanthropy Project, who uses mostly 2013 and 2014 data from the Food and Agriculture Organization of the United Nations.
5. Some of the antibiotics used on animal farms are called ionophores and not used for humans. There is a debate as to whether this group of antibiotics can lead to resistance in humans, but even if one removes ionophores from the calculation, the proportion of antibiotics used on animals is about 70%.
6. For a systematic analysis of cultured meat consumer acceptance research, see Bryant and Barnett 2018.

Further Reading

Belasco, W. (2006) *Meals to Come: A History of the Future of Food,* Berkeley and Los Angeles: University of California Press.

(An extended treatment of utopian/dystopian food futures and how people have envisioned synthetic biology changing food production.)

Shapiro, P. (2018) *Clean Meat: How Growing Meat Without Animals Will Revolutionize Dinner and the World,* New York: Gallery Books.

(A chronicle of the academics and entrepreneurs developing cultured meat technologies and businesses.)

van Mensvoort, K., and Grievink, H. (2014) *The in Vitro Meat Cookbook,* Amsterdam: BIS Publishers.

(Part fiction and part science, this "cookbook" explores what cultured meat might mean for our plates and our planet.)

Bibliography

Alexander, P., Brown, C., Arneth, A., Dias, C., Finnigan, J., Moran, D., and Rounsevell, M. (2017) "Could consumption of insects, cultured meat or imitation meat reduce global agriculture land use?" *Global Food Security* 15: 22–32.

Bramble, B. (2017) "Lab-grown meat could let humanity ignore a serious moral failing," *The Conversation,* http://theconversation.com/lab-grown-meat-could-let-humanity-ignore-a-serious-moral-failing-88909.

Bryant, C., and Barnett, J. (2018) "Consumer acceptance of cultured meat: A systematic review," *Meat Science* 143: 8–17.

Brown, E. (2015) "Is it possible to build 'meat' out of plant protein?" *PBS Newshour Interview,* www.pbs.org/newshour/show/is-it-possible-to-build-meat-out-of-plant-protein.

Churchill, W. (1932) "Fifty years hence," *Popular Mechanics*, pp. 390–397.

Dance, A. (2017) "Engineering the animal out of animal products," *Nature Biotechnology* 35: 704–707.

Datar, I., and Bolton, R. (2014) "The Carnery," in Van K. Mensvoort and H. J. Grievink (eds) *The In Vitro Meat Cookbook*, Amsterdam: BIS Publishers.

de Salcedo, A. M. (2015) *Combat-Ready Kitchen: How the U.S. Military Shapes the Way You Eat*, New York: Current.

FAO (2006) *Livestock's Long Shadow: Environmental Issues and Options,* Rome, www.fao.org/docrep/010/a0701e/a0701e00.HTM.

FAO (2012) *World Agriculture Towards 2030/2050: The 2012 Revision,* ESA Working Paper No. 12–03, Rome, www.fao.org/docrep/016/ap106e/ap106e.pdf.

FAO (2013) *World Livestock 2013—Changing Disease Landscapes,* Rome, www.fao.org/docrep/019/i3440e/i3440e.pdf.

FAO (2016) *The State of World Fisheries and Aquaculture,* Rome, www.fao.org/3/a-i5555e.pdf.

Foer, J. S. (2009) *Eating Animals*, New York: Little, Brown and Company.

Furnas, C. C. (1932) *America's Tomorrow: An Informal Excursion into the Era of the Two-Hour Working Day*, New York: Funk & Wagnalls.

Greig, K. (2017) "'Clean' meat or 'cultured' meat: A randomized trial evaluating the impact on self-reported purchasing preferences," *Animal Charity Evaluators*, December 10, https://animalcharityevaluators.org/blog/clean-meat-or-cultured-meat-a-randomized-trial-evaluating-the-impact-on-self-reported-purchasing-preferences/

Harari, Y. N. (2015) "Industrial farming is one of the worst crimes in history," *The Guardian,* December 17, 2017, www.theguardian.com/books/2015/sep/25/industrial-farming-one-worst-crimes-history-ethical-question.

Harris, A. J., and Walter, S. (2017) "All work. No pay," *Center for Investigative Reporting,* December 17, www.revealnews.org/article/they-thought-they-were-going-to-rehab-they-ended-up-in-chicken-plants/.

Haub, C. (2011) "How many people have ever lived on earth?" *Population Reference Bureau,* December 17, 2017, www.prb.org/Publications/Articles/2002/HowManyPeopleHaveEverLivedonEarth.aspx.

Hobbes, T. (2004) *Leviathan*, New York: Barnes & Noble Books.

Hopkins, P. D., and Dacey, A. (2008) "Vegetarian meat: Could technology save animals and satisfy meat eaters?" *Journal of Agricultural and Environmental Ethics* 21(6): 579–596.

Koneswaran, G., and Nierenberg, D. (2008) "Global farm animal production and global warming: Impacting and mitigating climate change," *Environmental Health Perspectives* 5(116): 578–582.

Kutern, M. (2014) "Death on the farm," *Newsweek,* December 17, 2017, www.newsweek.com/2014/04/18/death-farm-248127.html.

Lappé, F. M. (1991) *Diet for a Small Planet: Twentieth Anniversary Edition,* New York: Ballantine Books.

Lusk, J. (2016) *Unnaturally Delicious: How Science and Technology are Serving Up Super Foods to Save the World,* New York: St. Martin's Press.

Matticka, C., Landis, A., and Allenby, B. (2015) "A case for systemic environmental analysis of cultured meat," *Journal of Integrative Agriculture* 14(2): 249–254.

McWilliams, J. (2017) *Eating Promiscuously: Adventures in the Future of Food*, Berkeley: Counterpoint.

O'Neill, J. (2016) "Tackling drug-resistant infections globally: Final report and recommendations," *The Review on Antimicrobial Resistance.*

Pachirat, T. (2011) *Every Twelve Seconds: Industrialized Slaughter and the Politics of Sight,* New Haven: Yale University Press.

PETA (2017) *Glass Walls,* December 10, www.peta.org/videos/glass-walls-2/.

Ranganathan, J., Vennard, D., Waite, R., Dumas, P., Lipinski, B., Searchinger, T., and Model, G. (2016) "Shifting diets for a sustainable food future," *Working Paper, Installment 11 of Creating a Sustainable Food Future*, Washington, DC: World Resources Institute, www.worldresourcesreport.org.

Román, S., Sánchez-Siles, L., and Siegrist, M. (2017) "The importance of food naturalness for consumers: Results of a systematic review," *Trends in Food Science & Technology* 67: 44–57.

Rozin, P. (2005) "The meaning of 'natural': Process more important than content," *Psychological Science* 16(8): 652–628.

Searls, D. (2012) *The Intention Economy: When Customers Take Charge*, Boston: Harvard Business Review Press.

Sexton, A. (2016) "Alternative proteins and the (non)stuff of 'meat,'" *Gastronomica: The Journal of Critical Food Studies* 16(3): 66–78.

Shapiro, P. (2018) *Clean Meat: How Growing Meat Without Animals Will Revolutionize Dinner and the World,* New York: Gallery Books.

Siegrist, M. (2008) "Factors influencing public acceptance of innovative food technologies and products," *Trends in Food Science & Technology* 19: 603–608.
Sinclair, U. (1935) *I, Candidate for Governor: And How I Got Licked*, Berkeley: University of California Press.
Smith, F. E. (1930) *The World in 2030 A.D.*, London: Hodder and Stoughton.
Sneeringer, S., MacDonald, J., Key, N., McBride, W., and Mathews, K. (2015) "Economics of antibiotic use in U.S. livestock production," *Economic Research Report no. 200*, Washington, DC: USDA Economic Research Service.
Stephens, N. (2010) "In vitro meat: Zombies on the menu?" *SCRIPTed* 7(2): 394–400.
Tetrick, J. (2015) "What would it look like if we started over?" *Medium*, https://medium.com/@joshtetrick/what-would-it-look-like-if-we-started-over-716990050207.
Tuomisto, H. L., and Teixeira de Mattos, M. J. (2011) "Environmental impacts of cultured meat production," *Environmental Science and Technology* 45(14): 6117–6123.
Van der Weele, C., and Driessen, C. P. G. (2013) "Emerging profiles for cultured meat; ethics through and as design," *Animals* 3: 647–662.
Van der Weele, C., and Driessen, C. P. G. (2014) "In vitro meat as animal liberation," in Van K. Mensvoort and H. J. Grievink (eds) *The In Vitro Meat Cookbook*, Amsterdam: BIS Publishers.
Verbeke, W. (2011) "Consumer attitudes and communication challenges for agro-food technologies," *Agro-Food Industry Hi-Tech* 22(5): 34–36.
Verbeke, W., Marcu, A., Rutsaert, P., Gaspar, R., Seibt, B., Fletcher, D., and Barnett, J. (2015) "'Would you eat cultured meat?': Consumers' reactions and attitude formation in Belgium, Portugal and the United Kingdom," *Meat Science* 102: 49–58.
Verbeke, W., Sans, P., and Van Loo, E. (2015) "Challenges and prospects for consumer acceptance of cultured meat," *Journal of Integrative Agriculture* 14(2): 285–294.
Watson, E. (2018) "Memphis meats, NAMI: FDA and USDA both have roles to play in regulating cell-based meat and poultry," *FoodNavigator-USA*, www.foodnavigator-usa.com/Article/2018/08/23/Memphis-Meats-NAMI-FDA-and-USDA-both-have-roles-to-play-in-regulating-cell-based-meat-and-poultry#.
Welin, S. (2013) "Introducing the new meat. Problems and prospects," *Etikk i praksis: Nordic Journal of Applied Ethics* 7(1): 24–37.
Welin, S., and van der Weele, C. (2012) "Cultured meat: Will it separate us from nature?" in T. Potthast and S. Meisch (eds.) *Climate Change and Sustainable Development: Ethical Perspectives on Land Use and Food Production*, Wageningen Academic Publishers, Wageningen.
Wilks, M., and Phillips, C. J. C. (2017) "Attitudes to in vitro meat: A survey of potential consumers in the United States," *PLoS One* 12(2): e0171904.

43

VEGANISM, (ALMOST) HARM-FREE ANIMAL FLESH, AND NONMALEFICENCE

Navigating Dietary Ethics in an Unjust World

Cheryl Abbate

Arguments for ethical veganism draw attention to the ways in which animal agriculture harms non-human animals, the environment, and humans around the globe. Various philosophical approaches are used in defense of veganism, the dominant ones being utilitarian and rights-based approaches. Utilitarians, such as Peter Singer (1975) and Alastair Norcross (2004), argue that it is wrong to consume factory farmed products because doing so causes significant pain and suffering, which is not outweighed by the pleasures of meat-eating. Tom Regan (1983), who advances a theory of animal rights, argues that animals have the right to respectful treatment, which amounts to the right not to be viewed or treated as a resource. Because animals are viewed and treated like resources when they are raised and killed for food, the philosophy of animal rights declares that it is wrong to consume animal products when plant-based alternatives are available. James Rachels (2004), however, recommends that we consider the things that we eat one thing at a time, instead of supporting a sweeping prohibition against eating animal products. Dietary ethics is complicated, as we must consider thorny questions that present compelling challenges to veganism, such as: Which animals are sentient? Do plant-based diets cause harm? And if so, do they cause *more* harm than some animal-based diets?

Despite the moral complexities associated with food consumption, there is at least one well-grounded claim that helps us understand some of what morality demands when it comes to food ethics: it is wrong to raise and kill *sentient* animals for food *when plant-based alternatives are available*. This is the conclusion of the Nonmaleficence Argument, which I present and defend in this chapter.[1] But, as I argue, the truth of this claim does not entail that we should *only* harvest and eat plants. Rather, the truth of this claim suggests that we should consider ways to produce and consume animal products without harming sentient animals. In what follows, I propose how we might do just that.

Animal Welfare Considerations

Of animal products sold in supermarkets, 99% come from concentrated animal feeding operations (CAFOs), better known as "factory farms." Undercover videos taken at these farms reveal that factory farmed animals lead, as Singer puts it, "miserable lives from birth to slaughter" (Singer 1975: 97). Female pigs are often confined to farrowing and gestation crates, "veal" calves are penned in solitary crates, and 95% of laying hens are locked in battery cages, all of which render animals unable to turn around, let alone walk, *for their entire lives*. The movement of animals who are not confined to crates or cages is also severely restricted, as they are confined to enclosed, overcrowded sheds with thousands of other animals. The extreme crowding and unnatural living conditions induce aggressive and neurotic behaviors: chickens peck each other, often to death, pigs bite each other's tails, and cattle thrust their horns into each other. To prevent these aggressive behaviors, which result in financial loss to meat producers, farmers painfully mutilate these animals; they de-beak chickens, dehorn and brand cattle, clip the teeth of and tail-dock pigs, and castrate pigs and cattle, *all without pain relief.*

Factory farmed animals are surrounded by and continuously inhale overbearing fumes of urine and feces, and this often leads to painful illnesses, infections, and wounds that go untreated. The unnatural and restrictive conditions of factory farms cause animals to suffer from boredom, fear, stress, and anxiety. Babies are separated from their mothers before the time of natural weaning, which is emotionally traumatizing and stressful for both offspring and mother (Newberry and Swanson 2008).

Most factory farmed animals never see the light of day until they are hauled off to slaughter. The journey to slaughter is itself often long and painful. Some animals are on the road for days without food or water, and many die from thirst, hunger, wounds, and/or injury before reaching the slaughterhouse. Others arrive at the slaughterhouse tired, hungry, dehydrated, frightened, confused, and/or with broken bones. The cows, cattle, and pigs who are unable to walk because of injuries or illnesses, referred to as "downers," have ropes or chains tied around their legs and are dragged to slaughter (PETA 2019). The animals who can walk are kicked and smacked with iron pipes and electric prods as human handlers usher them into confinement pens, where they wait until it is their turn to walk single file to the "kill floor."

As they wait in line to be killed, animals smell the blood of their conspecifics, sense their fear, and listen to their terrifying screams. As one former slaughterhouse worker recalls,

> [T]he chickens are panicking. Many of them are squawking loudly, some are just sitting there trembling. Sometimes you catch one looking up at you, eye to eye, and you know it's terrified. . . . No one can convince me that that chicken did not know what was about to happen.
>
> *(Butler 2003)*

When it is time for their slaughter, cows, cattle, and pigs are restrained on conveyor belts, which bring them to a slaughterhouse employee who uses a stunning device, often a captive-bolt gun or electric shock gun, on the animals. If the stunning procedure is successful, the animals will be rendered unconscious before they are hung upside down, their throats are slit, and their bodies bleed out and are ripped apart. Yet because many slaughterhouse employees are poorly trained and the "kill line" moves so quickly, animals often are not stunned properly. Some slaughterhouse workers admit that they regularly cut the hocks off completely conscious cattle and that animals are often alive and fully conscious when they are mutilated by "tail cutters," "belly rippers," and "hide pullers." As Ramon Moreno describes it, "[y]hey blink. They make noises. . . . The head moves, the eyes are wide and looking around. . . . They die piece by piece" (Warrick 2001).

Legal Issues

It certainly would violate state anticruelty laws to treat dogs and cats the way that farmed animals are treated. But the terrible treatment of farmed animals just described is perfectly legal, as there are no federal laws governing the conditions in which farmed animals are raised (Wolfson 1996). Most acts that cause extreme pain and suffering to farmed animals, such as mutilations without anesthetic, are exempt from state anticruelty laws because they are considered "standard" or "commonly accepted" agriculture practices (Wolfson 1996). Although some cruel, nonstandard practices, such as bashing the heads of pigs into cement floors, are illegal, random acts of cruelty are nevertheless widespread on both factory farms and slaughterhouses, and those who commit such atrocious offenses are rarely punished (Singer and Mason 2006).

There are only two federal laws that apply to the treatment of farmed animals, and these laws pertain just to their slaughter and transport: *The Humane Slaughter Act*, which requires that (some) animals be rendered unconscious before slaughter, and *The Twenty-Eight Hour Law*, which requires that (some) animals be unloaded and given rest, water, and food if they are transported for more than 28 consecutive hours. However, these laws are rarely enforced and are routinely violated (Gail Eisnitz 1997). Moreover, neither law protects "poultry," despite that 95% of the land animals killed for food in the US are birds (Pew Charitable Trusts 2008). Thus, it is perfectly legal to grind up chickens alive or to suffocate them to death in trash bags, as is done to 260 million male chicks each year in the egg industry.

The Nonmaleficence Argument

While animal *rights* theorists claim that the abuse of animals on factory farms violates their rights to life, freedom, and bodily autonomy, one need not endorse an animal rights view to condemn factory farming. One might rather argue that factory farming is immoral because it causes *unnecessary harm*. This is the fundamental idea behind what I call the Nonmaleficence Argument:

P1. Raising and killing animals for food on factory farms causes unnecessary harm.[2]
P2. It is wrong to cause unnecessary harm.
Therefore, it is wrong to raise and kill animals for food on factory farms.

I expect there to be close to universal agreement with P2. This principle of nonmaleficence is perhaps the only moral principle that is beyond serious doubt, and it is accepted by virtually every minimally decent theory of ethics (DeGrazia 2002). But P1 needs further discussion.

P1 makes two assumptions: (1) raising and killing animals for food causes harm, and (2) this harm is unnecessary. Clearly, raising and killing animals for food on factory farms causes terrible pain and suffering to farmed animals. But sometimes causing harm is *necessary*. For instance, when children visit the doctor for standard immunizations, the injections usually cause pain to the children and thereby harms them. But it does not follow that immunizing children is wrong. After all, the harm is *necessary* for their health. Likewise, if it turns out that eating animals is necessary *and* animal factories are needed to feed a large and growing population of humans, then perhaps factory farming is not wrong. But not all acts characterized as "necessary" are morally permissible. It is *necessary* for me to torture kittens if I want to know what torturing kittens feels like, but clearly torturing kittens is impermissible. In general, when the term *necessary* is used in ethical discourse, the concern is with what is *morally necessary*.

For some harm X to be characterized as *morally necessary*, two conditions must be met:

1. The harm X must be caused in the name of an end Y that is worth the cost of the harm X.
2. We cannot achieve end Y unless we perform some activity that produces harm X.

One might argue that factory farms harm animals for the "worthy end" of human health, which is allegedly worth the cost of the harm done to animals. Yet, we must bear in mind that we are able to achieve this end without factory farming. After all, we can harvest plant-based foods. The position of the American Academy of Nutrition and Dietetics (2009), which is the world's largest organization of food and nutrition professionals devoted exclusively to nutrition and dietetics, is that an appropriately planned vegan diet is perfectly healthful and nutritionally adequate.[3] This is also the position of other reputable health organizations, including American Heart Association (2018), Kaiser

Permanente (2013), Mayo Clinic (2016), National Institutes of Health (2012), USDA Center for Nutrition Policy and Promotion (2011), and World Health Organization (2015). Because humans, at least in industrialized countries, do not need animal flesh to be healthy, the terrible suffering inflicted upon factory-farmed animals before they end up on our plates is done for mere *gustatory pleasure*. Although most humans value pleasure, gustatory pleasure surely is not worth the cost of the harm done to factory farmed animals.[4]

Ecological Considerations

The Nonmaleficence Argument maintains that animals endure terrible harms on factory farms and that these harms are *unnecessary*. But these are not the only harms produced by factory farming. After all, dietary choices impact both humans and wildlife through the intensive use of valuable resources, climate change, and deforestation (Sherriff 2017).

Climate Change: In 2006, the United Nations Food and Agriculture Organization (FAO) reported that animal agriculture is responsible for 18% of global greenhouse gas (GHG) emissions (Steinfeld et al. 2006). Moreover, a 2009 Worldwatch Institute study counters that animal agriculture is responsible for 51% of global emissions (Goodland and Anhang 2009). However, the methods and methodologies used in the Worldwatch study have been called into question by Mario Herrero et al. (2011) in a recent commentary in *Animal Feed Science and Technology*.[5]

But even if animal agriculture is responsible for 18%, and not 51%, of GHG emissions, it still stands that animal agriculture is responsible for more emissions than the *entire* transportation sector. Indeed, the United Nation's FAO report estimates that the animal agriculture sector is responsible for at least 9% of annual anthropogenic carbon dioxide (CO_2) emissions, 35% and 40% of annual anthropogenic methane emissions, and 65% of global nitrous oxide emissions (Steinfeld et al. 2006). Methane production and nitrous oxide production are especially worrisome because methane has 23 times the Global Warming Potential (GWP) as CO_2, and nitrous oxide has 296 times the GWP as CO_2 (Koneswaran and Nierenberg 2008).

The production of GHGs contributes to global warming, and an excessive amount of global warming causes climate change. Climate change, which refers to a change in global or regional climate patterns, causes an increase of storms, floods, heat waves, droughts, and desertification. Currently, *humans* around the globe are suffering from climate change, as their homes are destroyed by rising sea levels, floods, wildfires, tropical cyclones, and storms, leaving them displaced, often permanently, and forced to migrate (IPCC 2007). Wild animals, too, are vulnerable to the effects of climate change, as their homes are frequently washed or melted away due to excessive climate warming (Pagano et al. 2018).

Water Pollution: Animal manure and commercial fertilizers and pesticides, which are sprayed on animal-feed crops, frequently end up in oceans, lakes, and rivers, and this causes eutrophication. Manure and commercial fertilizers contain phosphorus and nitrogen, and excessive amounts cause algae to thrive. This sharply increases the occurrence of algal blooms, and algal blooms cause dead zones, which are areas in oceans, rivers, and lakes that do not have enough oxygen and sunlight to support marine life.

***Water Overuse*:** Agriculture accounts for 92% of the global freshwater footprint, and one third of this is used to grow grain that is eventually fed to livestock (Gerbens-Leenes et al. 2013; Mekonnen and Hoekstra 2012). Producing 1 kilogram of animal protein requires about 100 times more water than producing one kilogram of grain protein (Pimentel and Pimentel 2003). While one pound of beef requires approximately 2,400 gallons of water, one pound of Tofu requires approximately 219 gallons (Kreith 1991).

Deforestation: A third of the earth's arable land is used to grow crops to feed animals on CAFOs, and mass deforestation is needed to make way for fields of animal feed. Because an excessive amount of CO_2 is released into the atmosphere when forests are cleared, deforestation has a deleterious effect on climate change (Steinfeld et al. 2006). And since deforestation also destroys species-rich habitat, it is the biggest threat to biodiversity (Machovina et al. 2015).

Desertification: Thirty percent of the earth's territorial surface is used for livestock grazing, and 70% of the Amazon has been destroyed for livestock grazing (Steinfeld et al. 2006). Overgrazing leads to soil erosion, a process of desertification, which significantly reduces agricultural and forestry production. This impairs the livelihoods of the poor and vulnerable humans residing in these areas, as these areas can no longer be used for farming purposes (United Nations 2012).

Humanitarian Considerations

Factory farming directly impacts human well-being in a variety of ways. For one, those who work on slaughterhouses are subject to poor working conditions and serious physical health hazards, as evident by the number of job-related injuries. Approximately 25% of slaughterhouse workers suffer work-related illnesses or injuries each year (Woorall 2004). Slaughterhouse workers perform repetitive motions throughout the workday, such as cutting with knives every few seconds, which explains why the meatpacking industry has the highest rate of repetitive-motion injuries. Cumulative trauma disorders, such as carpal tunnel syndrome, "trigger finger," or back and shoulder problems, are common for slaughterhouse workers (Woorall 2004). Many injuries go unreported, as many workers are undocumented immigrants, who risk deportation if they report problems (Woorall 2004). Moreover, because slaughterhouse work is innately violent and workers gruesomely kill hundreds of animals each hour and thousands of animals each week, the workers likely suffer severe psychological trauma (Dillard 2008; Schlosser 2002; Eisnitz 1997). Slaughterhouse workers are susceptible to a post-traumatic stress disorder called perpetration-induced traumatic stress, which is the psychological stress one experiences in virtue of being the cause of another's trauma, such as the trauma that farmed animals endure (Dillard 2008; MacNair 2002). As a former slaughterhouse worker reveals,

> [a] lot of [slaughterhouse workers] have problems with alcohol. They have to drink, they have no other way of dealing with killing live, kicking animals all day long. If you stop and think about it, you're killing several thousand beings a day.
>
> *(Eisnitz 1997: 87–88)*

Researchers suspect that this psychological trauma has a "spillover" effect on nearby communities, as counties with slaughterhouses have noticeably more violent crime (Fitzgerald et al. 2009).

Moreover, CAFOs are usually located in impoverished, rural parts of America, and these communities are affected disproportionately by factory farm emissions (Pew Charitable Trusts 2008). For example, the tremendous quantities of animal waste produced on factory is disposed of in "cesspools" or manure lagoons, which are large open pits, often the size of four football fields. Eventually, cesspools must be emptied out, and this is done by spraying liquid waste, which often drifts downwind into neighboring communities (Toliver 2017). Because communities of color suffer disproportionately from asthma and other upper respiratory issues from the spraying of toxic waste, CAFOs are rightly charged with environmental racism (Carter 2016; Wallinga 2004; Kresge and Strochlic 2007).

The impacts of factory farming stretch well beyond neighboring communities and are felt by the global poor. Currently, millions of people go to bed hungry every night because they do not have enough to eat. Meanwhile, farmers feed, on average, 8 pounds of plant protein to pigs in return for 1 pound of pork, and they, on average, feed 21 pounds of plant protein to cattle in return for 1 pound of beef (DeGrazia 2002: 75). This inefficient use of plant protein is especially problematic because hunger is so widespread. Research shows that if animal-based products are replaced with plant-based foods, there would be enough food to feed 350 million additional people (Shepon et al. 2018).

Eating Animals Anyway: "Humane" Farms

Although factory farming clearly is impermissible, some argue that raising and killing sentient animals for food is not *inherently wrong*, as there are alternative, more "humane" forms of animal agriculture. This position, championed by Michael Pollan (2006), is known as "compassionate carnivorism," "benign carnivorism," or "welfarism," which contends that while animal agriculture ought to be *reformed*, it need not be *abolished*. On this view, so long as animals are given "happy" lives on farms and killed painlessly, it is permissible to use and kill them for food. As the argument goes, the lives of animals on "happy" farms contain more happiness than suffering, so "humane" farms do a good thing when they bring "happy" animals into existence, even if they are eventually killed for food.

The first problem with this argument is that it overlooks the tremendous suffering of animals on "happy" farms. For instance, animals on "free-range" and "cage-free" farms are often mutilated without anesthetic. Moreover, while chickens on "free-range" farms must be given outdoor access, the "outdoors" is usually a small patch of dirt, accessible only through a small hole in the wall of an overcrowded shed that only a handful of birds can ever reach, hence why "free-range" farms are said to provide only "phantom access" to the outdoors. Likewise, hens living in "cage-free" facilities are usually de-beaked without anesthetic, confined to overcrowded, filthy sheds with thousands of other birds, and denied outdoor access. Both "free-range" and "cage-free" chickens are eventually sent to slaughterhouses, where many are forced into scalding-hot de-feathering tanks while they are fully conscious. Male chicks, who have no value to the egg industry, are usually suffocated in trash bags or ground up alive.

There are no official federal guidelines for the definition of "humanely raised." Rather, animal producers determine their own definition of "humane," and the United States Department of Agriculture (USDA) Food Safety and Inspection Service (FSIS) approves this labelling by "verifying" that the company follows its own arbitrary standards (Farm Sanctuary 2009). To "verify" that meat producers adhere to their self-defined "humane" standards, the FSIS reviews "supporting documentation" that meat producers submit, such as testimonials and affidavits signed by *meat producers themselves* (USDA FSIS 2016). Neither third-party verification nor on-farm inspection is required for "humane" labeling, so the FSIS does not conduct inspections of these farms to determine whether they are really "humane" (USDA FSIS 2016).

Bob Fischer (2018) points out that even if there are farms that really do give animals good lives, it is unlikely that consumers know where to find these farms. While there are independent "humane" certifiers, the standards differ significantly from each other, and most, if not all, animal products that are labeled "certified humane" are products of serious pain and suffering (McWilliams 2017). For instance, animal products that come from farms that mutilate animals without anesthetic are certified as "humane" by the American Humane Association and the Humane Farm Animal Care organization, two popular independent "humane" certifiers. Moreover, most animals raised on "humane" farms are killed in the same slaughterhouses as factory-farmed animals (Stănescu 2016). As McWilliams (2017: 247) puts it, "even the most loving treatment of farm animals ends in a violent and unnecessary death." Although there may be less suffering on "humane" farms than there is on factory farms, the suffering still exists, and it is still unnecessary. Consequently, raising and killing sentient animals on "humane" farms is impermissible.

One problem with the rhetoric of "humane" farming is that it seems to endorse what I call the Inhumane Argument.

P1. Animals on "humane" farms suffer less than animals on factory farms.
C. Therefore, it is permissible to raise and kill animals for food on "humane" farms.

Note that there is a hidden premise of this argument: *For any act A, if act A causes less suffering than act B, then act A is permissible.* If it is not obvious why this is false, consider what it implies for human torture.

P1. A human who is tortured for 20 hours suffers less than a human who is tortured for 24 hours.
C. Therefore, it is permissible to torture a human for 20 hours.

While it might be more "humane" to torture a human for 20 hours than it is to torture a human for 24 hours, clearly torturing someone for 20 hours is impermissible. Likewise, while it might be more "humane" to raise and kill animals on nonfactory farms than it is to raise and kill animals on factory farms, it is still impermissible to harm animals on "humane" farms. Although the welfarist might insist that animals on "humane" farms suffer much less than animals on factory farms, to the point where they have lives worth living overall, the evidence strongly suggests that most animals raised and killed on "humane" farms *do not* have lives worth living. Just think about the hens living in "cage-free" facilities who are de-beaked without anesthetic and confined to overcrowded, filthy sheds with thousands of other birds for their entire lives before they are violently killed. Surely, that kind of life is not worth living. Given the tragic reality of "humane" farms, those who defend them *are* fundamentally committed to the Inhumane Argument, including its morally outrageous hidden premise.

But let us grant, for the sake of argument, that there exist *some* farms with animals who have lives worth living. Say there are farmers who provide their animals with appropriate veterinary care, refuse to mutilate them without anesthetic, protect them from inclement weather by offering them clean and comfortable housing, allow them to roam freely in the outdoors with their family members and conspecifics, and provide them with ample opportunities to engage in species-normal behavior. And say the animals are killed painlessly. Still, there are at least three reasons why animals should not be raised and killed on "ideal" farms.

First, approximately 56 billion land animals are killed for food each year and global animal flesh consumption is expected to double by 2050, which means that, if people continue to eat farmed animals (at the current rate of consumption), there will be approximately 120 billion animals raised and killed for food in 2050 (Steiner et al. 2006). There simply is not enough room on this planet for 120 billion animals to roam freely (Stănescu 2016). The "compassionate carnivore" ideal is thus *physically impossible* (Stănescu 2016).

Second, even if it were physically possible, it would be *ecologically devastating* (Stănescu 2016). For one, the amount of deforestation needed to make room for billions of free roaming animals would cause tremendous amounts of CO_2 to be released into the atmosphere, the massive degradation of wildlife habitat, and a sharp decrease in biodiversity (Stănescu 2016). Increased grazing of livestock will inevitably lead to increased killing of native wildlife, especially apex predators, by Wildlife Services, who, each year, kills millions of native animals, such as cougars, wolves, foxes, and coyotes, in the name of livestock protection (USDA Animal and Plant Health Inspection Service 2016). Moreover, since grass-fed cattle do not reach slaughter weight as quickly as do grain-fed cattle, grass-fed cattle live longer than grain-fed cattle, and with longer lives comes greater methane production (Capper 2012; Hayek and Garrett 2018). Although one might suggest that we could mitigate ecological impacts by raising the price of meat, which likely would decrease the demand for meat, this alternative "reform" would provide animal foods to only the privileged (McWilliams 2017: 249; Stănescu 2016: 138).

Third, other things equal, death *itself* harms animals, even when they are killed painlessly. To see why this is, imagine the case of a healthy, happy human—call her Sarah—who, while sleeping soundly in bed, is killed painlessly. Surely, other things equal, her killer acted wrongly because he harmed Sarah *herself*. But how could Sarah be harmed if she did not experience pain or suffering when killed?

One possible answer is that Sarah is harmed because she had a desire to continue living, and this desire was thwarted by her death. This explanation stems from the desire-satisfaction view, which holds that death harms the one who dies only when death thwarts a personal interest in continued life (Singer 1979). Some argue that, on this view, most nonhuman animals are not harmed by death because they are incapable of desiring to go on living (Singer 1979), a claim that is itself highly dubitable. As the argument goes, the only creatures who have the desire to live are self conscious creatures, that is, creatures with the ability to "conceive of themselves as distinct entities, existing over time with a past and a future" (Singer 1979). And as Singer (1979) argues, since most nonhuman animals are not *self-conscious*, they do not have a personal interest in continuing to live, so killing sentient animals for food is not always wrong.

The shortcomings of the desire-satisfaction view are revealed when we consider the case of a suicidal person who does not have the desire to live (Marquis 1989). Imagine someone—call him Sam—who is healthy, but momentarily depressed and thus intent on ending his life. Sam has no future oriented desires, besides the desire to kill himself, which would not be thwarted by his death. Arguably, it would be right to try to convince Sam not to kill himself. And presumably one would do so by pointing to the fact that, although Sam does not desire to live, his future is full of opportunities for pleasures and satisfactions. This suggests that the most plausible explanation why Sam's death is bad *for him* is that his death permanently forecloses his future opportunities for satisfaction—satisfactions he would enjoy if he continued to exist. Consequently, other things equal, *any* being who has future opportunities for satisfaction is harmed by death and because farmed animals who live decent lives have these opportunities, they, too, are harmed when they are killed prematurely, even if their deaths are painless (Regan 1983; DeGrazia 2002; Sapontzis 1987). And as Regan compellingly argues, an untimely death is the ultimate, irreversible harm "because it is the ultimate loss—the loss of life itself" (Regan 1983: 100). While, on the Nonmaleficence Argument, it *might* be permissible to raise animals on "ideal" farms, it surely is wrong to kill them, even painlessly.

Challenges for Veganism

So far, I have argued that it is wrong to raise and kill sentient animals on farms: factory or "humane" farms. But perhaps there is a gap between the ethics of *producing* animal products and the ethics of *purchasing and consuming* them. As one might argue, although it is wrong to *raise and kill* sentient animals on farms, it is not wrong to *buy and consume* animal products that come from these farms. As the argument goes, one person's consumption choices will not be noticed by farmers, so the purchasing decisions of individuals do not impact the number of animals raised and killed on farms. And since our purchases are causally inefficacious, we are morally permitted to consume meat. This is known as "the causal impotence problem."

There is a case to be made that our purchasing decisions *do* make a difference. As Norcross (2004) compellingly argues, while the market is insensitive to one person's purchasing choices, there is some threshold at which the number of vegans will influence the market and thereby reduce the number of sentient animals raised and killed for food. Because an individual's decision to stop purchasing traditional animal products may be the "tipping point," or it will at least reduce the amount of time it takes to reach this tipping point, it is impermissible for individuals to purchase farmed animal products.

Gaverick Matheny's (2002) provides a related and compelling argument. Say there are 20 million meat-eating customers per "threshold unit," and say if 20 million people stop eating animals, there will be one billion fewer animals raised and killed on farms each year. This means that each individual consumer has a 1-in-20 million chance of sparing 1 billion animals a life of misery. The expected "disutility" of purchasing meat, then, is the raising and slaughtering of 50 farmed animals (1/20-millionth times 1 billion). And since we know not the *actual* effects of our actions before we act, we ought to make consumption decisions based on their *expected* consequences. In this situation, the expected consequence of eating meat is the "disutility" of raising and killing 50 farmed animals per year. Surely that is substantial enough to conclude that buying the flesh of farmed animals is wrong.

Individuals such as you and me, who have access to less harmful alternatives, act wrongly when we purchase what Fischer (2018) calls "traditional animal products," that is, the flesh and products of sentient animals who were raised and killed for food. But this does not entail that it is always wrong to buy and consume animal products. And I do not intend for this argument to be used in defense of veganism. After all, ethical veganism is itself demonstrably arbitrary insofar as it dogmatically claims that although it is permissible to eat plants, it is *never* permissible to eat animals—sentient or insentient. For instance, when asked for his perspective on the ethics of bivalve consumption, popular vegan activist Gary Yourofsky responds:

> Clams and mussels and oysters are not plants and are not listed in any science book as plants. The fact they contain animal protein should let you know that they are off limits . . . being vegan means that you don't consume animal products . . . So, forget about the clams, forget about the scallops, stick to the fruits and vegetables.
>
> *(Barwick 2014)*

But what is the morally relevant difference between plants and insentient animals, such as bivalves (clams, mussels, oysters, and scallops)? What is the morally relevant feature or characteristic that *all* animals possess and *all* plants lack? If there is no such feature, veganism is guilty of what I call *kingdomism*—the view that an animal is entitled to serious moral consideration just because of its membership in the animal kingdom. This is not to suggest that it is inherently morally problematic to eat plants or to suggest that plants might be sentient, as some suggest (Smith 2016). Rather, this is a challenge to the vegan's claim that it is wrong to eat *insentient* animals. After all, the most compelling arguments against meat-eating ground serious moral status in *sentience*, but there are some animals that are not sentient, such as bivalves, which have no brains and only simple nervous systems. Moreover, farming bivalves has minimal ecological impacts and is relatively sustainable (Jacquet et al. 2017). Thus, some animal ethicists defend ostroveganism, the view that while it is impermissible to eat the flesh and product of *sentient* animals, it is permissible to eat the flesh of insentient bivalves (Cox 2010; Huemer 2019).

Relatedly, Chris Meyers (2013) and Fischer (2016) suggest that because it is highly unlikely that insects are conscious, we should move toward entomophagy—an insect-based diet. But some worry that insects may have the capacity for basic consciousness, as structures in their brains might function analogously to the mammalian cortex, and many insects engage in what appears to be intelligent behavior (Klein and Barron 2016). However, as DeGrazia (2019) points out, even if this hypothesis were true, it would not establish that insects are *sentient*; it would only establish that they have subjective awareness. Sentience requires awareness that involves *feeling*, and creatures such as insects might have basic consciousness without being sentient (DeGrazia 2019). Although to be safe, perhaps we should opt for eating bivalves instead of insects.

It is not morally problematic to claim that, other things equal, eating insentient animals is permissible. What is problematic is determining which beings are and are not sentient. Because we cannot definitively know whether bivalves are sentient, those who oppose the consumption of bivalves often appeal to a *precautionary principle*. The idea here is that we ought to give bivalves the benefit of the doubt and err on the side of caution by not eating them. But it is arbitrary to apply the precautionary principle to bivalves but not to plants. While one might argue that there is more doubt about whether bivalves are sentient than there is about plant sentience, justification for this claim is wanting. Although there is evidence that bivalves have opioids and opioid receptors, there is also evidence that plants excrete endogenous opioids when wounded or subjected to stress and that analgesics affect plant "response" (Smith 2016). While bivalves have simple nervous systems, some plant neurobiologists report that plants, too, have nervous systems (Baluška and Mancuso Stefano 2009). And even if there is slightly more evidence for bivalve sentience than there is for plant sentience, this does not imply that bivalves are off-limits. As Fischer (2016) compellingly argues, if we have the choice to harm either beings that we know to be sentient or beings that we do not know to be sentient, we should choose the latter. Since industrial plant production might cause harm to animals who are clearly conscious, such as field mice who are frequently run over by plow machines, we should consider reducing our consumption of plants by eating "maybe sentient" animals.

Perhaps, though, there is a better way to reduce our consumption of plants—a way that does not involve killing and eating "maybe sentient" animals. Jeff McMahan (2008: 9), for instance, proposes that it is morally permissible to raise sentient animals for food if (1) they are genetically programmed such that they die at an early age when "their meat would taste best" and (2) we eat them only after they die "naturally." Adam Shriver (2009) suggests using biotechnology to genetically engineer farmed animals to be insentient. Others suggest that it is permissible, if not obligatory, to eat scavenged flesh, such as roadkill or animal flesh obtained through dumpster diving (Milburn 2017; Bruckner 2015; Fischer 2018). While these proposed solutions are either controversial, unachievable fantasies, or unlikely to provide enough food for a growing population, there is a more realistic, less controversial, promising, and forthcoming "solution": growing animal flesh in laboratories. Cultured meat companies have successfully grown what is called "lab-grown," "in vitro," "cultured," or "clean" meat by harvesting, incubating, and feeding animal cells, and these products will soon be cost-effective and commercially available. The mass production of cultured animal products, *when certain energy sources and production systems are used*, has the potential to eliminate massive farmed animal suffering *and* minimize the ecological harms associated with (plant or animal) agriculture (Tuomisto and Joost Teixeira de Mattos 2011)

Some animal ethicists disapprove of the production of cultured meat on the grounds that it allegedly requires the harming of animals (Wrenn 2012). For instance, to grow meat in a lab, cells from animals must be harvested, through a biopsy procedure, from "donor" animals. The cells then must be "fed," and some companies use an animal-based blood serum (fetal bovine serum) to feed the harvested cells. There is thus a worry that lab grown meat still requires the raising and slaughtering of animals. Yet, others insist that animal use is not needed to produce cultured meat. For instance, the company Just used the feather of a chicken to harvest chicken cells, and the company grew these cells without animal-based blood serum.

Still, some worry that the normalization of cultured meat will perpetuate disrespectful attitudes towards animals or reinforce the view that humans have, while animals lack, dignity (Donaldson and Kymlicka 2011: 152). But it is unclear whether this concern is weightier than the alleged harms caused by industrial plant production. After all, growing plants for human consumption often involves deforestation and thus the destruction of wildlife habitat. In Indonesia and Malaysia, orangutans are often burned alive as forests are blazed in order to make room for palm oil plantations (Miles et al. 2007). Heavily processed meat substitutes are often made from plants raised in monoculture on formerly forested lands and require large amounts of pesticides and fertilizers, and this, as discussed earlier, causes serious ecological harm (Henning 2016). Many field animals are killed in the standard process of plowing, planting, harvesting, protecting crops, and pest management (Fischer and Lamey 2018). Because conventional plant agriculture allegedly involves animal suffering and death, Milburn (2017) and Fischer (2018) propose that the eating-animals discourse acknowledge that some animal-based diets are not simply permissible but, moreover, obligatory. The basic idea is that some animal-based diets may be less harmful than the average plant-based diet. If this is true, certain "non-traditional" animal-based diets may be morally preferable to strict veganism.

Elsewhere, I argue that eating animal bodies, including roadkill, is morally problematic, insofar as the act of eating animals expresses the disrespectful view that animals are mere consumables, at least in cultures that condemn the consumption of human corpses (Abbate 2019). I also question whether industrial plant production causes an *all-things-considered* harm to field animals. Hence, I remain skeptical of the claim that we are *obligated* to consume roadkill. I suggest that if we come across fresh roadkill and can transport it, we donate it to animal sanctuaries rather than consume it ourselves (Abbate 2019). But does this worry about the consumption of roadkill extend to the consumption of cultured meat?

Arguably, there is a moral distinction between eating lab-grown meat and consuming roadkill. Eating roadkill involves the consumption of animal bodies that once belonged to living, sentient animals, but, as Schaefer and Savulescu (2014) point out, cultured meat is essentially a group of cells and tissue in a petri dish, which are more like plants than animals. So, consuming cultured meat does not express the disrespectful view that animals are mere consumables; rather, it expresses the view that *a group of cells and tissues* are mere consumables. Surely that is not morally objectionable. So, *if* it turns out that industrial plant production is more harmful than the production of cultured meat, consumers ought to get some of their protein from cultured meat, when it becomes commercially available. But *roadkill* should be donated to animal sanctuaries.

There is a worry that at least some ways of producing cultured meat are as, or even more, harmful to the environment than cattle production (Lynch and Pierrehumbert 2019). Perhaps, then, if bivalve farming has minimal ecological impact, we ought to focus our efforts on increasing bivalve production instead of cultured meat. And for practical reasons, we ought to emphasize the reasons for eating bivalves in our conversations about food ethics, as ostroveganism arguably is manageable and less demanding than veganism. While we should emphasize the importance of pursuing a diet that causes *minimal harm*, we must be mindful of the importance of recommending diets that are *practical* in our growing world. Those who promote ostroveganism seem to understand the importance of doing just this.

Notes

1. Insofar as the argument presented here is not committed to a particular ethical theory, it has much in common with both David DeGrazia's (2002) and Rachels's (2004) commonsense arguments, which are supposed to rely on simple moral principles that any decent person will accept.
2. Throughout this chapter, when I use the phrase "unnecessary harm," I refer to harm that is either *intentional* or *foreseeable*.
3. Even if *some* people, such as those living in drought-stricken areas in the Horn of Africa, need to consume animal products, it doesn't follow that it justifies standard practices in the industry. If some people truly need animal products, their need could be supported by much more humane farms. Moreover, their *need* does not justify another's *choice* to harm animals.
4. Moreover, research shows that animal-based diets, which are high in cholesterol and saturated fat, are harmful (Harvard School of Public Health 2016; World Health Organization 2015). Appropriately planned vegan diets are considered optimal for human health because the alternative animal-based diets, bring about increased risks of developing heart disease, various cancers (including breast, liver, and prostate cancer), type 2 diabetes, and obesity. Hence, health experts from Kaiser Permanente (2013) urge physicians to recommend a plant-based diet to their patients. So, even if great gustatory pleasure is to be had from an animal-based diet, the long-term benefits of a healthier, longer life surely outweigh this pleasure (Garrett 2007).
5. Herrero, however, works for the International Livestock Research Institute, so given his invested interest in the livestock industry, this commentary is itself questionable.

Bibliography

Abbate, C. (2019) "Save the meat for cats: Why it's wrong to eat roadkill," *Journal of Agricultural and Environmental Ethics* 32(1): 165–182.

American Academy of Nutrition and Dietetics (2009) "Position of the American dietetic association: Vegetarian diets," *Journal of the American Dietetic Association* 109(7): 1266–1282.

American Heart Association (2018) "Vegetarian diets," *Healthy Living*, www.heart.org/HEARTORG/Healthy Living/HealthyEating/VegetarianDiets_UCM_306032_Article.jsp#.WvjDui0vyM9 (Accessed December 28, 2018).

Baluška, F., and Stefano, M. (2009) "Plant neurobiology: From stimulus perception to adaptive behavior of plants, via integrated chemical and electrical signaling," *Plant Signaling Behavior* 4(6): 475–476.

Barwick, E. (2014) "Are oysters vegan? Gary Yourofsky & The Vegan Black Metal Chef," *Bite Size Vegan*, www.bitesizevegan.org/bite-size-vegan-nuggets/main-nuggets/are-oysters-vegan-gary-yourofsky-the-vegan-black-metal-chef/ (Accessed December 28, 2018).

Bruckner, D. (2015) "Strict vegetarianism is immoral," in B. Bramble and B. Fischer (eds.) *The Moral Complexities of Eating Meat*, New York: Oxford University Press.

Butler, V. (2003) *The Cyberactivist: A Night in Tyson's Hell,* www.cyberactivist.blogspot.com/2003_09_01_archive. Html (Accessed December 28, 2018).

Capper, J. (2012) "Is the grass always greener? Comparing the environmental impact of conventional, natural and grass-fed beef production systems," *Animals* 2(2): 127–143.

Carter, C. (2016) "Vegan soul," in B. Donaldson and C. Carter (eds) *The Future of Meat Without Animals*, London: Rowman & Littlefield.

Cox, C. (2010) "Consider the oyster," *Slate,* www.slate.com/articles/life/food/2010/04/consider_the_oyster.html (Accessed December 28, 2018).

DeGrazia, D. (2002) *Animal Rights: A Very Short Introduction*, Oxford: Oxford University Press.

DeGrazia, D. (2019) "Sentience and consciousness as bases for interests and moral status: Considering the evidence and speculating about possibilities," in S. Johnson, A. Fenton, and A. Shriver (eds.) *Neuroethics and Nonhuman Animals*, New York: Springer.

Dillard, J. (2008) "A slaughterhouse nightmare: Psychological harm suffered by slaughterhouse employees and the possibility of redress through legal reform," *Georgetown Journal on Poverty Law & Policy* 15(2): 391–408.

Donaldson, S., and Will, K. (2011) *Zoopolis*, Oxford: Oxford University Press.

Eisnitz, G. (1997) *Slaughterhouse: The Shocking Story of Greed, Neglect, and Inhumane Treatment Inside the U.S. Meat Industry*, New York: Prometheus Books.

Farm Sanctuary (2009) *The truth behind the labels: Farm animal welfare standards and labeling practices: A farm sanctuary report*, https://faunalytics.org/wp-content/uploads/2015/05/Citation1051.pdf (Accessed March 1, 2019).

Fischer, B. (2016) "Bugging the strict vegan," *Journal of Agricultural and Environmental Ethics* 29(2): 255–263.

Fischer, B. (2018) "Arguments for consuming animal products," in A. Barnhill, M. Budolfson, and T. Doggett (eds) *The Oxford Handbook of Food Ethics*, Oxford: Oxford University Press.

Fischer, B., and Lamey, A. (2018) "Field deaths in plant agriculture," *Journal of Agricultural and Environmental Ethics* 31(4): 409–428.

Fitzgerald, A., Kalof, L., and Dietz, T. (2009) "Slaughterhouses and increased crime rates: An empirical analysis of the spillover from 'the jungle' into the surrounding community," *Organization & Environment* 22(2): 158–184.

Garrett, J. (2007) "Utilitarianism, vegetarianism, and human health: A response to the causal impotence objection," *Journal of Applied Philosophy* 24(3): 223–237.

Gerbens-Leenes, P., Mekonnen, M., and Hoekstra, A. (2013) "The water footprint of poultry, pork and beef: A comparative study in different countries and production systems," *Water Resources and Industry* 1–2: 25–36.

Goodland, R., and Anhang, J. (2009) "Livestock and climate change," *World Watch Magazine* 22(6): 10–19.

Harvard School of Public Health (2016) "Association of animal and plant protein intake with all-cause and cause-specific mortality," *JAMA Internal Medicine* 176(10): 1453–1463.

Hayek, M., and Garrett, D. (2018) "Nationwide shift to grass-fed beef requires larger cattle population," *Environmental Research Letters* 13(8): 1–8.

Henning, B. (2016) "Towards 2050: The projected costs of and possible alternatives to industrial livestock production," in B. Donaldson and C. Carter (eds) *The Future of Meat Without Animals*, London: Rowman & Littlefield.

Herrero, M., Gerber, P., Vellinga, T., Garnett, T., Leip, A., Opio, C., Westhoek, H., Thornton, P. Olesen, J., Hutchings, N., Montgomery, H., Soussana, J., Steinfeld, H., and McAllister, T. (2011) "Livestock and greenhouse gas emissions: The importance of getting the numbers right," *Animal Feed Science and Technology* 166–167: 779–782.

Huemer, M. (2019) *Dialogues on Ethical Vegetarianism,* New York: Routledge.

IPCC (2007) *Climate Change 2007: Synthesis Report. Contribution of Working Groups I, II and III to the Fourth Assessment Report of the Intergovernmental Panel on Climate Change*, Geneva: IPCC.

Jacquet, J., Sebo, J., and Elder, M. (2017) "Seafood in the future: Bivalves are better," *Solutions* 8(1): 27–32.

Kaiser Permanente (2013) "Nutritional update for physicians: Plant-based diets," *The Permanente Journal* 17(2): 61–66.

Klein, C., and Barron, A. (2016) "What insects can tell us about the origins of consciousness," *PNAS* 113: 4900–4908.

Kreith, M. (1991) *Water Inputs in California Food Production,* Sacramento: Water Education Foundation.

Kresge, L., and Strochlic, R. (2007) *Clearing the Air: Mitigating the Impact of Dairies on Fresno County's Air Quality and Public Health*, California Institute for Rural Studies.

Koneswaran, G., and Nierenberg, D. (2008) "Global farm animal production and global warming: Impacting and mitigating climate change," *Environmental Health Perspectives* 116(5): 578–582.

Lynch, J., and Pierrehumbert, R. (2019) "Climate impacts of cultured meat and beef cattle," *Frontiers in Sustainable Food Systems* 3(5): 1–11.
Machovina, B., Kenneth, F., and William, R. (2015) "Biodiversity conservation: The key is reducing meat consumption," *Science of The Total Environment* 536(1): 419–431.
MacNair, R. (2002) *Perpetration-Induced Traumatic Stress: The Psychological Consequences of Killing*, Westport: Greenwood Publishing.
Marquis, D. (1989) "Why abortion is immoral," *Journal of Philosophy* 86(4): 183–202.
Matheny, G. (2002) "Expected utility, contributory causation, and vegetarianism," *Journal of Applied Philosophy* 19(3): 293–297.
Mayo Clinic Staff (2016) "Vegetarian diet: How to get the best nutrition," *Mayo Clinic*, www.mayoclinic.org/healthy-lifestyle/nutrition-and-healthy-eating/in-depth/vegetarian-diet/art-20046446 (Accessed December 28, 2018).
McMahan, J. (2008) "Eating animals the nice way," *Dædalus*, Winter, 1–11.
McWilliams, J. (2017) "The ethics of humane animal agriculture," in M. Rawlinson and C. Ward (eds) *The Routledge Handbook of Food Ethics*, London: Routledge.
Mekonnen, M., and Hoekstra, A. (2012) "A global assessment of the water footprint of farm animal products," *Ecosystems* 15: 401–415.
Meyers, C. (2013) "Why it is morally good to eat (certain kinds of) meat: The case for entomophagy," *Southwest Philosophy Review* 29(1): 119–126.
Milburn, J. (2017) "Animal rights and food: Beyond Regan, beyond Regan," in M. Rawlinson and C. Ward (eds.) *The Routledge Handbook of Food Ethics*, London: Routledge.
Miles, L., Nellemann, C., Kaltenborn, B., Virtue, M., and Ahlenius, H. (2007) *The Last Stand of the Orangutan*, Norway: Grid-Arendal.
National Institute of Health (2012) "Digging a vegetarian diet plant-based eating can reap rewards," *NIH News in Health*, https://newsinhealth.nih.gov/2012/07/digging-vegetarian-diet (Accessed December 28, 2018).
Newberry, R., and Swanson, J. (2008) "Implications of breaking mother—young social bonds," *Applied Animal Behavior Science* 110: 3–23.
Norcross, A. (2004) "Puppies, pigs, and people: Eating meat and marginal cases," *Philosophical Perspectives* 18: 229–245.
Pagano, A., Durner, G. Rode, K., Atwood, T., Atkinson, S., Peacock, E., Costa, D., Owen, M., and William, T. (2018) "High-energy, high-fat lifestyle challenges an arctic apex predator, the polar bear," *Science* 359(6375): 568–572.
People for the Ethical Treatment of Animals (PETA). (2019) *Cow Transport and Slaughter* www.peta.org/issues/animals-used-for-food/factory- farming/cows/cow-transport-slaughter/ (Accessed March 1, 2019).
Pew Charitable Trusts (2008) *Putting Meat on the Table: Industrial Farm Animal Production in America*, www.pewtrusts.org/en/research-and- analysis/reports/2008/04/29/putting-meat-on-the-table-industrial-farm-animal-production-in-america (Accessed December 28, 2018).
Pimentel, D., and Pimentel, M. (2003) "Sustainability of meat-based and plant-based diets and the environment," *The American Journal of Clinical Nutrition* 78: 660s–663s.
Pollan, M. (2006) *The Omnivore's Dilemma: A Natural History of Four Meals*, London: Penguin Books.
Rachels, J. (2004) "The Basic arguments for vegetarianism," in S. Sapontzis (ed) *Food for Thought: The Debate Over Eating Meat*, New York: Prometheus Books.
Regan, T. (1983) *The Case for Animal Rights*, Berkeley: University of California.
Sapontzis, S. (1987) *Morals, Reasons, and Animals*, Philadelphia: Temple University Press.
Schaefer, G. O., and Savulescu, J. (2014) "The ethics of producing in vitro meat," *Journal of Applied Philosophy* 31: 188–202.
Schlosser, E. (2002) *Fast Food Nation: The Dark Side of the All-American Meal*, New York: Harper Perennial.
Shepon, A., Eshel, G., Noor, E, and Milo, R. (2018) "The opportunity cost of animal based diets exceeds all food losses," *Proceedings of the National Academy of Sciences*, 1–6.
Sherriff, G. (2017) "Food and environmental justice," in M. Rawlinson and C. Ward (eds) *The Routledge Handbook of Food Ethics*, London: Routledge.
Shriver, A. (2009) "Knocking out pain in livestock: Can technology succeed where morality has stalled?" *Neuroethics* 2(3): 115–124.
Singer, P. (1975) *Animal Liberation*, New York: Avon Books.
Singer, P. (1979) "Killing humans and killing animals," *Inquiry* 22: 145–156.
Singer, P., and Mason, J. (2006) *The Ethics of What We Eat: Why Our Food Choices Matter*, Emmaus: Rodale.
Smith, A. (2016) *A Critique of the Moral Defense of Vegetarianism*, New York: Palgrave: Macmillan.

Stănescu, V. (2016) "Beyond happy meat: The (im)possibilities of 'humane,' 'local' and 'compassionate' meat," in B. Donaldson and C. Carter (eds) *The Future of Meat without Animals*, London: Rowman & Littlefield.
Steinfeld, H., Gerber, P., Wassenaar, T., Castel, V., Rosales, M., and de Haan, C. (2006) *Livestock's Long Shadow: Environmental Issues and Options*, Rome: Food and Agriculture Organization of the United Nations.
Toliver, Z. (2017) "Communities of color take action after poisoning by pig urine, feces," *People for the Ethical Treatment of Animals,* www.peta.org/blog/communities-color-take-action-poisoning-pig-urine-feces/ (Accessed March 1, 2019).
Tuomisto, H., and Joost Teixeira de Mattos, M. (2011) "Environmental impacts of cultured meat production," *Environmental Science and Technology* 45(14): 6117–6123.
United Nations (2012) "Desertification, drought affect one third of planet, world's poorest people, second committee told as it continues debate on sustainable development," *United Nations Sixty-seventh General Assembly: Second Committee*, www.un.org/press/en/2012/gaef3352.doc.htm (Accessed December 28, 2018).
United States Department of Agriculture: Animal and Plant Health Inspection Service (2016) *Program Data Report G—2016: Animals Dispersed/Killed or Euthanized /Removed or Destroyed/Freed*, www.aphis.usda.gov/wildlife_damage/pdr/PDR- G_Report.php?fy=2016&fld=state&fld_val=ID (Accessed December 28, 2018).
United States Department of Agriculture Center for Nutrition Policy and Promotion (2011) "Healthy eating for vegetarians: 10 tips for vegetarians: DG tip-sheet No 8," https://choosemyplateprod.azureedge.net/sites/default/files/tentips/DGTipsheet8HealthyEatingForVegetarians.pdf (Accessed December 28, 2018).
United States Department of Agriculture Food Safety and Inspection Service (2016) *Food Safety and Inspection Service Labeling Guideline on Documentation Needed to Substantiate Animal Raising Claims for Label Submissions*, www.fsis.usda.gov/wps/wcm/connect/6fe3cd56-6809-4239-b7a2-bccb82a30588/RaisingClaims.pdf?MOD=AJPERES (Accessed December 28, 2018).
Wallinga, D. (2004) "Concentrated animal feeding operations: Health risks from air pollution," *Institute for Agriculture and Trade Policy*, www.iatp.org/documents/concentrated-animal-feeding-operations-health-risks-from-air-pollution (Accessed December 28, 2018).
Warrick, J. (2001) "They die piece by piece," *Washington Post,* www.washingtonpost.com/archive/politics/2001/04/10/they-die-piece-by-piece/f172dd3c-0383-49f8b6d8347e04b68da1/?utm_term=.c2241e2def31 (Accessed December 28, 2018).
Wolfson, D. (1996) "Beyond the law: Agribusiness and the systemic abuse of animals," *Lewis and Clark Animal Law Review* 2: 123.
World Health Organization (The International Agency for Research on Cancer). (2015) *ARC Monographs Evaluate Consumption of Red Meat and Processed Meat: Press Release N240,* www.iarc.fr/en/mediacentre/pr/2015/pdfs/pr240_E.pdf (Accessed December 28, 2018).
Woorall, M. (2004) "Meatpacking safety: Is OSHA enforcement adequate?" *Drake Journal of Agricultural Law* 9: 299–321.
Wrenn, C. (2012) *If You Care about Animals, In-Vitro Meat is Not the Answer,* www.coreyleewrenn.com/in-vitro/ (Accessed March 11, 2019).

44
ANIMAL SANCTUARIES

Elan Abrell

The Animal Sanctuary Movement

Animal sanctuaries are enclosed spaces created by humans to permanently house and care for animals that have been rescued from a variety of contexts in which they are typically killed, maltreated, exploited, or otherwise at risk of harm from human activities. These contexts span the many arenas of interaction that shape human–animal relations, including the use of animals for food, entertainment, scientific research, and companionship. Individual sanctuaries often focus their efforts on specific kinds of animals determined by the particular kinds of animal use from which they are rescued. For this reason, sanctuaries generally fall into several subcategories, including farm animal sanctuaries, which care for commonly farmed animals like cows, pigs, chickens, and goats; exotic animal sanctuaries, which care for animals used in entertainment or sold through the exotic pet trade, such as big cats, elephants, and monkeys; and companion animal sanctuaries, which generally care for companion animals, like cats and dogs, that are unlikely to be adopted from more temporary shelters. In practice, these categories are not static, and cats, dogs, and smaller farm animals can occasionally be found living in sanctuaries focused on other types of animals. There are also more specialized sanctuaries that focus on particular species, such as equine sanctuaries, wolf sanctuaries, and primate sanctuaries, which care for primates previously used in scientific research as well as those used in entertainment or previously kept as pets (see Fleury 2017; Fultz 2017; Fultz and Willis 2015; Gruen 2006, 2011: 158–162, 2013, 2014, 2016a; Hua and Ahuja 2013; Ross 2014).

Following Henry Bergh's founding of the American Society for the Prevention of Cruelty to Animals in 1866, advocates for more humane treatment of animals helped to create an unaffiliated system of shelters and pounds across the United States (see Winograd 2007). These facilities pioneered the practice of rescuing and caring for animals over the proceeding century. Extending these efforts to new categories of animals, the modern animal sanctuary movement started in the mid-1980s with the founding of organizations like the Performing Animal Welfare Society (PAWS) and Farm Sanctuary. In 1984, Pat Derby and her partner Ed Stewart created PAWS to care for captive wildlife rescued from the television and film industry and the exotic pet trade. Derby and Stewart based PAWS's first sanctuary in Galt, California. Derby may have even coined the contemporary usage of the term *sanctuary* to describe such facilities[1]:

> "sanctuary" was my descriptive designation of our attempt to properly house and provide care for the hundreds of exotic animals who were in need of refuge in the early 1980s. At

> that time, animal shelters were often as bad as roadside zoos, with handlers walking young lions and tigers on leashes and breeding animals to provide more homeless cubs for display and photo ops. I chose "sanctuary" to exemplify our mission which we hoped was different.
>
> *(Derby 2011; see also Doyle 2014, 2017)*

Two years later, Gene Baur and Lorri Houston used funds raised from selling veggie dogs at music concerts to establish Farm Sanctuary, one of the first sanctuaries for animals rescued from the agricultural industry, in Watkins Glen, New York (see Baur 2008). Since these first sanctuaries were established, their models have been replicated hundreds of times, and new models have been formed as well, as animal sanctuaries continue to proliferate both across the United States and around the world.

Growing public support for sanctuaries has raised concern about whether certain facilities are in fact legitimate sanctuaries or if they are simply using the name to "humane-wash" exploitative captive animal facilities, like small roadside zoos (Winders 2017). Accreditation organizations like the Global Federation of Animal Associations and the American Sanctuary Association have formed to ensure that facilities are, in fact, sanctuaries that conform to basic standards of care, but accreditation is voluntary, and many legitimate sanctuaries have chosen for various reasons not to seek accreditation.[2] Among the hundreds of actual sanctuaries, there are still exploitative facilities using the term to misrepresent their missions. Animal advocacy organizations such as People for the Ethical Treatment of Animals (PETA) encourage caution in determining if a facility is a legitimate sanctuary before supporting it (PETA n.d.).

While animal sanctuaries may be a relatively new form of animal advocacy, the idea of sanctuary as a form of political action or ethical practice goes back centuries. The etymological history of the word *sanctuary* can be traced to the Latin word *sanctuarium*, which means a place where sacred things are kept. Reflecting its more contemporary usage, "sanctuary" described the practice of churches in England granting protection to fugitives fleeing arrest or violence from the fourth to the early seventeenth century. This idea of providing protection to people at risk of oppression or violence has evolved into multiple contemporary forms, such as the legal practice now employed by many countries of granting political asylum to persecuted individuals and the New Sanctuary Movement, the present-day revival of the Sanctuary Movement of the eighties in which cities around the United States and Canada were declared "sanctuary cities" where undocumented immigrants (many of whom were fleeing civil wars in Latin America) could be partially shielded from the enforcement of harsh immigration laws.

Unlike these efforts to afford sanctuary to humans, however, animal sanctuaries share basic structural features with zoos and other animal care institutions, such as wildlife rehabilitation facilities. Most significantly, they keep animals in captivity. This necessary condition of sanctuary care raises several practical and ethical challenges (described in more detail later). Despite this commonality, however, sanctuaries generally differ significantly in their goals from wildlife preserves, zoos, and other modes of *in situ* animal care (see Alcampora 2010; Braverman 2013; Gruen 2016b; Jamieson 2003a, 2003b; Mullan and Marvin 1999). For example, in contrast to many modern zoos and aquaria, in which nature is often presented as a spectacle for public consumption (see Bulbeck 2005; Chris 2006; Davis 1997; Vivanco and Gordon 2006), animal sanctuaries foster care-based interspecies relationships that challenge ideas about how animals are used and treated in human society. Most sanctuaries function as spaces of interspecies sociality (Hua and Ahuja 2013: 634) in which humans interact with animals as subjects whose physical and emotional needs are the primary determinants of those interactions.

Some sanctuaries endeavor to create spaces in which animals can form social groups with each other with minimized interference from humans, especially sanctuaries that care for particular species that typically live in social groups in the wild, such as chimpanzees or elephants (see, e.g., Buckley and Bradshaw 2010). But even in these circumstances, the more limited human–animal interactions are still guided by an ethos that privileges serving the needs of the animals. These sanctuaries do share a goal with wildlife rehabilitation facilities that seek to minimize human contact with the animals in their care, but rehab facilities also seek to return animals to their native habitats (see Collard 2014). The majority of animals at these kinds of sanctuaries, on the other hand, were often born in captivity and have become habituated to humans through circumstances in which they were kept as pets or used in entertainment and thus cannot be safely released to the wild.

Although there is a variety of ethical frameworks informing care practices between different animal sanctuaries, the sanctuary movement, in general, is influenced by two main philosophical approaches to animal advocacy: welfarism and animal rights. The welfarist approach seeks to improve the conditions of animals used by humans by reducing suffering as much as possible. Current campaigns to improve the conditions of animals in agricultural and entertainment industries, for example, reflect a welfarist approach to animal ethics, whereas the rights approach seeks to end harmful treatment of animals by granting them certain legal rights. In understanding the two approaches' influence on sanctuaries, it is helpful to think about them as flexible and potentially complementary stances rather than the mutually exclusive positions they often appear to be in contemporary animal protection debates. Many of the caregivers and volunteers I encountered over two years of ethnographic fieldwork at several different sanctuaries expressed support for animal rights and the abolition of animal exploitation while also supporting welfarist policies that could help reduce animal suffering.[3] Others I spoke to criticized animal rights activism as unrealistically utopian, but still shared many of the rights approach's critiques of animal use.

Like the broader animal protection movement of which it is a part, the animal sanctuary movement is the product of a diverse range of philosophical perspectives and aspirations for the future of animals. Sanctuaries provided new models of rescue and care for animals that had not been served by the country's century-old shelter system. As mentioned earlier, however, the contradictions between sanctuaries' ethical visions and the practical limitations of everyday captive animal care present unique challenges to the sanctuary project.

Dilemmas of Care

The mission of most animal sanctuaries is twofold: to provide the best lives possible to the animals in their care and to educate the public about (in order to end) the mistreatment of animals that makes sanctuaries necessary. The first goal raises a number of challenges for sanctuaries. Providing permanent spaces in which animals can thrive entails trying to minimize the negative impacts of captivity as much as possible or, in other words, minimize human control over the animals' lives. But the everyday practicalities of animal care can make this goal difficult to achieve. Sanctuaries must decide how to balance animals' freedom from human control against concerns for their safety and well-being (Abrell 2017: 3; see Jones 2014). Captivity, by definition, involves a certain degree of control. As philosopher Lori Gruen observes, captivity is "a condition in which a being is confined and controlled and is reliant on those in control to satisfy her basic needs" (2011: 133).

Most sanctuaries, for example, have a strict policy against animal reproduction, as opposed to zoos, which intentionally breed animals as part of their conservation missions (see Braverman 2013). This is because sanctuary space and resources are limited, so bringing new baby animals into being would put a further burden on the sanctuary. Furthermore, sanctuaries' primary purpose is the rescue and care of maltreated animals, of which there is a boundless population. Allowing sanctuary animals to reproduce would limit sanctuaries' ability to rescue and care for already living animals.

Despite the practical reasons for limiting animal reproduction, this policy raises ethical concerns. As Gruen argues,

> [d]enying captive animals the possibility to reproduce strips them of the chance to engage fully in species-typical behaviors, and this is particularly detrimental to females who are,

> in most species, primarily responsible for rearing young. Having infants born in captivity allows individuals to experience a full range of social relations, and it serves as enrichment for captive groups.
>
> *(2011: 133)*

It is possible that many sanctuary animals do not have conceptually rich future-directed thought, such as anticipating or hoping to become mothers at some point in life. On the other hand, many (although not all) sanctuary birds do exhibit brooding behavior that suggests some form of interest in reproduction, such as sitting on their own eggs, the abandoned eggs of other birds, or even small balls or other egg-shaped objects. Furthermore, many sanctuary animals form close relationships with other animals, suggesting an interest in and desire for sociality. The concern restricting reproduction raises is not necessarily that it frustrates animals' desires for the future (although that is certainly possible) but that it deprives animals, in general, of the full range of social interactions that they would experience in the absence of humans and, in particular, of a significant and unique form of social enrichment that could improve their experience of captivity.

Limiting reproduction is just one of many ethical concerns raised by the necessities of captive animal care. Miriam Jones, one of the cofounders of a Vermont-based sanctuary for formerly farmed animals called VINE Animal Sanctuary, explains how her sanctuary wrestles with the dilemmas of animal control. At VINE, caregivers describe the sanctuary animals as "as free as possible" because "fences, enforced routines, involuntary medical procedures and regimes (including everything from forced sterilization to forced feeding), and other impositions certainly do not comprise a free state of being for those on the receiving end" (2014: 91). As Jones's description of the dilemmas of care demonstrates, these practices inevitably entail forms of violence or harm to animals (see Gruen 2014; van Dooren 2014). While spaying and neutering, for example, serve sanctuaries' larger rescue mission by ensuring as much space as possible for more rescues, they also inflict violence on the bodies of animals. Caregivers thus make decisions about care "as ethically as possible" within the context of a world "that requires the rescue of members of certain species because other members of our own species will hurt and kill them if we don't," making "survival on their own . . . an impossible goal" (Jones 2014: 92). I frequently encountered variations on this perspective among sanctuary caregivers I interviewed. As one caregiver at a sanctuary with several monkeys explained, "[t]here is no cage in the world big enough for a monkey, but they can't survive on their own. And as long they have to live here, we're going to give them as big and nice a space as we can" (Abrell 2016: 148).

Philosopher Karen S. Emmerman identifies another ethical concern related to the goal of providing animals the best lives possible. Emmerman points out that sanctuaries are incapable of providing restitution for the violence or exploitation to which rescued animals were subjected prior to coming to sanctuary, but they may appear to do precisely that to people who visit, learn about, or make donations to them:

> Once an animal is in sanctuary and people get to meet her, know her, hear her story of exploitation and trauma, a caring response ensues. The animal is romanticized, thought about, and held in awe in much the same way the animals in zoos are. We feel relief at seeing an end to her suffering and have a sense that things have gone well in the world.
>
> *(2014: 229)*

For this reason, Emmerman argues people must understand that sanctuaries are "one step in the work of moral repair rather than the final destination." They may provide "new beginnings," but they are also "wrapped in an inescapable past and captive present" (ibid.). Many, if not most, sanctuary animals live with the ongoing physiological and psychological effects and impacts of their treatment prior to rescue. Because they were selectively bred to grow tissue or produce eggs or milk as rapidly as possible, with no regard for how this could affect them later in life, animals rescued from industrial agricultural contexts also frequently develop chronic, painful medical conditions postrescue. For these animals, sanctuaries are spaces "of continued trauma" (ibid.). Emmerman argues that sanctuaries are "places where we get a glimpse of humans doing the very best kind of moral work," but "even the very best kind of moral work is tainted in some sense" by the fact that "lifelong captivity is the best we can offer animals" (230). Emmerman's argument highlights a fundamental paradox in the sanctuary movement. Despite their goal "of providing animals with the best living conditions and most freedom possible, both sanctuary animals and caregivers are still inextricably caught up in the broader power relations that buttress human-animal inequality across human societies" (Abrell 2017: 4).

These broader power relations also lead to dilemmas related to the second main goal of sanctuary work: educating the public about the harmful treatment of animals. The importance of this second goal is underscored by an article that appeared in *The Guardian* in August 2017 with the headline "Firefighters Eat Sausages Made of Piglets They Saved from Blaze," (Morris 2017). In the United Kingdom, a farmer tried to thank some firefighters for saving two-week-old piglets from a fire on her property by giving them sausages made from the same pigs six months later. The farmer is quoted in the story as explaining, "This is just what we do—we are not an animal sanctuary. We give the pigs the best opportunity and the best life they could have for six months" (ibid.). This story perfectly illustrates the contradiction between how sanctuaries understand their duties to formerly farmed animals and how the agriculture industry treats them, while jarringly exposing the cognitive dissonance shaping the way many people think about animals used for food. Exposing and challenging this contradiction is what sanctuaries seek to do in their public education role. Specifically, sanctuaries do this by removing animals from animal-based economic systems, such as agriculture, in which they are treated as property, and then modeling alternative ways of living with and relating to animals as subjects worthy of moral consideration.

Certain aspects of sanctuary work can complicate this mission, however. In challenging the objectification and exploitation of animals, sanctuaries form intersubjective relationships of care that are still embedded both in a larger political-economic context fueled by the use of animals to generate profits and in a larger social and legal contexts in which animals remain classified as property. So even in sanctuaries, animals can continue to operate as economic value generators in multiple ways, some directly resulting from their status as sanctuary animals. This raises multiple questions with which caregivers must contend.

First, can caring for sanctuary animals help to perpetuate animal-based economies? This is a dilemma that is unavoidable in a capitalist society, especially given that most of the resources used for caring for captive and farmed animal species are part of animal-based economies. For example, most, if not all, sanctuaries need to purchase vet medications and supplies, commercial animal food products, and fencing and equipment designed to facilitate animal captivity. Although sanctuaries successfully remove the animals in their care from certain forms of economic exploitation, their dependence on these resources ties sanctuaries and their animals to multiple circuits within the larger economy that still rely on animal bodies as a significant source of profit. There is not a separate or self-contained sanctuary care economy for these things, yet they are all important in caring for captive animals.

Second, should sanctuary animals play a role in generating the financial support that funds their own care? Take, for example, sanctuary donation solicitations, which often employ descriptions of animals' experiences before and after coming to a sanctuary. Practically, sharing animals' stories of mistreatment and rescue is an essential fundraising tool for many sanctuaries. But while their personal stories can help to make animals more relatable as individual subjects worthy of rescue and care, it could also be argued that monetizing their experiences through fundraising solicitations simultaneously recasts them as mechanisms for producing value, albeit a different kind of value than they may have generated in their previous lives. The concern this practice raises is not that animals may have a personal interest in their experiences and life stories not being monetized but that it potentially conflicts with or undermines the broader goal of many sanctuaries to transform cultural perspectives that see animals as valid sources of economic value in any context.

This kind of instrumentalization of animals' lives also relates to the use of animals as ambassadors for the animal protection movement. Many sanctuaries see their rescued animals as ambassadors who can educate the public about the abuse and exploitation of animals, in addition to motivating donations to help animals rescued from this mistreatment.[4] The organization Animal Charity Evaluators (ACE), which endeavors to advise potential donors on the most effective destinations for its donations, suggests that this is the only significant value of sanctuaries. A page about sanctuaries on its website states:

> The primary impact of sanctuaries does not occur from the direct number of animals that they save (though there is certainly value in each animal that is rescued), but rather from the effects of their promotion and education efforts to encourage others not to use animals . . . [A]nimal advocates should be mindful of the need to support sanctuaries enough that they can provide important symbolism as well as educational values to the public through tours and materials used by other advocacy groups. However, advocates should also be mindful of the potential for sanctuaries to absorb large amounts of funding through direct care of large numbers of animals—a need which is unlikely to be satiated.
>
> *(Bockman 2015)*

This passage directly links the use of animals as ambassadors to their ability to generate economic value for the sanctuaries, while also problematically suggesting that this purpose may be the only worthwhile reason to make donations to sanctuaries given that they are only able to save a tiny fraction of the animals subjected to harm by humans. It overlooks the possibility that sanctuary work has moral significance or strategic value to animal activism beyond the educational potential of rescued animals. While it is true that sanctuaries require ongoing financial support to care for a relatively small population of animals, it is debatable whether the number of lives saved per dollar should be the primary means of assessing the value of animal charities in all contexts. But even by this standard, it is difficult to measure the long-term impact of sanctuaries in influencing the kind of broad cultural shifts in the treatment and use of animals that ACE endeavors to effect. It is possible that growing public awareness of sanctuaries and the alternative ways of relating to animals they model could contribute as much or more to such cultural change as vegan advocacy and animal welfare initiatives, although such an impact may be indirect and gradual and thus not readily apparent.

Despite its advantages, enlisting animals as ambassadors also raises another moral dilemma for sanctuaries to navigate. One prominent way in which animals act as ambassadors at many sanctuaries is through their interaction with public tours. Allowing public access in the form of tours impacts animal autonomy by limiting animals' ability to regulate their proximity to humans. Take, for example, a tour I witnessed at a sanctuary I visited. The chickens at this sanctuary lived in several coops, each located in a small yard enclosed by wire fencing. When a group of a few dozen visitors stood inside one of these yards listening to a tour guide explain the typical conditions on an industrialized chicken farm, some chickens walked close to the visitors while others moved as far as they could to the opposite side of the yard. Any chickens who may have wanted to avoid humans could not put more than several meters distance between themselves and the visitors. Enlisting these chickens as ambassadors in the sanctuary's mission thus comes at a cost for these chickens, and other animals in similar circumstances, who must sacrifice some of their ability to move as far as they might want from unfamiliar humans in order to facilitate the educational goal of the tours. Sanctuaries that have made the decision to open their doors to the public in service of their educational goal necessarily must place further limits on animals' already-limited autonomy in exchange.

There is a third moral consideration regarding how sanctuaries are tied to larger economic processes, but this one affects humans: How should sanctuaries balance the needs of human caregivers with the needs of other sanctuary animals? Sanctuaries rely on various combinations of volunteer and paid human labor to operate. Understaffing is an ongoing problem at many sanctuaries, but the quality of labor is as much an issue as the quantity. One of the primary dilemmas sanctuaries face with employees is that they need skilled labor—people with a robust understanding of the care requirements of a range of species—but they often do not have the budget to adequately compensate that level of skill. Low wages can undermine sanctuaries' abilities to build and retain the kind of skilled and experienced caregivers they need. Increasing the employee budget would be one solution, but this would require shifting money away from other expenditures like feed and veterinary supplies. Limiting supplies would in turn place further pressure on the already difficult decisions sanctuaries must make about which and how many animals can be rescued. This dilemma reflects a paradoxical trend in animal rescue organizations more broadly in which the needs of animal subjects can eclipse the needs of the human subjects trying to help them.

Sanctuary as Political Action

Sanctuaries provide models for ways of relating to and living with animals as subjects rather than as living forms of property. As geographer Kathryn Gillespie observes, "[i]n a society so dedicated to teaching us that animals are here to be eaten, worn, experimented on, and kept as pets and entertainers, the sanctuary embodies an alternative conceptualization of how animals fit into multispecies social worlds" (2018: 143). But as long as these animals' new social lives can only exist within the walls of a sanctuary, captivity and the significant challenges to changing the way animals are treated outside those walls will continue to generate practical and moral dilemmas. In working through these dilemmas of care, caregivers and animals are co-creating new interspecies communities adapted to the current conditions that shape and constrain human–animal relations more broadly. Beyond the benefits this creates for the animals within these spaces, the political act of providing sanctuary also has a greater value to animal advocacy and potentially other social and environmental justice movements as well, despite the popular view articulated by ACE earlier.

Realizing this political potential would require not just a change in the cultural view of animals as resources to fulfill human desires but a shift in the same view of fellow humans as well, and such a shift would likely require a wide collaboration beyond the spaces of animal sanctuaries. It would require asking questions such as "What might it mean to rethink sanctuary more broadly as a site of resistance in the fight for global social justice, one that recognizes how the human-nonhuman divide authorizes violence against *all* who are deemed to be less than human?" (Pachirat 2018: 351) or "How can we work in solidarity with others who are creating refuge in other ways?"[5]

Political scientist Claire Jean Kim highlights one potential path toward collaborating outside of the sanctuary movement in her work on the intersection between racial politics and conflicts over the treatment of animals. Kim argues that by being attentive to how human–animal inequality intersects with other forms of domination, social justice activists and animal advocates could together create an "ethics of mutual avowal" that acknowledges the validity of each others' interests and provides a basis for collaboration and mutual support (2015: 20; see also Gruen 2015). Of course, forming such connections would eventually require a reciprocal effort on the part of other social justice activists, but spaces of sanctuary can begin to foster that reciprocity by continuing to extend their ethics of avowal beyond the politics of animal care and rescue toward the goals of the larger planetary movement for social and environmental justice, which has the capacity to make species, gender, racial, class, and sexual justice (among others) all central objectives.

Ultimately, the coalitional potential of sanctuary as a form of political resistance and transformation can be understood as a component of a much more expansive and historically longer, unfinished abolitionist political project that, as theorist Che Gosset observes, entails "ending anti-Black racism, racial capitalism, anti-trans, anti-queer, patriarchal policing, colonialism, and caging" of both human and nonhuman animals (Gosset and Filar 2016). Gosset explicitly ties this project to ecology, arguing that "abolition is always already about ecology and we continue to need . . . an abolitionist ecology" (ibid.). Providing sanctuary as a mode of political resistance has the potential to contribute to an abolitionist ecological politics equipped to address the mutually reinforcing processes of social, environmental, and species inequality that define and shape our current era. The realization of that potential may rely on the coalitional spaces it can foster.

Considering the range of dilemmas sanctuaries already face, such coalition building would not be a simple feat. And there are certainly limits on what individual sanctuaries could contribute to such a political project—for example, most sanctuaries do not have space to host the kinds of meetings that would facilitate building broad coalitions. Nonetheless, sanctuaries activists are no strangers to difficult circumstances. In trying to create on a small scale the world they want, sanctuaries face immense challenges, but in facing these challenges, they also build on the pragmatist tradition in animal activism of seeking to achieve immediate improvements for the animals living now while simultaneously endeavoring to bring about a more fundamental transformation in human–animal relations in the future. Sanctuaries are fundamentally experiments in community building. Broadening those communities further would be difficult work, creating potentially new dilemmas of care to navigate. However, given the transformative possibilities such work could create, those challenges may be worth facing.

Notes

1. Although see Fusari (2017) for a linguistic history of the development of the term "sanctuary" as it came to apply to protected spaces for animals in earlier contexts.
2. For example, some farm sanctuaries see these organizations' accreditation guidelines as better suited to address the needs of captive wild animals, like big cats or elephants, than the needs of formerly farmed animals.
3. This research was conducted with the support of the National Science Foundation under Grant Number 1322203. Any opinions, findings, and conclusions or recommendations expressed in this chapter are those of the author and do not necessarily reflect the views of the National Science Foundation.
4. Donaldson and Kymlicka critique this approach to sanctuary care, which they dub the "refuge + advocacy" model, proposing that sanctuaries instead adopt an "intentional community" model in which sanctuary animals are co-citizens with humans in the sanctuary space (2015).
5. This question was posed by pattrice jones at "Sanctuary: Reflecting on Refuge, a Conference about Care and Purpose at Farmed Animal Sanctuaries," held at Wesleyan University in Middletown, CT, from September 29 to October 1, 2017.

Bibliography

Abrell, E. (2016) "Lively sanctuaries: A shabbat of animal sacer," in I. Braverman (ed.) *Animals, Biopolitics, Law: Lively Legalities* (pp. 135–154), New York: Routledge.

Abrell, E. (2017) "Introduction: Interrogating captive freedom: The possibilities and limits of animal sanctuaries," *Animal Studies Journal* 6(2): 1–8.

Alcampora, R. (ed.) (2010) *Metamorphoses of the Zoo: Animal Encounter after Noah*, Lanham, CT: Lexington Books.

Baur, G. (2008) *Farm Sanctuary: Changing Hearts and Minds about Animals and Food*, New York: Touchstone.

Bockman, J. (2015) "The value of sanctuaries, and how to maximize their impact," *Animal Charity Evaluators*, September 24, https://animalcharityevaluators.org/blog/the-value-of-sanctuaries-and-how-to-maximize-their-impact/.

Braverman, I. (2013) *Zooland: The Institution of Captivity*, Stanford: Stanford University Press.

Buckley, C., and Bradshaw, G. A. (2010) "The art of cultural brokerage: Recreating elephant-human relationship and community," *Spring* 83: 37–63.

Bulbeck, C. (2005) *Facing the Wild: Ecotourism, Conservation, and Animal Encounters*, Los Angeles: Earthscan.

Chris, C. (2006) *Watching Wildlife*, Minneapolis, MN: University of Minnesota Press.

Collard, R. (2014) "Putting animals back together, taking commodities apart," *Annals of the Association of American Geographers* 104(1): 151–165.
Davis, S. (1997) *Spectacular Nature: Corporate Culture and the Sea World Experience*, Berkeley, CA: University of California Press.
Derby, P. (2011) *PAWS: An Introduction*, www.pawsweb.org/documents/PAWS_AN_INTRO.pdf.
Donaldson, S., and Kymlicka, W. (2015) "Farmed animal sanctuaries: The heart of the movement? A socio-political perspective," *Politics and Animals* 1(1): 50–74.
Doyle, C. (2014) "Captive elephants," in L. Gruen (ed.) *The Ethics of Captivity* (pp. 38–56), Oxford: Oxford University Press.
Doyle, C. (2017) "Captive wildlife sanctuaries: Definition, ethical considerations and public perception," *Animal Studies Journal* 6(2): 55–85.
Emmerman, K. (2014) "Sanctuary, not remedy: The problem of captivity and the need for moral repair," in L. Gruen (ed.) *The Ethics of Captivity*, Oxford: Oxford University Press.
Fleury, E. (2017) "Money for monkeys, and more: Ensuring sanctuary retirement of nonhuman primates," *Animal Studies Journal* 6(2): 30–54.
Fultz, A. (2017) "A guide for modern sanctuaries with examples from a captive chimpanzee sanctuary," *Animal Studies Journal* 6(2): 9–29.
Fultz, A., and Willis, C. (2015) "Foreword," in L. Gruen, *Entangled Empathy* (pp. xi–xv), New York: Lantern Books.
Fusari, S. (2017) "What is an animal sanctuary? Evidence from applied linguistics," *Animal Studies Journal* 6(2): 137–160.
Gillespie, K. (2018) *The Cow with Ear Tag #1389*, Chicago: University of Chicago Press.
Gosset, C., and Filar, R. (2016) "Cruising in the end times: An interview with Che Gossett," *VersoBooks.com*, December18,www.versobooks.com/blogs/3016-cruising-in-the-end-times-an-interview-with-che-gossett.
Gruen, L. (2006) *The First 100 Chimpanzees*, http://first100chimps.wesleyan.edu/.
Gruen, L. (2011) *Ethics and Animals: An Introduction*, Cambridge: Cambridge University Press.
Gruen, L. (2013) *The Last 1000*, http://last1000chimps.com/.
Gruen, L. (2014) "Dignity, captivity, and an ethics of sight," in L. Gruen (ed.) *The Ethics of Captivity*, Oxford: Oxford University Press.
Gruen, L. (2015) "Samuel Dubose, Cecil the Lion, and the ethics of avowal," *Al Jazeera*, http://america.aljazeera.com/opinions/2015/7/samuel-dubose-cecil-the-lion-and-the-ethics-of-avowal.html.
Gruen, L. (2016a) "Navigating difference (again): Animal ethics and entangled empathy," in J. Castricano and L. Corman (eds.) *Animal Subjects 2.0*, Waterloo, Canada: Wilfrid Laurier University Press.
Gruen, L. (2016b) "Shifting towards an ethics of sanctuary," www.humansandnature.org/shifting-toward-an-ethics-of-sanctuary
Hua, J., and Ahuja, N. (2013) "Chimpanzee sanctuary: 'Surplus' life and the politics of transspecies care," *American Quarterly* 65(3): 619–637.
Jamieson, D. (2003a) "Against zoos," in *Morality's Progress: Essays on Humans, Other Animals, and the Rest of Nature*, Oxford: Oxford University Press.
Jamieson, D. (2003b) "Zoos revisited," in *Morality's Progress: Essays on Humans, Other Animals, and the Rest of Nature*, Oxford: Oxford University Press.
Jones, M. (2014) "Captivity in the context of a sanctuary for formerly farmed animals," in L. Gruen (ed.) *The Ethics of Captivity*, Oxford: Oxford University Press.
Kim, C. (2015) *Dangerous Crossings: Race, Species, and Nature in Multicultural Age*. New York: Cambridge University Press.
Morris, S. (2017) "Firefighters eat sausages made of piglets they saved from blaze," *The Guardian*, August, www.theguardian.com/environment/2017/aug/23/firefighters-sausages-piglets-saved-blaze-wiltshire.
Mullan, B., and Marvin, G. (1999) *Zoo Culture*, London: Weidenfeld and Nicolson.
Pachirat, T. (2018) "Sanctuary," in L. Gruen (ed.) *Critical Terms for Animal Studies*, Chicago: Chicago University Press.
PETA (n.d.) "How to tell if a place is a REAL animal sanctuary," www.peta.org/features/real-animal-sanctuary-zoo/.
Ross, S. (2014) "Captive chimpanzees," in L. Gruen (ed.) *The Ethics of Captivity*, Oxford: Oxford University Press.
van Dooren, T. (2014) *Flight Ways: Life and Loss at the Edge of Extinction*. New York: Columbia University Press.
Vivanco, L., and Gordon, R. (eds.) (2006) *Tarzan was an Ecotourist . . . and Other Tales in the Anthropology of Adventure*, New York: Berghahn Books.
Winders, D. (2017) "Captive wildlife at a crossroads – Sanctuaries, accreditation, and humane-washing," *Animal Studies Journal* 6(2): 161–178.
Winograd, N. (2007) *Redemption: The Myth of Pet Overpopulation and the No Kill Revolution in America*, Los Angeles, CA: Almaden Books.

INDEX

Note: Page numbers in *italic* indicate a figure and page numbers in **bold** indicate a table or box on the corresponding page.

For Product Safety Concerns and Information please contact our
EU representative GPSR@taylorandfrancis.com Taylor & Francis
Verlag GmbH, Kaufingerstraße 24, 80331 München, Germany